Intermediate Algebra
Concepts with Applications

Charles P. McKeague

2ND PRINTING — December 2012
3RD PRINTING — March 2013
4TH PRINTING — June 2013

*xyz*textbooks

Intermediate Algebra:
Concepts with Applications
Charles P. McKeague

Project Manager: Staci Truelson

Developmental Editor: Katherine Heistand Shields

Illustrator and Cover Design: Kaela Soohoo

Cover Image: Katherine Heistand Shields

Text Designer: Devin Christ

Compositor: Donna Looper, Christina Machado

Reviewer: Jeffrey Saikali

ISBN-13: 978-1-936368-06-8   /   ISBN-10: 1-936368-06-4
Annotated Instructor's Edition: ISBN-13: 978-1-936368-07-5 / ISBN-10: 1-936368-07-2

For product information and technology assistance, contact us at
XYZ Textbooks, 1-877-745-3499

For permission to use material from this text or product,
e-mailed: **info@mathtv.com**

XYZ Textbooks
1339 Marsh St.
San Luis Obispo, CA 93401
USA

For your course and learning solutions, visit **www.xyztextbooks.com**

Printed in the United States of America

Brief Contents

Contents

3 Systems of Equations   249

4 Exponents and Polynomials   333

5 Rational Expressions, Equations, and Functions   415

Preface

XYZ Textbooks is on a mission to improve the quality and affordability of course materials for mathematics. Our 2013 Concepts with Applications Series brings a new level of technology innovation, which is certain to improve student proficiency.

As you can see, these books are presented in a format in which each section is organized by easy to follow learning objectives. Each objective includes helpful example problems, as well as additional practice problems in the margins, and are referenced in each exercise set. To help students relate what they are studying to the real world, each chapter and section begin with a real-life application of key concepts. You and your students will encounter extensions of these opening applications in the exercise sets.

We have put features in place that help your students stay on the trail to success. With that perspective in mind, we have incorporated this navigational theme into the following new and exciting features.

QR Codes

Unique technology embedded in each exercise set allows students with a smart phone or other internet-connected mobile device to view our video tutorials without being tied to a computer.

> **SCAN TO ACCESS**
>
> 10. Noelle recieves $17,000 in two loans. One loan charges 5% interest per year and the other 6.5%. If her total interest after 1 year is $970, how much was each loan?
>
> $9,000 at 5%, $8,000 at 6.5%

SAMPLE

Vocabulary Review

A list of fill-in-the-blank sentences appears at the beginning of each exercise set to help students better comprehend and verbalize concepts.

> **Vocabulary Review**
>
> Choose the correct words to fill in the blanks below.
>
> prime   composite   remainder   lowest terms
>
> 1. A _____ number is any whole number greater than 1 that has exactly two divisors: 1 and the number itself.
> 2. A number is a divisor of another number if it divides it without a _____.
> 3. Any whole number greater than 1 that is not a prime number is called a _____ number.
> 4. A fraction is said to be in _____ if the numerator and the denominator have no factors in common other than the number 1.

Getting Ready For Class

Simple writing exercises for students to answer after reading each section increases their understanding of key concepts.

> **GETTING READY FOR CLASS**
>
> *After reading through the preceding section, respond in your own words and in complete sentences.*
>
> A. What is a prime number?
> B. Why is the number 22 a composite number?
> C. Factor 120 into a product of prime factors.
> D. How would you reduce a fraction to lowest terms?

Key words

A list of important vocabulary appears at the beginning of each section to help students prepare for new concepts. These words can also be found in blue italics within the sections.

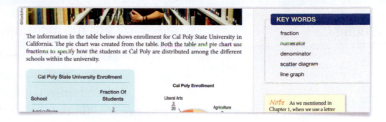

Navigation Skills

A discussion of important study skills appearing in each chapter anticipates students' needs as they progress through the course.

Navigation Skills: Prepare, Study, Achieve

Your instructor is a vital resource for your success in this class. Make note of your instructor's office hours and utilize them regularly. Compile a resource list that you keep with your class materials. This list should contain your instructor's office hours and contact information (e.g., office phone number or e-mail), as well as classmates' contact information that you can utilize outside of class. Communicate often with your classmates about how the course is going for you and any questions you may have. Odds are that someone else has the same question and you may be able to work together to find the answer.

Find the Mistake

Complete sentences that include common mistakes appear in each exercise set to help students identify errors and aid their comprehension of the concepts presented in that section.

Find the Mistake

Each sentence below contains a mistake. Circle the mistake and write the correct word or phrase on the line provided.

1. The set of all inputs for a function is called the range. _____
2. For the relation (1, 2), (3, 4), (5, 6), (7, 8), the domain is {2, 4, 6, 8}. _____
3. Graphing the relation $y = x^2 - 3$ will show that the domain is { x | x ≥ −3 }. _____
4. If a horizontal line crosses the graph of a relation in more than one place, the relation cannot be a function. _____

Landmark Review

A review appears in the middle of each chapter for students to practice key skills, check progress, and address difficulties.

Landmark Review: Checking Your Progress [7.1– 7.3]

Solve the following systems.

1. $y = 2x + 5$
   $y = -3x - 5$

2. $2x - y = -2$
   $2x - 3y$

3. $2x - y = 1$
   $y = -2x + 5$

4. $2x + 5y = 0$
   $x - 5y = -9$

5. $3x - 3y = 4$
   $x = 3y - 1$

6. $-6x + 2y = -9$
   $y = 3x + 4$

7. $x - y = 11$
   $x + y = 1$

8. $-3x + y = -15$
   $-2x + 3y = -17$

9. $2x + 5y = 17$
   $x - 4y = -24$

10. $x + 4y = 5$
    $3x + 8y = 14$

End of Chapter Reviews, Cumulative Reviews, and Tests

Problems at the end of each chapter provide students with a comprehensive review of each chapter's concepts.

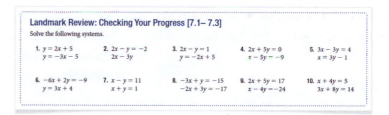

CHAPTER 5   Test

Write each ratio as a fraction in lowest terms. [5.1]

1. 48 to 18
2. $\frac{5}{8}$ to $\frac{3}{4}$
3. 6 to $2\frac{4}{7}$
4. 0.14 to 0.4

Find the unknown term in each proportion. [5.3, 5.4]

10. $\frac{4}{7} = \frac{24}{x}$
11. $\frac{2.5}{5} = \frac{1.5}{x}$
12. **Baseball** A baseball player gets 13 hits in his first 20 games of the season. If he continues at the same rate, how many hits will he get in 60 games? [5.5]

Trail Guide Projects

Individual or group projects appear at the end of each chapter for students to apply concepts learned to real life.

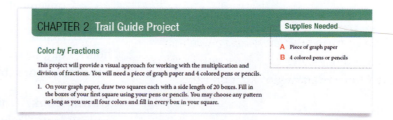

CHAPTER 2   Trail Guide Project   Supplies Needed

Color by Fractions

This project will provide a visual approach for working with the multiplication and division of fractions. You will need a piece of graph paper and 4 colored pens or pencils.

1. On your graph paper, draw two squares each with a side length of 20 boxes. Fill in the boxes of your first square using your pens or pencils. You may choose any pattern as long as you use all four colors and fill in every box in your square.

A Piece of graph paper
B 4 colored pens or pencils

XYZ Textbooks is committed to helping students achieve their goals of success. This new series of highly-developed, innovative books will help students prepare for the course ahead and navigate their way to a successful course completion.

Supplements

Instructor

Name: _____

Phone: _____

Email: _____

Office Hours: _____

Teacher Assistant

Name: _____

Phone: _____

Email: _____

Office Hours: _____

On-Campus Tutoring

Location: _____

Hours: _____

Study Group

Name: _____

Phone: _____

Email: _____

Name: _____

Phone: _____

Email: _____

Name: _____

Phone: _____

Email: _____

For the Instructor

Online Homework   XYZ Homework provides powerful online instructional tools for faculty and students. Randomized questions provide unlimited practice and instant feedback with all the benefits of automatic grading. Tools for instructors can be found at *www.xyzhomework.com* and include the following:

- Quick setup of your online class
- More than 3,000 randomized questions, similar to those in the textbook, for use in a variety of assessments, including online homework, quizzes, and tests
- Text and videos designed to supplement your instruction
- Automated grading of online assignments
- Flexible gradebook
- Message boards and other communication tools, enhanced with calculator-style input for proper mathematics notation

MathTV.com   With more than 8,000 videos, MathTV.com provides the instructor with a useful resource to help students learn the material. MathTV.com features videos of most examples in the book, explained by the author and a variety of peer instructors. If a problem can be completed more than one way, the peer instructors often solve it by different methods. Instructors can also use the site's Build a Playlist feature to create a custom list of videos for posting on their class blog or website.

For the Student

Online Homework   XYZ Homework provides powerful online instruction and homework practice for students. Benefits for the student can be found at *www.xyzhomework.com* and include the following:

- Unlimited practice with problems similar to those in the text
- Online quizzes and tests for instant feedback on performance
- Online video examples
- Convenient tracking of class programs

MathTV.com   Students have access to math instruction 24 hours a day, seven days a week, on MathTV.com. Assistance with any problem or subject is never more than a few clicks away.

Online Book   This text is available online for both instructors and students. Tightly integrated with MathTV.com, students can read the book and watch videos of the author and peer instructors explaining most examples. Access to the online book is available free with the purchase of a new book.

Additional Chapter Features

Learning Objectives

We have provided a list of the section's important concepts at the beginning of each section near the section title, noted with an orange capital letter, which is reiterated beside each learning objective header in the section.

Example and Practice Problems

Example problems, with work, explanations, and solutions are provided for every learning objective. Beside each example problem, we have also provided a corresponding practice problem and its answer in the margin.

Margin Notes

In many sections, you will find yellow notes in the margin that correspond to the section's current discussion. These notes contain important reminders, fun facts, and helpful advice for working through problems.

Colored Boxes

Throughout each section, we have highlighted important definitions, properties, rules, how to's, and strategies in colored boxes for easy reference.

Blueprint for Problem Solving

We provide students with a step-by-step guide for solving application problems in a clear and efficient manner.

Facts from Geometry

Throughout the book, we highlight specific geometric topics and show students how they apply to the concepts they are learning in the course.

Using Technology

Throughout the book, we provide students with step-by-step instructions for how to work some problems on a scientific or graphing calculator.

Descriptive Statistics

In some sections, we call out to the use of descriptive statistics as they are applied to the concepts students are learning, and thus, applied to real life.

Vocabulary Review

At the beginning of every exercise set, these fill-in-the-blank sentences help students comprehend and verbalize concepts from the corresponding section.

Problems

Each section contains problems ranging in difficulty that allow students to practice each learning objective presented in the section. The majority of the sections will also include problems in one or more of the following categories:

- Applying the Concepts
- Extending the Concepts or One Step Further
- Improving your Quantitative Literacy
- Getting Ready for the Next Section
- Extending the Concepts

Chapter Summary

At the end of each chapter, we provide a comprehensive summary of the important concepts from the chapter. Beside each concept, we also provide additional example problems for review.

Answers to Odd-Numbered Questions

At the end of the book, we have supplied answers for all odd-numbered problems that appear in the following features:

- Landmark Reviews
- Vocabulary Reviews
- Exercise Sets
- Find the Mistake Exercises

Navigation Skills at a Glance

- Compile a resource list for assistance.
- Set short- and long-term goals for the course.
- Complete homework completely and efficiently.
- Pay attention to instructions, and work ahead.
- Be an active listener while in class.
- Engage your senses when studying and memorizing concepts.
- Study with a partner or group.
- Maintain a positive academic self-image, and avoid burnout.
- Prepare early for the final exam, and reward yourself for a job well done!

Real Numbers and Algebraic Expressions

0

© 2010 Google
Griffith Park Observatory
Los Angeles, California

Griffith Park Observatory is one of the most visited public solar observatories in the world. Its solar telescope has enabled millions of people to safely view the sun. The telescope's copper-covered dome and three flat mirrors, called coelostats, work together to allow the sun to be viewed as stationary, despite Earth's rotation.

©iStockphoto.com/fotoVoyager

The distance from Earth to the sun is approximately 9.3×10^7 miles. If light travels 1.2×10^7 miles in one minute, how many minutes does it take the light from the sun to reach Earth and be viewed at the Griffith Observatory? As you work through this chapter, you will learn how to work with numbers written in scientific notation, as these are. You will also see this problem again in the exercise set of Section 4.

A Locate and label points on the number line.

B Change a fraction to an equivalent fraction with a new denominator.

C Simplify expressions containing absolute value.

D Identify the opposite of a number.

E Factor a number into a product of primes.

F Identify the reciprocal of a number.

G Find the perimeter and area of squares, rectangles, and triangles.

KEY WORDS

real number line

origin

coordinate

real numbers

fraction

numerator

denominator

equivalent fraction

absolute value

opposites

prime number

composite

factor

reciprocal

area

perimeter

Note: If there is no sign (+ or −) in front of a number, the number is assumed to be positive (+).

0.1 The Real Numbers

The cane toad was introduced to Australian agricultural fields in 1935 to help rid sugar cane crops of an invasive beetle. Instead, the toads became the invasive species, multiplying from an original number of 102 to a current population of more than 1.5 million. The toad preys on and competes with native species, and its skin contains a toxin that is deadly when ingested by native predators and even pets.

Scientists are studying how global warming may affect these amphibians. Cold-blooded animals, like toads, have a difficult time breathing in warmer temperatures, but the cane toad's heart and lungs appear to thrive in the heat. On the other hand, once temperatures drop below 15 degrees Celsius (59 degrees Fahrenheit), the toad's system shuts down and he can barely hop. This is good news for much of Southern Australia where weather conditions are too cold for the toad. However, as Earth warms, the toad population may invade those areas as well.

The following bar chart contains record low temperature readings for various cities in Australia. Notice that some of these temperatures are represented by negative numbers.

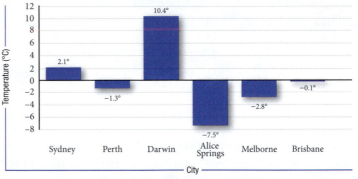

FIGURE 1

A The Number Line

In this section, we start our work with negative numbers. To represent negative numbers in algebra, we use what is called the *real number line*. Here is how we construct a real number line: First, we draw a straight line and label a convenient point on the line with 0. Then we mark off equally spaced distances in both directions from 0. We label the points to the right of 0 with the numbers 1, 2, 3, . . .(the dots mean "and so on"). The points to the left of 0 we label in order, −1, −2, −3, . . . Here is what it looks like:

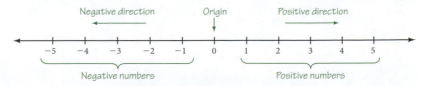

The numbers increase in value going from left to right. If we "move" to the right, we are moving in the positive direction. If we move to the left, we are moving in the negative direction. When we compare two numbers on the number line, the number on the left is always smaller than the number on the right. For instance, -3 is smaller than -1 because it is to the left of -1 on the number line.

EXAMPLE 1 Locate and label the points on the real number line associated with the numbers $-3.5, -1\frac{1}{4}, \frac{1}{2}, \frac{3}{4}, 2.5$.

Solution We draw a real number line from -4 to 4 and label the points in question.

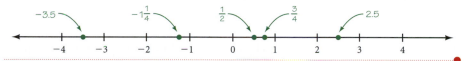

> **DEFINITION** coordinate
>
> The number associated with a point on the real number line is called the **coordinate** of that point.

In the preceding example, the numbers $\frac{1}{2}, \frac{3}{4}, 2.5, -3.5$, and $-1\frac{1}{4}$ are the coordinates of the points they represent.

> **DEFINITION** real numbers
>
> The numbers that can be represented with points on the real number line are called **real numbers**.

Real numbers include whole numbers, fractions, decimals, and other numbers that are not as familiar to us as these.

B Equivalent Fractions on the Number Line

As we proceed through the chapter, from time to time we will review some of the major concepts associated with fractions. To begin, here is the formal definition of a fraction:

> **DEFINITION** fraction
>
> If a and b are real numbers, then the expression
>
> $$\frac{a}{b} \qquad b \neq 0$$
>
> is called a **fraction**. The top number a is called the **numerator**, and the bottom number b is called the **denominator**. The restriction $b \neq 0$ keeps us from writing an expression that is undefined. (As you will see, division by zero is not allowed.)

The number line can be used to visualize fractions. Recall that for the fraction $\frac{a}{b}$, a is called the numerator and b is called the denominator. The denominator indicates the number of equal parts in the interval from 0 to 1 on the number line. The numerator indicates how many of those parts we have. If we take that part of the number line from 0 to 1 and divide it into *three equal parts*, we say that we have divided it into *thirds* (Figure 2). Each of the three segments is $\frac{1}{3}$ (one third) of the whole segment from 0 to 1.

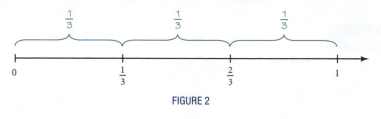

FIGURE 2

Practice Problems

1. Locate and label the points associated with $-2, -\frac{1}{2}, 0, 1.5, 2.75$.

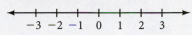

Note: There are other numbers on the number line that you may not be as familiar with. They are irrational numbers such as $\pi, \sqrt{2}, \sqrt{3}$. You will see these later in the book.

Answers

1.
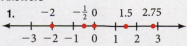

Two of these smaller segments together are $\frac{2}{3}$ (two thirds) of the whole segment. Three of them would be $\frac{3}{3}$ (three thirds), or the whole segment.

Let's do the same thing again with six equal divisions of the segment from 0 to 1 (Figure 3). In this case, we say each of the smaller segments has a length of $\frac{1}{6}$ (one sixth).

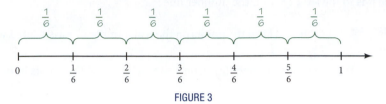

FIGURE 3

The same point we labeled with $\frac{1}{3}$ in Figure 2 is now labeled with $\frac{2}{6}$. Likewise, the point we labeled earlier with $\frac{2}{3}$ is now labeled $\frac{4}{6}$. It must be true then that

$$\frac{2}{6} = \frac{1}{3} \qquad \text{and} \qquad \frac{4}{6} = \frac{2}{3}$$

Actually, there are many fractions that name the same point as $\frac{1}{3}$. If we were to divide the segment between 0 and 1 into 12 equal parts, 4 of these 12 equal parts $\left(\frac{4}{12}\right)$ would be the same as $\frac{2}{6}$ or $\frac{1}{3}$; that is,

$$\frac{4}{12} = \frac{2}{6} = \frac{1}{3}$$

Even though these three fractions look different, each names the same point on the number line, as shown in Figure 4. All three fractions have the same value because they all represent the same number.

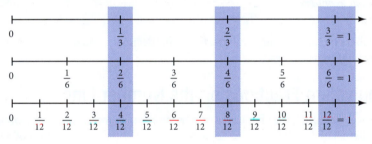

FIGURE 4

> **DEFINITION** equivalent fractions
>
> Fractions that represent the same relationship between the numerator and denominator are said to be *equivalent*. Equivalent fractions may look different, but they must have the same value.

It is apparent that every fraction has many different representations, each of which is equivalent to the original fraction. The next two properties give us a way of changing the terms of a fraction without changing its value.

> **PROPERTY** Multiplication of Equivalent Fractions
>
> Multiplying the numerator and denominator of a fraction by the same nonzero number never changes the value of the fraction.

> **PROPERTY** Division of Equivalent Fractions
>
> Dividing the numerator and denominator of a fraction by the same nonzero number never changes the value of the fraction.

EXAMPLE 2 Write $\frac{3}{4}$ as an equivalent fraction with denominator 20.

Solution The denominator of the original fraction is 4. The fraction we are trying to find must have a denominator of 20. We know that if we multiply 4 by 5, we get 20. Property 1 indicates that we are free to multiply the denominator by 5 as long as we do the same to the numerator.

$$\frac{3}{4} = \frac{3 \cdot 5}{4 \cdot 5} = \frac{15}{20}$$

The fraction $\frac{15}{20}$ is equivalent to the fraction $\frac{3}{4}$.

2. Write $\frac{5}{8}$ as an equivalent fraction with denominator 48.

C Absolute Values

Representing numbers on the number line lets us give each number two important properties: a direction from zero and a distance from zero. The direction from zero is represented by the sign in front of the number. (A number without a sign is understood to be positive.) The distance from zero is called the absolute value of the number, as the following definition indicates:

> **DEFINITION** absolute value
>
> The *absolute value* of a real number is its distance from zero on the number line. If x represents a real number, then the absolute value of x is written $|x|$.

This definition of absolute value is geometric in form because if defines absolute value in terms of the number line. Here is an alternative definition of absolute value that is algebraic in form because it involves only symbols.

> **ALTERNATIVE DEFINITION** absolute value
>
> If x represents a real number, then the *absolute value* of x is written $|x|$, and is given by
>
> $$|x| = \begin{cases} x & \text{if } x \geq 0 \\ -x & \text{if } x < 0 \end{cases}$$

Note It is important to recognize that if x is a real number, $-x$ is not necessarily negative. For example, if x is 5, then $-x$ is -5. On the other hand, if x were -5, then $-x$ would be $-(-5)$, which is 5.

If the original number is positive or 0, then its absolute value is the number itself. If the number is negative, its absolute value is its opposite, a term we will cover later in this section. For now, know that the opposite of a negative number must be positive.

EXAMPLE 3 Write each expression without absolute value symbols.

a. $|5|$　　**b.** $|-5|$　　**c.** $\left| -\frac{1}{2} \right|$

Solution

a. $|5| = 5$　　The number 5 is 5 units from zero.

b. $|-5| = 5$　　The number -5 is 5 units from zero.

c. $\left| -\frac{1}{2} \right| = \frac{1}{2}$　　The number $-\frac{1}{2}$ is $\frac{1}{2}$ unit from zero.

3. Write each expression without absolute value symbols.

a. $|7|$

b. $|-7|$

c. $\left| -\frac{3}{4} \right|$

The absolute value of a number is never negative. It is the distance the number is from zero without regard to which direction it is from zero. When working with the absolute value of sums and differences, we must simplify the expression inside the absolute value symbols first and then find the absolute value of the simplified expression.

Answers

2. $\frac{30}{48}$

3. a. 7　**b.** 7　**c.** $\frac{3}{4}$

4. Simplify each expression.
  a. $|7 - 2|$
  b. $|2 \cdot 3^2 + 5 \cdot 2^2|$
  c. $|10 - 4| - |11 - 9|$

5. Give the opposite of each number.
  a. 8
  b. -5
  c. $-\dfrac{2}{3}$
  d. -4.2

EXAMPLE 4 Simplify each expression.
a. $|8 - 3|$ **b.** $|3 \cdot 2^3 + 2 \cdot 3^2|$ **c.** $|9 - 2| - |8 - 6|$

Solution

a. $|8 - 3| = |5| = 5$

b. $|3 \cdot 2^3 + 2 \cdot 3^2| = |3 \cdot 8 + 2 \cdot 9| = |24 + 18| = |42| = 42$

c. $|9 - 2| - |8 - 6| = |7| - |2| = 7 - 2 = 5$

D Opposites

Another important concept associated with numbers on the number line is that of opposites. Here is the definition:

> **DEFINITION** opposites
>
> Any two real non-zero numbers the same distance from zero, but in opposite directions from zero on the number line, are called **opposites**, or *additive inverses*.

EXAMPLE 5 Give the opposite of each number.

a. 5   **b.** -3 **c.** $\dfrac{1}{4}$ **d.** -2.3

Solution

	Number	Opposite	
a.	5	-5	5 and -5 are opposites.
b.	-3	3	-3 and 3 are opposites.
c.	$\dfrac{1}{4}$	$-\dfrac{1}{4}$	$\dfrac{1}{4}$ and $-\dfrac{1}{4}$ are opposites.
d.	-2.3	2.3	-2.3 and 2.3 are opposites.

Each negative number is the opposite of some positive number, and each positive number is the opposite of some negative number. In symbols, if a represents a positive number, then

$$-(-a) = a$$

The negative sign in front of a number can be read in a few different ways. It can be read as "negative" or "the opposite of." We say -4 is the opposite of 4, or negative 4. The one we use will depend on the situation. For instance, the expression $-(-3)$ is best read "the opposite of negative 3." Because the opposite of -3 is 3, we have $-(-3) = 3$. In general, if a is any positive real number, then

$$-(-a) = a \quad \text{(The opposite of a negative is a positive)}$$

Opposites always have the same absolute value. And, when you add any two opposites, the result is always zero.

$$a + (-a) = 0$$

E Prime Numbers and Factoring

The following diagram shows the relationship between multiplication and factoring:

$$\text{Multiplication}$$
$$\text{Factors} \rightarrow 3 \cdot 4 = 12 \leftarrow \text{Product}$$
$$\text{Factoring}$$

When we read the problem from left to right, we say the product of 3 and 4 is 12. We can also say we multiply 3 and 4 to get 12. When we read the problem in the other direction, from right to left, we say we have *factored* 12 into 3 times 4, or 3 and 4 are *factors* of 12.

Answers

4. a. 5 **b.** 38 **c.** 4

5. a. -8 **b.** 5 **c.** $\dfrac{2}{3}$ **d.** 4.2

The number 12 can be factored still further:

$$12 = 4 \cdot 3$$
$$= 2 \cdot 2 \cdot 3$$
$$= 2^2 \cdot 3$$

The numbers 2 and 3 are called *prime* factors of 12 because neither can be factored any further.

> ### DEFINITION factor
>
> If a and b represent integers, then a is said to be a *factor* (or divisor) of b if a divides b evenly — that is, if a divides b with no remainder.

> ### DEFINITION prime, composite
>
> A *prime* number is any positive integer larger than 1 whose only positive factors (divisors) are itself and 1. An integer greater than 1 that is not prime is said to be *composite*.

Note Recall from your previous math classes that integers include whole numbers (0, 1, 2...) and their opposites (−1, −2, −3...).

Here is a list of the first few prime numbers:

Prime numbers $= \{2, 3, 5, 7, 11, 13, 17, 19, 23, 29, 31, 37, 41, \ldots\}$

When a number is not prime, we can factor it into the product of prime numbers. To factor a number into the product of primes, we simply factor it until it cannot be factored further.

EXAMPLE 6 Factor 525 into the product of primes.

Solution Because 525 ends in 5, it is divisible by 5.

$$525 = 5 \cdot 105$$
$$= 5 \cdot 5 \cdot 21$$
$$= 5 \cdot 5 \cdot 3 \cdot 7$$
$$= 3 \cdot 5^2 \cdot 7$$

6. Factor 420 into the product of primes.

EXAMPLE 7 Reduce $\dfrac{210}{231}$ to lowest terms.

Solution First we factor 210 and 231 into the product of prime factors. Then we reduce to lowest terms by dividing the numerator and denominator by any factors they have in common.

$$\frac{210}{231} = \frac{2 \cdot 3 \cdot 5 \cdot 7}{3 \cdot 7 \cdot 11} \qquad \text{Factor the numerator and denominator completely.}$$
$$= \frac{2 \cdot 3 \cdot 5 \cdot 7}{3 \cdot 7 \cdot 11} \qquad \text{Divide the numerator and denominator by } 3 \cdot 7.$$
$$= \frac{2 \cdot 5}{11}$$
$$= \frac{10}{11}$$

7. Reduce $\dfrac{154}{1,155}$ to lowest terms.

F Reciprocals

Before we go further with our study of the number line, we need to review multiplication with fractions. Recall that for the fraction $\frac{a}{b}$, a is called the numerator and b is called the denominator. To multiply two fractions, we simply multiply numerators and multiply denominators.

Answers

6. $2^2 \cdot 3 \cdot 5 \cdot 7$

7. $\dfrac{2}{15}$

8. Multiply.

a. $\dfrac{2}{3} \cdot \dfrac{7}{9}$

b. $4 \cdot \dfrac{1}{7}$

c. $\left(\dfrac{11}{12}\right)^2$

> **Note** In past math classes, you may have written fractions like $\dfrac{8}{5}$ (improper fractions) as mixed numbers, such as $1\dfrac{3}{5}$. In algebra, it is usually better to leave them as improper fractions.

9. Give the reciprocal of each number.

a. 6

b. 3

c. $\dfrac{1}{2}$

d. $\dfrac{2}{3}$

e. $\dfrac{1}{b}$

> **Note** The vertical line labeled h in the triangle is its height, or altitude. It extends from the top of the triangle down to the base, meeting the base at an angle of 90°. The altitude of a triangle is always perpendicular to the base. The small square shown where the altitude meets the base is used to indicate that the angle formed is 90°.

EXAMPLE 8   Multiply.

a. $\dfrac{3}{5} \cdot \dfrac{7}{8} = \dfrac{3 \cdot 7}{5 \cdot 8} = \dfrac{21}{40}$

b. $8 \cdot \dfrac{1}{5} = \dfrac{8}{1} \cdot \dfrac{1}{5} = \dfrac{8 \cdot 1}{1 \cdot 5} = \dfrac{8}{5}$

c. $\left(\dfrac{2}{3}\right)^4 = \dfrac{2}{3} \cdot \dfrac{2}{3} \cdot \dfrac{2}{3} \cdot \dfrac{2}{3} = \dfrac{16}{81}$

Multiplication of fractions is useful in understanding the concept of the reciprocal of a number. Here is a formal definition:

> **DEFINITION** reciprocals
>
> Any two real numbers whose product is 1 are called **reciprocals**, or *multiplicative inverses*.

EXAMPLE 9   Give the reciprocal of each number.

a. 5   b. 2   c. $\dfrac{1}{3}$ d. $\dfrac{3}{4}$ e. x

Solution

	Number	Reciprocal	
a.	5	$\dfrac{1}{5}$	Because $5\left(\dfrac{1}{5}\right) = \dfrac{5}{1}\left(\dfrac{1}{5}\right) = \dfrac{5}{5} = 1$
b.	2	$\dfrac{1}{2}$	Because $2\left(\dfrac{1}{2}\right) = \dfrac{2}{1}\left(\dfrac{1}{2}\right) = \dfrac{2}{2} = 1$
c.	$\dfrac{1}{3}$	3	Because $\dfrac{1}{3}(3) = \dfrac{1}{3}\left(\dfrac{3}{1}\right) = \dfrac{3}{3} = 1$
d.	$\dfrac{3}{4}$	$\dfrac{4}{3}$	Because $\dfrac{3}{4}\left(\dfrac{4}{3}\right) = \dfrac{12}{12} = 1$
e.	x	$\dfrac{1}{x}$	Because $x\left(\dfrac{1}{x}\right) = \dfrac{x}{1}\left(\dfrac{1}{x}\right) = \dfrac{x}{x} = 1, x \neq 0$

Although we will not develop multiplication with negative numbers until later in the chapter, you should know that the reciprocal of a negative number is also a negative number. For example, the reciprocal of -4 is $-\dfrac{1}{4}$.

G Formulas for Area and Perimeter

> **FACTS FROM GEOMETRY** Formulas for Area and Perimeter
>
> A square, rectangle, and triangle are shown in the following figures. Note that we have labeled the dimensions of each with variables. The formulas for the perimeter and area of each object are given in terms of its dimensions.
>
>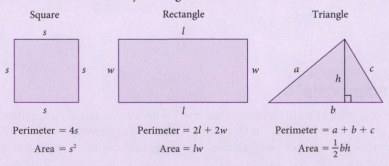

The formula for perimeter gives us the distance around the outside of the object along its sides, whereas the formula for area gives us a measure of the amount of surface the object has.

Answers

8. a. $\dfrac{14}{27}$ b. $\dfrac{4}{7}$ c. $\dfrac{121}{144}$

9. a. $\dfrac{1}{6}$ b. $\dfrac{1}{3}$ c. 2   d. $\dfrac{3}{2}$ e. b

EXAMPLE 10   Find the perimeter and area of each figure.

a.

5 ft

b.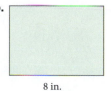

6 in.

8 in.

c.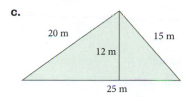

20 m   12 m   15 m

25 m

10. Find the perimeter and area of the figures in Example 10, with the following changes:

  a. $s = 4$ feet

  b. $l = 7$ inches
    $w = 5$ inches

  c. $a = 40$ meters
    $b = 50$ meters (base)
    $c = 30$ meters
    $h = 24$ meters

Solution   We use the preceding formulas to find the perimeter and the area. In each case, the units for perimeter are linear units, whereas the units for area are square units.

a. Perimeter $= 4s = 4 \cdot 5$ feet $= 20$ feet

   Area $= s^2 = (5 \text{ feet})^2 = 25$ square feet

b. Perimeter $= 2l + 2w = 2(8 \text{ inches}) + 2(6 \text{ inches}) = 28$ inches

   Area $= lw = (8 \text{ inches})(6 \text{ inches}) = 48$ square inches

c. Perimeter $= a + b + c = (20 \text{ meters}) + (25 \text{ meters}) + (15 \text{ meters})$

             $= 60$ meters

   Area $= \dfrac{1}{2}bh = \dfrac{1}{2}(25 \text{ meters})(12 \text{ meters}) = 150$ square meters

---

GETTING READY FOR CLASS

Each section of the book will end with some problems and questions like the ones below. They are for you to answer after you have read through the section but before you go to class. All of them require that you give written responses in complete sentences. Writing about mathematics is a valuable exercise. As with all problems in this course, approach these writing exercises with a positive point of view. You will get better at giving written responses to math questions as the course progresses. Even if you never feel comfortable writing about mathematics, just attempting the process will increase your understanding and ability in this course.

After reading through the preceding section, respond in your own words and in complete sentences.

A. What is a real number?

B. What is an equivalent fraction?

C. How do you find the reciprocal of a number?

D. Explain how you find the perimeter and the area of a rectangle.

Answers

10. a. Perimeter $= 16$ feet
     Area $= 16$ square feet
  b. Perimeter $= 24$ inches
     Area $= 35$ square inches
  c. Perimeter $= 120$ meters
     Area $= 600$ square meters

Vocabulary Review

Choose the correct words to fill in the blanks below.

prime	origin	numerator	real number line
factor	composite	denominator	equivalent
absolute value	opposites	reciprocals	coordinate

1. The _____ is used to represent positive and negative numbers in algebra.

2. Numbers to the left of the _____ on the real number line have negative values.

3. A _____ is a number associated with a point on the real number line.

4. The top number of a fraction is called the _____, and the bottom number is called the _____.

5. Fractions that represent the same number are said to be _____.

6. The notation $|x|$ is the definition for the _____ of a real number.

7. Any two real numbers the same distance from zero, but in opposite directions, are called _____.

8. If a divides b with no remainder, then a is a _____ of b.

9. A positive integer larger than 1 is _____ if its only positive factors are 1 and the integer itself.

10. A _____ is an integer greater than 1 that is not prime.

11. Any two real numbers whose product is 1 are called _____.

Problems

A Draw a number line that extends from -5 to 5. Label the points with the following coordinates.

1. 5

2. -2

3. -4

4. -3

5. 1.5

6. -1.5

7. $\dfrac{9}{4}$

8. $\dfrac{8}{3}$

B Write each of the following fractions as an equivalent fraction with denominator 24.

9. $\dfrac{3}{4}$

10. $\dfrac{5}{6}$

11. $\dfrac{1}{2}$

12. $\dfrac{1}{8}$

13. $\dfrac{5}{8}$

14. $\dfrac{7}{12}$

Write each fraction as an equivalent fraction with denominator 60.

15. $\dfrac{3}{5}$

16. $\dfrac{5}{12}$

17. $\dfrac{11}{30}$

18. $\dfrac{9}{10}$

19. $-\dfrac{5}{6}$

20. $-\dfrac{7}{12}$

C D F For each of the following numbers, give the opposite, the reciprocal, and the absolute value. (Assume all variables are nonzero.)

21. 10

22. 8

23. $\dfrac{3}{4}$

24. $\dfrac{5}{7}$

25. $\dfrac{11}{2}$

26. $\dfrac{16}{3}$

27. -3

28. -5

29. $-\dfrac{2}{5}$

30. $-\dfrac{3}{8}$

31. x

32. a

Write each of the following without absolute value symbols.

33. $|-2|$

34. $|-7|$

35. $\left|-\dfrac{3}{4}\right|$

36. $\left|\dfrac{5}{6}\right|$

37. $|\pi|$

38. $|-\sqrt{2}|$

39. $-|4|$

40. $-|5|$

41. $-|-2|$

42. $-|-10|$

43. $-\left|-\dfrac{3}{4}\right|$

44. $-\left|\dfrac{7}{8}\right|$

Place one of the symbols $<$ or $>$ between each of the following to make the resulting statement true.

45. $-5 \quad -3$

46. $-8 \quad -1$

47. $-3 \quad -7$

48. $-6 \quad 5$

49. $|-4| \quad -|-4|$

50. $3 \quad -|-3|$

51. $7 \quad -|-7|$

52. $-7 \quad |-7|$

53. $-\dfrac{3}{4} \quad -\dfrac{1}{4}$

54. $-\dfrac{2}{3} \quad -\dfrac{1}{3}$

55. $-\dfrac{3}{2} \quad -\dfrac{3}{4}$

56. $-\dfrac{8}{3} \quad -\dfrac{17}{3}$

C Simplify each expression.

57. $|8 - 2|$

58. $|6 - 1|$

59. $|5 \cdot 2^3 - 2 \cdot 3^2|$

60. $|2 \cdot 10^2 + 3 \cdot 10|$

61. $|7 - 2| - |4 - 2|$

62. $|10 - 3| - |4 - 1|$

63. $10 - |7 - 2(5 - 3)|$

64. $12 - |9 - 3(7 - 5)|$

65. $15 - |8 - 2(3 \cdot 4 - 9)| - 10$

66. $25 - |9 - 3(4 \cdot 5 - 18)| - 20$

E Factor each number into the product of prime factors.

67. 266   **68.** 385   **69.** 111   **70.** 735   **71.** 369   **72.** 1,155

Reduce each fraction to lowest terms.

73. $\dfrac{165}{385}$

74. $\dfrac{550}{735}$

75. $\dfrac{385}{735}$

76. $\dfrac{266}{285}$

77. $\dfrac{111}{185}$

78. $\dfrac{279}{310}$

79. $\dfrac{75}{135}$

80. $\dfrac{38}{30}$

81. $\dfrac{6}{8}$

82. $\dfrac{10}{25}$

83. $\dfrac{200}{5}$

84. $\dfrac{240}{6}$

G Find the perimeter and area of each figure.

85.
1 in.
1 in.

86.
15 mm
15 mm

87.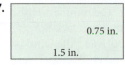
0.75 in.
1.5 in.

88.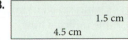
1.5 cm
4.5 cm

89.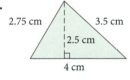
2.75 cm   3.5 cm
2.5 cm
4 cm

90.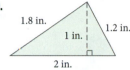
1.8 in.   1.2 in.
1 in.
2 in.

Applying the Concepts

91. Football Yardage A football team gains 6 yards on one play and then loses 8 yards on the next play. To what number on the number line does a loss of 8 yards correspond? The total yards gained or lost on the two plays corresponds to what negative number?

92. Checking Account Balance Nancy has a balance of $20 in her checking account. If she writes a check for $30, what negative number can be used to represent the new balance in her checking account?

Temperature In the United States, temperature is measured on the Fahrenheit temperature scale. On this scale, water boils at 212 degrees and freezes at 32 degrees. To denote a temperature of 32 degrees on the Fahrenheit scale, we write 32°F, which is read "32 degrees Fahrenheit."

Use this information for Problems 93 and 94.

93. Temperature and Altitude Marilyn is flying from Seattle to San Francisco on a Boeing 737 jet. When the plane reaches an altitude of 35,000 feet, the temperature outside the plane is 64 degrees below zero Fahrenheit. Represent the temperature with a negative number. If the temperature outside the plane gets warmer by 10 degrees, what will the new temperature be?

94. Temperature Change At 10:00 in the morning in White Bear Lake, Minnesota, John notices the temperature outside is 10 degrees below zero Fahrenheit. Write the temperature as a negative number. An hour later it has warmed up by 6 degrees. What is the temperature at 11:00 that morning?

Wind Chill The table to the right shows wind temperatures. The top column gives the air temperature, and the first row is wind speed in miles per hour. The numbers within the table indicate how cold the weather will feel. For example, if the thermometer reads 30°F and the wind is blowing at 15 miles per hour, the wind chill temperature is 9°F. Use Table 1 to answer Problems 95 and 96.

95. Reading Tables Find the wind chill temperature if the thermometer reads 20°F and the wind is blowing at 25 miles per hour.

96. Reading Tables Which will feel colder: a day with an air temperature of 10°F with a 25-mile-per-hour wind, or a day with an air temperature of 25°F and a 10-mile-per-hour wind?

Wind Chill Temperatures

Air Temperature (°F)	Wind Speed (mph)				
	10	15	20	25	30
30°	16°	9°	4°	1°	−2°
25°	10°	2°	−3°	−7°	−10°
20°	3°	−5°	−10°	−15°	−18°
15°	−3°	−11°	−17°	−22°	−25°
10°	−9°	−18°	−24°	−29°	−33°
5°	−15°	−25°	−31°	−36°	−41°
0°	−22°	−31°	−39°	−44°	−49°
−5°	−27°	−38°	−46°	−51°	−56°

97. Scuba Diving Steve is scuba diving near his home in Maui. At one point he is 100 feet below the surface. Assuming the surface is 0, represent this number with a negative number. If he descends another 5 feet, what negative number will represent his new position?

100 ft

98. eBook Memory The chart shows the memory capacity for various eBook readers. Write a mathematical statement using one of the symbols $=, <, >$ to compare the following capacities:

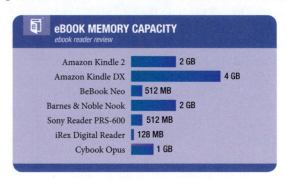

eBOOK MEMORY CAPACITY
ebook reader review

Amazon Kindle 2	2 GB
Amazon Kindle DX	4 GB
BeBook Neo	512 MB
Barnes & Noble Nook	2 GB
Sony Reader PRS-600	512 MB
iRex Digital Reader	128 MB
Cybook Opus	1 GB

a. Amazon Kindle 2 to the Amazon Kindle DX

b. Barnes and Noble Nook to the Cybook Opus

c. BeBook Neo to the Sony Reader PRS-600

99. Geometry Find the area and perimeter of an $8\frac{1}{2}$-by-11-inch piece of notebook paper.

100. Geometry Find the area and perimeter of an $8\frac{1}{2}$-by-$5\frac{1}{2}$-inch piece of paper.

Calories and Exercise The table here gives the amount of energy expended per hour for various activities for a person weighing 120, 150, or 180 pounds. Use the table to answer questions 101–104.

101. Suppose you weigh 120 pounds. How many calories will you burn if you play handball for 2 hours and then ride your bicycle for an hour?

102. How many calories are burned by a person weighing 150 pounds who jogs for $\frac{1}{2}$ hour and then goes bicycling for 2 hours?

103. Two people go skiing. One weighs 180 pounds and the other weighs 120 pounds. If they ski for 3 hours, how many more calories are burned by the person weighing 180 pounds?

104. Two people spend 3 hours bowling. If one weighs 120 pounds and the other weighs 150 pounds, how many more calories are burned during the evening by the person weighing 150 pounds?

Energy Expended from Exercising

Activity	Calories per Hour		
	120 lb	150 lb	180 lb
Bicycling	299	374	449
Bowling	212	265	318
Handball	544	680	816
Horseback trotting	278	347	416
Jazzercise	272	340	408
Jogging	544	680	816
Skiing (downhill)	435	544	653

Distance and Absolute Value These problems are based on absolute value $|x|$ representing the distance from 0 on a number line.

105. Find two numbers with an absolute value of 20.

106. Find two numbers x such that $|x| = 13$.

107. Is it possible for x to satisfy $|x| = -5$? Explain.

108. Is it possible for x to satisfy $|x| < 0$? Explain.

The exit for the Giffith Observatory is exit #141 from Interstate 5. The exit numbers increase by 1 for every mile as you travel north. Use this information for the next four problems.

109. If you are 16 miles north of the exit for the observatory, at what exit number are you located?

110. If you are 17 miles south of the exit for the observatory, at what exit number are you located?

111. If you are 20 miles away from the exit for the observatory, at what exit numbers could you be located? (Hint: there are two answers.)

112. If you are 35 miles away from the exit for the observatory, at what exit numbers could you be located? (Hint: there are two answers.)

113. iPad Apps The chart shows the top categories for iPad only applications. Use the chart to answer the following questions.

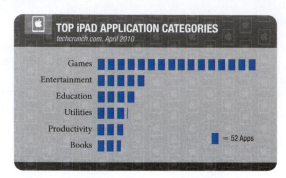

TOP iPAD APPLICATION CATEGORIES
techcrunch.com, April 2010

Games
Entertainment
Education
Utilities
Productivity
Books

■ = 52 Apps

a. How many entertainment applications are there?

b. True or false? There are more than 150 book applications.

c. True or false? There are less than 880 game applications.

114. Improving Your Quantitative Literacy Quantitative literacy is a subject discussed by many people involved in teaching mathematics. The person they are concerned with when they discuss it is you. We are going to work at improving your quantitative literacy, but before we do that we should answer the question, What is quantitative literacy? Lynn Arthur Steen, a noted mathematics educator, has stated that quantitative literacy is "the capacity to deal effectively with the quantitative aspects of life."

a. Give a definition for the word *quantitative*.

b. Give a definition for the word *literacy*.

c. Are there situations that occur in your life that you find distasteful, or that you try to avoid, because they involve numbers and mathematics? If so, list some of them here. (For example, some people find the process of buying a car particularly difficult because they feel that the numbers and details of the financing are beyond them.)

Find the Mistake

Each sentence below contains a mistake. Circle the mistake and write the correct word on the line provided.

1. On a real number line, the points to the right of the origin are labeled with negative numbers. _____

2. The top number in a fraction is called the denominator. _____

3. In the problem $4 \times 6 = 24$, the 24 is called a factor. _____

4. To find the perimeter of a triangle, we multiply $\frac{1}{2}$ by the length of the base times the height. _____

Navigation Skills: Prepare, Study, Achieve

At the end of the first exercise set of each chapter, we provide an important discussion of study skills that will help you succeed in this course. Pay special attention to these skills. Ponder each one and apply it to your life. Your success is in your hands.

Studying is the key to success in this course. However, many students have never learned effective skills for studying. Study skills include but are not limited to the following:

- Work done on problems for practice and homework
- Amount of time spent studying
- Time of day and location for studying
- Management of distractions during study sessions
- Material chosen to review
- Order and process of review

Let's begin our discussion with the topic of homework. From the first day of class, we recommend you spend two hours on homework for every hour you are scheduled to attend class. Any less may drastically impact your success in this course. To help visualize this commitment, map out a weekly schedule that includes your classes, work shifts, extracurriculars, and any additional obligations. Fill in the hours you intend to devote to completing assignments and studying for this class. Post this schedule at home and keep a copy with your study materials to remind you of your commitment to success.

0.2 Properties of Real Numbers

KEY WORDS

commutative property

associative property

grouping

distributive property

similar terms

LCD

algebraic expressions

identity

inverse

Each year, the coastal waters of Qingdao, China are infiltrated with a vibrant green algae bloom. The algae thrive in high temperatures and in polluted waters caused by runoff from agriculture land and fish farms. The bloom sucks oxygen out of the water, threatening the local marine life. As it washes onshore and dries in the hot sun, it gives off a noxious odor that smells like rotten eggs. Soldiers and other volunteers work tirelessly to clear the algae from more than 150 square miles of water and beaches before the bloom drastically damages the local ecosystem.

Consider the following figure as a general representation of the algae-affected coastline in China. The area of the large rectangle shown here can be found in two ways: We can multiply its width a by its length $b + c$, or we can find the areas of the two smaller rectangles and add those areas to find the total area.

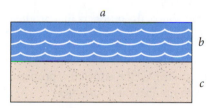

Area of large rectangle: $a(b + c)$
Sum of the areas of two smaller rectangles: $ab + ac$

Because the area of the large rectangle is the sum of the areas of the two smaller rectangles, we can write

$$a(b + c) = ab + ac$$

This expression is called the ***distributive property***. It is one of the properties we will be discussing in this section. Before we arrive at the distributive property, we need to review some basic properties.

A Properties of Real Numbers

We know that adding 3 and 7 gives the same answer as adding 7 and 3. The order of two numbers in an addition problem can be changed without changing the result. This fact about numbers and addition is called the *commutative property of addition*.

For all the properties listed in this section, a, b, and c represent real numbers.

PROPERTY	Commutative Property of Addition
In symbols:	$a + b = b + a$
In words:	The *order* of the numbers in a sum does not affect the result.

> **PROPERTY** Commutative Property of Multiplication
>
> *In symbols:* $a \cdot b = b \cdot a$
> *In words:*   The *order* of the numbers in a product does not affect the result.

1. Give an example of the commutative property of addition using 2 and *y*.

EXAMPLE 1

a. The statement $3 + 7 = 7 + 3$ is an example of the commutative property of addition.

b. The statement $3 \cdot x = x \cdot 3$ is an example of the commutative property of multiplication.

The other two basic operations (subtraction and division) are not commutative. If we change the order in which we are subtracting or dividing two numbers, we change the result.

Another property of numbers you have used many times has to do with grouping. When adding $3 + 5 + 7$, we can add the 3 and 5 first and then the 7, or we can add the 5 and 7 first and then the 3. Mathematically, it looks like this: $(3 + 5) + 7 = 3 + (5 + 7)$. Operations that behave in this manner are called *associative* operations.

> **PROPERTY** Associative Property of Addition
>
> *In symbols:* $a + (b + c) = (a + b) + c$
> *In words:*   The *grouping* of the numbers in a sum does not affect the result.

> **PROPERTY** Associative Property of Multiplication
>
> *In symbols:* $a(bc) = (ab)c$
> *In words:*   The *grouping* of the numbers in a product does not affect the result.

The following examples illustrate how the associative properties can be used to simplify expressions that involve both numbers and variables.

2. Simplify.

  a. $3 + (7 + x)$

  b. $3(4x)$

  c. $\frac{1}{3}(3x)$

  d. $3\left(\frac{1}{3}x\right)$

  e. $8\left(\frac{1}{2}x\right)$

EXAMPLE 2   Simplify by using the associative property.

a. $2 + (3 + y) = (2 + 3) + y$ Associative property

            $= 5 + y$ Add.

b. $5(4x) = (5 \cdot 4)x$ Associative property

         $= 20x$ Multiply.

c. $\frac{1}{4}(4a) = \left(\frac{1}{4} \cdot 4\right)a$ Associative property

         $= 1a$ Multiply.

         $= a$

d. $2\left(\frac{1}{2}x\right) = \left(2 \cdot \frac{1}{2}\right)x$ Associative property

         $= 1x$ Multiply.

         $= x$

e. $6\left(\frac{1}{3}x\right) = \left(6 \cdot \frac{1}{3}\right)x$ Associative property

         $= 2x$ Multiply.

Answers

1. $2 + y = y + 2$

2. a. $10 + x$ **b.** $12x$ **c.** x

    d. x **e.** $4x$

Our next property involves both addition and multiplication. It is called the *distributive property* and is stated as follows.

PROPERTY	**Distributive Property**
In symbols:	$a(b + c) = ab + ac$
In words:	Multiplication *distributes* over addition.

Note Although the properties we are listing are stated for only two or three real numbers, they hold for as many numbers as needed. For example, the distributive property holds for expressions like $3(x + y + z + 2)$. That is,

$3(x + y + z + 2) = 3x + 3y + 3z + 6$

You will see as we progress through the book that the distributive property is used very frequently in algebra. To see that the distributive property works, compare the following:

$$3(4 + 5) \qquad\qquad 3(4) + 3(5)$$
$$= 3(9) \qquad\qquad = 12 + 15$$
$$= 27 \qquad\qquad = 27$$

In both cases the result is 27. Because the results are the same, the original two expressions must be equal, or $3(4 + 5) = 3(4) + 3(5)$.

EXAMPLE 3 Apply the distributive property to each expression and then simplify the result.

a. $5(4x + 3) = 5(4x) + 5(3)$ Distributive property

$\qquad\qquad\quad = 20x + 15$ Multiply.

b. $6(3x + 2y) = 6(3x) + 6(2y)$ Distributive property

$\qquad\qquad\quad\ = 18x + 12y$ Multiply.

c. $\dfrac{1}{2}(3x + 6) = \dfrac{1}{2}(3x) + \dfrac{1}{2}(6)$ Distributive property

$\qquad\qquad\quad = \dfrac{3}{2}x + 3$ Multiply.

d. $2(3y + 4) + 2 = 2(3y) + 2(4) + 2$ Distributive property

$\qquad\qquad\qquad = 6y + 8 + 2$ Multiply.

$\qquad\qquad\qquad = 6y + 10$ Add.

3. Simplify

  a. $4(2x - 5)$

  b. $3(7x - 6y)$

  c. $\dfrac{1}{3}(3x + 6)$

  d. $5(7a - 3) + 2$

We can combine our knowledge of the distributive property with multiplication of fractions to manipulate expressions involving fractions. Here are some examples that show how we do this:

EXAMPLE 4 Apply the distributive property to each expression and then simplify the result.

a. $a\left(1 + \dfrac{1}{a}\right) = a \cdot 1 + a \cdot \dfrac{1}{a} = a + 1$

b. $3\left(\dfrac{1}{3}x + 5\right) = 3 \cdot \dfrac{1}{3}x + 3 \cdot 5 = x + 15$

c. $6\left(\dfrac{1}{3}x + \dfrac{1}{2}y\right) = 6 \cdot \dfrac{1}{3}x + 6 \cdot \dfrac{1}{2}y = 2x + 3y$

4. Apply the distributive property to each expression and then simplify the result.

  a. $x\left(3 + \dfrac{1}{x}\right)$

  b. $5\left(\dfrac{2}{5}x - 2\right)$

  c. $8\left(\dfrac{3}{2}x - \dfrac{3}{4}y\right)$

Combining Similar Terms

The distributive property can also be used to combine similar terms. (For now, a term is the product of a number with one or more variables.) Similar terms are terms with the same variable part. The terms $3x$ and $5x$ are similar, as are $2y$, $7y$, and $-3y$, because the variable parts are the same.

Answers

3. a. $8x - 20$ **b.** $21x - 18y$

    c. $x + 2$ **d.** $35a - 13$

4. a. $3x + 1$ **b.** $2x - 10$

    c. $12x - 6y$

5. Use the distributive property to combine similar terms.
  a. $5x - 2x$
  b. $6y + y$

EXAMPLE 5   Use the distributive property to combine similar terms.

a. $3x + 5x = (3 + 5)x$ Distributive property

             $= 8x$ Add.

b. $3y + y = (3 + 1)y$ Distributive property

             $= 4y$ Add.

B   Review of Addition with Fractions

To add fractions, each fraction must have the same denominator.

> **DEFINITION**   least common denominator
>
> The *least common denominator* (LCD) for a set of denominators is the smallest number divisible by *all* the denominators.

The first step in adding fractions is to find the least common denominator for all the denominators. We then rewrite each fraction (if necessary) as an equivalent fraction with the common denominator. Finally, we add the numerators and reduce to lowest terms if necessary.

6. Add $\dfrac{5}{18} + \dfrac{3}{14}$.

EXAMPLE 6   Add $\dfrac{5}{12} + \dfrac{7}{18}$.

Solution   The least common denominator for the denominators 12 and 18 must be the smallest number divisible by both 12 and 18. We can factor 12 and 18 completely and then build the LCD from these factors.

$$\left.\begin{array}{l} 12 = 2 \cdot 2 \cdot 3 \\ 18 = 2 \cdot 3 \cdot 3 \end{array}\right\} \quad \text{LCD} = 2 \cdot 2 \cdot 3 \cdot 3 = 36$$

12 divides the LCD

18 divides the LCD

Next, we rewrite our original fractions as equivalent fractions with denominators of 36. To do so, we multiply each original fraction by an appropriate form of the number 1.

$$\frac{5}{12} + \frac{7}{18} = \frac{5}{12} \cdot \frac{3}{3} + \frac{7}{18} \cdot \frac{2}{2}$$

$$= \frac{15}{36} + \frac{14}{36}$$

Finally, we add numerators and place the result over the common denominator, 36. (Remember, this is an application of the distributive property.)

$$\frac{15}{36} + \frac{14}{36} = \frac{15 + 14}{36} = \frac{29}{36}$$

C   Simplifying Expressions

We can use the commutative, associative, and distributive properties together to simplify algebraic expressions.

7. Simplify $3x + 9 + 4x - 7$.

EXAMPLE 7   Simplify $7x + 4 + 6x + 3$.

Solution   We begin by applying the commutative and associative properties to group similar terms.

$$7x + 4 + 6x + 3 = (7x + 6x) + (4 + 3) \qquad \text{Commutative and associative properties}$$

$$= (7 + 6)x + (4 + 3) \qquad \text{Distributive property}$$

$$= 13x + 7 \qquad \text{Add.}$$

Answers

5. a. $3x$ **b.** $7y$

6. $\dfrac{31}{63}$

7. $7x + 2$

EXAMPLE 8 Simplify $4 + 3(2y + 5) + 8y$.

Solution Because order of operations indicates that we are to multiply before adding, we must distribute the 3 across $2y + 5$ first.

$$
\begin{aligned}
4 + 3(2y + 5) + 8y &= 4 + 6y + 15 + 8y && \text{Distributive property} \\
&= (6y + 8y) + (4 + 15) && \text{Commutative property} \\
&&& \text{Associative property} \\
&= (6 + 8)y + (4 + 15) && \text{Distributive property} \\
&= 14y + 19 && \text{Add.}
\end{aligned}
$$

The numbers 0 and 1 are called the *additive identity* and *multiplicative identity*, respectively. Adding 0 to a number does not change the identity of that number. Likewise, multiplying a number by 1 does not alter the identity of that number. We see below that 0 is to addition what 1 is to multiplication.

> **PROPERTY** Additive Identity Property
>
> There exists a unique number 0 such that
> *In symbols:* $a + 0 = a$ and $0 + a = a$

> **PROPERTY** Multiplicative Identity Property
>
> There exists a unique number 1 such that
> *In symbols:* $a(1) = a$ and $1(a) = a$

> **PROPERTY** Additive Inverse Property
>
> For each real number a, there exists a unique real number $-a$ such that
> *In symbols:* $a + (-a) = 0$
> *In words:*   Opposites add to 0.

> **PROPERTY** Multiplicative Inverse Property
>
> For every real number a, except 0, there exists a unique real number $\frac{1}{a}$ such that
> *In symbols:* $a\left(\dfrac{1}{a}\right) = 1$
> *In words:*   Reciprocals multiply to 1.

EXAMPLE 9

a. $7(1) = 7$ Multiplicative identity property

b. $4 + (-4) = 0$ Additive inverse property

c. $6\left(\dfrac{1}{6}\right) = 1$ Multiplicative inverse property

d. $(5 + 0) + 2 = 5 + 2$ Additive identity property

> **GETTING READY FOR CLASS**
>
> *After reading through the preceding section, respond in your own words and in complete sentences.*
>
> **A.** Describe the commutative property of multiplication.
>
> **B.** Explain why subtraction and division are not commutative operations.
>
> **C.** What is the distributive property?
>
> **D.** Explain the additive inverse property.

---

8. Simplify $5 + 4(3y + 2) + 5y$.

9. State the property that justifies the given statement.

  a. $4\left(\dfrac{1}{4}\right) = 1$

  b. $8 + (-8) = 0$

  c. $(x + 2) + 3 = x + (2 + 3)$

  d. $5(1) = 5$

Answers

8. $17y + 13$

9. a. Multiplicative inverse property

  b. Additive inverse property

  c. Associative property of addition

  d. Multiplicative identity property

Vocabulary Review

Match the following definitions on the left with their property names on the right.

1. $a + b = b + a$

a. Multiplicative inverse property

2. $a \times b = b \times a$

b. Commutative property of multiplication

3. $a + (b + c) = (a + b) + c$

c. Distributive property

4. $a(bc) = (ab)c$

d. Associative property of addition

5. $a(b + c) = ab + ac$

e. Commutative property of addition

6. $a + 0 = a$ and $0 + a = a$

f. Additive identity property

7. $a(1) = a$ and $1(a) = a$

g. Associative property of multiplication

8. $a + (-a) = 0$

h. Additive inverse property

9. $a(\frac{1}{a}) = 1$

i. Multiplicative identity property

Problems

A Identify the property of real numbers that justifies each of the following.

1. $3 + 2 = 2 + 3$

2. $3(ab) = (3a)b$

3. $5x = x5$

4. $2 + 0 = 2$

5. $4 + (-4) = 0$

6. $1(6) = 6$

7. $x + (y + 2) = (y + 2) + x$

8. $(a + 3) + 4 = a + (3 + 4)$

9. $4(5 \cdot 7) = 5(4 \cdot 7)$

10. $6(xy) = (xy)6$

11. $4 + (x + y) = (4 + y) + x$

12. $(r + 7) + s = (r + s) + 7$

13. $3(4x + 2) = 12x + 6$

14. $5\left(\frac{1}{5}\right) = 1$

15. $-\frac{2}{3}\left(-\frac{3}{2}\right) = 1$

16. $3(2x - 6) = 6x - 18$

SCAN TO ACCESS

Use the associative property to rewrite each of the following expressions and then simplify the result.

17. $4 + (2 + x)$

18. $6 + (5 + 3x)$

19. $(a + 3) + 5$

20. $(4a + 5) + 7$

21. $5(3y)$

22. $7(4y)$

23. $\frac{1}{3}(3x)$

24. $\frac{1}{5}(5x)$

25. $4\left(\frac{1}{4}a\right)$

26. $7\left(\frac{1}{7}a\right)$

27. $\frac{2}{3}\left(\frac{3}{2}x\right)$

28. $\frac{4}{3}\left(\frac{3}{4}x\right)$

Apply the distributive property to each expression. Simplify when possible.

29. $3(x + 6)$

30. $5(x + 9)$

31. $2(6x + 4)$

32. $3(7x + 8)$

33. $5(3a + 2b)$

34. $7(2a + 3b)$

35. $\frac{1}{3}(4x + 6)$

36. $\frac{1}{2}(3x + 8)$

37. $\frac{1}{5}(10 + 5y)$

38. $\frac{1}{6}(12 + 6y)$

39. $(5t + 1)8$

40. $(3t + 2)5$

Apply the distributive property, then simplify if possible.

41. $3(3x + y - 2z)$

42. $2(2x - y + z)$

43. $10(0.3x + 0.7y)$

44. $10(0.2x + 0.5y)$

45. $100(0.06x + 0.07y)$

46. $100(0.09x + 0.08y)$

47. $3\left(x + \frac{1}{3}\right)$

48. $5\left(x - \frac{1}{5}\right)$

49. $2\left(x - \frac{1}{2}\right)$

50. $7\left(x + \frac{1}{7}\right)$

51. $x\left(1 + \frac{2}{x}\right)$

52. $x\left(1 - \frac{1}{x}\right)$

53. $a\left(1 - \frac{3}{a}\right)$

54. $a\left(1 + \frac{1}{a}\right)$

55. $8\left(\frac{1}{8}x + 3\right)$

56. $4\left(\frac{1}{4}x - 9\right)$

57. $6\left(\frac{1}{2}x - \frac{1}{3}y\right)$

58. $12\left(\frac{1}{4}x - \frac{1}{6}y\right)$

59. $12\left(\frac{1}{4}x + \frac{2}{3}y\right)$

60. $12\left(\frac{2}{3}x - \frac{1}{4}y\right)$

61. $20\left(\frac{2}{5}x + \frac{1}{4}y\right)$

62. $15\left(\frac{2}{3}x + \frac{2}{5}y\right)$

Apply the distributive property to each expression. Simplify when possible.

63. $3(5x + 2) + 4$

64. $4(3x + 2) + 5$

65. $4(2y + 6) + 8$

66. $6(2y + 3) + 2$

67. $5(1 + 3t) + 4$

68. $2(1 + 5t) + 6$

69. $3 + (2 + 7x)4$

70. $4 + (1 + 3x)5$

B Add the following fractions.

71. $\frac{2}{5} + \frac{1}{15}$

72. $\frac{5}{8} + \frac{1}{4}$

73. $\frac{17}{30} + \frac{11}{42}$

74. $\frac{19}{42} + \frac{13}{70}$

75. $\frac{9}{48} + \frac{3}{54}$

76. $\frac{6}{28} + \frac{5}{42}$

77. $\frac{25}{84} + \frac{41}{90}$

78. $\frac{23}{70} + \frac{29}{84}$

Simplify each expression.

79. $\left(\frac{3}{14} + \frac{7}{30}\right)$

80. $\left(\frac{3}{10} + \frac{11}{42}\right)$

81. $32\left(\frac{3}{4}\right) - 16\left(\frac{3}{4}\right)^2$

82. $32\left(\frac{3}{2}\right) - 16\left(\frac{3}{2}\right)^2$

C Use the commutative, associative, and distributive properties to simplify the following.

83. $5a + 7 + 8a + a$

84. $3y + y + 5 + 2y + 1$

85. $2(5x + 1) + 2x$

86. $7 + 2(4y + 2)$

87. $3 + 4(5a + 3) + 4a$

88. $5x + 2(3x + 8) + 4$

89. $5x + 3(x + 2) + 7$

90. $2a + 4(2a + 6) + 3$

91. $5(x + 2y) + 4(3x + y)$

92. $3x + 4(2x + 3y) + 7y$

93. $5b + 3(4b + a) + 6a$

94. $4 + 3(2x + 3y) + 6(x + 4)$

Applying the Concepts

95. Rhind Papyrus In approximately 1650 BC, a mathematical document was written in ancient Egypt called the Rhind Papyrus. An "exercise" in this document asked the reader to find "a quantity such that when it is added to one fourth of itself results in 15." Verify this quantity must be 12.

96. Clock Arithmetic In a normal clock with 12 hours on its face, 12 is the additive identity because adding 12 hours to any time on the clock will not change the hands of the clock. Also, if we think of the hour hand of a clock, the problem $10 + 4$ can be taken to mean: The hour hand is pointing at 10; if we add 4 more hours, it will be pointing at what number? Reasoning this way, we see that in clock arithmetic $10 + 4 = 2$ and $9 + 6 = 3$.

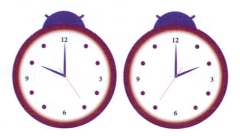

Find the following in clock arithmetic.

a. $10 + 5$ b. $10 + 6$

c. $10 + 1$ d. $10 + 12$

e. $x + 12$

Griffith Park Fires Three major fires have occurred since Griffith Park was formed. The acres burned are given in the table:

Year	Acres
1933	47
1961	814
2007	817

Source: www.nfpa.org

The total acreage of Griffith park is 4,217 acres.

97. How many more acres burned in 1961 as compared with 1933?

98. The acres burned in 2007 was 18 more than 17 times the acres burned in 1933. If the total acres burned in 2007 was 817, solve the equation and check your answer from the table.

99. What percent of the total acres was burned in 1933?

100. What percent of the total acres was burned in 1961?

101. Millionaires Use the chart to answer the questions below.

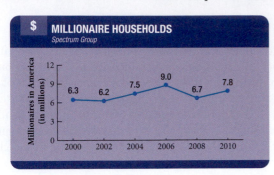

a. Between what years was the decrease in the number of millionaires the largest?

b. Is the rate of increase from 2000 to 2002 positive or negative?

c. What year was the number of millionaires the lowest?

102. Drinking and Driving The chart shows that if the risk of getting in an accident with no alcohol in your system is 1, then the risk of an accident with a blood-alcohol level of 0.24 is about 147 times as high.

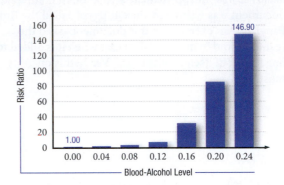

a. A person driving with a blood-alcohol level of 0.20 is how many times more likely to get in an accident than if she was driving with a blood-alcohol level of 0?

b. If the probability of getting in an accident while driving for an hour on surface streets in a certain city is 0.02%, what is the probability of getting in an accident in the same circumstances with a blood-alcohol level of 0.20?

One Step Further

Simplify.

103. $x(x + 8)$

104. $2a(a + 4)$

105. $x(y - 4)$

106. $3y(y - 2)$

Determine whether each statement is true (without exception) or false.

107. $-x$ is always a negative number.

108. $|x| = x$

109. $|-x| = x$

110. $|-x| = |x|$

111. $\dfrac{1}{-x} = \dfrac{-1}{x}$

112. $-(-x) = x$

113. $|x| > 0$ for all values of x.

114. $|x| \geq 0$ for all values of x.

115. $|x| \leq 0$ is never a true statement.

116. $|x| < 0$ is never a true statement.

Find the Mistake

Each sentence below contains a mistake. Circle the mistake and write the correct word on the line provided.

1. The value of the numbers in a sum does not affect the result. _____

2. The statement $2 \cdot 4 = 4 \cdot 2$ is an example of the associative property of multiplication. _____

3. The least common denominator for the fractions $\frac{2}{3}$ and $\frac{1}{7}$ is 14. _____

4. Reciprocals multiply to zero. _____

Landmark Review: Checking Your Progress

This feature is intended as a review of key skills from the preceding sections in the chapter. Each review will give you the opportunity to utilize important concepts and check your progress as you practice problems of different types. Take note of any problems in this review that you find difficult. There is no better time than now to revisit and master those difficult concepts in preparation for any upcoming exams or subsequent chapters.

Write each of the following fractions as an equivalent fraction with a denominator of 15.

1. $\frac{3}{5}$　　　　　**2.** $\frac{2}{3}$　　　　　**3.** 2　　　　　**4.** 1

For each of the following numbers give the opposite, the reciprocal, and the absolute value.

5. 5　　　　　　　　　　**6.** $-\frac{1}{2}$

Simplify each expression.

7. $|2 \times 3^2 - 4 \times 2^3|$　　　　　**8.** $14 - |4(3 - 4) - 13|$

Reduce each fraction to lowest terms.

9. $\frac{50}{75}$　　　　**10.** $\frac{52}{65}$　　　　**11.** $\frac{48}{80}$　　　　**12.** $\frac{42}{56}$

Add the following fractions.

13. $\frac{4}{10} + \frac{1}{15}$　　　**14.** $\frac{1}{12} + \frac{5}{16}$　　　**15.** $\frac{2}{5} + \frac{3}{8}$　　　**16.** $\frac{1}{2} + \frac{3}{4}$

Simplify the following expressions.

17. $4x + 3 + 2x + 5 + x$　　　　　**18.** $2y + 3(y - 3) + 8$

19. $2a + 3(4a + 3) + 5a + 4$　　　　　**20.** $5 + 2(3x + 2y) + 6(x + y)$

A Add and subtract real numbers.

B Multiply and divide real numbers.

C Divide fractions.

D Find the value of an expression.

sum

difference

product

quotient

undefined

algebraic expression

value

0.3 Arithmetic with Real Numbers

©iStockphoto.com/aldegonde

In 2010, 60 refurbished pianos were placed on street corners and in public parks of New York City. The idea is the brainchild of artist Luke Jerram, who has been installing street pianos in various cities since 2007. With the support of the non-profit arts group Sing for Hope, the city's population, musically trained or not, was encouraged to sit and play to passersby. Local artists had also volunteered their time to paint and decorate the pianos, which sat on the city streets for two weeks. The following table represents the number of pianos located in five boroughs of the city.

Borough	Number of Pianos
Manhattan	53
Brooklyn	11
Queens	6
The Bronx	4
Staten Island	4

Using the table, how many total pianos were in Queens and Brooklyn combined? How many more pianos were in Manhattan than The Bronx? In this section, we will review the rules for arithmetic with real numbers, which will allow us to answer questions similar to the preceding ones.

A Adding Real Numbers

This section reviews the rules for arithmetic with real numbers and the justification for those rules. We can justify the rules for addition of real numbers geometrically by use of the real number line. Consider the *sum* of -5 and 3:

$$-5 + 3$$

We can interpret this expression as meaning "start at the origin and move 5 units in the negative direction and then 3 units in the positive direction." With the aid of a number line, we can visualize the process.

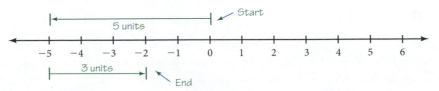

Because the process ends at -2, we say the sum of -5 and 3 is -2.

$$-5 + 3 = -2$$

We can use the real number line in this way to add any combination of positive and negative numbers.

The sum of -4 and -2, $-4 + (-2)$, can be interpreted as starting at the origin, moving 4 units in the negative direction, and then 2 more units in the negative direction.

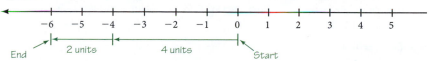

Because the process ends at -6, we say the sum of -4 and -2 is -6.

$$-4 + (-2) = -6$$

We can eliminate actually drawing a number line by simply visualizing it mentally. The following example gives the results of all possible sums of positive and negative 5 and 7.

EXAMPLE 1   Add all combinations of positive and negative 5 and 7.

Solution

$$5 + 7 = 12$$
$$-5 + 7 = 2$$
$$5 + (-7) = -2$$
$$-5 + (-7) = -12$$

Looking closely at the relationships in Example 1 (and trying other similar examples, if necessary), we can arrive at the following rule for adding two real numbers:

> **HOW TO**   Add Two Real Numbers
>
> With the *same* sign:
> **Step 1:** Add their absolute values.
> **Step 2:** Attach their common sign. If both numbers are positive, their sum is positive; if both numbers are negative, their sum is negative.
>
> With *opposite* signs:
> **Step 1:** Subtract the smaller absolute value from the larger.
> **Step 2:** Attach the sign of the number whose absolute value is larger.

Subtracting Real Numbers

To have as few rules as possible, we will not attempt to list new rules for the *difference* of two real numbers. We will instead define it in terms of addition and apply the rule for addition.

> **DEFINITION**   difference
>
> If a and b are any two real numbers, then the **difference** of a and b, written $a - b$, is given by
>
> $$a - b \qquad = \qquad a + (-b)$$
> To subtract b,   add the opposite of b.

We define the process of subtracting b from a as being equivalent to adding the opposite of b to a. In short, we say, "Subtraction is addition of the opposite."

2. Subtract.
   a. $3 - 9$
   b. $-9 - 6$
   c. $12 - (-5)$
   d. $-14 - (-3)$

EXAMPLE 2   Subtract.

a. $5 - 3 = 5 + (-3)$ *Subtracting 3 is equivalent to adding -3.*
$\quad\quad = 2$

b. $-7 - 6 = -7 + (-6)$ *Subtracting 6 is equivalent to adding -6.*
$\quad\quad\quad = -13$

c. $9 - (-2) = 9 + 2$ *Subtracting -2 is equivalent to adding 2.*
$\quad\quad\quad = 11$

d. $-6 - (-5) = -6 + 5$ *Subtracting -5 is equivalent to adding 5.*
$\quad\quad\quad\quad = -1$

3. Subtract -8 from -21.

EXAMPLE 3   Subtract -3 from -9.

Solution   Because subtraction is not commutative, we must be sure to write the numbers in the correct order. Because we are subtracting -3, the problem looks like this when translated into symbols:

$$-9 - (-3) = -9 + 3 \quad\quad \text{\textit{Change to addition of the opposite.}}$$
$$= -6 \quad\quad\quad \text{\textit{Add.}}$$

4. Add -6 to the difference of -2 and 8.

EXAMPLE 4   Add -4 to the difference of -2 and 5.

Solution   The difference of -2 and 5 is written $-2 - 5$. Adding -4 to that difference gives us

$$(-2 - 5) + (-4) = -7 + (-4) \quad\quad \text{\textit{Simplify inside parentheses.}}$$
$$= -11 \quad\quad\quad \text{\textit{Add.}}$$

B Multiplying Real Numbers

Multiplication with whole numbers is simply a shorthand way of writing repeated addition. For example, $3(-2)$ can be evaluated as follows:

$$3(-2) = -2 + (-2) + (-2) = -6$$

We can evaluate the product $-3(2)$ in a similar manner if we first apply the commutative property of multiplication.

$$-3(2) = 2(-3) = -3 + (-3)$$

From these results, it seems reasonable to say that the product of a positive and a negative is a negative number.

The last case we must consider is the product of two negative numbers, such as $-3(-2)$. To evaluate this product we will look at the expression $-3[2 + (-2)]$ in two different ways. First, since $2 + (-2) = 0$, we know the expression $-3[2 + (-2)]$ is equal to 0. On the other hand, we can apply the distributive property to get

$$-3[2 + (-2)] = -3(2) + (-3)(-2) = -6 + ?$$

Because we know the expression is equal to 0, it must be true that our ? is 6; 6 is the only number we can add to -6 to get 0. Therefore, we have

$$-3(-2) = 6$$

Here is a summary of what we have so far:

Original Numbers Have		Answer Is
The same sign	$3(2) = 6$	Positive
Different signs	$3(-2) = -6$	Negative
Different signs	$-3(2) = -6$	Negative
The same sign	$-3(-2) = 6$	Positive

Answers
2. a. -6 **b.** -15 **c.** 17 **d.** -11
3. -13
4. -16

> **HOW TO** Multiply Two Real Numbers
>
> **Step 1:** Multiply their absolute values.
>
> **Step 2:** If the two numbers have the *same* sign, the product is positive. If the two numbers have *opposite* signs, the product is negative.

EXAMPLE 5 Multiply all combinations of positive and negative 7 and 3.

Solution

$$7(3) = 21$$
$$7(-3) = -21$$
$$-7(3) = -21$$
$$-7(-3) = 21$$

Dividing Real Numbers

> **DEFINITION** quotient
>
> If a and b are any two real numbers, where $b \neq 0$, then the *quotient* of a and b, written $\frac{a}{b}$, is given by
>
> $$\frac{a}{b} = a \cdot \frac{1}{b}$$

Dividing a by b is equivalent to multiplying a by the reciprocal of b. In short, we say, "Division is multiplication by the reciprocal."

Because division is defined in terms of multiplication, the same rules hold for assigning the correct sign to a quotient as held for assigning the correct sign to a product. That is, the quotient of two numbers with like signs is positive, while the quotient of two numbers with unlike signs is negative.

EXAMPLE 6 Divide.

a. $\frac{6}{3} = 6 \cdot \left(\frac{1}{3}\right) = 2$

b. $\frac{6}{-3} = 6 \cdot \left(-\frac{1}{3}\right) = -2$

c. $\frac{-6}{3} = -6 \cdot \left(\frac{1}{3}\right) = -2$

d. $\frac{-6}{-3} = -6 \cdot \left(-\frac{1}{3}\right) = 2$

The second step in the preceding examples is written only to show that each quotient can be written as a product. It is not actually necessary to show this step when working problems.

In the examples that follow, we find a combination of operations. In each case, we use the rule for order of operations.

EXAMPLE 7 Simplify each expression as much as possible.

a. $(-2 - 3)(5 - 9) = (-5)(-4)$ Simplify inside parentheses.

$= 20$ Multiply.

b. $2 - 5(7 - 4) - 6 = 2 - 5(3) - 6$ Simplify inside parentheses.

$= 2 - 15 - 6$ Then, multiply.

$= -19$ Finally, subtract, left to right.

---

5. Multiply all combinations of positive and negative 9 and 4.

a. $9(4)$

b. $(9)(-4)$

c. $-9(4)$

d. $-9(-4)$

6. Divide.

a. $\frac{12}{4}$ **b.** $\frac{12}{-4}$

c. $\frac{-12}{4}$ **d.** $\frac{-12}{-4}$

> *Note* The problems in Example 6 indicate that if a and b are positive real numbers then $\frac{-a}{b} = \frac{a}{-b} = -\frac{a}{b}$ and $\frac{-a}{-b} = \frac{a}{b}$

7. Simplify each expression as much as possible.

a. $(-3 - 7)(6 - 2)$

b. $5 - 4(9 - 3) + 12$

c. $2(6 - 11)^2 - 4(3 - 9)^2$

d. $-6\left(\frac{2}{3}x\right)$

e. $5\left(\frac{x}{5} - 8\right)$

Answers

5. a. 36   **b.** -36 **c.** -36 **d.** 36

6. a. 3   **b.** -3 **c.** -3 **d.** 3

7. a. -40 **b.** -7 **c.** -94

    d. $-4x$ **e.** $x - 40$

c. $2(4-7)^3 + 3(-2-3)^2 = 2(-3)^3 + 3(-5)^2$ Simplify inside parentheses.

$$= 2(-27) + 3(25)$$ Evaluate numbers with exponents.

$$= -54 + 75$$ Multiply.

$$= 21$$ Add.

d. $6\left(\dfrac{t}{3}\right) = 6\left(\dfrac{1}{3}t\right)$ Dividing by 3 is the same as multiplying by $\frac{1}{3}$.

$$= \left(6 \cdot \dfrac{1}{3}\right)t$$ Associative Property

$$= 2t$$ Multiply.

e. $3\left(\dfrac{t}{3} - 2\right) = 3 \cdot \dfrac{t}{3} - 3 \cdot 2$ Distributive Property

$$= t - 6$$ Multiply.

Our next examples involve more complicated fractions. The fraction bar works like parentheses to separate the numerator from the denominator. Although we don't write expressions this way, here is one way to think of the fraction bar:

$$\frac{-8-8}{-5-3} = (-8-8) \div (-5-3)$$

As you can see, if we apply the rule for order of operations to the expression on the right, we would work inside each set of parentheses first, then divide. Applying this to the expression on the left, we work on the numerator and denominator separately, then we divide, or reduce the resulting fraction to the lowest terms.

8. Simplify.

a. $\dfrac{-14-7}{3-10}$

b. $\dfrac{-3(2)-6(4)}{3(-1)+5}$

c. $\dfrac{4^2+3^2}{2^2-3^2}$

EXAMPLE 8

a. $\dfrac{-8-8}{-5-3} = \dfrac{-16}{-8}$ Simplify numerator and denominator separately.

$$= 2$$ Divide.

b. $\dfrac{-5(-4)+2(-3)}{2(-1)-5} = \dfrac{20-6}{-2-5}$ Simplify numerator and denominator separately.

$$= \dfrac{14}{-7}$$

$$= -2$$ Divide.

c. $\dfrac{2^3+3^3}{2^2-3^2} = \dfrac{8+27}{4-9}$ Simplify numerator and denominator separately.

$$= \dfrac{35}{-5}$$

$$= -7$$ Divide.

Remember, since subtraction is defined in terms of addition, we can restate the distributive property in terms of subtraction. That is, if a, b, and c are real numbers, then $a(b-c) = ab - ac$.

9. Simplify $6(3y-2) + 2y$.

EXAMPLE 9   Simplify $3(2y-1) + y$.

Solution   We begin by multiplying the 3 and $2y-1$. Then, we combine similar terms.

$$3(2y-1)+y = 6y-3+y$$ Distributive property

$$= 7y-3$$ Combine similar terms.

Answers

8. a. 3   **b.** -15 **c.** -5

9. $20y - 12$

EXAMPLE 10 Simplify $8 - 3(4x - 2) + 5x$.

Solution First, we distribute the -3 across the $4x - 2$.

$$8 - 3(4x - 2) + 5x = 8 - 12x + 6 + 5x$$
$$= -7x + 14$$

EXAMPLE 11 Simplify $5(2a + 3) - (6a - 4)$.

Solution We begin by applying the distributive property to remove the parentheses. The expression $-(6a - 4)$ can be thought of as $-1(6a - 4)$. Thinking of it in this way allows us to apply the distributive property.

$$-1(6a - 4) = -1(6a) - (-1)(4) = -6a + 4$$

Here is the complete problem:

$$5(2a + 3) - (6a - 4) = 10a + 15 - 6a + 4 \qquad \text{Distributive property}$$
$$= 4a + 19 \qquad \text{Combine similar terms.}$$

C Dividing Fractions

In the first section of this book, we reviewed multiplication with fractions. Now we'll review division with fractions.

EXAMPLE 12 Divide and reduce to lowest terms.

a.
$$\frac{3}{4} \div \frac{6}{11} = \frac{3}{4} \cdot \frac{11}{6} \qquad \text{Definition of division}$$
$$= \frac{33}{24} \qquad \text{Multiply numerators; multiply denominators.}$$
$$= \frac{11}{8} \qquad \text{Divide numerator and denominator by 3.}$$

b.
$$10 \div \frac{5}{6} = \frac{10}{1} \cdot \frac{6}{5} \qquad \text{Definition of division}$$
$$= \frac{60}{5} \qquad \text{Multiply numerators; multiply denominators.}$$
$$= 12 \qquad \text{Divide.}$$

c.
$$-\frac{3}{8} \div 6 = -\frac{3}{8} \cdot \frac{1}{6} \qquad \text{Definition of division}$$
$$= -\frac{3}{48} \qquad \text{Multiply numerators; multiply denominators.}$$
$$= -\frac{1}{16} \qquad \text{Divide numerator and denominator by 3.}$$

Division With the Number 0

For every division problem an associated multiplication problem involving the same numbers exists. For example, the following two problems say the same thing about the numbers 2, 3, and 6:

$$\begin{array}{cc} \textit{Division} & \textit{Multiplication} \\ \dfrac{6}{3} = 2 & 6 = 2(3) \end{array}$$

We can use this relationship between division and multiplication to clarify division involving the number 0.

First, dividing 0 by a number other than 0 is allowed and always results in 0. To see this, consider dividing 0 by 5. We know the answer is 0 because of the relationship between multiplication and division. This is how we write it:

$$\frac{0}{5} = 0 \qquad \text{because} \qquad 0 = 0(5)$$

10. Simplify $9 - 2(5x + 3) + 4x$.

11. Simplify $3(6a + 4) - (9a + 7)$.

12. Divide and reduce to lowest terms.

a. $\dfrac{2}{3} \div \dfrac{7}{9}$

b. $16 \div \dfrac{4}{5}$

c. $\dfrac{6}{7} \div 9$

Answers

10. $-6x + 3$

11. $9a + 5$

12. a. $\dfrac{6}{7}$ **b.** 20   **c.** $\dfrac{2}{21}$

Note You may wonder why we can't divide 0 by 0, since $0 = (0)(0)$. Dividing zero by itself is a more complicated concept, one analyzed in calculus, not algebra. So for our purposes, we will still consider $\frac{0}{0}$ as undefined.

On the other hand, dividing a nonzero number by 0 is not allowed in the real numbers. Suppose we were attempting to divide 5 by 0. We don't know whether there is an answer to this problem, but if there is, let's say the answer is a number that we can represent with the letter n. If 5 divided by 0 is a number n, then

$$\frac{5}{0} = n \quad \text{and} \quad 5 = n(0)$$

But this is impossible, because no matter what number n is, when we multiply it by 0 the answer must be 0. It can never be 5. In algebra, we say expressions like $\frac{5}{0}$ are *undefined*, because there is no answer to them. That is, division by 0 is not allowed in the real numbers.

D Finding the Value of an Algebraic Expression

Recall that an algebraic expression is a combination of numbers, variables, and operation symbols. Each of the following is an algebraic expression:

$$7a \qquad x^2 - y^2 \qquad 2(3t - 4) \qquad \frac{2x - 5}{6}$$

An expression such as $2(3t - 4)$ will take on different values depending on what number we substitute for t. For example, if we substitute -8 for t then the expression $2(3t - 4)$ becomes $2[3(-8) - 4]$ which simplifies to -56. If we apply the distributive property to $2(3t - 4)$ we have

$$2(3t - 4) = 6t - 8$$

Substituting -8 for t in the simplified expression gives us $6(-8) - 8 = -56$, which is the same result we obtained previously. As you would expect, substituting the same number into an expression, and any simplified form of that expression, will yield the same result.

13. Evaluate $(x + 3)^2$, $x^2 + 9$, and $x^2 + 6x + 9$ when x is -3.

EXAMPLE 13   Evaluate the expressions $(a + 4)^2$, $a^2 + 16$, and $a^2 + 8a + 16$ when a is -2, 0, and 3.

Solution   Organizing our work with a table, we have

a	$(a + 4)^2$	$a^2 + 16$	$a^2 + 8a + 16$
-2	$(-2 + 4)^2 = 4$	$(-2)^2 + 16 = 20$	$(-2)^2 + 8(-2) + 16 = 4$
0	$(0 + 4)^2 = 16$	$0^2 + 16 = 16$	$0^2 + 8(0) + 16 = 16$
3	$(3 + 4)^2 = 49$	$3^2 + 16 = 25$	$3^2 + 8(3) + 16 = 49$

When we study polynomials later in the book, you will see that the expressions $(a + 4)^2$ and $a^2 + 8a + 16$ are equivalent, and that neither one is equivalent to $a^2 + 16$.

GETTING READY FOR CLASS

After reading through the preceding section, respond in your own words and in complete sentences.

A. Explain the steps you use to add two real numbers.

B. Use symbols to define the word *difference*.

C. Why do we define division as multiplication by the reciprocal?

D. Find the value of the expression $3(4t - 2)$ if $t = 2$.

Answers

13. 0, 18, 0

Vocabulary Review

Choose the correct words to fill in the blanks below.

undefined	positive	larger	algebraic expression
negative	quotient	difference	

1. The sum of two positive numbers will be a _____ number.

2. When adding two numbers with opposite signs, the sum will have the sign of the number whose absolute value is

   _____.

3. To subtract b from a, add the opposite of b to a to get the _____.

4. When multiplying two numbers with different signs, the answer is _____.

5. The _____ of a and b is given by $\frac{a}{b} = a \cdot \frac{1}{b}$.

6. In algebra, expressions like $\frac{x}{0}$ are _____.

7. An _____ is a combination of numbers, variables, and operation symbols.

Problems

A Find each of the following sums.

1. $6 + (-2)$ 2. $11 - 5$ 3. $-6 + 2$ 4. $-11 + 5$

Find each of the following differences.

5. $-7 - 3$ 6. $-6 - 9$ 7. $-7 - (-3)$ 8. $-6 - (-9)$

9. $\frac{3}{4} - \left(-\frac{5}{6}\right)$ 10. $\frac{2}{3} - \left(-\frac{7}{5}\right)$ 11. $\frac{11}{42} - \frac{17}{30}$ 12. $\frac{13}{70} - \frac{19}{42}$

13. Subtract 5 from -3.

14. Subtract -3 from 5.

15. Find the difference of -4 and 8.

16. Find the difference of 8 and -4.

17. Subtract $4x$ from $-3x$.

18. Subtract $-5x$ from $7x$.

19. What number do you subtract from 5 to get -8?

20. What number do you subtract from -3 to get 9?

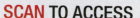

SCAN TO ACCESS

21. Add -7 to the difference of 2 and 9.

22. Add -3 to the difference of 9 and 2.

23. Subtract $3a$ from the sum of $8a$ and a.

24. Subtract $-3a$ from the sum of $3a$ and $5a$.

B Find the following products.

25. $3(-5)$

26. $-3(5)$

27. $-3(-5)$

28. $4(-6)$

29. $2(-3)(4)$

30. $-2(3)(-4)$

31. $-2(5x)$

32. $-5(4x)$

33. $-\dfrac{1}{3}(-3x)$

34. $-\dfrac{1}{6}(-6x)$

35. $-\dfrac{2}{3}\left(-\dfrac{3}{2}y\right)$

36. $-\dfrac{2}{5}\left(-\dfrac{5}{2}y\right)$

37. $-2(4x - 3)$

38. $-2(-5t + 6)$

39. $-\dfrac{1}{2}(6a - 8)$

40. $-\dfrac{1}{3}(6a - 9)$

Simplify each expression as much as possible.

41. $1(-2) - 2(-16) + 1(9)$

42. $6(1) - 1(-5) + 1(2)$

43. $1(1) - 3(-2) + (-2)(-2)$

44. $-2(-14) + 3(-4) - 1(-10)$

45. $-4(0)(-2) - (-1)(1)(1) - 1(2)(3)$

46. $1(0)(1) + 3(1)(4) + (-2)(2)(-1)$

47. $1[0 - (-1)] - 3(2 - 4) + (-2)(-2 - 0)$

48. $-3(-1 - 1) + 4(-2 + 2) - 5[2 - (-2)]$

49. $3(-2)^2 + 2(-2) - 1$

50. $4(-1)^2 + 3(-1) - 2$

51. $2(-2)^3 - 3(-2)^2 + 4(-2) - 8$

52. $5 \cdot 2^3 - 3 \cdot 2^2 + 4 \cdot 2 - 5$

53. $\dfrac{0-4}{0-2}$

54. $\dfrac{0+6}{0-3}$

55. $\dfrac{-4-4}{-4-2}$

56. $\dfrac{6+6}{6-3}$

57. $\dfrac{-6+6}{-6-3}$

58. $\dfrac{4-4}{4-2}$

59. $\dfrac{-2-4}{2-2}$

60. $\dfrac{3+6}{3-3}$

61. $\dfrac{3-(-1)}{-3-3}$

62. $\dfrac{-1-3}{3-(-3)}$

63. $\dfrac{-3^2+9}{-4-4}$

64. $\dfrac{-2^2+4}{-5-5}$

Simplify each expression.

65. $3(5x+4)-x$

66. $4(7x+3)-x$

67. $6-7(3-m)$

68. $3-5(5-m)$

69. $7-2(3x-1)+4x$

70. $8-5(2x-3)+4x$

71. $5(3y+1)-(8y-5)$

72. $4(6y+3)-(6y-6)$

73. $4(2-6x)-(3-4x)$

74. $7(1-2x)-(4-10x)$

75. $10-4(2x+1)-(3x-4)$

76. $7-2(3x+5)-(2x-3)$

77. $3x-5(x-3)-2(1-3x)$

78. $4x-7(x-3)+2(4-5x)$

B C Use the definition of division to write each division problem as a multiplication problem, then simplify.

79. $\dfrac{4}{0}$

80. $\dfrac{-7}{0}$

81. $\dfrac{0}{-3}$

82. $\dfrac{0}{5}$

83. $-\dfrac{3}{4} \div \dfrac{9}{8}$

84. $-\dfrac{2}{3} \div \dfrac{4}{9}$

85. $-8 \div \left(-\dfrac{1}{4}\right)$

86. $-12 \div \left(-\dfrac{2}{3}\right)$

87. $-40 \div \left(-\dfrac{5}{8}\right)$

88. $-30 \div \left(-\dfrac{5}{6}\right)$

89. $\dfrac{4}{9} \div (-8)$

90. $\dfrac{3}{7} \div (-6)$

Simplify as much as possible.

91. $\dfrac{3(-1) - 4(-2)}{8 - 5}$

92. $\dfrac{6(-4) - 5(-2)}{7 - 6}$

93. $8 - (-6)\left[\dfrac{2(-3) - 5(4)}{-8(6) - 4}\right]$

94. $-9 - 5\left[\dfrac{11(-1) - 9}{4(-3) + 2(5)}\right]$

95. $6 - (-3)\left[\dfrac{2 - 4(3 - 8)}{1 - 5(1 - 3)}\right]$

96. $8 - (-7)\left[\dfrac{6 - 1(6 - 10)}{4 - 3(5 - 7)}\right]$

D Complete each of the following tables.

97.

a	b	Sum $a + b$	Difference $a - b$	Product ab	Quotient $\dfrac{a}{b}$
3	12				
−3	12				
3	−12				
−3	−12				

98.

a	b	Sum $a + b$	Difference $a - b$	Product ab	Quotient $\dfrac{a}{b}$
8	2				
−8	2				
8	−2				
−8	−2				

99.

x	$3(5x - 2)$	$15x - 6$	$15x - 2$
−2			
−1			
0			
1			
2			

100.

x	$(x + 1)^2$	$x^2 + 1$	$x^2 + 2x + 1$
−2			
−1			
0			
1			
2			

101. Find the value of $-\dfrac{b}{2a}$ when

   a. $a = 3, b = -6$

   b. $a = -2, b = 6$

   c. $a = -1, b = -2$

   d. $a = -0.1, b = 27$

102. Find the value of $b^2 - 4ac$ when

   a. $a = 3, b = -2,$ and $c = 4$

   b. $a = 1, b = -3,$ and $c = -28$

   c. $a = 1, b = -6,$ and $c = 9$

   d. $a = 0.1, b = -27,$ and $c = 1700$

Use a calculator to simplify each expression. If rounding is necessary, round your answers to the nearest ten thousandth (4 places past the decimal point). You will see these types of problems later in the book.

103. $\dfrac{1.380}{0.903}$

104. $\dfrac{1.0792}{0.6690}$

105. $\dfrac{1}{2}(-0.1587)$

106. $\dfrac{1}{2}(-0.7948)$

107. $\dfrac{1}{2}\left(\dfrac{1.2}{1.4}-1\right)$

108. $\dfrac{1}{2}\left(\dfrac{1.3}{1.1}-1\right)$

109. $\dfrac{(6.8)(3.9)}{7.8}$

110. $\dfrac{(2.4)(1.8)}{1.2}$

111. $\dfrac{0.0005(200)}{(0.25)^2}$

112. $\dfrac{0.0006(400)}{(0.25)^2}$

113. $-500 + 27(100) - 0.1(100)^2$

114. $-500 + 27(170) - 0.1(170)^2$

115. $-0.05(130)^2 + 9.5(130) - 200$

116. $-0.04(130)^2 + 8.5(130) - 210$

Applying the Concepts

117. Convenience Stores The chart shows the states with the most convenience stores. Given the area of the state, find the number of convenience stores per square mile.

CONVENIENCE STORES AROUND THE U.S.
U.S. Census Bureau, 2002

Texas	9,143
California	5,768
Florida	5,676
Ohio	3,537
Illinois	3,280
Michigan	3,402
Pennsylvania	2,996
Georgia	4,050
North Carolina	4,081

*Convenience Store Establishments per State

a. Georgia: 59,424 square miles

b. Texas: 268,580 square miles

c. Which state has a higher concentration of stores, Georgia or Texas?

118. Oceans and Mountains The deepest ocean depth is 35,840 feet, found in the Pacific Ocean's Mariana Trench. The tallest mountain is Mount Everest, with a height of 29,028 feet. What is the difference between the highest point on Earth and the lowest point on Earth?

119. Downhill Skiing In our number system, everything is in terms of powers of 10. With minutes and seconds, we think in terms of 60's. The format for the times in the chart is

minutes:seconds:hundreths

Find the difference between each of the following downhill ski times:

a. Lindsey Vonn and Julia Mancuso

b. Elizabeth Goergl and Andrea Fischbacher

c. Lindsey Vonn and Andrea Fischbacher

Municipally-Owned City Parks The Trust for Public Land lists the size of Griffith Park as 4,217 acres.

120. Griffith Park is 17 acres more than five times the size of Central Park in New York. What is the acreage of Central Park?

121. South Mountain Preserve in Phoenix, Arizona is the largest municipally–owned city park at 16,094 acres. How many times larger is this park than Griffith Park? (Round to the nearest tenth.)

122. The largest state park within a city is Chugache State Park in Anchorage, Alaska at 490,125 acres. How many times would Griffith Park fit inside Chugache State Park?

123. Griffith Park is 109 acres larger than four times the size of Golden Gate Park in San Francisco, California. What is the size of Golden Gate Park?

124. Lincoln Park in Chicago, Illinois is 1,216 acres, while Cullen Park in Houston, Texas is 9,270 acres. How many times larger than Lincoln Park is Cullen Park?

Find the Mistake

Each problem below contains a mistake. Circle the mistake and explain the correction in the space provided.

1. $-3(-5) = -15$ _____

2. $-6 + 7 = -1$ _____

3. $-4 + 2 = -6$ _____

4. $-8 \div 4 = 2$ _____

5. The value of the expression $2x + 5$ when $x = -3$ is 11. _____

6. The value of the expression $x^2 - 4x$ when $x = 2$ is 12. _____

0.4 Exponents and Scientific Notation

Image © sxc.hu, Des1gn, 2005

OBJECTIVES

A Simplify expressions using the properties of exponents.

B Convert numbers between scientific notation and expanded form.

C Multiply and divide expressions written in scientific notation.

KEY WORDS

exponent

product property for exponents

power property for exponents

distributive property for exponents

negative exponent property

quotient property for exponents

identity property for exponents

zero exponent property

scientific notation

order of operations

In 1856, the steamboat Arabia (similar to the one pictured above) got stuck in sunken woody debris in the Missouri River. One of the dead trees ripped a hole in the ship's hull, which subsequently filled with water and sunk the ship. Over the years, the course of the river has shifted a half mile away, leaving the ship buried far below a Kansas City cornfield for more than a century.

In 1987, the local Hawley family set out to uncover the ship. After three weeks of digging using heavy equipment, the Hawleys finally reached the hull of the ship. The nutrient-rich mud had preserved much of the hull's cargo in pristine condition. The family uncovered artifacts such as elegant china, fancy clothes made of silk and beaver furs, medicines, perfumes, tools, weapons, eyeglasses, even jars of preserved food that are still edible nearly 150 years later!

The hole dug to find the Arabia was as large as a football field and was 45 feet deep. To calculate the volume of dirt excavated in order to unearth the Arabia, we need to multiply the square footage of a football field by the depth of the hole. A football field is 57,600 square feet (ft^2), therefore the hole's volume is 2,592,000 cubic feet (ft^3)! The notation for square feet and cubic feet uses exponents, which we will discuss further now.

The following figure shows a square and a cube, each with a side of length 1.5 centimeters. To find the area of the square, we raise 1.5 to the second power: 1.5^2. To find the volume of the cube, we raise 1.5 to the third power: 1.5^3.

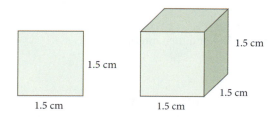

1.5 cm

1.5 cm

1.5 cm

1.5 cm

1.5 cm

1.5 cm

Because the area of the square is 1.5^2, we say second powers are *squares*; that is, x^2 is read "*x* squared." Likewise, because the volume of the cube is 1.5^3, we say third powers are *cubes*; that is, x^3 is read "*x* cubed." Exponents and the vocabulary associated with them are topics we will study in this section.

A Properties of Exponents

In this section, we will be concerned with the simplification of expressions that involve exponents. We begin by making some generalizations about exponents, based on specific examples.

Practice Problems

1. Write the product $x^3 \cdot x^7$ with a single exponent.

EXAMPLE 1   Write the product $x^3 \cdot x^4$ with a single exponent.

Solution

$$x^3 \cdot x^4 = (x \cdot x \cdot x)(x \cdot x \cdot x \cdot x)$$
$$= (x \cdot x \cdot x \cdot x \cdot x \cdot x \cdot x)$$
$$= x^7 \qquad \text{Note: } 3 + 4 = 7$$

We can generalize this result into the first property of exponents.

> **PROPERTY** Product Property for Exponents
>
> If a is a real number and r and s are integers, then
> $$a^r \cdot a^s = a^{r+s}$$

2. Write $(3^4)^5$ with a single exponent.

EXAMPLE 2   Write $(5^3)^2$ with a single exponent.

Solution

$$(5^3)^2 = 5^3 \cdot 5^3$$
$$= 5^6 \qquad \text{Note: } 3 \cdot 2 = 6$$

Generalizing this result, we have the second property of exponents.

> **PROPERTY** Power Property for Exponents
>
> If a is a real number and r and s are integers, then
> $$(a^r)^s = a^{r \cdot s}$$

A third property of exponents arises when we have the product of two or more numbers raised to an integer power.

3. Expand $(7x)^2$ and then multiply.

EXAMPLE 3   Expand $(3x)^4$ and then multiply.

Solution

$$(3x)^4 = (3x)(3x)(3x)(3x)$$
$$= (3 \cdot 3 \cdot 3 \cdot 3)(x \cdot x \cdot x \cdot x)$$
$$= 3^4 \cdot x^4 \qquad \text{Note: The exponent 4 distributes over the product } 3x.$$
$$= 81x^4$$

Generalizing Example 3, we have the distributive property for exponents.

> **PROPERTY** Distributive Property for Exponents
>
> If a and b are any two real numbers and r is an integer, then
> $$(ab)^r = a^r \cdot b^r$$

Answers

1. x^{10}
2. 3^{20}
3. $49x^2$

Here are some examples that use combinations of the first three properties of exponents to simplify expressions involving exponents.

EXAMPLE 4 Simplify each expression using the properties of exponents.

a. $(-3x^2)(5x^4) = (-3)(5)(x^2 \cdot x^4)$ *Commutative and associative properties*

$\qquad\qquad\qquad = -15x^6$ *Product property for exponents*

b. $(-2x^2)^3(4x^5) = (-2)^3(x^2)^3(4x^5)$ *Distributive property for exponents*

$\qquad\qquad\quad = (-8x^6) \cdot (4x^5)$ *Power property for exponents*

$\qquad\qquad\quad = (-8 \cdot 4)(x^6 \cdot x^5)$ *Commutative and associative properties*

$\qquad\qquad\quad = -32x^{11}$ *Product property for exponents*

c. $(x^2)^4(x^2y^3)^2(y^4)^3 = x^8 \cdot x^4 \cdot y^6 \cdot y^{12}$ *Power property and distributive property for exponents*

$\qquad\qquad\qquad\qquad = x^{12}y^{18}$ *Product property for exponents*

The next property of exponents deals with negative integer exponents.

PROPERTY **Negative Exponent Property**

If a is any nonzero real number and r is a positive integer, then

$$a^{-r} = \frac{1}{a^r}$$

EXAMPLE 5 Write with positive exponents, and then simplify.

a. $5^{-2} = \dfrac{1}{5^2} = \dfrac{1}{25}$

b. $(-2)^{-3} = \dfrac{1}{(-2)^3} = \dfrac{1}{-8} = -\dfrac{1}{8}$

c. $\left(\dfrac{3}{4}\right)^{-2} = \dfrac{1}{\left(\dfrac{3}{4}\right)^2} = \dfrac{1}{\dfrac{9}{16}} = \dfrac{16}{9}$

If we generalize the result in Example 5c, we have the following extension of the negative exponent property,

$$\left(\frac{a}{b}\right)^{-r} = \left(\frac{b}{a}\right)^{r}$$

which indicates that raising a fraction to a negative power is equivalent to raising the reciprocal of the fraction to the positive power.

The distributive property for exponents indicated that exponents distribute over products. Because division is defined in terms of multiplication, we can expect that exponents will distribute over quotients as well. The expanded distributive property for exponents is the formal statement of this fact.

PROPERTY **Expanded Distributive Property for Exponents**

If a and b are any two real numbers with $b \neq 0$, and r is an integer, then

$$\left(\frac{a}{b}\right)^{r} = \frac{a^r}{b^r}$$

---

4. Simplify.

  a. $(3y)^2(2y^3)$

  b. $(3x^3y^2)^2(2xy^4)^3$

  c. $(x^5)^2(x^3y^4)^2(y^3)^3$

Note The negative exponent property for exponents is actually a definition. That is, we are defining negative integer exponents as indicating reciprocals. Doing so gives us a way to write an expression with a negative exponent as an equivalent expression with a positive exponent.

5. Write with positive exponents, and then simplify.

  a. 4^{-3}

  b. $(-4)^{-3}$

  c. $\left(\dfrac{5}{7}\right)^{-2}$

Answers

4. a. $18y^5$ **b.** $72x^9y^{16}$ **c.** $x^{16}y^{17}$

5. a. $\dfrac{1}{64}$ **b.** $-\dfrac{1}{64}$ **c.** $\dfrac{49}{25}$

Proof of Expanded Distributive Property for Exponents

$$\left(\frac{a}{b}\right)^r = \underbrace{\left(\frac{a}{b}\right)\left(\frac{a}{b}\right)\left(\frac{a}{b}\right)\cdots\left(\frac{a}{b}\right)}_{r \text{ factors}}$$

$$= \frac{a \cdot a \cdot a \cdots a}{b \cdot b \cdot b \cdots b} \quad\begin{array}{l}\leftarrow \ r \text{ factors} \\ \leftarrow \ r \text{ factors}\end{array}$$

$$= \frac{a^r}{b^r}$$

Because multiplication with the same base resulted in addition of exponents, it seems reasonable to expect division with the same base to result in subtraction of exponents.

> **PROPERTY**   Quotient Property for Exponents
>
> If a is any nonzero real number, and r and s are any two integers, then
>
> $$\frac{a^r}{a^s} = a^{r-s}$$

Notice again that we have specified r and s to be any integers. Our definition of negative exponents is such that the properties of exponents hold for all integer exponents, whether positive or negative integers. Here is a proof of the quotient property for exponents.

Proof of The Quotient Property for Exponents

Our proof is centered on the fact that division by a number is equivalent to multiplication by the reciprocal of the number.

$$\frac{a^r}{a^s} = a^r \cdot \frac{1}{a^s} \qquad \text{Dividing by } a^s \text{ is equivalent to multiplying by } \frac{1}{a^s}.$$

$$= a^r a^{-s} \qquad \text{Negative exponent property}$$

$$= a^{r+(-s)} \qquad \text{Product property for exponents}$$

$$= a^{r-s} \qquad \text{Definition of subtraction}$$

EXAMPLE 6   Apply the quotient property for exponents to each expression, and then simplify the result. All answers that contain exponents should contain positive exponents only.

a. $\dfrac{2^8}{2^3} = 2^{8-3} = 2^5 = 32$

b. $\dfrac{x^2}{x^{18}} = x^{2-18} = x^{-16} = \dfrac{1}{x^{16}}$

c. $\dfrac{a^6}{a^{-8}} = a^{6-(-8)} = a^{14}$

d. $\dfrac{m^{-5}}{m^{-7}} = m^{-5-(-7)} = m^2$

Let's complete our list of properties by looking at how the numbers 0 and 1 behave when used as exponents.

We can use the original definition for exponents when the number 1 is used as an exponent.

$$a^1 = a$$

$$\uparrow$$

$$1 \text{ factor}$$

For 0 as an exponent, consider the expression $\dfrac{3^4}{3^4}$. Because $3^4 = 81$, we have

$$\frac{3^4}{3^4} = \frac{81}{81} = 1$$

---

6. Simplify.

a. $\dfrac{3^9}{3^7}$

b. $\dfrac{x^5}{x^{16}}$

c. $\dfrac{y^4}{y^{-5}}$

d. $\dfrac{n^{-6}}{n^{-13}}$

6. a. 9 **b.** $\dfrac{1}{x^{11}}$ **c.** y^9 **d.** n^7

However, because we have the quotient of two expressions with the same base, we can subtract exponents.

$$\frac{3^4}{3^4} = 3^{4-4} = 3^0$$

Hence, 3^0 must be the same as 1.

Summarizing these results, we have our last two properties for exponents.

PROPERTY Identity Property for Exponents

If a is any real number, then

$$a^1 = a$$

PROPERTY Zero Exponent Property

If a is any real number, then

$$a^0 = 1 \quad \text{(as long as } a \neq 0\text{)}$$

EXAMPLE 7 Simplify.

a. $(2x^2y^4)^0 = 1$

b. $(2x^2y^4)^1 = 2x^2y^4$

7. Simplify.

  a. $(4x^4y^9)^0$

  b. $(4x^4y^9)^1$

Here are some examples that use many of the properties of exponents. There are a number of ways to proceed on problems like these. You should use the method that works best for you.

EXAMPLE 8 Simplify.

a. $\dfrac{(x^3)^{-2}(x^4)^5}{(x^{-2})^7} = \dfrac{x^{-6}x^{20}}{x^{-14}}$ Power property for exponents

$= \dfrac{x^{14}}{x^{-14}}$ Product property for exponents

$= x^{28}$ Quotient property for exponents: $x^{14-(-14)} = x^{28}$

b. $\dfrac{6a^5b^{-6}}{12a^3b^{-9}} = \dfrac{6}{12} \cdot \dfrac{a^5}{a^3} \cdot \dfrac{b^{-6}}{b^{-9}}$ Write as separate fractions.

$= \dfrac{1}{2}a^2b^3$ Quotient property for exponents

c. $\dfrac{(4x^{-5}y^3)^2}{(x^4y^{-6})^{-3}} = \dfrac{16x^{-10}y^6}{x^{-12}y^{18}}$ Power property and distributive property for exponents

$= 16x^2y^{-12}$ Quotient property for exponents

$= 16x^2 \cdot \dfrac{1}{y^{12}}$ Negative exponent property

$= \dfrac{16x^2}{y^{12}}$ Multiply.

8. Simplify.

  a. $\dfrac{(x^3)^{-3}(x^9)^2}{(x^{-3})^5}$

  b. $\dfrac{3x^2y^{-4}}{15x^5y^{-6}}$

  c. $\dfrac{(-3x^{-6}y^7)^2}{(x^5y^{-4})^{-3}}$

Note The answer to Example 8(b) can also be written as $\frac{a^2b^3}{2}$. Either answer is correct.

We will complete our first look at exponents with the following discussion.

    Question: In what way are $(-5)^2$ and -5^2 different?

    Answer: In the first case, the base is -5. In the second case, the base is 5. The answer to the first is 25. The answer to the second is -25. Can you tell why? Would there be a difference in the answers if the exponent in each case were changed to 3? As you can see, the parentheses play an important role in determining the answer.

Answers

7. a. 1   **b.** $4x^4y^9$

8. a. x^{24} **b.** $\dfrac{y^2}{5x^3}$ **c.** $9x^3y^2$

B Scientific Notation

Scientific notation is a method for writing very large or very small numbers in a more manageable form. Here is the definition:

> **DEFINITION**　scientific notation
>
> A number is written in *scientific notation* if it is written as the product of a number between 1 and 10 and an integer power of 10. A number written in scientific notation has the form
>
> $$n \times 10^r$$
>
> where $1 \leq n < 10$ and $r =$ an integer.

9. Write 48,600,000 in scientific notation.

EXAMPLE 9　Write 376,000 in scientific notation.

Solution　We must rewrite 376,000 as the product of a number between 1 and 10 and a power of 10. To do so, we move the decimal point five places to the left so that it appears between the 3 and the 7. Then we multiply this number by 10^5. The number that results has the same value as our original number and is written in scientific notation.

$$376,000 = 3.76 \times 10^5$$

Move five places.　Decimal point originally here　Keeps track of the five places we moved the decimal point.

If a number written in expanded form is greater than or equal to 10, then when the number is written in scientific notation the exponent on 10 will be positive. A number that is less than one and greater than zero will have a negative exponent when written in scientific notation.

10. Write 3.05×10^5 in expanded form.

EXAMPLE 10　Write 4.52×10^3 in expanded form.

Solution　Because 10^3 is 1,000, we can think of this as simply a multiplication problem. That is,

$$4.52 \times 10^3 = 4.52 \times 1,000 = 4,520$$

On the other hand, we can think of the exponent 3 as indicating the number of places we need to move the decimal point to write our number in expanded form. Because our exponent is positive 3, we move the decimal point three places to the right.

$$4.52 \times 10^3 = 4,520$$

The following table lists some additional examples of numbers written in expanded form and in scientific notation. In each case, note the relationship between the number of places the decimal point is moved and the exponent on 10.

Number Written in Expanded Form		Number Written in Scientific Notation
376,000	=	3.76×10^5
49,500	=	4.95×10^4
3,200	=	3.2×10^3
591	=	5.91×10^2
46	=	4.6×10^1
8	=	8×10^0
0.47	=	4.7×10^{-1}
0.093	=	9.3×10^{-2}
0.00688	=	6.88×10^{-3}
0.0002	=	2×10^{-4}
0.000098	=	9.8×10^{-5}

Answers

9. 4.86×10^7

10. 305,000

Calculator Note Some scientific calculators have a key that allows you to enter numbers in scientific notation. The key is labeled

$$\boxed{\text{EXP}} \quad \text{or} \quad \boxed{\text{EE}} \quad \text{or} \quad \boxed{\text{SCI}}$$

To enter the number 3.45×10^6, you would enter the decimal number, press the scientific notation key, and then enter the exponent.

$$3.45 \,\boxed{\text{EXP}}\, 6$$

To enter 6.2×10^{-27}, you would use the following sequence:

$$6.2 \,\boxed{\text{EXP}}\, 27 \,\boxed{+/-}$$

C Simplifying Expressions with Scientific Notation

We can use the properties of exponents to do arithmetic with numbers written in scientific notation. Here are some examples:

EXAMPLE 11 Simplify each expression and write all answers in scientific notation.

a. $(2 \times 10^8)(3 \times 10^{-3}) = (2)(3) \times (10^8)(10^{-3})$

$$= 6 \times 10^5$$

b. $\dfrac{4.8 \times 10^9}{2.4 \times 10^{-3}} = \dfrac{4.8}{2.4} \times \dfrac{10^9}{10^{-3}}$

$$= 2 \times 10^{9-(-3)}$$

$$= 2 \times 10^{12}$$

c. $\dfrac{(6.8 \times 10^5)(3.9 \times 10^{-7})}{7.8 \times 10^{-4}} = \dfrac{(6.8)(3.9)}{7.8} \times \dfrac{(10^5)(10^{-7})}{10^{-4}}$

$$= 3.4 \times 10^2$$

●

Calculator Note On a scientific calculator with a scientific notation key, you would use the following sequence of keys to complete Example 11(b):

$$4.8 \,\boxed{\text{EXP}}\, 9 \,\boxed{\div}\, 2.4 \,\boxed{\text{EXP}}\, 3 \,\boxed{+/-}\,\boxed{=}$$

Order of Operations

It is important when evaluating arithmetic expressions in mathematics that each expression have only one answer in reduced form. Consider the expression

$$3 \cdot 7 + 2$$

If we find the product of 3 and 7 first, then add 2, the answer is 23. On the other hand, if we first combine the 7 and 2, then multiply by 3, we have 27. The problem seems to have two distinct answers depending on whether we multiply first or add first. To avoid this, we will decide that multiplication in a situation like this will always be done before addition. In this case, only the first answer, 23, is correct.

The complete set of rules for evaluating expressions follows.

> **RULE** Order of Operations
>
> When evaluating a mathematical expression, we will perform the operations in the following order:
>
> 1. Begin with the expression in the innermost parentheses or brackets and work your way out.
>
> 2. Simplify all numbers with exponents, working from left to right if more than one of these expressions is present.
>
> 3. Work all multiplications and divisions left to right.
>
> 4. Perform all additions and subtractions left to right.

11. Simplify.

a. $(4 \times 10^9)(5 \times 10^{-7})$

b. $\dfrac{4.8 \times 10^4}{1.6 \times 10^9}$

c. $\dfrac{(9.6 \times 10^9)(4.8 \times 10^{-7})}{7.2 \times 10^5}$

Answers

11. a. 2.0×10^3 **b.** 3.0×10^{-5}
  c. 6.4×10^{-3}

12. Simplify the expression.
$8 + 4(5 + 2)$

EXAMPLE 12　Simplify the expression $5 + 3(2 + 4)$.

Solution　$5 + 3(2 + 4) = 5 + 3(6)$ ⟶ Simplify inside parentheses.

$= 5 + 18$ ⟶ Then multiply.

$= 23$ ⟶ Add.

13. Simplify the expression.
$4 \cdot 2^4 - 5 \cdot 3^2$

EXAMPLE 13　Simplify the expression $5 \cdot 2^3 - 4 \cdot 3^2$.

Solution　$5 \cdot 2^3 - 4 \cdot 3^2 = 5 \cdot 8 - 4 \cdot 9$ ⟶ Simplify exponents left to right.

$= 40 - 36$ ⟶ Multiply left to right.

$= 4$ ⟶ Subtract.

14. Simplify the expression.
$90 - (3 \cdot 6^2 - 45)$

EXAMPLE 14　Simplify the expression $20 - (2 \cdot 5^2 - 30)$.

Solution　$20 - (2 \cdot 5^2 - 30) = 20 - (2 \cdot 25 - 30)$

$= 20 - (50 - 30)$

$= 20 - (20)$

$= 0$

Simplify inside parentheses, evaluating exponents first, then multiply, and finally subtract.

GETTING READY FOR CLASS

After reading through the preceding section, respond in your own words and in complete sentences.

A. Explain the difference between -2^4 and $(-2)^4$.

B. If a positive base is raised to a negative exponent, can the result be a negative number?

C. State the product property for exponents in your own words.

D. What is scientific notation?

Answers

12. 36
13. 19
14. 27

Vocabulary Review

Match the following definitions on the left with their property names on the right.

1. $a^r \cdot a^s = a^{r+s}$

2. $(a^r)^s = a^{r \cdot s}$

3. $(ab)^r = a^r \cdot b^r$

4. $a^{-r} = \dfrac{1}{a^r}$

5. $\left(\dfrac{a}{b}\right)^r = \dfrac{a^r}{b^r}$

6. $\dfrac{a^r}{a^s} = a^{r-s}$

7. $a^1 = a$

8. $a^0 = 1, a \neq 0$

9. $a \times 10^r, 1 \leq a \leq 10$

a. Scientific notation

b. Power property for exponents

c. Product property for exponents

d. Zero exponent property

e. Quotient property for exponents

f. Distributive property for exponents

g. Expanded distributive property for exponents

h. Negative exponent property

i. Identity property for exponents

The following is a list of steps for the order of operations. Write the correct step number in the blanks.

_____ Simplify all numbers with exponents, working from left to right if more than one of these expressions is present.

_____ Perform all additions and subtractions left to right.

_____ Work all multiplications and divisions left to right.

_____ Begin with the expressions in the innermost parentheses or brackets and work your way out.

Problems

A Evaluate each of the following.

1. 4^2

2. $(-4)^2$

3. -4^2

4. $-(-4)^2$

5. -0.3^3

6. $(-0.3)^3$

7. 2^5

8. 2^4

9. $\left(\dfrac{1}{2}\right)^3$

10. $\left(\dfrac{3}{4}\right)^2$

11. $\left(\dfrac{5}{6}\right)^2$

12. $\left(\dfrac{2}{3}\right)^3$

13. $\left(\dfrac{1}{10}\right)^4$

14. $\left(\dfrac{1}{10}\right)^5$

15. $\left(-\dfrac{5}{6}\right)^2$

16. $\left(-\dfrac{7}{8}\right)^2$

17. $\left(-\dfrac{3}{7}\right)^2$

18. $\left(-\dfrac{4}{5}\right)^3$

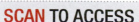

A Use the properties of exponents to simplify each of the following as much as possible.

19. $x^5 \cdot x^4$

20. $x^6 \cdot x^3$

21. $(2^3)^2$

22. $(3^2)^2$

23. $-3a^2(2a^4)$

24. $5a^7(-4a^6)$

25. $(4x^2)^2$

26. $(-3y^2)^2$

Write each of the following with positive exponents. Then simplify as much as possible.

27. 3^{-2}

28. $(-5)^{-2}$

29. $(-2)^{-5}$

30. 2^{-5}

31. $\left(\dfrac{3}{4}\right)^{-2}$

32. $\left(\dfrac{3}{5}\right)^{-2}$

33. $\left(\dfrac{1}{3}\right)^{-2} + \left(\dfrac{1}{2}\right)^{-3}$

34. $\left(\dfrac{1}{2}\right)^{-2} + \left(\dfrac{1}{3}\right)^{-3}$

Multiply.

35. $8x^3 \cdot 10y^6$

36. $5y^2 \cdot 4x^2$

37. $8x^3 \cdot 9y^3$

38. $4y^3 \cdot 3x^2$

39. $3x \cdot 5y$

40. $3xy \cdot 5z$

41. $4x^6y^6 \cdot 3x$

42. $16x^4y^4 \cdot 3y$

43. $27a^6c^3 \cdot 2b^2c$

44. $8a^3b^3 \cdot 5a^2b$

45. $12x^3y^4 \cdot 3xy^2$

46. $-8x^2y \cdot 2xy^4$

Divide. (Assume all variables are nonzero.)

47. $\dfrac{10x^5}{5x^2}$

48. $\dfrac{-15x^4}{5x^2}$

49. $\dfrac{20x^3}{5x^2}$

50. $\dfrac{25x^7}{-5x^2}$

51. $\dfrac{8x^3y^5}{-2x^2y}$

52. $\dfrac{-16x^2y^2}{-2x^2y}$

53. $\dfrac{4x^4y^3}{-2x^2y}$

54. $\dfrac{10a^4b^2}{4a^2b^2}$

Use the properties of exponents to simplify each expression. Write all answers with positive exponents only. (Assume all variables are nonzero.)

55. $\dfrac{x^{-1}}{x^9}$

56. $\dfrac{x^{-3}}{x^5}$

57. $\dfrac{a^4}{a^{-6}}$

58. $\dfrac{a^5}{a^{-2}}$

59. $\dfrac{t^{-10}}{t^{-4}}$

60. $\dfrac{t^{-8}}{t^{-5}}$

61. $\left(\dfrac{x^5}{x^3}\right)^6$

62. $\left(\dfrac{x^7}{x^4}\right)^5$

63. $\dfrac{(x^5)^6}{(x^3)^4}$

64. $\dfrac{(x^7)^3}{(x^4)^5}$

65. $\dfrac{(x^{-2})^3(x^3)^{-2}}{x^{10}}$

66. $\dfrac{(x^{-4})^3(x^3)^{-4}}{x^{10}}$

67. $\dfrac{5a^8b^3}{20a^5b^{-4}}$

68. $\dfrac{7a^6b^{-2}}{21a^2b^{-5}}$

69. $\dfrac{(3x^{-2}y^8)^4}{(9x^4y^{-3})^2}$

70. $\dfrac{(6x^{-3}y^{-5})^2}{(3x^{-4}y^{-3})^4}$

71. $\left(\dfrac{8x^2y}{4x^4y^{-3}}\right)^4$

72. $\left(\dfrac{5x^4y^5}{10xy^{-2}}\right)^3$

73. $\left(\dfrac{x^{-5}y^2}{x^{-3}y^5}\right)^{-2}$

74. $\left(\dfrac{x^{-8}y^{-3}}{x^{-5}y^6}\right)^{-1}$

75. $\left(\dfrac{ab^{-3}c^{-2}}{a^{-3}b^0c^{-5}}\right)^0$

76. $\left(\dfrac{a^3b^2c^1}{a^{-1}b^{-2}c^{-3}}\right)^0$

77. $\left(\dfrac{x^2}{x^{-3}}\right)^0$

78. $\left(\dfrac{2x^2y}{xy^5}\right)^0$

B Write each number in scientific notation.

79. 378,000

80. 3,780,000

81. 4,900

82. 490

83. 0.00037

84. 0.000037

85. 0.00495

86. 0.0495

Write each number in expanded form.

87. 5.34×10^3

88. 5.34×10^2

89. 7.8×10^6

90. 7.8×10^4

91. 3.44×10^{-3}

92. 3.44×10^{-5}

93. 4.9×10^{-1}

94. 4.9×10^{-2}

C Use the properties of exponents to simplify each of the following expressions. Write all answers in scientific notation.

95. $(4 \times 10^{10})(2 \times 10^{-6})$ **96.** $(3 \times 10^{-12})(3 \times 10^4)$ **97.** $\dfrac{8 \times 10^{14}}{4 \times 10^5}$ **98.** $\dfrac{6 \times 10^8}{2 \times 10^3}$

99. $\dfrac{(5 \times 10^6)(4 \times 10^{-8})}{8 \times 10^4}$ **100.** $\dfrac{(6 \times 10^{-7})(3 \times 10^9)}{5 \times 10^6}$ **101.** $\dfrac{(2.4 \times 10^{-3})(3.6 \times 10^{-7})}{(4.8 \times 10^6)(1 \times 10^{-9})}$ **102.** $\dfrac{(7.5 \times 10^{-6})(1.5 \times 10^9)}{(1.8 \times 10^4)(2.5 \times 10^{-2})}$

Simplify. Write answers without using scientific notation, rounding to the nearest whole number

103. $\dfrac{2.00 \times 10^8}{3.98 \times 10^6}$ **104.** $\dfrac{2.00 \times 10^8}{3.16 \times 10^5}$

Use a calculator to find each of the following. Write your answer in scientific notation with the first number in each answer rounded to the nearest tenth.

105. $10^{-4.1}$ **106.** $10^{-5.6}$

Simplify each expression.

107. a. $3 \cdot 5 + 4$ **108. a.** $3 \cdot 7 - 6$ **109. a.** $6 + 3 \cdot 4 - 2$ **110. a.** $8 + 2 \cdot 7 - 3$

 b. $3(5 + 4)$ **b.** $3(7 - 6)$ **b.** $6 + 3(4 - 2)$ **b.** $8 + 2(7 - 3)$

 c. $3 \cdot 5 + 3 \cdot 4$ **c.** $3 \cdot 7 - 3 \cdot 6$ **c.** $(6 + 3)(4 - 2)$ **c.** $(8 + 2)(7 - 3)$

111. a. $(5 + 7)^2$ **112. a.** $(8 - 3)^2$ **113. a.** $2 + 3 \cdot 2^2 + 3^2$ **114. a.** $3 + 4 \cdot 4^2 + 5^2$

 b. $5^2 + 7^2$ **b.** $8^2 - 3^2$ **b.** $2 + 3(2^2 + 3^2)$ **b.** $3 + 4(4^2 + 5^2)$

 c. $5^2 + 2 \cdot 5 \cdot 7 + 7^2$ **c.** $8^2 - 2 \cdot 8 \cdot 3 + 3^2$ **c.** $(2 + 3)(2^2 + 3^2)$ **c.** $(3 + 4)(4^2 + 5^2)$

Applying the Concepts

115. Large Numbers If you are 20 years old, you have been alive for more than 630,000,000 seconds. Write the number of seconds in scientific notation.

116. Our Galaxy The galaxy in which the Earth resides is called the Milky Way galaxy. It is a spiral galaxy that contains approximately 200,000,000,000 stars (our Sun is one of them). Write the number of stars in words and in scientific notation.

117. Light Year A light year, the distance light travels in 1 year, is approximately 5.9×10^{12} miles. The Andromeda galaxy is approximately 1.7×10^6 light years from our galaxy. Find the distance in miles between our galaxy and the Andromeda galaxy.

118. Distance to the Sun As mentioned in the chapter opener, the distance from the Earth to the Sun is approximately 9.3×10^7 miles and light travels 1.2×10^7 miles in 1 minute. How many minutes does it take the light from the Sun to reach the Griffith Observatory in Los Angeles?

119. TV Shows The chart shows the number of viewers for ABC's top television shows.

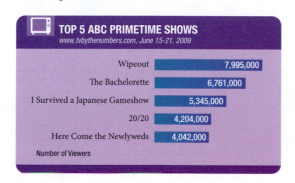

TOP 5 ABC PRIMETIME SHOWS
www.tvbythenumbers.com, June 15-21, 2009

Show	Number of Viewers
Wipeout	7,995,000
The Bachelorette	6,761,000
I Survived a Japanese Gameshow	5,345,000
20/20	4,204,000
Here Come the Newlyweds	4,042,000

For each of the following shows, write the number of viewers in scientific notation.

a. 20/20

b. Wipeout

c. The Bachelorette

120. Fingerprints The FBI has been collecting fingerprint cards since 1924. Their collection has grown to over 200 million cards. When digitized, each fingerprint card turns into about 10 MB of data. (A megabyte [MB] is $2^{20} \approx$ one million bytes.)

a. How many bytes of storage will they need for 200 million cards?

b. A compression routine called the WSQ method will compress the bytes by ratio of 12.9 to 1. Approximately how many bytes of storage will the FBI need for the compressed data from 200 million cards? (Hint: Divide by 12.9.)

Closest Stars to Earth The table below lists the distance (in light-years) of the five closest stars to our solar system:

Common Name	Distance (light-years)
Alpha Centauri C	4.2
Alpha Centauri A, B	4.3
Barnand's Star	5.9
Wolf 359	7.7
Lalande 21185	8.26

Source: space.about.com

Use the table and information given in problems 117 and 118 to help you answer each question.

121. How far (in miles) is Alpha Centauri C from our solar system? Answer using scientific notation.

122. How far (in miles) is Wolf 359 from our solar system? Answer using scientific notation.

123. It is estimated that our own galaxy, the Milky Way, is 4.13×10^{17} miles wide. How long would it take light to pass through the width of our galaxy?

124. Looking through a telescope in Griffith Observatory, you can see the Sombrero Galaxy, approximately 30 million light years away. How far (in miles) is this galaxy? Answer using scientific notation.

125. Credit Card Debt Outstanding credit-card debt in the United States is nearly $800 billion.

    a. Write the number 800 billion in scientific notation.

    b. If there are approximately 180 million credit card holders, find the average credit-card debt per holder, to the nearest dollar.

126. Cone Nebula Nebulas are collections of dust and gas in space that can be seen by some high-powered telescopes. One well-known nebula, the Cone Nebula, is 2.5 light years or 14,664,240,000,000 miles wide. Round this number to the nearest trillion and then write the result in scientific notation.

127. Computer Science We all use the language of computers to indicate how much memory our computers hold or how much information we can put on a storage device such as a flash drive. Scientific notation gives us a way to compare the actual numbers associated with the words we use to describe data storage in computers. The smallest amount of data that a computer can hold is measured in bits. A byte is the next largest unit and is equal to 8, or 2^3, bits. Fill in the following table:

Number of Bytes		
Unit	Exponential Form	Scientific Notation
Kilobyte	$2^{10} = 1,024$	
Megabyte	$2^{20} \approx 1,048,000$	
Gigabyte	$2^{30} \approx 1,074,000,000$	
Terabyte	$2^{40} \approx 1,099,500,000,000$	

One Step Further

Assume all variable exponents represent positive integers, and simplify each expression.

128. $x^{m+2} \cdot x^{-2m} \cdot x^{m-5}$

129. $x^{m-4}x^{m+9}x^{-2m}$

130. $(y^m)^2(y^{-3})^m(y^{m+3})$

131. $(y^m)^{-4}(y^3)^m(y^{m+6})$

132. $\dfrac{x^{n+2}}{x^{n-3}}$

133. $\dfrac{x^{n-3}}{x^{n-7}}$

134. $x^{a+3}x^{2a-5}x^{7-9a}$

135. $x^{2b-3}x^{5-8b}x^{-4-3b}$

136. $(x^{m+3})^2(x^{2m-5})^3$

137. $(x^{4m-1})^3(x^{m+2})^4$

138. $\dfrac{x^{a+5}}{x^{3a-7}}$

139. $\dfrac{x^{2b-6}}{x^{5b+2}}$

140. $\dfrac{(x^{m+3})^2}{(x^{2m-1})^3}$

141. $\dfrac{(x^{4m-3})^4}{(x^{2m+1})^3}$

142. $\left(\dfrac{x^{4a-3}}{x^{-2a+1}}\right)^2$

143. $\left(\dfrac{x^{-3a+2}}{x^{-5a-4}}\right)^3$

Find the Mistake

Each sentence below contains a mistake. Circle the mistake and write the correct word on the line provided.

  1. To find the volume of a cube with a side length of 3 cm, we raise 3 to the second power. _____

  2. The product of $x^4 \cdot x^5$ written with a single exponent is x^{20}. _____

  3. To write 3×10^{-4} in expanded form, move the decimal point 3 places to the left. _____

  4. According to the order of operations, we begin to simplify the expression $10 - 3 + 5^3 \cdot 2$ by finding the difference between 10 and 3. _____

Codebreakers

Codes have been used for centuries to communicate messages. A code could be as intricate as the sequences used to program computers, or as simple as the hand signals used to communicate direction by a coach to a baseball player. Regardless of the format, the process of writing a code, or decoding it for that matter, is conceptually similar to that of finding the value of an algebraic expression. If you do not know the value of the variables, you will not be able to understand the code, just as you would not be able to find a solution for the expression. Many codes are unbreakable unless you have a key; in other words, unless you have the sequence of values that when substituted for the coded variables will reveal the message.

1. For the first part of this project, we would like you and a partner to research different types of codes. Some examples are Morse code, binary code, tri code, or "leet" 1337. Choose three codes to examine. How do these codes work? Was any math used to create the keys for these codes? Explain.

2. Secondly, choose one type of code and work with your partner to compile a key for your code. Then each write a coded message. Make sure to keep your message a secret from your partner. Once you have coded your message, exchange the encrypted messages and use the key to decode them.

3. Lastly, algebraic expressions use letters as variables to represent numbers. Let's switch it up and use numbers to represent letters. Create a key for the English alphabet that assigns a number to each letter. The simple approach to this assignment would be to assign the numbers 1–26 to the letters A-Z respectively. However, we would like you to take it a step further and assign the numbers using an arithmetic or a geometric sequence. Recall that an arithmetic sequence is a sequence of numbers in which each number is derived by adding a constant value to the previous number (e.g., 1, 3, 5, 7 is an arithmetic sequence because we find each number by adding 2 to the number before it). A geometric sequence is a sequence of numbers in which each number is derived by multiplying a constant value to the previous number (e.g., 2, 6, 18, 54 is a geometric sequence because we find each number by multiplying the previous number by 3). To complete this last part of the project, each person should follow these steps:

    a. Choose a number to assign to the letter A. (We will be exchanging coded messages at the end, so be sure to keep your key a secret from your partner until then.)

    b. Decide whether to create an arithmetic sequence or a geometric sequence.

    c. Decide on the constant value you will add or multiply to create your sequence.

    d. Calculate your sequence for the entire alphabet.

    e. Write a message using your new code.

    f. Once each person has encoded a message, exchange messages. Give your partner the following three pieces of information to help him or her solve your code:

        i. The value for A

        ii. Whether you used an arithmetic or geometric sequence

        iii. The constant value

Chapter 0 Summary

The numbers in brackets refer to the section(s) in which the topic can be found.

Absolute Value [0.1]

1. $|5| = 5$
$|-5| = 5$

The *absolute value* of a real number is its distance from 0 on the number line. If $|x|$ represents the absolute value of x, then

$$|x| = \begin{cases} x & \text{if} & x \geq 0 \\ -x & \text{if} & x < 0 \end{cases}$$

The absolute value of a real number is never negative.

Opposites [0.1]

2. The numbers 5 and -5 are opposites; their sum is 0.
$5 + (-5) = 0$

Any two real numbers the same distance from 0 on the number line, but in opposite directions from 0, are called *opposites,* or *additive inverses.* Opposites always add to 0.

Reciprocals [0.1]

3. The numbers 3 and $\frac{1}{3}$ are reciprocals; their product is 1.
$3\left(\frac{1}{3}\right) = 1$

Any two real numbers whose product is 1 are called *reciprocals.* Every real number has a reciprocal except 0.

Properties of Real Numbers [0.2]

	For Addition	For Multiplication
Commutative	$a + b = b + a$	$ab = ba$
Associative	$a + (b + c) = (a + b) + c$	$a(bc) = (ab)c$
Identity	$a + 0 = a$	$a \cdot 1 = a$
Inverse	$a + (-a) = 0$	$a\left(\dfrac{1}{a}\right) = 1$
Distributive	$a(b + c) = ab + ac$	

Addition [0.3]

4. $5 + 3 = 8$
$5 + (-3) = 2$
$-5 + 3 = -2$
$-5 + (-3) = -8$

To add two real numbers with

1. *The same sign:* Simply add absolute values and use the common sign.

2. *Different signs:* Subtract the smaller absolute value from the larger absolute value. The answer has the same sign as the number with the larger absolute value.

Subtraction [0.3]

5. $6 - 2 = 6 + (-2) = 4$
$6 - (-2) = 6 + 2 = 8$

If a and b are real numbers,

$$a - b = a + (-b)$$

To subtract b, add the opposite of b.

Chapter 0 Summary

Multiplication [0.3]

To multiply two real numbers, simply multiply their absolute values. Like signs give a positive answer. Unlike signs give a negative answer.

6. $5(4) = 20$
$5(-4) = -20$
$-5(4) = -20$
$-5(-4) = 20$

Division [0.3]

If a and b are real numbers and $b \neq 0$, then

$$\frac{a}{b} = a \cdot \left(\frac{1}{b}\right)$$

To divide by b, multiply by the reciprocal of b.

7. $\frac{12}{-3} = -4$

$\frac{-12}{-3} = 4$

Properties of Exponents [0.4]

If a and b represent real numbers and r and s represent integers, then

1. $a^r \cdot a^s = a^{r+s}$ .... Product property for exponents

2. $(a^r)^s = a^{r \cdot s}$ .... Power property for exponents

3. $(ab)^r = a^r \cdot b^r$ .... Distributive property for exponents

4. $a^{-r} = \dfrac{1}{a^r}$ $\quad (a \neq 0)$ .... Negative exponent property

5. $\left(\dfrac{a}{b}\right)^r = \dfrac{a^r}{b^r}$ $\quad (b \neq 0)$ .... Expanded distributive property for exponents

6. $\dfrac{a^r}{a^s} = a^{r-s}$ $\quad (a \neq 0)$ .... Quotient property for exponents

7. $a^1 = a$ .... Identity property for exponents

8. $a^0 = 1$ $\quad (a \neq 0)$ .... Zero exponent property

8. These expressions illustrate the properties of exponents.

a. $x^2 \cdot x^3 = x^{2+3} = x^5$

b. $(x^2)^3 = x^{2 \cdot 3} = x^6$

c. $(3x)^2 = 3^2 \cdot x^2 = 9x^2$

d. $2^{-3} = \dfrac{1}{2^3} = \dfrac{1}{8}$

e. $\left(\dfrac{x}{5}\right)^2 = \dfrac{x^2}{5^2} = \dfrac{x^2}{25}$

f. $\dfrac{x^7}{x^5} = x^{7-5} = x^2$

g. $3^1 = 3$
$3^0 = 1$

Scientific Notation [0.4]

A number is written in scientific notation when it is written as the product of a number between 1 and 10 and an integer power of 10; that is, when it has the form

$$n \times 10^r$$

where $1 \leq n < 10$ and $r = $ an integer.

9. $49{,}800{,}000 = 4.98 \times 10^7$
$0.00462 = 4.62 \times 10^{-3}$

COMMON MISTAKES

1. Interpreting absolute value as changing the sign of the number inside the absolute value symbols. That is, $|-5| = +5$, $|+5| = -5$. To avoid this mistake, remember, absolute value is defined as a distance and distance is always measured in non-negative units.

2. Confusing $-(-5)$ with $-|-5|$ The first answer is $+5$, but the second answer is -5.

CHAPTER 0 Review

At the end of each chapter, we will provide a review of problems similar to those you have seen in the chapter's previous sections. These problems are meant to supplement, not replace, a thorough review of the concepts, examples, and exercise sets presented in the chapter.

Write each fraction as an equivalent fraction with denominator 36. [0.1]

1. $\dfrac{2}{3}$

2. $\dfrac{7}{9}$

Insert a $<$ or $>$ to make the statement true. [0.1]

3. $-4 \quad 9$

4. $-6 \quad -3$

5. $5 \quad 3$

6. $-\dfrac{2}{3} \quad -\dfrac{1}{6}$

Simplify each expression. [0.1], [0.3]

7. $|7 - 12 \div 3|$

8. $|9 - 12| - |3 - 11|$

9. $13 - |7(2 - 5)|$

10. $7 + |5 \cdot 3 - 7| - 6$

Reduce to lowest terms. [0.1]

11. $\dfrac{252}{468}$

12. $\dfrac{208}{496}$

Simplify. [0.2]

13. $3(2x + 5) - 4x$

14. $5x - 5 + 2x + 7$

15. $3x + 2(3x + 2y) + 4y$

16. $4a + 3(2b + 4) - 8$

17. $4y + 2(2x - 4) - 2y$

18. $16\left(\dfrac{5}{8}x + \dfrac{1}{4}\right) - 2$

Add the following fractions. [0.2]

19. $\dfrac{9}{14} + \dfrac{7}{24}$

20. $\dfrac{9}{16} + \dfrac{12}{36}$

Simplify each expression. [0.3]

21. $3 - (5 - x) + 3x$

22. $(6 - 3x) - (2x - 4)$

23. $3x - 4(2x - 3) - 12$

24. $\dfrac{3(-2) - 4(4)}{6 - 4}$

Simplify as much as possible. Write all answers with positive exponents. [0.4]

25. $(-2x^2)^{-2}$

26. $(3x^2y^2)^3$

27. $\left(\dfrac{2}{3}\right)^{-3}$

28. $2x^4y \cdot -2xy^4$

29. $-3x^2y \cdot 7x^5y^2$

30. $\dfrac{36a^5b}{6a^2b^4}$

31. $\dfrac{24a^6b^3}{-30ab^4}$

32. $\left(\dfrac{7a^4b}{14a^6b^5}\right)^{-2}$

Write in scientific notation. [0.4]

33. $482,000$

34. $7,280,000$

35. 0.00421

36. 0.0526

Write in expanded form. [0.4]

37. 6.29×10^{-3}

38. 3.29×10^3

39. 6.31×10^{-2}

40. 4.82×10^7

Simplify. Write in scientific notation. [0.4]

41. $\dfrac{(3.6 \times 10^7)(2 \times 10^{-4})}{2.4 \times 10^6}$

42. $\dfrac{(4 \times 10^{-6})(1.6 \times 10^4)}{(3.2 \times 10^9)(1 \times 10^{-7})}$

At the end of each chapter, we will provide a test that you should use as a helpful study tool for a chapter exam. Remember, it is still important to study every aspect of the material presented in the chapter, and not rely solely on the Chapter Test to study for an exam.

Write each fraction as an equivalent fraction with denominator 18. [0.1]

1. $\dfrac{1}{2}$

2. $\dfrac{5}{6}$

Insert a $<$ or $>$ to make the statement true. [0.1]

3. $-6 \quad -4$

4. $7 \quad 9$

5. $3 \quad -6$

6. $-\dfrac{6}{7} \quad -\dfrac{2}{5}$

Simplify each expression. [0.1], [0.3]

7. $|9 - 16 \div 8|$

8. $|7 - 10| - |3 - 5|$

9. $16 - |3(2 - 6)|$

10. $6 + |3 \cdot 2 - 4| - 3$

Reduce to lowest terms. [0.1]

11. $\dfrac{192}{312}$

12. $\dfrac{162}{459}$

Simplify. [0.2]

13. $4(3x + 2) - 6$

14. $6x - 3 + 4x + 5$

15. $4x + 3(2x + 4y) - 6y$

16. $3a + 2(5b + 4) - 1$

17. $6y + 3(x + 3) - 4y$

18. $18\left(\dfrac{5}{6}y - \dfrac{2}{3}x\right) - 5y$

Add the following fractions. [0.2]

19. $\dfrac{7}{18} + \dfrac{7}{9}$

20. $\dfrac{7}{12} + \dfrac{13}{34}$

Simplify the expression. [0.3]

21. $4 - (2x - 7) + 4x$

22. $(5 - 4x) - (3 - 2x)$

23. $5x - 3(2x + 4) + 15$

24. $\dfrac{(8)(-9) - (6)(-2)}{-(12 - 8)}$

Simplify as much as possible. Write all answers with positive exponents. [0.4]

25. $(-3x^4)^{-3}$

26. $(4x^2y)^3$

27. $\left(\dfrac{5}{9}\right)^{-2}$

28. $-5x^2y \cdot -7x^3y$

29. $-6xy^3 \cdot 3x^{-2}y^4$

30. $\dfrac{18a^4b}{3ab^3}$

31. $\dfrac{18a^5b^9}{24a^7b^4}$

32. $\left(\dfrac{3x^{-2}y}{9xy^2}\right)^{-3}$

Write in scientific notation. [0.4]

33. $12{,}530{,}000$

34. 0.0052

35. $6{,}320$

36. 0.00034

Write in expanded form. [0.4]

37. 5.26×10^{-3}

38. 4.9×10^5

39. 6.3×10^{-4}

40. 7.8×10^4

Simplify. Write in scientific notation. [0.4]

41. $\dfrac{(1.4 \times 10^7)(6.5 \times 10^{-4})}{(7.0 \times 10^2)}$

42. $\dfrac{(1.8 \times 10^7)(6.8 \times 10^6)}{(2.4 \times 10^9)(3.0 \times 10^{-4})}$

Linear Equations and Inequalities in One Variable

1.1 Linear Equations

1.2 Formulas

1.3 Applications

1.4 Interval Notation and Linear Inequalities

1.5 Sets and Compound Inequalities

1.6 Absolute Value Equations

1.7 Absolute Value Inequalities

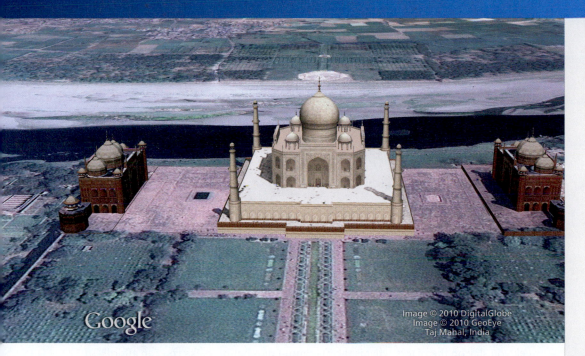

Image © 2010 DigitalGlobe
Image © 2010 GeoEye
Taj Mahal, India

The Taj Mahal stands prominently on the banks of River Yamuna in the city of Agra, India. Completed in 1653, it remains one of the most spectacular structures on Earth and is recognized as a symbol of love and romance. Legends tell the story of Mughal Emperor Shah Jahan. It is said that he ordered the construction of the Taj Mahal in memory of his wife Mumtaz Mahal, who died in 1631 while giving birth to their 14th child. It took approximately 22 years to complete, and construction required the services of 22,000 laborers and 1,000 elephants. The elaborate structure is built entirely out of white marble transported from all over India and Central Asia.

Celsius	Fahrenheit
0	32
7	45
46	115
100	212

Agra experiences extremely cold temperatures in the winter, hot temperatures in the summer, and monsoons in July. Therefore, if you visit this magnificent monument, you should first check the weather to ensure a pleasant viewing experience. Temperatures throughout the year range from 46°C down to 7°C. As shown by the table, this temperature range can be translated into the more familiar Fahrenheit, shown on the right column of the table. In order to complete this translation you must use the formula

$$F = \frac{9}{5}C + 32$$

where C is the temperature in degrees Celsius.

In this chapter, you will work with formulas like this one and also learn how to use them to work with inequalities.

1.1 Linear Equations

Image © Patrick Nugent, 2010

The newest craze in underwater exploration is by way of an aero-submarine called the Necker Nymph. This three-passenger, 15-foot-long submersible has an open cockpit and can dive nearly one hundred feet below the water's surface. Guests on Necker Island, a privately-owned tropical paradise in the Caribbean, have access to the aero-submarine. Passengers wear scuba gear as they maneuver the craft over beautiful coral reefs on the sea floor. The joys of this new thrill come at a cost; the Necker Nymph rents for $25,000 per week! Imagine for a moment that you have been given a generous gift of $60,000 to spend at your leisure while on the island. If you rent the aero-sub every week you stay on the island, what is the maximum number of weeks you can afford? To find a solution, we need to set up the following linear equation:

$$60,000 = 25,000x$$

Solving this type of equation is one of the things we will do in this section.

A Solutions to Linear Equations

> **DEFINITION** linear equation in one variable
>
> A *linear equation in one variable* is any equation that can be put in the form
>
> $$ax + b = c$$
>
> where a, b, and c are constants and $a \neq 0$.

For example, each of the equations

$$5x + 3 = 2 \qquad 2x = 7 \qquad 2x + 5 = 0$$

are linear because they can be put in the form $ax + b = c$. In the first equation, $5x$, 3, and 2 are called *terms* of the equation: $5x$ is a variable term; 3 and 2 are constant terms. Furthermore, we can find a solution for the equation by substituting a number for the variable.

> **DEFINITION** solution set
>
> The *solution set* for an equation is the set of all numbers that, when used in place of the variable, make the equation a true statement.

> **DEFINITION** equivalent equations
>
> Two or more equations with the same solution set are called *equivalent equations*.

The equations $2x - 5 = 9$, $x - 1 = 6$, and $x = 7$ are all equivalent equations because the solution set for each is {7}.

EXAMPLE 1   Determine whether 4 is a solution of $3x - 1 = 11$.

Solution   Substitute 4 for x in the equation to determine of it results in a true statement.

$$3(4) - 1 = 11 \qquad \text{\textcolor{green}{Substitute 4 for x.}}$$
$$12 - 1 = 11 \qquad \text{\textcolor{green}{Multiply 3(4).}}$$
$$11 = 11 \qquad \text{\textcolor{green}{Subtract 12 - 1.}}$$

Since the resulting statement is true, 4 is a solution to the equation $3x - 1 = 11$.

EXAMPLE 2   Determine whether -12 is a solution of $\frac{1}{3}x + 2 = -\frac{1}{4}x + 1$.

Solution   Substitute -12 for x in the equation to determine of it results in a true statement.

$$\frac{1}{3}(-12) + 2 = -\frac{1}{4}(-12) + 1 \qquad \text{\textcolor{green}{Substitute -12 for x.}}$$
$$-4 + 2 = 3 + 1 \qquad \text{\textcolor{green}{Multiply } \frac{1}{3}(-12) \text{ and } -\frac{1}{4}(-12).}$$
$$-2 = 4 \qquad \text{\textcolor{green}{Add -4 + 2 and 3 + 1.}}$$

Since the resulting statement is false, -12 is not a solution to the equation

$$\frac{1}{3}x + 2 = -\frac{1}{4}x + 1.$$

B Properties of Equality and Solving Linear Equations

The first property of equality states that adding the same quantity to both sides of an equation preserves equality. Or, more importantly, adding the same amount to both sides of an equation *never changes* the solution set. This property is called the *addition property of equality* and is stated in symbols as follows:

> **PROPERTY** Addition Property of Equality
>
> For any three algebraic expressions A, B, and C,
>
> $$\text{if} \qquad A = B$$
> $$\text{then} \qquad A + C = B + C$$
>
> *In words:*   Adding the same quantity to both sides of an equation will not change the solution set.

Our second property is called the *multiplication property of equality* and is stated as follows:

> **PROPERTY** Multiplication Property of Equality
>
> For any three algebraic expressions A, B, and C, where $C \neq 0$,
>
> $$\text{if} \qquad A = B$$
> $$\text{then} \qquad AC = BC$$
>
> *In words:*   Multiplying both sides of an equation by the same nonzero quantity will not change the solution set.

Practice Problems

1. Is -2 a solution to $3z + 4 = -2$?

2. Is 8 a solution to $\frac{x}{4} - 2 = \frac{x}{2} + 4$?

Note   Because subtraction is defined in terms of addition and division is defined in terms of multiplication, we do not need to introduce separate properties for subtraction and division. The solution set for an equation will never be changed by subtracting the same amount from both sides or by dividing both sides by the same nonzero quantity.

Answers

1. Yes
2. No

3. Solve the equation
$2x - 5 = -3x + 10.$

> *Note* We know that multiplication by a number and division by its reciprocal always produce the same result. Because of this fact, instead of multiplying each side of our equation by $\frac{1}{9}$, we could just as easily divide each side by 9. If we did so, the last two lines in our solution would look like this:
>
> $$\frac{9a}{9} = \frac{6}{9}$$
> $$a = \frac{2}{3}$$

4. Solve the equation
$\frac{3}{5}x + \frac{1}{2} = -\frac{7}{10}.$

EXAMPLE 3   Solve the equation $3a - 5 = -6a + 1$.

Solution   To solve for a, we must isolate it on one side of the equation. Let's decide to isolate a on the left side. We start by adding $6a$ to both sides of the equation.

$$3a - 5 = -6a + 1$$

$$3a + 6a - 5 = -6a + 6a + 1 \quad \text{Add } 6a \text{ to both sides.}$$

$$9a - 5 = 1$$

$$9a - 5 + 5 = 1 + 5 \qquad\qquad \text{Add 5 to both sides.}$$

$$9a = 6$$

$$\frac{1}{9}(9a) = \frac{1}{9}(6) \qquad\qquad \text{Multiply both sides by } \frac{1}{9}.$$

$$a = \frac{2}{3} \qquad\qquad\qquad \frac{1}{9}(6) = \frac{6}{9} = \frac{2}{3}$$

The solution is $\frac{2}{3}$.

The next example involves fractions. The least common denominator, which is the smallest expression that is divisible by each of the denominators, can be used with the multiplication property of equality to simplify equations containing fractions.

EXAMPLE 4   Solve the equation $\frac{2}{3}x + \frac{1}{2} = -\frac{3}{8}$.

Solution   We can solve this equation by applying our properties and working with fractions, or we can begin by eliminating the fractions. Let's work the problem using both methods.

Method 1: *Working with the fractions*

$$\frac{2}{3}x + \frac{1}{2} + \left(-\frac{1}{2}\right) = -\frac{3}{8} + \left(-\frac{1}{2}\right) \quad \text{Add } -\frac{1}{2} \text{ to each side.}$$

$$\frac{2}{3}x = -\frac{7}{8} \qquad\qquad -\frac{3}{8} + \left(-\frac{1}{2}\right) = -\frac{3}{8} + \left(-\frac{4}{8}\right)$$

$$\frac{3}{2}\left(\frac{2}{3}x\right) = \frac{3}{2}\left(-\frac{7}{8}\right) \qquad \text{Multiply each side by } \frac{3}{2}.$$

$$x = -\frac{21}{16}$$

The solution is $-\frac{21}{16}$.

Method 2: *Eliminating the fractions in the beginning*

Our original equation has denominators of 3, 2, and 8. The least common denominator, abbreviated LCD, for these three denominators is 24, and it has the property that all three denominators will divide it evenly. Therefore, if we multiply both sides of our equation by 24, each denominator will divide into 24, and we will be left with an equation that does not contain any denominators other than 1.

$$24\left(\frac{2}{3}x + \frac{1}{2}\right) = 24\left(-\frac{3}{8}\right) \qquad \text{Multiply each side by the LCD 24.}$$

$$24\left(\frac{2}{3}x\right) + 24\left(\frac{1}{2}\right) = 24\left(-\frac{3}{8}\right) \qquad \text{Distributive property on the left side}$$

$$16x + 12 = -9 \qquad\qquad \text{Multiply.}$$

$$16x = -21 \qquad\qquad \text{Add } -12 \text{ to each side.}$$

$$x = -\frac{21}{16} \qquad\qquad \text{Multiply each side by } \frac{1}{16}.$$

As the third line above indicates, multiplying each side of the equation by the LCD eliminates all the fractions from the equation. Both methods yield the same solution.

Answers

3. 3

4. -2

EXAMPLE 5 Solve the equation $0.06x + 0.05(10,000 - x) = 560$.

Solution We can solve the equation in its original form by working with the decimals, or we can eliminate the decimals first by using the multiplication property of equality and solve the resulting equation. Here are both methods:

Method 1: *Working with the decimals*

$$0.06x + 0.05(10,000 - x) = 560 \qquad \text{Original equation}$$
$$0.06x + 0.05(10,000) - 0.05x = 560 \qquad \text{Distributive property}$$
$$0.01x + 500 = 560 \qquad \text{Simplify the left side.}$$
$$0.01x + 500 + (-500) = 560 + (-500) \qquad \text{Add } -500 \text{ to each side.}$$
$$0.01x = 60$$
$$\frac{0.01x}{0.01} = \frac{60}{0.01} \qquad \text{Divide each side by 0.01.}$$
$$x = 6,000$$

Method 2: *Eliminating the decimals in the beginning.*
To move the decimal point two places to the right in $0.06x$ and 0.05, we multiply each side of the equation by 100.

$$0.06x + 0.05(10,000 - x) = 560 \qquad \text{Original equation}$$
$$0.06x + 500 - 0.05x = 560 \qquad \text{Distributive property}$$
$$100(0.06x) + 100(500) - 100(0.05x) = 100(560) \qquad \text{Multiply each side by 100.}$$
$$6x + 50,000 - 5x = 56,000 \qquad \text{Multiply.}$$
$$x + 50,000 = 56,000 \qquad \text{Simplify the left side.}$$
$$x = 6,000 \qquad \text{Add } -50,000 \text{ to each side.}$$

Using either method, the solution to our equation is 6,000. We check our work (to be sure we have not made a mistake in applying the properties or with our arithmetic) by substituting 6,000 into our original equation and simplifying each side of the result separately.

Check: Substituting 6,000 for x in the original equation, we have

$$0.06(6,000) + 0.05(10,000 - 6,000) = 0.06(6,000) + 0.05(4,000)$$
$$= 360 + 200$$
$$= 560 \qquad \text{A true statement}$$

5. Solve the equation
$0.06x + 0.04(x + 7,000) = 680$.

Note We are placing question marks over the equal signs because we don't know yet if the expressions on the left will be equal to the expressions on the right.

Here is a list of steps to use as a guideline for solving linear equations in one variable:

HOW TO Solve Linear Equations in One Variable

Step 1a: Use the distributive property to separate terms, if necessary.

1b: If fractions are present, consider multiplying both sides by the LCD to eliminate the fractions. If decimals are present, consider multiplying both sides by a power of 10 to clear the equation of decimals.

1c: Combine similar terms on each side of the equation.

Step 2: Use the addition property of equality to get all variable terms on one side of the equation and all constant terms on the other side. A variable term is a term that contains the variable. A constant term is a term that does not contain the variable (the number 3, for example).

Step 3: Use the multiplication property of equality to get the variable by itself on one side of the equation.

Step 4: Check your solution in the original equation to be sure that you have not made a mistake in the solution process.

Answer
5. 4,000

As you work through the problems in the problem set, you will see that it is not always necessary to use all four steps when solving equations. The number of steps used depends on the equation. In Example 6, there are no fractions or decimals in the original equation, so step 1(b) will not be used.

6. Solve the equation
$6 - 2(5x - 1) + 4x = 20$.

EXAMPLE 6   Solve the equation $8 - 3(4x - 2) + 5x = 35$.

Solution   We must begin by distributing the -3 across the quantity $4x - 2$. It would be a mistake to subtract 3 from 8 first, because the rule for order of operations indicates we are to do multiplication before subtraction. After we have simplified the left side of our equation, we apply the addition property and the multiplication property. In this example, we will show only the result.

$$8 - 3(4x - 2) + 5x = 35 \qquad \text{Original equation}$$

Step 1a: $\quad 8 - 12x + 6 + 5x = 35 \qquad$ Distributive property

Step 1c: $\quad\quad\quad\quad -7x + 14 = 35 \qquad$ Simplify.

Step 2: $\quad\quad\quad\quad\quad\quad -7x = 21 \qquad$ Add -14 to each side.

Step 3: $\quad\quad\quad\quad\quad\quad\quad x = -3 \qquad$ Multiply by $-\frac{1}{7}$.

Step 4: When x is replaced by -3 in the original equation, a true statement results. Therefore, -3 is the solution to our equation.

C   Identities and Equations With No Solution

There are two special cases associated with solving linear equations in one variable, which are illustrated in the following examples.

7. Solve the equation
$3(5x + 1) = 10 + 15x$.

EXAMPLE 7   Solve the equation $2(3x - 4) = 3 + 6x$.

Solution   Applying the distributive property to the left side gives us

$$6x - 8 = 3 + 6x \qquad \text{Distributive property}$$

Now, if we add $-6x$ to each side, we are left with $-8 = 3$ which is a false statement. This means that there is no solution to our equation. Any number we substitute for x in the original equation will lead to a similar false statement. We say the original equation is a *contradiction* because the left side is never equal to the right.

8. Solve the equation
$-4 + 8x = 2(4x - 2)$.

EXAMPLE 8   Solve the equation $-15 + 3x = 3(x - 5)$.

Solution   We start by applying the distributive property to the right side.

$$-15 + 3x = 3x - 15 \qquad \text{Distributive property}$$

If we add $-3x$ to each side, we are left with the true statement

$$-15 = -15$$

In this case, our result tells us that any number we use in place of x in the original equation will lead to a true statement. Therefore, all real numbers are solutions to our equation. We say the original equation is an *identity* because the left side is always identically equal to the right side.

Answers

6. -2

7. No solution

8. All real numbers are solutions.

Vocabulary Review

Choose the correct words to fill in the blanks below.

identity   solution   linear equation in one variable   contradiction

multiplication   equivalent   addition

1. An equation that can be put in the form $ax + b = c$ is called a _____ .

2. The _____ set for an equation is the set of all numbers that, when substituted in place of the variable, make the equation a true statement.

3. _____ equations have the same solution set.

4. The _____ property of equality says that adding the same quantity to both sides of an equation will not change the solution set.

5. The _____ property of equality says that multiplying both sides of an equation by the same nonzero quantity will not change the solution set.

6. An equation is called a(n) _____ if all real numbers are solutions, but if there is no solution, the equation is called a(n) _____ .

Problems

A Determine whether each number is a solution to the given equation.

1. $-3x + 2 = 5; -1$

2. $-4x + 3 = 11; -2$

3. $4x - 5 = -9; 1$

4. $5x + 1 = -6; -1$

5. $2x + 1 = 3x - 4; -1$

6. $2x + 1 = 3x - 4; 5$

7. $\frac{x}{2} + 1 = 4; 6$

8. $\frac{x}{3} - 1 = 1; -6$

9. $\frac{x}{3} + 1 = \frac{x}{4}; -12$

10. $\frac{x}{5} - 2 = \frac{x}{2} + 1; -10$

11. $3(x - 2) + 1 = 2x - 4; 2$

12. $-4(x + 3) + 2 = 5(x - 1) + 4; -1$

B Solve each of the following equations.

13. $7y - 4 = 2y + 11$

14. $5 - 2x = 3x + 1$

15. $-\frac{2}{5}x + \frac{2}{15} = \frac{2}{3}$

16. $\frac{1}{2}x + \frac{1}{4} = \frac{1}{3}x + \frac{5}{4}$

17. $0.14x + 0.08(10,000 - x) = 1220$

18. $-0.3y + 0.1 = 0.5$

19. $5(y + 2) - 4(y + 1) = 3$

20. $6(y - 3) - 5(y + 2) = 8$

Now that you have practiced solving a variety of equations, we can turn our attention to the type of equation you will see as you progress through the book.

Solve each equation.

21. $-3 - 4x = 15$

22. $-\dfrac{3}{5}a + 2 = 8$

23. $0 = 6400a + 70$

24. $0.07x = 1.4$

25. $5(2x + 1) = 12$

26. $50 = \dfrac{K}{48}$

27. $100P = 2{,}400$

28. $2x - 3(3x - 5) = -6$

29. $5\left(-\dfrac{19}{15}\right) + 5y = 9$

30. $2\left(-\dfrac{29}{22}\right) - 3y = 4$

31. $4x + (x - 2) \cdot 3 = 8$

32. $3x + (x - 2) \cdot 2 = 6$

33. $15 - 3(x - 1) = x - 2$

34. $2(2x - 3) + 2x = 45$

35. $2(20 + x) = 3(20 - x)$

36. $2x + 1.5(75 - x) = 127.5$

37. $0.08x + 0.09(9{,}000 - x) = 750$

38. $0.12x + 0.10(15{,}000 - x) = 1{,}600$

C Solve each equation, if possible.

39. $3x - 6 = 3(x + 4)$

40. $4y + 2 - 3y + 5 = 3 + y + 4$

41. $2(4t - 1) + 3 = 5t + 4 + 3t$

42. $7x - 3(x - 2) = -4(5 - x)$

43. $7(x + 2) - 4(2x - 1) = 18 - x$

44. $2x^2 + x - 1 = (2x + 3)(x - 1)$

45. Temperature and Altitude As an airplane gains altitude, the temperature outside the plane decreases. The relationship between temperature T and altitude A can be described with the formula

$$T = -0.0035A + 70$$

when the temperature on the ground is 70°F. Solve the equation below to find the altitude in feet at which the temperature outside the plane is -35°F.

$$-35 = -0.0035A + 70$$

46. Celsius and Fahrenheit Temperature As mentioned in the chapter opener, the relationship between Celsius and Fahrenheit temperature is

$$F = \frac{9}{5}C + 32$$

Solve the equation below to determine the temperature in which both scales are the same.

$$C = \frac{9}{5}C + 32$$

Pressure and Depth As a scuba diver descends deeper into the ocean, the water pressure increases. If x represents the depth (in feet) of the diver, then the relationship $P = 15 + 0.434x$ gives the water pressure P in psi (pounds per square inch). Use this formula to answer the following questions.

47. If the diver is 40 feet below the surface of the water, what is the water pressure (in psi)?

48. Recall the Necker Nymph from this section's opener. If the Necker Nymph dives to its maximum depth of 100 ft, what is the water pressure (in psi)?

49. Oxygen will become toxic at approximately 104.5 psi, resulting in death. At what diving depth will this pressure be reached? (Round to the nearest foot.)

50. The maximum scuba recreational diving pressure is 59.0 psi. What diving depth is this? (Round to the nearest foot.)

51. The deepest individual dive was made by a Navy diver in 2007 using a hardshell suit. His depth was 2,000 ft. What was the water pressure (in psi)?

52. In 2012, filmmaker James Cameron partnered with National Geographic and dove a submarine nearly 36,000 feet in the Pacific Ocean's Mariana Trench. What was the pressure (in psi) on the outside of the submarine?

Getting Ready for the Next Section

Problems under this heading, "Getting Ready for the Next Section", are problems that you must be able to work in order to understand the material in the next section. In this case, the problems below are variations on the type of problems you have already worked in this problem set. They are exactly the type of problems you will see in the explanations and examples in the next section.

Solve each equation.

53. $x \cdot 42 = 21$

54. $x \cdot 84 = 21$

55. $25 = 0.4x$

56. $35 = 0.4x$

57. $12 - 4y = 12$

58. $-6 - 3y = 6$

59. $525 = 900 - 300p$

60. $375 = 900 - 300p$

61. $486.7 = 78.5 + 31.4h$

62. $486.23 = 113.0 + 37.7h$

Find the Mistake

Each sentence below contains a mistake. Circle the mistake and write the correct word on the line provided.

1. A linear equation in one variable is any equation that can be put in the form $P = 2l + 2w$. _____

2. Using the multiplication property of equality to solve the equation $4x - 7 = 5$, we start by adding 7 to both sides of the equation. _____

3. The first step to solving linear equations in one variable is to use the addition property of equality to separate terms. _____

4. An equation is called an inverse if the left side is always identically equal to the right side. _____

Navigation Skills: Prepare, Study, Achieve

Your instructor is a vital resource for your success in this class. Make note of your instructor's office hours and utilize them regularly. Compile a resource list that you keep with your class materials. This list should contain your instructor's office hours and contact information (e.g., office phone number or e-mail), as well as classmates' contact information that you can utilize outside of class. Communicate often with your classmates about how the course is going for you and any questions you may have. Odds are that someone else has the same questions and you may be able to work together to find the answers.

1.2 Formulas

OBJECTIVES

A Solve a formula with numerical replacements for all but one of its variables.

B Solve formulas for the indicated variable.

©iStockphoto.com/chezzers

Three-time Olympic luge champion Georg Hackl uses a modified wok to compete in a lesser known sport called wok racing. The standard wok, imported directly from China, is reinforced with epoxy filling and lined with polyurethane to prevent injury. Hackl, wearing protective gear similar to a hockey uniform, sits in the wok and speeds feet first down an Olympic bobsled track.

Every year, the World Wok Racing Championships are held in Germany. In 2007, Hackl set a speed record of 91.70 km/h on a 1270-meter-long track. Let's suppose Hackl maintained this record speed for the last 100 meters of his run. How long did it take him to complete those last 100 meters?

To solve this problem, we need to find a formula that calculates time using the given quantities of distance and rate. In this section, we review formulas, such as the one needed for this problem. We want to evaluate formulas given specific values for some of its variables, and second, we must learn how to change the form of the formula without changing its meaning.

KEY WORDS

formula

area

perimeter

rate equation

average speed

A Solving Formulas

> **DEFINITION** formula
>
> A *formula* in mathematics is an equation that contains more than one variable.

Some formulas are probably already familiar to you — for example, the formula for the area A of a rectangle with length l and width w is $A = lw$.

To begin our work with formulas, we will consider some examples in which we are given numerical replacements for all but one of the variables.

EXAMPLE 1 Find y when x is 4 in the formula $3x - 4y = 2$.

Solution We substitute 4 for x in the formula and then solve for y.

When → $\qquad x = 4$

the formula → $3x - 4y = 2$

becomes → $\quad 3(4) - 4y = 2$

$\qquad 12 - 4y = 2$ Multiply 3 and 4.

$\qquad -4y = -10$ Add -12 to each side.

$\qquad y = \dfrac{5}{2}$ Divide each side by -4.

Practice Problems

1. Find x when y is 9 in the formula $3x + 2y = 6$.

Answer

1. -4

Note that, in the last line of Example 1, we divided each side of the equation by -4. Remember that this is equivalent to multiplying each side of the equation by $-\frac{1}{4}$. For the rest of the examples in this section, it will be more convenient to think in terms of division rather than multiplication.

2. For Example 2, what price should they charge if they want to sell 510 pads each week?

EXAMPLE 2   A store selling art supplies finds that they can sell x sketch pads each week at a price of p dollars each, according to the formula $x = 900 - 300p$. What price should they charge for each sketch pad if they want to sell 525 pads each week?

Solution   Here we are given a formula, $x = 900 - 300p$, and asked to find the value of p if x is 525. To do so, we simply substitute 525 for x and solve for p.

$$\text{When} \rightarrow \qquad x = 525$$
$$\text{the formula} \rightarrow \quad x = 900 - 300p$$
$$\text{becomes} \rightarrow \quad 525 = 900 - 300p$$
$$-375 = -300p \qquad \text{Add} -900 \text{ to each side.}$$
$$1.25 = p \qquad \text{Divide each side by } -300.$$

To sell 525 sketch pads, the store should charge \$1.25 for each pad.

3. For the boat in Example 3, find c if $d = 45$ miles, $r = 12$ miles per hour, and $t = 5$ hours.

EXAMPLE 3   A boat is traveling upstream against a current. If the speed of the boat in still water is r and the speed of the current is c, then the formula for the distance traveled by the boat is $d = (r - c) \cdot t$, where t is the length of time. Find c if $d = 52$ miles, $r = 16$ miles per hour, and $t = 4$ hours.

Solution   Substituting 52 for d, 16 for r, and 4 for t into the formula, we have

$$52 = (16 - c) \cdot 4$$
$$13 = 16 - c \qquad \text{Divide each side by 4.}$$
$$-3 = -c \qquad \text{Add} -16 \text{ to each side.}$$
$$3 = c \qquad \text{Divide each side by } -1.$$

The speed of the current is 3 miles per hour.

FACTS FROM GEOMETRY   Formulas for Area and Perimeter

To review, here are the formulas for the area and perimeter of some common geometric objects:

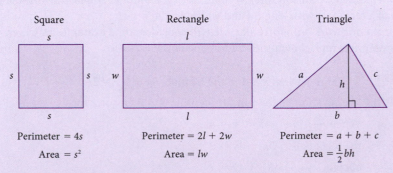

Square
Perimeter $= 4s$
Area $= s^2$

Rectangle
Perimeter $= 2l + 2w$
Area $= lw$

Triangle
Perimeter $= a + b + c$
Area $= \frac{1}{2}bh$

The formula for perimeter gives us the distance around the outside of the object along its sides, while the formula for area gives us a measure of the amount of surface the object covers.

Answers

2. \$1.30
3. $c = 3$

B Solving Formulas for an Indicated Variable

EXAMPLE 4 Given the formula $P = 2w + 2l$, solve for w.

Solution To solve for w, we must isolate it on one side of the equation. We can accomplish this if we delete the $2l$ term and the coefficient 2 from the right side of the equation.

To begin, we add $-2l$ to both sides.

$$P + (-2l) = 2w + 2l + (-2l)$$
$$P - 2l = 2w$$

To delete the 2 from the right side, we can multiply both sides by $\frac{1}{2}$.

$$\frac{1}{2}(P - 2l) = \frac{1}{2}(2w)$$
$$\frac{P - 2l}{2} = w$$

The two formulas

$$P = 2w + 2l \qquad \text{and} \qquad w = \frac{P - 2l}{2}$$

give the relationship between P, l, and w. They look different, but they both say the same thing about P, l, and w. The first formula gives P in terms of l and w, and the second formula gives w in terms of P and l.

EXAMPLE 5 Solve $ax - 3 = bx + 5$ for x.

Solution In this example, we must begin by collecting all the variable terms on the left side of the equation and all the constant terms on the other side (just like we did when we were solving linear equations in Section 1.1):

$$ax - 3 = bx + 5$$
$$ax - bx - 3 = 5 \qquad \text{Add } -bx \text{ to each side.}$$
$$ax - bx = 8 \qquad \text{Add 3 to each side.}$$

At this point, we need to apply the distributive property to write the left side as $(a - b)x$. After that, we divide each side by $a - b$:

$$(a - b)x = 8 \qquad \text{Distributive property}$$
$$x = \frac{8}{a - b} \qquad \text{Divide each side by } a - b.$$

EXAMPLE 6 Solve $\dfrac{y - b}{x - 0} = m$ for y.

Solution Although we will do more extensive work with formulas like this later in the book, we need to know how to solve this particular formula for y in order to understand some things in the next chapter. We begin by simplifying the denominator on the left side and then multiplying each side of the formula by x. Doing so makes the rest of the solution process simple.

$$\frac{y - b}{x - 0} = m \qquad \text{Original formula}$$
$$\frac{y - b}{x} = m \qquad x - 0 = x$$
$$x \cdot \frac{y - b}{x} = m \cdot x \qquad \text{Multiply each side by } x.$$
$$y - b = mx \qquad \text{Simplify each side.}$$
$$y = mx + b \qquad \text{Add } b \text{ to each side.}$$

4. Solve $P = 2w + 2l$ for l.

5. Solve $3x - 5 = bx + 7$ for x.

6. Solve the equation in Example 6 for x.

Answers

4. $l = \frac{P - 2w}{2}$

5. $x = \frac{12}{3 - b}$

6. $x = \frac{y - b}{m}$

This is our solution. If we look back to the first step, we can justify our result on the left side of the equation this way: Dividing by x is equivalent to multiplying by its reciprocal $\frac{1}{x}$. Here is what it looks like when written out completely:

$$x \cdot \frac{y-b}{x} = x \cdot \frac{1}{x} \cdot (y-b) = 1(y-b) = y-b$$

7. Solve $\dfrac{y-7}{x+5} = 4$ for y.

EXAMPLE 7   Solve $\dfrac{y-4}{x-5} = 3$ for y.

Solution   We proceed as we did in the previous example, but this time we clear the formula of fractions by multiplying each side of the formula by $x - 5$.

$$\frac{y-4}{x-5} = 3 \qquad\qquad \text{Original Formula}$$

$$(x-5) \cdot \frac{y-4}{x-5} = 3 \cdot (x-5) \qquad \text{Multiply each side by } (x-5).$$

$$y - 4 = 3x - 15 \qquad\qquad \text{Simplify each side.}$$

$$y = 3x - 11 \qquad\qquad \text{Add 4 to each side.}$$

We have solved for y. We can justify our result on the left side of the equation this way:

$$(x-5) \cdot \frac{y-4}{x-5} = (x-5) \cdot \frac{1}{x-5} \cdot (y-4) = 1(y-4) = y-4$$

Dividing by $x - 5$ is equivalent to multiplying by its reciprocal $\frac{1}{x-5}$.

8. Solve the formula
$V = lwh - \left(\dfrac{w}{2}\right)^2 \pi h$ for h.

EXAMPLE 8   Solve the formula $S = 2\pi rh + 2\pi r^2$ for h.

Solution   This is the formula for the surface area of a right circular cylinder, with radius r and height h, that is closed at both ends. To isolate h, we first add $-2\pi r^2$ to both sides.

$$S + (-2\pi r^2) = 2\pi rh + 2\pi r^2 + (-2\pi r^2)$$

$$S - 2\pi r^2 = 2\pi rh$$

Next, we divide each side by $2\pi r$.

$$\frac{S - 2\pi r^2}{2\pi r} = h$$

Exchanging the two sides of our equation, we have

$$h = \frac{S - 2\pi r^2}{2\pi r}$$

Rate Equation and Average Speed

Now we will look at some problems that use what is called the *rate equation*. You use this equation on an intuitive level when you are estimating how long it will take you to drive long distances. For example, if you drive at 50 miles per hour for 2 hours, you will travel 100 miles. Here is the rate equation:

$$\text{Distance} = \text{rate} \cdot \text{time, or } d = r \cdot t$$

The rate equation has two equivalent forms, one of which is obtained by solving for r, while the other is obtained by solving for t. Here they are:

$$r = \frac{d}{t} \quad \text{and} \quad t = \frac{d}{r}$$

The rate in this equation is also referred to as average speed.

Answers

7. $y = 4x + 27$

8. $h = \dfrac{V}{lw - \left(\frac{w}{2}\right)^2 \pi}$

The *average speed* of a moving object is defined to be the ratio of distance to time. If you drive your car for 5 hours and travel a distance of 200 miles, then your average rate of speed is

$$\text{Average speed} = \frac{200 \text{ miles}}{5 \text{ hours}} = 40 \text{ miles per hour}$$

Our next example involves both the formula for the circumference of a circle and the rate equation.

EXAMPLE 9 The first Ferris wheel was designed and built by George Ferris in 1893. The diameter of the wheel was 250 feet. It had 36 carriages, equally spaced around the wheel, each of which held a maximum of 40 people. One trip around the wheel took 20 minutes. Find the average speed of a rider on the first Ferris wheel. (Use 3.14 as an approximation for π.)

Solution Using 3.14 as an approximation for π the distance traveled is the circumference of the wheel, which is

$$C = \pi d = 250\pi = 250(3.14) = 785 \text{ feet}$$

To find the average speed, we divide the distance traveled by the amount of time it took to go once around the wheel.

$$r = \frac{d}{t} = \frac{785 \text{ feet}}{20 \text{ minutes}} = 39.3 \text{ feet per minute (to the nearest tenth)}$$

GETTING READY FOR CLASS

After reading through the preceding section, respond in your own words and in complete sentences.

A. What is a formula?

B. How would you solve the formula $2x + 4y = 10$ if $x = 3$

C. How would you solve for l in the perimeter formula for a rectangle?

D. Give two equivalent forms of the rate equation $d = rt$.

9. If the Ferris wheel in Example 9 has a diameter of 300 feet, what is the average speed of a rider. (Assume it takes the same amount of time for one rotation.)

Answer

9. 47.1 ft/min

EXERCISE SET 1.2

Vocabulary Review

Choose the correct words to fill in the blanks below.

perimeter   formula   area   average speed   rate equation   variable

1. An equation that contains more than one variable is called a _____.

2. The formula for the _____ of a triangle is given by $A = \dfrac{1}{2}bh$.

3. The formula for the _____ of a rectangle is given by $P = 2l + 2w$.

4. To solve for an indicated _____, we must isolate it on one side of the equation.

5. The three variable quantities in the _____ are distance, rate, and time.

6. The _____ of a moving object is the ratio of distance to time.

Problems

A Use the formula $3x - 4y = 12$ to find y if

1. x is 0

2. x is -2

3. x is 4

4. x is -4

Use the formula $y = 2x - 3$ to find x when

5. y is 0

6. y is -3

7. y is 5

8. y is -5

Problems 9 through 24 are problems that you will see later in the text.

9. If $x - 2y = 4$ and $y = -\dfrac{6}{5}$, find x.

10. If $x - 2y = 4$ and $x = \dfrac{8}{5}$, find y.

11. Let $x = 160$ and $y = 0$ in $y = a(x - 80)^2 + 70$ and solve for a.

12. Let $x = 0$ and $y = 0$ in $y = a(x - 80)^2 + 70$ and solve for a.

13. Find R if $p = 1.5$ and $R = (900 - 300p)p$.

14. Find R if $p = 2.5$ and $R = (900 - 300p)p$.

SCAN TO ACCESS

15. Find P if $P = -0.1x^2 + 27x + 1,700$ and

    a. $x = 100$

    b. $x = 170$

16. Find P if

    $P = -0.1x^2 + 27x + 1,820$ and

    a. $x = 130$

    b. $x = 140$

17. Find h if $h = 16 + 32t - 16t^2$ and

    a. $t = \dfrac{1}{4}$

    b. $t = \dfrac{7}{4}$

18. Find h if $h = 64t - 16t^2$ and

    a. $t = 1$

    b. $t = 3$

19. Use the formula $d = (r - c)t$ to find c if $d = 30$, $r = 12$, and $t = 3$.

20. Use the formula $d = (r - c)t$ to find r if $d = 49$, $c = 4$, and $t = 3.5$.

21. If $y = kx$, find k if $x = 5$ and $y = 15$.

22. If $d = kt^2$, find k if $t = 2$ and $d = 64$.

23. If $V = \dfrac{k}{P}$, find k if $P = 48$ and $V = 50$.

24. If $y = kxz^2$, find k if $x = 5$, $z = 3$, and $y = 180$.

Use the formula $5x - 3y = -15$ to find y if

25. $x = 2$

26. $x = -3$

27. $x = -\dfrac{1}{5}$

28. $x = 3$

B Solve each of the following formulas for the indicated variable.

29. $d = rt$ for r

30. $d = rt$ for t

31. $d = (r + c)t$ for t

32. $d = (r + c)t$ for r

33. $A = lw$ for l

34. $A = \dfrac{1}{2}bh$ for b

35. $I = prt$ for t

36. $I = prt$ for r

37. $PV = nRT$ for T

38. $PV = nRT$ for R

39. $y = mx + b$ for x

40. $A = P + Prt$ for t

41. $C = \dfrac{5}{9}(F - 32)$ for F

42. $F = \dfrac{9}{5}C + 32$ for C

43. $h = vt + 16t^2$ for v

44. $h = vt - 16t^2$ for v

45. $A = a + (n - 1)d$ for d

46. $A = a + (n - 1)d$ for n

47. $2x + 3y = 6$ for y

48. $2x - 3y = 6$ for y

49. $-3x + 5y = 15$ for y

50. $-2x - 7y = 14$ for y

51. $2x - 6y + 12 = 0$ for y

52. $7x - 2y - 6 = 0$ for y

53. $ax + 4 = bx + 9$ for x

54. $ax - 5 = cx - 2$ for x

55. $S = \pi r^2 + 2\pi rh$ for h

56. $A = P + Prt$ for P

57. $-3x + 4y = 12$ for x

58. $-3x + 4y = 12$ for y

59. $ax + 3 = cx - 7$ for x

60. $by - 9 = dy + 3$ for y

Problems 61 through 68 are problems that you will see later in the text. Solve each formula for y.

61. $x = 2y - 3$

62. $x = 4y + 1$

63. $y - 3 = -2(x + 4)$

64. $y - 1 = \dfrac{1}{4}(x - 3)$

65. $y - 3 = -\dfrac{2}{3}(x + 3)$

66. $y + 1 = -\dfrac{2}{3}(x - 3)$

67. $y - 4 = -\dfrac{1}{2}(x + 1)$

68. $y - 2 = \dfrac{1}{3}(x - 1)$

69. Solve for y.

   a. $\dfrac{y + 1}{x - 0} = 4$

   b. $\dfrac{y + 2}{x - 4} = -\dfrac{1}{2}$

   c. $\dfrac{y + 3}{x - 7} = 0$

70. Solve for y.

   a. $\dfrac{y - 1}{x - 0} = -3$

   b. $\dfrac{y - 2}{x - 6} = \dfrac{2}{3}$

   c. $\dfrac{y - 3}{x - 1} = 0$

Solve for y.

71. $\dfrac{x}{8} + \dfrac{y}{2} = 1$

72. $\dfrac{x}{7} + \dfrac{y}{9} = 1$

73. $\dfrac{x}{5} + \dfrac{y}{-3} = 1$

74. $\dfrac{x}{16} + \dfrac{y}{-2} = 1$

The next two problems are intended to give you practice reading, and paying attention to, the instructions that accompany the problems you are working.

75. Work each problem according to the instructions given.

   a. Solve: $-4x + 5 = 20$.

   b. Find the value of $-4x + 5$ when $x = 3$.

   c. Solve $-4x + 5y = 20$ for y.

   d. Solve $-4x + 5y = 20$ for x.

76. Work each problem according to the instructions given.

   a. Solve: $2x + 1 = -4$.

   b. Find the value of $2x + 1$ when $x = 8$.

   c. Solve $2x + y = 20$ for y.

   d. Solve $2x + y = 20$ for x.

Applying the Concepts

Estimating Vehicle Weight   If you can measure the area that the tires on your car contact the ground and know the air pressure in the tires, then you can estimate the weight of your car with the following formula:

$$W = \dfrac{APN}{2{,}000}$$

where W is the vehicle's weight in tons, A is the average tire contact area with a hard surface in square inches, P is the air pressure in the tires in pounds per square inch (psi, or lb/in^2), and N is the number of tires.

77. What is the approximate weight of a car if the average tire contact area is a rectangle 6 inches by 5 inches and if the air pressure in the tires is 30 psi?

78. What is the approximate weight of a car if the average tire contact area is a rectangle 5 inches by 4 inches and the tire pressure is 30 psi?

79. Current   It takes a boat 2 hours to travel 18 miles upstream against the current. If the speed of the boat in still water is 15 miles per hour, what is the speed of the current?

80. Current   It takes a boat 6.5 hours to travel 117 miles upstream against the current. If the speed of the current is 5 miles per hour, what is the speed of the boat in still water?

81. Wind   An airplane takes 4 hours to travel 864 miles while flying against the wind. If the speed of the airplane on a windless day is 258 miles per hour, what is the speed of the wind?

82. Wind   A cyclist takes 3 hours to travel 39 miles while pedaling against the wind. If the speed of the wind is 4 miles per hour, how fast would the cyclist be able to travel on a windless day?

For problems 83 and 84, use 3.14 as an approximation for π. Round answers to the nearest tenth.

83. Average Speed A person riding a Ferris wheel with a diameter of 65 feet travels once around the wheel in 30 seconds. What is the average speed of the rider in feet per second?

84. Average Speed A person riding a Ferris wheel with a diameter of 102 feet travels once around the wheel in 3.5 minutes. What is the average speed of the rider in feet per minute?

Digital Video Videos on the internet are compressed so they will be small enough for people to download. A formula for estimating the size, in kilobytes, of a compressed video is

$$S = \frac{height \cdot width \cdot fps \cdot time}{35{,}000}$$

where *height* and *width* are in pixels, *fps* is the number of frames per second the video is to play (television plays at 30 fps), and *time* is given in seconds.

85. Estimate the size in kilobytes of the *Star Wars* trailer that has a height of 480 pixels, a width of 216 pixels, plays at 30 fps, and runs for 150 seconds.

86. Estimate the size in kilobytes of the *Star Wars* trailer that has a height of 320 pixels, a width of 144 pixels, plays at 15 fps, and runs for 150 seconds.

87. The Taj Mahal is estimated to have cost 50 million rupees to build. If 1 gram of gold costs 1.4 rupees, find the cost of the Taj Mahal in grams of gold.

88. If the current price of gold is $43 per gram, use the information from problem 87 to find the cost of building the Taj Mahal in current dollars.

Fermat's Last Theorem Fermat's last theorem states that if n is an integer greater than 2, then there are no positive integers x, y, and z that will make the formula $x^n + y^n = z^n$ true. Use the formula $x^n + y^n = z^n$ to

89. Find x if $n = 1$, $y = 7$, and $z = 15$.

90. Find y if $n = 1$, $x = 23$, and $z = 37$.

Exercise Physiology In exercise physiology, a person's maximum heart rate, in beats per minute, is found by subtracting her age, in years, from 220. So, if A represents your age in years, then your maximum heart rate is

$$M = 220 - A$$

A person's training heart rate, in beats per minute, is her resting heart rate plus 60% of the difference between her maximum heart rate and her resting heart rate. If resting heart rate is R and maximum heart rate is M, then the formula that gives training heart rate is

$$T = R + 0.6(M - R)$$

91. Training Heart Rate Shar is 46 years old. Her daughter, Sara, is 26 years old. If they both have a resting heart rate of 60 beats per minute, find the training heart rate for each.

92. Training Heart Rate Shane is 30 years old and has a resting heart rate of 68 beats per minute. Her mother, Carol, is 52 years old and has the same resting heart rate. Find the training heart rate for Shane and for Carol.

Getting Ready for the Next Section

To understand all of the explanations and examples in the next section you must be able to work the problems below. Translate into symbols.

93. Three less than twice a number

94. Ten less than four times a number

95. The sum of x and y is 180.

96. The sum of a and b is 90.

Solve each equation.

97. $x + 2x = 90$

98. $x + 5x = 180$

99. $2(2x - 3) + 2x = 45$

100. $2(4x - 10) + 2x = 12.5$

101. $0.06x + 0.05(10{,}000 - x) = 560$

102. $x + 0.0725x = 17{,}481.75$

One Step Further

103. Solve $\dfrac{x}{a} + \dfrac{y}{b} = 1$ for x; $a, b \neq 0$.

104. Solve $\dfrac{x}{a} + \dfrac{y}{b} = 1$ for y; $a, b \neq 0$.

105. Solve $\dfrac{1}{a} + \dfrac{1}{b} = \dfrac{1}{c}$ for a; $a, b, c \neq 0$.

106. Solve $\dfrac{1}{a} + \dfrac{1}{b} = \dfrac{1}{c}$ for b; $a, b, c \neq 0$.

Find the Mistake

Each problem below contains a mistake. Circle the mistake and write the correct word or answer on the line provided.

1. A formula is an equation that has one variable. _____

2. Solving the formula $y - 3 = -2(x - 5)$ for y, you get $y = -6x + 30$. _____

3. The value of y in the formula $4y + 2x = 20$ when $x = 8$ is 9. _____

4. The rate equation is given by distance = rate · height. _____

1.3 Applications

An air-powered stomp rocket can be propelled over 200 feet using a blast of air. The harder you stomp on the launch pad, the farther the rocket flies.

If the rocket is launched straight up into the air with a velocity of 112 feet per second, then the formula

$$h = -16t^2 + 112t$$

gives the height h of the rocket t seconds after it is launched. We can use this formula to find the height of the rocket 3.5 seconds after launch by substituting $t = 3.5$.

$$h = -16(3.5)^2 + 112(3.5) = 196$$

At 3.5 seconds, the rocket reaches a height of 196 feet.

A Blueprint for Problem Solving

In this section, we begin our work with application problems and use the skills we have developed for solving equations to solve problems written in words. You may find that some of the examples and problems are more realistic than others. Because we are just beginning our work with application problems, even the ones that seem unrealistic are good practice. What is important in this section is the method we use to solve application problems, not the applications themselves. The method, or strategy, that we use to solve application problems is called the *Blueprint for Problem Solving*. It is an outline that will overlay the solution process we use on all application problems.

BLUEPRINT FOR PROBLEM SOLVING

Step 1: *Read* the problem, and then *list* the items that are known and the items that are unknown.

Step 2: *Assign a variable* to one of the unknown items. (In most cases, this will amount to letting $x =$ the item that is asked for in the problem.) Then *translate* the other *information* in the problem to expressions involving the variable.

Step 3: *Reread* the problem, and then *write an equation,* using the items and variable listed in steps 1 and 2, that describes the situation.

Step 4: *Solve the equation* found in step 3.

Step 5: *Write your answer* using a complete sentence.

Step 6: *Reread* the problem, and *check* your solution with the original words in the problem.

A number of substeps occur within each of the steps in our blueprint. For instance, with steps 1 and 2 it is always a good idea to draw a diagram or picture if it helps you visualize the relationship among the items in the problem.

EXAMPLE 1 The length of a rectangle is 3 inches less than twice the width. The perimeter is 45 inches. Find the length and width.

Solution When working problems that involve geometric figures, a sketch of the figure helps organize and visualize the problem.

Step 1: *Read and list.*
Known items: The figure is a rectangle. The length is 3 inches less than twice the width. The perimeter is 45 inches.
Unknown items: The length and the width

Step 2: *Assign a variable and translate information.*
Because the length is given in terms of the width (the length is 3 less than twice the width), we let x = the width of the rectangle. The length is 3 less than twice the width, so it must be $2x - 3$. The diagram in Figure 1 is a visual description of the relationships we have listed so far.

$2x - 3$

x

FIGURE 1

Step 3: *Reread and write an equation.*
The equation that describes the situation is

Twice the length + twice the width = perimeter
$$2(2x - 3) \quad + \quad 2x \quad = \quad 45$$

Step 4: *Solve the equation.*
$$2(2x - 3) + 2x = 45$$
$$4x - 6 + 2x = 45$$
$$6x - 6 = 45$$
$$6x = 51$$
$$x = 8.5$$

Step 5: *Write the answer.*
The width is 8.5 inches. The length is $2x - 3 = 2(8.5) - 3 = 14$ inches.

Step 6: *Reread and check.*
If the length is 14 inches and the width is 8.5 inches, then the perimeter must be $2(14) + 2(8.5) = 28 + 17 = 45$ inches. Also, the length, 14, is 3 less than twice the width.

Finding the solution to an application problem is a process; it doesn't happen all at once. The first step is to read the problem with a purpose in mind. That purpose is to mentally note the items that are known and the items that are unknown.

EXAMPLE 2 In April, Pat bought a Ford Mustang with a 5.0-liter engine. The total price, which includes the price of the used car plus sales tax, was $17,481.75. If the sales tax rate was 7.25%, what was the price of the car?

Solution

Step 1: *Read and list.*
Known items: The total price is $17,481.75. The sales tax rate is 7.25%, which is 0.0725 in decimal form.
Unknown item: The price of the car

Step 2: *Assign a variable and translate information.*
If we let x = the price of the car, then to calculate the sales tax we multiply the price of the car x by the sales tax rate:
Sales tax = (sales tax rate)(price of the car)
$$= 0.0725x$$

Step 3: *Reread and write an equation.*

Car price + sales tax = total price
$$x + 0.0725x = 17,481.75$$

Step 4: *Solve the equation.*

$$x + 0.0725x = 17,481.75$$
$$1.0725x = 17,481.75$$
$$x = \frac{17,481.75}{1.0725}$$
$$= 16,300.00$$

Step 5: *Write the answer.*
The price of the car is $16,300.00.

Step 6: *Reread and check.*
The price of the car is $16,300.00. The tax is 0.0725(16,300) = $1,181.75. Adding the retail price and the sales tax we have a total bill of $17,481.75.

FACTS FROM GEOMETRY Angles

An angle is formed by two rays with the same endpoint. The common endpoint is called the *vertex* of the angle, and the rays are called the *sides* of the angle.

In Figure 2, angle θ (theta) is formed by the two rays *OA* and *OB*. The vertex of θ is *O*. Angle θ is also denoted as angle *AOB*, where the letter associated with the vertex is always the middle letter in the three letters used to denote the angle.

Degree Measure
The angle formed by rotating a ray through one complete revolution about its endpoint (Figure 3) has a measure of 360 degrees, which we write as 360°.

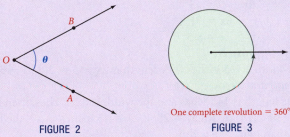

FIGURE 2   One complete revolution = 360°
                                   FIGURE 3

One degree of angle measure, written 1°, is $\frac{1}{360}$ of a complete rotation of a ray about its endpoint; there are 360° in one full rotation. (The number 360 was decided upon by early civilizations because it was believed that the Earth was at the center of the universe and the sun would rotate once around Earth every 360 days.) Similarly, 180° is half of a complete rotation, and 90° is a quarter of a full rotation. Angles that measure 90° are called *right angles*, and angles that measure 180° are called *straight angles*. If an angle measures between 0° and 90° it is called an *acute angle*, and an angle that measures between 90° and 180° is an *obtuse angle*. Figure 4 illustrates further.

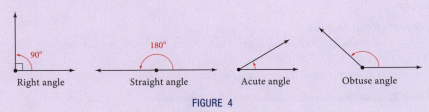

Right angle   Straight angle   Acute angle   Obtuse angle

FIGURE 4

FACTS FROM GEOMETRY Angles (continued)

Complementary Angles and Supplementary Angles
If two angles add up to 90°, we call them *complementary angles*, and each is called the *complement* of the other. If two angles have a sum of 180°, we call them *supplementary angles*, and each is called the *supplement* of the other. Figure 5 illustrates the relationship between angles that are complementary and angles that are supplementary.

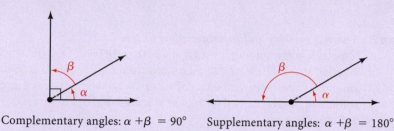

Complementary angles: $\alpha + \beta = 90°$ Supplementary angles: $\alpha + \beta = 180°$

FIGURE 5

EXAMPLE 3   Two complementary angles are such that one is twice as large as the other. Find the two angles.

Solution   Applying the Blueprint for Problem Solving, we have

Step 1: *Read and list.*
    Known items: Two complementary angles. One is twice as large as the other.
    Unknown items: The size of the angles

Step 2: *Assign a variable and translate information.*
    Let $x =$ the smaller angle. The larger angle is twice the smaller, so we represent the larger angle with $2x$.

Step 3: *Reread and write an equation.*
    Because the two angles are complementary, their sum is 90. Therefore,

$$x + 2x = 90$$

Step 4: *Solve the equation.*

$$x + 2x = 90$$
$$3x = 90$$
$$x = 30$$

Step 5: *Write the answer.*
    The smaller angle is 30°, and the larger angle is $2 \cdot 30 = 60°$.

Step 6: *Reread and check.*
    The larger angle is twice the smaller angle, and their sum is 90°.

Suppose we know that the sum of two numbers is 50. If we let x represent one of the two numbers, how can we represent the other? Let's suppose for a moment that x turns out to be 30. Then the other number will be 20, because their sum is 50. That is, if two numbers add up to 50, and one of them is 30, then the other must be $50 - 30 = 20$. Generalizing this to any number x, we see that if two numbers have a sum of 50, and one of the numbers is x, then the other must be $50 - x$.

The following table shows some additional examples:

If two numbers have a sum of	and one of them is	then the other must be
50	x	$50 - x$
10	y	$10 - y$
12	n	$12 - n$

3. Two supplementary angles are such that one is 24 degrees more than twice the other. Find the two angles.

4. Suppose a person put $3,000 more in an account that pays 5% annual interest than in an account that pays 4% annual interest. If the total interest earned from both accounts in a year is $510, how much was invested in each account?

EXAMPLE 4   Suppose a person invests a total of $10,000 in two accounts. One account earns 5% annually, and the other earns 6% annually. If the total interest earned from both accounts in a year is $560, how much is invested in each account?

Solution

Step 1:   *Read and list.*

Known items: Two accounts. One pays interest of 5%, and the other pays 6%. The total invested is $10,000.

Unknown items: The number of dollars invested in each individual account

Step 2:   *Assign a variable and translate information.*

If we let $x =$ the amount invested at 6%, then $10,000 - x$ is the amount invested at 5%. The total interest earned from both accounts is $560. The amount of interest earned on x dollars at 6% is $0.06x$, whereas the amount of interest earned on $10,000 - x$ dollars at 5% is $0.05(10,000 - x)$.

	Dollars at 6%	Dollars at 5%	Total
Number of	x	$10,000 - x$	10,000
Interest on	$0.06x$	$0.05(10,000 - x)$	560

Step 3:   *Reread and write an equation.*

The last line gives us the equation we are after:

$$0.06x + 0.05(10,000 - x) = 560$$

Step 4:   *Solve the equation.*

To make the equation a little easier to solve, we begin by multiplying both sides by 100 to move the decimal point two places to the right.

$$6x + 5(10,000 - x) = 56,000$$
$$6x + 50,000 - 5x = 56,000$$
$$x + 50,000 = 56,000$$
$$x = 6,000$$

Step 5:   *Write the answer.*

The amount of money invested at 6% is $6,000. The amount of money invested at 5% is $10,000 - $6,000 = $4,000.

Step 6:   *Reread and check.*

To check our results, we find the total interest from the two accounts:

The interest earned on $6,000 at 6% is $0.06(6,000) = $ 360
The interest earned on $4,000 at 5% is $0.05(4,000) = $ 200

The total interest $= \$560$

Answers

4. $4,000 at 4%, $7,000 at 5%

FACTS FROM GEOMETRY Special Triangles

An *isosceles triangle* as shown in Figure 6, is a triangle with two sides of equal length. Angles *A* and *B* in the isosceles triangle in Figure 6 are called the *base angles*: they are the angles opposite the two equal sides. In every isosceles triangle, the base angles are equal.

Isosceles Triangle
$a = b$

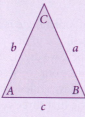

FIGURE 6

Equilateral Triangle
$a = b = c$

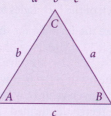

FIGURE 7

An equilateral triangle (Figure 7) is a triangle with three sides of equal length. If all three sides in a triangle have the same length, then the three interior angles in the triangle also must be equal. Because the sum of the interior angles in a triangle is always 180°, the three interior angles in any equilateral triangle must be 60°.

EXAMPLE 5 The base angles in an isosceles triangle are twice as large as the third angle. Find the measure of each angle.

Solution

Step 1: *Read and list.*
Known items: We have an isosceles triangle. The base angles are twice as large as the third angle.
Unknown items: The measure of each angle.

Step 2: *Assign a variable and translate information.*
Let x = the measure of the third angle, so the two base angles measure $2x$. The sum of all three angles is 180.

Step 3: *Reread and write an equation.*
$$x + 2x + 2x = 180$$

Step 4: *Solve the equation.*

$$5x = 180 \qquad \text{Combine like terms.}$$
$$x = 36 \qquad \text{Divide both sides by 5.}$$

Step 5: *Write the answer.*
The third angle measures 36. Two base angles measure $36 \cdot 2 = 72$.

Step 6: *Reread and check.*
The two base angles measure 72, twice the measure of the third angle. Their sum is 180.

5. The base angles in an isosceles triangle are four times as large as the third angle. Find the measure of each angle.

Answers

5. Base angles = 80°,
   third angle = 20°

6. If the string in Example 6 is 16 inches long, what is the length and width of a rectangle that will yield the largest area?

B Table Building

We can use our knowledge of formulas from Section 1.2 to build tables of paired data. As you will see, equations or formulas that contain exactly two variables produce pairs of numbers that can be used to construct tables.

EXAMPLE 6   A piece of string 12 inches long is to be formed into a rectangle. Build a table that gives the length of the rectangle if the width is 1, 2, 3, 4, or 5 inches. Then find the area of each of the rectangles formed.

Solution   Because the formula for the perimeter of a rectangle is $P = 2l + 2w$, and our piece of string is 12 inches long, the formula we will use to find the lengths for the given widths is $12 = 2l + 2w$. To solve this formula for l, we divide each side by 2 and then subtract w. The result is $l = 6 - w$. Table 1 organizes our work so that the formula we use to find l for a given value of w is shown, and we have added a last column to give us the areas of the rectangles formed. The units for the first three columns are inches, and the units for the numbers in the last column are square inches.

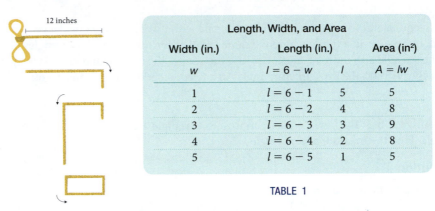

	Length, Width, and Area		
Width (in.)	Length (in.)		Area (in²)
w	$l = 6 - w$	l	$A = lw$
1	$l = 6 - 1$	5	5
2	$l = 6 - 2$	4	8
3	$l = 6 - 3$	3	9
4	$l = 6 - 4$	2	8
5	$l = 6 - 5$	1	5

TABLE 1

Figures 8 and 9 show two *bar charts* constructed from the information in Table 1.

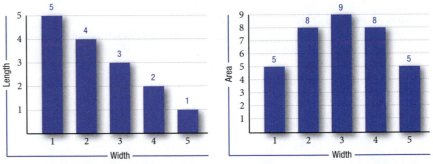

FIGURE 8 *Length and width of rectangles with perimeters fixed at 12 inches*

FIGURE 9 *Area and width of rectangles with perimeters fixed at 12 inches*

GETTING READY FOR CLASS

After reading through the preceding section, respond in your own words and in complete sentences.

A. What is the first step in solving an application problem?

B. What is the biggest obstacle between you and success in solving application problems?

C. Write an application problem for which the solution depends on solving the equation $2x + 2 \cdot 3 = 18$.

D. What is the last step in solving an application problem? Why is this step important?

Answers

6. $l = w = 4$ in

Vocabulary Review

Choose the correct words to fill in the blanks below.

supplementary   straight   equilateral   vertex   right
complementary   obtuse   acute   isosceles

1. The common endpoint of two rays of an angle is called the _____.

2. An angle that measures 90° is called a _____ angle, whereas an angle that measures 180° is called a _____ angle.

3. An angle that measures between 0° and 90° is an _____ angle, whereas an angle that measures between 90° and 180° is an _____ angle.

4. Two angles with a sum of 90° are considered _____ angles.

5. Two angles with a sum of 180° are considered _____ angles.

6. A triangle with two sides of equal length is an _____ triangle, whereas a triangle with three sides of equal length is an _____ triangle.

The following is an out-of-order list of steps for the Blueprint for Problem Solving. Write the correct step number in the blanks.

_____ Assign a variable to one of the unknown items, and then translate the remaining information into expressions involving the variable.

_____ Solve the equation.

_____ Read the problem, and then mentally list the known and unknown items.

_____ Reread the problem, and then write an equation.

_____ Reread the problem, and then check your solution.

_____ Write your answer using a complete sentence.

A Solve each application problem. Be sure to follow the steps in the Blueprint for Problem Solving.

Perimeter Problems

1. **Rectangle** A rectangle is twice as long as it is wide. The perimeter is 60 feet. Find the dimensions.

2. **Rectangle** The length of a rectangle is 5 times the width. The perimeter is 48 inches. Find the dimensions.

3. **Square** A square has a perimeter of 28 feet. Find the length of each side.

4. **Square** A square has a perimeter of 36 centimeters. Find the length of each side.

5. **Triangle** A triangle has a perimeter of 23 inches. The medium side is 3 inches more than the shortest side, and the longest side is twice the shortest side. Find the shortest side.

6. **Triangle** The longest side of a triangle is two times the shortest side, while the medium side is 3 meters more than the shortest side. The perimeter is 27 meters. Find the dimensions.

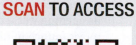

SCAN TO ACCESS

7. **Rectangle** The length of a rectangle is 3 meters less than twice the width. The perimeter is 18 meters. Find the width.

8. **Rectangle** The length of a rectangle is 1 foot more than twice the width. The perimeter is 20 feet. Find the dimensions.

9. **Livestock Pen** A livestock pen is built in the shape of a rectangle that is twice as long as it is wide. The perimeter is 48 feet. If the material used to build the pen is $1.75 per foot for the longer sides and $2.25 per foot for the shorter sides (the shorter sides have gates, which increase the cost per foot), find the cost to build the pen.

10. **Garden** A garden is in the shape of a square with a perimeter of 42 feet. The garden is surrounded by two fences. One fence is around the perimeter of the garden, whereas the second fence is 3 feet out from the first fence, as Figure 10 indicates. If the material used to build the two fences is $1.28 per foot, what was the total cost of the fences?

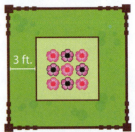

3 ft.

FIGURE 10

Percent Problems

11. **Money** Eric returned from a trip to Las Vegas with $300.00, which was 50% more money than he had at the beginning of the trip. How much money did Eric have at the beginning of his trip?

12. **Items Sold** Every item in the Just a Dollar store is priced at $1.00. When Mary Jo opens the store, there is $125.50 in the cash register. When she counts the money in the cash register at the end of the day, the total is $1,058.60. If the sales tax rate is 8.5%, how many items were sold that day?

13. **Textbook Price** Suppose a college bookstore buys a textbook from a publishing company and then marks up the price they paid for the book 33% and sells it to a student at the marked-up price. If the student pays $115.00 for the textbook, what did the bookstore pay for it? Round your answer to the nearest cent.

14. **Hourly Wage** A sheet metal worker earns $26.80 per hour after receiving a 4.5% raise. What was the sheet metal worker's hourly pay before the raise? Round your answer to the nearest cent.

15. **Movies** *Alice is Wonderland* grossed $116.3 million on its opening weekend and had the most successful 3-D movie launch in history at the time. If regular screenings accounted for approximately 30% of the revenue, how much of the box office receipts resulted from 3-D screenings?

16. **Fat Content in Milk** I was reading the information on a milk carton in at breakfast one morning when I was working on this book. According to the carton, this milk contains 70% less fat than whole milk. The nutrition label on the other side of the carton states that one serving of this milk contains 2.5 grams of fat. How many grams of fat are in an equivalent serving of whole milk?

Angle Problems

17. Angles Two supplementary angles are such that one is eight times larger than the other. Find the two angles.

18. Angles Two complementary angles are such that one is five times larger than the other. Find the two angles.

19. Angles One angle is 12° less than four times another. Find the measure of each angle if

    a. they are complements of each other.

    b. they are supplements of each other.

20. Angles One angle is 4° more than three times another. Find the measure of each angle if

    a. they are complements of each other.

    b. they are supplements of each other.

21. Triangles A triangle is such that the largest angle is three times the smallest angle. The third angle is 9° less than the largest angle. Find the measure of each angle.

22. Triangles The smallest angle in a triangle is half of the largest angle. The third angle is 15° less than the largest angle. Find the measure of all three angles.

23. Triangles The smallest angle in a triangle is one-third of the largest angle. The third angle is 10° more than the smallest angle. Find the measure of all three angles.

24. Triangles The third angle in an isosceles triangle is half as large as each of the two base angles. Find the measure of each angle.

25. Isosceles Triangles The third angle in an isosceles triangle is 8° more than twice as large as each of the two base angles. Find the measure of each angle.

26. Isosceles Triangles The third angle in an isosceles triangle is 4° more than one fifth of each of the two base angles. Find the measure of each angle.

Interest Problems

Answer the following interest problems assuming no payments were made in the first year.

27. Credit Cards Donna has two credit cards with a total balance of $9,000. One card has an 8% interest rate per year and the other charges 9%. If the interest charged in the first year is $750, what is the balance on each card?

	Dollars at 8%	Dollars at 9%	Total
Number of			
Interest on			

28. Credit Cards Matt has two credit cards with a total balance of $12,000. If one card charges 10% per year and the other charges 7% per year, what is the balance on each card if the total interest charged in the first year was $960?

	Dollars at 10%	Dollars at 7%	Total
Number of			
Interest on			

29. Loans Bill took out two loans that totaled $15,000. One of the loans charges 12% per year, and the other charges 10% per year. If the total interest charged in the first year is $1,600, how much was each loan?

30. Loans Nancy qualified for two loans that totalled $11,000. One loan charges 9% per year, and the other loan charges 11% per year. If the total interest charged in the first year is $1,150, how much was each loan?

31. Loans Stacy owes a total of $6,000 from two student loans. The total amount of interest she is charged from both accounts in the first year is $500. If one loan charges 8% interest per year and the other charges 9% interest per year, how much was each loan?

32. Loans Travis has a total of $6,000 from two auto loans. The total amount of interest charged in the first year is $410. If one loan charges 6% per year and the other charges 8% per year, how much is each loan?

Miscellaneous Problems

33. Tickets Tickets for the father-and-son breakfast were $5.00 for fathers and $3.50 for sons. If a total of 75 tickets were sold for $307.50, how many fathers and how many sons attended the breakfast?

34. Tickets A Girl Scout troop sells 62 tickets to their mother-and-daughter dinner, for a total of $216. If the tickets cost $4.00 for mothers and $3.00 for daughters, how many of each ticket did they sell?

35. Sales Tax Charlotte owns a small, cash-only business in a state that requires her to charge 6% sales tax on each item she sells. At the beginning of the day, she has $250 in the cash register. At the end of the day, she has $1,204 in the register. How much money should she send to the state government for the sales tax she collected?

36. Sales Tax A store is located in a state that requires 6% tax on all items sold. If the store brings in $3,392 in one day, how much of that total was sales tax?

B Table Building

37. Use $h = 32t - 16t^2$ to complete the table.

t	0	$\frac{1}{4}$	1	$\frac{7}{4}$	2
h					

38. Use $h = \dfrac{60}{t}$ to complete the table.

t	4	6	8	10
h				

39. Distance A search is being conducted for someone guilty of a hit-and-run felony. In order to set up roadblocks at appropriate points, the police must determine how far the guilty party might have traveled during the past half-hour. Use the formula $d = rt$ with $t = 0.5$ hour to complete the following table.

Speed (miles per hour)	Distance (miles)
20	
30	
40	
50	
60	
70	

40. Speed To determine the average speed of a bullet when fired from a rifle, the time is measured from when the gun is fired until the bullet hits a target that is 1,000 feet away. Use the formula $r = \dfrac{d}{t}$ with $d = 1,000$ feet to complete the following table.

Time (seconds)	Rate (feet per second)
1.00	
0.80	
0.64	
0.50	
0.40	
0.32	

41. Current A boat that can travel a rate r of 10 miles per hour in still water is traveling along a stream with a current c of 4 miles per hour. The distance d the boat will travel upstream is given by the formula $d = (r - c) \cdot t$, where t represents time. The distance it will travel downstream is given by the formula $d = (r + c) \cdot t$. Use these formulas with $r = 10$ and $c = 4$ to complete the following table.

Time (hours)	Distance upstream (miles)	Distance downstream (miles)
1		
2		
3		
4		
5		
6		

42. Wind A plane that can travel at a rate r of 300 miles per hour in still air is traveling in a wind stream with a speed of 20 miles per hour. The distance d the plane will travel against the wind is given by the formula $d = (r - w) \cdot t$, where t represents time. The distance it will travel with the wind is given by the formula $d = (r + w) \cdot t$. Use these formulas with $r = 300$ and $w = 20$ to complete the following table.

Time (hours)	Distance against the wind (miles)	Distance with the wind (miles)
0.50		
1.00		
1.50		
2.00		
2.50		
3.00		

Coffee Sales   Use the information to complete the following tables. Round to the nearest tenth of a billion dollars.

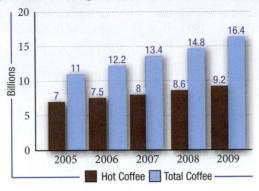

43.

Hot Coffee Sales	
Year	Sales (billions of dollars)
2005	
2006	
2007	
2008	
2009	

44.

Total Coffee Sales	
Year	Sales (billions of dollars)
2005	
2006	
2007	
2008	
2009	

45. Livestock Pen   A farmer buys 48 feet of fencing material to build a rectangular livestock pen. Fill in the second column of the table to find the length of the pen if the width is 2, 4, 6, 8, 10, or 12 feet. Then find the area of each of the pens formed.

Width (ft)	Length (ft)	Area (ft²)
2		
4		
6		
8		
10		
12		

46. Model Rocket   A small rocket is projected into the air with a velocity of 128 feet per second. The formula that gives the height h of the rocket t seconds after it is launched is

$$h = -16t^2 + 128t$$

Use this formula to find the height of the rocket after 1, 2, 3, 4, 5, and 6 seconds.

Time (seconds)	Height (feet)
1	
2	
3	
4	
5	
6	

Maximum Heart Rate In exercise physiology, a person's maximum heart rate, in beats per minute, is found by subtracting his age in years from 220. So, if A represents your age in years, then your maximum heart rate is

$$M = 220 - A$$

Use this formula to complete the following tables. Problems 47–48 may be solved using a graphing calculator.

47.

Age (years)	Maximum Heart Rate (beats per minute)
18	
19	
20	
21	
22	
23	

48.

Age (years)	Maximum Heart Rate (beats per minute)
15	
20	
25	
30	
35	
40	

Training Heart Rate A person's training heart rate, in beats per minute, is his resting heart rate plus 60% of the difference between his maximum heart rate and his resting heart rate. If resting heart rate is R and maximum heart rate is M, then the formula that gives training heart rate is

$$T = R + 0.6(M - R)$$

Use this formula along with the results of Problems 47 and 48 to fill in the following two tables.

49. For a 20-year-old person

Resting Heart Rate (beats per minute)	Training Heart Rate (beats per minute)
60	
62	
64	
68	
70	
72	

50. For a 40-year-old person

Resting Heart Rate (beats per minute)	Training Heart Rate (beats per minute)
60	
62	
64	
68	
70	
72	

Getting Ready for the Next Section

To understand all of the explanations and examples in the next section you must be able to work the problems below.

Graph each inequality.

51. $x < 2$

52. $x \leq 2$

53. $x \geq -3$

54. $x > -3$

Solve each equation.

55. $-2x - 3 = 7$

56. $3x + 3 = 2x - 1$

57. $3(2x - 4) - 7x = -3x$

58. $3(2x + 5) = -3x$

Find the Mistake

Each application problem below is followed by a corresponding equation. Write true or false if the equation accurately represents the problem. If false, write the correct equation on the line provided.

1. A swimming pool has a length that is 4 meters less than twice the width. The perimeter is 22 meters. Find the width.

   Equation: $2w - 4 = 22$ _____

2. Two complementary angles are such that one is four times larger than the other. Find the two angles.

   Equation: $5y = 90$ _____

3. A local company has decided to contribute 15.5% of every dollar donated to a local humane society during an annual fundraiser. At the end of the fundraiser, the humane society receives a total donation of $7,218.75. How much money was raised before the company's contribution?

   Equation: $15.5x = 7,218.75$ _____

4. The third angle in an isosceles triangle is 20 degrees more than twice as large as each of the two base angles. Find the measure of each angle.

   Equation: $4x + 20 = 180$ _____

Landmark Review: Checking Your Progress

Solve each of the following equations.

1. $4x + 3 = 15$

2. $3(y - 2) - 2(2y - 2) = 1$

3. $\frac{2}{3}x + \frac{3}{5} = \frac{2}{15}$

4. $0.2x + 0.3 = 0.5$

Use the formula $5x + 3y = 15$ to find y if

5. x is 0.

6. x is -3.

7. x is 1.

8. x is 4.

Solve the formula for the indicated variable.

9. $3x + 4y = 5$ for y

10. $d = (r+c)t$ for c

11. $A = p + prt$ for r

12. $5x - 3y = 15$ for x

13. Use $h = 64t - 16t^2$ to complete the table

t	0	$\frac{1}{4}$	1	$\frac{5}{4}$	2
h					

14. Use $h = \frac{70}{t}$ to complete the table

t	5	7	9	11
h				

OBJECTIVES

A Solve a linear inequality in one variable and graph the solution set.

B Write solutions to inequalities using interval notation.

C Solve application problems using inequalities.

The 'Apapane, a native Hawaiian bird, feeds mainly on the nectar of the 'ohi'a lehua blossom. The 'Apapane lives in high altitude regions where the 'ohi'a blossoms are found and where the birds are protected from mosquitoes which transmit avian malaria and avian pox. Predators of the 'Apapane include the rat, feral cat, mongoose, and owl. According to the *U.S. Geological Survey*:

> Annual survival rates based on 1,584 recaptures of 429 banded individuals: 0.72 ± 0.11 for adults and 0.13 ± 0.07 for juveniles.

The number following the plus or minus sign represents the margin of error for the data collection. Using this margin of error, we can write the survival rate for the adults as an inequality:

$$0.61 \leq r \leq 0.83$$

Inequalities are what we will study in this section.

KEY WORDS

linear inequality

addition property for inequalities

multiplication property for inequalities

set notation

interval notation

line graph

DEFINITION linear inequality

A *linear inequality* in one variable is any inequality that can be put in the form

$$ax + b < c \qquad (a, b, \text{ and } c \text{ constants}, a \neq 0)$$

where the inequality symbol ($<$) can be replaced with any of the other three inequality symbols ($\leq$, $>$, or $\geq$).

Some examples of *linear inequalities* are

$$3x - 2 \geq 7 \qquad -5y < 25 \qquad 3(x - 4) > 2x$$

A Solving Linear Inequalities

Our first property for inequalities is similar to the addition property we used when solving equations.

PROPERTY Addition Property for Inequalities

For any algebraic expressions, A, B, and C,

$$\text{if} \qquad A < B$$

$$\text{then} \qquad A + C < B + C$$

In words: Adding the same quantity to both sides of an inequality will not change the solution set.

> *Note* Because subtraction is defined as addition of the opposite, our new property holds for subtraction as well as addition. That is, we can subtract the same quantity from each side of an inequality and always be sure that we have not changed the solution.

Practice Problems

1. Solve $4x - 2 < 3x + 4$ and graph the solution.

EXAMPLE 1   Solve $3x + 3 < 2x - 1$ and graph the solution.

Solution   We use the addition property for inequalities to write all the variable terms on one side and all constant terms on the other side.

$$3x + 3 < 2x - 1$$
$$3x + (-2x) + 3 < 2x + (-2x) - 1 \qquad \text{Add } -2x \text{ to each side.}$$
$$x + 3 < -1$$
$$x + 3 + (-3) < -1 + (-3) \qquad \text{Add } -3 \text{ to each side.}$$
$$x < -4$$

The solution set is all real numbers that are less than -4. To show this we can use *set notation* and write

$$\{x \mid x < -4\}$$

We can graph the solution set on the number line using an open circle at -4 to show that -4 is not part of the solution set. This is the format you may have used when graphing inequalities in beginning algebra.

Here is an equivalent graph that uses a parenthesis opening left, instead of an open circle, to represent the end point of the graph.

This graph gives rise to the following notation, called *interval notation*, that is an alternative way to write the solution set.

$$(-\infty, -4)$$

The preceding expression indicates that the solution set is all real numbers from negative infinity up to, but not including, -4.

We have three equivalent representations for the solution set to our original inequality. Here are all three together:

Set Notation	Line Graph	Interval Notation
$\{x \mid x < -4\}$	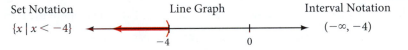	$(-\infty, -4)$

B   Interval Notation and Graphing

The following table shows the connection between set notation, interval notation, and number line graphs. We have included the graphs with open and closed circles for those of you who have used this type of graph previously. In this book, we will continue to show our graphs using the parentheses/brackets method.

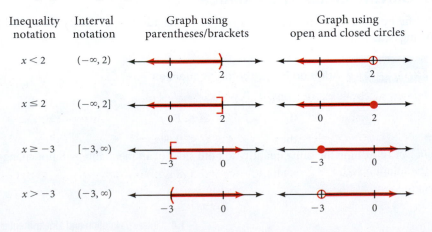

Inequality notation	Interval notation	Graph using parentheses/brackets	Graph using open and closed circles
$x < 2$	$(-\infty, 2)$		
$x \leq 2$	$(-\infty, 2]$		
$x \geq -3$	$[-3, \infty)$		
$x > -3$	$(-3, \infty)$		

Answer

1. $\{x \mid x < 6\}$

Before we state the multiplication property for inequalities, we will take a look at what happens to an inequality statement when we multiply both sides by a positive number and what happens when we multiply by a negative number.

We begin by writing three true inequality statements:

$$3 < 5 \qquad -3 < 5 \qquad -5 < -3$$

We multiply both sides of each inequality by a positive number, say, 4:

$$4(3) < 4(5) \qquad 4(-3) < 4(5) \qquad 4(-5) < 4(-3)$$
$$12 < 20 \qquad -12 < 20 \qquad -20 < -12$$

Notice in each case that the resulting inequality symbol points in the same direction as the original inequality symbol. Multiplying both sides of an inequality by a positive number preserves the *sense* of the inequality.

Let's take the same three original inequalities and multiply both sides by -4:

$$3 < 5 \qquad\qquad -3 < 5 \qquad\qquad -5 < -3$$

$$-4(3) > -4(5) \qquad -4(-3) > -4(5) \qquad -4(-5) > -4(-3)$$
$$-12 > -20 \qquad\quad 12 > -20 \qquad\qquad 20 > 12$$

Notice in this case that the resulting inequality symbol always points in the opposite direction from the original one. Multiplying both sides of an inequality by a negative number *reverses* the sense of the inequality. Keeping this in mind, we will now state the multiplication property for inequalities.

PROPERTY Multiplication Property for Inequalities

Let A, B, and C represent algebraic expressions.

$$\text{If} \qquad A < B$$
$$\text{then} \quad AC < BC \qquad \text{if} \qquad C \text{ is positive } (C > 0)$$
$$\text{or} \qquad AC > BC \qquad \text{if} \qquad C \text{ is negative } (C < 0)$$

In words: Multiplying both sides of an inequality by a positive number always produces an equivalent inequality. Multiplying both sides of an inequality by a negative number reverses the sense of the inequality.

The multiplication property for inequalities does not limit what we can do with inequalities. We are still free to multiply both sides of an inequality by any nonzero number we choose. If the number we multiply by happens to be *negative*, then we *must also reverse* the direction of the inequality.

EXAMPLE 2 Find the solution set for $-2y - 3 \leq 7$.

Solution We begin by adding 3 to each side of the inequality:

$$-2y - 3 \leq 7$$
$$-2y \leq 10 \qquad\qquad \text{Add 3 to both sides.}$$
$$-\frac{1}{2}(-2y) \geq -\frac{1}{2}(10) \qquad \text{Multiply by } -\frac{1}{2} \text{ and reverse the direction of}$$
$$\qquad\qquad\qquad\qquad\qquad \text{the inequality symbol.}$$
$$y \geq -5$$

The solution set is all real numbers that are greater than or equal to -5. The following are three equivalent ways to represent this solution set.

Set Notation	Line Graph	Interval Notation
$\{y \mid y \geq -5\}$		$[-5, \infty)$

Notice how a bracket is used with interval notation to show that -5 is part of the solution set.

Note Because division is defined as multiplication by the reciprocal, we can apply our new property to division as well as to multiplication. We can divide both sides of an inequality by any nonzero number as long as we reverse the direction of the inequality when the number we are dividing by is negative.

2. Find the solution set for $-3y - 2 < 7$.

Answer

2. $y > -3$

When our inequalities become more complicated, we use the same basic steps we used when we were solving equations. That is, we simplify each side of the inequality before we apply the addition property or multiplication property. When we have solved the inequality, we graph the solution on a number line.

3. Solve $4(3x - 2) - 2x \geq 4x$.

EXAMPLE 3   Solve $3(2x - 4) - 7x \leq -3x$.

Solution   We begin by using the distributive property to separate terms. Next, simplify both sides.

$$3(2x - 4) - 7x \leq -3x \qquad \text{Original inequality}$$
$$6x - 12 - 7x \leq -3x \qquad \text{Distributive property}$$
$$-x - 12 \leq -3x \qquad 6x - 7x = (6 - 7)x = -x$$
$$-12 \leq -2x \qquad \text{Add } x \text{ to both sides.}$$
$$-\frac{1}{2}(-12) \geq -\frac{1}{2}(-2x) \qquad \text{Multiply both sides by } -\frac{1}{2} \text{ and reverse the direction of the inequality symbol.}$$
$$6 \geq x$$

> **Note**   In Examples 2 and 3, notice that each time we multiplied both sides of the inequality by a negative number we also reversed the direction of the inequality symbol. Failure to do so would cause our graph to lie on the wrong side of the endpoint.

This last line is equivalent to $x \leq 6$. The solution set can be represented with any of the three following items.

Set Notation	Line Graph	Interval Notation
$\{x \mid x \leq 6\}$		$(-\infty, 6]$

4. Solve and graph
$3(1 - 2x) + 5 < 3x - 1$.

EXAMPLE 4   Solve and graph $2(1 - 3x) + 4 < 4x - 14$.

Solution
$$2 - 6x + 4 < 4x - 14 \qquad \text{Distributive property}$$
$$-6x + 6 < 4x - 14 \qquad \text{Simplify.}$$
$$-6x + 6 + (-6) < 4x - 14 + (-6) \qquad \text{Add } -6 \text{ to both sides.}$$
$$-6x < 4x - 20$$
$$-6x + (-4x) < 4x + (-4x) - 20 \qquad \text{Add } -4x \text{ to both sides.}$$
$$-10x < -20$$
$$\left(-\frac{1}{10}\right)(-10x) > \left(-\frac{1}{10}\right)(-20) \qquad \text{Multiply by } -\frac{1}{10}, \text{ reverse the direction of the inequality.}$$
$$x > 2$$

5. Solve $2x - 3y < 6$ for x.

EXAMPLE 5   Solve $2x - 3y < 6$ for y.

Solution   We can solve this formula for y by first adding $-2x$ to each side and then multiplying each side by $-\frac{1}{3}$. When we multiply by $-\frac{1}{3}$ we must reverse the direction of the inequality symbol. Because this is a formula, we will not graph the solution.

$$2x - 3y < 6 \qquad \text{Original formula}$$
$$2x + (-2x) - 3y < (-2x) + 6 \qquad \text{Add } -2x \text{ to each side.}$$
$$-3y < -2x + 6$$
$$-\frac{1}{3}(-3y) > -\frac{1}{3}(-2x + 6) \qquad \text{Multiply each side by } -\frac{1}{3}.$$
$$y > \frac{2}{3}x - 2 \qquad \text{Distributive property}$$

Answers

3. $x \geq \frac{4}{3}$

4. $x > 1$

5. $x < \frac{3}{2}y + 3$

C Applications

EXAMPLE 6   A company that manufactures ink cartridges for printers finds that they can sell x cartridges each week at a price of p dollars each, according to the formula $x = 1{,}300 - 100p$. What price should they charge for each cartridge if they want to sell at least 300 cartridges a week?

Solution   Because x is the number of cartridges they sell each week, an inequality that corresponds to selling at least 300 cartridges a week is

$$x \geq 300$$

Substituting $1{,}300 - 100p$ for x gives us an inequality in the variable p.

$$1{,}300 - 100p \geq 300$$
$$-100p \geq -1{,}000 \qquad \text{Add } -1{,}300 \text{ to each side.}$$

$$p \leq 10 \qquad \text{Divide each side by } -100, \text{ and reverse the direction of the inequality symbol.}$$

To sell at least 300 cartridges each week, the price per cartridge should be no more than $10. That is, selling the cartridges for $10 or less will produce weekly sales of 300 or more cartridges.

EXAMPLE 7   The formula $F = \frac{9}{5}C + 32$ gives the relationship between the Celsius and Fahrenheit temperature scales. If the temperature on a certain day is below 104° Fahrenheit, what is the temperature in degrees Celsius?

Solution   From the given information, we can write $F \leq 104$. However, because F is equal to $\frac{9}{5}C + 32$, we can also write

$$\frac{9}{5}C + 32 \leq 104$$
$$\frac{9}{5}C \leq 72 \qquad \text{Add } -32 \text{ to each number.}$$
$$\frac{5}{9}\left(\frac{9}{5}C\right) \leq \frac{5}{9}(72) \qquad \text{Multiply each number by } \frac{5}{9}.$$
$$C \leq 40$$

A temperature below 104° Fahrenheit corresponds to a temperature below 40° Celsius.

6. In Example 6, what price should they charge per cartridge if they want to sell at least 400 cartridges a week?

7. If the temperature range is below 77° Fahrenheit, what is the temperature range in Celsius?

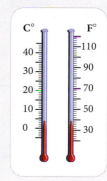

> ## GETTING READY FOR CLASS
>
> *After reading through the preceding section, respond in your own words and in complete sentences.*
>
> **A.** What is a linear inequality in one variable?
>
> **B.** What is the addition property for inequalities?
>
> **C.** When we use interval notation to denote a section of the real number line, when do we use parentheses ( ) and when do we use brackets [ ]?
>
> **D.** Explain the difference between the multiplication property of equality and the multiplication property for inequalities.

Answers

6. $p \leq 9$

7. $C \leq 25$

Vocabulary Review

Choose the correct words to fill in the blanks below.

set   linear inequality in one variable   addition property for inequalities

interval   multiplication property for inequalities

1. A _____ can be put in the form $ax + b < c$.

2. The _____ says that adding the same quantity to both sides of an inequality will not change the solution set.

3. Using _____ notation, we can write the solution set to the inequality $3x + 3 < 2x - 1$ as $\{x \mid x < -4\}$.

4. The solution set $[-2, \infty]$ for the inequality $-2y + 3 < 7$ is written using _____ notation.

5. The _____ says that multiplying both sides of an inequality by a negative number reverses the sense of the inequality.

Problems

A Graph the solution set for the following inequalities.

1. $2x \leq 3$

2. $5x \geq -115$

3. $\frac{1}{2}x > 2$

4. $\frac{1}{3}x > 4$

5. $-5x \leq 25$

6. $-7x \geq 35$

7. $-\frac{3}{2}x > -6$

8. $-\frac{2}{3}x < -8$

9. $-12 \leq 2x$

10. $-20 \geq 4x$

11. $-1 \geq -\frac{1}{4}x$

12. $-1 \leq -\frac{1}{5}x$

SCAN TO ACCESS

Graph the solution set for the following inequalities.

13. $-3x + 1 > 10$

14. $-2x - 5 \leq 15$

15. $\dfrac{1}{2} - \dfrac{m}{12} \leq \dfrac{7}{12}$

16. $\dfrac{1}{2} - \dfrac{m}{10} > -\dfrac{1}{5}$

17. $\dfrac{1}{2} \geq -\dfrac{1}{6} - \dfrac{2}{9}x$

18. $\dfrac{9}{5} > -\dfrac{1}{5} - \dfrac{1}{2}x$

19. $-40 \leq 30 - 20y$

20. $-20 > 50 - 30y$

21. $\dfrac{2}{3}x - 3 < 1$

22. $\dfrac{3}{4}x - 2 > 7$

23. $10 - \dfrac{1}{2}y \leq 36$

24. $8 - \dfrac{1}{3}y \geq 20$

25. $4 - \dfrac{1}{2}x < \dfrac{2}{3}x - 5$

26. $5 - \dfrac{1}{3}x > \dfrac{1}{4}x + 2$

27. $0.03x - 0.4 \leq 0.08x + 1.2$

28. $2.0 - 0.7x < 1.3 - 0.3x$

29. $3 - \dfrac{x}{5} < 5 - \dfrac{x}{4}$

30. $-2 + \dfrac{x}{3} \geq \dfrac{x}{2} - 5$

A B Simplify each side first, then solve the following inequalities. Write your answers with interval notation.

31. $2(3y + 1) \leq -10$

32. $3(2y - 4) > 0$

33. $-(a + 1) - 4a \leq 2a - 8$

34. $-(a - 2) - 5a \leq 3a + 7$

35. $\frac{1}{3}t - \frac{1}{2}(5 - t) < 0$

36. $\frac{1}{4}t - \frac{1}{3}(2t - 5) < 0$

37. $-2 \leq 5 - 7(2a + 3)$

38. $1 < 3 - 4(3a - 1)$

39. $-\frac{1}{3}(x + 5) \leq -\frac{2}{9}(x - 1)$

40. $-\frac{1}{2}(2x + 1) \leq -\frac{3}{8}(x + 2)$

41. $5(x - 2) - 7(x + 1) \leq -4x + 3$

42. $-3(1 - 2x) - 3(x - 4) < -3 - 4x$

43. $\frac{2}{3}x - \frac{1}{3}(4x - 5) < 1$

44. $\frac{1}{4}x - \frac{1}{2}(3x + 1) \geq 2$

45. $20x + 9,300 > 18,000$

46. $20x + 4,800 > 18,000$

47. $0.04x + 0.06(1,200 - x) \geq 54$

48. $0.08x + 0.05(900 - x) \geq 60$

The next two problems are intended to give you practice reading, and paying attention to, the instructions that accompany the problems you are working.

49. Work each problem according to the instructions.

    a. Evaluate when $x = 0$: $-\frac{1}{2}x + 1$

    b. Solve: $-\frac{1}{2}x + 1 = -7$

    c. Is 0 a solution to $-\frac{1}{2}x + 1 < -7$?

    d. Solve: $-\frac{1}{2}x + 1 < -7$

50. Work each problem according to the instructions.

    a. Evaluate when $x = 0$: $-\frac{2}{3}x - 5$.

    b. Solve: $-\frac{2}{3}x - 5 = 1$.

    c. Is 0 a solution to $-\frac{2}{3}x - 5 > 1$?

    d. Solve: $-\frac{2}{3}x - 5 > 1$.

Applying the Concepts

51. Geometry Problems The length of a rectangle is 3 times the width. If the perimeter is to be at least 48 meters, what are the possible values for the width? (If the perimeter is at least 48 meters, then it is greater than or equal to 48 meters.)

52. Geometry Problems The length of a rectangle is 3 more than twice the width. If the perimeter is to be at least 51 meters, what are the possible values for the width? (If the perimeter is at least 51 meters, then it is greater than or equal to 51 meters.)

53. Geometry Problems The numerical values of the three sides of a triangle are given by three consecutive even integers. If the perimeter is greater than 24 inches, what are the possibilities for the shortest side?

54. Geometry Problems The numerical values of the three sides of a triangle are given by three consecutive odd integers. If the perimeter is greater than 27 inches, what are the possibilities for the shortest side?

55. Car Heaters If you have ever sat in a cold car early in the morning, you know that the heater does not work until the engine warms up. This is because the heater relies on the heat coming off the engine. Write an equation using an inequality sign to express when the heater will work if the heater works only after the engine is 100°F.

56. Exercise When Kate exercises, she either swims or runs. She wants to spend a minimum of 8 hours a week exercising, and she wants to swim 3 times the amount she runs. What is the minimum amount of time she must spend doing each exercise?

57. Profit and Loss Movie theaters pay a certain price for the movies they show. Suppose a theater pays $1,500 for each showing of a popular movie. If they charge $7.50 for each ticket they sell, then they will lose money if ticket sales are less than $1,500. However, they will make a profit if ticket sales are greater than $1,500. What is the range of tickets they can sell and still lose money? What is the range of tickets they can sell and make a profit?

58. Stock Sales Suppose you purchase x shares of a stock at $12 per share. After 6 months you decide to sell all your shares at $20 per share. Your broker charges you $15 for the trade. If your profit is at least $4,015, how many shares did you purchase in the first place?

59. Art Supplies The owners of a store selling art supplies finds that they can sell x sketch pads each week at a price of p dollars each, according to the formula $x = 900 - 300p$. What price should they charge if they want to sell

a. at least 300 pads each week?

b. more than 600 pads each week?

c. less than 525 pads each week?

d. at most 375 pads each week?

60. Temperature Range Each of the following temperature ranges is in degrees Fahrenheit. Use the formula

$$F = \frac{9}{5}C + 32$$

to find the corresponding temperature range in degrees Celsius.

a. Below 113°

b. Above 86°

c. Below 14°

d. Above −4°

61. Fuel Efficiency The fuel efficiency (mpg rating) for cars has been increasing steadily since 1980. The formula for a car's fuel efficiency for a given year between 1980 and 1996 is

$$E = 0.36x + 15.9$$

where E is miles per gallon and x is the number of years after 1980 (U.S. Federal Highway Administration, *Highway Statistics,* annual).

a. In what years was the average fuel efficiency for cars less than 17 mpg?

b. In what years was the average fuel efficiency for cars more than 20 mpg?

62. Student Loan When considering how much debt to incur in student loans, you learn that it is wise to keep your student loan payment to 8% or less of your starting monthly income. Suppose you anticipate a starting annual salary of $36,000. Set up and solve an inequality that represents the amount of monthly debt for student loans that would be considered manageable.

63. Wage The annual pay for a mathematics major graduating from college is given by the formula $P = 46,400 + 2,800t$, where t represents years worked after college. How many years would a mathematics major expect to work to be earning at least $90,000 per year?

64. Wage The pay per year for a chemical engineering major graduating from college is given by the formula $P = 64,800 + 2,880t$, where t represents years worked after college. How many years would a chemical engineering major expect to work to be earning at least $108,000 per year?

65. Wage The pay for a government major graduating from college is given by the formula $P = 41,500 + 3,050t$, where t represents years worked after college. How many years would a government major expect to work to be earning more than the mathematics major from problem 63?

66. Wage Is it reasonable to expect that the government major from problem 65 will exceed the chemical engineering majors salary from problem 64? Solve an inequality to answer this question.

Getting Ready for the Next Section

Solve each inequality. Do not graph.

67. $2x - 1 \geq 3$

68. $3x + 1 \geq 7$

69. $-2x > -8$

70. $-3x > -12$

71. $-3 > 4x + 1$

72. $4x + 1 \leq 9$

73. $-4x + 3 < 15$

74. $-5x - 1 > -16$

One Step Further

Assume that a, b, and c are positive, and solve each formula for x.

75. $ax + b < c$

76. $ax - b > -c$

77. $\dfrac{x}{a} + \dfrac{y}{b} < 1$

78. $\dfrac{x}{a} - \dfrac{y}{b} \geq 1$

Assume that $0 < a < b < c < d$ and solve each formula for x.

79. $ax + b < cx + d$

80. $cx + b < ax + d$

81. $\dfrac{x}{a} + \dfrac{1}{b} < \dfrac{x}{c} + \dfrac{1}{d}$

82. $\dfrac{3}{a} - \dfrac{x}{b} \leq \dfrac{5}{c} - \dfrac{x}{d}$

Find the Mistake

Write true or false for each statement below. If false, circle what is false and write the correct answer on the line provided.

1. True or False: The graph of the solution for the inequality $-16 \leq 4x$ will use a parenthesis to show that -4 is included in the solution. _____

2. True or False: The inequality $-10 > 30 - 5y$ has a solution set that does not include the number 7. _____

3. True or False: 3 is a solution to $-\dfrac{2}{3}x - 5 < -7$. _____

4. True or False: The interval notation for the solution set to $5 > \dfrac{1}{2} + 3(4x - 1)$ begins with a parentheses and ends with a bracket. _____

KEY WORDS

set

elements

subset

empty set

union

intersection

set builder notation

compound inequality

open interval

closed interval

continued inequality

half-open interval

1.5 Sets and Compound Inequalities

Suppose a recent survey of web users asked to name a favorite search engine with the following four answer choices:

Ⓐ Google   Ⓑ Bing   Ⓒ Both   Ⓓ Neither

The results of this survey are displayed in an illustration called a Venn diagram, shown here:

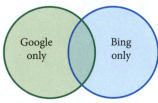

If we let $A = \{\text{all Google users}\}$ and $B = \{\text{all Bing users}\}$, then the notation

$A \cap B$ represents those who use both Google **and** Bing. This is the *intersection* of the two sets.

$A \cup B$ represents those who use Google **or** Bing or both. This is the *union* of the two sets.

We can use sets to represent everyday information like this survey. In this section, we will work with sets and use Venn diagrams to give visual representations of the relationships between different sets.

A Sets and Subsets

The concept of a set can be considered the starting point for all the branches of mathematics.

> **DEFINITION** set
>
> A *set* is a collection of objects or things. The objects in the set are called *elements*, or *members*, of the set.

Sets are usually denoted by capital letters, and elements of sets are denoted by lowercase letters. We use braces, { }, to enclose the elements of a set.

To show that an element is contained in a set, we use the symbol ∈. That is,

$x \in A$ is read "x is an element (member) of set A"

For example, if A is the set $\{1, 2, 3\}$, then $2 \in A$. However, $5 \notin A$ means 5 is not an element of set A.

DEFINITION subset

Set A is a *subset* of set B, written $A \subset B$, if every element in A is also an element of B. That is

$$A \subset B \qquad \text{if and only if} \qquad A \text{ is contained in } B$$

EXAMPLE 1 The set of numbers used to count things is $\{1, 2, 3, \ldots\}$. The dots mean the set continues indefinitely in the same manner. This is an example of an *infinite set*.

EXAMPLE 2 The set of all numbers represented by the dots on the faces of a regular die is $\{1, 2, 3, 4, 5, 6\}$. This set is a subset of the set in Example 1. It is an example of a *finite set* because it has a limited number of elements.

DEFINITION empty set

The set with no members is called the *empty,* or *null set.* It is denoted by the symbol $\varnothing$. The empty set is considered a subset of every set.

B Operations with Sets

As we discussed in the section opener, the two basic operations used to combine sets are *union* and *intersection*.

DEFINITION union

The *union* of two sets A and B, written $A \cup B$, is the set of all elements that are either in A or in B, or in both A and B. The key word here is *or*. For an element to be in $A \cup B$, it must be in A or B. These sets are also called *disjunctions*. In symbols, the definition looks like this:

$$x \in A \cup B \qquad \text{if and only if} \qquad x \in A \text{ or } x \in B$$

DEFINITION intersection

The *intersection* of two sets A and B, written $A \cap B$, is the set of elements in both A and B. The key word in this definition is the word *and*. These sets are also called *conjunctions*. For an element to be in $A \cap B$, it must be in both A and B. In symbols:

$$x \in A \cap B \qquad \text{if and only if} \qquad x \in A \text{ and } x \in B$$

EXAMPLE 3 Let $A = \{1, 3, 5\}$, $B = \{0, 2, 4\}$, and $C = \{1, 2, 3, \ldots\}$.

a. $A \cup B = \{0, 1, 2, 3, 4, 5\}$

b. $A \cap B = \varnothing$ (A and B have no elements in common.)

c. $A \cap C = \{1, 3, 5\} = A$

d. $B \cup C = \{0, 1, 2, 3\ldots\}$

Another notation we can use to describe sets is called *set-builder notation.* Here is how we write our definition for the union of two sets A and B using set-builder notation:

$$A \cup B = \{x \mid x \in A \text{ or } x \in B\}$$

The right side of this statement is read "the set of all x such that x is a member of A or x is a member of B." As you can see, the vertical line after the first x is read "such that."

Practice Problems

Identify each statement as either true or false.

1. The set $\{2, 4, 6, \ldots\}$ is an example of an infinite set.

2. The set $\{2, 4, 8\}$ is a subset of $\{2, 4, 6, \ldots\}$.

Note The symbol $\in$ means "is a member of a set."

3. For the following problems, let $A = \{1, 2, 3\}$, $B = \{0, 3, 7\}$, and $C = \{0, 1, 2, 3, \ldots\}$. Find the given set.

a. $A \cup B$

b. $A \cap B$

c. $A \cap C$

d. $B \cap C$

Answers

1. True

2. True

3. a. $\{0, 1, 2, 3, 7\}$ **b.** $\{3\}$ **c.** $\{1, 2, 3\}$ **d.** $\{0, 3, 7\}$

4. Let $A = \{2, 4, 6, 8, 10\}$ and find $B = \{x \mid x \in A \text{ and } x < 8\}$.

5. Use the Venn Diagram below to find the following sets.

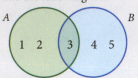

a. A
b. B
c. $A \cup B$
d. $A \cap B$

6. Graph $\{x \mid x < -3 \text{ or } x \geq 2\}$.

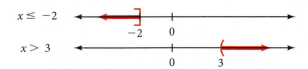

Note The square bracket indicates -2 is included in the solution set, and the parenthesis indicates 3 is not included.

EXAMPLE 4 If $A = \{1, 2, 3, 4, 5, 6\}$, find $C = \{x \mid x \in A \text{ and } x \geq 4\}$.

Solution We are looking for all the elements of A that are also greater than or equal to 4. They are 4, 5, and 6. Using set notation, we have

$$C = \{4, 5, 6\}$$

EXAMPLE 5 Use the information in the Venn diagram to find the following sets.

a. A
b. B
c. $A \cup B$
d. $A \cap B$

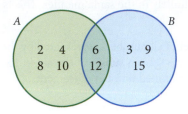

Solution

a. Set A includes the numbers in set A, as well as those common to both sets.
$$A = \{2, 4, 6, 8, 10, 12\}$$

b. Set B includes the numbers in set B, as well as those common to both sets.
$$B = \{3, 6, 9, 12, 15\}$$

c. The union of A and B includes all the numbers in either or both sets.
$$A \cup B = \{2, 3, 4, 6, 8, 9, 10, 12, 15\}$$

d. A intersect B includes only numbers common to both sets.
$$A \cap B = \{6, 12\}$$

C Graphing Compound Inequalities

Before we solve more inequalities, let's review some of the details involved with graphing more complicated inequalities.

Previously we defined the union of two sets A and B to be the set of all elements that are in either A or B. The word *or* is the key word in the definition. The intersection of two sets A and B is the set of all elements contained in both A and B, the key word here being *and*. We can use the words *and* and *or*, together with our methods of graphing inequalities, to graph some *compound inequalities*.

EXAMPLE 6 Graph: $\{x \mid x \leq -2 \text{ or } x > 3\}$.

Solution The two inequalities connected by the word *or* are referred to as a *compound inequality*. We begin by graphing each inequality separately.

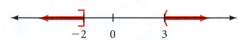

Because the two inequalities are connected by the word *or*, we graph their union. That is, we graph all points on either graph.

To represent this set of numbers with interval notation we use two intervals connected with the symbol for the union of two sets. Here is the equivalent set of numbers described with interval notation:

$$(-\infty, -2] \cup (3, \infty)$$

EXAMPLE 7   Graph: $\{x \mid x \geq -5 \text{ or } x \leq 2\}$.

Solution   We begin by graphing each inequality separately.

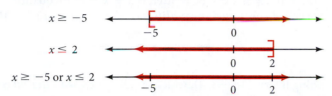

Since the third graph includes both inequalities and is completely shaded, the solution can be written as $(-\infty, \infty)$ in interval notation. In set builder notation, the solution uses the symbol $\mathbb{R}$ to represent the real numbers and is written $\{x \mid x = \mathbb{R}\}$.

EXAMPLE 8   Graph: $\{x \mid x > -1 \text{ and } x < 2\}$.

Solution   We first graph each inequality separately.

Because the two inequalities are connected by the word *and*, we graph their intersection — the part they have in common.

This graph corresponds to the interval $(-1, 2)$, which is called an ***open interval*** because neither endpoint is included in the interval.

EXAMPLE 9   Graph: $\{x \mid x > 6 \text{ and } x > -3\}$.

Solution   We begin by graphing each inequality separately.

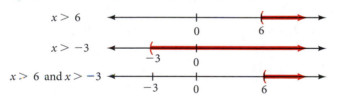

Since all the numbers greater than -3 are also greater than 6, the solution set it only those numbers greater than 6. We write the solution in set builder notation as $\{x \mid x > 6\}$.

EXAMPLE 10   Graph: $\{x \mid x \leq 2 \text{ and } x \geq 4\}$.

Solution   We begin by graphing each inequality separately.

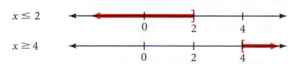

There are no numbers less than 2 and greater than 4.

$$\{x \mid x \leq 2\} \cap \{x \mid x \geq 4\} = \varnothing$$

There is no solution to this inequality. The blank number line represent the null set, $\varnothing$.

7. Graph $\{x \mid x \leq 1 \text{ or } x \geq -4\}$.

8. Graph $\{x \mid x < 3 \text{ and } x \geq -2\}$.

9. Graph $\{x \mid x \geq 3 \text{ and } x \geq -2\}$.

10. Graph $\{x \mid x \leq -3 \text{ and } x \geq -1\}$.

Answers
10. $\varnothing$ (No solution)

Notation   Sometimes compound inequalities that use the word *and* as the connecting word can be written in a shorter form. For example, the compound inequality $-3 \leq x$ and $x \leq 4$ can be written $-3 \leq x \leq 4$. Inequalities of the form $-3 \leq x \leq 4$ are called **continued inequalities** or *double inequalities*. The graph of $-3 \leq x \leq 4$ is

The corresponding interval is $[-3, 4]$, which is called a **closed interval** because both endpoints are included in the interval.

D Solving Compound Inequalities

EXAMPLE 11   Solve and graph $-3 \leq 2x - 5 \leq 3$.

Solution   We can extend our properties for addition and multiplication to cover this situation. If we add a number to the middle expression, we must add the same number to the outside expressions. We do the same for multiplication, remembering to reverse the direction of the inequality symbols if we multiply by a negative number.

$$-3 \leq 2x - 5 \leq 3$$
$$2 \leq 2x \leq 8 \qquad \text{Add 5 to all three members.}$$
$$1 \leq x \leq 4 \qquad \text{Multiply through by } \tfrac{1}{2}.$$

Here are three ways to write this solution set:

Set Notation	Line Graph	Interval Notation
$\{x \mid 1 \leq x \leq 4\}$	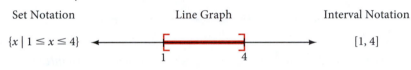	$[1, 4]$

EXAMPLE 12   Solve the compound inequality.

$$3t + 7 \leq -4 \qquad \text{or} \qquad 3t + 7 \geq 4$$

Solution   We solve each half of the compound inequality separately, then we graph the solution set.

$$3t + 7 \leq -4 \qquad \text{or} \qquad 3t + 7 \geq 4$$
$$3t \leq -11 \qquad \text{or} \qquad 3t \geq -3 \qquad \text{Add } -7.$$
$$t \leq -\frac{11}{3} \qquad \text{or} \qquad t \geq -1 \qquad \text{Multiply by } \tfrac{1}{3}.$$

The solution set can be written in any of the following ways:

Set Notation	Line Graph	Interval Notation
$\{t \mid t \leq -\tfrac{11}{3} \text{ or } t \geq -1\}$		$\left(-\infty, -\tfrac{11}{3}\right] \cup \left[-1, \infty\right)$

GETTING READY FOR CLASS

After reading through the preceding section, respond in your own words and in complete sentences.

A. What is a compound inequality?

B. Explain the shorthand notation that can be used to write two inequalities connected by the word *and*.

C. Write two inequalities connected by the word *and* that together are equivalent to $-1 < x < 2$.

D. Explain how you would graph the compound inequality $x < 2$ or $x > -3$.

---

11. Solve and graph
$-7 \leq 2x + 1 \leq 7$.

12. Graph the solution for the compound inequality
$3t - 6 \leq -3$ or $3t - 6 \geq 3$.

Note   As we have seen, an interval can appear open on one side and closed on the other side, such as $[-1, \infty)$. This interval is called a *half-open interval*.

Answers

11. $[-4, 3]$

12. $(-\infty, 1] \cup [3, \infty)$

Vocabulary Review

Choose the correct words to fill in the blanks below.

| set | union | subset | continued | compound | closed |
| empty set | set-builder | intersection | elements | open | |

1. A _____ is a collection of objects or things, also called _____.

2. If every element in A is also an element of B, then set A is a _____ of set B.

3. The set with no members is called the _____.

4. The _____ of two sets A and B with elements in either in A or B is written $A \cup B$.

5. The _____ of two sets A and B with elements in both A or B is written $A \cap B$.

6. Using _____ notation, we can write the union of two sets A and B as $A \cup B = \{x \mid x \in A \text{ or } x \in B\}$.

7. Two inequalities connected by the word *or* is called a _____ inequality.

8. A compound inequality written with the word *and* implied is called a _____ inequality.

9. A(n) _____ interval includes both endpoints, whereas a(n) _____ interval includes neither.

Problems

A Identify each statement as true or false.

1. The set of prime numbers is an example of an infinite set.

2. The set of odd numbers is an example of an infinite set.

3. The set of whole numbers less than 10 is an example of an infinite set.

4. The set of integers less than -20 is an example of an infinite set.

5. The set $\{1, 3, 5\}$ is a subset of $\{1, 3, 5 \ldots\}$.

6. The set $\{-2, 4, 6\}$ is a subset of $\{2, 4, 6 \ldots\}$.

7. The set of whole numbers is a subset of the set of real numbers.

8. The set of real numbers is a subset of the set of whole numbers.

B For the following problems, let $A = \{0, 2, 4, 6\}$, $B = \{1, 2, 3, 4, 5\}$, and $C = \{1, 3, 5, 7\}$.

9. $A \cup B$

10. $A \cup C$

11. $A \cap B$

12. $A \cap C$

13. $A \cup (B \cap C)$

14. $C \cup (A \cap B)$

15. $\{x \mid x \in A \text{ and } x < 4\}$

16. $\{x \mid x \in B \text{ and } x > 3\}$

17. $\{x \mid x \in A \text{ or } x \in C\}$

18. $\{x \mid x \in A \text{ or } x \in B\}$

19. $\{x \mid x \in B \text{ and } x \neq 3\}$

20. $\{x \mid x \in C \text{ and } x \neq 5\}$

SCAN TO ACCESS

Use the following Venn diagrams to find each set.

21. a. A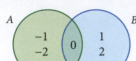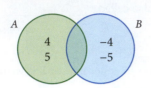

b. B

c. $A \cup B$

d. $A \cap B$

22. a. A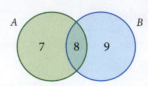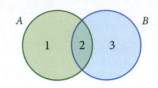

b. B

c. $A \cup B$

d. $A \cap B$

23. a. A

b. B

c. $A \cup B$

d. $A \cap B$

24. a. A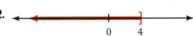

b. B

c. $A \cup B$

d. $A \cap B$

Construct a Venn diagram to represent the following sets.

25. $A = \{1, 3, 5, 7\}$
$B = \{3, 4, 5\}$

26. $A = \{2, 3, 4\}$
$B = \{4, 5, 6\}$

Use interval notation to describe each of the graphs.

27.

28.

29.

30.

31.

32.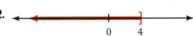

C Graph the following compound inequalities and write each solution set using interval notation.

33. $x < -1$ or $x > 5$

34. $x \le -2$ or $x \ge -1$

35. $x < 5$ and $x > 1$

36. $x \le 6$ and $x > -1$

37. $x \le 7$ and $x > 0$

38. $x > 2$ and $x < 4$

39. $x < 2$ or $x > 4$

40. $x \le 2$ or $x \ge 4$

41. $-1 < x < 3$

42. $-1 \le x \le 3$

43. $-3 < x \le -2$

44. $-5 \le x \le 0$

45. $-3 \le x < 2$

46. $-5 < x \le -1$

D Solve the following continued inequalities. Use both a line graph and interval notation to write each solution set.

47. $-2 \le m - 5 \le 2$

48. $-3 \le m + 1 \le 3$

49. $-40 < 20a + 20 < 20$

50. $-60 < 50a - 40 < 60$

51. $0.5 \le 0.3a - 0.7 \le 1.1$

52. $0.1 \le 0.4a + 0.1 \le 0.3$

53. $3 < \frac{1}{2}x + 5 < 6$

54. $5 < \frac{1}{4}x + 1 < 9$

55. $4 < 6 + \frac{2}{3}x < 8$

56. $3 < 7 + \frac{4}{5}x < 15$

57. $-2 < -\frac{1}{2}x + 1 < 1$

58. $-3 \le -\frac{1}{3}x - 1 < 2$

59. $-\frac{1}{2} \le \frac{3x + 1}{2} \le \frac{1}{2}$

60. $-\frac{5}{6} \le \frac{2x + 5}{3} \le \frac{5}{6}$

61. $-1.5 \le \frac{2x - 3}{4} \le 3.5$

62. $-1.25 \le \frac{2x + 3}{4} \le .75$

63. $-\frac{3}{4} \le \frac{4x - 3}{2} \le 1.5$

64. $-\frac{5}{6} \le \frac{5x + 2}{3} \le \frac{5}{6}$

Graph the solution sets for the following compound inequalities. Then write each solution set using interval notation.

65. $x + 5 \leq -2$ or $x + 5 \geq 2$

66. $3x + 2 < -3$ or $3x + 2 > 3$

67. $5y + 1 \leq -4$ or $5y + 1 \geq 4$

68. $7y - 5 \leq -2$ or $7y - 5 \geq 2$

69. $2x + 5 < 3x - 1$ or $x - 4 > 2x + 6$

70. $3x - 1 > 2x + 4$ or $5x - 2 < 3x + 4$

71. $3x + 1 < -8$ or $-2x + 1 \leq -3$

72. $2x - 5 \leq -1$ or $-3x - 6 < -15$

Translate each of the following phrases into an equivalent inequality statement.

73. x is greater than -2 and at most 4

74. x is less than 9 and at least -3

75. x is less than -4 or at least 1

76. x is at most 1 or more than 6

Applying the Concepts

77. Temperature Range Each of the following temperature ranges is in degrees Fahrenheit. Use the formula

$$F = \frac{9}{5}C + 32$$

to find the corresponding temperature range in degrees Celsius.

a. 95° to 113°

b. 68° to 86°

c. $-13°$ to 14°

d. $-4°$ to 23°

78. Calories Burned The chart shows the range of calories burned during the different events of an ironman triathlon.

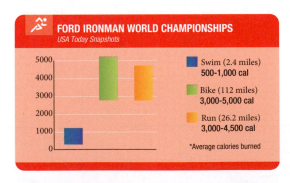

FORD IRONMAN WORLD CHAMPIONSHIPS
USA Today Snapshots

Swim (2.4 miles) 500–1,000 cal

Bike (112 miles) 3,000–5,000 cal

Run (26.2 miles) 3,000–4,500 cal

*Average calories burned

a. Write an inequality that describes the number of calories burned during the entire race.

b. If an athlete burned 750 calories during the swim, write an inequality that describes the number of calories burned during the entire race.

79. Here is what the United States Geological Survey has to say about the survival rates of the Apapane, one of the endemic birds of Hawaii.

> *Annual survival rates based on 1,584 recaptures of 429 banded individuals 0.72 ± 0.11 for adults and 0.13 ± 0.07 for juveniles.*

Write the survival rates using inequalities. Then give the survival rates in terms of percent.

80. Survival Rates for Sea Gulls Here is part of a report concerning the survival rates of Western Gulls that appeared on the web site of Cornell University.

> *Survival of eggs to hatching is 70%–80%; of hatched chicks to fledglings 50%–70%; of fledglings to age of first breeding <50%.*

Write the survival rates using inequalities without percent.

Getting Ready for the Next Section

To understand all of the explanations and examples in the next section you must be able to work the problems below.

Solve each equation.

81. $2a - 1 = -7$

82. $3x - 6 = 9$

83. $\frac{2}{3}x - 3 = 7$

84. $\frac{2}{3}x - 3 = -7$

85. $x - 5 = x - 7$

86. $x + 3 = x + 8$

87. $x - 5 = -x - 7$

88. $x + 3 = -x + 8$

One Step Further

Assume that a, b, and c are positive, and solve each formula for x.

89. $-c < ax + b < c$

90. $-1 < \dfrac{ax + b}{c} < 1$

Assume that $0 < a < b < c < d$ and solve each formula for x.

91. $-d < \dfrac{ax + b}{c} < d$

92. $\dfrac{x}{a} + \dfrac{1}{b} < \dfrac{x}{c}$

Find the Mistake

Each sentence below contains a mistake. Circle the mistake and write the correct word or phrase on the line provided.

1. The notation $x \in A$ is read "x is a set of element A." _____

2. The intersection of two sets A and B is written $A \cup B$. _____

3. The interval $[-6, 8]$ for the continued inequality $-6 \le x \le 8$ is called an open interval. _____

4. To solve the compound inequality $-2 \le 4x - 3 \le 9$, subtract 9 from the first two parts of the inequality.

_____

1.6 Absolute Value Equations

©iStockphoto.com/tarantas

At the U.S. National Handcar Races in Truckee, California, five-person teams face off on a railroad track. One team member gives the 1,000-pound handcar a push while the other four members, standing on its flatbed, use the hand pump to propel the handcar down a 1000-foot stretch of track. The team with the fastest time wins the race. The races are held each year in honor of California's great railroad history, which aside from a few local lines began in 1869 with the completion of the First Continental Railroad.

Today people traveling across the country by train use Amtrak, the government-established long-distance passenger train for inter-city travel in the United States. Amtrak's annual passenger revenue for the years 1985–1995 is modeled approximately by the formula

$$R = -60\,|x - 11| + 962$$

where R is the annual revenue in millions of dollars and x is the number of years after 1980 (Association of American Railroads, Washington, DC, *Railroad Facts, Statistics of Railroads of Class 1*, annual). In what year was the passenger revenue $722 million?

A Solve Equations with Absolute Values

In Chapter 0, we defined the *absolute value* of x, $|x|$, to be the distance between x and 0 on the number line. The absolute value of a number measures its distance from 0.

EXAMPLE 1   Solve $|x| = 5$ for x.

Solution   Using the definition of absolute value, we can read the equation as, "The distance between x and 0 on the number line is 5." If x is 5 units from 0, then x can be 5 or -5.

$$\text{If } |x| = 5 \quad \text{then } x = 5 \quad \text{or} \quad x = -5$$

In general, we can see that any equation of the form $|a| = b$ is equivalent to the equations $a = b$ or $a = -b$, as long as $b > 0$. Furthermore, $|a| = 0$ is equivalent to $a = 0$.

Practice Problems

1. Solve $|x| = 3$ for x.

Answer

1. $-3, 3$

EXAMPLE 2   Solve $|2a - 1| = 7$.

Solution   We can read this question as "$2a - 1$ is 7 units from 0 on the number line." The quantity $2a - 1$ must be equal to 7 or -7.

$$|2a - 1| = 7$$
$$2a - 1 = 7 \quad \text{or} \quad 2a - 1 = -7$$

We have transformed our absolute value equation into two equations that do not involve absolute value. We can solve each equation using the method in Section 1.1.

$2a - 1 = 7$	or $\quad 2a - 1 = -7$	
$2a = 8$	$2a = -6$	Add 1 to both sides.
$a = 4$	$a = -3$	Multiply by $\frac{1}{2}$.

The solutions are -3 and 4.

To check our solutions, we put them into the original absolute value equation.

When $\rightarrow$	$a = 4$	When $\rightarrow$	$a = -3$
the equation $\rightarrow$	$|2a - 1| = 7$	the equation $\rightarrow$	$|2a - 1| = 7$
becomes $\rightarrow$	$|2(4) - 1| = 7$	becomes $\rightarrow$	$|2(-3) - 1| = 7$
	$|7| = 7$		$|-7| = 7$
	$7 = 7$		$7 = 7$

EXAMPLE 3   Solve $\left|\frac{2}{3}x - 3\right| + 5 = 12$.

Solution   To use the definition of absolute value to solve this equation, we must isolate the absolute value on the left side of the equal sign. To do so, we add -5 to both sides of the equation to obtain

$$\left|\frac{2}{3}x - 3\right| = 7$$

Now that the equation is in the correct form, we can write

$\frac{2}{3}x - 3 = 7$	or $\quad \frac{2}{3}x - 3 = -7$	
$\frac{2}{3}x = 10$	$\frac{2}{3}x = -4$	Add 3 to both sides.
$x = 15$	$x = -6$	Multiply by $\frac{3}{2}$.

The solutions are -6 and 15.

EXAMPLE 4   Solve $|3a - 6| = -4$.

Solution   The solution set is $\varnothing$ because the right side is negative but the left side cannot be negative. No matter what we try to substitute for the variable a, the quantity $|3a - 6|$ will always be positive or zero. It can never be -4.

Consider the statement $|a| = |b|$. What can we say about a and b? We know they are equal in absolute value. By the definition of absolute value, they are the same distance from 0 on the number line. They must be equal to each other or opposites of each other. In symbols, we write

$$|a| = |b| \quad \Leftrightarrow \quad a = b \quad \text{or} \quad a = -b$$

$$\underset{\substack{\uparrow \\ \text{Equal in} \\ \text{absolute value}}}{} \qquad \underset{\substack{\uparrow \\ \text{Equals}}}{} \quad \text{or} \quad \underset{\substack{\uparrow \\ \text{Opposites}}}{}$$

2. Solve $|3x - 6| = 9$.

3. Solve $|4x - 3| + 2 = 3$.

4. Solve $|7a - 1| = -2$.

Note   The symbol $\Leftrightarrow$ means "if and only if" and "is equivalent to."

Answers

2. $-1, 5$

3. $1, \frac{1}{2}$

4. No solution

5. Solve $|4x - 5| = |3x + 9|$.

EXAMPLE 5   Solve $|3a + 2| = |2a + 3|$.

Solution   The quantities $3a + 2$ and $2a + 3$ have equal absolute values. They are, therefore, the same distance from 0 on the number line. They must be equals or opposites.

$$|3a + 2| = |2a + 3|$$

Equals		Opposites
$3a + 2 = 2a + 3$	or	$3a + 2 = -(2a + 3)$
$a + 2 = 3$		$3a + 2 = -2a - 3$
$a = 1$		$5a + 2 = -3$
		$5a = -5$
		$a = -1$

The solutions are -1 and 1.

It makes no difference in the outcome of the problem if we take the opposite of the first or second expression. It is very important, once we have decided which one to take the opposite of, that we take the opposite of both its terms and not just the first term. That is, the opposite of $2a + 3$ is $-(2a + 3)$, which we can think of as $-1(2a + 3)$. Distributing the -1 across *both* terms, we have

$$-1(2a + 3) = -2a - 3$$

6. Solve $|x + 3| = |x + 8|$.

EXAMPLE 6   Solve $|x - 5| = |x - 7|$.

Solution   As was the case in Example 5, the quantities $x - 5$ and $x - 7$ must be equal or they must be opposites, because their absolute values are equal.

Equals		Opposites
$x - 5 = x - 7$	or	$x - 5 = -(x - 7)$
$-5 = -7$		$x - 5 = -x + 7$
No solution here		$2x - 5 = 7$
		$2x = 12$
		$x = 6$

Because the first equation leads to a false statement, it will not give us a solution. (If either of the two equations were to reduce to a true statement, it would mean all real numbers would satisfy the original equation.) In this case, our only solution is 6.

GETTING READY FOR CLASS

After reading through the preceding section, respond in your own words and in complete sentences.

A. Why do some equations that involve absolute value have two solutions instead of one?

B. Translate $|x| = 6$ into words using the definition of absolute value.

C. Explain in words what the equation $|x - 3| = 4$ means with respect to distance on the number line.

D. When is the statement $|x| = x$ true?

Answers

5. $-\frac{4}{7}, 14$

6. $-\frac{11}{2}$

Vocabulary Review

Choose the correct words to fill in the blanks below.

negative $\Leftrightarrow$ same   positive

1. The absolute value of a number measures its distance from 0 on the number line and will never be _____.
2. The absolute value of a number will always be _____ or zero.
3. _____ means "if and only if" and "is equivalent to."
4. Two quantities with equal absolute values are the _____ distance from 0 on the number line.

Problems

A Use the definition of absolute value to solve each of the following equations.

1. $|x| = 4$

2. $|x| = 7$

3. $2 = |a|$

4. $5 = |a|$

5. $|x| = -3$

6. $|x| = -4$

7. $|a| + 2 = 3$

8. $|a| - 5 = 2$

9. $|y| + 4 = 3$

10. $|y| + 3 = 1$

11. $|a - 4| = \dfrac{5}{3}$

12. $|a + 2| = \dfrac{7}{5}$

13. $\left|\dfrac{3}{5}a + \dfrac{1}{2}\right| = 1$

14. $\left|\dfrac{2}{7}a + \dfrac{3}{4}\right| = 1$

15. $60 = |20x - 40|$

16. $800 = |400x - 200|$

17. $|2x + 1| = -3$

18. $|2x - 5| = -7$

19. $\left|\dfrac{3}{4}x - 6\right| = 9$

20. $\left|\dfrac{4}{5}x - 5\right| = 15$

21. $\left|1 - \dfrac{1}{2}a\right| = 3$

22. $\left|2 - \dfrac{1}{3}a\right| = 10$

23. $|2x - 5| = 3$

24. $|3x + 1| = 4$

25. $|4 - 7x| = 5$

26. $|9 - 4x| = 1$

27. $\left|3 - \dfrac{2}{3}y\right| = 5$

28. $\left|-2 - \dfrac{3}{4}y\right| = 6$

Solve each equation.

29. $|3x + 4| + 1 = 7$

30. $|5x - 3| - 4 = 3$

31. $|3 - 2y| + 4 = 3$

32. $|8 - 7y| + 9 = 1$

33. $3 + |4t - 1| = 8$

34. $2 + |2t - 6| = 10$

35. $\left|9 - \dfrac{3}{5}x\right| + 6 = 12$

36. $\left|4 - \dfrac{2}{7}x\right| + 2 = 14$

37. $5 = \left|\dfrac{2}{7}x + \dfrac{4}{7}\right| - 3$

38. $7 = \left|\dfrac{3}{5}x + \dfrac{1}{5}\right| + 2$

39. $2 = -8 + \left|4 - \dfrac{1}{2}y\right|$

40. $1 = -3 + \left|2 - \dfrac{1}{4}y\right|$

41. $|3(x + 1)| - 4 = -1$

42. $|2(2x + 3)| - 5 = -1$

43. $|1 + 3(2x - 1)| = 5$

44. $|3 + 4(3x + 1)| = 7$

45. $3 = -2 + \left|5 - \dfrac{2}{3}a\right|$

46. $4 = -1 + \left|6 - \dfrac{4}{5}a\right|$

47. $6 = |7(k + 3) - 4|$

48. $5 = |6(k - 2) + 1|$

Solve the following equations.

49. $|3a + 1| = |2a - 4|$

50. $|5a + 2| = |4a + 7|$

51. $\left|x - \dfrac{1}{3}\right| = \left|\dfrac{1}{2}x + \dfrac{1}{6}\right|$

52. $\left|\dfrac{1}{10}x - \dfrac{1}{2}\right| = \left|\dfrac{1}{5}x + \dfrac{1}{10}\right|$

53. $|y - 2| = |y + 3|$

54. $|y - 5| = |y - 4|$

55. $|3x - 1| = |3x + 1|$

56. $|5x - 8| = |5x + 8|$

57. $|0.03 - 0.01x| = |0.04 + 0.05x|$

58. $|0.07 - 0.01x| = |0.08 - 0.02x|$

59. $|x - 2| = |2 - x|$

60. $|x - 4| = |4 - x|$

61. $\left|\dfrac{x}{5} - 1\right| = \left|1 - \dfrac{x}{5}\right|$

62. $\left|\dfrac{x}{3} - 1\right| = \left|1 - \dfrac{x}{3}\right|$

63. $\left|\dfrac{2}{3}b - \dfrac{1}{4}\right| = \left|\dfrac{1}{6}b + \dfrac{1}{2}\right|$

64. $\left|-\dfrac{1}{4}x + 1\right| = \left|\dfrac{1}{2}x - \dfrac{1}{3}\right|$

65. $|0.1a - 0.04| = |0.3a + 0.08|$

66. $|-0.4a + 0.6| = |1.3 - 0.2a|$

67. Work each problem.

   a. Solve: $4x - 5 = 0$.

   b. Solve: $|4x - 5| = 0$.

   c. Solve: $4x - 5 = 3$.

   d. Solve: $|4x - 5| = 3$.

   e. Solve: $|4x - 5| = |2x + 3|$.

68. Work each problem.

   a. Solve: $3x + 6 = 0$.

   b. Solve: $|3x + 6| = 0$.

   c. Solve: $3x + 6 = 4$.

   d. Solve: $|3x + 6| = 4$.

   e. Solve: $|3x + 6| = |7x + 4|$.

Applying the Concepts

69. Amtrak Amtrak's annual passenger revenue for the years 1985–1995 is modeled approximately by the formula

$$R = -60|x - 11| + 962$$

where R is the annual revenue in millions of dollars and x is the number of years after 1980 (Association of American Railroads, Washington, DC, *Railroad Facts, Statistics of Railroads of Class 1*, annual). In what year was the passenger revenue $722 million?

70. Corporate Profits The corporate profits for various U.S. industries vary from year to year. An approximate model for profits of U.S. communications companies during a given year between 1990 and 1997 is given by

$$P = -3400\,|x - 5.5| + 36000$$

where P is the annual profits (in millions of dollars) and x is the number of years after 1990 (U.S. Bureau of Economic Analysis, Income and Product Accounts of the U.S. (1929 – 1994), *Survey of Current Business*, September 1998). Use the model to determine the years in which profits of communication companies were $31.5 billion ($31,500 million).

Getting Ready for the Next Section

To understand all of the explanations and examples in the next section you must be able to work the problems below.

Solve each inequality. Do not graph the solution set.

71. $2x - 5 < 3$

72. $-3 < 2x - 5$

73. $-4 \le 3a + 7$

74. $3a + 2 \le 4$

75. $4t - 3 \le -9$

76. $4t - 3 \ge 9$

One Step Further

Solve each formula for x. (Assume a, b, and c are positive.)

77. $|x - a| = b$

78. $|x + a| - b = 0$

79. $|ax + b| = c$

80. $|ax - b| - c = 0$

81. $\left| \dfrac{x}{a} + \dfrac{y}{b} \right| = 1$

82. $\left| \dfrac{x}{a} + \dfrac{y}{b} \right| = c$

Find the Mistake

Each sentence below contains a mistake. Circle the mistake and write the correct word or phrase on the line provided.

1. All absolute value equations have two solutions. _____

2. There are two solutions to the absolute value equation $|2x - 3| = |2x + 4|$. _____

3. There is no solution for the absolute equation $|y - 7| = |7 - y|$. _____

4. The absolute value equation $|6a - 4| = -9$ has a negative number for a solution. _____

1.7 Absolute Value Inequalities

Image © sxc.hu, emsago, 2003

A student survey conducted by the University of Minnesota, found that 30% of students were solely responsible for their finances. The survey was reported to have a margin of error plus or minus 3.74%. This means that the difference between the sample estimate of 30% and the actual percent of students who are responsible for their own finances is most likely less than 3.74%. We can write this as an inequality:

$$|x - 0.30| \leq 0.0374$$

where x represents the true percent of students who are responsible for their own finances.

A Solving Inequalities with Absolute Values

In this section, we will apply the definition of absolute value to solve inequalities involving absolute value. Again, the absolute value of x, which is denoted $|x|$, represents the distance that x is from 0 on the number line. We will begin by considering three absolute value expressions and their verbal translations:

Expression	In Words
$\|x\| = 7$	x is exactly 7 units from 0 on the number line.
$\|a\| < 5$	a is less than 5 units from 0 on the number line.
$\|y\| \geq 4$	y is greater than or equal to 4 units from 0 on the number line.

Once we have translated the expression into words, we can use the translation to graph the original equation or inequality. The graph is then used to write a final equation or inequality that does not involve absolute value.

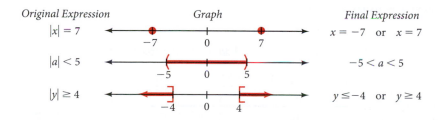

Original Expression	Graph	Final Expression
$\|x\| = 7$		$x = -7$ or $x = 7$
$\|a\| < 5$		$-5 < a < 5$
$\|y\| \geq 4$		$y \leq -4$ or $y \geq 4$

Although we will not always write out the verbal translation of an absolute value inequality, it is important that we understand the translation. Our second expression, $|a| < 5$, means a is within 5 units of 0 on the number line. The graph of this relationship is

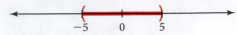

which can be written with the following continued inequality:

$$-5 < a < 5$$

We can follow this same kind of reasoning to solve more complicated absolute value inequalities.

EXAMPLE 1   Solve and graph $|2x - 5| < 3$.

Solution   The absolute value of $2x - 5$ is the distance that $2x - 5$ is from 0 on the number line. We can translate the inequality as, "$2x - 5$ is less than 3 units from 0 on the number line." That is, $2x - 5$ must appear between -3 and 3 on the number line.
    A picture of this relationship is

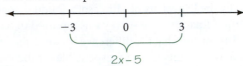

Using the picture, we can write an inequality without absolute value that describes the situation:

$$-3 < 2x - 5 < 3$$

Next, we solve the continued inequality by first adding 5 to all three members and then multiplying all three by $\frac{1}{2}$.

$$-3 < 2x - 5 < 3$$

$$2 < \quad 2x \quad < 8 \qquad \text{Add 5 to all three expressions.}$$

$$1 < \quad x \quad < 4 \qquad \text{Multiply each expression by } \frac{1}{2}.$$

The graph of the solution set is

We can see from the solution that for the absolute value of $2x - 5$ to be within 3 units of 0 on the number line, x must be between 1 and 4.

EXAMPLE 2   Solve and graph $|3a + 7| \le 4$.

Solution   We can read the inequality as, "The distance between $3a + 7$ and 0 is less than or equal to 4." Or, "$3a + 7$ is within 4 units of 0 on the number line." This relationship can be written without absolute value as

$$-4 \le 3a + 7 \le 4$$

Solving as usual, we have

$$-4 \le 3a + 7 \le 4$$

$$-11 \le \quad 3a \quad \le -3 \qquad \text{Add } -7 \text{ to all three members.}$$

$$-\frac{11}{3} \le \quad a \quad \le -1 \qquad \text{Multiply each expression by } \frac{1}{3}.$$

We can see from Examples 1 and 2 that to solve an inequality involving absolute value, we must be able to write an equivalent expression that does not involve absolute value.

EXAMPLE 3   Solve and graph $|x - 3| > 5$.

Solution   We interpret the absolute value inequality to mean that $x - 3$ is more than 5 units from 0 on the number line. The quantity $x - 3$ must be either above 5 or below -5. Here is a picture of the relationship:

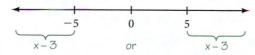

An inequality without absolute value that also describes this situation is

$$x - 3 < -5 \quad \text{or} \quad x - 3 > 5$$

Adding 3 to both sides of each inequality we have

$$x < -2 \quad \text{or} \quad x > 8$$

Here are three ways to write our result:

Set Notation      Interval Notation
$\{x \mid x < -2 \text{ or } x > 8\}$ $(-\infty, -2) \cup (8, \infty)$

We can use the results of our first few examples and the material in the previous section to summarize the information we have related to absolute value equations and inequalities in the table below. If c is a positive real number, then each of the following statements on the left is equivalent to the corresponding statement on the right.

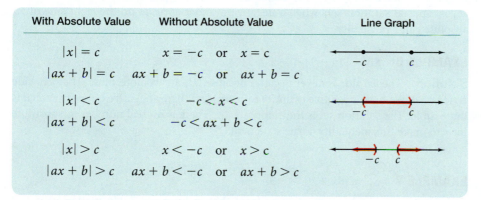

With Absolute Value	Without Absolute Value	Line Graph
$\|x\| = c$	$x = -c \quad \text{or} \quad x = c$	
$\|ax + b\| = c$	$ax + b = -c \quad \text{or} \quad ax + b = c$	
$\|x\| < c$	$-c < x < c$	
$\|ax + b\| < c$	$-c < ax + b < c$	
$\|x\| > c$	$x < -c \quad \text{or} \quad x > c$	
$\|ax + b\| > c$	$ax + b < -c \quad \text{or} \quad ax + b > c$	

EXAMPLE 4   Solve and graph $|2x + 3| + 4 < 9$.

Solution   Before we can apply the method of solution we used in the previous examples, we must isolate the absolute value on one side of the inequality. To do so, we add -4 to each side.

$$|2x + 3| + 4 < 9$$
$$|2x + 3| + 4 + (-4) < 9 + (-4)$$
$$|2x + 3| < 5$$

From this last line, we know that $2x + 3$ must be between -5 and 5.

$$-5 < 2x + 3 < 5$$
$$-8 < \quad 2x \quad < 2 \qquad \text{Add } -3x \text{ to each expression.}$$
$$-4 < \quad x \quad < 1 \qquad \text{Multiply each expression by } \tfrac{1}{2}.$$

Here are three equivalent ways to write our solution

Set Notation   Interval Notation
$\{x \mid -4 < x < 1\}$ $(-4, 1)$

3. Solve and graph $|x + 2| > 7$.

Note   The inequalities in the table may also be written using the symbols $\leq$ and $\geq$. Corresponding line graphs would then use a bracket at each value instead of a parenthesis.

4. Solve and graph $|2x + 5| - 2 < 9$.

Answers

3. $(-\infty, -9) \cup (5, \infty)$

4. $(-8, 3)$

5. Solve and graph $|5 - 2t| > 3$.

EXAMPLE 5   Solve and graph $|4 - 2t| > 2$.

Solution   The inequality indicates that $4 - 2t$ is less than -2 or greater than 2. Writing this without absolute value symbols, we have

$$4 - 2t < -2 \qquad \text{or} \qquad 4 - 2t > 2$$

To solve these inequalities we begin by adding -4 to each side.

$$4 + (-4) - 2t < -2 + (-4) \quad \text{or} \quad 4 + (-4) - 2t > 2 + (-4)$$

$$-2t < -6 \qquad\qquad\qquad -2t > -2$$

$$-\frac{1}{2}(-2t) > -\frac{1}{2}(-6) \qquad\qquad -\frac{1}{2}(-2t) < -\frac{1}{2}(-2)$$

$$t > 3 \qquad\qquad\qquad\qquad t < 1$$

Although in situations like this we are used to seeing the "less than" symbol written first, the meaning of the solution is clear. We want to graph all real numbers that are either greater than 3 or less than 1. Here is the graph:

Because absolute value always results in a nonnegative quantity, we sometimes come across special solution sets when a negative number appears on the right side of an absolute value inequality.

6. Solve $|8y + 3| \le -5$.

EXAMPLE 6   Solve $|7y - 1| < -2$.

Solution   The *left* side is never negative because it is an absolute value. The *right* side is negative. We have a positive quantity (or zero) less than a negative quantity, which is impossible. The solution set is the empty set, $\varnothing$. There is no real number to substitute for y to make this inequality a true statement.

7. Solve $|2x - 9| \ge -7$.

EXAMPLE 7   Solve $|6x + 2| > -5$.

Solution   This is the opposite case from that in Example 7. No matter what real number we use for x on the *left* side, the result will always be positive, or zero. The *right* side is negative. We have a positive quantity (or zero) greater than a negative quantity. Every real number we choose for x gives us a true statement. The solution set is the set of all real numbers.

GETTING READY FOR CLASS

After reading through the preceding section, respond in your own words and in complete sentences.

A. Write an inequality containing absolute value, the solution to which is all the numbers between -5 and 5 on the number line.

B. Translate $x \ge |3|$ into words using the definition of absolute value.

C. Explain in words what the inequality $|x - 5| < 2$ means with respect to distance on the number line.

D. Why is there no solution to the inequality $|2x - 3| < 0$?

Answers

5. $(-\infty, 1) \cup (4, \infty)$

6. No solution

7. All real numbers

Vocabulary Review

Match the following absolute value equations on the left to their equivalents without absolute values on the right.

1. $|x| = c$
2. $|x| < c$
3. $|x| > c$
4. $|ax + b| = c$
5. $|ax + b| < c$
6. $|ax + b| > c$

a. $ax + b < -c$ or $ax + b < c$
b. $x < -c$ or $x > c$
c. $-c < ax + b < c$
d. $ax + b = -c$ or $ax + b = c$
e. $x = -c$ or $x = c$
f. $-c < x < c$

Problems

A Solve each of the following inequalities using the definition of absolute value. Graph the solution set in each case and write the solution set in interval notation.

1. $|x| < 3$

2. $|x| \leq 7$

3. $|x| \geq 2$

4. $|x| > 4$

5. $|x| + 2 < 5$

6. $|x| - 3 < -1$

7. $|t| - 3 > 4$

8. $|t| + 5 > 8$

9. $|y| < -5$

10. $|y| > -3$

11. $|x| \geq -2$

12. $|x| \leq -4$

13. $|x - 3| < 7$

14. $|x + 4| < 2$

15. $|a + 5| \geq 4$

16. $|a - 6| \geq 3$

17. $|x - 5| < 3$

18. $|x - 3| \leq 5$

Solve each inequality and graph the solution set and write each solution set using interval notation.

19. $|a - 1| < -3$

20. $|a + 2| \geq -5$

21. $|2x - 4| < 6$

22. $|2x + 6| < 2$

23. $|3y + 9| \geq 6$

24. $|5y - 1| \geq 4$

25. $|2k + 3| \geq 7$

26. $|2k - 5| \geq 3$

27. $|x - 3| + 2 < 6$

28. $|x + 4| - 3 < -1$

29. $|2a + 1| + 4 \geq 7$

30. $|2a - 6| - 1 \geq 2$

31. $|3x + 5| - 8 < 5$

32. $|6x - 1| - 4 \leq 2$

Solve each inequality and write your answer using interval notation. Keep in mind that if you multiply or divide both sides of an inequality by a negative number you must reverse the sense of the inequality.

33. $|x - 3| \le 5$

34. $|a + 4| < 6$

35. $|3y + 1| < 5$

36. $|2x - 5| \le 3$

37. $|a + 4| \ge 1$

38. $|y - 3| > 6$

39. $|2x + 5| > 2$

40. $|-3x + 1| \ge 7$

41. $|-5x + 3| \le 8$

42. $|-3x + 4| \le 7$

43. $|-3x + 7| < 2$

44. $|-4x + 2| < 6$

Solve each inequality and graph the solution set.

45. $|5 - x| > 3$

46. $|7 - x| > 2$

47. $\left|3 - \dfrac{2}{3}x\right| \ge 5$

48. $\left|3 - \dfrac{3}{4}x\right| \ge 9$

49. $\left|2 - \dfrac{1}{2}x\right| > 1$

50. $\left|3 - \dfrac{1}{3}x\right| > 1$

51. $\left|\dfrac{1}{3}x - 2\right| > 4$

52. $\left|\dfrac{2}{5}x + 4\right| \ge 3$

Solve each inequality.

53. $|x - 1| < 0.01$

54. $|x + 1| < 0.01$

55. $|2x + 1| \ge \dfrac{1}{5}$

56. $|2x - 1| \ge \dfrac{1}{8}$

57. $|3x - 2| \le \dfrac{1}{3}$

58. $|2x + 5| < \dfrac{1}{2}$

59. $\left|\dfrac{3x + 1}{2}\right| > \dfrac{1}{2}$

60. $\left|\dfrac{2x - 5}{3}\right| \ge \dfrac{1}{6}$

61. $\left|\dfrac{4 - 3x}{2}\right| \geq 1$ **62.** $\left|\dfrac{2x - 3}{4}\right| < 0.35$ **63.** $\left|\dfrac{3x - 2}{5}\right| \leq \dfrac{1}{2}$ **64.** $\left|\dfrac{4x - 3}{2}\right| \leq \dfrac{1}{3}$

65. $\left|2x - \dfrac{1}{5}\right| < 0.3$ **66.** $\left|3x - \dfrac{3}{5}\right| < 0.2$ **67.** $\left|-2x + \dfrac{1}{2}\right| < \dfrac{1}{3}$ **68.** $\left|-4x - \dfrac{1}{3}\right| \leq \dfrac{1}{6}$

69. Write the continued inequality $-4 \leq x \leq 4$ as a single inequality involving absolute value.

70. Write the continued inequality $-8 \leq x \leq 8$ as a single inequality involving absolute value.

71. Write $-1 \leq x - 5 \leq 1$ as a single inequality involving absolute value.

72. Write $-3 \leq x + 2 \leq 3$ as a single inequality involving absolute value.

73. Work each problem according to the instructions given.
  a. Evaluate $|5x + 3|$ when $x = 0$.

  b. Solve: $|5x + 3| = 7$.

  c. Is 0 a solution to $|5x + 3| > 7$?

  d. Solve: $|5x + 3| > 7$.

74. Work each problem according to the instructions given.
  a. Evaluate $|-2x - 5|$ when $x = 0$.

  b. Solve: $|-2x - 5| = 1$.

  c. Is 0 a solution to $|-2x - 5| > 1$?

  d. Solve: $|-2x - 5| > 1$.

75. Wavelengths of Light When white light from the sun passes through a prism, it is broken down into bands of light that form colors. The wavelength v (in nanometers) of some common colors are

Blue:	$424 < v < 491$
Green:	$491 < v < 575$
Yellow:	$575 < v < 585$
Orange:	$585 < v < 647$
Red:	$647 < v < 700$

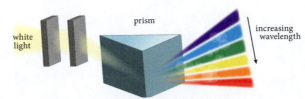

When a fireworks display made of copper is burned, it lets out light with wavelengths v that satisfy the relationship $|v - 455| < 23$. Write this inequality without absolute values, find the range of possible values for v, and then using the preceding list of wavelengths, determine the color of that copper fireworks display.

76. **Speed Limits** One of the highest interstate speed limits for cars in the United States is 75 miles per hour in Nebraska, Nevada, New Mexico, Oklahoma, South Dakota, Utah, and Wyoming. To discourage passing, minimum speeds are also posted, so that the difference between the fastest and slowest moving traffic is no more than 20 miles per hour. Write an absolute value inequality that describes the relationship between the minimum allowable speed and a maximum speed of 75 miles per hour.

One Step Further

Assume a, b, and c, are positive, and solve each inequality for x.

77. $x - a < b$

78. $x - a > b$

79. $ax - b > c$

80. $ax - b < c$

81. $ax + b \leq c$

82. $ax + b \geq c$

83. $\left| \dfrac{x}{a} + \dfrac{1}{b} \right| > c$

84. $\left| \dfrac{x}{a} + \dfrac{1}{b} \right| < c$

85. $\left| (x + a)^2 - x^2 \right| < 3a^2$

86. $\left| (x - a)^2 - x^2 \right| < 2a^2$

87. $\left| x - a^2 \right| \leq 2a^2$

88. $\left| x - 2b^2 \right| < 5b^2$

Find the Mistake

Write true or false for each statement below. If false, circle what is false and write the correct answer on the line provided.

1. True or False: The expression $|b| > 9$ is read as "b is less than or equal to 9 units from zero on the number line."

2. True or False: When solving $-3a < -9$ for a, we must reverse the direction of the inequality symbol.

3. True or False: When an absolute value expression is greater than a negative value, the solution set will be $\varnothing$.

4. True or False: When an absolute value expression is less than a negative value, the solution set will be $\varnothing$.

Saving for the Future

In this day and age, it is extremely important to have a savings account. Whether it be for acquiring a short term goal such as a new television or new car, or working toward a long-term goal such as retirement. Regardless of the goal, lending institutions compete readily for your investment. With all the choices out there, it is vital to do your research before choosing where to deposit your hard-earned cash.

For this project, choose four financial institutions in your area to compare. Then choose two savings account options from each institution: one for a short-term savings goal, and one for a long-term savings goal. Compile information in a table that compares details about the savings accounts. Include in your table answers to the following questions:

1. Does the account come with any fees? If so, what are they?

2. What is the interest rate on the account?

3. Does the account have a required minimum balance?

4. Am I required to make regular deposits into the account?

5. Am I allowed to make withdrawals from the account?

Now, decide which account you would choose for a short-term savings goal? Which account would you choose for a long-term savings goal?

For the short-term savings account, answer the following questions:

1. What is the interest rate on your account? When do you receive interest accrued?

2. What is the minimum amount of money needed to open your account? If you don't have a minimum amount, use the quantity of $100 to answer the next three questions.

3. If you don't make any deposits after your opening deposit, how much money will you have in your account after 5 years?

4. If you don't make any deposits after your opening deposit, how long will it take for your account balance to double?

5. If you make regular deposits of $50/month, how long will it take for your account balance to reach 5 times your opening balance?

For the long-term savings account, answer the following questions:

1. What is the interest rate on your account? When do you receive interest accrued?

2. What is the minimum amount of money needed to open your account? If you don't have a minimum amount, use the quantity of $1000 to answer the next three questions.

3. If you don't make any deposits after your opening deposit, how much money will you have in your account after 15 years?

4. If you don't make any deposits after your opening deposit, how long will it take for your account balance to double?

5. If you make regular deposits of $500/month, how long will it take for your account balance to reach 10 times your opening balance?

Chapter 1 Summary

The numbers in brackets refer to the section(s) in which the topic can be found.

Addition Property of Equality [1.1]

For algebraic expressions A, B, and C,

$$\text{if} \qquad A = B$$
$$\text{then} \qquad A + C = B + C$$

This property states that we can add the same quantity to both sides of an equation without changing the solution set.

1. We can solve

$$x + 3 = 5$$

by adding -3 to both sides:

$$x + 3 + (-3) = 5 + (-3)$$
$$x = 2$$

Multiplication Property of Equality [1.1]

For algebraic expressions A, B, and C,

$$\text{if} \qquad A = B$$
$$\text{then} \qquad AC = BC \quad (C \neq 0)$$

Multiplying both sides of an equation by the same nonzero quantity never changes the solution set.

2. We can solve $3x = 12$ by multiplying both sides by $\frac{1}{3}$.

$$3x = 12$$
$$\frac{1}{3}(3x) = \frac{1}{3}(12)$$
$$x = 4$$

Strategy for Solving Linear Equations in One Variable [1.1]

Step 1: **a.** Use the distributive property to separate terms, if necessary.

    b. If fractions are present, consider multiplying both sides by the LCD to eliminate the fractions. If decimals are present, consider multiplying both sides by a power of 10 to clear the equation of decimals.

    c. Combine similar terms on each side of the equation.

Step 2: Use the addition property of equality to get all variable terms on one side of the equation and all constant terms on the other side. A variable term is a term that contains the variable (for example, $5x$). A constant term is a term that does not contain the variable (the number 3, for example).

Step 3: Use the multiplication property of equality to get the variable by itself on one side of the equation.

Step 4: Check your solution in the original equation to be sure that you have not made a mistake in the solution process.

3. Solve: $3(2x - 1) = 9$.

$$3(2x - 1) = 9$$
$$6x - 3 = 9$$
$$6x - 3 + 3 = 9 + 3$$
$$6x = 12$$
$$\frac{1}{6}(6x) = \frac{1}{6}(12)$$
$$x = 2$$

Formulas [1.2]

A *formula* in algebra is an equation involving more than one variable. To solve a formula for one of its variables, simply isolate that variable on one side of the equation.

4. Solve for w:

$$P = 2l + 2w$$
$$P - 2l = 2w$$
$$\frac{P - 2l}{2} = w$$

5. The perimeter of a rectangle is 32 inches. If the length is 3 times the width, find the dimensions.

Step 1: This step is done mentally.

Step 2: Let x = the width. Then the length is $3x$.

Step 3: The perimeter is 32; therefore

$$2x + 2(3x) = 32$$

Step 4: $8x = 32$
$$x = 4$$

Step 5: The width is 4 inches. The length is $3(4) = 12$ inches.

Step 6: The perimeter is $2(4) + 2(12)$, which is 32. The length is 3 times the width.

Blueprint for Problem Solving [1.3]

Step 1: *Read* the problem, and then mentally *list* the items that are known and the items that are unknown.

Step 2: *Assign a variable* to one of the unknown items. (In most cases this will amount to letting x = the item that is asked for in the problem.) Then *translate* the other *information* in the problem to expressions involving the variable.

Step 3: *Reread* the problem, and then *write an equation,* using the items and variables listed in steps 1 and 2, that describes the situation.

Step 4: *Solve the equation* found in step 3.

Step 5: *Write your answer* using a complete sentence.

Step 6: *Reread* the problem, and *check* your solution with the original words in the problem.

Addition Property for Inequalities [1.4]

6. Adding 5 to both sides of the inequality $x - 5 < -2$ gives

$$x - 5 + 5 < -2 + 5$$

$$x < 3$$

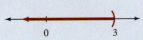

For expressions A, B, and C,

$$\text{if} \qquad A < B$$
$$\text{then} \qquad A + C < B + C$$

Adding the same quantity to both sides of an inequality never changes the solution set.

Multiplication Property for Inequalities [1.4]

7. Multiplying both sides of $-2x \geq 6$ by $-\frac{1}{2}$ gives

$$-2x \geq 6$$

$$-\frac{1}{2}(-2x) \leq -\frac{1}{2}(6)$$

$$x \leq -3$$

For expressions A, B, and C,

$$\text{if} \qquad A < B$$
$$\text{then} \qquad AC < BC \qquad \text{if} \qquad C > 0 \ (C \text{ is positive})$$
$$\text{or} \qquad AC > BC \qquad \text{if} \qquad C < 0 \ (C \text{ is negative})$$

We can multiply both sides of an inequality by the same nonzero number without changing the solution set as long as each time we multiply by a negative number we also reverse the direction of the inequality symbol.

Compound Inequalities [1.5]

8. Graph $x < -3$ or $x > 1$

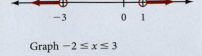

Graph $-2 \leq x \leq 3$

Two inequalities connected by the word *and* or *or* form a compound inequality. If the connecting word is *or*, we graph all points that are on either graph. If the connecting word is *and*, we graph only those points that are common to both graphs. The inequality $-2 \leq x \leq 3$ is equivalent to the compound inequality $-2 \leq x$ and $x \leq 3$.

Absolute Value Equations [1.6]

To solve an equation that involves absolute value, we isolate the absolute value on one side of the equation and then rewrite the absolute value equation as two separate equations that do not involve absolute value. In general, if b is a positive real number, then

$$|a| = b \quad \text{is equivalent to} \quad a = b \quad \text{or} \quad a = -b$$

9. To solve

$$|2x - 1| + 2 = 7$$

we first isolate the absolute value on the left side by adding -2 to each side to obtain

$$|2x - 1| = 5$$

$$\begin{aligned} 2x - 1 = 5 \quad &\text{or} \quad 2x - 1 = -5 \\ 2x = 6 \quad\quad & \quad\quad 2x = -4 \\ x = 3 \quad\quad & \quad\quad\; x = -2 \end{aligned}$$

Absolute Value Inequalities [1.7]

To solve an inequality that involves absolute value, we first isolate the absolute value on the left side of the inequality symbol. Then we rewrite the absolute value inequality as an equivalent continued or compound inequality that does not contain absolute value symbols. In general, if b is a positive real number, then

$$|a| < b \text{ is equivalent to } -b < a < b$$

and

$$|a| > b \text{ is equivalent to } a < -b \text{ or } a > b$$

10. To solve

$$|x - 3| + 2 < 6$$

we first add -2 to both sides to obtain

$$|x - 3| < 4$$

which is equivalent to

$$\begin{aligned} -4 < x - 3 < 4 \\ -1 < x \;\;\; < 7 \end{aligned}$$

COMMON MISTAKE

A very common mistake in solving inequalities is to forget to reverse the direction of the inequality symbol when multiplying both sides by a negative number. When this mistake occurs, the graph of the solution set is always drawn on the wrong side of the endpoint.

CHAPTER 1 Review

Solve the following equations. [1.1]

1. $x - 4 = 7$ **2.** $3y = -5$

3. $2(x + 2) - 4(2x - 3) = 4(2x - 1)$

4. $-0.05x - 0.08 = 0.03 - 0.04(2x - 1)$

Use the formula $6x - 9y = 15$ to find y when [1.2]

5. $x = 3$ **6.** $x = -2$

Use the formula $-5x + 7y = 35$ to find x when [1.2]

7. $y = 0$ **8.** $y = -5$

Solve for the indicated variable. [1.2]

9. $5x - 10y = 20$ for y **10.** $ax + by = cx + dy$ for x

Solve for y. [1.2]

11. $5x - y = 5$ **12.** $6x - 3y = 15$

13. $y - 2 = \frac{2}{3}(x + 6)$ **14.** $y - 5 = \frac{1}{4}(12x + 8)$

Solve each geometry problem. [1.3]

15. Find the dimensions of the rectangle below if its length is twice its width and the perimeter is 102 inches.

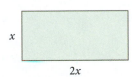

16. The perimeter of the triangle below is 24 meters. Find the value of x.

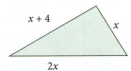

17. Find the measures of the two supplementary angles.

18. Find the measures of the two acute angles in the given right triangle.

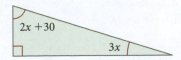

Solve each of the following. [1.3]

19. Sales Tax At the beginning of the day, the cash register at a coffee shop contains $75. At the end of the day, it contains $1,021.40. If the sales tax rate is 8.2%, how much of the total is sales tax?

20. Investing Suppose you invest a certain amount of money in an account that pays 9% interest annually, and $2,000 more than that in an account that pays 11% annually. How much money do you have in each account if the total interest for a year is $320?

	Dollars at 9%	Dollars at 11%
Number of		x
Interest on		

Solve the following inequalities. Write the solution set using interval notation, then graph the solution set. [1.4]

21. $-3x < 12$ **22.** $-6 - \frac{4}{3}x \geq 2$

23. $1.5x - 2.6 \geq 0.5x + 1.4$ **24.** $3(5x - 6) < 6(3x + 5)$

If $A = \{1, 2, 3, 4\}$, $B = \{-2, 0, 2, 4\}$, and $C = \{0, 3, 5\}$, find each of the following. [1.5]

25. $A \cap C$ **26.** $\{x \mid x \in A \text{ and } x > 2\}$

Solve the following compound inequalities. [1.5]

27. $-2 \leq \frac{2}{3}x - 4 \leq 8$ **28.** $x - 5 > 3$ or $2x - 7 < 3$

Solve the following equations. [1.6]

29. $\left| \frac{1}{8}x - 4 \right| = \frac{1}{2}$ **30.** $\left| \frac{3}{2}a + 7 \right| = 4$

31. $|4y - 5| - 3 = 10$ **32.** $\left| \frac{5}{6} - \frac{1}{9}x \right| + 5 = 2$

Solve the following inequalities and graph the solutions. [1.7]

33. $|3y - 4| \geq 2$ **34.** $|5x + 3| - 4 \leq 9$

35. $|4x - 6| > -8$ **36.** $|2 - 5t| < -3$

At the end of each chapter from here on, we will also provide a review that consists of problems similar to those found in the current and previous chapters. Use these reviews as a study tool for exams, and as a way to check your progress, attending to any difficulties you come across. Remember, that math concepts build on each other. It is important to master the skills that come before in order to find success as you proceed through the book.

Simplify each of the following.

1. 6^2

2. $|-5|$

3. $4 \cdot (-3) + 7$

4. $3(4 - 5)$

5. $9 - 2 \cdot 4 + 1$

6. $(7 - 1)(5 + 4)$

7. $8 + 2(3 + 2)$

8. $45 - 24 \div 8 + 3$

9. $-(4 + 1) - (8 - 11)$

10. $3(-4)^2 + 2(-3)^3$

11. $\dfrac{-4 - 4}{-1 - 3}$

12. $\dfrac{7(-2) + 4}{9(-2) + 6(2)}$

Reduce the following fractions to lowest terms.

13. $\dfrac{57}{76}$

14. $\dfrac{129}{387}$

Add the following fractions.

15. $\dfrac{7}{28} + \dfrac{5}{12}$

16. $\dfrac{45}{56} + \dfrac{17}{21}$

17. Find the difference of 4 and -5.

18. Subtract -3 from the product of 4 and -7.

Let $A = \{1, 3, 5\}$; $B = \{3, 6, 9, 12\}$; $C = \{5, 6, 7, 8\}$. Find the following.

19. $A \cup C$

20. $A \cap B$

21. $\{x \mid x \in B \text{ and } x > 6\}$

22. $\{x \mid x \in C \text{ and } x \leq 7\}$

Simplify each of the following expressions.

23. $14\left(\dfrac{x}{7} + 1\right)$

24. $4(3x - 2) + 7$

25. $x\left(1 + \dfrac{3}{x}\right)\left(1 + \dfrac{3}{x}\right)$

26. $1,000(0.03x + 0.01y)$

Solve each of the following equations.

27. $3 - 6y = -9$

28. $3(2 - x) = 4(1 + 3x)$

29. $6 - \dfrac{4}{3}a = -2$

30. $300x^3 = 100x^2$

31. $|x| + 3 = 5$

32. $|5y + 3| = 18$

Substitute the given values in each formula, and then solve for the variable that does not have a numerical replacement.

33. $A = P + Prt$: $A = 1{,}550$, $P = 50$, $r = 0.3$

34. $A = a + (n - 1)d$: $A = 15$, $a = 3$, $d = 6$

Solve each formula for the indicated variable.

35. $y = mx + b$ for m

36. $S = \dfrac{a}{1 - r}$ for r

37. $C = \dfrac{5}{9}(F - 32)$ for F

38. $ax - 2 = bx + 3$ for x

Solve each application. In each case, be sure to show the equation that describes the situation.

39. Angles Two angles are complementary. If the larger angle is 5° more than four times the smaller angle, find the measure of each angle.

40. Rectangle A rectangle is three times as long as it is wide. The perimeter is 64 feet. Find the dimensions.

Solve each inequality. Write you answer using interval notation.

41. $500 + 200x > 300$

42. $-\dfrac{1}{8} \leq \dfrac{1}{16}x \leq \dfrac{1}{4}$

Solve each inequality and graph the solution set.

43. $|3x - 1| > 7$

44. $|4t - 3| + 2 < 4$

The chart shows the average hours per day spent commuting in different cities in the United States. Use the information to answer the following questions.

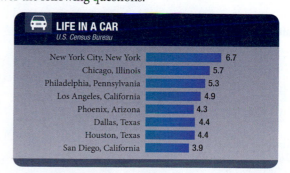

LIFE IN A CAR
U.S. Census Bureau

New York City, New York	6.7
Chicago, Illinois	5.7
Philadelphia, Pennsylvania	5.3
Los Angeles, California	4.9
Phoenix, Arizona	4.3
Dallas, Texas	4.4
Houston, Texas	4.4
San Diego, California	3.9

45. What is the difference between New York City and San Diego in daily commuting hours?

46. What is the difference between Philadelphia and Houston in daily commuting hours?

Solve the following equations. [1.1]

1. $x + 4 = 3$

2. $5y = -2$

3. $4(3x - 1) + 3(x - 2) = 7x + 2$

4. $-0.07x - 0.02 = 0.05 - 0.03(4x + 2)$

Use the formula $8x + 4y = 16$ to find y when [1.2]

5. $x = 0$

6. $x = 2$

Use the formula $-3x + 9y = 21$ to find x when [1.2]

7. $y = 0$

8. $y = 3$

Solve for the indicated variable. [1.2]

9. $y = mx + b$ for m

10. $C = \dfrac{5}{9}(F - 32)$ for F

Solve for y. [1.2]

11. $7x - y = 2$

12. $4x - 5y = 10$

13. $y + 9 = \dfrac{1}{5}(x + 10)$

14. $y - 7 = \dfrac{4}{3}(9x - 6)$

Solve each geometry problem. [1.3]

15. Find the dimensions of the rectangle below if its length is three times the width and the perimeter is 72 inches.

16. The perimeter of the triangle below is 15 meters. Find the value of x.

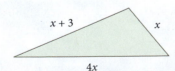

17. Find the measures of the two supplementary angles.

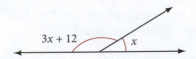

18. Find the measures of the two acute angles in the given right triangle.

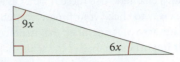

Solve each of the following. [1.3]

19. Sales Tax At the beginning of the day, the cash register at a coffee shop contains $55. At the end of the day, it contains $928.30. If the sales tax rate is 6.5%, how much of the total is sales tax?

20. Investing Suppose you invest a certain amount of money in an account that pays 13% interest annually, and $1,000 more than that in an account that pays 15% annually. How much money do you have in each account if the total interest for a year is $990?

	Dollars at 13%	Dollars at 15%
Number of	x	
Interest on		

Solve the following inequalities. Write the solution set using interval notation, then graph the solution set. [1.4]

21. $-4t \le 16$

22. $-7 - \dfrac{3}{4}x \le 2$

23. $1.2x + 1.5 < 0.6x + 3.9$

24. $2(3y - 2) > 5(y + 3)$

Solve the following equations. [1.6]

25. $\left| \dfrac{1}{6}x - 2 \right| = \dfrac{1}{3}$

26. $\left| \dfrac{4}{7}a + 8 \right| = 4$

27. $|3y + 9| - 2 = 7$

28. $\left| \dfrac{3}{4} - \dfrac{1}{17}x \right| + 4 = 3$

Solve the following inequalities and graph the solutions. [1.7]

29. $|4y + 7| < 3$

30. $|8x + 4| - 3 \ge 9$

31. $|11x - 8| \ge -6$

32. $|2 - 9t| < -7$

Graphs of Equations, Inequalities, and Functions

2

© Google 2010
Altamont Pass, California

Wind power generation has been extensively studied as a replacement for traditional power produced from fossil fuels. Offshore wind has been viewed as a potentially limitless source of power since the ocean has stronger and more constant winds compared to land. Recently scientists analyzed five years of wind data from eleven meteorological stations off the Atlantic coast of the United States from Florida to Maine. They found that combining power from all stations with a transmission cable could prevent large power fluctuations. In a simulation, underwater transmission cable stretched more than 1,550 miles and connected all eleven stations. Although individual sites showed erratic patterns, the total power output changed very little. For

example, the power output of an individual station would regularly drop to zero and could fluctuate by more than 50 percent in an hour. The output of the entire grid, however, did not change more than 10 percent in any given hour. Grid power never dropped to zero during the entire five-year period.

If you consider each individual wind station, the power generated by that station would be its output of electricity as a function of the input of the wind. If you consider all the stations together, the composition of those functions will give you the output of the entire grid. Functions and combinations of functions will be one of the topics we will cover in this chapter.

KEY WORDS

rectangular coordinate system

x-axis

y-axis

bar chart

line graph

ordered pair

x-coordinate

y-coordinate

origin

intercepts

2.1 Graphs of Equations

In this section, we place our work with charts and graphs in a more formal setting. Our foundation will be the *rectangular coordinate system*, because it gives us a link between algebra and geometry. With it we notice relationships between certain equations and different lines and curves.

Table 1 gives the wholesale price of a popular intermediate algebra text at the beginning of each year in which a new edition was published. (The wholesale price is the price the bookstore pays for the book, not the price you pay for it.)

Table 1	Price of a Textbook	
Edition	Year Published	Wholesale Price ($)
First	1991	30.50
Second	1995	39.25
Third	1999	47.50
Fourth	2003	55.00
Fifth	2007	65.75

The information in Table 1 is represented visually in Figures 1 and 2. The diagram in Figure 1 is called a *bar chart*. The diagram in Figure 2 is called a *line graph*. The data in Table 1 is called *paired data* because each number in the year column is paired with a specific number in the price column.

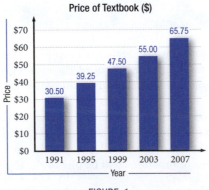

FIGURE 1

FIGURE 2

A Ordered Pairs

Paired data play an important role in equations that contain two variables. Working with these equations is easier if we standardize the terminology and notation associated with paired data. So here is a definition that will do just that:

> **DEFINITION** ordered pair, *x*-coordinate, *y*-coordinate
>
> A pair of numbers enclosed in parentheses and separated by a comma, such as $(-2, 1)$, is called an ***ordered pair*** of numbers. The first number in the pair is called the ***x-coordinate*** of the ordered pair; the second number is called the ***y-coordinate.*** For the ordered pair $(-2, 1)$, the *x*-coordinate is -2 and the *y*-coordinate is 1.

Rectangular Coordinate System

A rectangular coordinate system allows us to connect algebra and geometry by associating geometric shapes (the lines and curves shown in the diagrams) with algebraic equations. The French philosopher and mathematician René Descartes (1596–1650) is usually credited with the invention of the rectangular coordinate system, which is often referred to as the Cartesian coordinate system in his honor. As a philosopher, Descartes is responsible for the statement, "I think, therefore, I am." Until Descartes invented his coordinate system in 1637, algebra and geometry were treated as separate subjects.

A *rectangular coordinate system* is made by drawing two real number lines at right angles to each other. The two number lines, called *axes*, cross each other at 0. This point is called the *origin*. Positive directions are to the right and up. Negative directions are to the left and down. The rectangular coordinate system is shown in Figure 3.

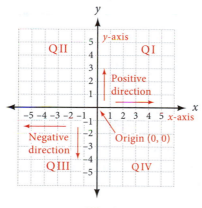

FIGURE 3

> **Note** Any point in quadrant I has positive $(+, +)$ *x*- and *y*-coordinates. Points in quadrant II have negative *x*-coordinates and positive *y*-coordinates $(-, +)$. In quadrant III, both coordinates are negative $(-, -)$. In quadrant IV, the form is $(+, -)$. If a point lies on an axis or at the origin, then it is not considered in any quadrant.

The horizontal number line is called the *x-axis*, and the vertical number line is called the *y-axis*. The two number lines divide the coordinate system into four quadrants, which we number, Q I through Q IV, in a counterclockwise direction. Points on the axes are not considered as being in any quadrant.

Graphing Ordered Pairs

To graph the ordered pair (a, b) on a rectangular coordinate system, we start at the origin and move a units right or left (right if a is positive, left if a is negative). Then we move b units up or down (up if b is positive, down if b is negative). The point where we end up is the graph of the ordered pair (a, b).

1. Graph the ordered pairs (2, 3), (2, −3), (−2, 3), and (−2, −3).

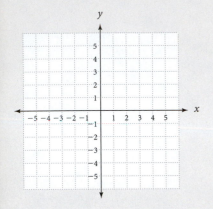

2. Graph the ordered pairs (−2, 1), (3, 5), (0, 2), and (−3, −3).

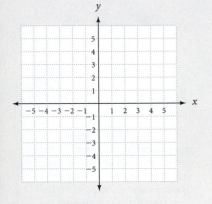

3. Graph the equation $y = \frac{3}{2}x$.

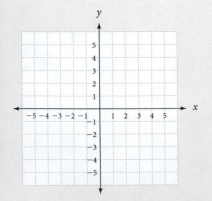

EXAMPLE 1   Plot (graph) the ordered pairs (2, 5), (−2, 5), (−2, −5), and (2, −5).

Solution   To graph the ordered pair (2, 5), we start at the origin and move 2 units to the right, then 5 units up. We are now at the point whose coordinates are (2, 5). We graph the other three ordered pairs in a similar manner (see Figure 4).

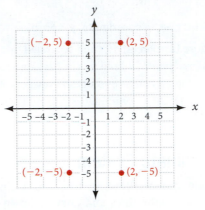

FIGURE 4

EXAMPLE 2   Graph the ordered pairs (1, −3), $\left(\frac{1}{2}, 2\right)$, (3,0), (0, −2), (−1, 0), and (0, 5).

Solution

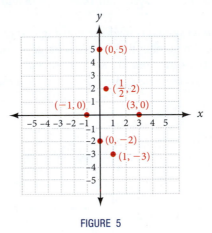

FIGURE 5

From Figure 5, we see that any point on the *x*-axis has a *y*-coordinate of 0 (it has no vertical displacement), and any point on the *y*-axis has an *x*-coordinate of 0 (no horizontal displacement).

B Graphing Equations

We can plot a single point from an ordered pair, but to draw a line or a curve, we need more points.

   To graph an equation in two variables, we simply graph its solution set. That is, we draw a line or smooth curve through all the points whose coordinates satisfy the equation.

EXAMPLE 3   Graph the equation $y = -\frac{1}{3}x$.

Solution   The graph of this equation will be a straight line. We need to find three ordered pairs that satisfy the equation. To do so, we can let *x* equal any numbers we choose and find corresponding values of *y*. However, because every value of *x* we substitute into the equation is going to be multiplied by $-\frac{1}{3}$, let's use numbers for *x* that are divisible by 3, like −3, 0, and 3. That way, when we multiply them by $-\frac{1}{3}$, the result will be an integer.

Let $x = -3$; $\quad y = -\frac{1}{3}(-3) = 1$

The ordered pair $(-3, 1)$ is one solution.

Let $x = 0$; $\quad y = -\frac{1}{3}(0) = 0$

The ordered pair $(0, 0)$ is a second solution.

Let $x = 3$; $\quad y = -\frac{1}{3}(3) = -1$

The ordered pair $(3, -1)$ is a third solution.

In table form

x	y
−3	1
0	0
3	−1

Plotting the ordered pairs $(-3, 1)$, $(0, 0)$, and $(3, -1)$ and drawing a straight line through their graphs, we have the graph of the equation $y = -\frac{1}{3}x$ as shown in Figure 6.

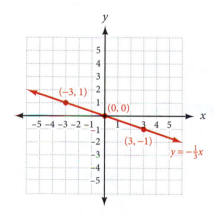

FIGURE 6

Example 3 illustrates again the connection between algebra and geometry that we mentioned earlier in this section. Descartes's rectangular coordinate system allows us to associate the equation $y = -\frac{1}{3}x$ (an algebraic concept) with a specific straight line (a geometric concept). The study of the relationship between equations in algebra and their associated geometric figures is called *analytic geometry*.

Note It takes only two points to determine a straight line. We have included a third point for "insurance." If all three points do not line up in a straight line, we have made a mistake.

Lines Through the Origin

As you can see from Figure 6, the graph of the equation $y = -\frac{1}{3}x$ is a straight line that passes through the origin. The same will be true of the graph of any equation that has the same form as $y = -\frac{1}{3}x$. Here are three more equations that have that form, along with their graphs.

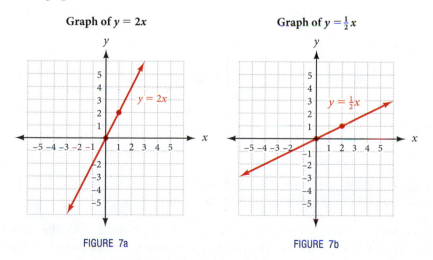

Graph of $y = 2x$

Graph of $y = \frac{1}{2}x$

FIGURE 7a

FIGURE 7b

Graph of $y = x$

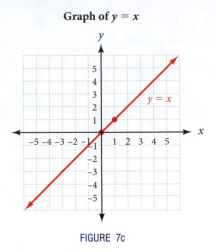

FIGURE 7c

Here is a summary of this discussion:

> **RULE** Lines Through the Origin
>
> The graph of any equation of the form
>
> $$y = mx$$
>
> where m is a real number, will be a straight line through the origin.

4. Graph the equation $y = \dfrac{3}{2}x - 3$.

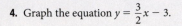

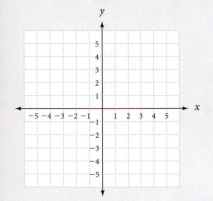

EXAMPLE 4 Graph the equation $y = -\dfrac{1}{3}x + 2$.

Solution Again, we need ordered pairs that are solutions to our equation. Noticing the similarity of this equation to our previous equation, we choose the same values of x for our inputs.

Input x	Calculate using the equation	Output y	Form ordered pairs
-3	$y = -\frac{1}{3}(-3) + 2 = 1 + 2 =$	3	$(-3, 3)$
0	$y = -\frac{1}{3}(0) + 2 = 0 + 2 =$	2	$(0, 2)$
3	$y = -\frac{1}{3}(3) + 2 = -1 + 2 =$	1	$(3, 1)$

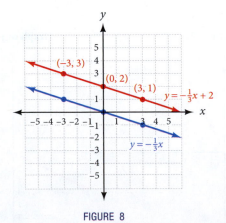

FIGURE 8

Notice that we have included the graph of $y = -\frac{1}{3}x$ along with the graph of $y = -\frac{1}{3}x + 2$. We can see that the graph of $y = -\frac{1}{3}x + 2$ looks just like the graph of $y = -\frac{1}{3}x$, but all points are moved up vertically 2 units.

EXAMPLE 5 Graph the equation $y = -\frac{1}{3}x - 4$.

Solution We create a table from ordered pairs, then we graph the information in the table. However, we are expecting the graph of $y = -\frac{1}{3}x - 4$ to be 4 units below the graph of $y = -\frac{1}{3}x$.

Input x	Calculate using the equation	Output y	Form ordered pairs
-3	$y = -\frac{1}{3}(-3) - 4 = 1 - 4 =$	-3	$(-3, -3)$
0	$y = -\frac{1}{3}(0) - 4 = 0 - 4 =$	-4	$(0, -4)$
3	$y = -\frac{1}{3}(3) - 4 = -1 - 4 =$	-5	$(3, -5)$

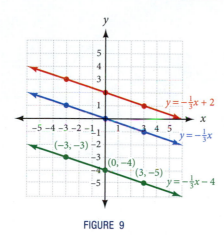

FIGURE 9

Vertical Translations

We know that the graph of $y = mx$ is a line that passes through the origin. From our previous two examples we can generalize as follows:

	If K is a positive number, then:	
The graph of	Is the graph of $y = mx$ translated	
$y = mx + K$	K units up	
$y = mx - K$	K units down	

EXAMPLE 6 Find an equation for the blue line.

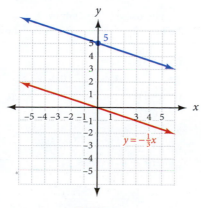

FIGURE 10

5. Graph the equation $y = \frac{3}{2}x + 4$.

5. Graph the equation $y = \frac{3}{2}x + 4$.

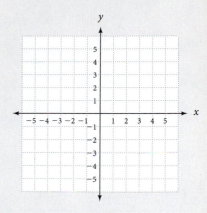

6. Find the equation of the blue line.

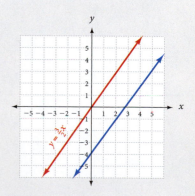

Answer

6. $y = \frac{3}{2}x - 4$

Solution   The blue line is parallel to the line $y = -\frac{1}{3}x$, but translated up 5 units from $y = -\frac{1}{3}x$. According to what we have done up to this point, the equation for the blue line is

$$y = -\frac{1}{3}x + 5$$

EXAMPLE 7   Graph the equation $y = x^2$.

Solution   We input values of x then calculate the values of y using the equation. The result is a set of ordered pairs that we plot and then connect with a smooth curve.

Input x	Calculate using the equation	Output y	Form ordered pairs
-3	$y = (-3)^2 =$	9	$(-3, 9)$
-2	$y = (-2)^2 =$	4	$(-2, 4)$
-1	$y = (-1)^2 =$	1	$(-1, 1)$
0	$y = (0)^2 =$	0	$(0, 0)$
1	$y = (1)^2 =$	1	$(1, 1)$
2	$y = (2)^2 =$	4	$(2, 4)$
3	$y = (3)^2 =$	9	$(3, 9)$

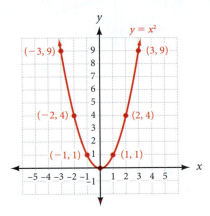

FIGURE 11

EXAMPLE 8   Graph $y = x^2 - 4$.

Solution   We make a table as we did in the previous example. If the vertical translation idea works for this type of equation as it did with our straight lines, we expect this graph to be the graph in Example 7 translated down 4 units.

Input x	Calculate using the equation	Output y	Form ordered pairs
-3	$y = (-3)^2 - 4 =$	5	$(-3, 5)$
-2	$y = (-2)^2 - 4 =$	0	$(-2, 0)$
-1	$y = (-1)^2 - 4 =$	-3	$(-1, -3)$
0	$y = (0)^2 - 4 =$	-4	$(0, -4)$
1	$y = (1)^2 - 4 =$	-3	$(1, -3)$
2	$y = (2)^2 - 4 =$	0	$(2, 0)$
3	$y = (3)^2 - 4 =$	5	$(3, 5)$

---

7. Graph the equation $y = 2x^2$.

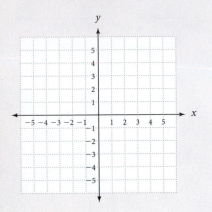

Note   Recall our discussion earlier in the book about raising a negative number to a specific power. It is very important to enclose the negative number in a set of parentheses before working the exponent.

8. Graph $y = 2x^2 - 2$.

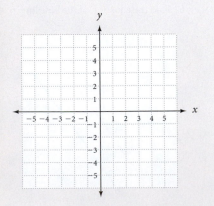

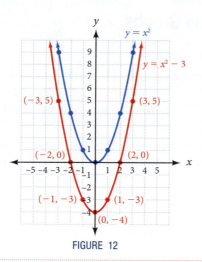

FIGURE 12

As you can see, the graph of $y = x^2 - 4$ is the graph of $y = x^2$ translated down 4 units. We generalize this as follows:

If K is a positive number, then:	
The graph of	**is the graph of $y = x^2$ translated**
$y = x^2 + K$	K units up
$y = x^2 - K$	K units down

EXAMPLE 9 Graph $y = |x|$ and $y = |x| - 4$.

Solution We let x take on values of -4, -3, -2, -1, 0, 1, 2, 3, and 4. The corresponding values of y are shown in the tables.

$y = \lvert x \rvert$	
Input x	Output y
-4	4
-3	3
-2	2
-1	1
0	0
1	1
2	2
3	3
4	4

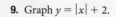

FIGURE 13

$y = \lvert x \rvert - 4$	
Input x	Output y
-4	0
-3	-1
-2	-2
-1	-3
0	-4
1	-3
2	-2
3	-1
4	0

9. Graph $y = |x| + 2$.

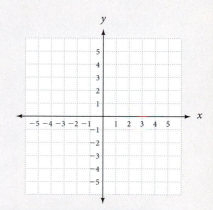

As you can see, the graph of $y = x - 4$ is the graph of $y = x$ translated down 4 units. We will do more with these vertical translations later in the chapter. For now, we simply want to notice how the relationship between the equations can be used to predict how one graph will look when the other graph is given.

C Intercepts and Graphs

DEFINITION intercepts

An *x-intercept* of the graph of an equation is the coordinates of a point where the graph intersects the *x*-axis. The *y-intercept* of the graph of an equation is the coordinates of a point where the graph intersects the *y*-axis.

Because any point on the *x*-axis has a *y*-coordinate of 0, we can find the *x*-intercept by letting $y = 0$ and solving the equation for *x*. We find the *y*-intercept by letting $x = 0$ and solving for *y*.

EXAMPLE 10   Find the *x*- and *y*-intercepts for $2x + 3y = 6$; then graph the equation.

Solution   To find the *y*-intercept, we let $x = 0$.

When → $\qquad\qquad x = 0$

we have → $\quad 2(0) + 3y = 6$

$$3y = 6$$

$$y = 2$$

y-intercept = $(0, 2)$. To find the *x*-intercept, we let $y = 0$.

When → $\qquad\qquad y = 0$

we have → $\quad 2x + 3(0) = 6$

$$2x = 6$$

$$x = 3$$

x-intercept = $(3, 0)$. We use these results to graph the solution set for $2x + 3y = 6$. The graph is shown in Figure 14.

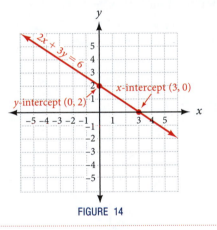

FIGURE 14

10. Graph the solution set for $2x - 4y = 8$.

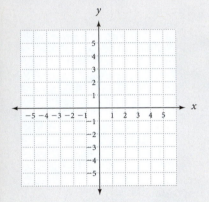

> *Note* Graphing straight lines by finding the intercepts works best when the coefficients of *x* and *y* are factors of the constant term.

11. Find the intercepts of $y = -2x^2 + 8$.

EXAMPLE 11   Find the intercepts for $y = x^2 - 4$.

Solution   We drew the graph of this equation in Example 8. Looking back to that graph we see that the *x*-intercepts are $(-2, 0)$ and $(2, 0)$, and the *y*-intercept is $(0, -4)$. Let's see if we obtain the same results using the algebraic method shown in Example 10.

x-intercept	*y-intercept*
When → $\quad y = 0$,	When → $\qquad x = 0$,
we have → $\quad 0 = x^2 - 4$	we have → $\quad y = 0^2 - 4 = -4$
$\qquad\qquad 0 = (x + 2)(x - 2)$	The *y*-intercept is $(0, -4)$.
$\qquad x + 2 = 0 \quad \text{or} \quad x - 2 = 0$	
$\qquad\quad x = -2 \qquad\qquad x = 2$	
The *x*-intercepts are $(-2, 0)$ and $(2, 0)$.	

Answers

11. *y*-intercept: $(0, 8)$,
$\quad$ *x*-intercepts: $(-2, 0)$, $(2, 0)$

D Horizontal and Vertical Lines

EXAMPLE 12   Graph each of the following lines.

a. $x = 3$ **b.** $y = -2$

Solution

a. The line $x = 3$ is the set of all points whose x-coordinate is 3. The variable y does not appear in the equation, so the y-coordinate can be any number. Note that we can write our equation as a linear equation in two variables by writing it as $x + 0y = 3$. Because the product of 0 and y will always be 0, y can be any number. The graph of $x = 3$ is the vertical line shown in Figure 15a.

b. The line $y = -2$ is the set of all points whose y-coordinate is -2. The variable x does not appear in the equation, so the x-coordinate can be any number. Again, we can write our equation as a linear equation in two variables by writing it as $0x + y = -2$. Because the product of 0 and x will always be 0, x can be any number. The graph of $y = -2$ is the horizontal line shown in Figure 15b.

12. Graph.

a. $x = -4$

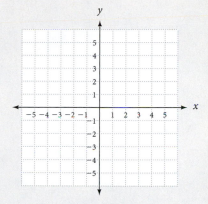

b. $y = 3$

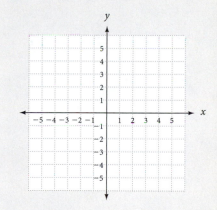

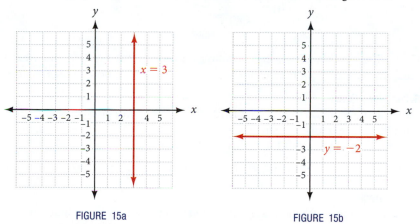

FIGURE 15a   FIGURE 15b

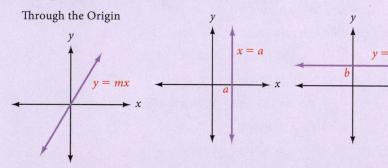

FACTS FROM GEOMETRY   Special Equations and Their Graphs

For the equations below, m, a, and b are real numbers.

FIGURE 16a *Any equation of the form $y = mx$ has a graph that passes through the origin.*

FIGURE 16b *Any equation of the form $x = a$ has a vertical line for its graph.*

FIGURE 16c *Any equation of the form $y = b$ has a horizontal line for its graph.*

USING TECHNOLOGY   Graphing Calculators

Graphing With Trace and Zoom

All graphing calculators have the ability to graph a function and then trace over the points on the graph, giving their coordinates. Furthermore, all graphing calculators can zoom in and out on a graph that has been drawn.

To graph an equation on a graphing calculator, we first set the graph window. The counterpart values of y are Ymin and Ymax. We will use the notation

$$\text{Window:}\quad -5 \le x \le 4 \text{ and } -3 \le y \le 2$$

to stand for a window in which Xmin $= -5$, Xmax $= 4$, Ymin $= -3$, and Ymax $= 2$

Set your calculator to the following window:

$$\text{Window:}\quad -10 \le x \le 10 \text{ and } -10 \le y \le 10$$

Graph the equation $Y = -X + 8$ and compare your results with this graph:

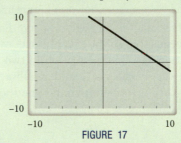

FIGURE 17

Use the Trace feature of your calculator to name three points on the graph. Next, use the Zoom feature of your calculator to zoom out so your window is twice as large.

Solving for y First
To graph the equation from Example 10, $2x + 3y = 6$, on a graphing calculator, you must first solve it for y. When you do so, you will get $y = -\frac{2}{3}x + 2$, which results with the graph in Figure 14. Use this window:

$$\text{Window:}\quad -6 \le x \le 6 \text{ and } -6 \le y \le 6$$

Hint on Tracing
If you are going to use the Trace feature and you want the x-coordinates to be exact numbers, set your window so the range of X inputs is a multiple of the number of horizontal pixels on your calculator screen. On the TI-83/84, the screen is 94 pixels wide. Here are a few convenient trace windows:

To trace with x to the nearest tenth use $-4.7 \le x \le 4.7$ or $0 \le x \le 9.4$

To trace with x to the nearest integer use $-47 \le x \le 47$ or $0 \le x \le 94$

Graph each equation using the indicated window.

1. $y = \frac{1}{2}x - 3$ $-10 \le x \le 10$ and $-10 \le y \le 10$

2. $y = \frac{1}{2}x^2 - 3$ $-10 \le x \le 10$ and $-10 \le y \le 10$

3. $y = \frac{1}{2}x^2 - 3$ $-4.7 \le x \le 4.7$ and $-10 \le y \le 10$

4. $y = x^3$ $-10 \le x \le 10$ and $-10 \le y \le 10$

5. $y = x^3 - 5$ $-4.7 \le x \le 4.7$ and $-10 \le y \le 10$

GETTING READY FOR CLASS

After reading through the preceding section, respond in your own words and in complete sentences.

A. Explain how you would construct a rectangular coordinate system from two real number lines.

B. Explain in words how you would graph the ordered pair $(2, -3)$.

C. How can you tell if an ordered pair is a solution to the equation $y = \frac{1}{3}x + 5$?

D. Explain the difference between the x- and y-intercepts of the graph $y = x^2 - 3$.

Vocabulary Review

Choose the correct words to fill in the blanks below.

y-intercept	horizontal	x-coordinate	y-axis	vertical
x-intercept	y-coordinate	origin	x-axis	

1. The horizontal number line on a rectangular coordinate system is called the _____ , whereas the vertical number line is called the _____ .
2. For the ordered pair $(-9, 6)$, the _____ is -9 and the _____ is 6.
3. The graph of an equation $y = mx$ will be a straight line through the _____ .
4. The _____ is the pair of coordinates of a point where a graph crosses the x-axis.
5. The _____ is the pair of coordinates of a point where a graph crosses the y-axis.
6. Any equation of the form $x = a$ has a _____ line for its graph.
7. Any equation of the form $y = b$ has a _____ line for its graph.

Problems

A Graph each of the following ordered pairs on a rectangular coordinate system.

1. a. $(-1, 2)$
b. $(-1, -2)$
c. $(5, 0)$
d. $(0, 2)$
e. $(-5, -5)$
f. $\left(\frac{1}{2}, 2\right)$

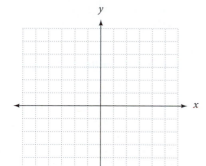

2. a. $(-1, 2)$
b. $(1, -2)$
c. $(0, -3)$
d. $(4, 0)$
e. $(-4, -1)$
f. $\left(3, \frac{1}{4}\right)$

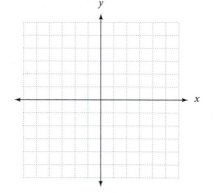

Give the coordinates of each point.

3.

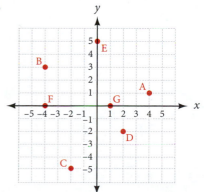

4.

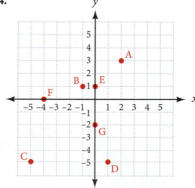

5. Which of the following tables could be produced from the equation $y = 2x - 6$?

a.

x	y
0	6
1	4
2	2
3	0

b.

x	y
0	-6
1	-4
2	-2
3	0

c.

x	y
0	-6
1	-5
2	-4
3	-3

6. Which of the following tables could be produced from the equation $3x - 5y = 15$?

d.

x	y
0	5
-3	0
10	3

e.

x	y
0	-3
5	0
10	3

f.

x	y
0	-3
-5	0
10	-3

7. The graph shown here is the graph of which of the following equations?

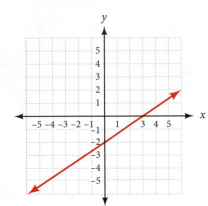

a. $y = \dfrac{3}{2}x - 3$

b. $y = \dfrac{2}{3}x - 2$

c. $y = -\dfrac{2}{3}x + 2$

8. The graph shown here is the graph of which of the following equations?

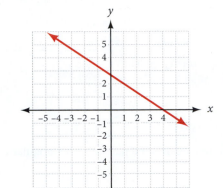

a. $3x - 2y = 8$

b. $2x - 3y = 8$

c. $2x + 3y = 8$

For each problem below, the equation of the red graph is given. Find the equation for the blue graph.

9.

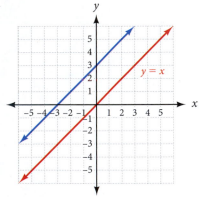

10.

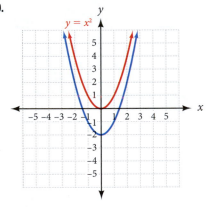

11.

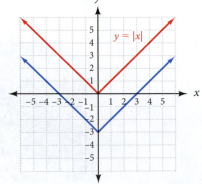

12.

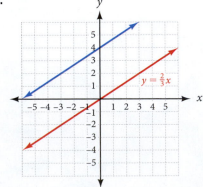

Graph each of the following. Use one coordinate system for each problem.

13. a. $y = 2x$

b. $y = 2x + 3$

c. $y = 2x - 5$

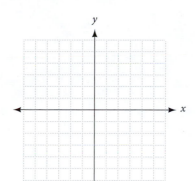

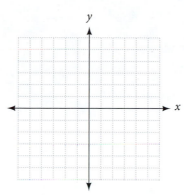

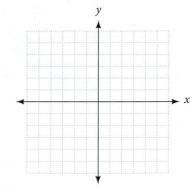

14. a. $y = \dfrac{1}{3}x$

b. $y = \dfrac{1}{3}x + 1$

c. $y = \dfrac{1}{3}x - 3$

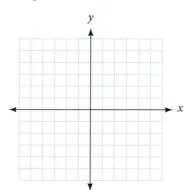

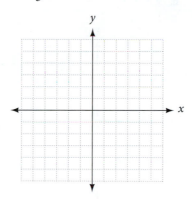

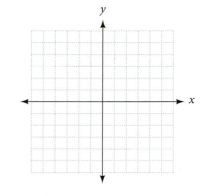

15. a. $y = \dfrac{1}{2}x^2$

b. $y = \dfrac{1}{2}x^2 - 2$

c. $y = \dfrac{1}{2}x^2 + 2$

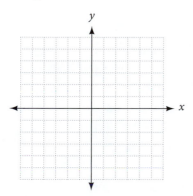

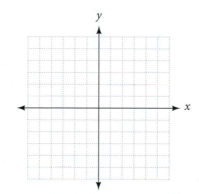

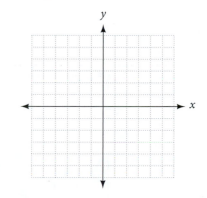

16. a. $y = 2x^2$

b. $y = 2x^2 - 5$

c. $y = 2x^2 + 1$

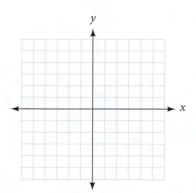

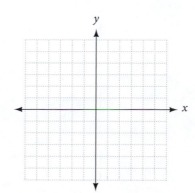

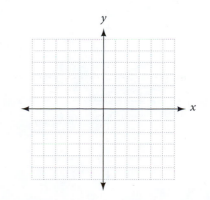

17. Graph the straight line $0.02x + 0.03y = 0.06$.

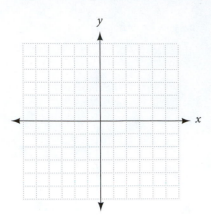

18. Graph the straight line $0.05x - 0.03y = 0.15$.

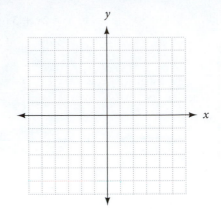

B D Graph each of the following lines.

19. a. $y = 2x$

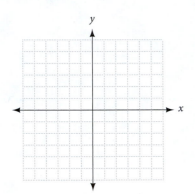

b. $x = -3$

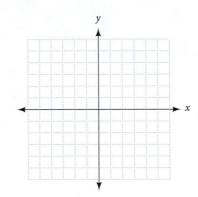

c. $y = 2$

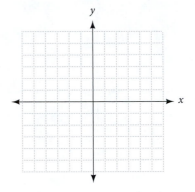

20. a. $y = 3x$

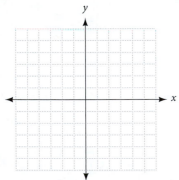

b. $x = -2$

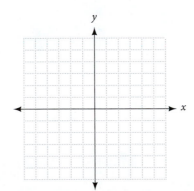

c. $y = 4$

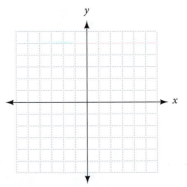

21. a. $y = -\dfrac{1}{2}x$

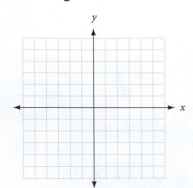

b. $x = 4$

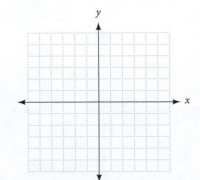

c. $y = -3$

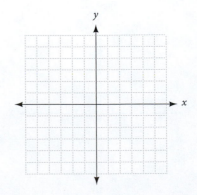

22. a. $y = -\dfrac{1}{3}x$

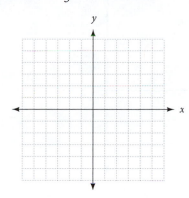

b. $x = 1$

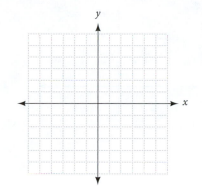

c. $y = -5$

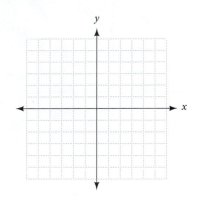

C Find the intercepts for each graph. Then use them to help sketch the graph.

23. $y = x^2 - 9$

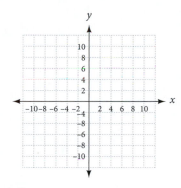

24. $y = x^2$

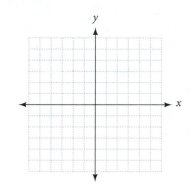

25. $y = 2x - 4$

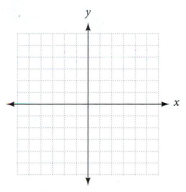

26. $y = 4x - 2$

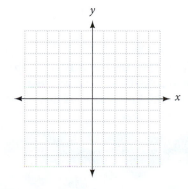

27. $y = \dfrac{1}{2}x + 1$

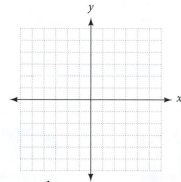

28. $y = -\dfrac{1}{2}x + 1$

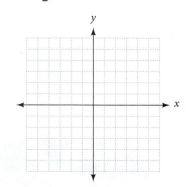

29. $y = 3x$

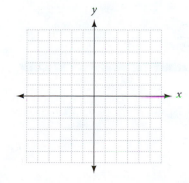

30. $y = -\dfrac{1}{3}x$

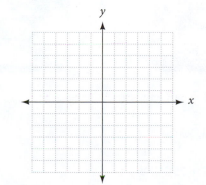

31. $y = x^2 - x$

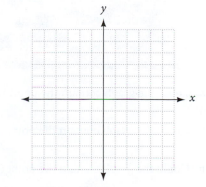

32. $y = x^2 + 3$

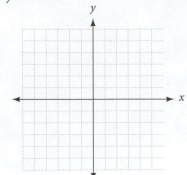

33. $y = x - 3$

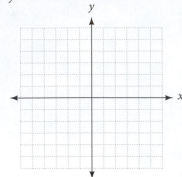

34. $y = x + 2$

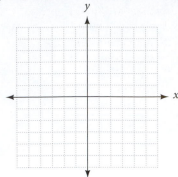

35. a. Solve: $4x + 12 = -16$.

   b. Find x when y is 0: $4x + 12y = -16$.

   c. Find y when x is 0: $4x + 12y = -16$.

   d. Graph: $4x + 12y = -16$.

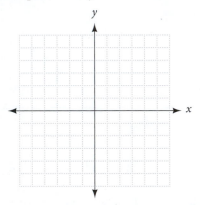

   e. Solve for y: $4x + 12y = -16$.

36. a. Solve: $3x - 8 = -12$.

   b. Find x when y is 0: $3x - 8y = -12$.

   c. Find y when x is 0: $3x - 8y = -12$.

   d. Graph: $3x - 8y = -12$.

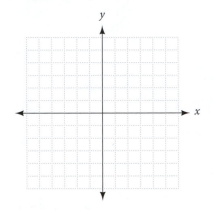

   e. Solve for y: $3x - 8y = -12$.

Applying the Concepts

37. Movie Tickets The graph shows the rise in movie ticket prices from 2000 to 2009. Use the chart to answer the following questions.

   a. Could the graph contain the point (2002, 5.75)?

   b. Could the graph contain the point (2005, 5.5)?

   c. Could the graph contain the point (2007, 6.75)?

38. Gym Memberships The chart shows the rise in health club memberships from 2000 to 2005. Using years as x and millions of members as y, write 5 ordered pairs that describe the information in the chart.

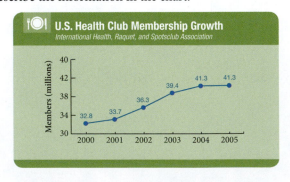

39. Hourly Wages Suppose you have a job that pays $9.50 per hour, and you work anywhere from 0 to 40 hours per week. Table 2 gives the amount of money you will earn in 1 week for working various hours. Construct a line graph from the information in Table 2.

Table 2

Weekly Wages

Hours worked	Pay ($)
0	0
10	95
20	190
30	285
40	380

40. Softball Toss Chaudra is tossing a softball into the air with an underhand motion. It takes exactly 2 seconds for the ball to come back to her. Table 3 shows the distance the ball is above her hand at quarter-second intervals. Construct a line graph from the information in the table.

Table 3

Tossing a softball into the air

Time (sec)	Distance (ft)
0	0
0.25	7
0.5	12
0.75	15
1	16
1.25	15
1.5	12
1.75	7
2	0

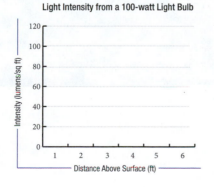

Weekly Wages

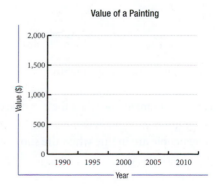

Tossing a Softball Into the Air

41. Intensity of Light Table 4 gives the intensity of light that falls on a surface at various distances from a 100-watt light bulb. Construct a bar chart from the information in Table 4.

Table 4 Light intensity from a 100-watt light bulb

Distance above surface (ft)	Intensity (lumens/sq ft)
1	120.0
2	30.0
3	13.3
4	7.5
5	4.8
6	3.3

42. Value of a Painting Presley purchased a piece of abstract art in 1990 for $125. Table 5 shows the value of the painting at various times, assuming that it doubles in value every 5 years. Construct a bar chart from the information in the table.

Table 5 Value of a Painting

Year	Value ($)
1990	125
1995	250
2000	500
2005	1,000
2010	2,000

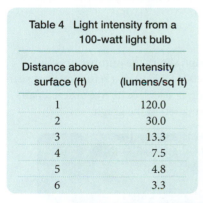

Light Intensity from a 100-watt Light Bulb

Value of a Painting

43. Wind Power Table 6 shows the power output generated by different sized wind turbines. The data is based on power output generated at 33 mph winds, which is considered the ideal power generation speed. Construct a line graph from the information in the table.

Table 6 Power Output from a Wind Turbine	
Blade diameter (m)	Power output(kw)
10	25
17	100
27	225
33	300
40	500

Source: Danish Wind Industry Association, American Wind Energy Association

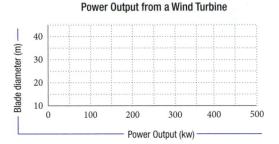

Power Output from a Wind Turbine

44. Solar Power Table 7 shows the voltage and current from various power resistors for a typical PV Solar Panel.

Table 7 Power Output from a PV Solar Panel		
Current (Amps)	Voltage (Volts)	Power
0.00	30.9	
0.29	28.8	
0.54	27.0	
0.90	22.5	
1.03	3.1	
1.10	0.0	

Source: Danish Wind Industry Association, American Wind Energy Association

a. Construct a line graph from the information in the table for current and voltage.

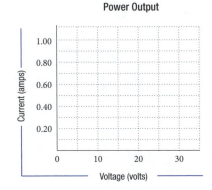

Power Output

b. The power output is the current multiplied by the volts. Compute the power output for each resistance.

c. Identify the point on your line graph with the longest power output. This represents the peak power generated from the solar panel.

45. Reading Graphs The graph shows the number of people in line at a theater box office to buy tickets for a movie that starts at 7:30. The box office opens at 6:45.

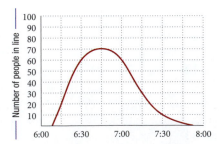

a. How many people are in line at 6:30?

b. How many people are in line when the box office opens?

c. How many people are in line when the show starts?

d. At what times are there 60 people in line?

e. How long after the show starts is there no one left in line?

46. Facebook Users The graph shows the number of active Facebook users from 2004 to 2010. If *x* represents the year and *y* represents million of users, write five ordered pairs that describe the information in the graph.

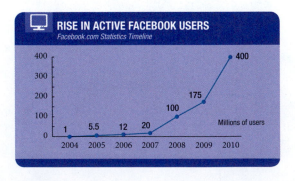

RISE IN ACTIVE FACEBOOK USERS
Facebook.com Statistics Timeline

Getting Ready for the Next Section

Complete each table using the given equation.

47. $y = 7.5x$

x	y
0	
10	
20	

48. $h = 32t - 16t^2$

t	h
0	
1	
-1	

49. $y = 7.5x$

x	y
0	
$\frac{1}{2}$	
1	

50. $h = 32t - 16t^2$

t	h
-3	
0	
3	

Find the Mistake

Write true or false for each statement below. If false, circle what is false and write the correct answer on the line provided.

1. True or False: On a rectangular coordinate system, the vertical axis is called the x-axis. _____

2. True or False: Any point on the x-axis has a y-coordinate of 0. _____

3. True or False: The graph of a line that passes through the origin will have the ordered pair (0, 0) as a solution.
   _____

4. True or False: The x-intercept for the graph of the equation $4x + 7y = 12$ is 3. _____

Navigation Skills: Prepare, Study, Achieve

Completing homework assignments in full is a key piece to succeeding in this class. To do this effectively, you must pay special attention to each set of instructions. When you do your homework, you usually work a number of similar problems at a time. But the problems may vary on a test. It is very important to make a habit of paying attention to the instructions to elicit correct answers on a test. Secondly, to complete an assignment efficiently, you will need to memorize various definitions, properties, and formulas. Reading the definition in the book alone is not enough. There are many techniques for successful memorization. Here are a few:

- Spend some time rereading the definition.
- Say the definition out loud.
- Explain the definition to another person.
- Write the definition down on a separate sheet of notes.
- Create a mnemonic device using key words from the definition.
- Analyze how the definition applies to your homework problems.

The above suggestions are ways to engage your senses when memorizing an abstract concept. This will help anchor it in your memory. For instance, it is easier to remember explaining to your friend a difficult math formula, than it is to simply recall it from a single read of the chapter. Lastly, once you've completed an assignment, take any extra time you've allotted for studying to work more problems, and if you feel ready, read ahead and work problems you will encounter in the next section.

OBJECTIVES

A Construct a table or a graph from a function rule.

B Identify the domain and range of a function.

C Determine whether a relation is also a function.

D Use the vertical line test to determine whether a graph is a function.

KEY WORDS

relation

function

domain

range

function map

vertical line test

Before the harvest each fall, a Nebraskan farmer climbs onto his riding lawn mower and rides through acres of corn. He uses the mower to carve an intricate maze in his cornfield. One of his most impressive designs took the shape of a Star Wars starship. He is known to fly in a plane over his freshly cut maze to ensure its perfection. Thousands flock to Grandpa John's Amazing Maze; each visitor paying $5.00 to walk through it.

In this section, we will introduce you to functions. Using Grandpa John's Amazing Maze as an example, the amount of money the maze earns for the farmer each season depends on the number of people who visit it. In mathematics, we say that his seasonal earnings are a *function* of how many people visit the maze.

Relations

Before we begin our discussion of functions, let's examine what happens when a coordinate from a set of inputs is paired with one or more elements from a set of outputs.

Table 1 shows the prices of used Ford Mustangs that were listed in the local newspaper. The diagram in Figure 1 is called a *scatter diagram*. It gives a visual representation of the data in Table 1.

Table 1	
Used Mustang Prices	
Year *x*	Price ($) *y*
2009	17,995
2009	16,945
2009	25,985
2008	15,995
2008	14,867
2007	16,960
2006	14,999
2006	19,150
2005	11,352

FIGURE 1

Ordered Pairs

(2009, 17,995)

(2009, 16,945)

(2009, 25,985)

(2008, 15,995)

(2008, 14,867)

(2007, 16,960)

(2006, 14,999)

(2006, 19,150)

(2005, 11,352)

Looking at the graph, you can see that the year 2009 is paired with three different prices: $17,995, $16,945, and $25,985. This data is considered a *relation*. Furthermore, all sets of paired data are considered relations where each number in the domain is paired with one or more numbers in the range. The following are two formal definitions of a relation.

A *relation* is a rule that pairs each element in one set, called the domain, with **one or more elements** from a second set, called the **range**.

A *relation* is a set of ordered pairs. The set of all first coordinates is the *domain* of the relation. The set of all second coordinates is the *range* of the relation.

A Constructing Tables and Graphs

A special case occurs when each number in a domain is paired with exactly one number in a range. This is called a *function*.

Suppose you have a job that pays $9.50 per hour and that requires you to work anywhere from 0 to 40 hours per week. If we let the variable x represent hours and the variable y represent the money you make, then the relationship between x and y can be written as

$$y = 9.5x \qquad \text{for} \qquad 0 \le x \le 40$$

EXAMPLE 1 Construct a table and graph for the function

$$y = 9.5x \qquad \text{for} \qquad 0 \le x \le 40$$

Solution Table 2 gives some of the paired data that satisfy the equation $y = 9.5x$. Figure 2 is the graph of the equation with the restriction $0 \le x \le 40$.

Table 2 Weekly Wages

Hours Worked	Rule	Pay
x	$y = 9.5x$	y
0	$y = 9.5(0)$	0
10	$y = 9.5(10)$	95
20	$y = 9.5(20)$	190
30	$y = 9.5(30)$	285
40	$y = 9.5(40)$	380

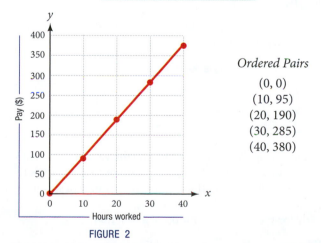

Ordered Pairs

$(0, 0)$
$(10, 95)$
$(20, 190)$
$(30, 285)$
$(40, 380)$

FIGURE 2

The equation $y = 9.5x$ with the restriction $0 \le x \le 40$, Table 2, and Figure 2 are three ways to describe the same relationship between the number of hours you work in one week and your gross pay for that week. In all three, we *input* values of x, and then use the function rule to *output* values of y.

Practice Problems

1. Construct a table and graph for $y = 8x, 0 \le x \le 40$.

x	y
0	
10	
20	
30	
40	

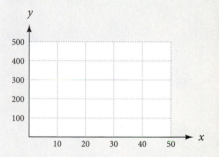

B Domain and Range of a Function

We began this discussion by saying that the number of hours worked during the week was from 0 to 40, so these are the values that x can assume. From the line graph in Figure 2, we see that the values of y range from 0 to 380. We call the complete set of values that x can assume the *domain* of the function. The values that are assigned to y are called the *range* of the function.

Domain: The set of all inputs	**The Function Rule** →	Range: The set of all outputs

EXAMPLE 2   State the domain and range for the function

$$y = 9.5x, \quad 0 \le x \le 40$$

Solution   From the previous discussion, we have

$$\text{Domain} = \{x \,|\, 0 \le x \le 40\} = [0, 40]$$
$$\text{Range} = \{y \,|\, 0 \le y \le 380\} = [0, 380]$$

We are familiar with writing domain and range using set-builder notation and interval notation, shown in Example 2. We can also write the domain and range using *roster notation*. This notation lists all the elements in a set one by one. In the case of Example 1, our set consists of the coordinates

$$(0, 0), (10, 95), (20, 190), (30, 285), (40, 380)$$

Therefore, using roster notation, we give the domain as {0, 10, 20, 30, 40}, and the range as {0, 95, 190, 285, 380}. Here is a summary of the three types of notation using the values from our first example:

	Set-Builder Notation	Roster Notation	Interval Notation
Domain	$\{x \,|\, 0 \le x \le 40\}$	$\{0, 10, 20, 30, 40\}$	$[0, 40]$
Range	$\{y \,|\, 0 \le y \le 380\}$	$\{0, 95, 190, 285, 380\}$	$[0, 380]$

Function Maps

Another way to visualize the relationship between x and y is with the diagram in Figure 2, which we call a *function map*.

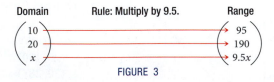

FIGURE 3

Although the diagram in Figure 3 does not show all the values that x and y can assume, it does give us a visual description of how x and y are related. It shows that values of y in the range come from values of x in the domain according to a specific rule (multiply by 9.5 each time).

We are now ready for the formal definition of a *function*.

> **DEFINITION** function
>
> A *function* is a rule that pairs each element in one set, called the *domain,* with exactly one element from a second set, called the *range.* In other words, a function is a rule for which each input is paired with exactly one output.

2. State the domain and range for $y = 8x, 0 \le x \le 40$.

Answers

2. Domain = $\{x \,|\, 0 \le x \le 40\}$
Range = $\{y \,|\, 0 \le y \le 320\}$

EXAMPLE 3 Determine the domain and range of the following relations.

a. $\{(5, 7), (0, -3), (2, 1), (x, y)\}$

b.

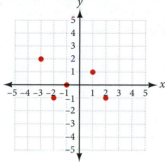

c.

Input	Rule: Population in millions	Output

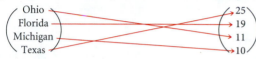

Ohio, Florida, Michigan, Texas → 25, 19, 11, 10

3. Determine the domain and range for the following relations.

a. $\{(3, 3), (1, -1), (0, -3), (a, b)\}$

b.

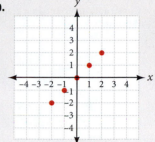

c.

Input	Output

5, 7, 3, 2 → 0, 1, 4, 2

Solution

a. For the set of ordered pairs, the first coordinate is the domain and the second coordinate is the range.

Domain $= \{5, 0, 2, x\}$

Range $= \{7, -3, 1, y\}$

b. The points on the graph represent ordered pairs. The set of x-coordinates are the domain and the set of y-coordinates are the range.

The relation is $\{(-3, 2), (-2, -1), (-1, 0), (1, 1), (2, -1)\}$.

Domain $= \{-3, -2, -1, 1, 2\}$

Range $= \{2, -1, 0, 1\}$

c. The domain is the set of states, and the range is the set of numbers that correspond to each state.

Domain $= \{$Ohio, Florida, Michigan, Texas$\}$

Range $= \{10, 11, 19, 25\}$

EXAMPLE 4 Kendra tosses a softball into the air with an underhand motion. The distance of the ball above her hand is given by the function

$$h = 32t - 16t^2 \qquad \text{for} \qquad 0 \le t \le 2$$

where h is the height of the ball in feet and t is the time in seconds. Construct a table that gives the height of the ball at quarter-second intervals, starting with $t = 0$ and ending with $t = 2$, then graph the function.

Solution We construct Table 3 using the following values of t: 0, $\frac{1}{4}$, $\frac{1}{2}$, $\frac{3}{4}$, 1, $\frac{5}{4}$, $\frac{3}{2}$, $\frac{7}{4}$, 2. Then we construct the graph in Figure 4 from the table. The graph appears only in the first quadrant because neither t nor h can be negative.

Answers

3. a. Domain $= \{3, 1, 0, a\}$

Range $= \{-3, -1, 3, b\}$

b. Domain $= \{-2, -1, 0, 1, 2\}$

Range $= \{-2, -1, 0, 1, 2\}$

c. Domain $= \{-2, 3, 5, 7\}$

Range $= \{0, 1, 2, 4\}$

4. Kendra is tossing a softball into the air so that the distance h the ball is above her hand t seconds after she begins the toss is given by $h = 48t - 16t^2$, $0 \le t \le 3$. Construct a table and line graph for this function.

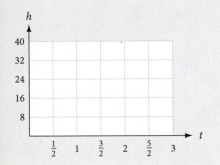

Table 3 Tossing a Softball into the Air		
Input		**Output**
Time (sec) t	Function Rule $h = 32t - 16t^2$	Distance (ft) h
0	$h = 32(0) - 16(0)^2 = 0 - 0 = 0$	0
$\frac{1}{4}$	$h = 32\left(\frac{1}{4}\right) - 16\left(\frac{1}{4}\right)^2 = 8 - 1 = 7$	7
$\frac{1}{2}$	$h = 32\left(\frac{1}{2}\right) - 16\left(\frac{1}{2}\right)^2 = 16 - 4 = 12$	12
$\frac{3}{4}$	$h = 32\left(\frac{3}{4}\right) - 16\left(\frac{3}{4}\right)^2 = 24 - 9 = 15$	15
1	$h = 32(1) - 16(1)^2 = 32 - 16 = 16$	16
$\frac{5}{4}$	$h = 32\left(\frac{5}{4}\right) - 16\left(\frac{5}{4}\right)^2 = 40 - 25 = 15$	15
$\frac{3}{2}$	$h = 32\left(\frac{3}{2}\right) - 16\left(\frac{3}{2}\right)^2 = 48 - 36 = 12$	12
$\frac{7}{4}$	$h = 32\left(\frac{7}{4}\right) - 16\left(\frac{7}{4}\right)^2 = 56 - 49 = 7$	7
2	$h = 32(2) - 16(2)^2 = 64 - 64 = 0$	0

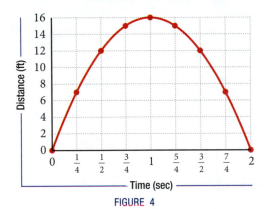

FIGURE 4

Here is a summary of what we know about functions as it applies to this example: We input values of t and output values of h according to the function rule

$$h = 32t - 16t^2 \qquad \text{for} \qquad 0 \le t \le 2$$

The domain is given by the inequality that follows the equation; it is

$$\text{Domain} = \{t \,|\, 0 \le t \le 2\} = [0, 2]$$

The range is the set of all outputs that are possible by substituting the values of t from the domain into the equation. From our table and graph, it seems that the range is

$$\text{Range} = \{h \,|\, 0 \le h \le 16\} = [0, 16]$$

As you can see from the examples we have done to this point, the function rule produces ordered pairs of numbers. We use this result to write an alternative definition for a function.

ALTERNATE DEFINITION   function

A *function* is a set of ordered pairs in which no two different ordered pairs have the same first coordinate. The set of all first coordinates is called the *domain* of the function. The set of all second coordinates is called the *range* of the function. The restriction on first coordinates keeps us from assigning a number in the domain to more than one number in the range.

C Determining Relations and Functions

Here are some facts that will help clarify the distinction between relations and functions:

1. Any rule that assigns numbers from one set to numbers in another set is a relation. If that rule makes the assignment so no input has more than one output, then it is also a function.
2. Any set of ordered pairs is a relation. If none of the first coordinates of those ordered pairs is repeated, the set of ordered pairs is also a function.
3. Every function is a relation.
4. Not every relation is a function.

EXAMPLE 5   Sketch the graph of $x = y^2$.

Solution   Without going into much detail, we graph the equation $x = y^2$ by finding a number of ordered pairs that satisfy the equation, plotting these points, then drawing a smooth curve that connects them. A table of values for x and y that satisfy the equation follows, along with the graph of $x = y^2$ shown in Figure 5.

x	y
0	0
1	1
1	−1
4	2
4	−2
9	3
9	−3

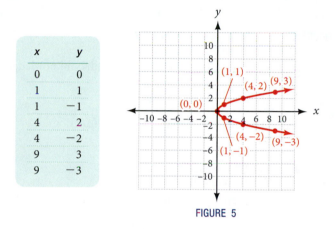

FIGURE 5

As you can see from looking at the table and the graph in Figure 5, several ordered pairs whose graphs lie on the curve have repeated first coordinates, for instance (1, 1) and (1, −1), (4, 2) and (4, −2), as well as (9, 3) and (9, −3). Therefore, the graph is not the graph of a function.

D Vertical Line Test

Look back at the scatter diagram for used Mustang prices shown in Figure 1. Notice that some of the points on the diagram lie above and below each other along vertical lines. This is an indication that the data do not constitute a function. Two data points that lie on the same vertical line must have come from two ordered pairs with the same first coordinates.

Now, look at the graph shown in Figure 1. The reason this graph is the graph of a relation, but not of a function, is that some points on the graph have the same first coordinates, for example, the points (4, 2) and (4, −2). Furthermore, any time two points on a graph have the same first coordinates, those points must lie on a vertical line. [To convince yourself, connect the points (4, 2) and (4, −2) with a straight line. You will see that it must be a vertical line.] This allows us to write the following test that uses the graph to determine whether a relation is also a function.

> **RULE** Vertical Line Test
>
> If a vertical line crosses the graph of a relation in more than one place, the relation cannot be a function. If no vertical line can be found that crosses a graph in more than one place, then the graph represents a function.

5. Use the equation $x = y^2 - 4$ to fill in the following table. Then use the table to sketch the graph.

x	y
	−3
	−2
	−1
	0
	1
	2
	3

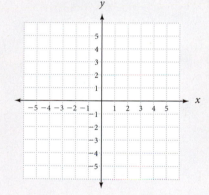

6. Graph $y = |x| - 3$.

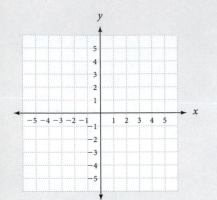

If we look back to the graph of $h = 32t - 16t^2$ as shown in Figure 4, we see that no vertical line can be found that crosses this graph in more than one place. The graph shown in Figure 4 is therefore the graph of a function.

EXAMPLE 6   Match each relation with its graph, then indicate which relations are functions.

a. $y = |x| - 4$ **b.** $y = x^2 - 4$ **c.** $y = 2x + 2$

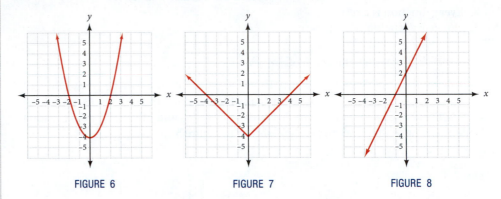

FIGURE 6   FIGURE 7   FIGURE 8

Solution   Using the basic graphs for a guide along with our knowledge of translations, we have the following:

a. Figure 7   **b.** Figure 6   **c.** Figure 8

And, since all graphs pass the vertical line test, all are functions.

GETTING READY FOR CLASS

After reading through the preceding section, respond in your own words and in complete sentences.

A. What is a relation?

B. What is a function?

C. Which variable is usually associated with the domain of a function? Which variable is associated with the range?

D. What is the vertical line test?

Vocabulary Review

Choose the correct words to fill in the blanks below.

    function   domain   right   left

    range   relation   vertical line test

1. The _____ of a function is the complete set of inputs, whereas the _____ is the set of all outputs.

2. On a function map, the values for the domain of the function appear on the _____ and the values for the range appear on the _____ .

3. A _____ is a rule that pairs each element in one set with exactly one element from a second set.

4. A _____ is a rule that pairs each element in one set with one or more elements from a second set.

5. The _____ can be used to determine whether the graph of a relation is a function.

Problems

A For each of the following functions, construct a table and graph.

1. $y = 4x$ for $-2 \leq x \leq 5$

x	y
−2	
0	
2	
5	

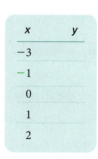

2. $y = -3x$ for $-3 \leq x \leq 2$

x	y
−3	
−1	
0	
1	
2	

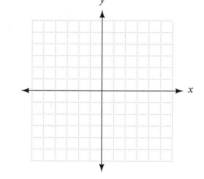

3. $y = 100 - 4x$ for $0 \leq x \leq 25$

x	y
0	
5	
10	
15	
20	
25	

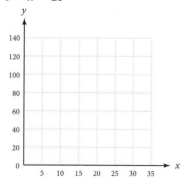

4. $y = 250 - 10x$ for $0 \leq x \leq 25$

x	y
0	
5	
10	
15	
20	
25	

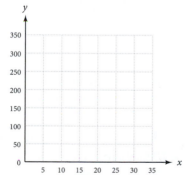

5. $y = 1600 - 10x$ for $0 \le x \le 160$

x	y
0	
40	
80	
120	
160	

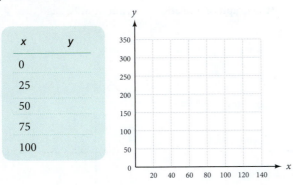

6. $y = 1200 - 40x$ for $0 \le x \le 30$

x	y
0	
10	
20	
30	

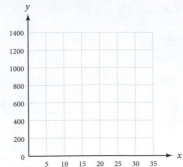

7. $y = 40 + 2.5x$ for $0 \le x \le 100$

x	y
0	
25	
50	
75	
100	

8. $y = 100 + 0.5x$ for $0 \le x \le 40$

x	y
0	
10	
20	
30	
40	

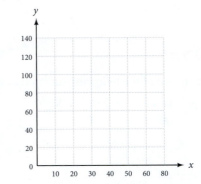

B C For each of the following relations, give the domain and range using roster notation, and indicate which are also functions.

9. $(1, 2), (3, 4), (5, 6), (7, 8)$

10. $(2, 1), (4, 3), (6, 5), (8, 7)$

11. $(2, 5), (3, 4), (1, 4), (0, 6)$

12. $(0, 4), (1, 6), (2, 4), (1, 5)$

13. $(a, 3), (b, 4), (c, 3), (d, 5)$

14. $(a, 5), (b, 5), (c, 4), (d, 5)$

15. $(a, 1), (a, 2), (a, 3), (a, 4)$

16. $(a, 1), (b, 1), (c, 1), (d, 1)$

Give the domain and range for the following function maps.

17.

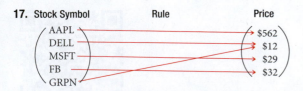

18.

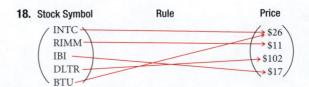

19.

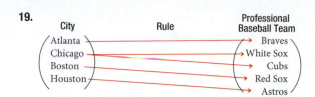

20.

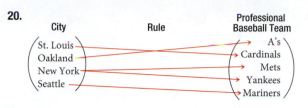

C D State whether each of the following graphs represents the graph of a function.

21.

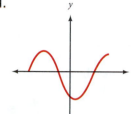

22.

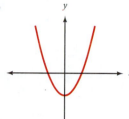

23.

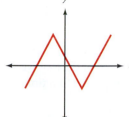

24.

25.

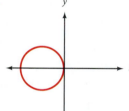

26.

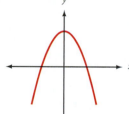

27.

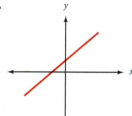

28.

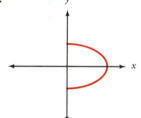

29.

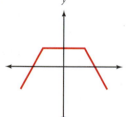

30.

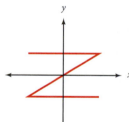

31.

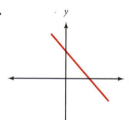

32.

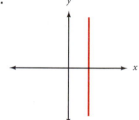

Determine the domain and range of the following functions. Assume the *entire* function is shown. Write your answers in set-builder notation and interval notation.

33.

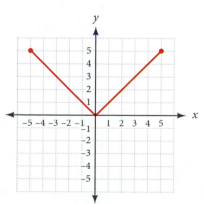

34.

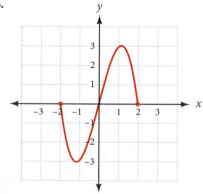

35.

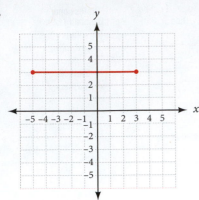

36.

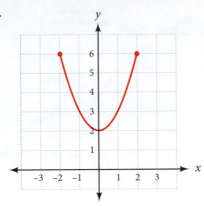

Graph each of the following relations. In each case, use the graph to find the domain and range, and indicate whether the graph is the graph of a function.

37. $y = x^2 - 1$

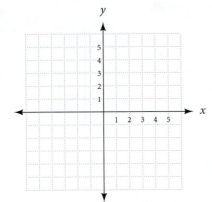

38. $y = x^2 + 1$

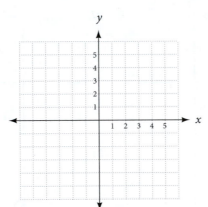

39. $y = x^2 + 4$

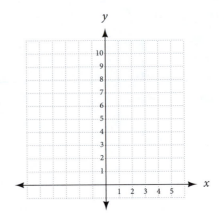

40. $y = x^2 - 9$

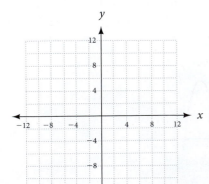

41. $x = y^2 - 1$

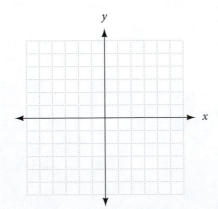

42. $x = y^2 + 1$

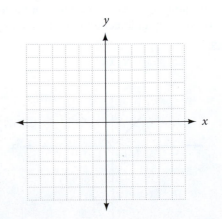

43. $y = (x + 2)^2$

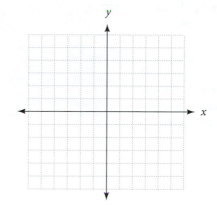

44. $y = (x - 3)^2$

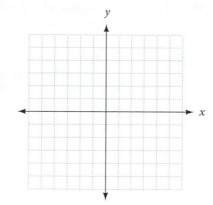

45. $x = (y + 1)^2$

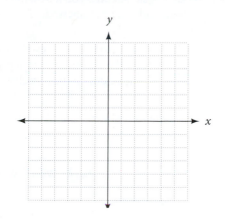

46. $x = 3 - y^2$

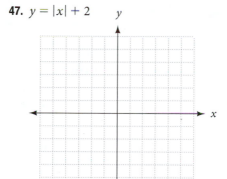

47. $y = |x| + 2$

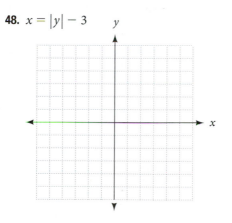

48. $x = |y| - 3$

Applying the Concepts

49. iPod Memory The chart shows the memory capacity for different generations of iPods from 2001 to 2009. Using the chart, list all the values in the domain and range for the memory capacity.

50. Red Box Kiosks The chart shows the number of Redbox DVD rental kiosks installed from 2005 to 2010. Use the chart to state the domain and range of the function for Redbox kiosk installation.

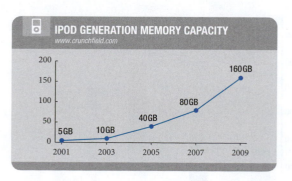

IPOD GENERATION MEMORY CAPACITY
www.crunchfield.com

5GB 10GB 40GB 80GB 160GB
2001 2003 2005 2007 2009

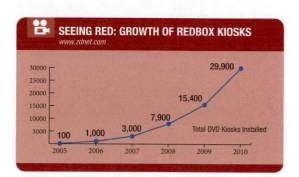

SEEING RED: GROWTH OF REDBOX KIOSKS
www.zdnet.com

100 1,000 3,000 7,900 15,400 29,900
Total DVD Kiosks Installed
2005 2006 2007 2008 2009 2010

51. Weekly Wages Suppose you have a job that pays $8.50 per hour and you work anywhere from 10 to 40 hours per week.

 a. Write an equation, with a restriction on the variable x, that gives the amount of money, y, you will earn for working x hours in one week.

 b. Use the function rule you have written in part **a.** to complete Table 4.

Table 4   Weekly Wages

Hours Worked	Function Rule	Gross Pay ($)
x		y
10		
20		
30		
40		

 c. Construct a line graph from the information in Table 4.

Weekly Wages

 d. State the domain and range of this function.

 e. What is the minimum amount you can earn in a week with this job? What is the maximum amount?

52. Weekly Wages The ad shown here was in the local newspaper. Suppose you are hired for the job described in the ad.

 a. If x is the number of hours you work per week and y is your weekly gross pay, write the equation for y. (Be sure to include any restrictions on the variable x that are given in the ad.)

 b. Use the function rule you have written in part a to complete Table 5.

Table 5   Weekly Wages

Hours Worked	Function Rule	Gross Pay ($)
x		y
15		
20		
25		
30		

 c. Construct a line graph from the information in Table 5.

Weekly Wages

 d. State the domain and range of this function.

 e. What is the minimum amount you can earn in a week with this job? What is the maximum amount?

53. Global Solar Energy Demand The chart shows the global solar power demand from 2008 to 2012. Using the chart:

 a. List the values for the domain and for the range.

 b. Did demand increase more in 2011 or 2012 (projected)?

 c. In what year did demand decrease?

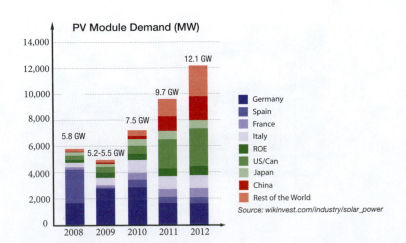

PV Module Demand (MW)

Source: wikinvest.com/industry/solar_power

54. Global Wind Power Capacity The chart shows the global wind power capacity from 1996 to 2010:

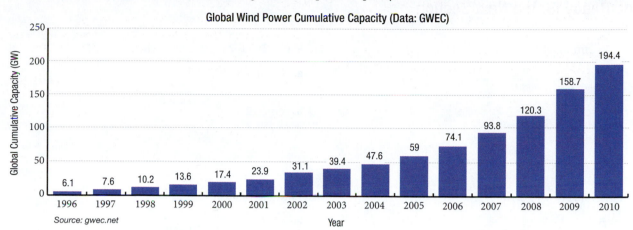

Global Wind Power Cumulative Capacity (Data: GWEC)

Source: gwec.net

Use the information from the chart to answer the following questions.

a. List the values for the domain and for the range.

b. Did capacity increase more in 2006 or 2007?

c. In which year did capacity increase the most?

55. Profits Match each of the following statements to the appropriate graph indicated by labels I-IV.

a. Sarah works 25 hours to earn $250.

b. Justin works 35 hours to earn $560.

c. Rosemary works 30 hours to earn $360.

d. Marcus works 40 hours to earn $320.

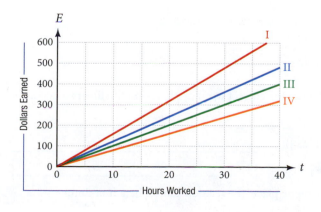

FIGURE 9

56. Find an equation for each of the functions shown in the Figure 9. Show dollars earned, E, as a function of hours worked, t. Then indicate the domain and range of each function.

a. Graph I: $E =$

   Domain = $\{ t \mid \qquad \}$

   Range = $\{ E \mid \qquad \}$

b. Graph II: $E =$

   Domain = $\{ t \mid \qquad \}$

   Range = $\{ E \mid \qquad \}$

c. Graph III: $E =$

   Domain = $\{ t \mid \qquad \}$

   Range = $\{ E \mid \qquad \}$

d. Graph IV: $E =$

   Domain = $\{ t \mid \qquad \}$

   Range = $\{ E \mid \qquad \}$

Getting Ready for the Next Section

57. If $s = \dfrac{60}{t}$, find s when

   a. $t = 10$

   b. $t = 8$

58. If $y = 3x^2 + 2x - 1$, find y when

   a. $x = 0$

   b. $x = -2$

59. Find the value of $x^2 + 2$ for

   a. $x = 5$

   b. $x = -2$

60. Find the value of $125 \cdot 2^t$ for

   a. $t = 0$

   b. $t = 1$

For the equation $y = x^2 - 3$:

61. Find y if x is 2.

62. Find y if x is -2.

63. Find y if x is 0.

64. Find y if x is -4.

65. For which value of x is $y = -3$?

66. For which values of x is $y = 6$? (There are two answers)

The problems that follow review some of the more important skills you have learned in previous sections and chapters.

67. If $x - 2y = 4$, and $x = \dfrac{8}{5}$ find y.

68. If $\dfrac{x^2}{25} + \dfrac{y^2}{9} = 1$, find y when x is -4.

69. Let $x = 0$ and $y = 0$ in $y = a(x - 8)^2 + 70$ and solve for a.

70. Find R if $p = 2.5$ and $R = (900 - 300p)p$.

Find the Mistake

Each sentence below contains a mistake. Circle the mistake and write the correct word or phrase on the line provided.

1. The set of all inputs for a function is called the range. _____

2. For the relation (1, 2), (3, 4), (5, 6), (7, 8), the domain is {2, 4, 6, 8}. _____

3. Graphing the relation $y = x^2 - 3$ will show that the domain is $\{\, x \mid x \geq -3 \,\}$. _____

4. If a horizontal line crosses the graph of a relation in more than one place, the relation cannot be a function.

_____

2.3 Function Notation

©iStockphoto.com/tap10

OBJECTIVES

A Use function notation to find the value of a function for a given value of the variable.

B Use graphs to visualize the relationship between a function and a variable.

C Use function notation in formulas.

KEY WORDS

dependent variable

independent variable

function notation

inputs

outputs

A Evaluate a Function at a Point

Pogopalooza is an annual world championship for stunt pogo stick athletes. Pogo jumpers compete in various events, such as the most or the least jumps in a minute, the highest jump, and numerous exhibitions of acrobatic stunts. Imagine a group of local stunt pogo enthusiasts are holding a pogo stick competition in your city. They have gathered a variety a sponsors, and their budget allows them to hire you to market and promote the event. If the job pays $8.50 per hour for working from 0 to 40 hours a week, then the amount of money y earned in one week is a function of the number of hours worked x. The exact relationship between x and y is written

$$y = 8.5x \quad \text{for} \quad 0 \le x \le 40$$

Because the amount of money earned y depends on the number of hours worked x, we call y the *dependent variable* and x the *independent variable*. Furthermore, if we let f represent all the ordered pairs produced by the equation, then we can write

$$f = \{(x, y) \,|\, y = 8.5x \quad \text{and} \quad 0 \le x \le 40\}$$

Once we have named a function with a letter, we can use an alternative notation to represent the dependent variable y. The alternative notation for y is $f(x)$. It is read "f of x" and can be used instead of the variable y when working with functions. The notation y and the notation $f(x)$ are equivalent. That is,

$$y = 8.5x \Leftrightarrow f(x) = 8.5x$$

When we use the notation $f(x)$ we are using *function notation*. The benefit of using function notation is that we can write more information with fewer symbols than we can by using just the variable y. For example, asking how much money a person will make for working 20 hours is simply a matter of asking for $f(20)$. Without function notation, we would have to say, "Find the value of y that corresponds to a value of $x = 20$." To illustrate further, using the variable y, we can say "y is 170 when x is 20." Using the notation $f(x)$, we simply say "$f(20) = 170$." Each expression indicates that you will earn $170 for working 20 hours.

> **Note** An important thing to remember is that f is the function name, it is not a variable and cannot be replaced with a number. Furthermore, $f(x)$ is read "f of x" but does not imply multiplication.

EXAMPLE 1 If $f(x) = 7.5x$, find $f(0)$, $f(10)$, and $f(20)$.

Solution To find $f(0)$, we substitute 0 for x in the expression $7.5x$ and simplify. We find $f(10)$ and $f(20)$ in a similar manner — by substitution.

$$\text{If} \qquad f(x) = 7.5x$$
$$\text{then} \qquad f(0) = 7.5(0) = 0$$
$$f(10) = 7.5(10) = 75$$
$$f(20) = 7.5(20) = 150$$

Practice Problems

1. If $f(x) = 8x$, find
  a. $f(0)$
  b. $f(5)$
  c. $f(10.5)$

Answers

1. a. 0 **b.** 40 **c.** 84

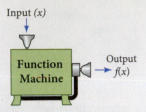

Input (x)

Function Machine

Output $f(x)$

If we changed the example in the discussion that opened this section so the hourly wage was $6.50 per hour, we would have a new equation to work with, namely,

$$y = 6.5x \qquad \text{for} \qquad 0 \le x \le 40$$

Suppose we name this new function with the letter g. Then

$$g = \{(x, y) \,|\, y = 6.5x \quad \text{and} \quad 0 \le x \le 40\}$$

and

$$g(x) = 6.5x$$

If we want to talk about both functions in the same discussion, having two different letters, f and g, makes it easy to distinguish between them. For example, since $f(x) = 7.5x$ and $g(x) = 6.5x$, asking how much money a person makes for working 20 hours is simply a matter of asking for $f(20)$ or $g(20)$, avoiding any confusion over which hourly wage we are talking about.

The diagrams shown in Figure 1 further illustrate the similarities and differences between the two functions we have been discussing.

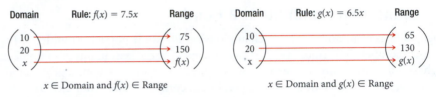

FIGURE 1

B Function Notation and Graphs

We can visualize the relationship between x and $f(x)$ on the graph of the function. Figure 2 shows the graph of $f(x) = 7.5x$ along with two additional line segments. The horizontal line segment corresponds to $x = 20$, and the vertical line segment corresponds to $f(20)$. (Note that the domain is restricted to $0 \le x \le 40$.)

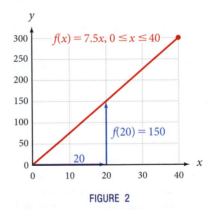

FIGURE 2

We can use functions and function notation to talk about numbers in the following chart showing the number of millionaire households in the U.S. Let's let x represent one of the years in the chart.

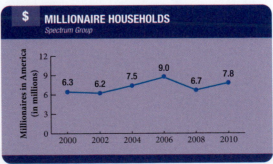

If the function f pairs each year in the chart with the number of millionaire households for that year, then each statement below is true:

$$f(2006) = 9.0$$

The domain of $f = \{2000, 2002, 2004, 2006, 2008, 2010\}$

In general, when we refer to the function f we are referring to the domain, the range, and the rule that takes elements in the domain and outputs elements in the range. When we talk about $f(x)$ we are talking about the rule itself, or an element in the range, or the variable y.

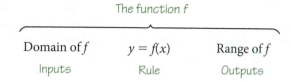

The function f

Domain of f	$y = f(x)$	Range of f
Inputs	Rule	Outputs

C Using Function Notation

The remaining examples in this section show a variety of ways to use and interpret function notation.

EXAMPLE 2   If Lorena takes t minutes to run a mile, then her average speed s, in miles per hour, is given by the formula

$$s(t) = \frac{60}{t} \qquad \text{for} \qquad t > 0$$

Find $s(10)$ and $s(8)$, and then explain what they mean.

Solution   To find $s(10)$, we substitute 10 for t in the equation and simplify.

$$s(10) = \frac{60}{10} = 6$$

In words: When Lorena runs a mile in 10 minutes, her average speed is 6 miles per hour. We calculate $s(8)$ by substituting 8 for t in the equation. Doing so gives us

$$s(8) = \frac{60}{8} = 7.5$$

In words: Running a mile in 8 minutes is running at a rate of 7.5 miles per hour.

EXAMPLE 3   A painting is purchased as an investment for $125. If its value increases continuously so that it doubles every 5 years, then its value is given by the function

$$V(t) = 125 \cdot 2^{t/5} \qquad \text{for} \qquad t \geq 0$$

where t is the number of years since the painting was purchased, and V is its value (in dollars) at time t. Find $V(5)$ and $V(10)$, and explain what they mean.

©Marissa McKeague, 2011

Solution   The expression $V(5)$ is the value of the painting when $t = 5$ (5 years after it is purchased). We calculate $V(5)$ by substituting 5 for t in the equation $V(t) = 125 \cdot 2^{t/5}$. Here is our work:

$$V(5) = 125 \cdot 2^{5/5} = 125 \cdot 2^1 = 125 \cdot 2 = 250$$

In words: After 5 years, the painting is worth $250.

2. When Lorena runs a mile in t minutes, then her average speed in feet per second is given by
$$s(t) = \frac{88}{t}, t > 0$$
   a. Find $s(8)$ and explain what it means.
   b. Find $s(11)$ and explain what it means.

3. A medication has a half-life of 5 days. If the concentration of the medication in a patient's system is 80 ng/mL, and the patient stops taking it, then t days later the concentration will be
$$C(t) = 80\left(\frac{1}{2}\right)^{t/5}$$
Find each of the following, and explain what they mean.
   a. $C(5)$
   b. $C(10)$

Note   Remember, as stated in the order of operations, work the exponents before any multiplication the problem requires.

Answers

2. **a.** $s(8) = 11$; runs a mile in 8 minutes, average speed is 11 feet per second.
   b. $s(11) = 8$; runs a mile in 11 minutes, average speed is 8 feet per second.
3. **a.** $C(5) = 40$ ng/mL; after 5 days the concentration is 40 ng per mL.
   b. $C(10) = 20$ ng/mL; after 10 days the concentration is 20 ng per mL.

The expression $V(10)$ is the value of the painting after 10 years. To find this number, we substitute 10 for t in the equation:

$$V(10) = 125 \cdot 2^{10/5} = 125 \cdot 2^2 = 125 \cdot 4 = 500$$

In words: The value of the painting 10 years after it is purchased is $500.

4. The following formulas give the circumference and area of a circle with radius r. Use the formulas to find the circumference and area of a circular plate if the radius is 5 inches.
  a. $C(r) = 2\pi r$
  b. $A(r) = \pi r^2$

EXAMPLE 4   A balloon has the shape of a sphere with a radius of 3 inches. Use the following formulas to find the volume and surface area of the balloon.

$$V(r) = \frac{4}{3}\pi r^3 \qquad S(r) = 4\pi r^2$$

Solution   As you can see, we have used function notation to write the formulas for volume and surface area, because each quantity is a function of the radius. To find these quantities when the radius is 3 inches, we evaluate $V(3)$ and $S(3)$.

$$V(3) = \frac{4}{3}\pi \cdot 3^3 = \frac{4}{3}\pi \cdot 27$$

$$= 36\pi \text{ cubic inches, or } 113 \text{ cubic inches}$$
$$\text{(to the nearest whole number)}$$

$$S(3) = 4\pi \cdot 3^2$$

$$= 36\pi \text{ square inches, or } 113 \text{ square inches}$$
$$\text{(to the nearest whole number)}$$

The fact that $V(3) = 36\pi$ means that the ordered pair $(3, 36\pi)$ belongs to the function V. Likewise, the fact that $S(3) = 36\pi$ tells us that the ordered pair $(3, 36\pi)$ is a member of function S.

We can generalize the discussion at the end of Example 4 this way:

$$(a, b) \in f \qquad \text{if and only if} \qquad f(a) = b$$

USING TECHNOLOGY More About Example 4

If we look at Example 4, we see that when the radius of a sphere is 3, the numerical values of the volume and surface area are equal. How unusual is this? Are there other values of r for which $V(r)$ and $S(r)$ are equal? We can answer this question by looking at the graphs of both V and S.
  To graph the function $V(r) = \frac{4}{3}\pi r^3$, set $Y_1 = 4\pi X^3/3$. To graph $S(r) = 4\pi r^2$, set $Y_2 = 4\pi X^2$. Graph the two functions in each of the following windows:

    Window 1:   X from -4 to 4, Y from -2 to 10

    Window 2:   X from 0 to 4, Y from 0 to 50

    Window 3:   X from 0 to 4, Y from 0 to 150

Then use the Trace and Zoom features of your calculator to locate the point in the first quadrant where the two graphs intersect. How do the coordinates of this point compare with the results in Example 4?

5. If $f(x) = 4x^2 - 3$, find
  a. $f(0)$
  b. $f(3)$
  c. $f(-2)$

EXAMPLE 5   If $f(x) = 3x^2 + 2x - 1$, find $f(0)$, $f(3)$, and $f(-2)$.

Solution   Since $f(x) = 3x^2 + 2x - 1$, we have

$$f(0) = 3(0)^2 + 2(0) - 1 = 0 + 0 - 1 = -1$$

$$f(3) = 3(3)^2 + 2(3) - 1 = 27 + 6 - 1 = 32$$

$$f(-2) = 3(-2)^2 + 2(-2) - 1 = 12 - 4 - 1 = 7$$

Answers
4. $C(5) = 10\pi$
    $A(5) = 25\pi$
5. a. -3 **b.** 33 **c.** 13

In Example 5, the function f is defined by the equation $f(x) = 3x^2 + 2x - 1$. We could just as easily have said $y = 3x^2 + 2x - 1$. That is, $y = f(x)$. Saying $f(-2) = 7$ is exactly the same as saying y is 7 when x is -2.

EXAMPLE 6 If $f(x) = 4x - 1$ and $g(x) = x^2 + 2$, find the value of each for $x = 5, -2, 0, z, a,$ and $a + 3$.

Solution

$$f(5) = 4(5) - 1 = 19 \quad \text{and} \quad g(5) = 5^2 + 2 = 27$$

$$f(-2) = 4(-2) - 1 = -9 \quad \text{and} \quad g(-2) = (-2)^2 + 2 = 6$$

$$f(0) = 4(0) - 1 = -1 \quad \text{and} \quad g(0) = 0^2 + 2 = 2$$

$$f(z) = 4z - 1 \quad \text{and} \quad g(z) = z^2 + 2$$

$$f(a) = 4a - 1 \quad \text{and} \quad g(a) = a^2 + 2$$

$$f(a + 3) = 4(a + 3) - 1 \qquad g(a + 3) = (a + 3)^2 + 2$$

$$= 4a + 12 - 1 \qquad = (a^2 + 6a + 9) + 2$$

$$= 4a + 11 \qquad = a^2 + 6a + 11$$

USING TECHNOLOGY More About Example 6

Most graphing calculators can use tables to evaluate functions. To work Example 6 using a graphing calculator table, set Y_1 equal to $4X - 1$ and Y_2 equal to $X^2 + 2$. Then set the independent variable in the table to Ask instead of Auto. Go to your table and input 5, -2, and 0. Under Y_1 in the table, you will find $f(5)$, $f(-2)$, and $f(0)$. Under Y_2, you will find $g(5)$, $g(-2)$, and $g(0)$.

Plot1 Plot2 Plot3
\Y₁ ■ 4X − 1
\Y₂ ■ X² + 2
\Y₃ =
\Y₄ =
\Y₅ =
\Y₆ =
\Y₇ =

TABLE SETUP
TblStart = 0
ΔTbl = 1
Indpnt: Auto Ask
Depend: Auto Ask

The table will look like this:

X	Y_1	Y_2
5	19	27
−2	−9	6
0	−1	2

Although the calculator asks us for a table increment, the increment doesn't matter because we are inputting the X values ourselves.

---

6. If $f(x) = 2x + 1$ and $g(x) = x^2 - 3$, find

a. $f(5)$

b. $g(5)$

c. $f(-2)$

d. $g(-2)$

e. $f(a)$

f. $g(a)$

Answers

6. a. 11   b. 22   c. -3
   d. 1   e. $2a + 1$ f. $a^2 - 3$

7. If $f = \{(-4, 1), (2, -3), (7, 9)\}$, find
  a. $f(-4)$
  b. $f(2)$
  c. $f(7)$

EXAMPLE 7   If the function f is given by $f = \{(-2, 0), (3, -1), (2, 4), (7, 5)\}$, find $f(-2), f(3), f(2)$, and $f(7)$.

Solution $f(-2) = 0, f(3) = -1, f(2) = 4$, and $f(7) = 5$.

8. If $f(x) = 3x^2$ and $g(x) = 4x + 1$, find
  a. $f(g(2))$
  b. $g(f(2))$

EXAMPLE 8   If $f(x) = 2x^2$ and $g(x) = 3x - 1$, find
  a. $f(g(2))$ **b.** $g(f(2))$

Solution   The expression $f(g(2))$ is read "f of g of 2."

  a. Because $g(2) = 3(2) - 1 = 5$,
$$f(g(2)) = f(5) = 2(5)^2 = 50$$

  b. Because $f(2) = 2(2)^2 = 8$,
$$g(f(2)) = g(8) = 3(8) - 1 = 23$$

GETTING READY FOR CLASS

After reading through the preceding section, respond in your own words and in complete sentences.

A. Explain what you are calculating when you find $f(2)$ for a given function f.

B. If $s(t) = \frac{60}{t}$ how do you find $s(10)$?

C. If $f(2) = 3$ for a function f, what is the relationship between the numbers 2 and 3 and the graph of f?

D. If $f(6) = 0$ for a particular function f, then you can immediately graph one of the intercepts. Explain.

Answers
7. a. 1   **b.** −3   **c.** 9
8. a. 243   **b.** 49

Vocabulary Review

Choose the correct words to fill in the blanks below.

dependent   independent $f(x)$ function f

1. The variable that represents the input of a function is called the _____ variable.
2. The variable that represents the output of a function is called the _____ variable.
3. When we use the notation _____ , we are using function notation.
4. The _____ refers to the domain, the range, and the rule that takes elements in the domain and outputs elements in the range.

Problems

A Let $f(x) = 2x - 5$ and $g(x) = x^2 + 3x + 4$. Evaluate the following.

1. $f(2)$

2. $f(3)$

3. $f(-3)$

4. $g(-2)$

5. $g(-1)$

6. $f(-4)$

7. $g(-3)$

8. $g(2)$

9. $g(a)$

10. $f(a)$

11. $f(a + 6)$

12. $g(a + 6)$

Let $f(x) = 3x^2 - 4x + 1$ and $g(x) = 2x - 1$. Evaluate the following.

13. $f(0)$

14. $g(0)$

15. $g(-4)$

16. $f(1)$

17. $f(-1)$

18. $g(-1)$

19. $g\left(\dfrac{1}{2}\right)$

20. $g\left(\dfrac{1}{4}\right)$

21. $f(a)$

22. $g(a)$

23. $f(a + 2)$

24. $g(a + 2)$

SCAN TO ACCESS

If $f = \{(1, 4), (-2, 0), \left(3, \frac{1}{2}\right), (\pi, 0)\}$ and $g = \{(1, 1),(-2, 2), \left(\frac{1}{2}, 0\right)\}$, find each of the following values of f and g.

25. $f(1)$

26. $g(1)$

27. $g\left(\frac{1}{2}\right)$

28. $f(3)$

29. $g(-2)$

30. $f(\pi)$

31. $f(g(1))$

32. $f(-2)-g(-2)$

33. For which value of x is $g(x) = f(\pi)$?

34. Is $f(1) = g(1)$?

Let $f(x) = x^2 - 2x$ and $g(x) = 5x - 4$. Evaluate the following.

35. $f(-4)$

36. $g(-3)$

37. $f(-2) + g(-1)$

38. $f(-1) + g(-2)$

39. $2f(x) - 3g(x)$

40. $f(x) - g(x^2)$

41. $f(g(3))$

42. $g(f(3))$

43. $f(g(2)) - g(f(2))$

44. $g(f(1)) - f(g(1))$

45. $[f(-3)]^2$

46. $[g(3)]^2$

Let $f(x) = \dfrac{1}{x + 3}$ and $g(x) = \dfrac{1}{x} + 1$. Evaluate the following.

47. $f\left(\frac{1}{3}\right)$

48. $g\left(\frac{1}{3}\right)$

49. $f\left(-\frac{1}{2}\right)$

50. $g\left(-\frac{1}{2}\right)$

51. $f(-3)$

52. $g(0)$

53. $g(-2)$

54. $f(-1)$

55. For the function $f(x) = x^2 - 4$, evaluate each of the following expressions.

a. $f(a) - 3$

b. $f(a - 3)$

c. $f(x) + 2$

d. $f(x + 2)$

e. $f(a + b)$

f. $f(x + h)$

56. For the function $f(x) = 3x^2$, evaluate each of the following expressions.

a. $f(a) - 2$

b. $f(a - 2)$

c. $f(x) + 5$

d. $f(x + 5)$

e. $f(a + b)$

f. $f(x + h)$

B

57. Graph the function $f(x) = \frac{1}{2}x + 2$. Then draw and label the line segments that represent $x = 4$ and $f(4)$.

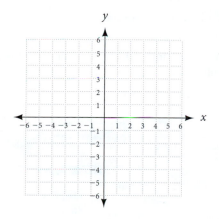

58. Graph the function $f(x) = -\frac{1}{2}x + 6$. Then draw and label the line segments that represent $x = 4$ and $f(4)$.

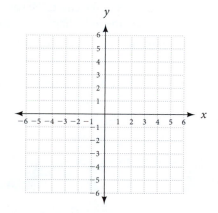

59. For the function $f(x) = \frac{1}{2}x + 2$, find the value of x for which $f(x) = x$.

60. For the function $f(x) = -\frac{1}{2}x + 6$, find the value of x for which $f(x) = x$.

61. Graph the function $f(x) = x^2$. Then draw and label the line segments that represent $x = 1$ and $f(1)$, $x = 2$ and $f(2)$ and, finally, $x = 3$ and $f(3)$.

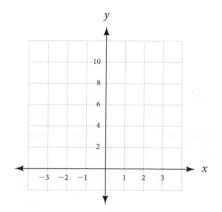

62. Graph the function $f(x) = x^2 - 2$. Then draw and label the line segments that represent $x = 2$ and $f(2)$ and the line segments corresponding to $x = 3$ and $f(3)$.

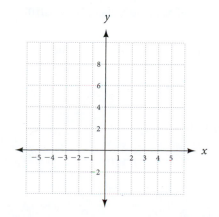

Applying the Concepts

63. Investing in Art Charlie purchased a painting as an investment for $150. If its value increases continuously so that it doubles every 3 years, then its value is given by the function

$$V(t) = 150 \cdot 2^{t/3} \quad \text{for} \quad t \geq 0$$

where t is the number of years since the painting was purchased, and $V(t)$ is its value (in dollars) at time t. Find $V(3)$ and $V(6)$, and then explain what they mean.

64. Average Speed If Minke takes t minutes to run a mile, then her average speed $s(t)$, in miles per hour, is given by the formula

$$s(t) = \frac{60}{t} \quad \text{for} \quad t > 0$$

Find $s(4)$ and $s(5)$, and then explain what they mean.

65. iPhone Sales The chart below shows the percentage of respondents who purchased the iPhone on each of its first day of sales, shown by their current carrier. Suppose x represents one of the carriers in the chart. We have three functions f, g, and h that do the following:

f pairs each year with the percentage of respondents who were AT&T customers for that year.

g pairs each year with the percentage of respondents who were Verizon customers for that year.

h pairs each year with the percentage of respondents who were T-Mobile customers for that year.

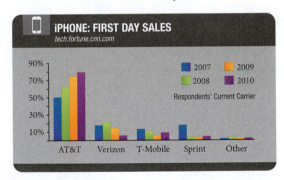

For each statement below, indicate whether the statement is true or false.

a. The domain of g is {2007, 2008, 2009, 2010}

b. The domain of f is $\{x \mid 2008 \leq x \leq 2009\}$

c. $f(2008) > g(2008)$

d. $h(2009) > 30\%$

e. $h(2008) < g(2008) < f(2010)$

66. Newspaper Sales Suppose x represents one of the publications in the chart. We have three functions f, g, and h that do the following:

f pairs each publication with the number of paper subscribers for that publication.

g pairs each publication with the number of online viewers for that publication.

h is such that $h(x) = f(x) + g(x)$.

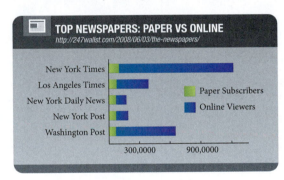

For each statement below, indicate whether the statement is true or false.

a. The domain of f is {New York Times, Los Angeles Times, New York Daily News, New York Post, Washington Post}

b. $h(\text{Los Angeles Times}) = 600,000$

c. $g(\text{New York Times}) > f(\text{New York Times})$

d. $g(\text{New York Daily News}) < g(\text{Washington Post})$

e. $h(\text{New York Times}) > h(\text{Los Angeles Times}) > h(\text{New York Post})$

Straight-Line Depreciation Straight-line depreciation is an accounting method used to help spread the cost of new equipment over a number of years. It takes into account both the cost when new and the salvage value, which is the value of the equipment at the time it gets replaced.

67. Value of a Copy Machine The function
$V(t) = -3,300t + 18,000$, where V is value and t is time in years, can be used to find the value of a large copy machine during the first 5 years of use.

a. What is the value of the copier after 3 years and 9 months?

b. What is the salvage value of this copier if it is replaced after 5 years?

c. State the domain of this function.

d. Sketch the graph of this function.

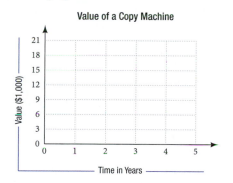

Value of a Copy Machine

e. What is the range of this function?

f. After how many years will the copier be worth only $10,000?

68. Step Function Figure 3 shows the graph of the step function C that was used to calculate the first-class postage on a large envelope weighing x ounces in 2010. Use this graph to answer the following questions.

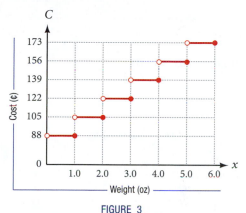

FIGURE 3

a. Fill in the following table:

Weight (ounces)	0.6	1.0	1.1	2.5	3.0	4.8	5.0	5.3
Cost (cents)								

b. If a letter cost 122 cents to mail, how much does it weigh? State your answer in words and as an inequality.

c. If the entire function is shown in Figure 3, state the domain.

d. State the range of the function shown in Figure 3.

69. Wind Turbine The annual output from a wind turbine with an average wind velocity of 10mph is given by the function $f(d) = 13.28d^2$, where d is the diameter (in feet) of the blade and f is kWh/year.

a. Complete the table for various diameter blades.

d (feet)	f(d) (kWh/year)
0	
5	
10	
15	
20	

Source: builditsolar.com/Projects/Wind/Wind.htm

b. What is the range of this function for $0 \le d \le 20$?

70. Wind Turbine The annual output from a wind turbine with a 10–foot–diameter blade is given by the function $f(v) = 1.328v^3$, where v is the average velocity of the wind (in mph) and f is kWh/year.

a. Complete the table for various diameter blades.

v (mph)	f(v) (kWh/year)
0	
10	
20	
30	
40	

Source: builditsolar.com/Projects/Wind/Wind.htm

b. What is the range of this function for $0 \le v \le 40$?

Getting Ready for the Next Section

Multiply.

71. $x(35 - 0.1x)$ **72.** $0.6(M - 70)$ **73.** $(4x - 3)(x - 1)$ **74.** $(4x - 3)(4x^2 - 7x + 3)$

Simplify.

75. $(35x - 0.1x^2) - (8x + 500)$ **76.** $(4x - 3) + (4x^2 - 7x + 3)$

77. $(4x^2 + 3x + 2) - (2x^2 - 5x - 6)$ **78.** $(4x^2 + 3x + 2) + (2x^2 - 5x - 6)$

79. $4(2)^2 - 3(2)$ **80.** $4(-1)^2 - 7(-1)$

Simplify.

81. $2(x - 3)^2 - 5(x - 3)$ **82.** $5(3x - 7) - 4$

83. $-8(-3x + 1) + 2$ **84.** $4(x + 1)^2 - 6(x + 1)$

Find the Mistake

Each sentence below contains a mistake. Circle the mistake and write the correct word or phrase on the line provided.

1. To evaluate the function $f(x) = 10x + 3$ for $f(1)$ we begin by setting $10x + 3$ equal to 1. _____

2. For the average speed function $s(t) = \dfrac{60}{t}$, finding $s(6)$ shows an average speed of 6 miles per hour. _____

3. If the volume of a ball is given by $V(r) = \dfrac{4}{3}\pi r^3$ where r is the radius, we see that the volume decreases as the radius of the ball increases. _____

4. If $f(x) = x^2$ and $g(x) = 2x + 4$, evaluating $f(g(2))$ is the same as multiplying $f(2) \cdot g(2)$. _____
_____

OBJECTIVES

A Use algebra to combine functions.

B Use composition of functions to create a new function.

KEY WORDS

composition of functions

Insects are an important part of the diet for many of the world's cultures. Some scientists suggest eating insects as a solution to potential world hunger once conventional meat sources become scarce. Still, it is rare to find insects on the dinner table in Western households, which is why a certain candy company in Pismo Beach, California remains such a novelty. The company produces and sells lollipops and other sugary treats with real insects inside! You can choose from a wide selection of worms, crickets, scorpions, ants, or butterflies in hard candy or chocolate. Suppose the price $p(x)$ they charge for their scorpion lollipops is related to the number of units sold by the demand function

$$p(x) = 35 - 0.1x$$

We find the revenue for this business by multiplying the number of lollipops sold by the price per lollipop. When we do so, we are forming a new function by combining two existing functions. That is, if $n(x) = x$ is the number of lollipop sold and $p(x) = 35 - 0.1x$ is the price per lollipop in dollars, then revenue is

$$R(x) = n(x) \cdot p(x) = x(35 - 0.1x) = 35x - 0.1x^2$$

In this case, the revenue function is the product of two functions. When we combine functions in this manner, we are applying our rules for algebra to functions.

To carry this situation further, we know the profit function is the difference between two functions. If the cost function for producing x lollipops is $C(x) = 8x + 500$, then the profit function is

$$P(x) = R(x) - C(x) = (35x - 0.1x^2) - (8x + 500) = -500 + 27x - 0.1x^2$$

The relationship between these last three functions is represented visually in Figure 1.

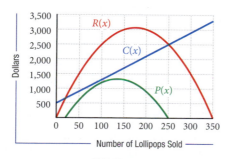

FIGURE 1

A Algebra with Functions

Again, when we combine functions in the manner shown, we are applying our rules for algebra to functions. To begin this section, we take a formal look at addition, subtraction, multiplication, and division with functions.

If we are given two functions $f(x)$ and $g(x)$ with a common domain, in other words, the domain of f and the domain of g consist of common elements, we can define four other functions as follows:

DEFINITION *f* and *g* functions

$(f + g)(x) = f(x) + g(x)$ The function $(f + g)(x)$ is the sum of the functions $f(x)$ and $g(x)$.

$(f - g)(x) = f(x) - g(x)$ The function $(f - g)(x)$ is the difference of the functions $f(x)$ and $g(x)$.

$(f \cdot g)(x) = f(x) \cdot g(x)$ The function $(f \cdot g)(x)$ is the product of the functions $f(x)$ and $g(x)$.

$\left(\dfrac{f}{g}\right)(x) = \dfrac{f(x)}{g(x)}$ The function $\left(\dfrac{f}{g}\right)(x)$ is the quotient of the functions $f(x)$ and $g(x)$, where $g(x) \neq 0$.

Practice Problems

1. If $f(x) = x^2 - 4$ and $g(x) = x + 2$, find formulas for

   a. $(f + g)(x)$

   b. $(f - g)(x)$

   c. $(f \cdot g)(x)$

   d. $\left(\dfrac{f}{g}\right)(x)$

EXAMPLE 1   If $f(x) = 4x^2 + 3x + 2$ and $g(x) = 2x^2 - 5x - 6$, write the formulas for the functions $(f + g)(x)$, $(f - g)(x)$, $(f \cdot g)(x)$, and $\left(\dfrac{f}{g}\right)(x)$.

Solution   The function $(f + g)(x)$ is defined by

$$(f + g)(x) = f(x) + g(x)$$
$$= (4x^2 + 3x + 2) + (2x^2 - 5x - 6)$$
$$= 6x^2 - 2x - 4$$

The function $(f - g)(x)$ is defined by

$$(f - g)(x) = f(x) - g(x)$$
$$= (4x^2 + 3x + 2) - (2x^2 - 5x - 6)$$
$$= 4x^2 + 3x + 2 - 2x^2 + 5x + 6$$
$$= 2x^2 + 8x + 8$$

The function $(f \cdot g)(x)$ is defined by

$$(f \cdot g)(x) = f(x) \cdot g(x)$$
$$= (4x^2 + 3x + 2)(2x^2 - 5x - 6)$$
$$= 8x^4 - 20x^3 - 24x^2 + 6x^3 - 15x^2 - 18x + 4x^2 - 10x - 12$$
$$= 8x^4 - 14x^3 - 35x^2 - 28x - 12$$

The function $\left(\dfrac{f}{g}\right)(x)$ is defined by

$$\left(\frac{f}{g}\right)(x) = \frac{f(x)}{g(x)}$$
$$= \frac{4x^2 + 3x + 2}{2x^2 - 5x - 6}$$

Answers

1. a. $x^2 + x - 2$ b. $x^2 - x - 6$
   c. $x^3 + 2x^2 - 4x - 8$ d. $x - 2$

EXAMPLE 2 Let $f(x) = 4x - 3$, $g(x) = 4x^2 - 7x + 3$, and $h(x) = x - 1$. Find $(f + g)(x)$, $(f \cdot h)(x)$, $(f \cdot g)(x)$ and $\left(\frac{g}{f}\right)(x)$.

Solution The function $(f + g)(x)$, the sum of functions $f(x)$ and $g(x)$, is defined by

$$(f + g)(x) = f(x) + g(x)$$
$$= (4x - 3) + (4x^2 - 7x + 3)$$
$$= 4x^2 - 3x$$

The function $(f \cdot h)(x)$, the product of functions $f(x)$ and $h(x)$, is defined by

$$(f \cdot h)(x) = f(x) \cdot h(x)$$
$$= (4x - 3)(x - 1)$$
$$= 4x^2 - 7x + 3$$
$$= g(x)$$

The function $(f \cdot g)(x)$, the product of the functions $f(x)$ and $g(x)$, is defined by

$$(f \cdot g)(x) = f(x) \cdot g(x)$$
$$= (4x - 3)(4x^2 - 7x + 3)$$
$$= 16x^3 - 28x^2 + 12x - 12x^2 + 21x - 9$$
$$= 16x^3 - 40x^2 + 33x - 9$$

The function $\left(\frac{g}{f}\right)(x)$, the quotient of the functions $g(x)$ and $f(x)$, is defined by

$$\left(\frac{g}{f}\right)(x) = \frac{g(x)}{f(x)}$$
$$= \frac{4x^2 - 7x + 3}{4x - 3}$$

Factoring the numerator, we can reduce to lowest terms.

$$\left(\frac{g}{f}\right)(x) = \frac{(4x - 3)(x - 1)}{4x - 3}$$
$$= x - 1$$
$$= h(x)$$

EXAMPLE 3 Let $f(x) = 2x - 4$, $g(x) = \frac{2}{x}$, and $h(x) = x^2 - 4$. Find $(f + h)(x)$, $(f \cdot g)(x)$, and $\left(\frac{f}{g}\right)(x)$.

Solution The function $(f + h)(x)$, the sum of functions $f(x)$ and $h(x)$, is defined by

$$(f + h)(x) = f(x) + h(x)$$
$$= (2x - 4) + (x^2 - 4)$$
$$= x^2 + 2x - 8$$

The function $(f \cdot g)(x)$, the product of functions $f(x)$ and $g(x)$, is defined by

$$(f \cdot g)(x) = f(x) \cdot g(x)$$
$$= (2x - 4)\left(\frac{2}{x}\right)$$
$$= 4 - \frac{8}{x}$$

The function $\left(\frac{f}{g}\right)(x)$, the quotient of the functions $f(x)$ and $g(x)$, is defined by

$$\left(\frac{f}{g}\right)(x) = \frac{f(x)}{g(x)}$$
$$= \frac{2x - 4}{\frac{2}{x}}$$
$$= \frac{(2x - 4)(x)}{2}$$
$$= x^2 - 2x$$

2. Let $f(x) = 3x + 2$, $g(x) = 3x^2 - 10x - 8$, and $h(x) = x - 4$. Find

a. $(f + g)(x)$

b. $(f \cdot h)(x)$

c. $(f \cdot g)(x)$

d. $\left(\frac{g}{f}\right)(x)$

3. Let $f(x) = 3x - 6$, $g(x) = \frac{6}{x}$, and $h(x) = x^2 - 6$.

a. $(f + h)(x)$

b. $(g \cdot h)(x)$

c. $\left(\frac{h}{g}\right)(x)$

Answers

2. a. $3x^2 - 7x - 6$ b. $g(x)$
   c. $9x^3 - 24x^2 - 44x - 16$ d. $h(x)$

3. a. $x^2 + 3x - 12$ b. $6x - \frac{36}{x}$
   c. $\frac{x^3 - 6x}{6}$

4. Use the functions from Practice Problem 2 to find

a. $(f + g)(3)$

b. $(f \cdot h)(-1)$

c. $(f \cdot g)(0)$

d. $\left(\dfrac{g}{f}\right)(10)$

EXAMPLE 4   If $f(x)$, $g(x)$, and $h(x)$ are the same functions defined in Example 2, evaluate $(f + g)(2)$, $(f \cdot h)(-1)$, $(f \cdot g)(0)$, and $\left(\dfrac{g}{f}\right)(5)$.

Solution   We use the formulas for $f + g, f \cdot h, f \cdot g$ and $\dfrac{g}{f}$ found in Example 2.

$$(f + g)(2) = 4(2)^2 - 3(2)$$
$$= 16 - 6$$
$$= 10$$

$$(f \cdot h)(-1) = 4(-1)^2 - 7(-1) + 3$$
$$= 4 + 7 + 3$$
$$= 14$$

$$(f \cdot g)(0) = 16(0)^3 - 40(0)^2 + 33(0) - 9$$
$$= 0 - 0 + 0 - 9$$
$$= -9$$

$$\left(\dfrac{g}{f}\right)(5) = 5 - 1$$
$$= 4$$

B   Composition of Functions

In addition to the four operations used to combine functions shown so far in this section, there is a fifth way to combine two functions to obtain a new function. It is called *composition of functions*. To illustrate the concept, recall from Chapter 1 the definition of training heart rate: training heart rate, in beats per minute, is resting heart rate plus 60% of the difference between maximum heart rate and resting heart rate. If your resting heart rate is 70 beats per minute, then your training heart rate is a function of your maximum heart rate M.

$$T(M) = 70 + 0.6(M - 70) = 70 + 0.6M - 42 = 28 + 0.6M$$

But your maximum heart rate is found by subtracting your age in years from 220. So, if x represents your age in years, then your maximum heart rate is

$$M(x) = 220 - x$$

Therefore, if your resting heart rate is 70 beats per minute and your age in years is x, then your training heart rate can be written as a function of x.

$$T(x) = 28 + 0.6(220 - x)$$

This last line is the composition of functions T and M. We input x into function M, which outputs $M(x)$. Then we input $M(x)$ into function T, which outputs $T(M(x))$. This is the training heart rate as a function of age x. Here is a diagram, called a function map, of the situation:

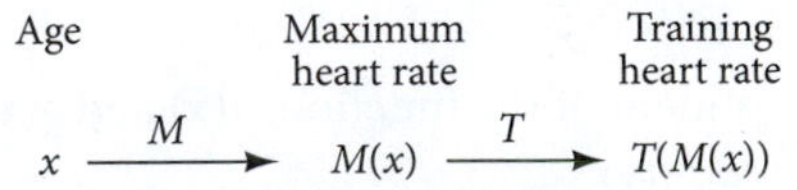

FIGURE 2

Now let's generalize the preceding ideas into a formal development of composition of functions. To find the composition of two functions f and g, we first require that the range of g have numbers in common with the domain of f. Then the composition of f with g, is defined this way:

$$(f \circ g)(x) = f(g(x))$$

Answers

4. a. 0   **b.** 5   **c.** −16   **d.** 6

To understand this new function, we begin with a number x, and we operate on it with g, giving us $g(x)$. Then we take $g(x)$ and operate on it with f, giving us $f(g(x))$. The only numbers we can use for the domain of the composition of f with g are numbers x in the domain of g, for which $g(x)$ is in the domain of f. The diagrams in Figure 3 illustrate the composition of f with g.

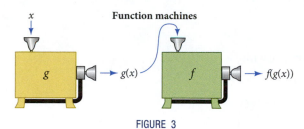

Function machines

FIGURE 3

Composition of functions is not commutative. The composition of f with g, $f \circ g$, may therefore be different from the composition of g with f, $g \circ f$.

$$(g \circ f)(x) = g(f(x))$$

Again, the only numbers we can use for the domain of the composition of g with f are numbers in the domain of f, for which $f(x)$ is in the domain of g. The diagrams in Figure 4 illustrate the composition of g with f.

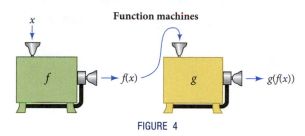

Function machines

FIGURE 4

EXAMPLE 5 If $f(x) = x + 5$ and $g(x) = x^2 - 2x$, find $(f \circ g)(x)$ and $(g \circ f)(x)$.

Solution The composition of f with g is

$$(f \circ g)(x) = f(g(x))$$
$$= f(x^2 - 2x)$$
$$= (x^2 - 2x) + 5$$
$$= x^2 - 2x + 5$$

The composition of g with f is

$$(g \circ f)(x) = g(f(x))$$
$$= g(x + 5)$$
$$= (x + 5)^2 - 2(x + 5)$$
$$= (x^2 + 10x + 25) - 2x - 10$$
$$= x^2 + 8x + 15$$

5. If $f(x) = x - 4$ and $g(x) = x^2 + 3x$, find

a. $(f \circ g)(x)$
b. $(g \circ f)(x)$

Answers

5. a. $x^2 + 3x - 4$ **b.** $x^2 - 5x + 4$

6. If $f(x) = x^2 - 4$ and $g(x) = \frac{2}{x}$, find

   a. $(f \circ g)(x)$

   b. $(g \circ f)(x)$

EXAMPLE 6   If $f(x) = x + 7$ and $g(x) = \frac{1}{x}$, find $(f \circ g)(x)$ and $(g \circ f)(x)$.

Solution   The composition of f with g is

$$(f \circ g)(x) = f(g(x))$$
$$= f\left(\frac{1}{x}\right)$$
$$= \frac{1}{x} + 7$$

The composition of g with f is

$$(g \circ f)(x) = g(f(x))$$
$$= g(x + 7)$$
$$= \frac{1}{x + 7}$$

GETTING READY FOR CLASS

Respond in your own words and in complete sentences.

A. Use function notation to show how profit, revenue, and cost are related.

B. Explain in words the definition for $(f + g)(x) = f(x) + g(x)$.

C. For functions f and g, how do you find the composition of f with g? Of g with f?

D. For functions f and g, how do you determine the domain of f composed with g?

Answers

6. a. $\frac{4}{x^2} - 4$ **b.** $\frac{2}{x^2 - 4}$

Vocabulary Review

Match the functions on the left to their definitions on the right.

1. $(f - g)(x) = f(x) - g(x)$

2. $(f + g)(x) = f(x) + g(x)$

3. $(f \circ g)(x) = f(g(x))$

4. $(f \cdot g)(x) = f(x)g(x)$

5. $\left(\dfrac{f}{g}\right)(x) = \dfrac{f(g)}{g(x)}$

a. The function $(f \cdot g)(x)$ is the product of the functions $f(x)$ and $g(x)$.

b. The function $\left(\dfrac{f}{g}\right)(x)$ is the quotient of the functions $f(x)$ and $g(x)$, where $g(x) \neq 0$.

c. The function $(f - g)(x)$ is the difference of the functions $f(x)$ and $g(x)$.

d. The composition of the functions $f(x)$ with $g(x)$.

e. The function $(f + g)(x)$ is the sum of the functions $f(x)$ and $g(x)$.

Problems

A Let $f(x) = 4x - 3$ and $g(x) = 2x + 5$. Write a formula for each of the following functions.

1. $(f + g)(x)$

2. $(f - g)(x)$

3. $(g - f)(x)$

4. $(g + f)(x)$

5. $(f \cdot g)(x)$

6. $\left(\dfrac{f}{g}\right)(x)$

7. $\left(\dfrac{g}{f}\right)(x)$

8. $(f \cdot f)(x)$

Let $f(x) = 3x - 5$ and $g(x) = \dfrac{5}{x}$. Write a formula for each of the following functions.

9. $(f + g)(x)$

10. $(f - g)(x)$

11. $(g - f)(x)$

12. $(g + f)(x)$

13. $(f \cdot g)(x)$

14. $\left(\dfrac{f}{g}\right)(x)$

15. $\left(\dfrac{g}{f}\right)(x)$

16. $(g \cdot g)(x)$

A If the functions f, g, and h are defined by $f(x) = 3x - 5$, $g(x) = x - 2$ and $h(x) = 3x^2 - 11x + 10$, write a formula for each of the following functions.

17. $(g + f)(x)$

18. $(f + h)(x)$

19. $(g + h)(x)$

20. $(f - g)(x)$

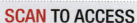

SCAN TO ACCESS

21. $(g - f)(x)$

22. $(h - g)(x)$

23. $(f \cdot g)(x)$

24. $(g \cdot f)(x)$

25. $(f \cdot h)(x)$

26. $(g \cdot h)(x)$

27. $\left(\dfrac{h}{f}\right)(x)$

28. $\left(\dfrac{h}{g}\right)(x)$

29. $\left(\dfrac{f}{h}\right)(x)$

30. $\left(\dfrac{g}{h}\right)(x)$

31. $(f + g + h)(x)$

If the functions f, g, and h are defined by $f(x) = 3x - 5$, $g(x) = x - 2$ and $h(x) = 3x^2 - 11x + 10$, write a formula for each of the following functions. Also, find the domain for each function.

32. $(h - g + f)(x)$

33. $(h + f \cdot g)(x)$

34. $(h - f \cdot g)(x)$

35. $\left(\dfrac{f + g}{h}\right)(x)$

36. $\left(\dfrac{fg}{h}\right)(x)$

37. $(f^2)(x)$

38. $(g^2)(x)$

39. $(f^2 - g^2)(x)$

40. $(f + g)(f - g)(x)$

Let $f(x) = 2x + 1$, $g(x) = 4x + 2$, and $h(x) = 4x^2 + 4x + 1$, and find the following:

41. $(f + g)(2)$

42. $(f - g)(-1)$

43. $(f \cdot g)(3)$

44. $\left(\dfrac{f}{g}\right)(-3)$

45. $\left(\dfrac{h}{g}\right)(1)$

46. $(h \cdot g)(1)$

47. $(f \cdot h)(0)$

48. $(h - g)(-4)$

49. $(f + g + h)(2)$

50. $(h - f + g)(0)$

51. $(h + f \cdot g)(3)$

52. $(h - f \cdot g)(5)$

B

53. Let $f(x) = x^2$ and $g(x) = x + 4$, and find
  a. $(f \circ g)(5)$

  b. $(g \circ f)(5)$

  c. $(f \circ g)(x)$

  d. $(g \circ f)(x)$

54. Let $f(x) = 3 - x$ and $g(x) = x^3 - 1$, and find
  a. $(f \circ g)(0)$

  b. $(g \circ f)(0)$

  c. $(f \circ g)(x)$

  d. $(g \circ f)(x)$

55. Let $f(x) = x^2 + 3x$ and $g(x) = 4x - 1$, and find

    a. $(f \circ g)(0)$

    b. $(g \circ f)(0)$

    c. $(f \circ g)(x)$

    d. $(g \circ f)(x)$

56. Let $f(x) = (x - 2)^2$ and $g(x) = x + 1$, and find

    a. $(f \circ g)(-1)$

    b. $(g \circ f)(-1)$

    c. $(f \circ g)(x)$

    d. $(g \circ f)(x)$

57. Let $f(x) = x^2$ and $g(x) = \frac{3}{x}$, and find

    a. $(f \circ g)(3)$

    b. $(g \circ f)(3)$

    c. $(f \circ g)(x)$

    d. $(g \circ f)(x)$

58. Let $f(x) = x - 10$ and $g(x) = \frac{5}{x}$, and find

    a. $(f \circ g)(5)$

    b. $(g \circ f)(5)$

    c. $(f \circ g)(x)$

    d. $(g \circ f)(x)$

For each of the following pairs of functions f and g, show that $(f \circ g)(x) = (g \circ f)(x) = x$.

59. $f(x) = 5x - 4$ and $g(x) = \dfrac{x + 4}{5}$

60. $f(x) = \dfrac{x}{6} - 2$ and $g(x) = 6x + 12$

61. $f(x) = 3x + 6$ and $g(x) = \dfrac{x}{3} - 2$

62. $f(x) = \dfrac{x + 2}{x - 1}$ and $g(x) = \dfrac{x + 2}{x - 1}$

For the following problems, a fixed point of a function is defined as a value of x for which $f(x) = x$. For each function given, find the fixed point.

63. $f(x) = 3x + 2$

64. $f(x) = 5x - 8$

65. $f(x) = 7x + 8$

66. $f(x) = -2x + 3$

67. $f(x) = -4x + 2$

68. $f(x) = -5x - 7$

69. $f(x) = \frac{1}{2}x - 1$

70. $f(x) = \frac{1}{3}x - 2$

71. $f(x) = -\frac{3}{4}x + \frac{1}{2}$

72. $f(x) = -\frac{1}{4}x + \frac{1}{3}$

Applying the Concepts

73. Profit, Revenue, and Cost A company manufactures and sells DVDs. Here are the equations they use in connection with their business:

Number of DVDs sold each day: $n(x) = x$
Selling price for each DVD: $p(x) = 11.5 - 0.05x$
Daily fixed costs: $f(x) = 200$
Daily variable costs: $v(x) = 2x$
Find the following functions:

a. Revenue $= R(x) =$ the product of the number of DVDs sold each day and the selling price of each DVD.

b. Cost $= C(x) =$ the sum of the fixed costs and the variable costs.

c. Profit $= P(x) =$ the difference between revenue and cost.

d. Average cost $= \overline{C}(x) =$ the quotient of cost and the number of DVDs sold each day.

74. Profit, Revenue, and Cost A company manufactures and sells flash drives for home computers. Here are the equations they use in connection with their business:

Number of flash drives sold each day: $n(x) = x$
Selling price for each flash drive: $p(x) = 3 - \frac{1}{300}x$
Daily fixed costs: $f(x) = 200$
Daily variable costs: $v(x) = 2x$
Find the following functions:

a. Revenue $= R(x) =$ the product of the number of flash drives sold each day and the selling price of each drive.

b. Cost $= C(x) =$ the sum of the fixed costs and the variable costs.

c. Profit $= P(x) =$ the difference between revenue and cost.

d. Average cost $= \overline{C}(x) =$ the quotient of cost and the number of flash drives sold each day.

Recall our discussion of heart rate and composition of functions from the end of the section. Use this information to answer the following two questions.

75. Training Heart Rate Find the training heart rate function, $T(M)$ for a person with a resting heart rate of 62 beats per minute, then find the following:

  a. Find the maximum heart rate function, $M(x)$, for a person x years of age.

  b. What is the maximum heart rate for a 24-year-old person?

  c. What is the training heart rate for a 24-year-old person with a resting heart rate of 62 beats per minute?

  d. What is the training heart rate for a 36-year-old person with a resting heart rate of 62 beats per minute?

  e. What is the training heart rate for a 48-year-old person with a resting heart rate of 62 beats per minute?

76. Training Heart Rate Find the training heart rate function, $T(M)$ for a person with a resting heart rate of 72 beats per minute, then find the following to the nearest whole number.

  a. Find the maximum heart rate function, $M(x)$, for a person x years of age.

  b. What is the maximum heart rate for a 20-year-old person?

  c. What is the training heart rate for a 20-year-old person with a resting heart rate of 72 beats per minute?

  d. What is the training heart rate for a 30-year-old person with a resting heart rate of 72 beats per minute?

  e. What is the training heart rate for a 40-year-old person with a resting heart rate of 72 beats per minute?

77. Cost of Energy Production One of the problems with alternative energy production is the relatively high costs associated with renewable sources. For example, in 2009 the cost per mWh (megawatt hour) for conventional coal was $94.80, while that of wind was $97.00.

  a. If x represents the mWh of production, create functions for the cost of coal and the cost of wind.

  b. Sketch a graph of these two functions on the same axis, using a domain of $0 \le x \le 100$.

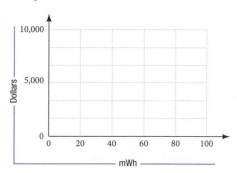

  c. Find a function representing the increased cost of wind as compared with coal.

78. Cost of Energy Production In 2009, the cost per mWh for nuclear power was $113.90, while that of solar photo-voltaic cells was $210.70.

  a. If x represents the mWh of production, create functions for the cost of nuclear and the cost of solar.

  b. Sketch a graph of these two functions on the same axis, using a domain of $0 \le x \le 100$.

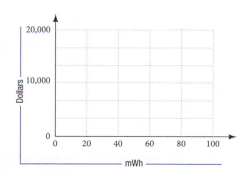

  c. Find a function that represents the increased cost of solar as compared with nuclear.

79. Cost of Energy Production Newer coal-fired generation plants utilize carbon control and sequestration (css) which reduces the amount of carbon dioxide emissions. The projected costs per mWh of such coal plants is $136.20. Repeat problem 67 using this new function to compare the costs of coal and wind energy production

80. Cost of Energy Production The cost per mWh for geothermal energy production is $101.70. Repeat problem 68 comparing the cost of nuclear and geothermal energy production.

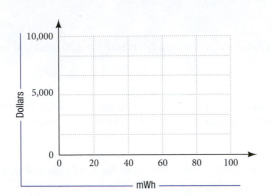

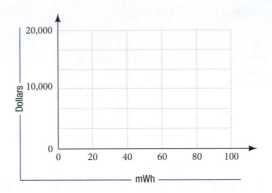

Maintaining Your Skills

The problems that follow review some of the more important skills you have learned in previous sections and chapters.

Solve the following equations.

81. $x - 5 = 7$

82. $3y = -4$

83. $5 - \dfrac{4}{7}a = -11$

84. $\dfrac{1}{5}x - \dfrac{1}{2} - \dfrac{1}{10}x + \dfrac{2}{5} = \dfrac{3}{10}x + \dfrac{1}{2}$

85. $5(x - 1) - 2(2x + 3) = 5x - 4$

86. $0.07 - 0.02(3x + 1) = -0.04x + 0.01$

Solve for the indicated variable.

87. $P = 2l + 2w$ for w

88. $A = \dfrac{1}{2}h(b + B)$ for B

Solve the following inequalities. Write the solution set using interval notation, then graph the solution set.

89. $-5t \leq 30$

90. $5 - \dfrac{3}{2}x > -1$

91. $1.6x - 2 < 0.8x + 2.8$

92. $3(2y + 4) \geq 5(y - 8)$

Solve the following equations.

93. $\left| \dfrac{1}{4}x - 1 \right| = \dfrac{1}{2}$ **94.** $\left| \dfrac{2}{3}a + 4 \right| = 6$ **95.** $|3 - 2x| + 5 = 2$ **96.** $5 = |3y + 6| - 4$

Find the Mistake

Each sentence below contains a mistake. Circle the mistake and write the correct word or phrase on the line provided.

1. The function $(f + g)(x)$ is the product of the functions $f(x)$ and $g(x)$. _____

2. The function $\left(\dfrac{f}{g} \right)(x) = f(x) - g(x)$. _____

3. If $f(x) = 3x + 4$ and $g(x) = x - 3$, the function $(f + g)(x) = (3x + 4)(x - 3)$. _____

4. For $(f \circ g)(x)$ the numbers in the domain of the composite function cannot be numbers in the domain of f. _____

Landmark Review: Checking Your Progress

Graph each of the following.

1. $y = 3x$

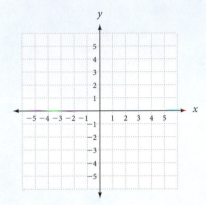

2. $y = -3$

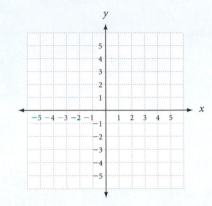

3. $y = 2x + 1$

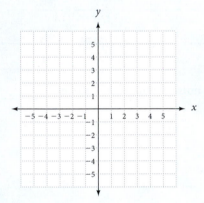

4. $y = \dfrac{1}{4}x^2$

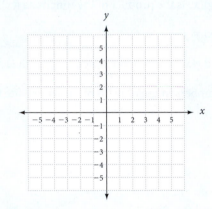

State the domain and range for each of the following relations and indicate which are also functions.

5. $(4,3), (3,2), (1,3), (7,9)$

6. $(0,3), (1,4), (3,2), (1,3)$

7. $(a,0), (4,3), (b,3), (5,1)$

8. $(1,3), (1,5), (1,2), (1,0)$

Let $f(x) = 3x - 4$ and $g(x) = x^2 + 3x + 2$. Evaluate the following.

9. $f(3)$ **10.** $g(2)$ **11.** $f(a + 1)$ **12.** $g(a + 3)$

Let $f(x) = 2x + 1$ and $g(x) = 3x - 5$. Write an expression for each of the following functions.

13. $f + g$ **14.** $g - f$ **15.** fg **16.** $\dfrac{g}{f}$ **17.** $(f \circ g)(1)$ **18.** $(g \circ f)(3)$

2.5 Slope and Average Rate of Change

©iStockphoto.com/bbuong

OBJECTIVES

A Find the slope of a line from its graph.

B Find the slope of a line given two points on the line.

C Interpret the slope as the average rate of change of a function.

KEY WORDS

slope

rise

run

parallel

perpendicular

rate of change

San Francisco, California is known for its steep streets, as shown in the photo above. Let's look at the following table and compare some of San Francisco's streets with others found around the world.

Steep Streets Around the World	
Street	**Grade**
Fargo Street, San Francisco	32%
Eldred Street, Los Angeles	33.33%
Ffordd Penllech, Wales	34%
Baldwin Street, New Zealand	35%
Canton Avenue, Pennsylvania	37%

Unofficially, Canton Avenue in Pittsburgh, Pennsylvania is the steepest public street in the United States. However, the Guinness Book of World Records recognizes Baldwin Street, in southern New Zealand, as the world's steepest street. The road's incline begins at 30 meters above sea level and rises to 100 meters above sea level. Its maximum grade is 35%, which means that every 2.86 meters travelled horizontally is accompanied by a 1-meter drop in elevation.

In mathematics, we would say the slope of Baldwin Street is $-0.35 = -\frac{35}{100} = -\frac{7}{20}$. The slope is the ratio of the vertical change to the accompanying horizontal change.

In defining the slope of a straight line, we want to associate a number with the line. This does two things. First, we want the slope of a line to measure the "steepness" of the line. That is, in comparing two lines, the slope of the steeper line should have the larger numerical value. Second, we want a line that rises going from left to right to have a *positive* slope. We want a line that falls going from left to right to have a *negative* slope. (A line that neither rises nor falls going from left to right must, therefore, have 0 slope.)

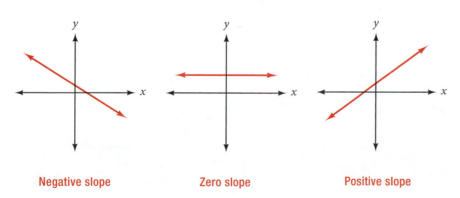

Negative slope   **Zero slope**   **Positive slope**

Geometrically, we can define the *slope* of a line as the ratio of the vertical change to the horizontal change encountered when moving from one point to another on the line. The vertical change is sometimes called the *rise*. The horizontal change is called the *run*.

A   Finding the Slope From Graphs

EXAMPLE 1   Find the slope of the line $y = 2x - 3$.

Solution   To use our geometric definition, we first graph $y = 2x - 3$ (Figure 1). We then pick any two convenient points and find the ratio of rise to run. By convenient points we mean points with integer coordinates. If we let $x = 2$ in the equation, then $y = 1$. Likewise, if we let $x = 4$, then y is 5.

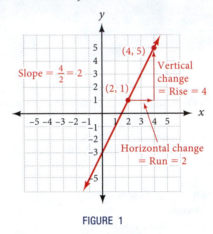

FIGURE 1

The ratio of vertical change to horizontal change is 4 to 2, giving us a slope of $\frac{4}{2} = 2$. Our line has a slope of 2.

Notice that we can measure the signed vertical change (rise) by subtracting the y-coordinates of the two points shown in Figure 1: $5 - 1 = 4$. The signed horizontal change (run) is the difference of the x-coordinates: $4 - 2 = 2$. This gives us a second way of defining the slope of a line.

B   Finding Slopes Using Two Points

> **DEFINITION**   slope
>
> The *slope* of the line between two points (x_1, y_1) and (x_2, y_2) is given by
>
> $$\text{Slope} = m = \frac{\text{Rise}}{\text{Run}} = \frac{y_2 - y_1}{x_2 - x_1}$$
>
> Geometric Form   Algebraic Form
>
>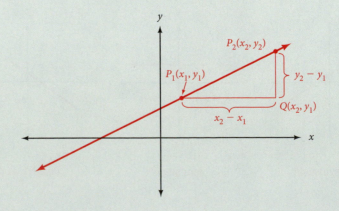

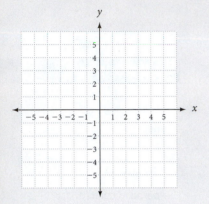

EXAMPLE 2 Find the slope of the line through $(-2, -3)$ and $(-5, 1)$.

Solution

$$m = \frac{y_2 - y_1}{x_2 - x_1} = \frac{1 - (-3)}{-5 - (-2)} = \frac{4}{-3} = -\frac{4}{3}$$

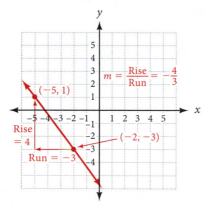

FIGURE 2

Looking at the graph of the line between the two points (Figure 2), we can see our geometric approach does not conflict with our algebraic approach.

We should note here that it does not matter which ordered pair we call (x_1, y_1) and which we call (x_2, y_2). If we were to reverse the order of subtraction of both the x- and y-coordinates in the preceding example, we would have

$$m = \frac{-3 - 1}{-2 - (-5)} = \frac{-4}{3} = -\frac{4}{3}$$

which is the same as our previous result.

EXAMPLE 3 Find the slope of the line containing $(3, -1)$ and $(3, 4)$.

Solution Using the definition for slope, we have

$$m = \frac{-1 - 4}{3 - 3} = \frac{-5}{0}$$

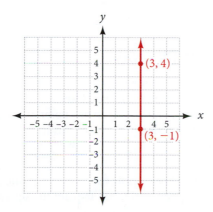

FIGURE 3

The expression $\frac{-5}{0}$ is undefined. That is, there is no real number to associate with it. In this case, we say the line has *an undefined slope*.

The graph of our line is shown in Figure 3. Our line with an undefined slope is a vertical line. All vertical lines have an undefined slope. (All horizontal lines, as we mentioned earlier, have 0 slope.)

2. Find the slope of the line through $(3, 4)$ and $(1, -2)$.

> *Note* The two most common mistakes students make when first working with the formula for the slope of a line are
>
> **1.** Putting the difference of the x-coordinates over the difference of the y-coordinates.
>
> **2.** Subtracting in one order in the numerator and then subtracting in the opposite order in the denominator. You would make this mistake in Example 2 if you wrote $1 - (-3)$ in the numerator and then $-2 - (-5)$ in the denominator.

3. Find the slope of the line through $(2, -3)$ and $(-1, -3)$.

Answers

2. 3

3. 0

Slopes of Parallel and Perpendicular Lines

In geometry, we call lines in the same plane that never intersect *parallel*. For two lines to be nonintersecting, they must rise or fall at the same rate. In other words, two non-vertical lines are parallel if and only if they have the *same slope*. Furthermore, two vertical lines are automatically parallel.

Although it is not as obvious, it is also true that two nonvertical lines are *perpendicular* if and only if the *product of their slopes is* -1. This is the same as saying their slopes are negative reciprocals.

We can state these facts with symbols as follows: If line l_1 has slope m_1 and line l_2 has slope m_2, then

$$l_1 \text{ is parallel to } l_2 \Leftrightarrow m_1 = m_2$$

and

$$l_1 \text{ is perpendicular to } l_2 \Leftrightarrow m_1 \cdot m_2 = -1 \text{ or } \left(m_1 = \frac{-1}{m_2} \right)$$

For example, if a non-vertical line has a slope of $\frac{2}{3}$, then any non-vertical line parallel to it has a slope of $\frac{2}{3}$. Any line perpendicular to it has a slope of $-\frac{3}{2}$ (the negative reciprocal of $\frac{2}{3}$).

Although we cannot give a formal proof of the relationship between the slopes of perpendicular lines at this level of mathematics, we can offer some justification for the relationship. Figure 4 shows the graphs of two lines. One of the lines has a slope of $\frac{2}{3}$; the other has a slope of $-\frac{3}{2}$. As you can see, the lines are perpendicular.

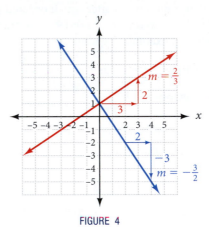

FIGURE 4

Slope and Rate of Change

So far, the slopes we have worked with represent the ratio of the change in y to the corresponding change in x, or, on the graph of the line, the slope is the ratio of vertical change to horizontal change in moving from one point on the line to another. However, when our variables represent quantities from the world around us, slope can have additional interpretations.

EXAMPLE 4   On the chart below, find the slope of the line connecting the first point (1981, 10.75) with the last point (2011, 80). Explain the significance of the result.

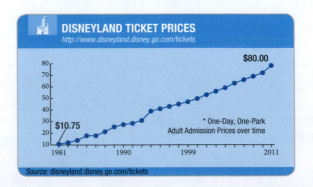

4. Using the graph from Example 4, find the slope of the line connecting ticket prices in 1985, (1985, 18) to 2005, (2005, 62).

Answer

4. $2.20/year

Solution The slope of the line connecting the first point (1981, 10.75) with the last point (2011, 80), is

$$m = \frac{80 - 10.75}{2011 - 1981} = \frac{69.25}{30} = 2.3083$$

The units are dollars/year so we have

$$m = \$2.31/\text{year}$$

which is the average change in price of a ticket to Disneyland over a 30-year period of time.

Likewise, if we connect the points (1999, 47) and (2011, 80), the line that results has a slope of

$$m = \frac{80 - 47}{2011 - 1999} = \frac{25}{12} = \$2.08/\text{year}$$

which is the average change in the price of admission over a 12-year period. If you were to summarize this information for an article in the newspaper, we could say, "The price of admission to Disneyland has increased $2.31 per year over the last 30 years and $2.08 per year over the last 12 years."

C Slope and Average Rate of Change

Previously we introduced the rate equation $d = rt$. Suppose that a boat is traveling at a constant speed of 15 miles per hour in still water. The following table shows the distance the boat will have traveled in the specified number of hours. The graph of this data shown in Figure 5. Notice that the points all lie along a line.

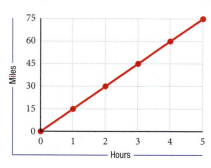

t (Hours)	d (Miles)
0	0
1	15
2	30
3	45
4	60
5	75

FIGURE 5

We can calculate the slope of this line using any two points from the table. Notice we have graphed the data with t on the horizontal axis and d on the vertical axis. Using the points (2, 30) and (3, 45), the slope will be

$$m = \frac{\text{Rise}}{\text{Run}} = \frac{45 - 30}{3 - 2} = \frac{15}{1} = 15$$

The units of the rise are miles and the units of the run are hours, so the slope will be in units of miles per hour. We see that the slope is simply the change in distance divided by the change in time, which is how we compute the average speed. Since the speed is constant, the slope of the line represents the speed of 15 miles per hour.

5. What would be the speed of the car if it traveled 330 miles in 6 hours?

EXAMPLE 5 A car is traveling at a constant speed. A graph (Figure 6) of the distance the car has traveled over time is shown below. Use the graph to find the speed of the car.

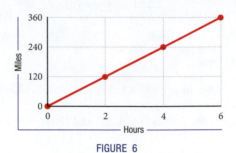

FIGURE 6

Solution Using the second and third points, we see the rise is $240 - 120 = 120$ miles, and the run is $4 - 2 = 2$ hours. The speed is given by the slope, which is

$$m = \frac{\text{rise}}{\text{run}}$$

$$= \frac{120 \text{ miles}}{2 \text{ hours}}$$

$$= 60 \text{ miles per hour}$$

In general, the slope between any two points of a function represents the *average rate of change* of the function.

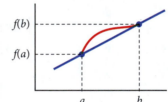

$$m = \frac{f(b) - f(a)}{b - a}$$

Specifically, the average rate of change of $f(x)$ on the interval $[a, b]$ is given by the ratio $\frac{f(b) - f(a)}{b - a}$.

6. Find the average rate of change of $f(x) = x^2 + 2x$ on the interval $[2, 4]$.

EXAMPLE 6 Let $f(x) = x^2 - 3x$. Find the average rate of change of $f(x)$ on the interval $[1, 3]$.

Solution The average rate of change is given by the relationship

$$\frac{f(b) - f(a)}{b - a}$$

where $a = 1$ and $b = 3$. Therefore

$$m = \frac{f(3) - f(1)}{3 - 1} \qquad f(3) = 3^2 - 3(3) = 9 - 9 = 0$$
$$f(1) = 1^2 - 3(1) = 1 - 3 = -2$$

$$= \frac{0 - (-2)}{3 - 1}$$

$$= \frac{0 + 2}{2}$$

$$= \frac{2}{2}$$

$$= 1$$

Answers

5. 55 miles/hour

6. 8

USING TECHNOLOGY Family of Curves

We can use a graphing calculator to investigate the effects of the numbers a and b on the graph of $y = ax + b$. To see how the number b affects the graph, we can hold a constant and let b vary. Doing so will give us a *family of curves*. Suppose we set $a = 1$ and then let b take on integer values from -3 to 3.

We will give three methods of graphing this set of equations on a graphing calculator.

Method 1: Y-Variables List

To use the Y-variables list, enter each equation at one of the Y variables, set the graph window, then graph. The calculator will graph the equations in order, starting with Y_1 and ending with Y_7. Following is the Y-variables list, an appropriate window, and a sample of the type of graph obtained (Figure 7).

$Y_1 = X - 3$
$Y_2 = X - 2$
$Y_3 = X - 1$
$Y_4 = X$
$Y_5 = X + 1$
$Y_6 = X + 2$
$Y_7 = X + 3$

Window: X from -4 to 4, Y from -4 to 4

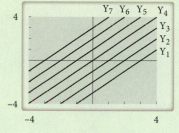

FIGURE 7

Method 2: Programming

The same result can be obtained by programming your calculator to graph $Y = X + B$ for $B = -3, -2, -1, 0, 1, 2,$ and 3. Here is an outline of a program that will do this. Check the manual that came with your calculator to find the commands for your calculator.

Step 1: Clear screen
Step 2: Set window for X from -4 to 4 and Y from -4 to 4
Step 3: $-3 \rightarrow B$
Step 4: Label 1
Step 5: Graph $Y = X + B$
Step 6: $B + 1 \rightarrow B$
Step 7: If $B < 4$, Go to 1
Step 8: End

Method 3: Using Lists

On the TI-83/84 you can set Y_1 as follows

$$Y_1 = X + \{-3, -2, -1, 0, 1, 2, 3\}$$

When you press GRAPH, the calculator will graph each line from $y = x + (-3)$ to $y = x + 3$.

Each of the three methods will produce graphs similar to those in Figure 7.

GETTING READY FOR CLASS

After reading through the preceding section, respond in your own words and in complete sentences.

A. Use symbols to define the slope of a line.

B. Describe the behavior of a line with a negative slope, a line with an undefined slope, and a line with 0 slope.

C. Would you rather climb a hill with a slope of $\frac{1}{2}$ or a slope of 3? Explain why.

D. Describe how to obtain the slope of a line if you know the coordinates of two points on the line.

Problems

A Find the slope of each of the following lines from the given graph.

1.

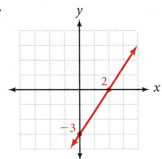

2.

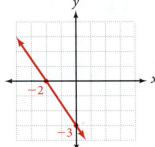

3.

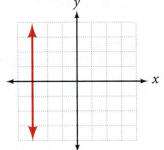

4.

5.

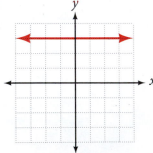

6.

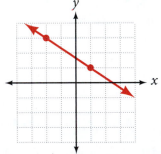

B Find the slope of the line through each of the following pairs of points. Then, plot each pair of points, draw a line through them, and indicate the rise and run in the graph in the manner shown in Example 2.

7. $(2, 1), (4, 4)$

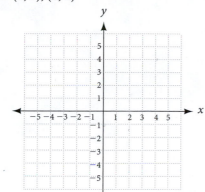

8. $(3, 1), (5, 4)$

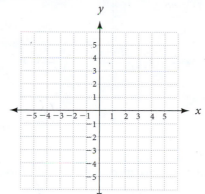

9. $(1, 4), (5, 2)$

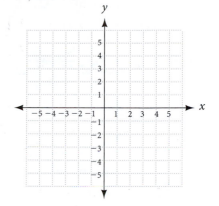

10. $(1, 3), (5, 2)$

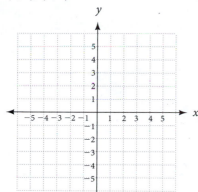

11. $(1, -3), (4, 2)$

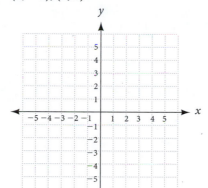

12. $(2, -3), (5, 2)$

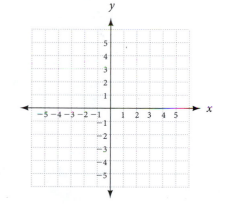

13. $(2, -4), (5, -9)$

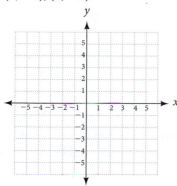

14. $(-3, 2), (-1, 6)$

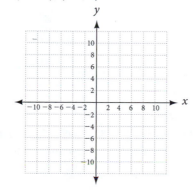

15. $(-3, 5), (1, -1)$

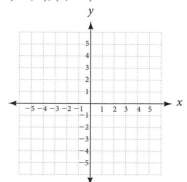

16. $(-2, -1), (3, -5)$

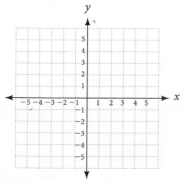

17. $(-4, 6), (2, 6)$

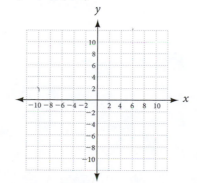

18. $(2, -3), (2, 7)$

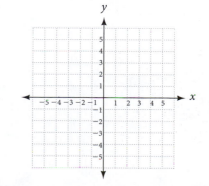

19. $(a, -3), (a, 5)$

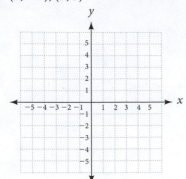

20. $(x, 2y), (4x, 8y)$

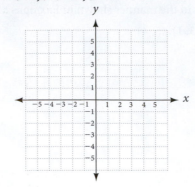

Solve for the indicated variable if the line through the two given points has the given slope.

21. $(a, 3)$ and $(2, 6)$, $m = -1$

22. $(a, -2)$ and $(4, -6)$, $m = -3$

23. $(2, b)$ and $(-1, 4b)$, $m = -2$

24. $(-4, y)$ and $(-1, 6y)$, $m = 2$

25. $(2, 4)$ and (x, x^2), $m = 5$

26. $(3, 9)$ and (x, x^2), $m = -2$

27. $(1, 3)$ and $(x, 2x^2 + 1)$, $m = -6$

28. $(3, 7)$ and $(x, x^2 - 2)$, $m = -4$

For each of the equations in Problems 29–32, complete the table, and then use the results to find the slope of the graph of the equation.

29. $2x + 3y = 6$

x	y
0	
	0

30. $3x - 2y = 6$

x	y
0	
	0

31. $f(x) = \dfrac{2}{3}x - 5$

x	f(x)
0	
3	

32. $f(x) = -\dfrac{3}{4}x + 2$

x	f(x)
0	
4	

33. Finding Slope From Intercepts Graph the line that has an x-intercept of 3 and a y-intercept of -2. What is the slope of this line?

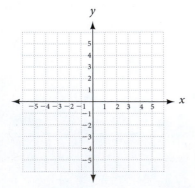

34. Finding Slope From Intercepts Graph the line with x-intercept -4 and y-intercept -2. What is the slope of this line?

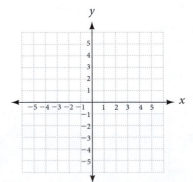

35. Parallel Lines Find the slope of any line parallel to the line through $(2, 3)$ and $(-8, 1)$.

36. Parallel Lines Find the slope of any line parallel to the line through $(2, 5)$ and $(5, -3)$.

37. Perpendicular Lines Line l contains the points $(5, -6)$ and $(5, 2)$. Give the slope of any line perpendicular to l.

38. Perpendicular Lines Line l contains the points $(3, 4)$ and $(-3, 1)$. Give the slope of any line perpendicular to l.

39. Parallel Lines Line l contains the points $(-2, 1)$ and $(4, -5)$. Find the slope of any line parallel to l.

40. Parallel Lines Line l contains the points $(3, -4)$ and $(-2, -6)$. Find the slope of any line parallel to l.

41. Perpendicular Lines Line l contains the points $(-2, -5)$ and $(1, -3)$. Find the slope of any line perpendicular to l.

42. Perpendicular Lines Line l contains the points $(6, -3)$ and $(-2, 7)$. Find the slope of any line perpendicular to l.

43. Determine if each of the following tables could represent ordered pairs from an equation of a line.

a.

x	y
0	5
1	7
2	9
3	11

b.

x	y
-2	-5
0	-2
2	0
4	1

44. The following lines have slope $2, \frac{1}{2}, 0$, and -1. Match each line to its slope value.

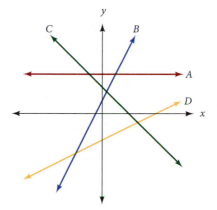

An object is traveling at a constant speed. The distance and time data are shown on the given graph. Use the graph to find the speed of the object.

45.

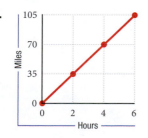

46.

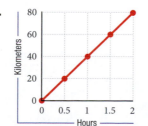

47.

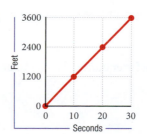

48.

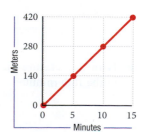

C Find the average rate of change of each function on the given interval.

49. $f(x) = x^2 + 5x, [1, 4]$ **50.** $f(x) = 2x^2 - 4, [-2, 2]$ **51.** $f(x) = 5x + 6, [-3, -1]$ **52.** $f(x) = -4x + 7, [-2, 5]$

53. $f(x) = x^2 - 3x, [-2, 3]$ **54.** $f(x) = x^2 + x, [-4, 0]$ **55.** $f(x) = 2x^2 + 3, [1, 5]$ **56.** $f(x) = 3x^2 + 7, [0, 2]$

Applying the Concepts

57. Heating a Block of Ice A block of ice with an initial temperature of $-20°C$ is heated at a steady rate. The graph shows how the temperature changes as the ice melts to become water and the water boils to become steam and water.

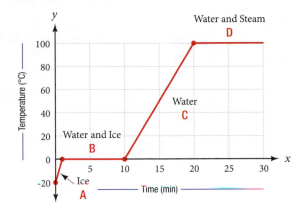

a. How long does it take all the ice to melt?

b. From the time the heat is applied to the block of ice, how long is it before the water boils?

c. Find the slope of the line segment labeled A. What units would you attach to this number?

d. Find the slope of the line segment labeled C. Be sure to attach units to your answer.

e. Is the temperature changing faster during the 1st minute or the 16th minute?

58. Slope of a Highway A sign at the top of the Cuesta Grade, outside of San Luis Obispo, reads "7% downgrade next 3 miles." The following diagram is a model of the Cuesta Grade that illustrates the information on that sign.

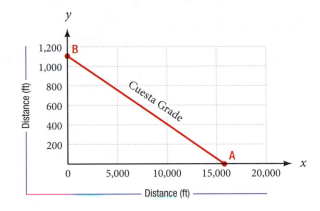

a. At point B, the graph crosses the y-axis at 1,106 feet. How far is it from the origin to point A?

b. What is the slope of the Cuesta Grade?

59. Facebook Users The graph below shows the number of Facebook users over a six year period. Using the graph, find the slope of the line connecting the first (2004, 1) and last (2010, 400) points on the line. Explain in words what the slope represents.

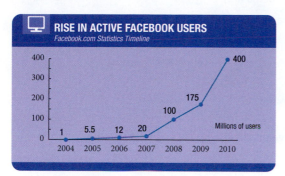

60. Mac Users The graph shows the average increase in the percentage of computer users who use Macs since 2003. Find the slope of the line connecting the first (2003, 2.2) and last (2010, 6.7) points on the line. Explain in words what the slope represents.

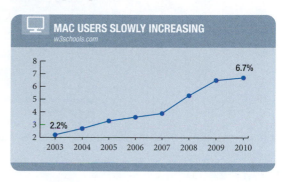

61. Baseball Attendance The following chart shows the seasons' attendance for Major League Baseball from 1999 to 2009.

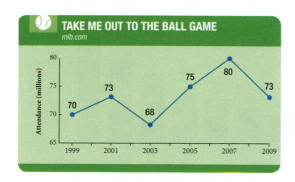

a. Find the slope of the line from 1999 to 2009. Explain in words what the slope represents.

b. Find the slope of the line from 2007 to 2009. Explain in words what the slope represents.

62. Nike Women's Marathon The following chart shows the number of finishers at the Nike Women's Marathon in San Francisco, CA from 2005 to 2010.

a. Find the slope of the line from 2005 to 2010. Explain in words what the slope represents.

b. Find the slope of the line from 2009 to 2010. Explain in words what the slope represents.

63. Recall the following table from a previous section:

Power Output from a Wind Turbine	
Blade diameter (m)	**Power output(kw)**
10	25
17	100
27	225
33	300
40	500

Source: Danish Wind Industry Association, American Wind Energy Association

Use the power output from the wind turbine table to find the average rate of change (slope) for the following intervals:

a. $10 \le d \le 17$

b. $17 \le d \le 27$

c. $27 \le d \le 33$

d. $33 \le d \le 40$

e. Based on your answers, is it advantageous to build longer wind turbines or smaller ones?

64. Recall the following chart from a previous section:

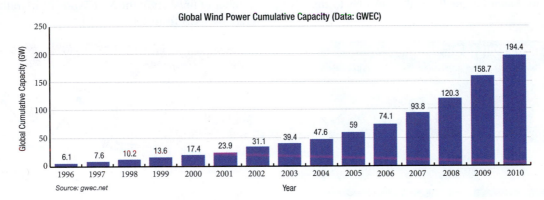

Use the global wind power capacity chart to find the average rate of change (slope) for the following intervals:

a. $1996 \le t \le 2000$

b. $2000 \le t \le 2005$

c. $2005 \le t \le 2010$

d. Based on your answers, would you say that global wind power capacity is remaining constant, increasing at a steady rate, or increasing at faster and faster rates?

65. Recall the following chart from a previous section:
Use the global solar energy demand chart to find the
average rate of change (slope) for the following intervals.

    a. $2009 \le t \le 2010$ (Use 5.35 as the 2009 value)

    b. $2010 \le t \le 2011$

    c. $2011 \le t \le 2012$

    d. Based on your answers, would you say that global
solar energy demand is remaining constant, increasing
at a steady rate, or increasing at faster and faster rates?

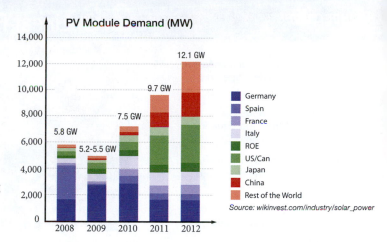

PV Module Demand (MW)

Source: wikinvest.com/industry/solar_power

Getting Ready for the Next Section

Simplify.

66. $2\left(-\dfrac{1}{2}\right)$

67. $\dfrac{3-(-1)}{-3-3}$

68. $-\dfrac{5-(-3)}{2-6}$

69. $3\left(-\dfrac{2}{3}x+1\right)$

Solve for y.

70. $\dfrac{y-b}{x-0}=m$

71. $2x+3y=6$

72. $y-3=-2(x+4)$

73. $y+1=-\dfrac{2}{3}(x-3)$

74. If $y=-\dfrac{4}{3}x+5$, find y when x is 0.

75. If $y=-\dfrac{4}{3}x+5$, find y when x is 3.

Find the Mistake

Each sentence below contains a mistake. Circle the mistake and write the correct word or phrase on the line provided.

1. A line that rises from left to right has a negative slope. _____

2. All horizontal lines have a slope that equals one. _____

3. To find the slope of a line between two points (x_1, y_1) and (x_2, y_2) divide the expression $y_1 - y_2$ by the expression $x_2 - x_1$. _____

4. The graphs of two non-vertical lines are parallel if and only if the product of their slopes is -1. _____

OBJECTIVES

A Find the equation of a linear function given its slope and *y*-intercept.

B Find the slope and *y*-intercept from the equation of a linear function.

C Find the equation of a linear function given the slope and a point on the line.

D Find the equation of a linear function given two points on the line.

KEY WORDS

linear function

slope-intercept form

point-slope form

standard form

Every winter, brave swimmers around the country dive into frigid waters during events called polar bear plunges. Plungefest, sponsored by the Maryland State Police, draws thousands of eager participants each January. For each swimmer to participate, he or she must have raised at least $50 in pledges benefiting the Special Olympics. During the plunge, the swimmers run, skip, leap, and inch their way into the 30°F water of Chesapeake Bay.

In the United States, it is more common to see temperature readings in Fahrenheit like the temperature in the above paragraph, as opposed to Celsius. However, we can determine a Celsius temperature given a Fahrenheit temperature, and vice versa, using a linear function.

The table and illustrations below show some corresponding temperatures on the Fahrenheit and Celsius temperature scales. For example, water freezes at 32°F and 0°C, and boils at 212°F and 100°C.

Degrees Celsius	Degrees Fahrenheit
0	32
25	77
50	122
75	167
100	212

If we plot all the points in the table using the *x*-axis for temperatures on the Celsius scale and the *y*-axis for temperatures on the Fahrenheit scale, we see that they line up in a straight line (Figure 1).

This means that a linear function in two variables will give a perfect description of the relationship between the two scales. That function is

$$F(C) = \frac{9}{5}C + 32$$

The techniques we use to find the equation of a linear function from a set of points is what this section is all about.

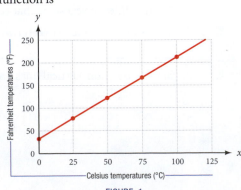

FIGURE 1

A Finding Linear Functions Using the Slope and Intercepts

Suppose non-vertical line l has slope m and y-intercept $(0, b)$. What is the equation of l? Because the y-intercept is b, we know the point $(0, b)$ is on the line. If (x, y) is any other point on l, then using the definition for slope, we have

$$\frac{y - b}{x - 0} = m \qquad \text{Definition of slope}$$

$$y - b = mx \qquad \text{Multiply both sides by } x.$$

$$y = mx + b \qquad \text{Add } b \text{ to both sides.}$$

$$f(x) = mx + b \qquad \text{Using function notation}$$

This last equation is known as the *slope-intercept form* of the equation of a straight line. Written as a function, $f(x) = mx + b$ is known as a *linear function*

> **PROPERTY** Slope-Intercept Form and Linear Functions
>
> The equation of any non-vertical line with slope m and y-intercept b is given by
>
> $$y = mx + b$$
>
> Slope $\qquad$ y-intercept
>
> Using function notation, this can be written as $f(x) = mx + b$.

When the equation is in this form, the *slope* of the line is always the *coefficient* of x and the y-intercept is always the *constant term*.

EXAMPLE 1 Find the equation of the line with slope $-\frac{4}{3}$ and y-intercept 5. Then graph the line.

Solution Substituting $m = -\frac{4}{3}$ and $b = 5$ into the equation $y = mx + b$, we have

$$y = -\frac{4}{3}x + 5, \text{ or } f(x) = -\frac{4}{3}x + 5$$

Finding the equation from the slope and y-intercept is just that easy. If the slope is m and the y-intercept is b, then the equation is always $y = mx + b$. Now, let's graph the line.

Because the y-intercept is 5, the graph goes through the point $(0, 5)$. To find a second point on the graph, we can use the given slope of $-\frac{4}{3}$. We start at $(0, 5)$ and move 4 units down (that's a rise of -4) and 3 units to the right (a run of 3). The point we end up at is $(3, 1)$. Drawing a line that passes through $(0, 5)$ and $(3, 1)$, we have the graph of our equation. (Note that we could also let the rise $= 4$ and the run $= -3$ and obtain the same graph.) The graph is shown in Figure 2.

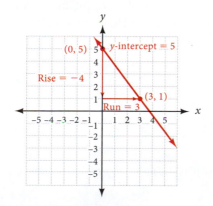

FIGURE 2

Practice Problems

1. Find the equation of the line with slope $\frac{2}{3}$ and y-intercept 1. Write your answer as a linear function. Then graph the line.

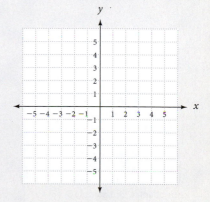

Answer

1. $f(x) = \frac{2}{3}x + 1$

2. Give the slope and *y*-intercept for the line $4x - 5y = 7$.

B Finding Slopes and Intercepts

EXAMPLE 2 Give the slope and *y*-intercept for the line $2x - 3y = 5$.

Solution To use the slope-intercept form, we must solve the equation for *y* in terms of *x*.

$$2x - 3y = 5$$

$$-3y = -2x + 5 \qquad \text{Add } -2x \text{ to both sides.}$$

$$y = \frac{2}{3}x - \frac{5}{3} \qquad \text{Divide by } -3.$$

The last equation has the form $y = mx + b$. The slope must be $m = \frac{2}{3}$ and the *y*-intercept is $b = -\frac{5}{3}$.

3. Graph the linear function $f(x) = \frac{3}{2}x - 3$ using the slope and *y*-intercept.

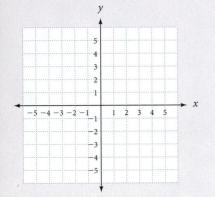

EXAMPLE 3 Graph the linear function $f(x) = -\frac{2}{3}x + 2$ using the slope and *y*-intercept.

Solution The slope is $m = -\frac{2}{3}$ and the *y*-intercept is $b = 2$. Therefore, the point $(0, 2)$ is on the graph, and the ratio of rise to run going from $(0, 2)$ to any other point on the line is $-\frac{2}{3}$. If we start at $(0, 2)$ and move 2 units up (that's a rise of 2) and 3 units to the left (a run of -3), we will be at another point on the graph. (We could also go down 2 units and right 3 units and still be assured of ending up at another point on the line because $\frac{2}{-3}$ is the same as $\frac{-2}{3}$.)

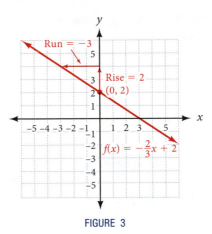

FIGURE 3

C Finding Linear Functions Using Slopes and a Point

A second useful form of the equation of a straight line is the *point-slope form*.

Let line *l* contain the point (x_1, y_1) and have slope *m*. If (x, y) is any other point on *l*, then by the definition of slope we have

$$\frac{y - y_1}{x - x_1} = m$$

Multiplying both sides by $(x - x_1)$ gives us

$$(x - x_1) \cdot \frac{y - y_1}{x - x_1} = m(x - x_1)$$

$$y - y_1 = m(x - x_1)$$

This last equation is known as the *point-slope form* of the equation of a straight line.

Note As we mentioned earlier in this chapter, the rectangular coordinate system is the tool we use to connect algebra and geometry. Example 3 illustrates this connection, as do the many other examples in this chapter. In Example 3, Descartes's rectangular coordinate system allows us to associate the linear function $f(x) = -\frac{2}{3}x + 2$ (an algebraic concept) with the straight line (a geometric concept) shown in Figure 3.

PROPERTY Point-Slope Form of the Equation of a Line

The equation of the non-vertical line through (x_1, y_1) with slope *m* is given by

$$y - y_1 = m(x - x_1)$$

Answers

2. $m = \frac{4}{5}, b = -\frac{7}{5}$

This form of the equation of a straight line is used to find the equation of a line, either given one point on the line and the slope, or given two points on the line.

EXAMPLE 4 Find the equation of the line with slope -2 that contains the point $(-4, 3)$. Write the answer in slope-intercept form using function notation.

Solution

Using $\quad (x_1, y_1) = (-4, 3) \quad$ and $\quad m = -2$

in $\quad y - y_1 = m(x - x_1) \qquad$ Point-slope form

gives us $\quad y - 3 = -2(x + 4) \qquad$ Note: $x - (-4) = x + 4$

$\qquad\qquad y - 3 = -2x - 8 \qquad$ Multiply out right side.

$\qquad\qquad\quad y = -2x - 5 \qquad$ Add 3 to each side.

$\qquad\qquad f(x) = -2x - 5 \qquad$ Using function notation

Figure 4 is the graph of the line that contains $(-4, 3)$ and has a slope of -2. Notice that the y-intercept on the graph matches that of the equation we found.

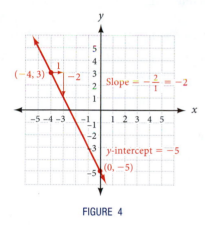

FIGURE 4

D Finding the Linear Function Using Two Points

EXAMPLE 5 Find the equation of the line that passes through the points $(-3, 3)$ and $(3, -1)$. Write your answer in slope-intercept form using function notation.

Solution We begin by finding the slope of the line.

$$m = \frac{3 - (-1)}{-3 - 3} = \frac{4}{-6} = -\frac{2}{3}$$

Using $(x_1, y_1) = (3, -1)$ and $m = -\frac{2}{3}$ in $y - y_1 = m(x - x_1)$ yields

$$y + 1 = -\frac{2}{3}(x - 3)$$

$$y + 1 = -\frac{2}{3}x + 2 \qquad \text{Multiply out right side.}$$

$$y = -\frac{2}{3}x + 1 \qquad \text{Add } -1 \text{ to each side.}$$

$$f(x) = -\frac{2}{3}x + 1 \qquad \text{Using function notation}$$

Figure 5 shows the graph of the line that passes through the points $(-3, 3)$ and $(3, -1)$. As you can see, the slope and y-intercept are $-\frac{2}{3}$ and 1, respectively

4. Find the equation of the linear function with slope 3 that contains the point $(-1, 2)$. Write the answer in slope-intercept form using function notation.

5. Find the equation of the linear function through $(2, 5)$ and $(6, -3)$. Write you answer in slope-intercept form using function notation.

Answer

4. $f(x) = 3x + 5$

5. $f(x) = -2x + 9$

Note We could have used the point $(-3, 3)$ instead of $(3, -1)$ and obtained the same equation. That is, using $(x_1, y_1) = (-3, 3)$ and $m = -\frac{2}{3}$ in

$$y - y_1 = m(x - x_1) \text{ gives us}$$

$$y - 3 = -\frac{2}{3}(x + 3)$$

$$y - 3 = -\frac{2}{3}x - 2$$

$$y = -\frac{2}{3}x + 1$$

$$f(x) = -\frac{2}{3}x + 1$$

which is the same result we obtained using $(3, -1)$.

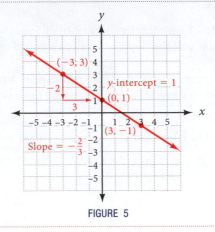

FIGURE 5

The last form of the equation of a line that we will consider in this section is called the *standard form*. It is used mainly to write equations in a form that is free of fractions and is easy to compare with other equations. Note that this form represents a line as an equation, rather than as a function.

> **PROPERTY** Standard Form for the Equation of a Line
>
> If A, B, and C are real numbers and A and B are not both zero, then the equation of a non-vertical line is in standard form when it has the form
>
> $$Ax + By = C$$

If we were to write the equation

$$y = -\frac{2}{3}x + 1$$

in standard form, we would first multiply both sides by 3 to obtain

$$3y = -2x + 3$$

Then we would add $2x$ to each side, yielding

$$2x + 3y = 3$$

which is a linear equation in standard form.

EXAMPLE 6 Give the equation of the line through $(-1, 4)$ whose graph is perpendicular to the graph of $2x - y = -3$. Write the answer in standard form.

Solution   To find the slope of $2x - y = -3$, we solve for y.

$$2x - y = -3$$

$$y = 2x + 3$$

The slope of this line is 2. The line we are interested in is perpendicular to the line with slope 2 and must, therefore, have a slope of $-\frac{1}{2}$.

Using $(x_1, y_1) = (-1, 4)$ and $m = -\frac{1}{2}$, we have

$$y - y_1 = m(x - x_1)$$

$$y - 4 = -\frac{1}{2}(x + 1)$$

Because we want our answer in standard form, we multiply each side by 2.

$$2y - 8 = -1(x + 1)$$

$$2y - 8 = -x - 1$$

$$x + 2y - 8 = -1$$

$$x + 2y = 7$$

The last equation is in standard form.

6. Find the equation of the line through $(3, 2)$ that is perpendicular to the graph of $3x - y = 2$. Write your answer in standard form.

Answer

6. $x + 3y = 9$

As a final note, the following summary reminds us that all horizontal lines have equations of the form $y = b$, and slopes of 0. Since they cross the y-axis at b, the y-intercept is b; there is no x-intercept. Written as $f(x) = b$, they are referred to as constant functions. Vertical lines have no slope, and equations of the form $x = a$. Each will have an x-intercept at a, and no y-intercept. Also note that vertical lines, by definition, fail the vertical line test, so they do not represent a function. Finally, equations of the form $y = mx$ have graphs that pass through the origin. The slope is always m and both the x-intercept and the y-intercept are 0.

FACTS FROM GEOMETRY Special Equations: Their Graphs, Slopes and Intercepts

For the equations below, m, a, and b are real numbers.

Through the Origin	*Vertical Line*	*Horizontal Line*
Equation: $y = mx$	Equation: $x = a$	Equation: $y = b$
Slope $= m$	Undefined slope	Slope $= 0$
x-intercept $= 0$	x-intercept $= a$	No x-intercept
y-intercept $= 0$	No y-intercept	y-intercept $= b$
Function	Not a function	Function

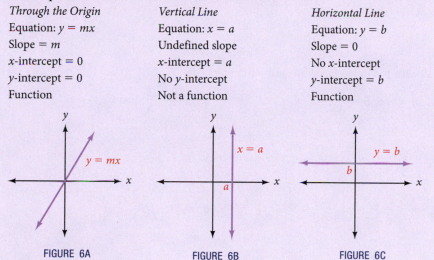

FIGURE 6A   FIGURE 6B   FIGURE 6C

USING TECHNOLOGY Graphing Calculators

One advantage of using a graphing calculator to graph lines is that a calculator does not care whether the equation has been simplified or not. To illustrate, in Example 5 we found that the equation of the line with slope $-\frac{2}{3}$ that passes through the point $(3, -1)$ is

$$y + 1 = -\frac{2}{3}(x - 3)$$

Normally, to graph this equation we would simplify it first. With a graphing calculator, we add -1 to each side and enter the function this way:

$$Y_1 = -(2/3)(X - 3) - 1$$

No simplification is necessary. We can graph the function in this form, and the graph will be the same as the simplified form of the equation, which is $y = -\frac{2}{3}x + 1$. To convince yourself that this is true, graph both the simplified form for the equation and the unsimplified form in the same window. As you will see, the two graphs coincide.

GETTING READY FOR CLASS

After reading through the preceding section, respond in your own words and in complete sentences.

A. How would you graph the linear function $f(x) = \frac{1}{2}x + 3$?

B. What is the slope-intercept form of the equation of a line?

C. Describe how you would find the equation of a line if you knew the slope and a point on the line.

D. What is standard form for the equation of a line?

EXERCISE SET 2.6

Problems

A Give the equation of the linear function $f(x)$ with the following slope and y-intercept.

1. $m = -4, b = -3$

2. $m = -6, b = \dfrac{4}{3}$

3. $m = -\dfrac{2}{3}, b = 0$

4. $m = 0, b = \dfrac{3}{4}$

5. $m = -\dfrac{2}{3}, b = \dfrac{1}{4}$

6. $m = \dfrac{5}{12}, b = -\dfrac{3}{2}$

B Find the slope of a line **a.** parallel and **b.** perpendicular to the given line.

7. $f(x) = 3x - 4$

8. $f(x) = -4x + 1$

9. $3x + y = -2$

10. $2x - y = -4$

11. $2x + 5y = -11$

12. $3x - 5y = -4$

SCAN TO ACCESS

13. $2y = 6$

14. $\dfrac{1}{2}y = 4$

Give the slope and *y*-intercept for each of the following functions or equations. Sketch the graph using the slope and *y*-intercept. Also, give the slope of any line perpendicular to the given line.

15. $f(x) = 3x - 2$

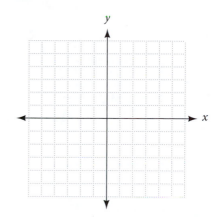

16. $f(x) = 2x + 3$

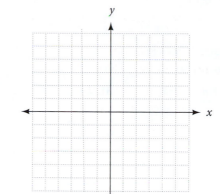

17. $2x - 3y = 12$

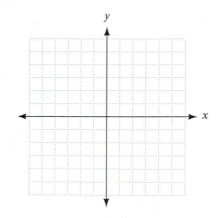

18. $3x - 2y = 12$

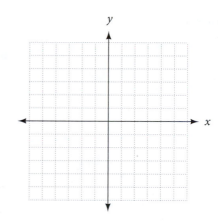

19. $4x + 5y = 20$

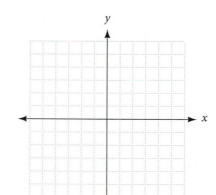

20. $5x - 4y = 20$

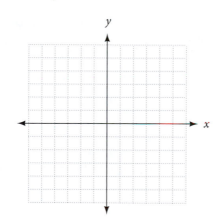

For each of the following lines, name the slope and *y*-intercept. Then write the equation of the linear function.

21.

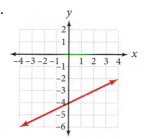

22.

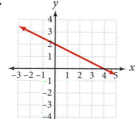

23.

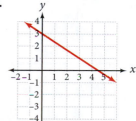

24.

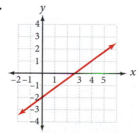

C For each of the following problems, the slope and one point on the line are given. In each case, find the equation of the linear function.

25. $(-2, -5)$; $m = 2$ **26.** $(-1, -5)$; $m = 2$ **27.** $(-4, 1)$; $m = -\dfrac{1}{2}$ **28.** $(-2, 1)$; $m = -\dfrac{1}{2}$

29. $\left(-\dfrac{1}{3}, 2\right)$; $m = -3$ **30.** $\left(-\dfrac{2}{3}, 5\right)$; $m = -3$ **31.** $(-4, 2)$, $m = \dfrac{2}{3}$ **32.** $(3, -4)$, $m = -\dfrac{1}{3}$

33. $(-5, -2)$, $m = -\dfrac{1}{4}$ **34.** $(-4, -3)$, $m = \dfrac{1}{6}$ **35.** $(-4, 0)$, $m = -\dfrac{2}{3}$ **36.** $(6, 2)$, $m = 0$

D Find the equation of the line that passes through each pair of points. Write your answers in standard form.

37. $(3, -2)$, $(-2, 1)$ **38.** $(-4, 1)$, $(-2, -5)$ **39.** $\left(-2, \dfrac{1}{2}\right)$, $\left(-4, \dfrac{1}{3}\right)$ **40.** $(-6, -2)$, $(-3, -6)$

41. $\left(\dfrac{1}{3}, -\dfrac{1}{5}\right)$, $\left(-\dfrac{1}{3}, -1\right)$ **42.** $\left(-\dfrac{1}{2}, -\dfrac{1}{2}\right)$, $\left(\dfrac{1}{2}, \dfrac{1}{10}\right)$ **43.** $\left(-\dfrac{1}{2}, -\dfrac{1}{2}\right)$, $\left(\dfrac{1}{4}, \dfrac{1}{4}\right)$ **44.** $\left(-\dfrac{2}{3}, \dfrac{2}{3}\right)$, $\left(\dfrac{1}{6}, -\dfrac{1}{6}\right)$

For each of the following lines, name the coordinates of any two points on the line. Then use those two points to find the linear function represented by the graph.

45.
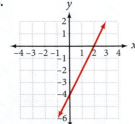

46.
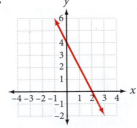

47.
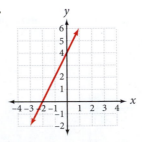

48.
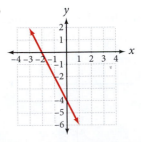

49. The equation $3x - 2y = 10$ is a linear equation in standard form. From this equation, answer the following:

a. Find the x- and y-intercepts.

b. Find a solution to this equation other than the intercepts in part a.

c. Write this equation as a linear function.

d. Is the point $(2, 2)$ a solution to the equation?

50. The equation $4x + 3y = 8$ is a linear equation in standard form. From this equation, answer the following:

a. Find the x- and y-intercepts.

b. Find a solution to this equation other than the intercepts in part a.

c. Write this equation as a linear function.

d. Is the point $(-3, 2)$ a solution to the equation?

The next two problems are intended to give you practice reading, and paying attention to the instructions that accompany the problems you are working. Working these problems is an excellent way to get ready for a test or a quiz.

51. Work each problem according to the instructions given.

a. Solve: $-2x + 1 = -3$.

b. Find x when y is 0: $-2x + y = -3$.

c. Find y when x is 0: $-2x + y = -3$.

d. Graph: $-2x + y = -3$.

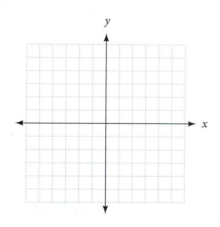

e. Solve for y: $-2x + y = -3$.

52. Work each problem according to the instructions given.

a. Solve: $\dfrac{x}{3} + \dfrac{1}{4} = 1$.

b. Find x when y is 0: $\dfrac{x}{3} + \dfrac{y}{4} = 1$.

c. Find y when x is 0: $\dfrac{x}{3} + \dfrac{y}{4} = 1$.

d. Graph: $\dfrac{x}{3} + \dfrac{y}{4} = 1$.

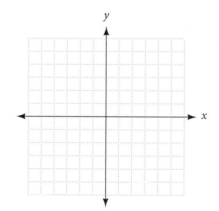

e. Solve for y: $\dfrac{x}{3} + \dfrac{y}{4} = 1$.

Graph each of the following lines. In each case, name the slope, the x-intercept, and the y-intercept.

53. a. $y = \dfrac{1}{2}x$

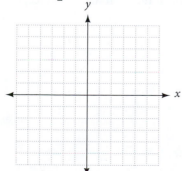

b. $x = 3$

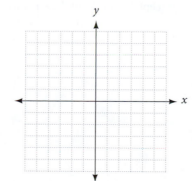

c. $y = -2$

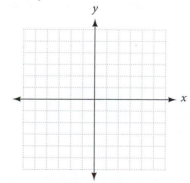

54. a. $y = -2x$

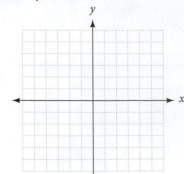

b. $x = 2$

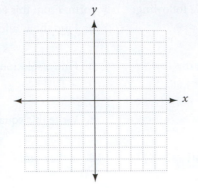

c. $y = -4$

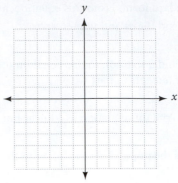

55. Find the equation of the linear function parallel to the graph of $3x - y = 5$ that contains the point $(-1, 4)$.

56. Find the equation of the linear function parallel to the graph of $2x - 4y = 5$ that contains the point $(0, 3)$.

57. Line l is perpendicular to the graph of the equation $2x - 5y = 10$ and contains the point $(-4, -3)$. Find the equation for l.

58. Line l is perpendicular to the graph of the equation $-3x - 5y = 2$ and contains the point $(2, -6)$. Find the equation for l.

59. Give the equation of the linear function perpendicular to the graph of $y = -4x + 2$ that has an x-intercept of -1.

60. Write the equation of the linear function parallel to the graph of $7x - 2y = 14$ that has an x-intercept of 5.

61. Find the equation of the linear function with x-intercept 3 and y-intercept 2.

62. Find the equation of the linear function with x-intercept 2 and y-intercept 3.

Applying the Concepts

63. Deriving the Temperature Equation The table below resembles the table from the introduction to this section. The rows of the table give us ordered pairs (C, F).

Degrees Celsius	Degrees Fahrenheit
°C	°F
0	32
25	77
50	122
75	167
100	212

a. Use any two of the ordered pairs from the table to derive the equation $F(C) = \frac{9}{5}C + 32$.

b. Use the equation from part **a.** to find the Fahrenheit temperature that corresponds to a Celsius temperature of 30°.

64. Maximum Heart Rate The table below gives the maximum heart rate for adults 30, 40, 50, and 60 years old. Each row of the table gives us an ordered pair (A, M).

Age (years)	Maximum Heart Rate (beats per minute)
A	M
30	190
40	180
50	170
60	160

a. Use any two of the ordered pairs from the table to derive the equation $M(A) = 220 - A$, which gives the maximum heart rate M for an adult whose age is A.

b. Use the equation from part **a.** to find the maximum heart rate for a 25-year-old adult.

65. Textbook Cost To produce this textbook, suppose the publisher spent $125,000 for typesetting and $6.50 per book for printing and binding. The total cost to produce and print n books can be written as

$$C(n) = 125{,}000 + 6.5n$$

a. Suppose the number of books printed in the first printing is 10,000. What is the total cost?

b. If the average cost is the total cost divided by the number of books printed, find the average cost per book of producing 10,000 textbooks.

c. Find the cost to produce one more textbook when you have already produced 10,000 textbooks.

66. Exercise Heart Rate In an aerobics class, the instructor indicates that her students' exercise heart rate is 60% of their maximum heart rate, where maximum heart rate is 220 minus their age.

a. Determine the equation that gives exercise heart rate E in terms of age A.

b. Use the equation to find the exercise heart rate of a 22-year-old student.

c. Sketch the graph of the equation for students from 18 to 80 years of age.

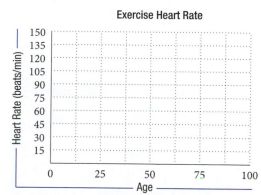

Exercise Heart Rate

67. World Cup The graph shows the total number of goals scored by the United States during several FIFA World Cup soccer tournaments. Find the equation for the linear function between 1934 and 1950. Write the equation in slope-intercept form. Round the slope and the intercept to the nearest tenth.

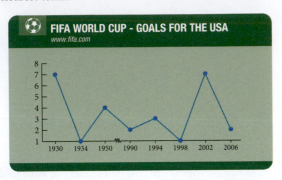

68. Disneyland The graph shows the Disneyland attendance from 1955 to 2010. Find the equation for the linear function that connects the points (2005, 14.5) and (2010, 15.9). Write the equation in slope-intercept form.

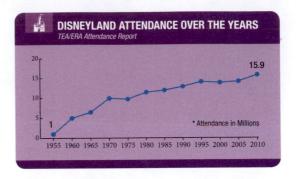

69. Computers A computer is purchased for $1,800, and four years later its value is estimated to be $400. Write a linear function for its value t years after purchase.

70. Cars A car is purchased for $22,500, and eight years later its value is estimated to be $2,500. Write a linear function for its value t years after purchase.

71. Using the data points (10 m, 25 kW) and (40 m, 500 kW) representing the power output from a wind turbine:

   a. Find a linear function relating the blade diameter d with the power output p.

   b. Using your function, predict the power output from a wind turbine with a blade diameter of 50 m.

   c. Using your function, what blade diameter would have a power output of 400 kW?

72. Using the data points (2008, 120.36 W) and (2010, 194.4 GW) representing the global wind power capacity:

   a. Find a linear function relating the year x with the power capacity p.

   b. Using your function, predict the global wind power capacity in the years 2011 and 2012.

   c. Using your function, in what year will the global wind power capacity reach 1,000 GW?

Getting Ready for the Next Section

73. Which of the following are solutions to $x + y \leq 4$?

$(0, 0)$ $(4, 0)$ $(2, 3)$

74. Which of the following are solutions to $y < 2x - 3$?

$(0, 0)$ $(3, -2)$ $(-3, 2)$

75. Which of the following are solutions to $y \leq \frac{1}{2}x$?

$(0, 0)$ $(2, 0)$ $(-2, 0)$

76. Which of the following are solutions to $y > -2x$?

$(0, 0)$ $(2, 0)$ $(-2, 0)$

Find the Mistake

Each sentence below contains a mistake. Circle the mistake and write the correct word or phrase on the line provided.

1. The equation of a line with slope 8 and y-intercept $-\frac{2}{3}$ is given by $y = -\frac{2}{3}x + 8$. _____

2. If you are given a point and the slope of a line, you can find the equation of the line using slope-intercept form. _____

3. The standard form for the equation of a line is $y - y_1 = m(x - x_1)$. _____

4. The graphs of two non-vertical lines are parallel if and only if the product of their slopes is -1. _____

KEY WORDS

linear inequality in two variables

boundary

strict inequalities

2.7 Linear Inequalities

A small movie theater holds 100 people. The owner charges more for adults than for children, so it is important to know the different combinations of adults and children that can be seated at one time. The shaded region in Figure 1 contains all the seating combinations. The line $x + y = 100$ shows the combinations for a full theater. The y-intercept corresponds to a theater full of adults, and the x-intercept corresponds to a theater full of children. In the shaded region below the line $x + y = 100$ are the combinations that occur if the theater is not full.

Shaded regions like the one shown in Figure 1 are produced by linear inequalities in two variables, which is the topic of this section.

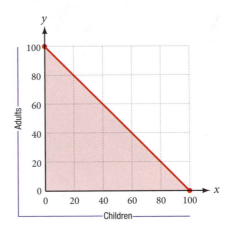

FIGURE 1

A Graphing Inequalities

> **DEFINITION** linear inequality in two variables
>
> A *linear inequality in two variables* is any expression that can be put in the form
>
> $$Ax + By < C$$
>
> where A, B, and C are real numbers (A and B not both 0). The inequality symbol can be any one of the following four: $<, \leq, >, \geq$.

Some examples of linear inequalities are

$$2x + 3y < 6 \qquad y \geq 2x + 1 \qquad x - y \leq 0$$

Although not all of these examples have the form $Ax + By < C$, each one can be put in that form.

The *solution* to a linear inequality is an ordered pair whose coordinates satisfy the inequality. The *solution set* for a linear inequality is the set of all solutions and represented by a *section of the coordinate plane*. The *boundary* for the section is found by replacing the inequality symbol with an equal sign and graphing the resulting equation. The boundary is included in the solution set (and is represented with a *solid line*) if the inequality symbol used originally is $\leq$ or $\geq$. The boundary is not included (and is represented with a *broken line*) if the original symbol is $<$ or $>$. Inequalities that consist of the symbol $<$ or $>$ are also called *strict inequalities*.

EXAMPLE 1   Graph the solution set for $x + y \leq 4$.

Solution   The boundary for the graph is the graph of $x + y = 4$. The boundary is included in the solution set because the inequality symbol is $\leq$. Figure 2 is the graph of the boundary.

Practice Problems

1. Graph the solution set for
   $x - y \geq 3$.

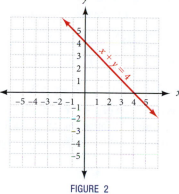

FIGURE 2

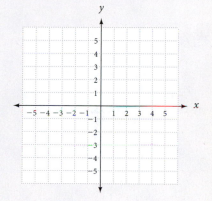

The boundary separates the coordinate plane into two regions: the region above the boundary and the region below it. The solution set for $x + y \leq 4$ is one of these two regions along with the boundary. To find the correct region, we simply choose any convenient point that is *not* on the boundary. We then substitute the coordinates of the point into the original inequality $x + y \leq 4$. If the point we choose satisfies the inequality, then it is a member of the solution set, and we can assume that all points on the same side of the boundary as the chosen point are also in the solution set. If the coordinates of our point do not satisfy the original inequality, then the solution set lies on the other side of the boundary.

In this example, a convenient point that is not on the boundary is the origin.

$$\text{Substituting} \rightarrow \quad (0, 0)$$
$$\text{into} \rightarrow \quad x + y \leq 4$$
$$\text{gives us} \rightarrow \quad 0 + 0 \leq 4$$
$$0 \leq 4 \qquad \text{A true statement}$$

Because the origin is a solution to the inequality $x + y \leq 4$ and the origin is below the boundary, all other points below the boundary are also solutions.

Figure 3 is the graph of $x + y \leq 4$.

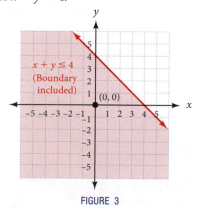

FIGURE 3

The region above the boundary is described by the inequality $x + y > 4$.

Here is a list of steps to follow when graphing the solution sets for linear inequalities in two variables:

> **HOW TO**   Graph a Linear Inequality in Two Variables
>
> **Step 1:** Replace the inequality symbol with an equals sign. The resulting equation represents the boundary for the solution set.
>
> **Step 2:** Graph the boundary found in step 1. Use a *solid line* if the boundary is included in the solution set (i.e., if the original inequality symbol was either $\leq$ or $\geq$). Use a *broken line* to graph the boundary if it is *not* included in the solution set. (It is not included if the original inequality was either $<$ or $>$.)
>
> **Step 3:** Choose any convenient point not on the boundary and substitute the coordinates into the *original* inequality. If the resulting statement is *true,* the solution set lies on the *same* side of the boundary as the chosen point. If the resulting statement is *false,* the solution set lies on the *opposite* side of the boundary.

2. Graph the solution set for

$$y < \frac{1}{2}x + 3.$$

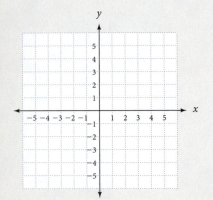

EXAMPLE 2   Graph the solution set for $y < 2x - 3$.

Solution   The boundary is the graph of $y = 2x - 3$, a line with slope 2 and y-intercept -3. The boundary is not included because the original inequality symbol is $<$. We therefore use a broken line to represent the boundary in Figure 4.

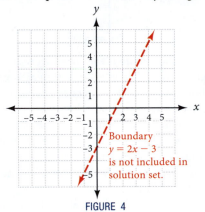

FIGURE 4

A convenient test point is again the origin.

$$\begin{aligned}
\text{Substituting} \rightarrow &\quad (0, 0) \\
\text{into} \rightarrow &\quad y < 2x - 3 \\
\text{gives us} \rightarrow &\quad 0 < 2(0) - 3 \\
&\quad 0 < -3 \qquad \text{A false statement}
\end{aligned}$$

Because our test point gives us a false statement and it lies above the boundary, the solution set must lie on the other side of the boundary (Figure 5).

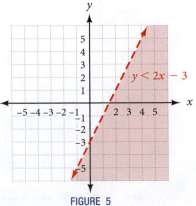

FIGURE 5

USING TECHNOLOGY Graphing Calculators

Most graphing calculators have a Shade command that allows a portion of a graphing screen to be shaded. With this command we can visualize the solution sets to linear inequalities in two variables. Because most graphing calculators cannot draw a dotted line, however, we are not actually "graphing" the solution set, only visualizing it.

Strategy for Visualizing a Linear Inequality in Two Variables on a Graphing Calculator

Step 1: Solve the inequality for y.

Step 2: Replace the inequality symbol with an equal sign. The resulting equation represents the boundary for the solution set.

Step 3: Graph the equation in an appropriate viewing window.

Step 4: Use the Shade command to indicate the solution set:
For inequalities having the $<$ or $\leq$ sign, use Shade (Xmin, Y_1).
For inequalities having the $>$ or $\geq$ sign, use Shade (Y_1, Xmax). Figures 6 and 7 show the graphing calculator screens that help us visualize the solution set to the inequality $y < 2x - 3$ that we graphed in Example 2.

Windows: X from −5 to 5, Y from −5 to 5

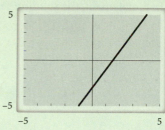 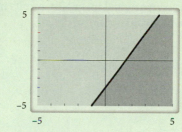

FIGURE 6 $Y1 = 2X − 3$ **FIGURE 7** *Shade (Xmin, Y1)*

EXAMPLE 3 Graph the solution set for $x \leq 5$.

3. Graph $y > -2$.

Solution The boundary is $x = 5$, which is a vertical line. All points in Figure 8 to the left have x-coordinates less than 5 and all points to the right have x-coordinates greater than 5.

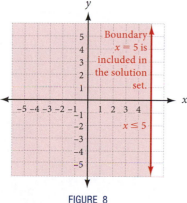

Boundary $x = 5$ is included in the solution set.

$x \leq 5$

FIGURE 8

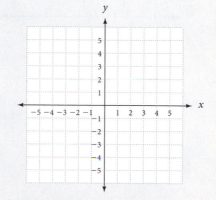

GETTING READY FOR CLASS

After reading through the preceding section, respond in your own words and in complete sentences.

A. When graphing a linear inequality in two variables, how do you find the equation of the boundary line?

B. What is the significance of a broken line in the graph of an inequality?

C. When graphing a linear inequality in two variables, how do you know which side of the boundary line to shade?

D. Describe the set of ordered pairs that are solutions to $x + y < 6$.

EXERCISE SET 2.7

Problems

A Graph the solution set for each of the following.

1. $x + y < 5$

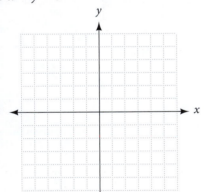

2. $x + y \leq 5$

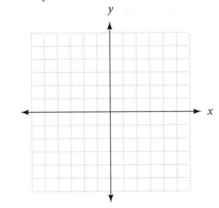

3. $x - y \geq -3$

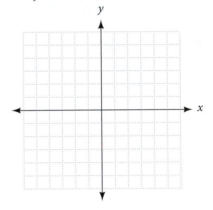

4. $x - y > -3$

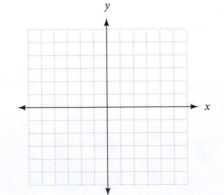

5. $2x + 3y < 6$

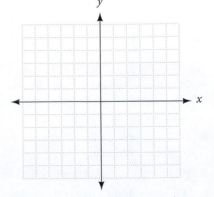

SCAN TO ACCESS

6. $2x - 3y > -6$

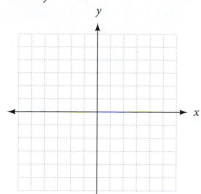

7. $-x + 2y > -4$

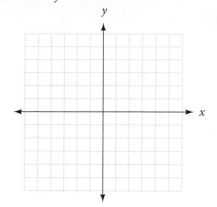

8. $-x - 2y < 4$

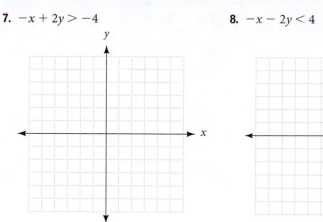

9. $2x + y < 5$

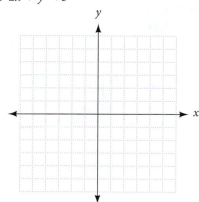

10. $2x + y < -5$

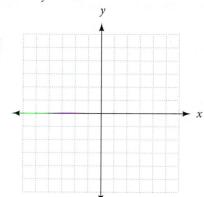

11. $y < 2x - 1$

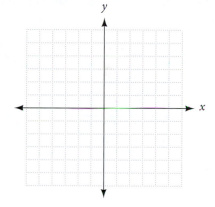

12. $y \leq 2x - 1$

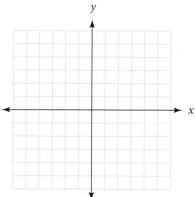

13. $3x - 4y < 12$

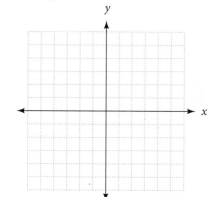

14. $-2x + 3y < 6$

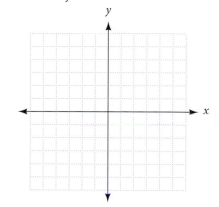

15. $-5x + 2y \leq 10$

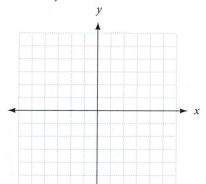

16. $4x - 2y \leq 8$

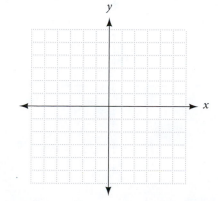

17. $x + y \geq 0$

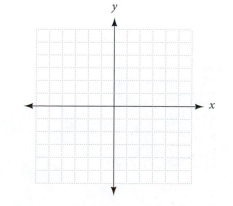

18. $x - y < 0$

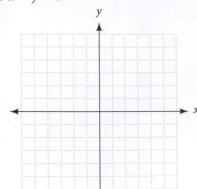

19. $3x + 2y \geq 0$

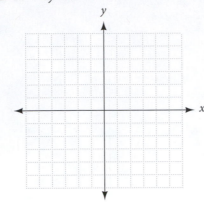

20. $2x - 3y < 0$

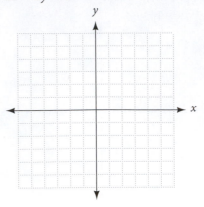

For each graph shown here, name the linear inequality in two variables that is represented by the shaded region.

21.

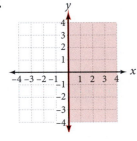

22.

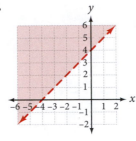

23.

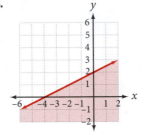

24.

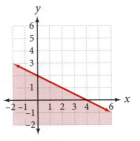

25.

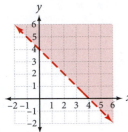

26.

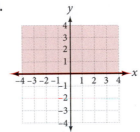

27.

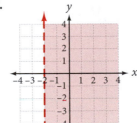

28.

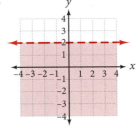

Graph each inequality.

29. $x \geq 3$

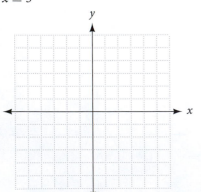

30. $x > -2$

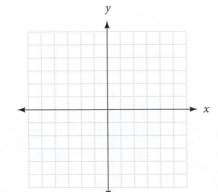

31. $2y \leq 8$

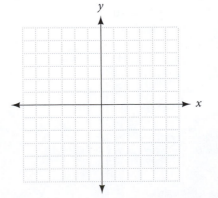

Graph each inequality.

32. $6y > 4y - 10$

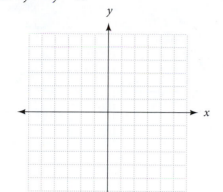

33. $y < 2x$

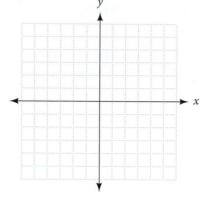

34. $y > -3x$

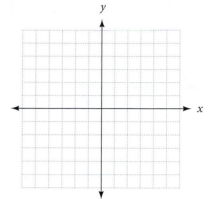

35. $y \geq \dfrac{1}{2}x$

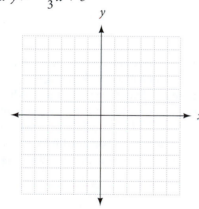

36. $y \leq \dfrac{1}{3}x$

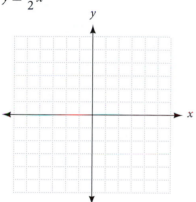

37. $y \geq \dfrac{3}{4}x - 2$

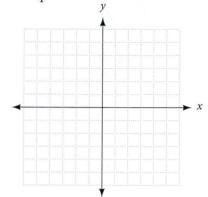

38. $y > -\dfrac{2}{3}x + 3$

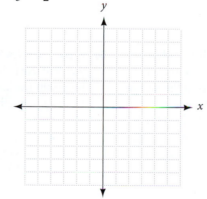

39. $\dfrac{x}{3} + \dfrac{y}{2} > 1$

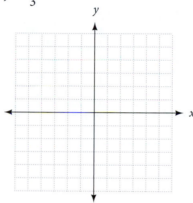

40. $\dfrac{x}{5} + \dfrac{y}{4} < 1$

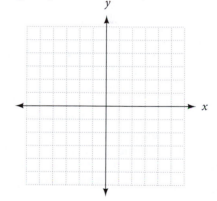

41. $\dfrac{x}{3} - \dfrac{y}{2} > 1$

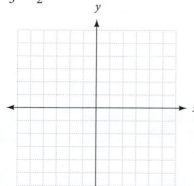

42. $-\dfrac{x}{4} - \dfrac{y}{3} > 1$

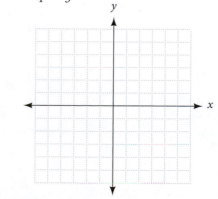

43. $y \leq -\dfrac{2}{3}x$

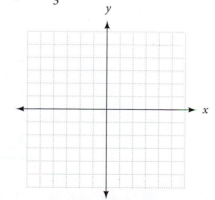

44. $y \geq \frac{1}{4}x - 1$

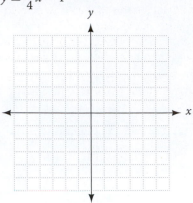

45. $5x - 3y < 0$

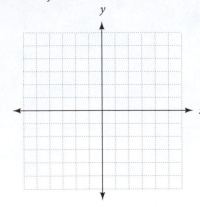

46. $2x + 3y > 0$

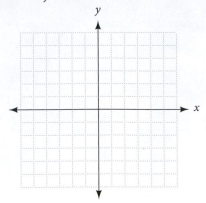

47. $\frac{x}{4} + \frac{y}{5} \leq 1$

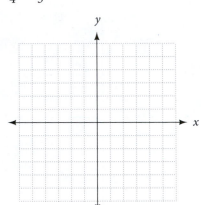

48. $\frac{x}{2} + \frac{y}{3} < 1$

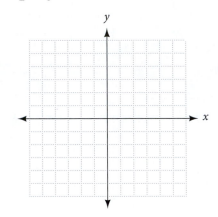

Applying the Concepts

49. Number of People in a Dance Club A dance club holds a maximum of 200 people. The club charges one price for students and a higher price for nonstudents. If the number of students in the club at any time is x and the number of nonstudents is y, sketch the graph and shade the region in the first quadrant that contains all combinations of students and nonstudents that are in the club at any time.

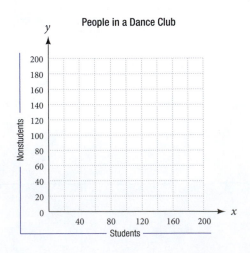

50. Many Perimeters Suppose you have 500 feet of fencing that you will use to build a rectangular livestock pen. Let x represent the length of the pen and y represent the width. Sketch the graph and shade the region in the first quadrant that contains all possible values of x and y that will give you a rectangle from 500 feet of fencing. (You don't have to use all of the fencing, so the perimeter of the pen could be less than 500 feet.)

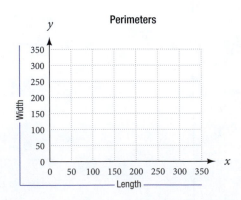

51. Gas Mileage You have two cars. The first car travels an average of 12 miles on a gallon of gasoline, and the second averages 22 miles per gallon. Suppose you can afford to buy up to 30 gallons of gasoline this month. If the first car is driven x miles this month, and the second car is driven y miles this month, sketch the graph and shade the region in the first quadrant that gives all the possible values of x and y that will keep you from buying more than 30 gallons of gasoline this month.

52. Student Loan Payments When considering how much debt to incur in student loans, it is advisable to keep your student loan payment after graduation to 8% or less of your starting monthly income. Let x represent your starting monthly salary and let y represent your monthly student loan payment, and write an inequality that describes this situation. Sketch the graph and shade the region in the first quadrant that is a solution to your inequality.

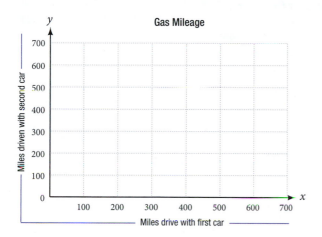

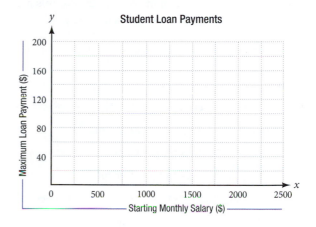

Find the Mistake

Each sentence below contains a mistake. Circle the mistake and write the correct word or phrase on the line provided.

1. The equation found when replacing an inequality symbol with an equals sign represents the solution set for the inequality. _____

2. The boundary for the solution set for the inequality $y > 5x + 3$ is represented by a solid line. _____

3. The solution set for $y \leq 8$ includes all points above the boundary $y = 8$. _____

4. Use $(0, 1)$ as a convenient test point to find the solution set of an inequality that does not pass through the origin.

_____

Supplies Needed

A Piece of graph paper

B Access to internet for research

Product Sales

One of the simplest ways for a company to judge whether a new product is popular is to track its sales. Choose a product that has been in production for at least one year. Go online and search the sales history of your chosen product. Search using the product's name in addition to phrases such as "sales figures" or "sales history." Then create a graph of the sales data in the form of a scatter plot. You may find some sales tracked by year or by quarter within a single year. You may also encounter enough data from one source, or you may need to reference multiple sources. For this project, you'll need a graph that includes four points (e.g., quarterly sales for one year, or annual sales for four years).

Based on your graph, what is the domain and range of the data? Answer the questions below for the first and second points, the second and third points, and the third and fourth points.

1. What are the ordered pairs of the two points?

2. Graph the line between the two points.

3. Find the slope between the two points.

4. Does the line have *x*- and *y*-intercepts? If so, what are they?

5. Find the equation that represents the line between the two points.

6. Create two inequalities from your equation, one using < and a second using >. What does the solution set for the inequality that uses < represent? What does the solution set for the other inequality represent?

Chapter 2 Summary

The numbers in brackets refer to the section(s) in which the topic can be found.

Linear Equations in Two Variables [2.1, 2.6]

A *linear equation in two variables* is any equation that can be put in *standard form* $ax + by = c$. The graph of every linear equation is a straight line.

EXAMPLES

1. The equation $3x + 2y = 6$ is an example of a linear equation in two variables.

Intercepts [2.1]

The *x-intercept* of an equation is the *x-coordinate* of the point where the graph crosses the *x-axis*. The *y-intercept* is the *y-coordinate* of the point where the graph crosses the *y-axis*. We find the *y*-intercept by substituting $x = 0$ into the equation and solving for *y*. The *x*-intercept is found by letting $y = 0$ and solving for *x*.

2. To find the *x*-intercept for $3x + 2y = 6$, we let $y = 0$ and get

$$3x = 6$$
$$x = 2$$

In this case the *x*-intercept is 2, and the graph crosses the *x*-axis at $(2, 0)$.

Relations and Functions [2.2]

A *relation* is any set of ordered pairs. The set of all first coordinates is called the *domain* of the relation, and the set of all second coordinates is the *range* of the relation. A *function* is a rule that pairs each element in one set, called the *domain,* with exactly one element from a second set, called the *range.* A function is a relation in which no two different ordered pairs have the same first coordinates.

3. The relation $\{(8, 1), (6, 1), (-3, 0)\}$ is also a function because no ordered pairs have the same first coordinates. The domain is $\{8, 6, -3\}$ and the range is $\{1, 0\}$.

Vertical Line Test [2.2]

If a vertical line crosses the graph of a relation in more than one place, the relation cannot be a function. If no vertical line can be found that crosses the graph in more than one place, the relation must be a function.

4. The graph of $x = y^2$ shown in Figure 5 in Section 2.2 fails the vertical line test. It is not the graph of a function.

Function Notation [2.3]

The alternative notation for *y* is $f(x)$. It is read "*f* of *x*" and can be used instead of the variable *y* when working with functions. The notation *y* and the notation $f(x)$ are equivalent; that is, $y = f(x)$.

5. If $f(x) = 5x - 3$, then
$$f(0) = 5(0) - 3 = -3$$
$$f(1) = 5(1) - 3 = 2$$
$$f(-2) = 5(-2) - 3 = -13$$
$$f(a) = 5a - 3$$

Algebra with Functions [2.4]

If *f* and *g* are any two functions with a common domain, then

$(f + g)(x) = f(x) + g(x)$ The function $(f + g)(x)$ is the sum of the functions $f(x)$ and $g(x)$.

$(f - g)(x) = f(x) - g(x)$ The function $(f - g)(x)$ is the difference of the functions $f(x)$ and $g(x)$.

$(f \cdot g)(x) = f(x) \cdot g(x)$ The function $(f \cdot g)(x)$ is the product of the functions $f(x)$ and $g(x)$.

$\left(\dfrac{f}{g}\right)(x) = \dfrac{f(x)}{g(x)}$ The function $\left(\dfrac{f}{g}\right)(x)$ is the quotient of the functions $f(x)$ and $g(x)$, where $g(x) \neq 0$.

Composition of Functions [2.4]

If $f(x)$ and $g(x)$ are two functions for which the range of each has numbers in common with the domain of the other, then we have the following definitions:

$$\text{The composition of } f(x) \text{ with } g(x): (f \circ g)(x) = f(g(x))$$
$$\text{The composition of } g(x) \text{ with } f(x): (g \circ f)(x) = g(f(x))$$

The Slope of a Line [2.5]

6. The slope of the line through $(6, 9)$ and $(1, -1)$ is

$$m = \frac{9 - (-1)}{6 - 1} = \frac{10}{5} = 2$$

The *slope* of the line containing points (x_1, y_1) and (x_2, y_2) is given by

$$\text{Slope} = m = \frac{\text{Rise}}{\text{Run}} = \frac{y_2 - y_1}{x_2 - x_1}$$

Horizontal lines have 0 slope, and vertical lines have no slope. Parallel lines have equal slopes, and perpendicular lines have slopes that are negative reciprocals.

Linear Functions [2.6]

7. The linear function with slope 5 and y-intercept 3 is

$$f(x) = 5x + 3$$

A *linear function* with slope m and y-intercept b is given by

$$f(x) = mx + b$$

The Point-Slope Form of a Line [2.6]

8. The equation of the line through $(3, 2)$ with slope -4 is
$$y - 2 = -4(x - 3)$$

which can be written as the linear function $f(x) = -4x + 14$.

The equation of the line through (x_1, y_1) that has slope m can be written as

$$y - y_1 = m(x - x_1)$$

Linear Inequalities [2.7]

9. The graph of $x - y \leq 3$ is

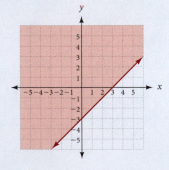

An inequality of the form $ax + by < c$ is a *linear inequality in two variables*. The equation for the boundary of the solution set is given by $ax + by = c$. (This equation is found by simply replacing the inequality symbol with an equal sign.)

To graph a linear inequality, first graph the boundary, using a solid line if the boundary is included in the solution set and a broken line if the boundary is not included in the solution set. Next, choose any point not on the boundary and substitute its coordinates into the original inequality. If the resulting statement is true, the graph lies on the same side of the boundary as the test point. A false statement indicates that the solution set lies on the other side of the boundary.

> ### COMMON MISTAKE
>
> **1.** When graphing ordered pairs, the most common mistake is to associate the first coordinate with the y-axis and the second with the x-axis. Remember, the first coordinate is always associated with the horizontal axis, and the second coordinate is always associated with the vertical axis.
>
> **2.** The two most common mistakes students make when first working with the formula for the slope of a line are the following:
>
> **a.** Putting the difference of the x-coordinates over the difference of the y-coordinates.
>
> **b.** Subtracting in one order in the numerator and then subtracting in the opposite order in the denominator.

Graph each line. [2.1]

1. $4x - 8y = 16$

2. $y = -\dfrac{5}{3}x + 4$

3. $y = 5$

State the domain and range of each relation, and then indicate which relations are also functions. [2.2]

4. $\{(3, 0), (2, -4), (2, 1)\}$

5. $\{(3, 0), (4, 2), (5, -3)\}$

If $f = \{(2, 3), (-4, 4), (5, 0), (-3, -4)\}$ and $g = \{(\pi, 3), (-2, 0), (0, -4)\}$, find the following. [2.3]

6. $f(2)$

7. $f(5) + g(-2)$

Let $f(x) = 3x^2 - 4x + 1$ and $g(x) = 4x - 3$, and evaluate each of the following. [2.4]

8. $f(0)$

9. $g(3)$

10. $f(g(1))$

11. $f(g(0))$

Let $f(x) = 4x - 3$, $g(x) = x^2 + 4$, and $h(x) = x + 5$, and find the following. [2.4]

12. $(g \circ h)(-2)$

13. $(f + g)(-4)$

14. $\left(\dfrac{g}{f}\right)(2)$

15. $(g - fh)(-5)$

Find the slope of the line through the following pairs of points. [2.5]

16. $(6, 3), (-4, 8)$

17. $(-3, 2), (-5, 7)$

Find x if the line through the two given points has the given slope. [2.5]

18. $(5, y), (4, 4); m = -1$

19. $(-3, 7), (x, 9); m = \dfrac{1}{4}$

20. Find the slope of any line parallel to the line through $(5, -3)$ and $(-7, 4)$. [2.5]

21. The line through $(4, 2y)$ and $(6, 3y)$ is parallel to a line with slope $-\frac{1}{2}$. What is the value of y? [2.5]

Give the equation of the line with the following slope and y-intercept. [2.6]

22. $m = -5, b = 4$

23. $m = -7, b = -3$

Give the slope and y-intercept of each equation. [2.6]

24. $4x - y = 3$

25. $4x - 5y = 10$

Find the equation of the line that contains the given point and has the given slope. [2.6]

26. $(3, 5), m = 3$

27. $(-2, 5), m = \dfrac{3}{2}$

Find the equation of the line that contains the given pair of points. [2.6]

28. $(3, 0), (-2, 4)$

29. $(-4, 2), (3, -5)$

30. $(-2, 3), (4, 1)$

31. Find the equation of the line that is parallel to $3x + y = 5$ and contains the point $(0, 4)$. [2.6]

32. Find the equation of the line perpendicular to $y = -\frac{2}{3}x + 2$ that has an x-intercept of -4. [2.6]

Graph each linear inequality. [2.7]

33. $4y \le 2x + 8$

34. $-2x < 6$

Simplify each of the following.

1. -6^2

2. $-|-4|$

3. $3^3 + 4^2$

4. $(4 + 9)^2$

5. $36 \div 12 \cdot 6$

6. $108 \div 6 \cdot 3$

7. $36 - 24 \div 6 + 2$

8. $36 - 24 \div (6 + 2)$

Find the value of each expression when x is 5.

9. $x^2 + 3x - 7$

10. $x^2 - 36$

Reduce the following fractions to lowest terms.

11. $\dfrac{108}{126}$

12. $\dfrac{252}{468}$

Perform the indicated operation.

13. $\dfrac{7}{18} + \dfrac{7}{24}$

14. $\dfrac{7}{32} - \dfrac{5}{24}$

15. Add $-\dfrac{5}{4}$ to the product of 7 and $\dfrac{1}{4}$.

16. Subtract $\dfrac{7}{9}$ from the product of 8 and $\dfrac{5}{6}$.

Simplify each of the following expressions.

17. $12\left(\dfrac{2}{3}x - \dfrac{5}{6}y\right)$

18. $6\left(\dfrac{1}{3}x - \dfrac{3}{2}y\right)$

Solve the following equations.

19. $\dfrac{2}{3}x + 3 = -9$

20. $\dfrac{1}{2}(4x - 7) + \dfrac{1}{6} = -1$

21. $-3 + 4(2x - 5) = 9$

22. $4x - (2x - 7) = -9$

23. $|3x - 5| = 4$

24. $|2x - 5| = 2x + 7$

Solve the following expressions for y and simplify.

25. $y - 3 = -4(x - 2)$

26. $y + 4 = \dfrac{3}{2}(4x - 6)$

Solve each inequality and graph the solution set.

27. $|2x - 5| \le 9$

28. $|3x + 5| - 3 > 7$

Graph on a rectangular coordinate system.

29. $4x - 3y = -12$

30. $y = -3$

31. $y = \dfrac{2}{3}x - 2$

32. $x = 2$

33. Find the slope of the line through $(-6, 4)$ and $(5, -7)$.

34. Find y if the line through $(-6, 2)$ and $(3, y)$ has a slope of $\dfrac{2}{3}$.

35. Find the slope of the line $x = -3$.

36. Give the equation of a line with slope $-\dfrac{3}{5}$ and y-intercept $= 7$.

37. Find the slope and y-intercept of $3x - 6y = 12$.

38. Find the equation of the line with x-intercept -3 and y-intercept 6.

39. Find the equation of the line that is parallel to $3x - 6y = 4$ and contains the point $(6, 7)$.

40. Find the equation of a line with slope $\dfrac{5}{7}$ that contains the point $(14, -3)$. Write the answer in slope-intercept form.

If $f(x) = x^2 + 5x$, $g(x) = 2x + 3$ and $h(x) = 3x - 1$, find the following.

41. $g(0)$

42. $f(-1) + h(2)$

43. $(f \circ h)(3)$

44. $(f - g)(x)$

45. Specify the domain and range for the relation $\{(-2, 3), (-5, 3), (1, 4)\}$. Is the relation also a function?

46. Sarah left a \$2.25 tip for a \$15 dinner. What percent of the cost of dinner was the tip?

47. If the sales tax rate is 8.2% of the purchase price, how much sales tax will Randy pay on a \$200 iPod?

Use the information in the illustration to work Problems 48–49.

CONVENIENCE STORES AROUND THE U.S.
U.S. Census Bureau, 2002

State	Stores
Texas	9,143
California	5,768
Florida	5,676
Ohio	3,537
Illinois	3,280
Michigan	3,402
Pennsylvania	2,996
Georgia	4,050
North Carolina	4,081

*Convenience Store Establishments per State

48. How many more convenience stores are there in Texas as compared to California? Express your answer as a percent rounded to the nearest tenth.

49. How many more convenience stores are there in Michigan as compared to Pennsylvania? Express your answer as a percent rounded to the nearest tenth.

For each of the following straight lines, identify the x-intercept, y-intercept, and slope, and sketch the graph. [2.1, 2.5]

1. $3x - y = 9$

2. $y = -\dfrac{1}{2}x + 4$

3. $y = \dfrac{3}{5}x - 6$

4. $x = -4$

State the domain and range for the following relations, and indicate which relations are also functions. [2.2]

5. $\{(0, 3), (2, 5), (6, -4)\}$

6. $y = 3x^2 + 2$

7. $\{(0, 5), (-6, 5), (0, 7)\}$

8. $y = x - 5$

Determine the domain and range of the following functions. Assume the *entire* function is shown. [2.2]

9.

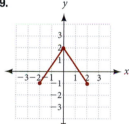

10.

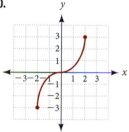

Let $f(x) = x - 5$, $g(x) = 2x - 5$ and $h(x) = x^2 - 5x + 6$, and find the following. [2.3]

11. $f(0) + g(-3)$

12. $h(0) - g(0)$

13. $f(g(-1))$

14. $g(f(3))$

Use the graph to answer Questions 15–18.

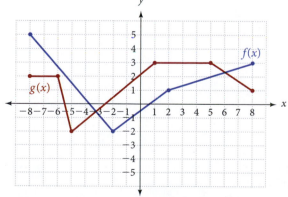

Evaluate. [2.4]

15. $f(-3) + g(4)$

16. $(f - g)(5)$

17. $(g \circ f)(2)$

18. $f(g(-7))$

Find the slope of each graph. [2.5]

19.

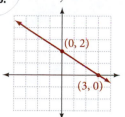

20.

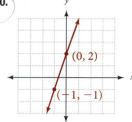

Use slope-intercept form to write the equation of each line below. [2.6]

21.

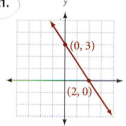

22.

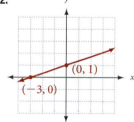

Find the equation for each line. [2.6]

23. The line through $(-6, 1)$ that has slope $m = -3$

24. The line through $(6, 8)$ and $(-3, 2)$

25. The line which contains the point $(-4, 1)$ and is perpendicular to the line $y = -\dfrac{2}{3}x + 4$

26. The line which contains the point $(0, -6)$ and is parallel to the line $4x - 2y = 3$

For each graph below, state the inequality represented by the shaded region. [2.7]

27.

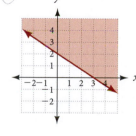

28.

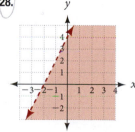

Graph the following linear inequalities. [2.7]

29. $4x - 8y \leq 16$

30. $y > \dfrac{2}{5}x - 4$

Systems of Equations

Denver International Airport in Denver, Colorado is the largest international airport in the United States. The facility can accommodate up to 50 million passengers per year, and the four busiest airports in the country could fit inside its 53 square miles of land. It was built to resemble the nearby Rocky Mountains, and the unique architecture makes it one of the most recognizable airports in the world. The roof of the main terminal building, the Jeppesen Terminal, is made of a Teflon-coated fiberglass that is as thin as a credit card and weighs less than two pounds per square foot. This material is held up by 34 masts and 10 miles of steel cable, which allows the roof to withstand the roughest of weather in its very windy location. It was one of the first airports in the U.S. to incorporate art into its public spaces, permanent displays and temporary exhibitions continue to be a feature of the airport.

Suppose a plane completes a trip from Denver International Airport to an airport 600 miles away flying with the wind in 2 hours. The return trip against the wind takes $2\frac{1}{2}$ hours. What is the speed of the wind and what would the speed of the plane be in still air? To begin solving this problem, review the table below. Later in this chapter, you will see how we complete this table and you will use new knowledge of systems of equations to solve the problem.

	Distance d (miles)	Rate r (mph)	Time t (hours)
Against the wind	600	$x - y$	2.5
With the wind	600	$x + y$	2

OBJECTIVES

A Solve a system of linear equations in two variables by graphing.

B Use the substitution method to solve a system of linear equations in two variables.

C Use the elimination method to solve a system of linear equations in two variables.

KEY WORDS

linear system

solution set

solution

consistent

inconsistent

dependent

independent

point of intersection

substitution method

elimination method

coefficients

3.1 Solving Linear Systems

© iStockphoto/cstewart

Art and physical endurance enthusiasts compete in an annual race called the Kinetic Sculpture Grand Championship. Each team engineers an amphibious work of art that can travel over any terrain. The vehicle must be able to carry its pilots, who power it over sand dunes, across a harbor, up and down steep city streets, and through a swamp. The vehicles are built and decorated as if they were elaborate parade floats, many with animated parts and their pilots dressed in costume.

Let's suppose a kinetic sculpture team races its vehicle on a river. It takes 5 minutes for the vehicle to race down the river going with a current of 3 miles per hour, and 10 minutes to travel the same distance back up the river against the current. After working through this section, you will be able to set up a *linear system of equations* for this problem and use an algebraic method, called the substitution method, to solve the system. But first, let's examine how graphing is a useful method for finding the solution set to a system of equations.

A Solving Systems by Graphing

Previously, we found the graph of an equation of the form $ax + by = c$ to be a straight line. Because the graph is a straight line, the equation is said to be a linear equation. Two linear equations considered together form a *linear system* of equations. For example, the following is a linear system.

$$3x - 2y = 6$$
$$2x + 4y = 20$$

If we graph each equation on the same set of axes, we can see the solution set (see Figure 1).

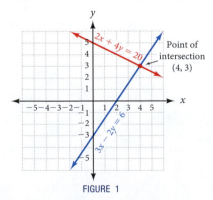

FIGURE 1

The point (4, 3) lies on both lines and therefore must satisfy both equations. It is obvious from the graph that it is the only point that does so. The solution set for the system is {(4, 3)}.

> **DEFINITION** solution set
>
> The *solution set* to the system is the set of all ordered pairs that satisfy both equations.

More generally, if $a_1x + b_1y = c_1$ and $a_2x + b_2y = c_2$ are linear equations, then the solution set for the system

$$a_1x + b_1y = c_1$$
$$a_2x + b_2y = c_2$$

can be illustrated through one of the graphs in Figure 2.

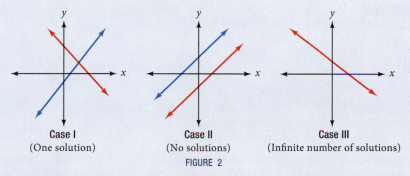

Case I	Case II	Case III
(One solution)	(No solutions)	(Infinite number of solutions)

FIGURE 2

Case I The two lines intersect at one and only one point. The coordinates of the point give the *solution* to the system. We call this a *consistent system* with *independent* equations.

Case II The lines are parallel and therefore have no points in common. The solution set to the system is the empty set, $\varnothing$. In this case, we say the system is *inconsistent* and it has independent equations.

Case III The lines coincide. That is, their graphs represent the same line. The solution set consists of all ordered pairs that satisfy either equation. In this case, the equations are said to be *dependent* and the system is considered consistent.

EXAMPLE 1 Solve the following system by graphing.

$$x + y = 4$$
$$x - y = -2$$

Solution On the same set of coordinate axes we graph each equation separately. Figure 3 shows both graphs, without showing the work necessary to get them. We can see from the graphs that they intersect at the point $(1, 3)$. Therefore, the point $(1, 3)$ must be the solution to our system because it is the only ordered pair whose graph lies on both lines. Its coordinates satisfy both equations.

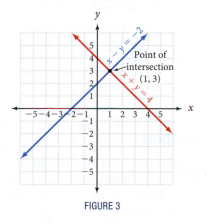

FIGURE 3

Practice Problems

1. Solve this system by graphing.

$$x + y = 3$$
$$x - y = 5$$

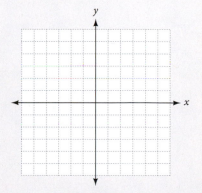

Answer

1. $(4, -1)$

We can check our results by substituting the coordinates $x = 1$, $y = 3$ into both equations to see if they work.

When →	$x = 1$	When →	$x = 1$
and →	$y = 3$	and →	$y = 3$
the equation →	$x + y = 4$	the equation →	$x - y = -2$
becomes →	$1 + 3 \overset{?}{=} 4$	becomes →	$1 - 3 \overset{?}{=} -2$
	$4 = 4$		$-2 = -2$

The point $(1, 3)$ satisfies both equations. This is a Case I system.

Here are some steps to follow in solving linear systems by graphing.

> **HOW TO**   Solving a Linear System by Graphing
>
> **Step 1:** Graph the first equation.
>
> **Step 2:** Graph the second equation on the same set of axes used for the first equation.
>
> **Step 3:** Read the coordinates of the point of intersection of the two graphs.
>
> **Step 4:** Check the solution in both equations.

As previously discussed, we sometimes use special vocabulary to describe systems that are categorized as Case II or Case III systems. When a system of equations has no solution because the lines are parallel, we say the system is *inconsistent*. When the lines coincide, we say the equations are *dependent*. These two special cases do not happen often. Usually, a system has a single ordered pair as a solution (Case I). Solving a system of linear equations by graphing is useful only when the ordered pair in the solution set has integers for coordinates.

Here is a summary of three possible types of solutions to a system of equations in two variables:

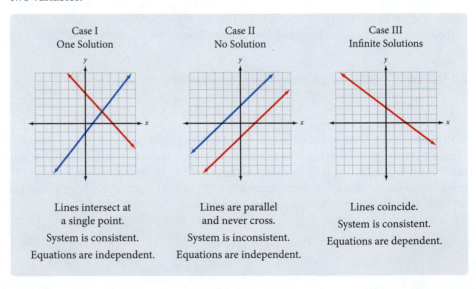

Case I One Solution	Case II No Solution	Case III Infinite Solutions
Lines intersect at a single point.	Lines are parallel and never cross.	Lines coincide.
System is consistent.	System is inconsistent.	System is consistent.
Equations are independent.	Equations are independent.	Equations are dependent.

Two other solution methods work well in all cases. We will develop the other two methods now.

B Substitution Method

As you may have guessed, solving a system of linear equations by graphing is actually the least accurate method. If the coordinates of the point of intersection are not integers, it can be difficult to read the solution set from the graph. There are other methods of solving a linear system that do not depend on the graph. The *substitution method* is one of them.

Note   Make sure to graph the equations as accurately as possible so you may correctly identify the solution set.

EXAMPLE 2 Solve the following system using the substitution method.

$$x + y = 2$$
$$y = 2x - 1$$

2. Solve by substitution.

$$x + y = 3$$
$$y = x + 5$$

Solution The second equation tells us that y is $2x - 1$. We can replace the y variable in the first equation with the expression $2x - 1$ from the second equation; that is, we substitute $2x - 1$ from the second equation for y in the first equation. Here is what it looks like:

$$x + (2x - 1) = 2$$

The equation we end up with contains only the variable x. The y variable has been eliminated by substitution.

Solving the resulting equation, we have

$$x + (2x - 1) = 2$$
$$3x - 1 = 2$$
$$3x = 3$$
$$x = 1$$

This is the x-coordinate of the solution to our system. To find the y-coordinate, we substitute $x = 1$ into the second equation of our system. (We could substitute $x = 1$ into the first equation also and have the same result.)

$$y = 2(1) - 1$$
$$y = 2 - 1$$
$$y = 1$$

The solution to our system is the ordered pair $(1, 1)$. It satisfies both of the original equations. Figure 1 provides visual evidence that the substitution method yields the correct solution.

> *Note* Sometimes this method of solving systems of equations is confusing the first time you see it. If you are confused, you may want to read through this first example more than once and try it on your own before attempting the practice problem.

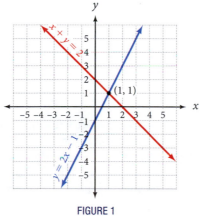

FIGURE 1

EXAMPLE 3 Solve the following system by solving the first equation for x and then using the substitution method.

$$x - 3y = -1$$
$$2x - 3y = 4$$

3. Solve the first equation for x, and then substitute the result into the second equation to solve the system by substitution.

$$x - 4y = -5$$
$$3x - 2y = 5$$

Solution We solve the first equation for x by adding $3y$ to both sides to get

$$x = 3y - 1$$

Using this value of x in the second equation, we have

$$2(3y - 1) - 3y = 4$$
$$6y - 2 - 3y = 4$$
$$3y - 2 = 4$$
$$3y = 6$$
$$y = 2$$

Answers

2. $(-1, 4)$
3. $(3, 2)$

Next, we find x.

$$\text{When} \rightarrow \qquad y = 2$$
$$\text{the equation} \rightarrow x = 3y - 1$$
$$\text{becomes} \rightarrow \qquad x = 3(2) - 1$$
$$x = 6 - 1$$
$$x = 5$$

The solution to our system is $(5, 2)$.

Here are the steps to use in solving a system of equations by the substitution method:

> **HOW TO** **Strategy for Solving a System of Linear Equations by the Substitution Method**
>
> **Step 1:** Solve either one of the equations for x or y. (This step is not necessary if one of the equations is already in the correct form, as in Examples 1 and 2.)
>
> **Step 2:** Substitute the expression for the variable obtained in step 1 into the other equation and solve it.
>
> **Step 3:** Substitute the solution from step 2 into any equation in the system that contains both variables and solve it.
>
> **Step 4:** Check your results, if necessary.

4. Solve by substitution.

$$2x + 6y = 10$$
$$x = -3y + 5$$

EXAMPLE 4   Solve by substitution.

$$4x + 2y = 8$$
$$y = -2x + 4$$

Solution   Substituting the expression $-2x + 4$ from the second equation for y in the first equation, we have

$$4x + 2(-2x + 4) = 8$$
$$4x - 4x + 8 = 8$$
$$8 = 8 \qquad \text{\color{green}A true statement}$$

Both variables have been eliminated, and we are left with a true statement. Recall from the last section that a true statement in this situation tells us the lines coincide; that is, the equations $4x + 2y = 8$ and $y = -2x + 4$ have exactly the same graph. Any point on that graph has coordinates that satisfy both equations and is a solution to the system. Therefore, this system is considered consistent with dependent equations.

5. Solve by substitution.

$$6x + 3y = 1$$
$$y = -2x - 5$$

EXAMPLE 5   Solve by substitution.

$$2x + 6y = 7$$
$$x = -3y + 5$$

Solution   Substituting the expression $-3y + 5$ from the second equation for x in the first equation, we have

$$2(-3y + 6) + 6y = 8$$
$$-6x + 10 + 6y = 8$$
$$10 = 8 \qquad \text{\color{green}A false statement}$$

Both variables have been eliminated, and we are left with a false statement. The lines are parallel. The system is considered inconsistent because there are no solutions.

Answers

4. Dependent (lines coincide)

5. Inconsistent (no solution)

C The Elimination Method

We will now turn our attention to another method of solving systems of linear equations in two variables, called the *elimination method*. This method is also called the addition method and makes use of the addition property of equality. The addition property of equality states that if equal quantities are added to both sides of an equation, the solution set is unchanged. In the past, we have used this property to help solve equations in one variable. We will now use it to solve systems of linear equations. Here is another way to state the addition property of equality.

$$\text{If} \qquad A = B$$
$$\text{and} \qquad C = D$$
$$\text{then} \quad A + C = B + D$$

where A, B, C, and D are algebraic expressions. Because C and D are equal (that is, they represent the same number), what we have done is added the same amount to both sides of the equation $A = B$. Let's see how we can use this form of the addition property of equality to solve a system of linear equations.

EXAMPLE 6   Solve the following system.

$$x + y = 4$$
$$x - y = 2$$

Solution   The system is written in the form of the addition property of equality as written in this section. It looks like this:

$$A = B$$
$$C = D$$

where A is $x + y$, B is 4, C is $x - y$, and D is 2.

We use the addition property of equality to add the left sides together and the right sides together.

$$\begin{array}{r} x + y = 4 \\ \underline{x - y = 2} \\ 2x + 0 = 6 \end{array}$$

We now solve the resulting equation for x.

$$2x + 0 = 6$$
$$2x = 6$$
$$x = 3$$

The value we get for x is the value of the x-coordinate of the point of intersection of the two lines $x + y = 4$ and $x - y = 2$. To find the y-coordinate, we simply substitute $x = 3$ into either of the two original equations. Using the first equation, we get

$$3 + y = 4$$
$$y = 1$$

The solution to our system is the ordered pair $(3, 1)$. It satisfies both equations.

When →	$x = 3$	When →	$x = 3$
and →	$y = 1$	and →	$y = 1$
the equation →	$x + y = 4$	the equation →	$x - y = 2$
becomes →	$3 + 1 \overset{?}{=} 4$	becomes →	$3 - 1 \overset{?}{=} 2$
	$4 = 4$		$2 = 2$

6. Solve the following system.

$$x + y = 3$$
$$x - y = 5$$

Answer

6. $(4, -1)$

The most important part of this method of solving linear systems is eliminating one of the variables when we add the left and right sides together. In our first example, the equations were written so that the y variable was eliminated when we added the left and right sides together. If the equations are not set up this way to begin with, we have to work on one or both of them separately before we can add them together to eliminate one variable.

EXAMPLE 7 Solve the following system.

$$2x - y = 6$$
$$x + 3y = 3$$

Solution Let's eliminate the y variable from the two equations. We can do this by multiplying the first equation by 3 and leaving the second equation unchanged.

$$2x - y = 6 \xrightarrow[]{\text{Multiply by 3.}} 6x - 3y = 18$$
$$x + 3y = 3 \xrightarrow[\text{No change}]{} x + 3y = 3$$

The important thing about our system now is that the coefficients (the numbers in front) of the y variables are opposites. When we add the terms on each side of the equal sign, then the terms in y will add to zero and be eliminated.

$$\begin{array}{r} 6x - 3y = 18 \\ x + 3y = 3 \\ \hline 7x = 21 \end{array} \quad \text{Add corresponding terms.}$$

This gives us $x = 3$. Using this value of x in the second equation of our original system, we have

$$3 + 3y = 3$$
$$3y = 0$$
$$y = 0$$

We could substitute $x = 3$ into any of the equations with both x and y variables and also get $y = 0$. The solution to our system is the ordered pair $(3, 0)$.

EXAMPLE 8 Solve the system.

$$\frac{1}{2}x - \frac{1}{3}y = 2$$
$$\frac{1}{4}x + \frac{2}{3}y = 6$$

Solution Although we could solve this system without clearing the equations of fractions, there is probably less chance for error if we have only integer coefficients to work with. So let's begin by multiplying both sides of the top equation by 6 and both sides of the bottom equation by 12, to clear each equation of fractions.

$$\frac{1}{2}x - \frac{1}{3}y = 2 \xrightarrow[]{\text{Multiply by 6.}} 3x - 2y = 12$$
$$\frac{1}{4}x + \frac{2}{3}y = 6 \xrightarrow[\text{Multiply by 12.}]{} 3x + 8y = 72$$

Now we can eliminate x by multiplying the top equation by -1 and leaving the bottom equation unchanged.

$$3x - 2y = 12 \xrightarrow[]{\text{Multiply by } -1.} -3x + 2y = -12$$
$$3x + 8y = 72 \xrightarrow[\text{No change}]{} \begin{array}{r} 3x + 8y = 72 \\ \hline 10y = 60 \\ y = 6 \end{array}$$

---

7. Solve the following system.

$$3x - y = 7$$
$$x + 2y = 7$$

> **Note** If you are having trouble understanding this method of solution, it is probably because you can't see why we chose to multiply by 3 in the first step of Example 7. Look at the result of doing so: the $3y$ and $-3y$ will add to 0. We chose to multiply by 3 because they produce $3y$ and $-3y$, which will add to 0.

8. Solve the system.

$$\frac{1}{3}x + \frac{1}{2}y = 1$$
$$x + \frac{3}{4}y = 0$$

Answers

7. $(3, 2)$

8. $(-3, 4)$

We can substitute $y = 6$ into any equation that contains both x and y. Let's use $3x - 2y = 12$.

$$3x - 2(6) = 12$$
$$3x - 12 = 12$$
$$3x = 24$$
$$x = 8$$

The solution to the system is $(8, 6)$.

Our next two examples will show what happens when we apply the elimination method to a system of equations consisting of parallel lines and to a system in which the lines coincide.

EXAMPLE 9 Solve the system.

$$2x - y = 2$$
$$4x - 2y = 12$$

Solution Let us choose to eliminate y from the system. We can do this by multiplying the first equation by -2 and leaving the second equation unchanged.

$$2x - y = 2 \xrightarrow{\text{Multiply by } -2.} -4x + 2y = -4$$
$$4x - 2y = 12 \xrightarrow{\text{No change}} 4x - 2y = 12$$

If we add both sides of the resulting system, we have

$$\begin{array}{r} -4x + 2y = -4 \\ \underline{4x - 2y = 12} \\ 0 + 0 = 8 \end{array}$$
$$0 = 8 \qquad \text{A false statement}$$

Both variables have been eliminated and we end up with the false statement $0 = 8$. We have tried to solve a system that consists of two parallel lines. There is no solution, and the system is inconsistent.

EXAMPLE 10 Solve the system.

$$4x - 3y = 2$$
$$8x - 6y = 4$$

Solution Multiplying the top equation by -2 and adding, we can eliminate the variable x.

$$4x - 3y = 2 \xrightarrow{\text{Multiply by } -2.} -8x + 6y = -4$$
$$8x - 6y = 4 \xrightarrow{\text{No change}} \begin{array}{r} 8x - 6y = 4 \\ \hline 0 = 0 \end{array}$$

Both variables have been eliminated, and the resulting statement $0 = 0$ is true. In this case the lines coincide because the equations are equivalent. The solution set consists of all ordered pairs that satisfy either equation, and the system is dependent.

9. Solve the system.
$$x - 3y = 2$$
$$-3x + 9y = 2$$

10. Solve the system.
$$5x - y = 1$$
$$10x - 2y = 2$$

Answers

9. Inconsistent (no solution)
10. Dependent (lines coincide)

The preceding two examples illustrate the two special cases in which the graphs of the equations in the system either coincide or are parallel.

> Here is a summary of our results from these two examples:
>
> Both variables are eliminated and ↔ The lines are parallel and there
> the resulting statement is false.   is no solution to the system.
>
> Both variables are eliminated and ↔ The lines coincide and there
> the resulting statement is true.   is an infinite number of
>   solutions to the system.

The main idea in solving a system of linear equations by the elimination method is to use the multiplication property of equality on one or both of the original equations, if necessary, to make the coefficients of either variable opposites. The following box shows some steps to follow when solving a system of linear equations by the elimination method.

HOW TO   **Solving a System of Linear Equations by the Elimination Method**

Step 1: Decide which variable to eliminate. (In some cases one variable will be easier to eliminate than the other. With some practice you will notice which one it is.)

Step 2: Use the multiplication property of equality on each equation separately to make the coefficients of the variable that is to be eliminated opposites.

Step 3: Add the respective left and right sides of the system together.

Step 4: Solve for the variable remaining.

Step 5: Substitute the value of the variable from step 4 into an equation containing both variables and solve for the other variable.

Step 6: Check your solution in both equations, if necessary.

GETTING READY FOR CLASS

After reading through the preceding section, respond in your own words and in complete sentences.

A. What is a system of two linear equations in two variables?

B. What is a solution to a system of linear equations?

C. What does it mean when we solve a system of linear equations by the substitution method and we end up with the statement $8 = 8$?

D. How do you use the addition property of equality in the elimination method of solving a system of linear equations?

Vocabulary Review

Choose the correct words to fill in the blanks below.

intersect   dependent   system   inconsistent   solution set

1. Two linear equations considered together form a linear _____ of equations.

2. For a linear system, Case I occurs when two lines _____ at one and only one point.

3. If there are no solutions for a linear system, the system is said to be _____.

4. If the lines of a linear system coincide, the equations of those lines are said to be _____.

5. The _____ for a system of linear equations must satisfy all equations in the system.

Use the graphs to fill in the blanks below.

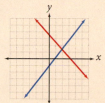 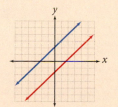 

6. Give the number of solutions for each system.

_____   _____   _____

7. Label each system as consistent or inconsistent.

_____   _____   _____

8. Label each system as dependent or independent.

_____   _____   _____

Problems

A Solve each system by graphing both equations on the same set of axes and then reading the solution from the graph.

1. $3x - 2y = 6$
$x - y = 1$

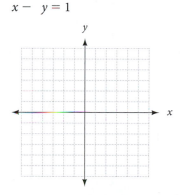

2. $5x - 2y = 10$
$x - y = -1$

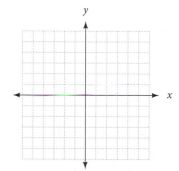

3. $y = \dfrac{3}{5}x - 3$
$2x - y = -4$

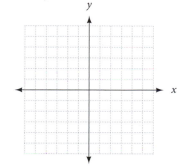

4. $y = \dfrac{1}{2}x - 2$
$2x - y = -1$

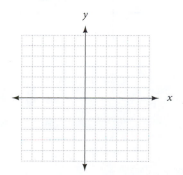

5. $y = \dfrac{1}{2}x$
$y = -\dfrac{3}{4}x + 5$

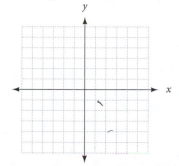

SCAN TO ACCESS

6. $y = \dfrac{2}{3}x$

$y = -\dfrac{1}{3}x + 6$

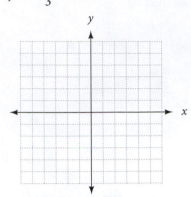

7. $3x + 3y = -2$
$y = -x + 4$

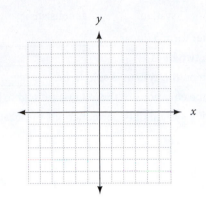

8. $2x - 2y = 6$
$y = x - 3$

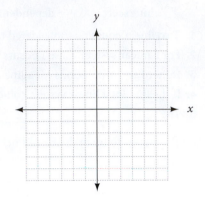

9. $x + y = 3$
$x - y = 1$

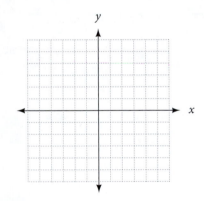

10. $x + y = 2$
$x - y = 4$

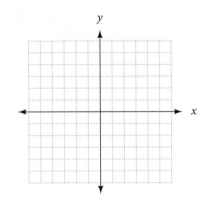

11. $x + 2y = 0$
$2x - y = 0$

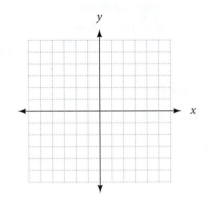

12. $3x + y = 0$
$5x - y = 0$

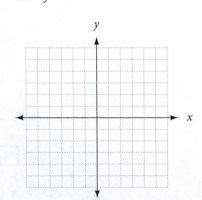

13. $x = -4$
$y = 6$

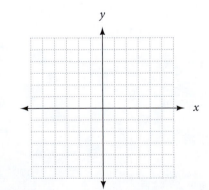

14. $x = 5$
$y = -1$

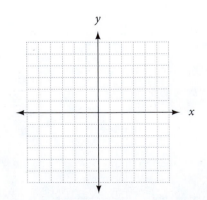

B Solve the following systems by solving one of the equations for x or y and then using the substitution method.

15. $x + y = 11$
$y = 2x - 1$

16. $x - y = -3$
$y = 3x + 5$

17. $x + y = 20$
$y = 5x + 2$

18. $3x - y = -1$
$x = 2y - 7$

19. $-2x + y = -1$
$y = -4x + 8$

20. $4x - y = 5$
$y = -4x + 1$

21. $5x + 4y = 7$
$y = -3x$

22. $10x + 2y = -6$
$y = -5x$

23. $x + 3y = 4$
$x - 2y = -1$

24. $x - y = 5$
$x + 2y = -1$

25. $2x + y = 1$
$x - 5y = 17$

26. $2x - 2y = 2$
$x - 3y = -7$

27. $2x - y = 5$
$4x - 2y = 10$

28. $-10x + 8y = -6$
$y = \dfrac{5}{4}x$

29. $-3x - 9y = 7$
$x + 3y = 12$

30. $2x + 6y = -18$
$x + 3y = -9$

31. $0.05x + 0.10y = 1.70$
$y = 22 - x$

32. $0.20x + 0.50y = 3.60$
$y = 12 - x$

C Solve the following systems by elimination.

33. $x + y = 3$
$x - y = 1$

34. $x - y = 4$
$2x + y = 8$

35. $x + y = -1$
$3x - y = -3$

36. $2x - y = -2$
$-2x - y = 2$

37. $3x + 2y = 1$
$-3x - 2y = -1$

38. $-3x - 2y = 5$
$9x - 6y = -14$

39. $\dfrac{1}{2}x + \dfrac{1}{3}y = 13$
$\dfrac{2}{5}x + \dfrac{1}{4}y = 10$

40. $\dfrac{1}{2}x + \dfrac{1}{3}y = \dfrac{2}{3}$
$\dfrac{2}{3}x + \dfrac{2}{5}y = \dfrac{14}{15}$

41. $\dfrac{2}{3}x + \dfrac{2}{5}y = -4$
$\dfrac{1}{3}x - \dfrac{1}{2}y = -\dfrac{1}{3}$

42. $\dfrac{1}{2}x - \dfrac{1}{3}y = \dfrac{5}{6}$
$-\dfrac{2}{5}x + \dfrac{1}{2}y = -\dfrac{9}{10}$

Solve each of the following systems by eliminating the *y* variable.

43. $3x - y = 4$
   $2x + 2y = 24$

44. $2x + y = 3$
   $3x + 2y = 1$

45. $5x - 3y = -2$
   $10x - y = 1$

46. $4x - y = -1$
   $2x + 4y = 13$

47. $11x - 4y = 11$
   $5x + y = 5$

48. $3x - y = 7$
   $10x - 5y = 25$

49. $3x + y = 5$
   $2x - 3y = -2$

50. $4x - y = -6$
   $3x + 2y = 7$

Solve each of the following systems by eliminating the *x* variable.

51. $3x - 5y = 7$
   $-x + y = -1$

52. $4x + 2y = 32$
   $x + y = -2$

53. $-x - 8y = -1$
   $-2x + 4y = 13$

54. $-x + 10y = 1$
   $-5x + 15y = -9$

55. $-3x - y = 7$
   $6x + 7y = 11$

56. $-5x + 2y = -6$
   $10x + 7y = 34$

57. $2x - 3y = 5$
   $-6x + y = 4$

58. $x - 3y = -4$
   $-3x + 9y = 12$

Solve each of the following systems by substitution or by elimination.

59. $x - 3y = 7$
   $2x + y = -6$

60. $2x - y = 9$
   $x + 2y = -11$

61. $y = \dfrac{1}{2}x + \dfrac{1}{3}$

   $y = -\dfrac{1}{3}x + 2$

62. $y = \dfrac{3}{4}x - \dfrac{4}{5}$

   $y = \dfrac{1}{2}x - \dfrac{1}{2}$

63. $3x - 4y = 12$

   $x = \dfrac{2}{3}y - 4$

64. $-5x + 3y = -15$

   $x = \dfrac{4}{5}y - 2$

65. $4x - 3y = -7$
   $-8x + 6y = -11$

66. $3x - 4y = 8$

   $y = \dfrac{3}{4}x - 2$

67. $3y + z = 17$
   $5y + 20z = 65$

68. $x + y = 850$
   $1.5x + y = 1,100$

69. $\dfrac{3}{4}x - \dfrac{1}{3}y = 1$

   $y = \dfrac{1}{4}x$

70. $-\dfrac{2}{3}x + \dfrac{1}{2}y = -1$

   $y = -\dfrac{1}{3}x$

71. $\dfrac{1}{4}x - \dfrac{1}{2}y = \dfrac{1}{3}$

   $\dfrac{1}{3}x - \dfrac{1}{4}y = -\dfrac{2}{3}$

72. $\dfrac{1}{5}x - \dfrac{1}{10}y = -\dfrac{1}{5}$

   $\dfrac{2}{3}x - \dfrac{1}{2}y = -\dfrac{1}{6}$

73. $\dfrac{2}{3}x - \dfrac{3}{4}y = -\dfrac{1}{6}$

   $\dfrac{3}{4}x - \dfrac{1}{5}y = -\dfrac{1}{10}$

74. $-\dfrac{1}{2}x + \dfrac{1}{6}y = -\dfrac{1}{3}$

   $-\dfrac{1}{8}x + \dfrac{1}{2}y = \dfrac{3}{4}$

Applying the Concepts

75. Gas Mileage Daniel is trying to decide whether to buy a car or a truck. The truck he is considering will cost him $150 a month in loan payments, and it gets 20 miles per gallon in gas mileage. The car will cost $180 a month in loan payments, but it gets 35 miles per gallon in gas mileage. Daniel estimates that he will pay $4.00 per gallon for gas. This means that the monthly cost to drive the truck x miles will be $y = \frac{4.00}{20}x + 150$. The total monthly cost to drive the car x miles will be $y = \frac{4.00}{35}x + 180$. The following figure shows the graph of each equation.

a. At how many miles do the car and the truck cost the same to operate?

b. If Daniel drives more than 600 miles in a month, which will be cheaper?

c. If Daniel drives fewer than 100 miles in a month, which will be cheaper?

d. Why do the graphs appear in the first quadrant only?

76. Truck Rental You need to rent a moving truck for two days. U-Haul charges $50 per day and $0.50 per mile. Budget Truck Rental charges $45 per day and $0.75 per mile. The following figure represents the cost of renting each of the trucks for two days.

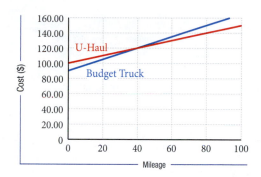

a. From the figure, after how many miles would the trucks cost the same?

b. Which company will give you a better deal if you drive less than 30 miles?

c. Which company will give you a better deal if you drive more than 60 miles?

77. Cost of Solar Panels Suppose the average cost of electricity is $0.15/kWh. Solar panels cost $100/sq ft and you want to install 400 sq ft of solar panels. Assume each square foot of solar panels produces 20 kWh every year. Further, assume government tax incentives pay for 50% of the initial costs.

a. Find the cost to install the 400 ft² of solar panels. Remember to reduce this cost for the government tax incentives.

b. Find a function for the amount of money saved by using the solar panels, where t represents the number of years.

c. Find a break-even point for when the solar panels pay off their original cost.

78. Cost of Solar Panels Suppose the average cost of electricity is $0.25/kWh. Solar panels cost $100/sq ft and you want to install 400 sq ft of solar panels. Assume each square foot of solar panels produces 20 kWh every year. Further, assume government tax incentives pay for 50% of the initial costs.

a. Find the cost to install the 400 ft² of solar panels. Remember to reduce this cost for the government tax incentives.

b. Find a function for the amount of money saved by using the solar panels, where t represents the number of years.

c. Find a break-even point for when the solar panels pay off their original cost.

79. Charter Jets A company is considering using a charter jet to fly employees from Los Angeles to New York. The charter rents for $5,200 per hours for the five hour flight. Alternatively, the company can purchase business class seats on a commercial airliner for $1,300 per ticket.

a. Express each cost as a linear equation.

b. Graph each equations on the same set of axes.

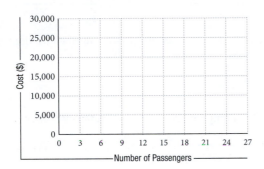

c. If the company wants to send 15 employees to New York, which option is cheaper?

d. How many passengers would be needed for the charter jet to be the cheaper option?

80. Speed of Planes A Boeing 747 has a cruising speed of 570 mph. Forty-five minutes after the 747 has taken off, a F18 Hornet with a speed of 1,190 mph is dispatched to intercept the 747.

1,190 mph

a. Express each distance traveled as a linear equation. Let t represent the time for the 747 jet.

b. Graph each equation on the same set of axes.

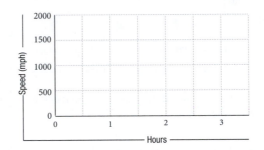

c. By solving a system, find the time at which the F18 will intercept the Boeing 747. Answer to the closest minute.

81. Cost of a Hybrid Vehicle You may have wondered what the break-even point is for purchasing a hybrid vehicle if you paid in cash (no interest). Below is data for two models of Honda Civic bases on $4.00 per gallon gas prices.

	Base Price ($)	Cost/Mile ($)
Civic Hybrid	$22,600	$0.095
Civic Ex	$18,710	$0.14

Source: greencar.com

a. Find the cost function for each car, as a function of miles driven x.

b. Determine the break-even miles. This represents the number of miles you would need to drive the hybrid before it becomes the cheaper alternative.

82. Kinetic Sculpture In the introduction to this section, using d for distance and x for the kinetic sculpture rate, the system of equations is:

$$d = 5(x + 3)$$
$$d = 10(x - 3)$$

Find the rate of the kinetic sculpture.

Getting Ready for the Next Section

Simplify the expression.

83. $2 - 2(6)$ **84.** $2(1) - 2 + 3$ **85.** $(x + 3y) - 1(x - 2z)$ **86.** $(x + y + z) + (2x - y + z)$

Solve for the variable.

87. $-9y = -9$ **88.** $30x = 38$ **89.** $3(1) + 2z = 9$ **90.** $4\left(\dfrac{19}{15}\right) - 2y = 4$

Apply the distributive property, then simplify if possible.

91. $2(5x - z)$ **92.** $-1(x - 2z)$ **93.** $3(3x + y - 2z)$ **94.** $2(2x - y + z)$

Find the Mistake

Each sentence below contains a mistake. Circle the mistake and write the correct word or phrase on the line provided.

1. When two lines are parallel, the equations of those lines are dependent. _____

2. Equations are said to be dependent if their lines intersect. _____

3. True or False: To solve the following system, it would be easiest to add $2x$ to both sides of the first equation and then divide both sides by 6 to isolate y.

$$-2x + 6y = -6$$
$$x - 3y = 3$$
_____

4. The solution to the following system is $(1, 2)$.

$$4x - 3y = -2$$
$$-8x + 6y = 20$$
_____

Navigation Skills: Prepare, Study, Achieve

Preparation is another key component for success in this course. What does it mean to be prepared for this class? Before you come to class, make sure to do the following:

- Complete the homework from the previous section.
- Read the upcoming section.
- Answer the Getting Ready for Class questions.
- Rework example problems, and work practice problems.
- Make a point to attend every class session and arrive on time.
- Prepare a list of questions to ask your instructor or fellow students that will help you work through difficult problems.
- Commit to being an engaged, attentive, and active listener while in class.
- Take notes as the instructor speaks and don't hesitate to ask questions if you need clarification.
- Research note-taking techniques online if you find that you aren't quite sure how to take notes or your current method of taking notes is not helpful.

You shouldn't expect to master a new topic the first time you read about it or learn about it in class. Mastering mathematics takes a lot of practice, so make the commitment to come to class prepared and practice, practice, practice.

3.2 Systems of Linear Equations in Three Variables

Image © sxc.hu, johnnyberg. 2008

If a human faced off against an elephant in a hot dog bun eating contest, who would win? Each July, three brave humans sign up to find out firsthand. Three elephants, Susie, Minnie, and Bunny, from Ringling Bros. and Barnum & Bailey Circus stand behind a table on which a tall pile of hot dog buns sits. A few feet away, three competitive eating contestants stand in front of their own table outfitted with stacks of buns and large cups of water in which they can dunk the buns to help their cause. When the announcer yells "Go!", the two teams begin devouring the buns. They have 6 minutes to eat as many as possible. The 2010 final tally: 40 dozen buns for the elephants, 15 dozen for the humans.

We can use this scenario to help us introduce the topic of this section: systems of linear equations in three variables. Up until now, we have worked with systems in two variables. Now we will work with those with three unknowns. For instance, let's assume the individual totals for the amount of buns Susie, Minnie, and Bunny ate are x, y, and z, respectively and the total number of buns consumed by the elephants was 401 dozen. Suppose Minnie ate twice as many buns as Susie plus 10 more buns, and Bunny ate as much as Minnie plus one third the amount the humans ate. After reading through this section, you will be able to assign three linear equations and solve for the three variables in this problem.

A Solving Systems in Three Variables

A solution to an equation in three variables such as

$$2x + y - 3z = 6$$

is an ordered triple of numbers (x, y, z). For example, the ordered triples $(0, 0, -2)$, $(2, 2, 0)$, and $(0, 9, 1)$ are solutions to the equation $2x + y - 3z = 6$, because they produce a true statement when their coordinates are substituted for x, y, and z in the equation.

> **DEFINITION** solution set
>
> The *solution set* for a system of three linear equations in three variables is the set of ordered triples that satisfies all three equations.

Practice Problems

1. Solve the system.

$$x + 2y + z = 2$$
$$x + y - z = 6$$
$$x - y + 2z = -7$$

Answer

1. $(1, 2, -3)$

EXAMPLE 1 Solve the system.

$$x + y + z = 6 \qquad \text{Equation (1)}$$
$$2x - y + z = 3 \qquad \text{Equation (2)}$$
$$x + 2y - 3z = -4 \qquad \text{Equation (3)}$$

Solution We want to find the ordered triple (x, y, z) that satisfies all three equations. We have numbered the equations so it will be easier to keep track of where they are and what we are doing.

There are many ways to proceed. The main idea is to take two different pairs of equations and eliminate the same variable from each pair. We begin by adding equations (1) and (2) to eliminate the y-variable. The resulting equation is numbered (4).

$$
\begin{array}{ll}
x + y + z = 6 & (1) \\
2x - y + z = 3 & (2) \\
\hline
3x \quad\; + 2z = 9 & \text{Equation (4)}
\end{array}
$$

Adding twice equation (2) to equation (3) will also eliminate the variable y. The resulting equation is numbered (5):

$$
\begin{array}{ll}
4x - 2y + 2z = 6 & \text{Twice (2)} \\
x + 2y - 3z = -4 & (3) \\
\hline
5x \qquad\;\; - z = 2 & \text{Equation (5)}
\end{array}
$$

Equations (4) and (5) form a linear system in two variables. By multiplying equation (5) by 2 and adding the result to equation (4), we succeed in eliminating the variable z from the new pair of equations.

$$
\begin{array}{ll}
3x + 2z = 9 & (4) \\
10x - 2z = 4 & \text{Twice (5)} \\
\hline
13x \qquad = 13 & \\
x \qquad = 1 &
\end{array}
$$

Substituting $x = 1$ into equation (4), we have

$$
\begin{aligned}
3(1) + 2z &= 9 \\
2z &= 6 \\
z &= 3
\end{aligned}
$$

Using $x = 1$ and $z = 3$ in equation (1) gives us

$$
\begin{aligned}
1 + y + 3 &= 6 \\
y + 4 &= 6 \\
y &= 2
\end{aligned}
$$

The solution is the ordered triple $(1, 2, 3)$.

EXAMPLE 2 Solve the system.

$$
\begin{array}{ll}
2x + y - z = 3 & (1) \\
3x + 4y + z = 6 & (2) \\
2x - 3y + z = 1 & (3)
\end{array}
$$

Solution It is easiest to eliminate z from the equations using the elimination method. The equation produced by adding (1) and (2) is

$$
5x + 5y = 9 \qquad (4)
$$

The equation that results from adding (1) and (3) is

$$
4x - 2y = 4 \qquad (5)
$$

2. Solve the system.

$$
\begin{array}{l}
3x - 2y + z = 2 \\
3x + y + 3z = 7 \\
x + 4y - z = 4
\end{array}
$$

Answer

2. $(1, 1, 1)$

Equations (4) and (5) form a linear system in two variables. We can eliminate the variable y from this system as follows:

$$5x + 5y = 9 \xrightarrow{\text{Multiply by 2.}} 10x + 10y = 18$$

$$4x - 2y = 4 \xrightarrow[\text{Multiply by 5.}]{} 20x - 10y = 20$$

$$\overline{30x = 38}$$

$$x = \frac{38}{30}$$

$$= \frac{19}{15}$$

Substituting $x = \frac{19}{15}$ into equation (5) or equation (4) and solving for y gives

$$y = \frac{8}{15}$$

Using $x = \frac{19}{15}$ and $y = \frac{8}{15}$ in equation (1), (2), or (3) and solving for z results in

$$z = \frac{1}{15}$$

The ordered triple that satisfies all three equations is $\left(\frac{19}{15}, \frac{8}{15}, \frac{1}{15} \right)$.

EXAMPLE 3   Solve the system.

$$2x + 3y - z = 5 \qquad (1)$$

$$4x + 6y - 2z = 10 \qquad (2)$$

$$x - 4y + 3z = 5 \qquad (3)$$

Solution   Multiplying equation (1) by -2 and adding the result to equation (2) looks like this:

$$-4x - 6y + 2z = -10 \qquad -2 \text{ times } (1)$$

$$\underline{4x + 6y - 2z = 10 \qquad (2)}$$

$$0 = 0$$

All three variables have been eliminated, and we are left with a true statement. This implies that the two equations are dependent. With a system of three equations in three variables, however, a dependent system can have no solution or an infinite number of solutions. After we have concluded the examples in this section, we will discuss the geometry behind these systems. Doing so will give you some additional insight into dependent systems.

EXAMPLE 4   Solve the system.

$$x - 5y + 4z = 8 \qquad (1)$$

$$3x + y - 2z = 7 \qquad (2)$$

$$-9x - 3y + 6z = 5 \qquad (3)$$

Solution   Multiplying equation (2) by 3 and adding the result to equation (3) produces

$$9x + 3y - 6z = 21 \qquad 3 \text{ times } (2)$$

$$\underline{-9x - 3y + 6z = 5 \qquad (3)}$$

$$0 = 26$$

In this case, all three variables have been eliminated, and we are left with a false statement. The two equations are inconsistent; there are no ordered triples that satisfy both equations. There is no solution to the system. If equations (2) and (3) have no ordered triples in common, then certainly (1), (2), and (3) do not either.

---

3. Solve the system.

$$3x + 5y - 2z = 1$$
$$4x - 3y - z = 5$$
$$x - 8y + z = 4$$

4. Solve the system.

$$3x - y + 2z = 4$$
$$-2x + 2y - 5z = 1$$
$$5x - 3y + 7z = 5$$

Answers

3. Dependent (no unique solution)

4. Inconsistent (no solution)

EXAMPLE 5 Solve the system.

$$x + 3y = 5 \quad (1)$$
$$6y + z = 12 \quad (2)$$
$$x - 2z = -10 \quad (3)$$

Solution It may be helpful to rewrite the system as

$$x + 3y \qquad = 5 \quad (1)$$
$$6y + z = 12 \quad (2)$$
$$x \qquad - 2z = -10 \quad (3)$$

Equation (2) does not contain the variable x. If we multiply equation (3) by -1 and add the result to equation (1), we will be left with another equation that does not contain the variable x.

$$\begin{array}{ll} x + 3y \qquad = 5 & (1) \\ \underline{-x \qquad + 2z = 10} & -1 \text{ times (3)} \\ 3y + 2z = 15 & (4) \end{array}$$

Equations (2) and (4) form a linear system in two variables. Multiplying equation (2) by -2 and adding the result to equation (4) eliminates the variable z.

$$\begin{array}{l} 6y + z = 12 \xrightarrow{\text{Multiply by } -2.} -12y - 2z = -24 \\ 3y + 2z = 15 \xrightarrow{\text{No Change}} \underline{\quad 3y + 2z = 15} \\ \qquad\qquad\qquad\qquad\qquad -9y + \quad = -9 \\ \qquad\qquad\qquad\qquad\qquad\qquad y = 1 \end{array}$$

Using $y = 1$ in equation (4) and solving for z, we have

$$z = 6$$

Substituting $y = 1$ into equation (1) gives

$$x = 2$$

The ordered triple that satisfies all three equations is $(2, 1, 6)$.

The Geometry Behind Linear Equations in Three Variables

We can graph an ordered triple on a coordinate system with three axes. The graph will be a point in space. The coordinate system is drawn in perspective; you have to imagine that the x-axis comes out of the paper and is perpendicular to both the y-axis and the z-axis. To graph the point $(3, 4, 5)$, we move 3 units in the x-direction, 4 units in the y-direction, and then 5 units in the z-direction, as shown in Figure 1.

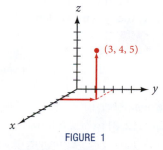

FIGURE 1

Although in actual practice it is sometimes difficult to graph equations in three variables, if we were to graph a linear equation in three variables, we would find that the graph was a plane in space. A system of three equations in three variables is represented by three planes in space.

5. Solve the system.

$$x + 2y = 0$$
$$3y + z = -3$$
$$2x - z = 5$$

Answer

5. $(4, -2, 3)$

There are a number of possible ways in which these three planes can intersect, some of which are shown below. And there are still other possibilities that are not among these shown.

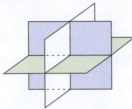

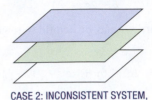

CASE 1: ONE SOLUTION
The three planes have exactly one point in common. In this case we get one solution to our system, as in Examples 1, 2, and 5.

CASE 2: INCONSISTENT SYSTEM, THREE PARALLEL PLANES
The three planes have no points in common because they are all parallel to one another. The system they represent is an inconsistent system.

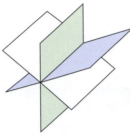

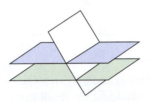

CASE 3: DEPENDENT SYSTEM
The three planes intersect in a line. Any point on the line is a solution to the system of equations represented by the planes, so there is an infinite number of solutions to the system. This is an example of a dependent system.

CASE 4: INCONSISTENT SYSTEM, TWO PARALLEL LINES
Two of the planes are parallel; the third plane intersects each of the parallel planes. In this case, the three planes have no points in common. There is no solution to the system; it is an inconsistent system.

In Example 3, we found that equations (1) and (2) were dependent equations. They represent the same plane. That is, they have all their points in common. But the system of equations that they came from has either no solution or an infinite number of solutions. It all depends on the third plane. If the third plane coincides with the first two, then the solution to the system is a plane. If the third plane is parallel to the first two, then there is no solution to the system. Finally, if the third plane intersects the first two but does not coincide with them, then the solution to the system is that line of intersection.

In Example 4, we found that trying to eliminate a variable from the second and third equations resulted in a false statement. This means that the two planes represented by these equations are parallel. It makes no difference where the third plane is; there is no solution to the system in Example 4. (If we were to graph the three planes from Example 4, we would obtain a diagram similar to Case 2 or Case 4 above.)

If, in the process of solving a system of linear equations in three variables, we eliminate all the variables from a pair of equations and are left with a false statement, we will say the system is inconsistent. If we eliminate all the variables and are left with a true statement, then we will say the system is a dependent one.

GETTING READY FOR CLASS

After reading through the preceding section, respond in your own words and in complete sentences.

A. What is an ordered triple of numbers?

B. Explain what it means for (1, 2, 3) to be a solution to a system of liner equations in three variables.

C. Explain in a general way the procedure you would use to solve a system of three linear equations in three variables.

D. How do you know when a system of linear equations in three variables has no solution?

Vocabulary Review

Choose the correct words to fill in the blanks below.

dependent   triples   infinite   inconsistent

1. The solution set for a system of three linear equations in three variables is the set of ordered _____ that satisfies all three equations.

2. If three variables are eliminated and we are left with a true statement, we consider the system to be _____.

3. With a system of three equations in three variables, a dependent system can have no solution or an _____ number of solutions.

4. If three variables are eliminated and we are left with a false statement, we consider the system to be _____ with no solutions.

Problems

A Solve the following systems.

1. $x + y + z = 4$
$x - y + 2z = 1$
$x - y - 3z = -4$

2. $x - y - 2z = -1$
$x + y + z = 6$
$x + y - z = 4$

3. $x + y + z = 6$
$x - y + 2z = 7$
$2x - y - 4z = -9$

4. $x + y + z = 0$
$x + y - z = 6$
$x - y + 2z = -7$

5. $x + 2y + z = 3$
$2x - y + 2z = 6$
$3x + y - z = 5$

6. $2x + y - 3z = -14$
$x - 3y + 4z = 22$
$3x + 2y + z = 0$

7. $2x + 3y - 2z = 4$
$x + 3y - 3z = 4$
$3x - 6y + z = -3$

8. $4x + y - 2z = 0$
$2x - 3y + 3z = 9$
$-6x - 2y + z = 0$

9. $-x + 4y - 3z = 2$
$2x - 8y + 6z = 1$
$3x - y + z = 0$

10. $4x + 6y - 8z = 1$
$-6x - 9y + 12z = 0$
$x - 2y - 2z = 3$

11. $\frac{1}{2}x - y + z = 0$
$2x + \frac{1}{3}y + z = 2$
$x + y + z = -4$

12. $\frac{1}{3}x + \frac{1}{2}y + z = -1$
$x - y + \frac{1}{5}z = -1$
$x + y + z = -5$

13. $2x - y - 3z = 1$
$x + 2y + 4z = 3$
$4x - 2y - 6z = 2$

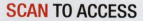

14. $3x + 2y + z = 3$
$x - 3y + z = 4$
$-6x - 4y - 2z = 1$

15. $2x - y + 3z = 4$
$x + 2y - z = -3$
$4x + 3y + 2z = -5$

16. $6x - 2y + z = 5$
$3x + y + 3z = 7$
$x + 4y - z = 4$

17. $x + y = 9$
$y + z = 7$
$x - z = 2$

18. $x - y = -3$
$x + z = 2$
$y - z = 7$

19. $2x + y = 2$
$y + z = 3$
$4x - z = 0$

20. $2x + y = 6$
$3y - 2z = -8$
$x + z = 5$

21. $2x - 3y = 0$
$6y - 4z = 1$
$x + 2z = 1$

22. $3x + 2y = 3$
$y + 2z = 2$
$6x - 4z = 1$

23. $x + y - z = 2$
$2x + y + 3z = 4$
$x - 2y + 2z = 6$

24. $x + 2y - 2z = 4$
$3x + 4y - z = -2$
$2x + 3y - 3z = -5$

25. $2x + 3y = -\dfrac{1}{2}$
$4x + 8z = 2$
$3y + 2z = -\dfrac{3}{4}$

26. $3x - 5y = 2$
$4x + 6z = \dfrac{1}{3}$
$5y - 7z = \dfrac{1}{6}$

27. $\dfrac{1}{3}x + \dfrac{1}{2}y - \dfrac{1}{6}z = 4$
$\dfrac{1}{4}x - \dfrac{3}{4}y + \dfrac{1}{2}z = \dfrac{3}{2}$
$\dfrac{1}{2}x - \dfrac{2}{3}y - \dfrac{1}{4}z = -\dfrac{16}{3}$

28. $-\dfrac{1}{4}x + \dfrac{3}{8}y + \dfrac{1}{2}z = -1$
$\dfrac{2}{3}x - \dfrac{1}{6}y - \dfrac{1}{2}z = 2$
$\dfrac{3}{4}x - \dfrac{1}{2}y - \dfrac{1}{8}z = 1$

29. $x - \dfrac{1}{2}y - \dfrac{1}{3}z = -\dfrac{4}{3}$
$\dfrac{1}{3}x - \dfrac{1}{2}z = 5$
$-\dfrac{1}{4}x + \dfrac{2}{3}y - z = -\dfrac{3}{4}$

30. $x + \dfrac{1}{3}y - \dfrac{1}{2}z = -\dfrac{3}{2}$
$\dfrac{1}{2}x - y + \dfrac{1}{3}z = 8$
$\dfrac{1}{3}x - \dfrac{1}{4}y - z = -\dfrac{5}{6}$

Applying the Concepts

31. Electric Current In the following diagram of an electrical circuit, x, y, and z represent the amount of current (in amperes) flowing across the 5-ohm, 20-ohm, and 10-ohm resistors, respectively. (In circuit diagrams, resistors are represented by —W— and cells that provide electrical energy are ⊣⊢.)

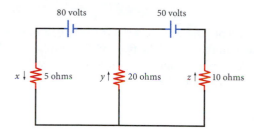

The system of equations used to find the three currents x, y, and z is

$$x - y - z = 0$$
$$5x + 20y = 80$$
$$20y - 10z = 50$$

Solve the system for all variables.

32. Cost of a Rental Car If a car rental company charges $10 a day and 8¢ a mile to rent one of its cars, then the cost z, in dollars, to rent a car for x, days and drive y miles can be found from the equation

$$z = 10x + 0.08y$$

a. How much does it cost to rent a car for 2 days and drive it 200 miles under these conditions?

b. A second company charges $12 a day and 6¢ a mile for the same car. Write an equation that gives the cost z, in dollars, to rent a car from this company for x days and drive it y miles.

c. A car is rented from each of the companies mentioned in a. and b. for 2 days. To find the mileage at which the cost of renting the cars from each of the two companies will be equal, solve the following system for y:

$$z = 10x + 0.08y$$
$$z = 12x + 0.06y$$
$$x = 2$$

33. Hot Dogs Refer to the hot dog eating contest at the beginning of this section. Set up a system of three equations in three unknowns, then solve it, to find the number of hot dog buns Susie, Minnie, and Bunny each ate.

34. Hot Dogs If the organizers of the hot dog eating contest paid $1.20 for each dozen hot dog buns, how much did they spend on the buns consumed by the elephants and the humans?

Getting Ready for the Next Section

Translate into symbols.

35. Two more than 3 times a number

36. One less than twice a number

Apply the distributive property, then simplify.

37. $10(0.2x + 0.5y)$

38. $100(0.09x + 0.08y)$

Solve.

39. $x + (3x + 2) = 26$

40. $5x = 2,500$

Solve each system.

41. $-2y - 4z = -18$
$\quad -7y + 4z = 27$

42. $-x + 2y = 200$
$\quad 4x - 2y = 1{,}300$

43. The sum of three numbers is 20. The largest number equals the sum of the two smaller numbers, and the middle number is one more than twice the smaller number. Find these numbers.

44. The sum of three numbers is 34. The largest number is six more than the sum of the two smaller numbers, and the middle number is one less than twice the smaller number. Find the three numbers.

45. A 20 foot board is cut into three pieces. The longest piece is 2 ft. more than the sum of the two smaller pieces, and the middle piece is twice the length of the shortest piece. Find the length of each piece.

46. A 10 foot board is cut into three pieces. The longest piece is the sum of the lengths of the two smaller pieces, and the middle piece is one foot longer than the shortest piece. Find the length of each piece.

Find the Mistake

Each sentence below contains a mistake. Circle the mistake and write the correct word or phrase on the line provided.

1. The solution to the following system is (3, 3, 2). _____
$\quad x + y + z = -2$
$\quad 2x + 3y + z = -3$
$\quad 2x - y - z = 11$

2. It is easiest to solve the following system by eliminating y from each equation. _____
$\quad x + 2y + z = -7$
$\quad 2x + y - z = 4$
$\quad 2x - 2y + z = 2$

3. The following system has no solution because the equations represent parallel lines. _____
$\quad x + 2y + 2 = -8$
$\quad 2x + y + 2z = -7$
$\quad 4x + 2y + 4z = -14$

4. The following system has an infinite number of solutions. _____
$\quad x + 3y + z = 14$
$\quad 2x - y - 3z = 2$
$\quad 2x + 2y + 2z = 16$

3.3 Matrix Solutions to Linear Systems

OBJECTIVES

A Find the dimensions of a matrix.

B Find the coefficient matrix, the constant matrix, and the augmented matrix for a system of equations.

C Solve a system of linear equations using an augmented matrix.

KEY WORDS

matrix

rectangular array

coefficient matrix

constant matrix

augmented matrix

row operation

Suppose you decide to sell homegrown zucchini and artichokes at two local farmer's market. For the first market, you price the zucchini at $0.75 each and the artichokes at $1.50 each, and you make a total of $21. For the second market, you increase your prices by a quarter each and end up selling the same quantity of each vegetables as you did at the first market. You make a total of $26 at the second market. The following system of equations shows how many of each vegetable you sold at the markets.

$$0.75x + 1.5y = 21$$
$$x + 1.75y = 26$$

We can use a matrix to display a system of equations. For example, the above system can be written as a *rectangular array* of $m \times n$ real numbers with m rows and n columns. The coefficients of the variables are called the elements of the matrix and are enclosed by brackets.

$$\text{Matrix } A = \begin{bmatrix} 0.75 & 1.5 & 21 \\ 1 & 1.75 & 26 \end{bmatrix} \; m \text{ rows}$$

n columns

Matrix A above has 2 rows and 3 columns, with elements from the coefficients of the equations in the given system. The dimension of matrix A is written 2×3 and read "two by three."

A Matrix Dimensions

In this section, we will learn how to solve a system of equations using a matrix.

EXAMPLE 1 Give the dimensions of each of the following matrices.

a. $\begin{bmatrix} -2 & 1 \\ 5 & 3 \end{bmatrix}$ **b.** $\begin{bmatrix} 1 & 0 \\ 4 & -2 \\ -3 & 7 \end{bmatrix}$ **c.** $\begin{bmatrix} 5 \\ -2 \\ 1 \end{bmatrix}$ **d.** $[4 \; -2]$

Solutions

a. $A = \begin{bmatrix} -2 & 1 \\ 5 & 3 \end{bmatrix}$ Matrix A is a 2×2 matrix and is called a square matrix because $m = n$.

b. $B = \begin{bmatrix} 1 & 0 \\ 4 & -2 \\ -3 & 7 \end{bmatrix}$ Matrix B is a 3×2 matrix as there are thee rows and two columns.

c. $C = \begin{bmatrix} 5 \\ -2 \\ 1 \end{bmatrix}$ Matrix C is a 3×1 matrix as there are three rows but one column.

d. $D = [4 \; -2]$ Matrix D is a 1×2 matrix as there is one row and two columns.

Practice Problems

1. Give the dimensions of each matrix.

a. $\begin{bmatrix} 7 & 0 & -5 \\ -1 & 2 & -3 \\ 5 & -3 & 4 \end{bmatrix}$

b. $[2 \; 4 \; 6]$

c. $\begin{bmatrix} 4 \\ -2 \end{bmatrix}$

Answers

1. a. 3×3 **b.** 1×3 **c.** 2×1

B Augmented Matrices

Matrices are used to solve systems of equations. When a system of equations is written so that the variables appear in the same order in each equation, the matrix of coefficients if the variables is called the *coefficient matrix.* The constant terms on the right side of the equations form what is *constant matrix.* The combination of the coefficients and constants written in one matrix is called the *augmented matrix* of the system.

EXAMPLE 2   Find the coefficient matrix, constant matrix and augmented matrix of the system

$$x + 5y - 3z = 4$$
$$-x + 2y = -4$$

Solution   The coefficient matrix comes from the coefficients of the variables written in the same order. The constant matrix comes from the terms on the right side of each equation. Finally the augmented matrix combines these two. They are each written as

Coefficient   Constant   Augmented
Matrix   Matrix   Matrix

$$\begin{bmatrix} 1 & 5 & -3 \\ -1 & 2 & 0 \end{bmatrix} \qquad \begin{bmatrix} 4 \\ -4 \end{bmatrix} \qquad \left[\begin{array}{ccc|c} 1 & 5 & -3 & 4 \\ -1 & 2 & 0 & -4 \end{array}\right]$$

C Row Operations

Now that we know how a matrix can represent a system of equations, we will now see that operations on that matrix can be used to find solutions to the system. As you will see, these matrix methods are similar to the elimination method used to solve systems of equations. We will apply these methods to rows, columns, and elements rather than equations, variables, and coefficients.

We can use the following *row operations* to transform an augmented matrix into an equivalent system:

1. We can interchange any two rows of a matrix.

2. We can multiply any row by a nonzero constant.

3. We can add to any row a constant multiple of another row.

The three row operations are simply a list of the properties we use to solve systems of linear equations, translated to fit an augmented matrix. For instance, the second operation in our list is actually just another way to state the multiplication property of equality.

We solve a system of linear equations by first transforming the augmented matrix into a matrix that has 1's down the diagonal of the coefficient matrix, and 0's below it. For instance, we will solve the system

$$2x + 5y = -4$$
$$x - 3y = 9$$

by transforming the matrix

$$\left[\begin{array}{cc|c} 2 & 5 & -4 \\ 1 & -3 & 9 \end{array}\right]$$

using the row operations listed earlier to get a matrix of the form

$$\left[\begin{array}{cc|c} 1 & \square & \square \\ 0 & 1 & \square \end{array}\right]$$

To accomplish this, we begin with the first column and try to produce a 1 in the first position and a 0 below it. Interchanging rows 1 and 2 gives us a 1 in the top position of the first column:

$$\left[\begin{array}{cc|c} 1 & -3 & 9 \\ 2 & 5 & -4 \end{array}\right] \qquad \text{Interchange rows 1 and 2.}$$

Multiplying row 1 by -2 and adding the result to row 2 gives us a 0 where we want it.

$$\begin{bmatrix} 1 & -3 & 9 \\ 0 & 11 & -22 \end{bmatrix}$$ Multiply row 1 by –2 and add the result to row 2.

Continue to produce 1's down the diagonal by multiplying row 2 by $\frac{1}{11}$.

$$\begin{bmatrix} 1 & -3 & | & 9 \\ 0 & 1 & | & -2 \end{bmatrix}$$ Multiply row 2 by $\frac{1}{11}$.

Taking this last matrix and writing the system of equations it represents, we have

$$x - 3y = 9$$
$$y = -2$$

Substituting -2 for y in the top equation gives us

$$x = 3$$

The solution to our system is $(3, -2)$.

EXAMPLE 3   Solve the following system using an augmented matrix.

$$x + y - z = 2$$
$$2x + 3y - z = 7$$
$$3x - 2y + z = 9$$

Solution   We begin by writing the system as an augmented matrix.

$$\begin{bmatrix} 1 & 1 & -1 & | & 2 \\ 2 & 3 & 1 & | & 7 \\ 3 & -2 & 1 & | & 9 \end{bmatrix}$$

Next, we want to produce 0's in the second two positions of column 1.

$$\begin{bmatrix} 1 & 1 & -1 & | & 2 \\ 0 & 1 & 1 & | & 3 \\ 3 & -2 & 1 & | & 9 \end{bmatrix}$$ Multiply row 1 by –2 and add the result to row 2.

$$\begin{bmatrix} 1 & 1 & -1 & | & 2 \\ 0 & 1 & 1 & | & 3 \\ 0 & -2 & 4 & | & 3 \end{bmatrix}$$ Multiply row 1 by –3 and add the result to row 3.

Note that we could have done these two steps in one single step. As you become more familiar with this method of solving systems of equations, you will do just that.

$$\begin{bmatrix} 1 & 1 & -1 & | & 2 \\ 0 & 1 & 1 & | & 3 \\ 0 & 0 & 9 & | & 18 \end{bmatrix}$$ Multiply row 2 by 5 and add the result to row 3.

$$\begin{bmatrix} 1 & 1 & -1 & | & 2 \\ 0 & 1 & 1 & | & 3 \\ 0 & 0 & 1 & | & 2 \end{bmatrix}$$ Multiply row 3 by $\frac{1}{9}$.

Converting back to a system of equations, we have

$$x + y - z = 2$$
$$y + z = 3$$
$$z = 2$$

3. Solve the following system using an augmented matrix.

$$x - y + z = -6$$
$$2x + y - 2z = 12$$
$$2x - 2y + 2z = -12$$

Answer

3. $(1, 4, -3)$

This system is equivalent to our first one, but much easier to solve. Substituting $z = 2$ into the second equation, we have

$$y = 1$$

Substituting $z = 2$ and $y = 1$ into the first equation, we have

$$x = 3$$

The solution to our original system is (3, 1, 2). It satisfies each of our original equations. You can check this, if you want.

GETTING READY FOR CLASS

After reading through the preceding section, respond in your own words and in complete sentences.

A. What are the dimensions of a matrix?

B. What advantage do augmented matrices have in solving systems over Cramer's rule?

C. What are the row operations that can be applied to an augmented matrix?

D. The form for an augmented matrix that solves systems is called upper triangular. Why do you think this term is used?

Vocabulary Review

Choose the correct words to fill in the blanks below.

coefficients	constant	rectangular array	augmented
dimensions	elements	matrix	

1. A _____ is used to display and solve a system of equations.

2. When creating a _____ for a system of equations, the coefficients of the variables are called the _____ of the matrix.

3. The _____ of a matrix are written as number of rows m by number of columns n, or $m \times n$.

4. A _____ matrix comes from the coefficients of the variables in a system written in the same order, whereas a _____ matrix comes from the terms on the right side of each equation.

5. The combination of the coefficients and constants written in one matrix is called the _____ matrix of the system of equations.

Problems

A Solve the following systems of equations by using matrices.

1. $x + y = 5$
$3x - y = 3$

2. $x + y = -2$
$2x - y = -10$

3. $3x - 5y = 7$
$-x + y = -1$

4. $2x - y = 4$
$x + 3y = 9$

5. $2x - 8y = 6$
$3x - 8y = 13$

6. $3x - 6y = 3$
$-2x + 3y = -4$

7. $2x - y = -10$
$4x + 3y = 0$

8. $3x - 7y = 36$
$5x - 4y = 14$

9. $5x - 3y = 27$
$6x + 2y = -18$

10. $3x + 4y = 2$
$5x + 3y = 29$

11. $5x + 2y = -14$
$y = 2x + 11$

12. $3x + 5y = 3$
$x = 4y + 1$

13. $2x + 3y = 11$
$-x - y = -2$

14. $5x + 2y = -25$
$-3x + 2y = -1$

15. $3x - 2y = 16$
$4x + 3y = -24$

16. $6x + y = 3$
$x = 4y + 13$

17. $3x - 2y = 16$
$y = 2x - 12$

18. $4x - 3y = 28$
$y = -x - 7$

SCAN TO ACCESS

19. $x + y + z = 4$
$x - y + 2z = 1$
$x - y - z = -2$

20. $x - y - 2z = -1$
$x + y + z = 6$
$x + y - z = 4$

21. $x + 2y + z = 3$
$2x - y + 2z = 6$
$3x + y - z = 5$

22. $x - 3y + 4z = -4$
$2x + y - 3z = 14$
$3x + 2y + z = 10$

23. $x - 2y + z = -4$
$2x + y - 3z = 7$
$5x - 3y + z = -5$

24. $3x - 2y + 3z = -3$
$x + y + z = 4$
$x - 4y + 2z = -9$

25. $5x - 3y + z = 10$
$x - 2y - z = 0$
$3x - y + 2z = 10$

26. $2x - y - z = 1$
$x + 3y + 2z = 13$
$4x + y - z = 7$

27. $2x - 5y + 3z = 2$
$3x - 7y + z = 0$
$x + y + 2z = 5$

28. $3x - 4y + 2z = -2$
$2x + y + 3z = 13$
$x - 3y + 2z = -3$

29. $x + 2y = 3$
$y + z = 3$
$4x - z = 2$

30. $x + y = 2$
$3y - 2z = -8$
$x + z = 5$

31. $x + 3y = 7$
$3x - 4z = -8$
$5y - 2z = -5$

32. $x + 4y = 13$
$2x - 5z = -3$
$4y - 3z = 9$

33. $x + 4y = 13$
$2x - 5z = -3$
$4y - 3z = 9$

34. $x - 2y = 5$
$4x + 3z = 11$
$5y + 4z = -12$

35. $x - 2y + z = -5$
$2x + 3y - 2z = -9$
$2x - y + 2z = -1$

36. $-4x - 3y - z = -7$
$3x + 2y + 2z = 7$
$-x - y + 2z = 2$

37. $4x - 2y - z = -5$
$x + 3y - 4z = 13$
$3x - y - 3z = 0$

38. $3x - 5y + z = 15$
$2x + 6y - 4z = 10$
$x - 5y - 3z = -5$

39. $5y + z = 11$
$7x - 2y = 1$
$5x + 2z = -3$

40. $x - 2y - z = 1$
$3x - 2y + 3z = 3$
$2x + y + 4z = 5$

Solve each system using matrices. Remember, multiplying a row by a nonzero constant will not change the solution to a system.

41. $\frac{1}{3}x + \frac{1}{5}y = 2$

$\frac{1}{3}x - \frac{1}{2}y = -\frac{1}{3}$

42. $\frac{1}{2}x + \frac{1}{3}y = 13$

$\frac{1}{5}x + \frac{1}{8}y = 5$

43. $\frac{1}{3}x - \frac{1}{4}y = 1$

$\frac{1}{3}x + \frac{1}{4}y = 3$

44. $\frac{1}{3}x - \frac{5}{6}y = 16$

$-\frac{1}{2}x + \frac{3}{4}y = -18$

The systems that follow are inconsistent systems. In both cases, the lines are parallel. Try solving each system using matrices and see what happens.

45. $2x - 3y = 4$
$4x + 6y = 4$

46. $10x - 15y = 5$
$-4x + 6y = -4$

The systems that follow are dependent systems. In each case, the lines coincide. Try solving each system using matrices and see what happens

47. $-6x + 4y = 8$
$-3x + 2y = 4$

48. $x + 2y = 5$
$-x - 2y = -5$

Getting Ready for the Next Section

Graph the following equations.

49. $y = 3x - 5$

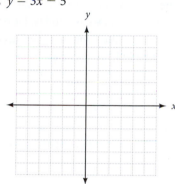

50. $y = -\dfrac{3}{4}x$

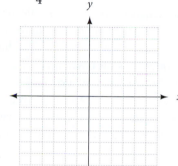

51. $x = -3$

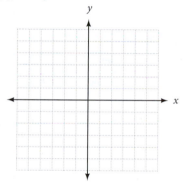

Find the Mistake

Each sentence below contains a mistake. Circle the mistake and write the correct word or phrase on the line provided.

1. The dimensions of the matrix $[5 \quad 2 \quad -1]$ is 3×1. _____

2. When transforming an augmented matrix into an equivalent system, we can interchange any two rows of a matrix, multiply any row by zero, and add to any row a constant multiple of another row. _____

3. The following matrices represent the given system, but they are mislabeled. Write the correct labels on the lines provided.

$$x + 3y - 4z = 2$$
$$2x + 4y + 5z = -6$$
$$-3x - 2y + 8z = 4$$

$$\begin{bmatrix} 1 & -3 & -4 \\ -2 & -4 & -5 \\ -3 & -2 & -8 \end{bmatrix}$$
Augmented matrix
_____

$$\begin{bmatrix} 2 \\ -6 \\ 4 \end{bmatrix}$$
Coefficient matrix
_____

$$\begin{bmatrix} 1 & 3 & -4 & | & 2 \\ 2 & 4 & 5 & | & -6 \\ -3 & -2 & 8 & | & 4 \end{bmatrix}$$
Constant matrix
_____

4. The following system of equations should represent the given matrix, but it is written incorrectly. Circle the mistakes, and write the correct system on the lines provided.

$$\begin{bmatrix} 1 & 0 & -1 & | & 4 \\ 0 & 1 & 0 & | & 8 \\ 1 & 1 & 1 & | & 5 \end{bmatrix}$$

$x + y - z = 4$ _____
$x + y = 8$ _____
$x = 5$ _____

A Find the value of a 2 × 2 determinant.

B Solve a system of linear equations in two variables using Cramer's rule.

C Find the value of a 3 × 3 determinant.

D Solve a system of linear equations in three variables using Cramer's rule.

KEY WORDS

2 × 2 determinant

two-variable system

Cramer's rule

3 × 3 determinant

expansion of minors

sign array

three-variable system

3.4 Determinants and Cramer's Rule

© iStockphoto/Marcco73

In 1714, Daniel Gabriel Fahrenheit invented the first mercury thermometer. Ten years later, he introduced the scale to measure temperature that we use today. He based this scale on the temperature of the human body, which at the time was measured to be 100°F, although has since been adjusted to 98.6°F. In 1742, Anders Celsius invented the centigrade scale based on the freezing point of water at 0°C and the boiling point at 100°C. Since there were now two prominent scales to measure temperature, the linear formula $F = \frac{9}{5}C + 32$ was derived to convert Celsius temperatures to Fahrenheit temperatures. In this section, you will learn how a linear formula can also be written as a *determinant* equation. For example, $F = \frac{9}{5}C + 32$ is equivalent to the following determinant:

$$\begin{vmatrix} C & F & 1 \\ 5 & 41 & 1 \\ -10 & 14 & 1 \end{vmatrix} = 0$$

But first, let's learn how to expand and evaluate determinants. The purpose of this section is simply to be able to find the value of a given determinant. As we will see, determinants are very useful in solving systems of linear equations. Before we apply determinants to systems of linear equations; however, we must practice calculating the value of some determinants.

A 2 × 2 Determinants

> **DEFINITION** 2 × 2 Determinant
>
> The value of the **2 × 2** (read as "2 by 2") *determinant* is given by
> $$\begin{vmatrix} a & c \\ b & d \end{vmatrix} = ad - bc.$$

From the preceding definition we see that a determinant is simply a square array of numbers with two vertical lines enclosing it. The value of a 2 × 2 determinant is found by cross-multiplying on the diagonals and then subtracting, a diagram that looks like

$$\begin{vmatrix} a & c \\ b & d \end{vmatrix} = ad - bc$$

EXAMPLE 1 Find the value of the following 2 × 2 determinants.

a. $\begin{vmatrix} 1 & 2 \\ 3 & 4 \end{vmatrix} = 1(4) - 3(2) = 4 - 6 = -2$

b. $\begin{vmatrix} 3 & -2 \\ 5 & 7 \end{vmatrix} = 3(7) - (-2)5 = 21 + 10 = 31$

Practice Problems

1. Find the value of each determinant.

a. $\begin{vmatrix} 2 & 1 \\ 4 & 3 \end{vmatrix}$

b. $\begin{vmatrix} 4 & -2 \\ 0 & 3 \end{vmatrix}$

Answers

1. **a.** 2 **b.** 12

EXAMPLE 2 Solve for x if $\begin{vmatrix} x & 2 \\ x & 4 \end{vmatrix} = 8$

Solution We expand the determinant on the left side to get

$$x(4) - x(2) = 8$$
$$4x - 2x = 8$$
$$2x = 8$$
$$x = 4$$

We can now look at how determinants can be used to solve a system of linear equations in two variables. We will use Cramer's rule to do so, but first we state it here as a theorem without proof.

B Solving Two-Variable Systems with Cramer's Rule

DEFINITION Cramer's Rule I

The solution to the system

$$a_1x + b_1y = c_1$$
$$a_2x + b_2y = c_2$$

is given by

$$x = \frac{D_x}{D}, \ y = \frac{D_y}{D}$$

where

$$D = \begin{vmatrix} a_1 & b_1 \\ a_2 & b_2 \end{vmatrix} \qquad D_x = \begin{vmatrix} c_1 & b_1 \\ c_2 & b_2 \end{vmatrix} \qquad D_y = \begin{vmatrix} a_1 & c_1 \\ a_2 & c_2 \end{vmatrix} \qquad (D \neq 0)$$

The determinant D is made up of the coefficients of x and y in the original system. The determinants D_x and D_y are found by replacing the coefficients of x or y by the constant terms in the original system. Notice also that Cramer's rule does not apply if $D = 0$. In this case the equations are dependent, or the system is inconsistent.

EXAMPLE 3 Use Cramer's rule to solve

$$2x - 3y = 4$$
$$4x + 5y = 3$$

Solution We begin by calculating the determinants D, D_x, and D_y.

$$D = \begin{vmatrix} 2 & -3 \\ 4 & 5 \end{vmatrix} = 2(5) - 4(-3) = 22$$

$$D_x = \begin{vmatrix} 4 & -3 \\ 3 & 5 \end{vmatrix} = 4(5) - 3(-3) = 29$$

$$D_y = \begin{vmatrix} 2 & 4 \\ 4 & 3 \end{vmatrix} = 2(3) - 4(4) = -10$$

$$x = \frac{D_x}{D} = \frac{29}{22} \qquad \text{and} \qquad y = \frac{D_y}{D} = \frac{-10}{22} = -\frac{5}{11}$$

The solution set for the system is $\left\{ \left(\dfrac{29}{22}, -\dfrac{5}{11} \right) \right\}$.

---

2. Solve for x if $\begin{vmatrix} -3 & x \\ 2 & x \end{vmatrix} = 20$.

3. Use Cramer's rule to solve

$$3x - 5y = 2$$
$$2x + 4y = 1$$

Answers

2. -4

3. $\left(\dfrac{13}{22}, -\dfrac{1}{22} \right)$

C 3 × 3 Determinants

We now turn our attention to 3 × 3 determinants. A 3 × 3 determinant is also a square array of numbers enclosed by a vertical line, the value of which is given by the following definition.

DEFINITION   3 × 3 Determinant

The value of the **3 × 3 determinant** is given by

$$\begin{vmatrix} a_1 & b_1 & c_1 \\ a_2 & b_2 & c_2 \\ a_3 & b_3 & c_3 \end{vmatrix} = a_1 b_2 c_3 + a_3 b_1 c_2 + a_2 b_3 c_1 - a_3 b_2 c_1 - a_1 b_3 c_2 - a_2 b_1 c_3$$

At first glance, the expansion of a 3 × 3 determinant looks a little complicated. There are actually two different methods used to find the six products in the preceding definition, which simplifies matters somewhat.

Method 1   We begin by writing the determinant with the first two columns repeated on the right.

$$\begin{vmatrix} a_1 & b_1 & c_1 \\ a_2 & b_2 & c_2 \\ a_3 & b_3 & c_3 \end{vmatrix} \begin{matrix} a_1 & b_1 \\ a_2 & b_2 \\ a_3 & b_3 \end{matrix}$$

The positive products in the definition come from multiplying down the three full diagonals:

The negative products come from multiplying up the three full diagonals.

4. Find the value of

$$\begin{vmatrix} 2 & 0 & -1 \\ 3 & 1 & 2 \\ 5 & -2 & 1 \end{vmatrix}$$

EXAMPLE 4   Find the value of

$$\begin{vmatrix} 1 & 3 & -2 \\ 2 & 0 & 1 \\ 4 & -1 & 1 \end{vmatrix}$$

Solution   Repeating the first two columns and then finding the products down the diagonals and the products up the diagonals as given in Method 1, we have

$$= 1(0)(1) + 3(1)(4) + (-2)(2)(-1) - 4(0)(-2) - (-1)(1)(1) - 1(2)(3)$$

$$= 0 + 12 + 4 - 0 + 1 - 6$$

$$= 11$$

Answer

4. 21

Method 2 The second method of evaluating a 3×3 determinant is called *expansion of minors.*

> **DEFINITION**   Expansion of Minors
>
> The *minor* for an element in a 3×3 determinant is the determinant consisting of the elements remaining when the row and column to which the element belongs are deleted. For example, in the determinant
>
> $$\begin{vmatrix} a_1 & b_1 & c_1 \\ a_2 & b_2 & c_2 \\ a_3 & b_3 & c_3 \end{vmatrix}$$
>
> Minor for element $a_1 = \begin{vmatrix} b_2 & c_2 \\ b_3 & c_3 \end{vmatrix}$
>
> Minor for element $b_2 = \begin{vmatrix} a_1 & c_1 \\ a_3 & c_3 \end{vmatrix}$
>
> Minor for element $c_3 = \begin{vmatrix} a_1 & b_1 \\ a_2 & b_2 \end{vmatrix}$

Before we can evaluate a 3×3 determinant by Method 2, we must first define what is known as the *sign array* for a 3×3 determinant.

> **DEFINITION**   Sign Array
>
> The *sign array* for a 3×3 determinant is a 3×3 array of signs in this pattern:
>
> $$\begin{vmatrix} + & - & + \\ - & + & - \\ + & - & + \end{vmatrix}$$
>
> The sign array begins with a plus sign in the upper left-hand corner. The signs then alternate between plus and minus across every row and down every column.

Note   If you have read this far and are confused, hang on. After you have done a couple of examples you will find expansion by minors to be a fairly simple process. It just takes a lot of writing to explain it.

To Evaluate a 3 × 3 Determinant by Expansion of Minors

> **HOW TO**   Evaluate a 3 × 3 Determinant by Expansion of Minors
>
> We can evaluate a 3×3 determinant by expanding across any row or down any column as follows:
>
> **Step 1:** Choose a row or column to expand.
>
> **Step 2:** Write the product of each element in the row or column chosen in Step 1 with its minor.
>
> **Step 3:** Connect the three products in Step 2 with the signs in the corresponding row or column in the sign array.

To illustrate the procedure, we will use the same determinant we used in Example 3.

EXAMPLE 5   Expand across the first row.

$$\begin{vmatrix} 1 & 3 & -2 \\ 2 & 0 & 1 \\ 4 & -1 & 1 \end{vmatrix}$$

Solution   The products of the three elements in row 1 with their minors are

$$1\begin{vmatrix} 0 & 1 \\ -1 & 1 \end{vmatrix} \qquad 3\begin{vmatrix} 2 & 1 \\ 4 & 1 \end{vmatrix} \qquad (-2)\begin{vmatrix} 2 & 0 \\ 4 & -1 \end{vmatrix}$$

5. Expand across the first row.

$$\begin{vmatrix} 2 & 0 & -1 \\ 3 & 1 & 2 \\ 5 & -2 & 1 \end{vmatrix}$$

Answer

5. 21

Connecting these three products with the signs from the first row of the sign array, we have

$$+1\begin{vmatrix} 0 & 1 \\ -1 & 1 \end{vmatrix} -3\begin{vmatrix} 2 & 1 \\ 4 & 1 \end{vmatrix} +(-2)\begin{vmatrix} 2 & 0 \\ 4 & -1 \end{vmatrix}$$

We complete the problem by evaluating each of the three 2×2 determinants and then simplifying the resulting expression.

$$+1[0 - (-1)] - 3(2 - 4) + (-2)(-2 - 0)$$
$$= 1(1) - 3(-2) + (-2)(-2)$$
$$= 1 + 6 + 4$$
$$= 11$$

The results of Examples 4 and 5 match. It makes no difference which method we use; the value of a 3×3 determinant is unique.

6. Expand down column 2.

$$\begin{vmatrix} 0 & 4 & -2 \\ 3 & 1 & 1 \\ 1 & -2 & 0 \end{vmatrix}$$

EXAMPLE 6　Expand down column 2.

$$\begin{vmatrix} 2 & 3 & -2 \\ 1 & 4 & 1 \\ 1 & 5 & -1 \end{vmatrix}$$

Solution　We connect the products of elements in column 2 and their minors with the signs from the second column in the sign array.

$$\begin{vmatrix} 2 & 3 & -2 \\ 1 & 4 & 1 \\ 1 & 5 & -1 \end{vmatrix} = -3\begin{vmatrix} 1 & 1 \\ 1 & -1 \end{vmatrix} + 4\begin{vmatrix} 2 & -2 \\ 1 & -1 \end{vmatrix} - 5\begin{vmatrix} 2 & -2 \\ 1 & 1 \end{vmatrix}$$
$$= -3(-1 - 1) + 4[-2 - (-2)] - 5[2 - (-2)]$$
$$= -3(-2) + 4(0) - 5(4)$$
$$= 6 + 0 - 20$$
$$= -14$$

D Solving Three-Variable Systems with Cramer's Rule

Cramer's rule can also be used to solve systems of linear equations in three variables.

DEFINITION　Cramer's Rule II

The solution set to the system

$$a_1 x + b_1 y + c_1 z = d_1$$
$$a_2 x + b_2 y + c_2 z = d_2$$
$$a_3 x + b_3 y + c_3 z = d_3$$

is given by

$$x = \frac{D_x}{D}, \quad y = \frac{D_y}{D}, \quad \text{and} \quad y = \frac{D_z}{D},$$

where

$$D = \begin{vmatrix} a_1 & b_1 & c_1 \\ a_2 & b_2 & c_2 \\ a_3 & b_3 & c_3 \end{vmatrix} \qquad D_x = \begin{vmatrix} d_1 & b_1 & c_1 \\ d_2 & b_2 & c_2 \\ d_3 & b_3 & c_3 \end{vmatrix} \qquad (D \neq 0)$$

$$D_y = \begin{vmatrix} a_1 & d_1 & c_1 \\ a_2 & d_2 & c_2 \\ a_3 & d_3 & c_3 \end{vmatrix} \qquad D_z = \begin{vmatrix} a_1 & b_1 & d_1 \\ a_2 & b_2 & d_2 \\ a_3 & b_3 & d_3 \end{vmatrix}$$

Answer

6. 18

Again, the determinant D consists of the coefficients of x, y, and z in the original system. The determinants D_x, D_y, and D_z are found by replacing the coefficients of x, y, and z, respectively, with the constant terms from the original system. If $D = 0$, there is no unique solution to the system.

EXAMPLE 7   Use Cramer's rule to solve.

$$x + y + z = 6$$
$$2x - y + z = 3$$
$$x + 2y - 3z = -4$$

Solution   We begin by setting up and evaluating D, D_x, D_y, and D_z. (Recall that there are a number of ways to evaluate a 3×3 determinant. Since we have four of these determinants, we can use both Methods 1 and 2 from earlier in the section.) We evaluate D using Method 1.

$$D = \begin{vmatrix} 1 & 1 & 1 \\ 2 & -1 & 1 \\ 1 & 2 & -3 \end{vmatrix} \begin{matrix} 1 & 1 \\ 2 & -1 \\ 1 & 2 \end{matrix}$$

$$= 3 + 1 + 4 - (-1) - (2) - (-6)$$
$$= 13$$

We evaluate D_x using Method 2 from this section and expanding across row 1.

$$D_x = \begin{vmatrix} 6 & 1 & 1 \\ 3 & -1 & 1 \\ -4 & 2 & -3 \end{vmatrix}$$

$$= 6 \begin{vmatrix} -1 & 1 \\ 2 & -3 \end{vmatrix} - 1 \begin{vmatrix} 3 & 1 \\ -4 & -3 \end{vmatrix} + 1 \begin{vmatrix} 3 & -1 \\ -4 & 2 \end{vmatrix}$$

$$= 6(1) - 1(-5) + 1(2)$$
$$= 13$$

Find D_y by expanding across row 2.

$$D_y = \begin{vmatrix} 1 & 6 & 1 \\ 2 & 3 & 1 \\ 1 & -4 & -3 \end{vmatrix}$$

$$= -2 \begin{vmatrix} 6 & 1 \\ -4 & -3 \end{vmatrix} + 3 \begin{vmatrix} 1 & 1 \\ 1 & -3 \end{vmatrix} - 1 \begin{vmatrix} 1 & 6 \\ 1 & -4 \end{vmatrix}$$

$$= -2(-14) + 3(-4) - 1(-10)$$
$$= 26$$

Find D_z by expanding down column 1.

$$D_z = \begin{vmatrix} 1 & 1 & 6 \\ 2 & -1 & 3 \\ 1 & 2 & -4 \end{vmatrix}$$

$$= 1 \begin{vmatrix} -1 & 3 \\ 2 & -4 \end{vmatrix} - 2 \begin{vmatrix} 1 & 6 \\ 2 & -4 \end{vmatrix} + 1 \begin{vmatrix} 1 & 6 \\ -1 & 3 \end{vmatrix}$$

$$= 1(-2) - 2(-16) + 1(9)$$
$$= 39$$

Now find x, y, and z.

$$x = \frac{D_x}{D} = \frac{13}{13} = 1 \qquad y = \frac{D_y}{D} = \frac{26}{13} = 2 \qquad z = \frac{D_z}{D} = \frac{39}{13} = 3$$

The solution set is $\{(1, 2, 3)\}$.

7. Use Cramer's rule to solve
$$x + 2y + z = -2$$
$$x + 2y - z = -6$$
$$x - 2y + z = -4$$

Note   When we are solving a system of linear equations by Cramer's rule, it is best to find the determinant D first. If $D = 0$, then there is no unique solution to the system and we may not want to go further.

Note   We are finding each of these determinants by expanding about different rows or columns just to show the different ways these determinants can be evaluated.

Answer

7. $\left(-5, \frac{1}{2}, 2\right)$

8. Use Cramer's rule to solve

$x + y = 3$

$2x - z = 3$

$y + 2z = 9$

EXAMPLE 8 Use Cramer's rule to solve.

$x + y = -1$

$2x - z = 3$

$y + 2z = -1$

Solution It is helpful to rewrite the system using zeros for the coefficients of those variables not shown.

$x + y + 0z = -1$

$2x + 0y - z = 3$

$0x + y + 2z = -1$

The four determinants used in Cramer's rule are

$$D = \begin{vmatrix} 1 & 1 & 0 \\ 2 & 0 & -1 \\ 0 & 1 & 2 \end{vmatrix} = -3$$

$$D_x = \begin{vmatrix} -1 & 1 & 0 \\ 3 & 0 & -1 \\ -1 & 1 & 2 \end{vmatrix} = -6$$

$$D_y = \begin{vmatrix} 1 & -1 & 0 \\ 2 & 3 & -1 \\ 0 & -1 & 2 \end{vmatrix} = 9$$

$$D_z = \begin{vmatrix} 1 & 1 & -1 \\ 2 & 0 & 3 \\ 0 & 1 & -1 \end{vmatrix} = -3$$

$$x = \frac{D_x}{D} = \frac{-6}{-3} = 2 \qquad y = \frac{D_y}{D} = \frac{9}{-3} = -3 \qquad z = \frac{D_z}{D} = \frac{-3}{-3} = 1$$

The solution set is $\{(2, -3, 1)\}$.

Finally, we should mention the possible situations that can occur when the determinant D is 0 and we are using Cramer's rule. If $D = 0$ and at least one of the other determinants, D_x or D_y (or D_z), is not 0, then the system is inconsistent. In this case, there is no solution to the system.

However, if $D = 0$ and both D_x and D_y (and D_z in a system of three equations in three variables) are 0, then the equations are dependent.

GETTING READY FOR CLASS

After reading through the preceding section, respond in your own words and in complete sentences.

A. Why is Method II (Expansion of Minors) a better method for finding determinants than Method I?

B. What happens if $D = 0$ while using Cramer's rule?

C. If you are solving a 3 × 3 system of equations, how many determinants will you need to find?

D. What is the advantage of using Cramer's rule over either substitution or elimination methods?

Answers

8. $(4, -1, 5)$

Vocabulary Review

Choose the correct words to fill in the blanks below.

3 × 3   2 × 2   constant   minor   sign array   coefficients

1. The value of a _____ determinant $\begin{vmatrix} a & c \\ b & d \end{vmatrix}$ is given by $\begin{vmatrix} a & c \\ b & d \end{vmatrix} = ad - bc$.

2. For a two-variable system, the determinants D_x and D_y are found using Cramer's rule and replacing the _____ of x or y by the constant terms in the original system.

3. The value of a _____ determinant is given by

$$\begin{vmatrix} a_1 & b_1 & c_1 \\ a_2 & b_2 & c_2 \\ a_3 & b_3 & c_3 \end{vmatrix} = a_1 b_2 c_3 + a_3 b_1 c_2 + a_2 b_3 c_1 - a_3 b_2 c_1 - a_1 b_3 c_2 - a_2 b_1 c_3.$$

4. The _____ for an element in a 3 × 3 determinant consists of the elements remaining when the row and column to which the element belongs are deleted.

5. A _____ for a 3 × 3 determinant begins with a plus sign in the upper left hand corner, and then the signs alternate between plus and minus across every row and down every column.

6. For a three-variable system, Cramer's rule allows us to find the determinants D_x, D_y, and D_z by replacing the coefficients of x, y, and z with the _____ terms from the original system.

Problems

A Find the value of the following 2 × 2 determinants.

1. $\begin{vmatrix} 1 & 0 \\ 2 & 3 \end{vmatrix}$

2. $\begin{vmatrix} 5 & 3 \\ 3 & 2 \end{vmatrix}$

3. $\begin{vmatrix} 1 & 2 \\ 3 & 4 \end{vmatrix}$

4. $\begin{vmatrix} 4 & 1 \\ 5 & 2 \end{vmatrix}$

5. $\begin{vmatrix} 5 & 4 \\ 3 & 2 \end{vmatrix}$

6. $\begin{vmatrix} 2 & 1 \\ 3 & 4 \end{vmatrix}$

7. $\begin{vmatrix} 0 & 1 \\ 1 & 0 \end{vmatrix}$

8. $\begin{vmatrix} 1 & 0 \\ 0 & 1 \end{vmatrix}$

9. $\begin{vmatrix} -3 & 2 \\ 6 & -4 \end{vmatrix}$

10. $\begin{vmatrix} 8 & -3 \\ -2 & -5 \end{vmatrix}$

11. $\begin{vmatrix} -3 & -1 \\ 4 & -2 \end{vmatrix}$

12. $\begin{vmatrix} 5 & 3 \\ 7 & -6 \end{vmatrix}$

Solve each of the following for x.

13. $\begin{vmatrix} 2x & 1 \\ x & 3 \end{vmatrix} = 10$

14. $\begin{vmatrix} 3x & -2 \\ 2x & 3 \end{vmatrix} = 26$

15. $\begin{vmatrix} 1 & 2x \\ 2 & -3x \end{vmatrix} = 21$

16. $\begin{vmatrix} -5 & 4x \\ 1 & -x \end{vmatrix} = 27$

17. $\begin{vmatrix} 2x & -4 \\ x & 2 \end{vmatrix} = -16$

18. $\begin{vmatrix} 3x & -2 \\ x & 4 \end{vmatrix} = 21$

SCAN TO ACCESS

19. $\begin{vmatrix} 11x & -7x \\ 3 & -2 \end{vmatrix} = 3$ **20.** $\begin{vmatrix} -3x & -5x \\ 4 & 6 \end{vmatrix} = -14$ **21.** $\begin{vmatrix} 2x & -4 \\ 2 & x \end{vmatrix} = -8x$ **22.** $\begin{vmatrix} 3x & 2 \\ 2 & x \end{vmatrix} = -11x$

23. $\begin{vmatrix} x^2 & 3 \\ x & 1 \end{vmatrix} = 10$ **24.** $\begin{vmatrix} x^2 & -2 \\ x & 1 \end{vmatrix} = 35$ **25.** $\begin{vmatrix} x^2 & -4 \\ x & 1 \end{vmatrix} = 32$ **26.** $\begin{vmatrix} x^2 & 6 \\ x & 1 \end{vmatrix} = 72$

27. $\begin{vmatrix} x & 5 \\ 1 & x \end{vmatrix} = 4$ **28.** $\begin{vmatrix} 3x & 4 \\ 2 & x \end{vmatrix} = 10x$

B Solve each of the following systems using Cramer's rule.

29. $2x - 3y = 3$
$\quad 4x - 2y = 10$

30. $3x + y = -2$
$\quad -3x + 2y = -4$

31. $5x - 2y = 4$
$\quad -10x + 4y = 1$

32. $-4x + 3y = -11$
$\quad 5x + 4y = 6$

33. $4x - 7y = 3$
$\quad 5x + 2y = -3$

34. $3x - 4y = 7$
$\quad 6x - 2y = 5$

35. $9x - 8y = 4$
$\quad 2x + 3y = 6$

36. $4x - 7y = 10$
$\quad -3x + 2y = -9$

37. $3x + 2y = 6$
$\quad 4x - 5y = 8$

38. $-4x + 3y = 12$
$\quad 6x - 7y = 14$

39. $12x - 13y = 16$
$\quad 11x + 15y = 18$

40. $-13x + 15y = 17$
$\quad 12x - 14y = 19$

C Find the value of the following 3 × 3 determinants by using Method 1 of this section.

41. $\begin{vmatrix} 1 & 2 & 0 \\ 0 & 2 & 1 \\ 1 & 1 & 1 \end{vmatrix}$

42. $\begin{vmatrix} -1 & 0 & 2 \\ 3 & 0 & 1 \\ 0 & 1 & 3 \end{vmatrix}$

43. $\begin{vmatrix} 1 & 2 & 3 \\ 3 & 2 & 1 \\ 1 & 1 & 1 \end{vmatrix}$

44. $\begin{vmatrix} -1 & 2 & 0 \\ 3 & -2 & 1 \\ 0 & 5 & 4 \end{vmatrix}$

Find the value of the following 3 × 3 determinants by using Method 2 and expanding across the first row.

45. $\begin{vmatrix} 0 & 1 & 2 \\ 1 & 0 & 1 \\ -1 & 2 & 0 \end{vmatrix}$

46. $\begin{vmatrix} 3 & -2 & 1 \\ 0 & -1 & 0 \\ 2 & 0 & 1 \end{vmatrix}$

47. $\begin{vmatrix} 3 & 0 & 2 \\ 0 & -1 & -1 \\ 4 & 0 & 0 \end{vmatrix}$

48. $\begin{vmatrix} 1 & 1 & 1 \\ 1 & -1 & 1 \\ 1 & 1 & -1 \end{vmatrix}$

Find the value of each of the following determinants by expanding across any row or down any column.

49. $\begin{vmatrix} 2 & -1 & 0 \\ 1 & 0 & -2 \\ 0 & 1 & 2 \end{vmatrix}$

50. $\begin{vmatrix} 5 & 0 & -4 \\ 0 & 1 & 3 \\ -1 & 2 & -1 \end{vmatrix}$

51. $\begin{vmatrix} 1 & 3 & 7 \\ -2 & 6 & 4 \\ 3 & 7 & -1 \end{vmatrix}$

52. $\begin{vmatrix} 2 & 1 & 5 \\ 6 & -3 & 4 \\ 8 & 9 & -2 \end{vmatrix}$

53. $\begin{vmatrix} -2 & 0 & 1 \\ 0 & 3 & 2 \\ 1 & 0 & -5 \end{vmatrix}$

54. $\begin{vmatrix} -1 & 1 & 1 \\ -2 & 2 & 2 \\ 5 & 7 & -4 \end{vmatrix}$

55. $\begin{vmatrix} -2 & 4 & -1 \\ 0 & 3 & 1 \\ -5 & -2 & 3 \end{vmatrix}$

56. $\begin{vmatrix} -3 & 2 & 4 \\ 1 & 2 & 3 \\ -1 & 1 & 5 \end{vmatrix}$

D Solve each of the following systems using Cramer's rule.

57. $x + y + z = 4$
$x - y - z = 2$
$2x + 2y - z = 2$

58. $-x + y + 3z = 6$
$x + y + 2z = 7$
$2x + 3y + z = 4$

59. $x + y - z = 2$
$-x + y + z = 3$
$x + y + z = 4$

60. $-x - y + z = 1$
$x - y + z = 3$
$x + y - z = 4$

61. $3x - y + 2z = 4$
$6x - 2y + 4z = 8$
$x - 5y + 2z = 1$

62. $2x - 3y + z = 1$
$3x - y - z = 4$
$4x - 6y + 2z = 3$

63. $2x - y + 3z = 4$
$x - 5y - 2z = 1$
$-4x - 2y + z = 3$

64. $4x - y + 5z = 1$
$2x + 3y + 4z = 5$
$x + y + 3z = 2$

65. $x + 2y - z = 4$
$2x + 3y + 2z = 5$
$x - 3y + z = 6$

66. $3x + 2y + z = 6$
$2x + 3y - 2z = 4$
$x - 2y + 3z = 8$

67. $3x - 4y + 2z = 5$
$2x - 3y + 4z = 7$
$4x + 2y - 3z = 6$

68. $5x - 3y - 4z = 3$
$4x - 5y + 3z = 5$
$3x + 4y - 5z = -4$

69. $x - 3z = 1$
$y + 2z = 8$
$x + z = 10$

70. $x - 5y = -6$
$y - 4z = -5$
$2x + 3z = -6$

71. $-x - 7y = 1$
$x + 3z = 11$
$2y + z = 0$

72. $x + y = 2$
$-x + 3z = 0$
$2y + z = 3$

73. $x - y = 12$
$3x + z = 11$
$y - 2z = -3$

74. $4x + 5y = -1$
$2y + 3z = -5$
$x + 2z = -1$

Applying the Concepts

75. Slope-Intercept Form Show that the following determinant equation is another way to write the slope-intercept form of the equation of a line.

$$\begin{vmatrix} y & x \\ m & 1 \end{vmatrix} = b$$

76. Temperature Conversions Show that the following determinant equation is another way to write the equation $F = \frac{9}{5}C + 32$.

$$\begin{vmatrix} C & F & 1 \\ 5 & 41 & 1 \\ -10 & 14 & 1 \end{vmatrix} = 0$$

77. Amusement Park Income From 1986 to 1990, the annual income of amusement parks was linearly increasing, after which time it remained fairly constant. The annual income y, in billions of dollars, may be found for one of these years by evaluating the following determinant equation, in which x represents the number of years past January 1, 1986.

$$\begin{vmatrix} x & -1.7 \\ 2 & 0.3 \end{vmatrix} = y$$

a. Write the determinant equation in slope-intercept form.

b. Use the equation from part a to find the approximate income for amusement parks in the year 1988.

78. Military Enlistment From 1981, the enlistment of women in the United States armed forces was linearly increasing until 1990, after which it declined. The approximate number of women, w, enlisted in the armed forces from 1981 to 1990 may be found by evaluating the following determinant equation, in which x represents the number of years past January 1, 1981.

$$\begin{vmatrix} 6{,}509 & -2 \\ 85{,}709 & x \end{vmatrix} = w$$

Use this equation to determine the number of women enlisted in the armed forces in 1985.

79. Per Capita Income From 1990 to 1998, the per capita income in California was linearly increasing, according to the following determinant equation

$$\begin{vmatrix} x & -3 \\ 7121 & 767.5 \end{vmatrix} = I$$

where I is income and x is the number of years since January 1, 1990 (U.S. Bureau of Economic Analysis, *Survey of Current Business*, May 1999 and unpublished data.).

a. Write the determinant equation in slope-intercept form.

b. Use the equation from part a to find the approximate personal income per capita in the year 1994.

80. Median Income For the years 1990 through 1998, the U.S. median family income I, in dollars, can be estimated by the following determinant (x is the number of years since January 1, 1990) (U.S. Census Bureau).

$$\begin{vmatrix} x & 3535.3 \\ -10 & 1264.89 \end{vmatrix} = I$$

a. Use the determinant equation to determine the median family income for 1994.

b. Use the determinant equation to predict the median family income for 1999.

81. School Enrollment Enrollment in higher education has been increasing from 1990 to 1999. The higher education enrollment E, in millions, may be found by evaluating the following determinant equation (U.S. Census Bureau). If x is the number of years since January 1, 1990, determine the higher education enrollment for 1996.

$$\begin{vmatrix} 0.1 & 6.9 \\ -2 & x \end{vmatrix} = y$$

82. Earnings The median income for women I has increased throughout the years, according to the determinant equation below (x is the number of years since January 1, 1990) (Current Population Reports, U.S. Census Bureau.). Use the equation to find the median income for women in 2001.

$$\begin{vmatrix} 457.5 & -10 \\ 1007 & x \end{vmatrix} = I$$

83. Organ Transplants The number of procedures for heart transplants H have increased during the years 1985 through 1994, according to the equation

$$H = 164.2x + 719$$

where x is the number of years since January 1, 1985 (*U.S. Department of Health and Human Services, Public Health Services, Division of Organ Transplantation and United Network of Organ Sharing*). Solving the system below for H will give the number of heart transplants in the year 1990. Solve this system for H using Cramer's rule.

$$-164.2x + H = 719$$
$$x = 5$$

84. Automobile Air Bags The percent of automobiles with air bags P since 1990 can be modeled by the following equation.

$$P = 5.6x - 3.6$$

where x is the number of years since January 1, 1990 (*National Highway Traffic Safety Administration*). Solving the system below for x will allow you to find the year in which 80.4% of all automobiles will have air bags. Use Cramer's rule to solve this system for x.

$$5.6x - P = 3.6$$
$$P = 80.4$$

85. Break-Even Point If a company has fixed costs of $100 per week and each item it produces costs $10 to manufacture, then the total cost y per week to produce x items is

$$y = 10x + 100$$

If the company sells each item it manufactures for $12, then the total amount of money y the company brings in for selling x items is

$$y = 12x$$

Use Cramer's rule to solve the system

$$y = 10x + 100$$
$$y = 12x$$

for x to find the number of items the company must sell per week to break even.

86. Break-Even Point Suppose a company has fixed costs of $200 per week and each item it produces costs $20 to manufacture.

a. Write an equation that gives the total cost per week y to manufacture x items.

b. If each item sells for $25, write an equation that gives the total amount of money y the company brings in for selling x items.

c. Use Cramer's rule to find the number of items the company must sell each week to break even.

87. Health Insurance For years between 1980 and 1991, the number (in millions) of U.S. residents without health insurance, y, may be approximated by the equation

$$y = 0.98x - 1,915.8$$

where x represents the year, and $1980 \le x \le 1991$. To determine the year in which 30 million U.S. residents were without health insurance, we solve the system of equations made up of the equation above and the equation $y = 30$. Solve this system using Cramer's rule. (When you obtain an answer, you will need to round it to the nearest year.)

88. Price Index From 1970 to 1990, the price index of dental care, d, may be closely approximated by the equation

$$d = 6x - 11,780$$

where x is the year, and $1970 \le x \le 1990$. Determine when the price index for dental care reached 120 by forming a system of equations using the equation above along with the equation $d = 120$. Solve this system using Cramer's rule.

Extending the Concepts

A 4×4 determinant can be evaluated only by using Method 2, expansion by minors; Method 1 will not work. Below is a 4×4 determinant and its associated sign array.

$$\begin{vmatrix} 2 & 0 & 1 & -3 \\ -1 & 2 & 0 & 1 \\ -3 & 0 & 1 & 0 \\ 1 & 1 & 0 & 0 \end{vmatrix} \qquad \begin{vmatrix} + & - & + & - \\ - & + & - & + \\ + & - & + & - \\ - & + & - & + \end{vmatrix}$$

4×4 determinant $\qquad\qquad$ 4×4 sign array

89. Use expansion by minors to evaluate the preceding 4×4 determinant by expanding it across row 1.

90. Evaluate the preceding determinant by expanding it down column 4.

91. Use expansion by minors down column 3 to evaluate the preceding determinant.

92. Evaluate the preceding determinant by expanding it across row 4.

Find the value of the following determinants.

93. $\begin{vmatrix} 1 & 3 & 2 & -4 \\ 0 & 4 & 1 & 0 \\ -2 & 1 & 3 & 0 \\ 2 & 3 & 4 & -1 \end{vmatrix}$

94. $\begin{vmatrix} 2 & 4 & -2 & -3 \\ 1 & 2 & 0 & 2 \\ -1 & 2 & 3 & -2 \\ 3 & 2 & 1 & -3 \end{vmatrix}$

Use Cramer's rule to solve the following systems.

95. $x + 2y - z + 3w = 4$
$2x + y + 2z - 2w = 9$
$x - 3y + z - w = 1$
$-2x + y - z + 3w = -3$

96. $3x - 2y - z - w = -10$
$2x + y + 2z - 2w = -19$
$x - 3y + 3z + w = -9$
$-2x + 3y - z + 3w = 25$

97. $ax + y + z = 1$
$x + ay + z = 1$
$x + y + az = 1$

98. $ax + y + z = a$
$x + ay + z = a$
$x + y + az = a$

Getting Ready for the Next Section

Translate into symbols.

99. Two more than 3 times a number

100. One less than twice a number

Simplify.

101. $25 - \frac{385}{9}$

102. $0.30(12)$

103. $0.08(4,000)$

104. $500(1.5)$

Apply the distributive property, then simplify.

105. $10(0.2x + 0.5y)$

106. $100(0.09x + 0.08y)$

107. $x + (3x + 2) = 26$

108. $5x = 2,500$

Solve each system.

109. $3y + z = 17$
   $5y + 20z = 65$

110. $x + y = 850$
   $1.5x + y = 1,100$

Find the Mistake

Each sentence below contains a mistake. Circle the mistake and write the correct word or phrase on the line provided.

1. The 2×2 determinant $\begin{vmatrix} 4 & 9 \\ -3 & 6 \end{vmatrix}$ is equal to $9(-3) - 4(6) = -51$. _____

2. The second step when evaluating a 3×3 determinant by expansion of minors is to connect the product with the signs in the corresponding row or column in the sign array. _____
_____

3. For a two-variable system, Cramer's rule states that there is no unique solution to the system if $D = 0$.
_____

4. When using Cramer's rule to solve the following system of equations $\begin{aligned} x + 0y + 2z &= 4 \\ 2x + 3y - z &= 7 \\ 0x + y + 3z &= 2 \end{aligned}$, the value for the determinant D_y is found by evaluating $\begin{vmatrix} 4 & 0 & 2 \\ 7 & 3 & -1 \\ 2 & 1 & 3 \end{vmatrix}$.
_____

Landmark Review: Checking Your Progress

Solve the following systems by graphing.

1. $3x + 2y = 7$
$x + 4y = 9$

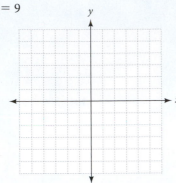

2. $3x - 4y = 16$
$x - 2y = 8$

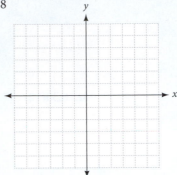

3. $x + 3y = 11$
$2x - y = 1$

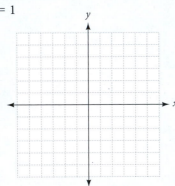

4. $y = 2x - 2$
$y = 2x + 1$

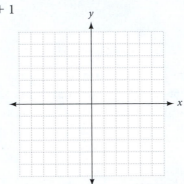

Solve the following systems.

5. $2x - 3y = -6$
$y = x + 1$

6. $x - 4y = 1$
$y = 3x - 14$

7. $x - 5y = -4$
$x = 2y - 1$

8. $x - y = 1$
$x = 4y - 5$

9. $x - y = 0$
$x + y = 2$

10. $3x - 2y = 4$
$-3x + 2y = -4$

11. $4x + 2y = 6$
$6x + 4y = 2$

12. $10x - 4y = -2$
$5x - 3y = 1$

13. $x + y + z = 2$
$x - y + 2z = 0$
$x - y + 3z = 2$

14. $x + y + z = 0$
$x - y + 2z = -5$
$x - y + 3z = -6$

15. $\frac{1}{4}x - \frac{1}{4}y - \frac{1}{4}z = \frac{9}{4}$
$-\frac{1}{2}x - \frac{1}{4}y - z = \frac{3}{4}$
$\frac{1}{2}x - y + \frac{1}{2}z = -2$

16. $8x - 4y = 0$
$-2y + 4z = 1$
$12x - 2z = 2$

3.5 Applications of Systems of Equations

OBJECTIVES

A Solve application problems whose solutions are found through systems of linear equations.

Boise's Mad Hatter Relay Race is the epitome of a "fun run." Two-person teams from across the state participate in 4-mile combined races, 2 miles for each team member. Teams can enter the women's, men's, or co-ed races. All participants must wear silly or theme hats while running, and are strongly encouraged to wear costumes as well. For the 2010 event, pre-registration for each team cost $28, with an additional $10 for race-day registration.

Let's suppose x teams pre-registered for the co-ed relay race and y teams registered on race day for a total of 29 teams. Let's also suppose that the total registration fees paid for this race were $922. Can you use this information to create a linear system and solve for its variables? Read through this section and practice the application problems. Then return to the relay race problem and see if you can solve it.

Many times word problems involve more than one unknown quantity. If a problem is stated in terms of two unknowns and we represent each unknown quantity with a different variable, then we must write the relationships between the variables with two equations. The two equations written in terms of the two variables form a system of linear equations that we solve using the methods developed in this chapter. If we find a problem that relates three unknown quantities, then we need three equations to form a linear system we can solve.

A Application Problems

Here is our Blueprint for Problem Solving, modified to fit the application problems that you will find in this section.

> **BLUEPRINT Using a System of Equations**
>
> **Step 1:** *Read* the problem, and then mentally *list* the items that are known and the items that are unknown.
>
> **Step 2:** *Assign variables* to each of the unknown items. That is, let $x =$ one of the unknown items and $y =$ the other unknown item (and $z =$ the third unknown item, if there is a third one). Then *translate* the other *information* in the problem to expressions involving the two (or three) variables.
>
> **Step 3:** *Reread* the problem, and then *write a system of equations*, using the items and variables listed in steps 1 and 2, that describes the situation.
>
> **Step 4:** *Solve the system* found in step 3.
>
> **Step 5:** *Write your answers* using complete sentences.
>
> **Step 6:** *Reread* the problem, and *check* your solution with the original words in the problem.

Note When working an application problem, it is useful to draw a sketch or make a table to represent the knowns and the unknowns. This will help you better visualize how to solve the problem.

EXAMPLE 1   One number is 2 more than 3 times another. Their sum is 26. Find the two numbers.

Solution   Applying the steps from our Blueprint, we have

> **Step 1:** *Read and list.*
> We know that we have two numbers, whose sum is 26. One of them is 2 more than 3 times the other. The unknown quantities are the two numbers.

> **Step 2:** *Assign variables and translate information.*
> Let x = one of the numbers and y = the other number.

> **Step 3:** *Write a system of equations.*
> The first sentence in the problem translates into $y = 3x + 2$. The second sentence gives us a second equation $x + y = 26$. Together, these two equations give us the following system of equations:
>
> $$x + y = 26$$
> $$y = 3x + 2$$

> **Step 4:** *Solve the system.*
> Substituting the expression for y from the second equation into the first and solving for x yields
>
> $$x + (3x + 2) = 26$$
> $$4x + 2 = 26$$
> $$4x = 24$$
> $$x = 6$$
>
> Using $x = 6$ in $y = 3x + 2$ gives the second number.
>
> $$y = 3(6) + 2$$
> $$y = 20$$

> **Step 5:** *Write answers.*
> The two numbers are 6 and 20.

> **Step 6:** *Reread and check.*
> The sum of 6 and 20 is 26, and 20 is 2 more than 3 times 6.

EXAMPLE 2   Suppose 850 tickets were sold for a game for a total of $4,725. If adult tickets cost $7.00 and children's tickets cost $3.50, how many of each kind of ticket were sold?

Solution

> **Step 1:** *Read and list.*
> The total number of tickets sold is 850. The total income from tickets is $4,725. Adult tickets are $7.00 each. Children's tickets are $3.50 each. We don't know how many of each type of ticket have been sold.

> **Step 2:** *Assign variables and translate information.*
> We let x = the number of adult tickets and y = the number of children's tickets.

> **Step 3:** *Write a system of equations.*
> The total number of tickets sold is 850, giving us our first equation.
>
> $$x + y = 850$$

Because each adult ticket costs $7.00, and each children's ticket costs $3.50, and the total amount of money paid for tickets was $4,725, a second equation is

$$7.00x + 3.50y = 4,725$$

The same information can also be obtained by summarizing the problem with a table. One such table follows. Notice that the two equations we obtained previously are given by the two rows of the table.

	Adult tickets	Children's tickets	Total
Number	x	y	850
Value	$7.00x$	$3.50y$	4,725

Whether we use a table to summarize the information in the problem or just talk our way through the problem, the system of equations that describes the situation is

$$x + y = 850$$

$$7.00x + 3.50y = 4,725$$

Step 4: *Solve the system.*

If we multiply the second equation by 10 to clear it of decimals, we have the system

$$x + y = 850$$
$$70x + 35y = 47,250$$

Multiplying the first equation by -35 and adding the result to the second equation eliminates the variable y from the system.

$$
\begin{aligned}
-35x - 35y &= -29,750 \\
70x + 35y &= 47,250 \\
\hline
35x &= 17,500 \\
x &= 500
\end{aligned}
$$

The number of adult tickets sold was 500. To find the number of children's tickets, we substitute $x = 500$ into $x + y = 850$ to get

$$500 + y = 850$$

$$y = 350$$

Step 5: *Write answers.*

The number of children's tickets is 350, and the number of adult tickets is 500.

Step 6: *Reread and check.*

The total number of tickets is $350 + 500 = 850$. The amount of money from selling the two types of tickets is

$$
\begin{aligned}
&350 \text{ children's tickets at \$3.50 each is } 350(3.50) = \$1,225 \\
&\phantom{350 \text{ children's}}500 \text{ adult tickets at \$7.00 each is } 500(7.00) = \$3,500 \\
\hline
&\phantom{350 \text{ children's tic}}\text{The total income from ticket sales is } \$4,725.
\end{aligned}
$$

3. Amy has a total balance of $10,000 on two credit cards. One card has a 6% annual interest rate, and the other has 7%. If she is charged $630 in interest for the year, what was the initial balance on the credit card?

> *Note* Remember that most problems that involve percents require you to change each percent to a decimal (or fraction) before working with them, as you see in step 3 of Example 3.

EXAMPLE 3 Suppose a person has a total balance of $10,000 on two credit cards. One card has an 8% annual interest rate and the toher has a 9% annual interest rate. If the total interest charged from both accounts in a year is $860, what was the balance of each card at the beginning of the year? (Assume this person did not make any payments or accrue any penalty fees during this year).

Solution

Step 1: *Read and list.*

The total balance is $10,000 split between two accounts. One account charges 8% in interest annually, and the other charges 9% annually. The interest from both accounts is $860 in 1 year. We don't know how much is in each account.

Step 2: *Assign variables and translate information.*

We let x equal the amount at 9% and y be the amount at 8%.

Step 3: *Write a system of equations.*

Because the total balance is $10,000, one relationship between x and y can be written as

$$x + y = 10{,}000$$

The total interest charged from both accounts is $860. The amount of interest charged on x dollars at 9% is $0.09x$, while the amount of interest charged on y dollars at 8% is $0.08y$. This relationship is represented by the equation

$$0.09x + 0.08y = 860$$

The two equations we have just written can also be found by first summarizing the information from the problem in a table. Again, the two rows of the table yield the two equations just written. Here is the table:

	Dollars at 9%	Dollars at 8%	Total
Number	x	y	10,000
Interest	$0.09x$	$0.08y$	860

The system of equations that describes this situation is given by

$$x + y = 10{,}000$$
$$0.09x + 0.08y = 10{,}860$$

Step 4: *Solve the system.*

Multiplying the second equation by 100 will clear it of decimals. The system that results after doing so is

$$x + 8y = 10{,}000$$
$$9x + 8y = 86{,}000$$

We can eliminate y from this system by multiplying the first equation by -8 and adding the result to the second equation.

$$
\begin{aligned}
-8x - 8y &= -80{,}000 \\
9x + 8y &= 86{,}000 \\
\hline
x &= 6{,}000
\end{aligned}
$$

The amount of money on the card at 9% interest annually is $6,000. Because the total balance was $10,000, the amount on the card with 8% must be $4,000.

Answers

3. $3,000 at 7%, $7,000 at 6%

Step 5: *Write answers.*

The amount on the card with 8% interest is $4,000, and the amount on the card with 9% interest is $6,000.

Step 6: *Reread and check.*

The total balance is $4,000 + $6,000 = $10,000. The amount of interest charged from the two accounts is

In 1 year, $4,000 spent at 8% is charged 0.08(4,000) = $320
In 1 year, $6,000 spent at 9% is charged 0.09(6,000) = $540

The total interest from the two accounts is $860.

EXAMPLE 4   How much 20% alcohol solution and 50% alcohol solution must be mixed to obtain 12 gallons of 30% alcohol solution?

Solution   To solve this problem, we must first understand that a 20% alcohol solution is 20% alcohol and 80% water.

Step 1: *Read and list.*

We will mix two solutions to obtain 12 gallons of solution that is 30% alcohol. One of the solutions is 20% alcohol and the other 50% alcohol. We don't know how much of each solution we need.

Step 2: *Assign variables and translate information.*

Let x = the number of gallons of 20% alcohol solution needed, and y = the number of gallons of 50% alcohol solution needed.

Step 3: *Write a system of equations.*

Because we must end up with a total of 12 gallons of solution, one equation for the system is

$$x + y = 12$$

The amount of alcohol in the x gallons of 20% solution is $0.20x$, while the amount of alcohol in the y gallons of 50% solution is $0.50y$. Because the total amount of alcohol in the 20% and 50% solutions must add up to the amount of alcohol in the 12 gallons of 30% solution, the second equation in our system can be written as

$$0.20x + 0.50y = 0.30(12)$$

Again, let's make a table that summarizes the information we have to this point in the problem.

	20% Solution	50% Solution	Final Solution
Total number of gallons	x	y	12
Gallons of alcohol	$0.20x$	$0.50y$	$0.30(12)$

Our system of equations is

$$x + y = 12$$
$$0.20x + 0.50y = 0.30(12) = 3.6$$

4. How much 30% alcohol solution and 60% alcohol solution must be mixed to get 25 gallons of 48% solution?

Answers

4. 10 gallons of 30%, 15 gallons of 60%

Step 4: *Solve the system.*

Multiplying the second equation by 10 gives us an equivalent system.

$$x + y = 12$$
$$2x + 5y = 36$$

Multiplying the top equation by -2 to eliminate the x-variable, we have

$$-2x - 2y = -24$$
$$\underline{2x + 5y = 36}$$
$$3y = 12$$
$$y = 4$$

Substituting $y = 4$ into $x + y = 12$, we solve for x.

$$x + 4 = 12$$
$$x = 8$$

Step 5: *Write answers.*

It takes 8 gallons of 20% alcohol solution and 4 gallons of 50% alcohol solution to produce 12 gallons of 30% alcohol solution.

Step 6: *Reread and check.*

If we mix 8 gallons of 20% solution and 4 gallons of 50% solution, we end up with a total of 12 gallons of solution. To check the percentages we look for the total amount of alcohol in the two initial solutions and in the final solution.

The amount of alcohol in 8 gallons of 20% solution is $0.20(8) = 1.6$ gallons
The amount of alcohol in 4 gallons of 50% solution is $0.50(4) = 2.0$ gallons

The total amount of alcohol in the initial solutions is 3.6 gallons.

The amount of alcohol in 12 gallons of 30% solution is $0.30(12) = 3.6$ gallons.

EXAMPLE 5   It takes 2 hours for a boat to travel 28 miles downstream (with the current). The same boat can travel 18 miles upstream (against the current) in 3 hours. What is the speed of the boat in still water, and what is the speed of the current of the river?

Solution

Step 1: *Read and list.*

A boat travels 18 miles upstream and 28 miles downstream. The trip upstream takes 3 hours. The trip downstream takes 2 hours. We don't know the speed of the boat or the speed of the current.

Step 2: *Assign variables and translate information.*

Let $x =$ the speed of the boat in still water and let $y =$ the speed of the current. The average speed (rate) of the boat upstream is $x - y$, because it is traveling against the current. The rate of the boat downstream is $x + y$, because the boat is traveling with the current.

Step 3: *Write a system of equations.*

Putting the information into a table, we have

	Distance d (miles)	Rate r (mph)	Time t (hours)
Upstream	18	$x - y$	3
Downstream	28	$x + y$	2

5. A boat can travel 20 miles downstream in 2 hours. The same boat can travel 18 miles upstream in 3 hours. What is the speed of the boat in still water, and what is the speed of the current?

Answers

5. Boat 8 miles per hour; current 2 miles per hour

The formula for the relationship between distance d, rate r, and time t is $d = rt$ (the rate equation). Because $d = r \cdot t$, the system we need to solve the problem is

$$18 = (x - y) \cdot 3$$
$$28 = (x + y) \cdot 2$$

which is equivalent to

$$6 = x - y$$
$$14 = x + y$$

Step 4: *Solve the system.*

Adding the two equations, we have

$$20 = 2x$$
$$x = 10$$

Substituting $x = 10$ into $14 = x + y$, we see that

$$y = 4$$

Step 5: *Write answers.*

The speed of the boat in still water is 10 miles per hour; the speed of the current is 4 miles per hour.

Step 6: *Reread and check.*

The boat travels at $10 + 4 = 14$ miles per hour downstream, so in 2 hours it will travel $14 \cdot 2 = 28$ miles. The boat travels at $10 - 4 = 6$ miles per hour upstream, so in 3 hours it will travel $6 \cdot 3 = 18$ miles.

EXAMPLE 6   A coin collection consists of 14 coins with a total value of $1.35. If the coins are nickels, dimes, and quarters, and the number of nickels is 3 less than twice the number of dimes, how many of each coin is there in the collection?

Solution   This problem will require three variables and three equations.

Step 1: *Read and list.*

We have 14 coins with a total value of $1.35. The coins are nickels, dimes, and quarters. The number of nickels is 3 less than twice the number of dimes. We do not know how many of each coin we have.

Step 2: *Assign variables and translate information.*

Because we have three types of coins, we will have to use three variables. Let's let $x =$ the number of nickels, $y =$ the number of dimes, and $z =$ the number of quarters.

Step 3: *Write a system of equations.*

Because the total number of coins is 14, our first equation is

$$x + y + z = 14$$

Because the number of nickels is 3 less than twice the number of dimes, a second equation is

$$x = 2y - 3 \qquad \text{which is equivalent to} \qquad x - 2y = -3$$

Our last equation is obtained by considering the value of each coin and the total value of the collection. Let's write the equation in terms of cents, so we won't have to clear it of decimals later.

$$5x + 10y + 25z = 135$$

Here is our system, with the equations numbered for reference:

$$
\begin{aligned}
x + y + z &= 14 & (1) \\
x - 2y &= -3 & (2) \\
5x + 10y + 25z &= 135 & (3)
\end{aligned}
$$

6. A collection of nickels, dimes, and quarters consist of 15 coins with a total value of $1.10. If the number of nickels is one less than 4 times the number of dimes, how many of each coin is contained in the collection?

Note   Writing our third equation in terms of cents is a helpful tip. Recall from the multiplication property of equality that doing so will not change the equation as long as we apply the same quantity to each term on both sides.

Step 4: *Solve the system.*

Let's begin by eliminating x from the first and second equations, and the first and third equations. Adding -1 times the second equation to the first equation gives us an equation in only y and z. We call this equation (4).

$$3y + z = 17 \qquad (4)$$

Adding -5 times equation (1) to equation (3) gives us

$$5y + 20z = 65 \qquad (5)$$

We can eliminate z from equations (4) and (5) by adding -20 times (4) to (5). Here is the result:

$$-55y = -275$$
$$y = 5$$

Substituting $y = 5$ into equation (4) gives us $z = 2$. Substituting $y = 5$ and $z = 2$ into equation (1) gives us $x = 7$.

Step 5: *Write answers.*

The collection consists of 7 nickels, 5 dimes, and 2 quarters.

Step 6: *Reread and check.*

The total number of coins is $7 + 5 + 2 = 14$. The number of nickels, 7, is 3 less than twice the number of dimes, 5. To find the total value of the collection, we have

$$\text{The value of the 7 nickels is } 7(0.05) = \$0.35$$
$$\text{The value of the 5 dimes is } 5(0.10) = \$0.50$$
$$\text{The value of the 2 quarters is } 2(0.25) = \$0.50$$

$$\text{The total value of the collection is } \$1.35$$

If you go on to take a chemistry class, you may see the next example (or one much like it).

EXAMPLE 7   In a chemistry lab, students record the temperature of water at room temperature and find that it is 77° on the Fahrenheit temperature scale and 25° on the Celsius temperature scale. The water is then heated until it boils. The temperature of the boiling water is 212°F and 100°C. Assume that the relationship between the two temperature scales is a linear one, then use the preceding data to find the formula that gives the Celsius temperature C in terms of the Fahrenheit temperature F.

Solution   The data is summarized in the following table.

Corresponding Temperatures	
In Degrees Fahrenheit	In Degrees Celsius
77	25
212	100

If we assume the relationship is linear, then the formula that relates the two temperature scales can be written in slope-intercept form as

$$C = mF + b$$

Substituting $C = 25$ and $F = 77$ into this formula gives us

$$25 = 77m + b$$

Substituting $C = 100$ and $F = 212$ into the formula yields

$$100 = 212m + b$$

7. Solve the equation from Example 7 for F and then find what temperature is the same in both Celsius and Fahrenheit.

Answer

7. -40

Together, the two equations form a system of equations, which we can solve using the elimination method.

$$25 = 77m + b \xrightarrow{\text{Multiply by } -1.} -25 = -77m - b$$

$$100 = 212m + b \xrightarrow{\text{No Change}} \frac{100 = 212m + b}{75 = 135m}$$

$$m = \frac{75}{135} = \frac{5}{9}$$

To find the value of b, we substitute $m = \frac{5}{9}$ into $25 = 77m + b$ and solve for b.

$$25 = 77\left(\frac{5}{9}\right) + b$$

$$25 = \frac{385}{9} + b$$

$$b = 25 - \frac{385}{9} = \frac{225}{9} - \frac{385}{9} = -\frac{160}{9}$$

The equation that gives C in terms of F is

$$C = \frac{5}{9}F - \frac{160}{9}$$

GETTING READY FOR CLASS

After reading through the preceding section, respond in your own words and in complete sentences.

A. To apply the Blueprint for Problem Solving to the examples in this section, what is the first step?

B. When would you write a system of equations while working a problem in this section using the Blueprint for Problem Solving?

C. When working application problems involving boats moving in rivers, how does the current of the river affect the speed of the boat?

D. Write an application problem for which the solution depends on solving the following system of equations:

$$x + y = 1,000$$
$$0.05x + 0.06y = 55$$

Number Problems

1. One number is 3 more than twice another. The sum of the numbers is 18. Find the two numbers.

2. The sum of two numbers is 32. One of the numbers is 4 less than 5 times the other. Find the two numbers.

3. The difference of two numbers is 6. Twice the smaller is 4 more than the larger. Find the two numbers.

4. The larger of two numbers is 5 more than twice the smaller. If the smaller is subtracted from the larger, the result is 12. Find the two numbers.

5. The sum of three numbers is 8. Twice the smallest is 2 less than the largest, while the sum of the largest and smallest is 5. Use a linear system in three variables to find the three numbers.

6. The sum of three numbers is 14. The largest is 4 times the smallest, while the sum of the smallest and twice the largest is 18. Use a linear system in three variables to find the three numbers.

7. One number is eight more than five times another; their sum is 26. Find the numbers.

8. One number is three less than four times another; their sum is 27. Find the numbers.

9. The difference of two positive numbers is nine. The larger number is six less than twice the smaller number. Find the numbers.

10. The difference of two positive numbers is 17. The larger number is one more than twice the smaller number. Find the numbers.

Ticket and Interest Problems

11. Linda sold 925 tickets for her choir's concert for a total of $6,000. If adult tickets sold for $8.00 and children's tickets sold for $6.00, how many of each kind of ticket did Linda sell?

12. Jarrett was selling tickets to ride the Ferris wheel at a local carnival. The tickets cost $2.00 for adults and $1.50 for children. How many of each kind of ticket did he sell if he sold a total of 300 tickets for $525?

13. Mr. Jones has $20,000 in credit on two credit cards. He charges part at 6% annual interest and the rest at 7%. If he accululates $1,280 in interest on his balances after 1 year, how much did he initially charge at each rate? (Assume he makes no payments on either card during that year, nor does he accrue any fees.)

14. Noelle recieves $17,000 in two loans. One loan charges 5% interest per year and the other 6.5%. If her total interest after 1 year is $970, how much was each loan?

15. Susan charges twice as much money on a credit card with a 7.5% annual interest rate as she does on a card with 6% interest. If her total interest after 1 year is $840, how much did she charge on each card? (Again, assume she does not make any payments or accrue any fees on her cards during the year.)

16. Katherine owes $1,350 in interest from two loan accounts in 1 year. If she has a balance that is three times as much at 7% interest as she does at 6%, how much does she have in each account?

17. William has withdrawn $2,200 from three loan accounts that charge 6%, 8%, and 9% in annual interest, respectively. He has three times as much from the account with 9% interest as he does from the account with 6%. If his total interest for the year is $178, how much was withdrwan from each account?

18. Mary has balances in three accounts that charge 5%, 7%, and 8% in annual interest. She has three times as much in the account with 8% as she does with 5%. If the total balance from all three accounts is $1,600 and her interest for the year comes to $115, how much was the loan for each account?

Mixture Problems

19. How many gallons of 20% alcohol solution and 50% alcohol solution must be mixed to get 9 gallons of 30% alcohol solution?

20. How many ounces of 30% hydrochloric acid solution and 80% hydrochloric acid solution must be mixed to get 10 ounces of 50% hydrochloric acid solution?

21. A mixture of 16% disinfectant solution is to be made from 20% and 14% disinfectant solutions. How much of each solution should be used if 15 gallons of the 16% solution are needed?

22. How much 25% antifreeze and 50% antifreeze should be combined to give 40 gallons of 30% antifreeze?

23. Paul mixes nuts worth $1.55 per pound with oats worth $1.35 per pound to get 25 pounds of trail mix worth $1.45 per pound. How many pounds of nuts and how many pounds of oats did he use?

24. A chemist has three different acid solutions. The first acid solution contains 20% acid, the second contains 40%, and the third contains 60%. He wants to use all three solutions to obtain a mixture of 60 liters containing 50% acid, using twice as much of the 60% solution as the 40% solution. How many liters of each solution should be used?

Rate Problems

25. It takes a boat 2 hours to travel 24 miles downstream and 3 hours to travel 18 miles upstream. What is the speed of the boat in still water? What is the speed of the current of the river?

26. A boat on a river travels 20 miles downstream in only 2 hours. It takes the same boat 6 hours to travel 12 miles upstream. What are the speed of the boat and the speed of the current?

27. Recall the airplane from the chapter introduction that flies from Denver International Airport to an airport 600 miles away in 2 hours with the wind and $2\frac{1}{2}$ hours against the wind. How fast is the plane and what is the speed of the wind?

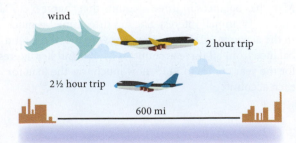

wind

2 hour trip

2½ hour trip

600 mi

28. An airplane covers a distance of 1,500 miles in 3 hours when it flies with the wind and $3\frac{1}{3}$ hours when it flies against the wind. What is the speed of the plane in still air?

wind

3 hour trip

3⅓ hour trip

1,500 mi

Coin Problems

29. Bob has 20 coins totaling $1.40. If he has only dimes and nickels, how many of each coin does he have?

30. If Amy has 15 coins totaling $2.70, and the coins are quarters and dimes, how many of each coin does she have?

31. A collection of nickels, dimes, and quarters consists of 9 coins with a total value of $1.20. If the number of dimes is equal to the number of nickels, find the number of each type of coin.

32. A coin collection consists of 12 coins with a total value of $1.20. If the collection consists only of nickels, dimes, and quarters, and the number of dimes is two more than twice the number of nickels, how many of each type of coin are in the collection?

33. Kaela has a collection of nickels, dimes, and quarters that amounts to $10.00. If there are 140 coins in all and there are twice as many dimes as there are quarters, find the number of nickels.

34. A cash register contains a total of 95 coins consisting of pennies, nickels, dimes, and quarters. There are only 5 pennies and the total value of the coins is $12.05. Also, there are 5 more quarters than dimes. How many of each coin is in the cash register?

35. John has $1.70 in nickels and dimes in his pocket. He has four more nickels than he does dimes. How many of each does he have?

36. Jamie has $2.65 in dimes and quarters in his pocket. He has two more dimes than she does quarters. How many of each does she have?

Additional Problems

37. Price and Demand A manufacturing company finds that they can sell 300 items if the price per item is $2.00, and 400 items if the price is $1.50 per item. If the relationship between the number of items sold x and the price per item p is a linear one, find a formula that gives x in terms of p. Then use the formula to find the number of items they will sell if the price per item is $3.00.

38. Price and Demand A company manufactures and sells bracelets. They have found from past experience that they can sell 300 bracelets each week if the price per bracelet is $2.00, but only 150 bracelets are sold if the price is $2.50 per bracelet. If the relationship between the number of bracelets sold x and the price per bracelet p is a linear one, find a formula that gives x in terms of p. Then use the formula to find the number of bracelets they will sell at $3.00 each.

39. The length of a rectangle is five inches more than three times the width. The perimeter is 58 inches. Find the length and width.

40. The length of a rectangle is three inches less than twice the width. The perimeter is 36 inches. Find the length and width.

41. Height of a Ball Lisa tosses a ball into the air. The height of the ball after 1, 3, and 5 seconds is as given in the following table.

t (sec)	h (ft)
1	128
3	128
5	0

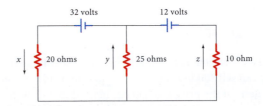

If the relationship between the height of the ball h and the time t is quadratic, then the relationship can be written as

$$h = at^2 + bt + c$$

Use the information in the table to write a system of three equations in three variables a, b, and c. Solve the system to find the exact relationship between h and t.

42. Height of a Ball A ball is tossed into the air and its height above the ground after 1, 3, and 4 seconds is recorded as shown in the following table.

t (sec)	h (ft)
1	96
3	64
4	0

The relationship between the height of the ball h and the time t is quadratic and can be written as

$$h = at^2 + bt + c$$

Use the information in the table to write a system of three equations in three variables a, b, and c. Solve the system to find the exact relationship between the variables h and t.

43. Relay Race If 29 teams entered a relay race, and pre-registration costs $28 while on-site registration costs $38, find the amount of teams preregistered and on-site registered if the total fees paid were $922.

44. Electric Current In the following diagram of an electrical circuit, x, y, and z represent the amount of current (in amperes) flowing across the 20-ohm, 25-ohm, and 10-ohm resistors, respectively. (In circuit diagrams, resistors are represented by ⌇⌇⌇ and potential differences by ⊣⊢.)

The system of equations used to find the three currents x, y, and z is

$$x - y - z = 0$$
$$20x - 25y = 32$$
$$25y - 10z = 12$$

Solve the system for all variables.

Getting Ready for the Next Section

45. Does the graph of $x + y < 4$ include the boundary line?

46. Does the graph of $-x + y \leq 3$ include the boundary line?

47. Where do the graphs of the lines $x + y = 4$ and $x - 2y = 4$ intersect?

48. Where do the graphs of the line $x = -1$ and $x - 2y = 4$ intersect?

Solve.

49. $20x + 9{,}300 > 18{,}000$

50. $20x + 4{,}800 > 18{,}000$

Find the Mistake

Each application problem below is followed by a corresponding system that contains mistakes. Circle the mistakes and write the correct equations on the lines provided.

1. Christina sold 950 tickets for a school sporting event for a total of $1,875. If adult tickets were $2.50 and student tickets $1.50, how many of each kind of ticket were sold? If $a =$ adult tickets and $s =$ student tickets, then

$$a + s = 1875$$
$$2.5a + 1.5s = 950$$

2. Cameron invested three times as much money at 8% as he does at 6%. If his total interest after a year is $900, how much did he originally invest at each rate? If $x = \$$ at 8% and $y = \$$ at 6%, then

$$y = 3x$$
$$8x + 6y = 900$$

3. Nicole invested $1,700 in three accounts. Account x pays 6% in annual interest, account y pays 7%, and account z pays 8%. Nicole has 2 times as much invested at 8% than she does at 6%. If her total interest for the year is $123, how much did she invest at each rate?

$$x + y + z = 123$$
$$2x = z$$
$$.06z + .07x + .08y = 1700$$

4. Lisa has a collection of nickels, dimes, and quarters. She has 55 coins worth a total of $5.75. If she has twice as many dimes as quarters, how many of each coin does she have?

$$n + d + q = 5.75$$
$$2q + d = 55$$
$$.05n + .10d + .25q = 5.75$$

3.6 Systems of Linear Inequalities and Applications

A Graph the solution to a system of linear inequalities in two variables.

B Solve application problems using systems of inequalities, including break-even analysis and supply-demand equilibrium equations.

KEY WORDS

system of linear inequalities

fixed cost

variable cost

revenue

profit

break even point

equilibrium point

supply

demand

©iStockphoto.com/monkeybusinessimages

Imagine you work at a local coffee house and your boss wants you to create a new house blend using the three most popular coffees: the dark roast, the medium roast, and the flavored blend. The dark roast coffee costs $20.00 per pound, the medium roast costs $14.50 per pound, and the flavored blend costs $16.00 per pound. Your boss wants the cost of the new blend to be at most $17.00 per pound and is requesting to see possible combinations to serve customers. As we have discussed, words like "at most" imply the use of inequalities. Working through this section will help us visualize solutions to systems of linear inequalities in two and three variables, such as the coffee house example.

A Graphing Solutions to Linear Inequalities

In Section 2.7, we graphed linear inequalities in two variables. To review, we graph the boundary line, using a solid line if the boundary is part of the solution set and a broken line if the boundary is not part of the solution set. Then we test any point that is not on the boundary line in the original inequality. A true statement tells us that the point lies in the solution set, a false statement tells us the solution set is the other region.

Figure 1 shows the graph of the inequality $x + y < 4$. Note that the boundary is not included in the solution set, and is therefore drawn with a broken line. Figure 2 shows the graph of $-x + y \leq 3$. Note that the boundary is drawn with a solid line, because it is part of the solution set.

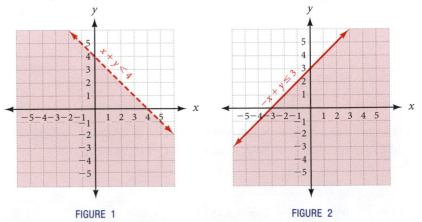

FIGURE 1   FIGURE 2

If we form a system of inequalities with the two inequalities, the solution set will be all the points common to both solution sets shown in the two figures above. It is the intersection of the two solution sets.

Therefore, the solution set for the system of inequalities from Figure 1 and 2 is all the ordered pairs that satisfy both inequalities. It is the set of points that are below the line $x + y = 4$, and also below (and including) the line $-x + y = 3$. The graph of the solution set to this system is shown in Figure 3. We have written the system in Figure 3 with the word *and* just to remind you that the solution set to a system of equations or inequalities is all the points that satisfy both equations or inequalities.

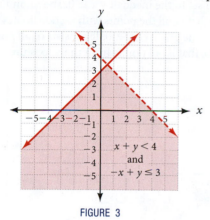

FIGURE 3

EXAMPLE 1   Graph the solution to the system of linear inequalities.

$$y < \frac{1}{2}x + 3$$

$$y \geq \frac{1}{2}x - 2$$

Solution   Figures 4 and 5 show the solution set for each of the inequalities separately.

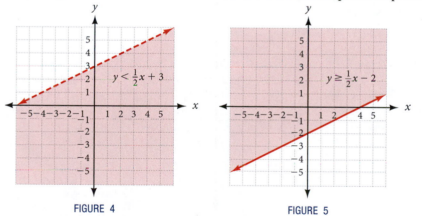

FIGURE 4   **FIGURE 5**

Figure 6 is the solution set to the system of inequalities. It is the region consisting of points whose coordinates satisfy both inequalities.

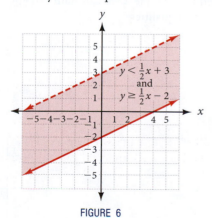

FIGURE 6

1. Graph the solution set to the system.

$$y \leq \frac{1}{3}x + 1$$

$$y > \frac{1}{3}x - 3$$

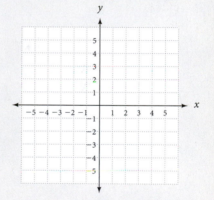

2. Graph the solution set to the system.

$$x + y < 3$$
$$x \geq 0$$
$$y \geq 0$$

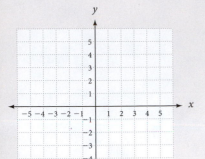

EXAMPLE 2   Graph the solution to the system of linear inequalities.

$$x + y < 4$$
$$x \geq 0$$
$$y \geq 0$$

Solution   We graphed the first inequality, $x + y < 4$, in Figure 1 at the beginning of this section. The solution set to the inequality $x \geq 0$, shown in Figure 7, is all the points to the right of the y-axis; that is, all the points with x-coordinates that are greater than or equal to 0. Figure 8 shows the graph of $y \geq 0$. It consists of all points with y-coordinates greater than or equal to 0; that is, all points from the x-axis up.

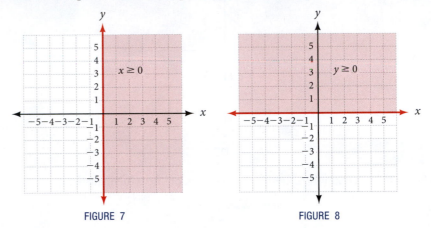

FIGURE 7   FIGURE 8

The regions shown in Figures 7 and 8 overlap in the first quadrant. Therefore, putting all three regions together we have the points in the first quadrant that are below the line $x + y = 4$. This region is shown in Figure 9, and it is the solution to our system of inequalities.

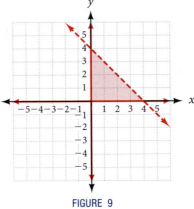

FIGURE 9

Extending the discussion in Example 2 we can name the points in each of the four quadrants using systems of inequalities.

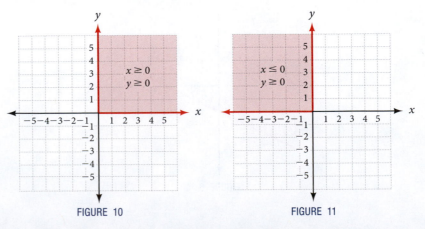

FIGURE 10   FIGURE 11

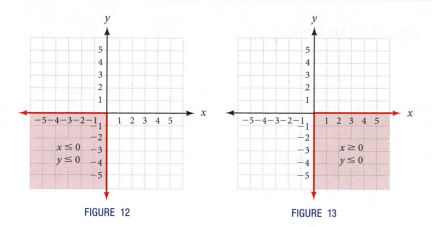

FIGURE 12   FIGURE 13

EXAMPLE 3   Graph the solution to the system of linear inequalities.

$$x \le 4$$
$$y \ge -3$$

Solution   The solution to this system will consist of all points to the left of and including the vertical line $x = 4$ that intersect with all points above and including the horizontal line $y = -3$. The solution set is shown in Figure 14.

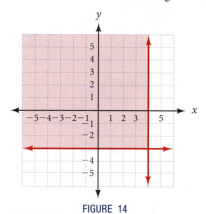

FIGURE 14

EXAMPLE 4   Graph the solution set for the following system.

$$x - 2y \le 4$$
$$x + y \le 4$$
$$x \ge -1$$

Solution   We have three linear inequalities, representing three sections of the coordinate plane. The graph of the solution set for this system will be the intersection of these three sections. The graph of $x - 2y \le 4$ is the section above and including the boundary $x - 2y = 4$. The graph of $x + y \le 4$ is the section below and including the boundary line $x + y = 4$. The graph of $x \ge -1$ is all the points to the right of, and including, the vertical line $x = -1$. The intersection of these three graphs is shown in Figure 15.

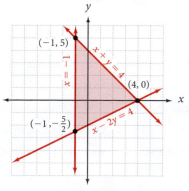

FIGURE 15

3. Graph the solution to the system
$$x > -3$$
$$y < 4$$

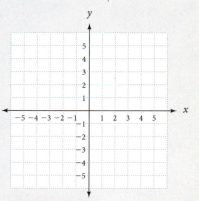

4. Graph the solution set for the following system.
$$x - y < 5$$
$$x + y < 5$$
$$x > 1$$

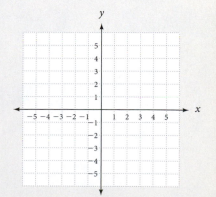

5. How would the graph for Example 5 change if they had reserved only 300 tickets at the $15 rate?

B Application Problems

EXAMPLE 5   A college basketball arena plans on charging $20 for certain seats and $15 for others. They want to bring in more than $18,000 from all ticket sales and have reserved at least 500 tickets at the $15 rate. Find a system of inequalities describing all possibilities and sketch the graph. If 620 tickets are sold for $15, at least how many tickets are sold for $20?

Solution   Let x = the number of $20 tickets and y = the number of $15 tickets. We need to write a list of inequalities that describe this situation. That list will form our system of inequalities. First of all, we note that we cannot use negative numbers for either x or y. So, we have our first inequalities:

$$x \geq 0$$
$$y \geq 0$$

Next, we note that they are selling at least 500 tickets for $15, so we can replace our second inequality with $y \geq 500$. Now our system is

$$x \geq 0$$
$$y \geq 500$$

Now the amount of money brought in by selling $20 tickets is $20x$, and the amount of money brought in by selling $15 tickets is $15y$. It the total income from ticket sales is to be more than $18,000, then $20x + 15y$ must be greater than 18,000. This gives us our last inequality and completes our system.

$$20x + 15y > 18,000$$
$$x \geq 0$$
$$y \geq 500$$

We have used all the information in the problem to arrive at this system of inequalities. The solution set contains all the values of x and y that satisfy all the conditions given in the problem. Figure 16 is the graph of the solution set.

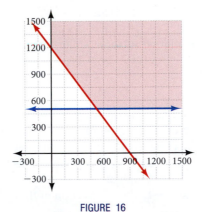

FIGURE 16

If 620 tickets are sold for $15, then we substitute 620 for y in our first inequality to obtain

$$20x + 15(620) > 18000 \qquad \text{Substitute 620 for } y.$$
$$20x + 9300 > 18000 \qquad \text{Multiply.}$$
$$20x > 8700 \qquad \text{Add } -9300 \text{ to each side.}$$
$$x > 435 \qquad \text{Divide each side by 20.}$$

If they sell 620 tickets for $15 each, then they need to sell more than 435 tickets at $20 each to bring in more than $18,000.

Answer

5. The horizontal line would move down to $y = 300$ from $y = 500$.

EXAMPLE 6 Make-A-Sale is a new company designed to assist clients selling household items over the internet. Start-up costs for the company are $40,000, and they expect variable costs to be $6.00 for each item sold. They plan to charge $14 per item.

a. Find the total cost, revenue, and profit for selling x items.

b. If they sell 3,000 items, find their cost, revenue, and profit.

c. If they sell 12,000 items, find their cost, revenue, and profit.

d. Find the break-even point (when revenue equals cost). Include a graph of the inequalities $R(x) \leq C(x)$ and $R(x) \geq C(x)$ to show intervals of profit and loss.

Solution

a. The total cost includes the fixed cost ($40,000) plus the variable cost ($6 per item). So the total cost is given by $C(x) = 40,000 + 6x$.

The revenue is the price times the number sold, so the revenue is $R(x) = 14x$.

The profit is the revenue minus the cost, so the profit is given by

$$P(x) = R(x) - C(x)$$
$$= (14x) - (40,000 + 6x)$$
$$= 8x - 40,000$$

b. Substituting $x = 3,000$ into the cost, revenue, and profit functions:

$$C(3,000) = 40,000 + 6(3,000) = \$58,000$$
$$R(3,000) = 14(3,000) = \$42,000$$
$$P(3,000) = 8(3,000) - 40,000 = -\$16,000$$

Note that $P(3,000) = R(3,000) - C(3,000)$

c. Substituting $x = 12,000$ into the cost, revenue, and profit functions:

$$C(12,000) = 40,000 + 6(12,000) = \$112,000$$
$$R(12,000) = 14(12,000) = \$168,000$$
$$P(12,000) = 8(12,000) - 40,000 = -\$56,000$$

Again, note that $P(12,000) = R(12,000) - C(12,000)$

d. Set the revenue equal to the cost to find the break-even point.

$$R(x) = C(x)$$
$$14x = 40,000 + 6x$$
$$8x = 40,000$$
$$x = 5,000$$

Figure 17 indicates areas of profit ($x > 5,000$) and areas of loss ($x < 5,000$).

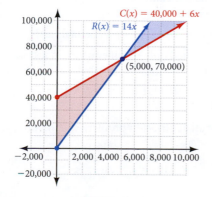

FIGURE 17

6. Suppose a company's fixed costs are $70,000 and variable costs are $8 per item. They plan to charge $15 per item.

a. Find the total cost, revenue, and profit for selling x items.

b. If they sell 7,500 items, find their cost, revenue, and profit.

c. If they sell 20,000 items, find their cost, revenue, and profit.

d. Find the break-even point and sketch the areas of profit and loss.

Answers

6 a. $C(x) = 70,000 + 8x$
$R(x) = 15x$
$P(x) = 7x - 70,000$

b. $C(7,500) = \$130,000$
$R(7,500) = \$112,500$
$P(7,500) = \$17,500$

c. $C(20,000) = \$230,000$
$R(20,000) = \$300,000$
$P(20,000) = -\$70,000$

d. 10,000 items

As the price (p) of an item increases, the supply (s) will increase since producers will want to sell at the higher price. However, the demand (d) will decrease since consumers will not want to purchase the item at a higher price. The equilibrium point represents the price and quantity that the consumer and the producer agree.

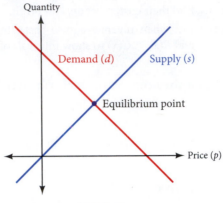

FIGURE 18

7. Find the equilibrium price and quantity for the following supply and demand equations.

$$d(p) = 7,000 - 12p$$
$$s(p) = 1,000 + 18p$$

EXAMPLE 7   Find the equilibrium price and quantity for the following supply and demand equations:

$$d(p) = 5,000 - 4p$$
$$s(p) = 1,000 + 6p$$

Solution   The equilibrium point occurs when $d(p) = s(p)$

$$5,000 - 4p = 1,000 + 6p$$
$$5,000 - 10p = 1,000$$
$$-10p = -4,000$$
$$p = 400$$

The equilibrium price is $400 per unit. Substituting this price into either the supply or demand function, we get

$$d(400) = 5,000 - 4(400) = 3,400$$
$$p(400) = 1,000 + 6(400) = 3,400$$

The equilibrium quantity is 3,400.

GETTING READY FOR CLASS

After reading through the preceding section, respond in your own words and in complete sentences.

A. What is the solution set for a system of inequalities?

B. If an ordered pair satisfied only one inequality in a system of inequalities, is the ordered pair part of the solution set for that system? Explain.

C. Explain how you would graph the solution set for a system of three linear inequalities.

D. Give an example of an application problem that involves a system of linear inequalities.

Answers

7. Price = $200;
   Quantity = 4,600 units

Vocabulary Review

Choose the correct words to fill in the blanks below.

shaded   first   second

third   fourth   common

1. The solution set to a system of inequalities will be all the points _____ to both solution sets.

2. When graphing a system of inequalities, the solution set is represented by a _____ portion of the graph.

3. For the system $x \leq 0$ and $y \geq 0$, the solution set consists of points in the _____ quadrant.

4. For the system $x \geq 0$ and $y \geq 0$, the solution set consists of points in the _____ quadrant.

5. For the system $x \geq 0$ and $y \leq 0$, the solution set consists of points in the _____ quadrant.

6. For the system $x \leq 0$ and $y \leq 0$, the solution set consists of points in the _____ quadrant.

Problems

A Graph the solution set for each system of linear inequalities.

1. $x + y < 5$
   $2x - y > 4$

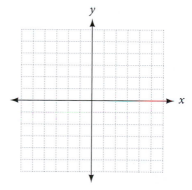

2. $x + y < 5$
   $2x - y < 4$

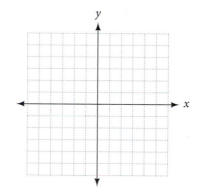

3. $y < \dfrac{1}{3}x + 4$

   $y \geq \dfrac{1}{3}x - 3$

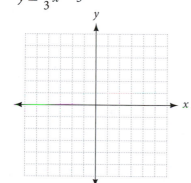

4. $y < 2x + 4$
   $y \geq 2x - 3$

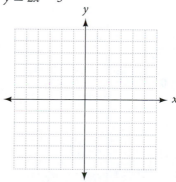

5. $x \geq -3$
   $y < -2$

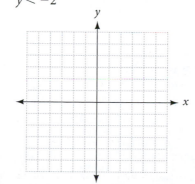

6. $x \leq 4$
  $y \geq -2$

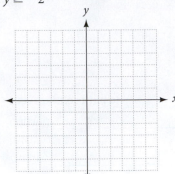

7. $1 \leq x \leq 3$
  $2 \leq y \leq 4$

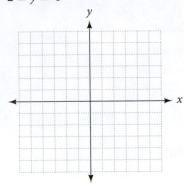

8. $-4 \leq x \leq -2$
  $1 \leq y \leq 3$

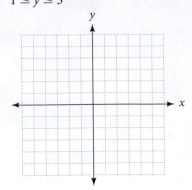

9. $x + y \leq 4$
  $x \geq 0$
  $y \geq 0$

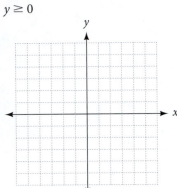

10. $x - y \leq 2$
  $x \geq 0$
  $y \leq 0$

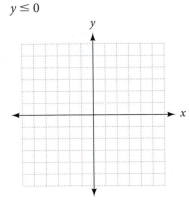

11. $x + y \leq 3$
  $x - 3y \leq 3$
  $x \geq -2$

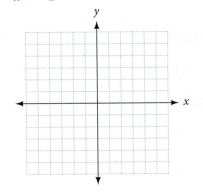

12. $x - y \leq 4$
  $x + 2y \leq 4$
  $x \geq -1$

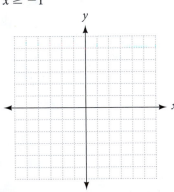

13. $x + y \leq 2$
  $-x + y \leq 2$
  $y \geq -2$

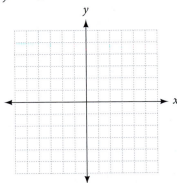

14. $x - y \leq 3$
  $-x - y \leq 3$
  $y \leq -1$

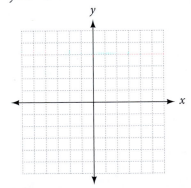

15. $x + y < 5$
  $y > x$
  $y \geq 0$

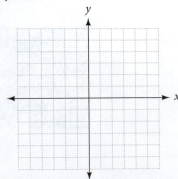

16. $x + y < 5$
  $y > x$
  $x \geq 0$

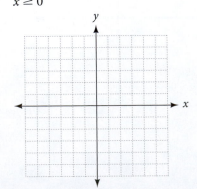

17. $2x + 3y \leq 6$
  $x \geq 0$
  $y \geq 0$

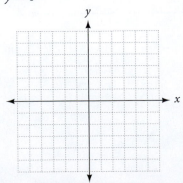

18. $x + 2y \le 10$
$3x + 2y \le 12$
$x \ge 0$
$y \ge 0$

19. $x + 4y > 12$
$2x + y > 6$
$x > 0$
$y > 0$

20. $3x + y > 6$
$2x + 3y > 12$
$x > 0$
$y > 0$

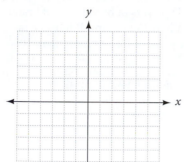

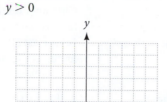

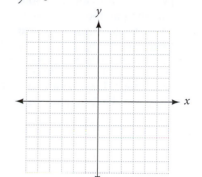

For each figure below, find a system of inequalities that describes the shaded region.

21.

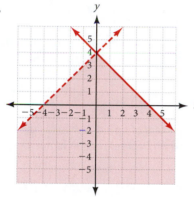

FIGURE 19

22.

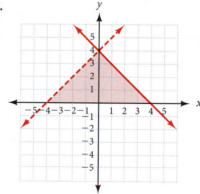

FIGURE 20

For each figure below, find a system of inequalities that describes the shaded region.

23.

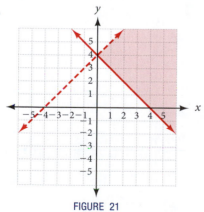

FIGURE 21

24.

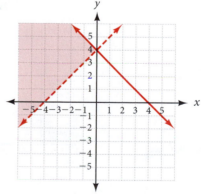

FIGURE 22

Applying the Concepts

25. Office Supplies An office worker wants to purchase some $0.55 postage stamps and also some $0.65 postage stamps totaling no more than $40. She also wants to have at least twice as many $0.55 stamps and more than 15 $0.55 stamps.

a. Find a system of inequalities describing all the possibilities and sketch the graph.

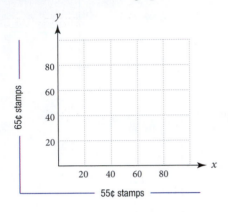

b. If she purchases 20 $0.55 stamps, what is the maximum number of $0.65 stamps she can purchase?

26. Inventory A store sells two brands of DVD players. Customer demand indicates that it is necessary to stock at least twice as many DVD players of brand A as of brand B. At least 30 of brand A and 15 of brand B must be on hand. In the store, there is room for not more than 100 DVD players.

a. Find a system of inequalities describing all possibilities, then sketch the graph.

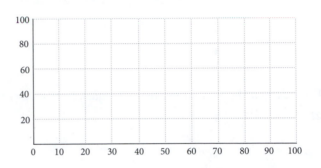

b. If there are 35 DVD players of brand A, what is the most number of brand B DVD players on hand?

Find the break-even point for the following fixed costs, variable costs, and price. Include a graph showing areas of profit and loss.

27. Fixed cost = $24,000
Variable cost = $14
Price = $22

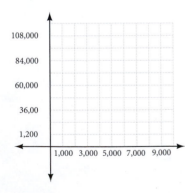

28. Fixed cost = $50,000
Variable cost = $24
Price = $44

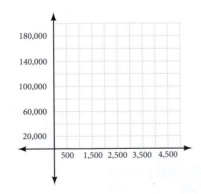

29. Fixed cost = $120,000
Variable cost = $175
Price = $225

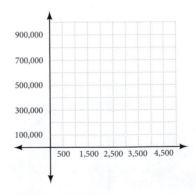

30. Fixed cost = $175,000
Variable cost = $215
Price = $315

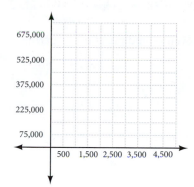

31. Fixed cost = $19,500
Variable cost = $6.50
Price = $9.50

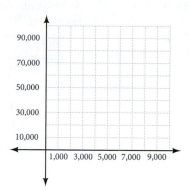

32. Fixed cost = $21,300
Variable cost = $9.75
Price = $12.25

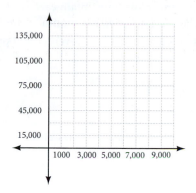

33. Fixed cost = $1,200
Variable cost = $8.25
Price = $9.75

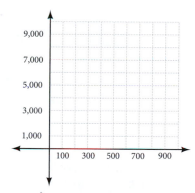

34. Fixed cost = $6,000
Variable cost = $11.80
Price = $14.30

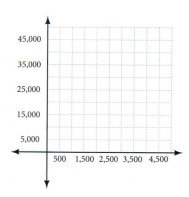

Find the equilibrium price and quantity for the following supply and demand equations.

35. $d(p) = 4,200 - 6p$
$s(p) = 4p$

36. $d(p) = 6,700 - 9p$
$s(p) = 2,200 + 6p$

37. $d(p) = 10,000 - 11p$
$s(p) = 2,000 + 5p$

38. $d(p) = 12,500 - 17p$
$s(p) = 2,500 + 3p$

39. $d(p) = 24,000 - 24p$
$s(p) = 6,000 + 6p$

40. $d(p) = 300,000 - 47p$
$s(p) = 3p$

41. Websites View-Me is a company which assists clients in creating and maintaining website addresses. Start-up costs for the company are $75,000, and they expect monthly costs to maintain each website to be $24. They plan to charge $40 per month for their service.

   a. Find the total cost, revenue, and profit for maintaining x websites.

   b. If they maintain 2,000 websites, find their cost, revenue, and profit.

   c. If they maintain 8,000 websites, find their cost, revenue, and profit.

   d. Find the break-even point. Include a graph to show intervals of profit and loss.

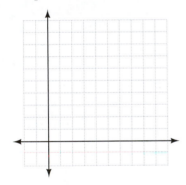

42. Tax Returns Give-It-Back is a company which prepares tax returns for clients. Start-up costs for the company are $24,000, and each return costs the company $48 to process and file. They charge $80 to prepare an individual tax return.

   a. Find the total cost, revenue, and profit for preparing x tax returns.

   b. If they prepare 500 tax returns, find their cost, revenue, and profit.

   c. Find the break-even point. Include a graph to show intervals of profit and loss.

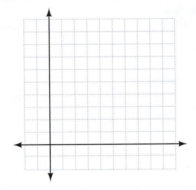

   d. If the company wants to make a net profit of $60,000, how many tax returns would they need to prepare?

Maintaining Your Skills

For each of the following straight lines, identify the x-intercept, y-intercept, and slope, and sketch the graph.

43. $2x + y = 6$

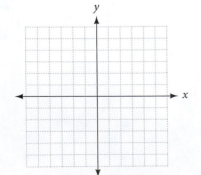

44. $y = \dfrac{3}{2}x + 4$

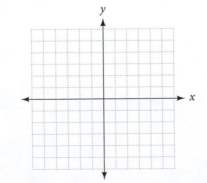

45. $x = -2$

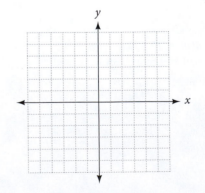

Find the equation for each line.

46. Give the equation of the line through $(-1, 3)$ that has slope $m = 2$.

47. Give the equation of the line through $(-3, 2)$ and $(4, -1)$.

48. Line l contains the point $(5, -3)$ and has a graph parallel to the graph of $2x - 5y = 10$. Find the equation for l.

49. Give the equation of the vertical line through $(4, -7)$.

State the domain and range for the following relations, and indicate which relations are also functions.

50. $\{(-2, 0), (-3, 0), (-2, 1)\}$

51. $y = x^2 - 9$

Let $f(x) = x - 2$, $g(x) = 3x + 4$ and $h(x) = 3x^2 - 2x - 8$, and find the following.

52. $f(3) + g(2)$

53. $h(0) + g(0)$

54. $f(g(2))$

55. $g(f(2))$

56. $h(f(x))$

57. $f(g(x))$

58. $\dfrac{h(x)}{f(x)}$

59. $\dfrac{g(x)}{h(x)}$

Find the Mistake

Each sentence below contains a mistake. Circle the mistake and write the correct word or phrase on the line provided.

1. The solution set for the system of inequalities $x \leq 0$ and $y \geq 0$ is all the points in the first quadrant only.

_____

2. The solution set for the system of inequalities $x \leq 0$ and $y \leq 0$ is all the points in the second quadrant.

_____

Write true or false for each statement below. Graph each system to support your answer.

3. True or False: The solution for the following system of inequalities contains the point $(-1, 2)$.

_____

$$x - y \leq 2$$
$$x + 3y \leq -2$$
$$x \geq -2$$

4. True or False: The solution for the following system of inequalities contains the point $(3, 3)$.

_____

$$3x + 2y \geq 12$$
$$3x - y \geq -6$$
$$y \geq 0$$

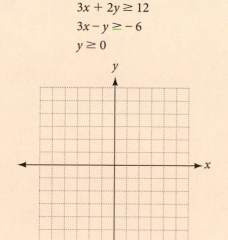

Applied Mathematics

The field of study that extends much of what we do in the classroom for use in the real world is called applied mathematics. Research the field of applied mathematics and then answer the following questions:

1. Give examples of how applied mathematics is used in the real world.

2. How does this field use concepts taught in this course?

3. What is the difference between the field of applied mathematics and a field of study involving applications of mathematics? In other words, how might a professional use applied mathematics in their work as opposed to doing applied mathematics for their work?

4. Would a statistician be considered an applied mathematician? Explain.

The numbers in brackets refer to the section(s) in which the topic can be found.

Systems of Linear Equations [3.1–3.2]

A system of linear equations consists of two or more linear equations considered simultaneously. The solution set to a linear system in two variables is the set of ordered pairs that satisfy both equations. The solution set to a linear system in three variables consists of all the ordered triples that satisfy each equation in the system.

EXAMPLES

1. The solution to the system

$$x + 2y = 4$$
$$x - y = 1$$

is the ordered pair (2, 1). It is the only ordered pair that satisfies both equations.

Strategy for Solving a System by Graphing [3.1]

Step 1: Graph the first equation.

Step 2: Graph the second equation on the same set of axes.

Step 3: Read the coordinates of the point where the graphs cross each other (the coordinates of the point of intersection).

Step 4: Check the solution to see that it satisfies *both* equations.

2. Solving the system in Example 1 by graphing looks like

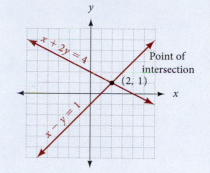

To Solve a System by the Substitution Method [3.1]

Step 1: Solve either of the equations for one of the variables (this step is not necessary if one of the equations has the correct form already).

Step 2: Substitute the results of step 1 into the other equation, and solve.

Step 3: Substitute the results of step 2 into an equation with both *x*-and *y*-variables, and solve. (The equation produced in step 1 is usually a good one to use.)

Step 4: Check your solution if necessary.

3. We can apply the substitution method to the system in Example 1 by first solving the second equation for *x* to get

$$x = y + 1$$

Substituting this expression for *x* into the first equation we have

$$(y + 1) + 2y = 4$$
$$3y + 1 = 4$$
$$3y = 3$$
$$y = 1$$

Using $y = 1$ in either of the original equations gives $x = 2$.

To Solve a System by the Elimination Method [3.1]

Step 1: Look the system over to decide which variable will be easier to eliminate.

Step 2: Use the multiplication property of equality on each equation separately, if necessary, to ensure that the coefficients of the variable to be eliminated are opposites.

Step 3: Add the left and right sides of the system produced in step 2, and solve the resulting equation.

Step 4: Substitute the solution from step 3 back into any equation with both *x*- and *y*-variables, and solve.

Step 5: Check your solution in both equations if necessary.

4. We can eliminate the *y*-variable from the system in Example 1 by multiplying both sides of the second equation by 2 and adding the result to the first equation.

$$x + 2y = 4 \xrightarrow{\text{No Change}} x + 2y = 4$$
$$x - y = 1 \xrightarrow{\text{Multiply by 2}} \underline{2x - 2y = 2}$$
$$3x = 6$$
$$x = 2$$

Substituting $x = 2$ into either of the original two equations gives $y = 1$. The solution is (2, 1).

5. If the two lines are parallel, then the system will be inconsistent and the solution is ∅. If the two lines coincide, then the equations are dependent.

Inconsistent and Dependent Equations [3.1–3.2]

A system of two linear equations that have no solutions in common is said to be an *inconsistent* system, whereas two linear equations that have all their solutions in common are said to be *dependent* equations.

Matrix Solutions [3.3]

A *matrix* is written as a rectangular array of $m \times n$ real numbers with m rows and n columns, and is used to display and solve a system of equations.

Determinants [3.4]

2×2 determinant:
$$\begin{vmatrix} a & c \\ b & d \end{vmatrix} = ad - bc$$

3×3 determinant:
$$\begin{vmatrix} a_1 & b_1 & c_1 \\ a_2 & b_2 & c_2 \\ a_3 & b_3 & c_3 \end{vmatrix} = a_1 b_2 c_3 + a_3 b_1 c_2 + a_2 b_3 c_1 - a_3 b_2 c_1 - a_1 b_3 c_2 - a_2 b_1 c_3$$

Applications [3.5]

Step 1: *Read* the problem, and then mentally *list* the items that are known and the items that are unknown.

Step 2: *Assign variables* to each of the unknown items. That is, let $x =$ one of the unknown items and $y =$ the other unknown item (and $z =$ the third unknown item, if there is a third one). Then *translate* the other *information* in the problem to expressions involving the two (or three) variables.

Step 3: *Reread* the problem, and then *write a system of equations*, using the items and variables listed in steps 1 and 2, that describes the situation.

Step 4: *Solve the system* found in step 3.

Step 5: *Write your answers* using complete sentences.

Step 6: *Reread* the problem, and *check* your solution with the original words in the problem.

Break-even Point $R(x) = C(x)$ [revenue = cost]

Equilibrium Point $s(p) = d(p)$ [supply = demand]

Systems of Linear Inequalities [3.6]

6. The solution set for the system

$$x + y < 4$$
$$-x + y \leq 3$$

is shown below.

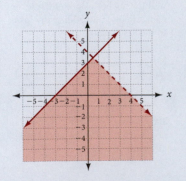

A system of linear inequalities is two or more linear inequalities considered at the same time. To find the solution set to the system, we graph each of the inequalities on the same coordinate system. The solution set is the region that is common to all the regions graphed.

Solve the following systems by the substitution method. [3.1]

1. $4x + 2y = -2$
   $-2x + y = 11$

2. $x - 5y = 10$
   $5x + 3y = -6$

Solve the following systems by the elimination method. [3.1]

3. $6x - y = 7$
   $3x + 2y = 16$

4. $\dfrac{1}{5}x + \dfrac{3}{5}y = 2$
   $\dfrac{1}{4}x + \dfrac{1}{2}y = 2$

5. Solve the system using the elimination method. [3.1]
   $-2x + 4y + 4z = 7$
   $-2x + 3y - z = 1$
   $-4x + 2y - 4z = -4$

6. Solve the system using augmented matrices. [3.3]
   $5x - 2y - 2z = 3$
   $3x - y - 4z = -7$
   $x + 3y + 2z = 1$

7. Find the determinant. [3.4]
   $\begin{vmatrix} 4 & 7 \\ -2 & -3 \end{vmatrix}$

8. Find the determinant. [3.4]
   $\begin{vmatrix} 1 & -2 & -3 \\ 2 & 1 & 3 \\ 1 & -3 & 2 \end{vmatrix}$

Solve the system using Cramer's rule. [3.4]

9. $2x + 3y = 7$
   $4x - 5y = 11$

Solve each word problem. [3.5]

10. Number Problem A number is 4 more than three times another. Their sum is 16. Find the two numbers.

11. Investing Ralph owes twice as much money at a 10% annual interest rate as he does at 12%. If his loan charges a total of $560 in 1 year, how much does he owe at each rate?

12. Ticket Cost There were 905 tickets sold for a basketball game for a total of $9,050. If adult tickets cost $12.00 and children's tickets cost $7.00, how many of each kind were sold?

13. Speed of a Boat A boat can travel 80 miles downstream in 8 hours. The same boat can travel 20 miles upstream in 5 hours. What is the speed of the boat in still water, and what is the speed of the current?

14. Coin Problem A collection of nickels, dimes, and quarters consists of 22 coins with a total value of $1.75. If the number of nickels is three times the number of dimes, how many of each coin are contained in the collection?

Graph the solution set for each system of linear inequalities. [3.6]

15. $x - 2y \geq 4$
   $2x + 3y < 6$

16. $y < -\dfrac{3}{4}x + 5$
   $x \geq 0$
   $y \geq 0$

Find a system of inequalities that describes the shaded region. [3.6]

17.

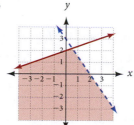

18.

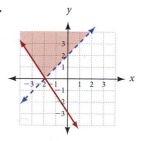

Simplify each of the following.

1. $3^2 - 7^2$

2. $(3 - 7)^2$

3. $16 - 9 \div 3 - 6 \cdot 2$

4. $3(8 - 11)^3 + 2(2 - 6)^2$

5. $6\left(\dfrac{2}{3}\right) - 18\left(\dfrac{2}{3}\right)^2$

6. $18\left(\dfrac{5}{6}\right) - 72\left(\dfrac{5}{6}\right)^2$

Find the value of each expression when x is -3.

7. $x^2 + 10x - 25$

8. $(x + 5)(x - 5)$

Simplify each of the following expressions.

9. $3(2x - 4) + 5(3x - 1)$

10. $7 - 2[4x - 5(x + 3)]$

11. $-6\left(\dfrac{2}{3}x + \dfrac{5}{2}y\right)$

12. $8\left(\dfrac{1}{4}x - \dfrac{3}{8}y\right)$

Solve.

13. $-3y + 6 = 7y - 4$

14. $-30 + 6(7x + 5) = 0$

15. $|7x + 2| + 4 = 3$

16. $\left|\dfrac{3}{2}x - 1\right| + 2 = 1$

Solve the following equations for y.

17. $y - 4 = -2(x + 1)$

18. $y + 2 = \dfrac{3}{4}(x + 8)$

Solve the following equations for x.

19. $ax + 1 = bx - 3$

20. $ax - 3 = cx - 4$

Solve each inequality and write your answers using interval notation.

21. $-4t \le 12$

22. $|3x - 1| \ge 5$

Let $A = \{0, 5, 10, 15\}$ and $B = \{1, 3, 5, 7\}$. Find the following.

23. $A \cup B$

24. $A \cap B$

Solve each of the following systems.

25. $-2x + 7y = -27$
     $-x + 3y = -11$

26. $6x + 3y = 13$
     $10x + 5y = 7$

27. $3x + y = 1$
     $y = -3x + 1$

28. $5x + 2y = -z$
     $3y - 4z = -22$
     $7x + 2z = 8$

29. Graph on a rectangular coordinate system.
$$3x + y < -2$$

30. Find the determinant: $\begin{vmatrix} -2 & 0 & 1 \\ 1 & 2 & 1 \\ 3 & -1 & 0 \end{vmatrix}$

31. Graph the solution set to the system.
$$x + y < -3$$
$$y > -5$$

32. Graph the solution set to the system.
$$y < \dfrac{1}{4}x + 3$$
$$y \ge \dfrac{1}{4}x - 2$$

33. Find the slope of the line through $\left(\dfrac{3}{2}, -\dfrac{1}{5}\right)$ and $\left(\dfrac{3}{5}, \dfrac{7}{10}\right)$.

34. Find the slope and y-intercept of $2x - 7y = 14$.

35. Give the equation of a line with slope $-\dfrac{1}{2}$ and y-intercept $= 13$.

36. Find the equation of the line that is perpendicular to $4x - 3y = 17$ and contains the point $(0, -3)$.

37. Find the equation of the line through $(5, -6)$ and $(9, 2)$.

Let $f(x) = x - 1$ and $g(x) = 3 - x^2$. Find the following.

38. $f(-3)$

39. $g(-2) + 5$

40. $(f - g)(x)$

41. $(g \circ f)(x)$

Use the graph below to work Problems 41–44.

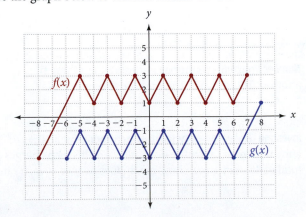

42. Find $f(2)$.

43. Find $(f + g)(-3)$.

44. Find $(g \circ f)(-8)$.

45. Find x if $g(x) = 1$.

Solve the following systems by the substitution method. [3.1]

1. $4x - 2y = 9$
$6x + y = 8$

2. $x - 6y = -50$
$-2x + 8y = 64$

Solve the following systems by the elimination method. [3.1]

3. $x - y = -1$
$14x - 2y = 70$

4. $\frac{1}{7}x + \frac{3}{7}y = 2$
$\frac{1}{5}x - \frac{1}{3}y = 0$

5. Solve the system using the elimination method. [3.2]
$x + 4y + 2z = 5$
$x - 2y - 4z = -3$
$4x - y - 2z = 2$

6. Solve the system using augmented matrices. [3.3]
$4x - y - 2z = -12$
$-3x - 6y + z = -5$
$-x + 5y + z = 13$

7. Find the determinant. [3.4]
$$\begin{vmatrix} 3 & -5 \\ -2 & 4 \end{vmatrix}$$

8. Find the determinant. [3.4]
$$\begin{vmatrix} 1 & 2 & 3 \\ 4 & 5 & 6 \\ 7 & 8 & 9 \end{vmatrix}$$

Solve the system using Cramer's rule. [3.4]

9. $4x - 2y = 5$
$3x + 5y = 11$

Solve each word problem. [3.5]

10. **Number Problem** A number is 2 more than half another. Their sum is 8. Find the two numbers.

11. **Investing** Ralph owes three times as much money on a credit card with a 17% annual interest rate as he does at 13%. If he owes a total of $768 in interest for 1 year, how much does he owe at each rate?

12. **Ticket Cost** There were 890 tickets sold for a baseball game for a total of $11,750. If adult tickets cost $15.00 and children's tickets cost $5.00, how many of each kind were sold?

13. **Speed of a Boat** A boat can travel 36 miles downstream in 4 hours. The same boat can travel 33 miles upstream in 11 hours. What is the speed of the boat in still water, and what is the speed of the current?

14. **Coin Problem** A collection of nickels, dimes, and quarters consists of 8 coins with a total value of $1.05. If the number of nickels is equal to the number of quarters and dimes together, how many of each coin are in the collection?

Graph the solution set for each system of linear inequalities. [3.6]

15. $x + 2y < 6$
$6x + 3y \geq 6$

16. $y \leq \frac{1}{4}x + 2$
$x \geq 0$
$y \geq 0$

Find a system of inequalities that describes the shaded region. [3.6]

17.

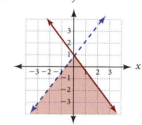

18.

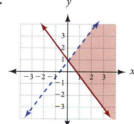

Exponents and Polynomials

Image © 2010 DigitalGlobe
Gray Buildings © 2008 Sanborn
Statue of Liberty, New York

The Statue of Liberty National Monument has stood tall and proud on Liberty Island in New York Harbor for over 120 years. The people of France gave the monument as a gift to the people of the United States. However, construction of the statue was a combined effort between the two countries. The pedestal for the monument was built in New York, and the statue itself was completed in France in 1884. Then the statue was reduced to 350 individual pieces and packed in 214 crates for transport by ship across the Atlantic Ocean to New York Harbor. Under the direction of the French, the statue was reassembled on her new pedestal and dedicated on October 28, 1886 before a crowd of thousands.

Suppose you were responsible for packing the pieces of Lady Liberty for her transatlantic trip. You would need to know the volume of the crates for which to pack the pieces, and the volume of the ship's hold to ensure there would be enough room for all of the crates. Knowing how to calculate volumes of different sized crates requires that you know how to multiply polynomials, which is one of the topics we will discuss in this chapter.

A Give the degree of a polynomial.

B Add and subtract polynomials.

C Evaluate a polynomial for a given value of its variable.

KEY WORDS

polynomial

term

monomial

numerical coefficient

degree

similar terms

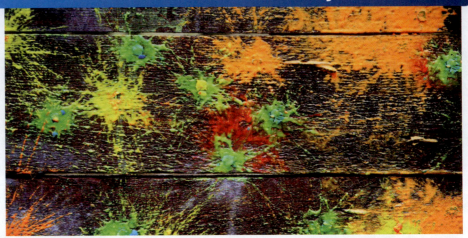

Let's assume the chart is from a company that sells paintballs. It shows the revenue and cost to sell 2000-count boxes of paintballs. From the chart, you can see that 60 boxes will bring $3,600 in revenue, with a cost of $2,100. The profit is the difference between revenue and cost, and amounts to $1,500.

Revenue and Cost for 2000-Count Boxes of Paintballs

Cost (vertical axis): $3,500 — $3,000 — $2,500 — $2,000 — $1,500 — $1,000 — $500

Number of Paintball Boxes (horizontal axis): 10, 20, 30, 40, 50, 60

Revenue values: 600, 1,200, 1,800, 2,400, 3,000, 3,600
Cost values: 350, 700, 1,050, 1,400, 1,750, 2,100

■ = Revenue ■ = Cost

FIGURE 1

The relationship between profit, revenue, and cost is one application of the polynomials we will study in this section. Let's begin with a definition that we will use to build polynomials.

Polynomials in General

> **DEFINITION** term, or monomial
>
> A *term*, or *monomial*, is a constant or the product of a constant and one or more variables raised to whole-number exponents.

The following are monomials, or terms:

$$-16 \qquad 3x^2y \qquad -\frac{2}{5}a^3b^2c \qquad xy^2z$$

The numerical part of each monomial is called the *numerical coefficient*, or just *coefficient*. For the preceding terms, the coefficients are -16, 3, $-\frac{2}{5}$, and 1. Notice that the coefficient for xy^2z is understood to be 1.

DEFINITION polynomial

A *polynomial* is any finite sum of terms. Because subtraction can be written in terms of addition, finite differences are also included in this definition.

The following are polynomials:

$$2x^2 - 6x + 3 \qquad -5x^2y + 2xy^2 \qquad 4a - 5b + 6c + 7d$$

Polynomials can be classified further according to the number of terms present. If a polynomial consists of two terms, it is said to be a *binomial*. If it has three terms, it is called a *trinomial*. And, as stated, a polynomial with only one term is said to be a *monomial*.

A Degree of a Polynomial

DEFINITION degree

The *degree* of a polynomial with one variable is the highest power to which the variable is raised in any one term.

EXAMPLE 1

a. $6x^2 + 2x - 1$ *A trinomial of degree 2*

b. $5x - 3$ *A binomial of degree 1*

c. $7x^6 - 5x^3 + 2x - 4$ *A polynomial of degree 6*

d. $-7x^4$ *A monomial of degree 4*

e. 15 *A monomial of degree 0*

Polynomials in one variable are usually written in decreasing powers of the variable. When this is the case, the coefficient of the first term is called the *leading coefficient*. In Example 1a, the leading coefficient is 6. In Example 1b, it is 5. The leading coefficient in Example 1c is 7. However, if a polynomial is not written in decreasing powers of the variable, the leading coefficient is attached to the variable with the highest degree.

DEFINITION similar, or like, terms

Two or more terms that differ only in their numerical coefficients are called *similar*, or *like*, terms. Since similar terms can differ only in their coefficients, they have identical variable parts, which includes identical powers on the variables. For example, $2x$ and $5x$ are like terms, as is x^2 and $4x^2$. However, x and x^2 are not similar.

B Addition and Subtraction of Polynomials

To add two polynomials, we simply apply the commutative and associative properties to group similar terms together, and then use the distributive property as we have in the following example:

EXAMPLE 2   Add $5x^2 - 4x + 2$ and $3x^2 + 9x - 6$.

Solution $(5x^2 - 4x + 2) + (3x^2 + 9x - 6)$

$$= (5x^2 + 3x^2) + (-4x + 9x) + (2 - 6) \qquad \text{Commutative and associative properties}$$

$$= (5 + 3)x^2 + (-4 + 9)x + (2 - 6) \qquad \text{Distributive property}$$

$$= 8x^2 + 5x + (-4)$$

$$= 8x^2 + 5x - 4$$

Practice Problems

1. Identify each expression as monomial, binomial, or trinomial, and give the degree of each.
   a. $3x + 1$
   b. $4x^2 + 2x + 5$
   c. -17
   d. $4x^5 - 7x^3$
   e. $4x^3 - 5x^2 + 2x$

Note In practice it is not necessary to show all the steps shown in Example 2. It is important to understand that addition of polynomials is equivalent to combining similar terms.

2. Add $3x^2 + 2x - 5$ and $2x^2 - 7x + 3$.

Answers

1. **a.** Binomial, 1   **b.** Trinomial, 2
   c. Monomial, 0   **d.** Binomial, 5
   e. Trinomial, 3
2. $5x^2 - 5x - 2$

3. Find the sum of $x^3 + 7x^2 + 3x + 2$ and $-3x^3 - 2x^2 + 3x - 1$.

EXAMPLE 3   Find the sum of $-8x^3 + 7x^2 - 6x + 5$ and $10x^3 + 3x^2 - 2x - 6$.

Solution   We can add the two polynomials using the method of Example 2, or we can arrange similar terms in columns and add vertically. Using the column method, we have

$$-8x^3 + 7x^2 - 6x + 5$$
$$\underline{10x^3 + 3x^2 - 2x - 6}$$
$$2x^3 + 10x^2 - 8x - 1$$

To find the difference of two polynomials, we need to use the fact that the opposite of a sum is the sum of the opposites; that is,

$$-(a + b) = -a + (-b)$$

One way to remember this is to observe that $-(a + b)$ is equivalent to $-1(a + b) = (-1)a + (-1)b = -a + (-b)$.

If a negative sign directly precedes the parentheses surrounding a polynomial, we may remove the parentheses and the preceding negative sign by changing the sign of each term within the parentheses. For example,

$$-(3x + 4) = -3x + (-4) = -3x - 4$$
$$-(5x^2 - 6x + 9) = -5x^2 + 6x - 9$$
$$-(-x^2 + 7x - 3) = x^2 - 7x + 3$$

4. Let $p(x) = (4x^2 - 2x + 7)$ and $q(x) = (7x^2 - 3x + 1)$. Find $p(x) - q(x)$.

EXAMPLE 4   Let $p(x) = (9x^2 - 3x + 5)$ and $q(x) = (4x^2 + 2x - 3)$. Find $p(x) - q(x)$.

Solution   We subtract by adding the opposite of each term in the polynomial that follows the subtraction sign.

$$p(x) - q(x) = (9x^2 - 3x + 5) - (4x^2 + 2x - 3)$$
$$= 9x^2 - 3x + 5 + (-4x^2) + (-2x) + 3 \quad \text{The opposite of a sum is the sum of the opposites.}$$
$$= (9x^2 - 4x^2) + (-3x - 2x) + (5 + 3) \quad \text{Commutative and associative property}$$
$$= 5x^2 - 5x + 8 \quad \text{Combine similar terms.}$$

5. Let $p(x) = 3x + 5$ and $q(x) = 7x - 4$. Find $q(x) - p(x)$.

EXAMPLE 5   Let $p(x) = 4x^2 - 9x + 1$ and $q(x) = -3x^2 + 5x - 2$. Find $q(x) - p(x)$.

Solution   Again, to subtract, we add the opposite.

$$q(x) - p(x) = (-3x^2 + 5x - 2) - (4x^2 - 9x + 1)$$
$$= -3x^2 + 5x - 2 - 4x^2 + 9x - 1$$
$$= (-3x^2 - 4x^2) + (5x + 9x) + (-2 - 1)$$
$$= -7x^2 + 14x - 3$$

6. Simplify $2x - 4[6 - (5x + 3)]$.

EXAMPLE 6   Simplify $4x - 3[2 - (3x + 4)]$.

Solution   Removing the innermost parentheses first, we have

$$4x - 3[2 - (3x + 4)] = 4x - 3(2 - 3x - 4)$$
$$= 4x - 3(-3x - 2)$$
$$= 4x + 9x + 6$$
$$= 13x + 6$$

Answers
3. $-2x^3 + 5x^2 + 6x + 1$
4. $-3x^2 + x + 6$
5. $4x - 9$
6. $22x - 12$

EXAMPLE 7 Simplify $(2x + 3) - [(3x + 1) - (x - 7)]$.

Solution $(2x + 3) - [(3x + 1) - (x - 7)] = (2x + 3) - (3x + 1 - x + 7)$

$$= (2x + 3) - (2x + 8)$$

$$= 2x + 3 - 2x - 8$$

$$= -5$$

7. Simplify.
$(9x - 4) - [(2x + 5) - (x + 3)]$

C Evaluating Polynomials

In the example that follows we will find the value of a polynomial for a given value of the variable.

EXAMPLE 8 Find the value of $5x^3 - 3x^2 + 4x - 5$ when x is 2.

Solution We begin by substituting 2 for x in the original polynomial.

When $\rightarrow$ $x = 2$

the polynomial $\rightarrow$ $5x^3 - 3x^2 + 4x - 5$

becomes $\rightarrow$ $5 \cdot 2^3 - 3 \cdot 2^2 + 4 \cdot 2 - 5 = 5 \cdot 8 - 3 \cdot 4 + 4 \cdot 2 - 5$

$$= 40 - 12 + 8 - 5$$

$$= 31$$

8. Find the value of
$2x^3 - 3x^2 + 4x - 8$
when x is -2.

Polynomials and Function Notation

Example 8 can be restated using function notation by calling the polynomial $P(x)$ and asking for $P(2)$. The solution would look like this:

If $P(x) = 5x^3 - 3x^2 + 4x - 5$

then $P(2) = 5 \cdot 2^3 - 3 \cdot 2^2 + 4 \cdot 2 - 5$

$$= 31$$

Three functions that occur very frequently in business and economics classes are profit $P(x)$, revenue $R(x)$, and cost functions $C(x)$. If a company manufactures and sells x items, then the revenue $R(x)$ is the total amount of money obtained by selling all x items. The cost $C(x)$ is the total amount of money it costs the company to manufacture the x items. The profit $P(x)$ obtained by selling all x items is the difference between the revenue and the cost and is given by the equation

$$P(x) = R(x) - C(x)$$

You will use this function to calculate profit in the exercise set for this section.

GETTING READY FOR CLASS

After reading through the preceding section, respond in your own words and in complete sentences.

A. Is $3x^2 + 2x - \frac{1}{x}$ a polynomial? Explain.

B. What is the degree of a polynomial?

C. What are similar terms?

D. Explain in words how you subtract one polynomial from another.

Answers

7. $8x - 6$

8. -44

Vocabulary Review

Choose the correct words to fill in the blanks below.

| monomial | binomial | trinomial |
| polynomial | degree | similar |

1. A constant or the product of a constant and one or more variables raised to whole-number exponents is called a _____ .

2. The numerical part of a monomial is called a _____ .

3. A _____ is a polynomial that contains two terms.

4. $4x^3 + 3x^2 - 2x + 6$ is an example of a _____ .

5. The trinomial $2x^3 + 4x + 5$ has a _____ of 3.

6. _____ terms have identical variable parts and differ only in their numerical coefficients.

Problems

A Identify those of the following that are monomials, binomials, or trinomials. Give the degree of each, and name the leading coefficient.

1. $5x^2 - 3x + 2$

2. $2x^2 + 4x - 1$

3. $3x - 5$

4. $5y + 3$

5. $8a^2 + 3a - 5$

6. $9a^2 - 8a - 4$

7. $4x^3 - 6x^2 + 5x - 3$

8. $9x^4 + 4x^3 - 2x^2 + x$

9. $-\dfrac{3}{4}$

10. -16

11. $4x - 5 + 6x^3$

12. $9x + 2 + 3x^3$

B Simplify each of the following by combining similar terms.

13. $(4x + 2) + (3x - 1)$

14. $(8x - 5) + (-5x + 4)$

15. $2x^2 - 3x + 10x - 15$

16. $6x^2 - 4x - 15x + 10$

17. $12a^2 + 8ab - 15ab - 10b^2$

18. $28a^2 - 8ab + 7ab - 2b^2$

19. $(5x^2 - 6x + 1) - (4x^2 + 7x - 2)$

20. $(11x^2 - 8x) - (4x^2 - 2x - 7)$

21. $\left(\frac{1}{2}x^2 - \frac{1}{3}x - \frac{1}{6}\right) - \left(\frac{1}{4}x^2 + \frac{7}{12}x\right) + \left(\frac{1}{3}x - \frac{1}{12}\right)$

22. $\left(\frac{2}{3}x^2 - \frac{1}{2}x\right) - \left(\frac{1}{4}x^2 + \frac{1}{6}x\right) + \frac{1}{12} - \left(\frac{1}{2}x^2 + \frac{1}{4}\right)$

23. $(y^3 - 2y^2 - 3y + 4) - (2y^3 - y^2 + y - 3)$

24. $(8y^3 - 3y^2 + 7y + 2) - (-4y^3 + 6y^2 - 5y - 8)$

25. $(5x^3 - 4x^2) - (3x + 4) + (5x^2 - 7) - (3x^3 + 6)$

26. $(x^3 - x) - (x^2 + x) + (x^3 - 1) - (-3x + 2)$

27. $\left(\frac{4}{7}x^2 - \frac{1}{7}xy + \frac{1}{14}y^2\right) - \left(\frac{1}{2}x^2 - \frac{2}{7}xy - \frac{9}{14}y^2\right)$

28. $\left(\frac{1}{5}x^2 - \frac{1}{2}xy + \frac{1}{10}y^2\right) - \left(-\frac{3}{10}x^2 + \frac{2}{5}xy - \frac{1}{2}y^2\right)$

29. $(3a^3 + 2a^2b + ab^2 - b^3) - (6a^3 - 4a^2b + 6ab^2 - b^3)$

30. $(a^3 - 3a^2b + 3ab^2 - b^3) - (a^3 + 3a^2b + 3ab^2 + b^3)$

31. Let $p(x) = 2x^2 - 7x$ and $q(x) = 2x^2 - 4x$. Find $p(x) - q(x)$.

32. Let $p(x) = -3x + 9$ and $q(x) = -3x + 6$. Find $p(x) - q(x)$.

33. Find the sum of $x^2 - 6xy + y^2$ and $2x^2 - 6xy - y^2$.

34. Find the sum of $9x^3 - 6x^2 + 2$ and $3x^2 - 5x + 4$.

35. Let $p(x) = -8x^5 - 4x^3 + 6$ and $q(x) = 9x^5 - 4x^3 - 6$. Find $q(x) - p(x)$.

36. Let $p(x) = 4x^4 - 3x^3 - 2x^2$ and $q(x) = 2x^4 + 3x^3 + 4x^2$. Find $q(x) - p(x)$.

37. Find the sum of $11a^2 + 3ab + 2b^2$, $9a^2 - 2ab + b^2$, and $-6a^2 - 3ab + 5b^2$.

38. Find the sum of $a^2 - ab - b^2$, $a^2 + ab - b^2$, and $a^2 + 2ab + b^2$.

Simplify each of the following. Begin by working within the innermost parentheses.

39. $-[2 - (4 - x)]$

40. $-[-3 - (x - 6)]$

41. $-5[-(x - 3) - (x + 2)]$

42. $-6[(2x - 5) - 3(8x - 2)]$

43. $4x - 5[3 - (x - 4)]$

44. $x - 7[3x - (2 - x)]$

45. $-(3x - 4y) - [(4x + 2y) - (3x + 7y)]$

46. $(8x - y) - [-(2x + y) - (-3x - 6y)]$

47. $4a - (3a + 2[a - 5(a + 1) + 4])$

48. $6a - (-2a - 6[2a + 3(a - 1) - 6])$

C Evaluate the following polynomials.

49. Find the value of $2x^2 - 3x - 4$ when x is 2.

50. Find the value of $4x^2 + 3x - 2$ when x is -1.

51. If $P(x) = \dfrac{3}{2}x^2 - \dfrac{3}{4}x + 1$, find

   a. $P(12)$

   b. $P(-8)$

52. If $P(x) = \dfrac{2}{5}x^2 - \dfrac{1}{10}x + 2$, find

   a. $P(10)$

   b. $P(-10)$

53. If $Q(x) = x^3 - x^2 + x - 1$, find

   a. $Q(4)$

   b. $Q(-2)$

54. If $Q(x) = x^3 + x^2 + x - 1$, find

   a. $Q(5)$

   b. $Q(-2)$

55. If $R(x) = 11.5x - 0.05x^2$, find

   a. $R(10)$

   b. $R(-10)$

56. If $R(x) = 11.5x - 0.01x^2$, find

   a. $R(10)$

   b. $R(-10)$

57. If $P(x) = 600 + 1{,}000x - 100x^2$, find

   a. $P(-4)$

   b. $P(4)$

58. If $P(x) = 500 + 800x - 100x^2$, find

   a. $P(-6)$

   b. $P(8)$

Applying the Concepts

59. Caffeine The chart shows the number of milligrams of caffeine in different sodas. If Mary drinks three Dr. Peppers, a Mountain Dew, and two Barq's Root Beers, write an expression that describes the amount of caffeine she drinks, and solve it.

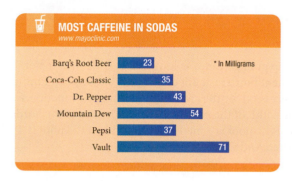

60. Sports Fields The chart shows the minimum acreage needed for different sports. A school has twice as many basketball courts as it does baseball fields and half as many soccer fields as it does baseball fields. Write an expression that describes the minimum acreage of sports fields, then solve the expression if the school has two baseball fields.

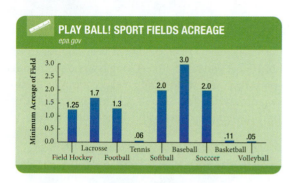

Problems 61–66 may be solved using a graphing calculator.

61. Height of an Object If an object is thrown straight up into the air with a velocity of 128 feet/second, then its height $h(t)$ above the ground t seconds later is given by the formula

$$h(t) = -16t^2 + 128t$$

Find the height after 3 seconds and after 5 seconds. [Find $h(3)$ and $h(5)$.]

62. Height of an Object The formula for the height of an object that has been thrown straight up with a velocity of 64 feet/second is

$$h(t) = -16t^2 + 64t$$

Find the height after 1 second and after 3 seconds. [Find $h(1)$ and $h(3)$.]

63. Profits The total cost (in dollars) for a company to manufacture and sell x items per week is $C(x) = 60x + 300$. If the revenue brought in by selling all x items is $R(x) = 100x - 0.5x^2$, find the weekly profit. How much profit will be made by producing and selling 60 items each week?

64. Profits The total cost (in dollars) for a company to produce and sell x items per week is $C(x) = 200x + 1,600$. If the revenue brought in by selling all x items is $R(x) = 300x - 0.6x^2$, find the weekly profit. How much profit will be made by producing and selling 50 items each week?

65. Profits Suppose it costs a company selling sewing patterns $C(x) = 800 + 6.5x$ dollars to produce and sell x patterns a month. If the revenue obtained by selling x patterns is $R(x) = 10x - 0.002x^2$, what is the profit equation? How much profit will be made if 1,000 patterns are produced and sold in May?

66. Profits Suppose a company manufactures and sells x picture frames each month with a total cost of $C(x) = 1,200 + 3.5x$ dollars. If the revenue obtained by selling x frames is $R(x) = 9x - 0.003x^2$, find the profit equation. How much profit will be made if 1,000 frames are manufactured and sold in June?

67. A pool 10 feet wide and 40 feet long is to be surrounded by a pathway x feet wide. Find an expression for the area of the pathway as a function of x. What area will the pathway have if it is 5 feet wide? What about 8 feet wide?

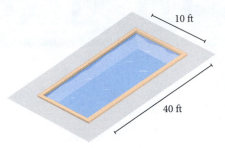

10 ft

40 ft

68. A lap pool 10 times as long as it is wide. A 2-meter cement border is built around the entire pool. Find an expression for the area of the border as a function w, the width of the pool. What will the area of the border be if the pool is 3 meters wide? What if the pool is 4 meters wide?

69. A rectangular box is built by cutting out square corners from a 9" by 12" piece of cardboard, then folding the resulting flaps up to form the height. Let x represent the sides of the square corners being cut out. Express the volume of the box as a function of x. What will the volume be if 2" squares are cut out?

x

12 in

9 in

70. A rectangular box is built by cutting out square corners from a 9 cm by 12 cm piece of cardboard, then folding the resulting flaps up to form the height. Let x represent the sides of the square corners being cut out. Express the volume of the box as a function of x. What will the volume be if 3 cm squares are cut out? What is the domain of the volume function you found? (Hint: What is the largest square that can be cut out from the sides?)

Getting Ready for the Next Section

Simplify.

71. $2x^2 - 3x + 10x - 15$

72. $12a^2 + 8ab - 15ab - 10b^2$

73. $(6x^3 - 2x^2y + 8xy^2) + (-9x^2y + 3xy^2 - 12y^3)$

74. $(3x^3 - 15x^2 + 18x) + (2x^2 - 10x + 12)$

75. $4x^3(-3x)$

76. $5x^2(-4x)$

77. $4x^3(5x^2)$

78. $5x^2(3x^2)$

79. $(a^3)^2$

80. $(a^4)^2$

81. $11.5(130) - 0.05(130)^2$

82. $-0.05(130)^2 + 9.5(130) - 200$

Maintaining Your Skills

Simplify each expression.

83. $-1(5 - x)$

84. $-1(a - b)$

85. $-1(7 - x)$

86. $-1(6 - y)$

87. $5\left(x - \dfrac{1}{5}\right)$

88. $7\left(x + \dfrac{1}{7}\right)$

89. $x\left(1 - \dfrac{1}{x}\right)$

90. $a\left(1 + \dfrac{1}{a}\right)$

91. $12\left(\dfrac{1}{4}x + \dfrac{2}{3}y\right)$

92. $20\left(\dfrac{2}{5}x + \dfrac{1}{4}y\right)$

93. $x^2\left(1 - \dfrac{4}{x^2}\right)$

94. $4a^2\left(\dfrac{1}{a^2} - \dfrac{3}{4a}\right)$

One Step Further

95. The graphs of two polynomial functions are given in Figures 2 and 3. Use the graphs to find the following.

a. $f(-3)$ **b.** $f(0)$ **c.** $f(1)$ **d.** $g(-1)$

e. $g(0)$ **f.** $g(2)$ **g.** $f(g(2))$ **h.** $g(f(2))$

i. $(f(2))^2$ **j.** $f(f(2))$ **k.** $f(f(1))$ **l.** $g(g(1))$

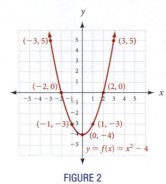

FIGURE 2

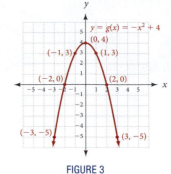

FIGURE 3

m. For which values of x is $f(x) = 0$?

n. For which values of x is $g(x) = -5$

o. For which values of x is $f(x) = g(x)$?

Find the Mistake

Each sentence below contains a mistake. Circle the mistake and write the correct word, phrase, or number on the line provided.

1. The numerical coefficient for the monomial $4x^2y^3z^5$ is 5. _____

2. The degree of the polynomial $\frac{1}{2}x^6 + 2x^2 - 4$ is $\frac{1}{2}$. _____

3. Similar terms of a polynomial have the same coefficients but different variable parts.

_____

4. The sum of the polynomials $2x^2 + 4x - 3$ and $3x + 5$ is $9x^2 + 2$. _____

Navigation Skills: Prepare, Study, Achieve

When taking this course, expect to encounter problems that you find difficult. Also expect to make mistakes. Mistakes highlight possible difficulties you are having and help you learn how to overcome them. We suggest making a list of problems you find difficult. As the course progresses, add new problems to the list, rework the problems on your list, and use the list to study for exams. Be aware of the mistakes you make and what you need to do to ensure you will not make that same mistake twice.

4.2 Multiplication of Polynomials

Image © Katherine Heistand Shields, 2010

In the previous section, we found the relationship between profit, revenue, and cost to be

$$P(x) = R(x) - C(x)$$

Revenue itself can be broken down further by another formula common in the business world. The revenue obtained from selling all x items is the product of the number of items sold and the price per item; that is,

Revenue = (Number of items sold)(Price of each item)

$$R = xp$$

Many times, x and p are polynomials, which means that the expression xp is the product of two polynomials. In this section, we learn how to multiply polynomials, and in so doing, increase our understanding of the equations and formulas that describe business applications.

A Multiplying Polynomials

EXAMPLE 1 Find the product of $4x^3$ and $5x^2 - 3x + 1$.

Solution To multiply, we apply the distributive property.

$4x^3(5x^2 - 3x + 1)$

$\quad = 4x^3(5x^2) + 4x^3(-3x) + 4x^3(1)$ Distributive property

$\quad = 20x^5 - 12x^4 + 4x^3$

Notice that we multiply coefficients and add exponents.

EXAMPLE 2 Multiply $2x - 3$ and $x + 5$.

Solution Distributing the $2x - 3$ across the sum $x + 5$ gives us

$(2x - 3)(x + 5)$

$\quad = (2x - 3)x + (2x - 3)5$ Distributive property

$\quad = 2x(x) + (-3)x + 2x(5) + (-3)5$ Distributive property

$\quad = 2x^2 - 3x + 10x - 15$

$\quad = 2x^2 + 7x - 15$ Combine like terms.

Notice the third line in this example. It consists of all possible products of terms in the first binomial and those of the second binomial. We can generalize this into a rule for multiplying two polynomials.

Practice Problems

1. Multiply $5x^2(3x^2 - 4x + 2)$.

2. Multiply $4x - 1$ and $x + 3$.

Answers

1. $15x^4 - 20x^3 + 10x^2$
2. $4x^2 + 11x - 3$

RULE   Multiplication of Polynomials

To multiply two polynomials, multiply each term in the first polynomial by each term in the second polynomial.

Multiplying polynomials can be accomplished by a method that looks very similar to long multiplication with whole numbers.

3. Multiply using the vertical method.

$$(3x + 2)(x^2 - 5x + 6)$$

Note   The vertical method of multiplying polynomials does not directly show the use of the distributive property. It is, however, very useful since it always gives the correct result and is easy to remember.

EXAMPLE 3   Multiply $(2x - 3y)$ and $(3x^2 - xy + 4y^2)$ vertically.

Solution $3x^2 - \quad xy + \quad 4y^2$

$$\underline{\qquad\qquad 2x - \quad 3y}$$

$\underline{6x^3 - \quad 2x^2y + \quad 8xy^2}$ Multiply $(3x^2 - xy + 4y^2)$ by $2x$.

$\underline{\qquad - \quad 9x^2y + \quad 3xy^2 - 12y^3}$ Multiply $(3x^2 - xy + 4y^2)$ by $-3y$.

$6x^3 - 11x^2y + 11xy^2 - 12y^3$ Add similar terms.

B Multiplying Binomials—The FOIL Method

Consider the product of $(2x - 5)$ and $(3x - 2)$. Distributing $(3x - 2)$ over $2x$ and -5, we have

$$(2x - 5)(3x - 2) = (2x)(3x - 2) + (-5)(3x - 2)$$
$$= (2x)(3x) + (2x)(-2) + (-5)(3x) + (-5)(-2)$$
$$= 6x^2 - 4x - 15x + 10$$
$$= 6x^2 - 19x + 10$$

Looking closely at the second and third lines, we notice the following:

Note   The FOIL method does not show the properties used in multiplying two binomials. It is simply a way of finding products of binomials quickly. Remember, the FOIL method applies only to products of two binomials. The vertical method applies to all products of polynomials with two or more terms.

1. $6x^2$ comes from multiplying the *first* terms in each binomial.

$(2x - 5)(3x - 2)$ $2x(3x) = 6x^2$ *First terms*

2. $-4x$ comes from multiplying the *outside* terms in the product.

$(2x - 5)(3x - 2)$ $2x(-2) = -4x$ *Outside terms*

3. $-15x$ comes from multiplying the *inside* terms in the product.

$(2x - 5)(3x - 2)$ $-5(3x) = -15x$ *Inside terms*

4. 10 comes from multiplying the *last* two terms in the product.

$(2x - 5)(3x - 2)$ $-5(-2) = 10$ *Last terms*

Once we know where the terms in the answer come from, we can reduce the number of steps used in finding the product.

$$(2x - 5)(3x - 2) = 6x^2 - 4x - 15x + 10 = 6x^2 - 19x + 10$$
$$\qquad\qquad\qquad\quad \uparrow \qquad \uparrow \qquad \uparrow \qquad \uparrow$$
$$\qquad\qquad\qquad \text{First Outside Inside Last}$$

4. Multiply using the FOIL method.

$$(2a - 3b)(5a - b)$$

EXAMPLE 4   Multiply $(4a - 5b)(3a + 2b)$ using the FOIL method.

Solution $(4a - 5b)(3a + 2b) = 12a^2 + 8ab - 15ab - 10b^2$

$$\qquad\qquad\qquad\qquad\qquad \text{F} \qquad \text{O} \qquad \text{I} \qquad \text{L}$$

$$= 12a^2 - 7ab - 10b^2$$

Answers

3. $3x^3 - 13x^2 + 8x + 12$

4. $10a^2 - 17ab + 3b^2$

EXAMPLE 5   Multiply $(3 - 2t)(4 + 7t)$ using the FOIL method.

Solution $(3 - 2t)(4 + 7t) = 12 + 21t - 8t - 14t^2$

F   O   I   L

$$= 12 + 13t - 14t^2$$

EXAMPLE 6   Let $p(x) = \left(2x + \frac{1}{2}\right)$ and $q(x) = \left(4x - \frac{1}{2}\right)$. Find $p(x) \cdot q(x)$.

Solution $p(x) \cdot q(x) = \left(2x + \frac{1}{2}\right)\left(4x - \frac{1}{2}\right) = 8x^2 - x + 2x - \frac{1}{4} = 8x^2 + x - \frac{1}{4}$

F   O   I   L

EXAMPLE 7   Multiply $(a^5 + 3)(a^5 - 7)$ using the FOIL method.

Solution $(a^5 + 3)(a^5 - 7) = a^{10} - 7a^5 + 3a^5 - 21$

F   O   I   L

$$= a^{10} - 4a^5 - 21$$

EXAMPLE 8   Multiply $(2x + 3)(5y - 4)$ using the FOIL method.

Solution $(2x + 3)(5y - 4) = 10xy - 8x + 15y - 12$

F   O   I   L

C The Square of a Binomial

Consider Figure 1 as a representation of $(a + b)^2$.

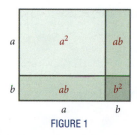

FIGURE 1

This example is the square of a binomial. This type of product occurs frequently enough in algebra that we have special formulas for it. Here are the formulas for binomial squares:

$$(a + b)^2 = (a + b)(a + b) = a^2 + ab + ab + b^2 = a^2 + 2ab + b^2$$
$$(a - b)^2 = (a - b)(a - b) = a^2 - ab - ab + b^2 = a^2 - 2ab + b^2$$

Observing the results in both cases, we have the following rule:

> **RULE**   Squaring of a Binomial
>
> The square of a binomial is the sum of the square of the first term, twice the product of the two inside terms, and the square of the last term. Or:
>
> | $(a + b)^2 =$ | a^2 | $+$ | $2ab$ | $+$ | b^2 |
> |---|---|---|---|---|---|
> | | Square of first term | | Twice the product of the two terms | | Square of last term |
> | $(a - b)^2 =$ | a^2 | $-$ | $2ab$ | $+$ | b^2 |

5. Multiply using the FOIL method.
$$(6 - 3t)(2 + 5t)$$

6. Let $p(x) = \left(3x + \frac{1}{4}\right)$ and $q(x) = \left(4x - \frac{1}{3}\right)$. Find $p(x) \cdot q(x)$.

7. Multiply using the FOIL method.
$$(a^4 - 2)(a^4 + 5)$$

8. Multiply using the FOIL method.
$$(7x - 2)(3y + 8)$$

Note   Be careful not to mistake $(a + b)^2$ as the same as $a^2 + b^2$. We must multiply both terms of the first binomial by both terms of the second to get $a^2 - 2ab + b^2$. Again, $(a + b)^2 \neq a^2 + b^2$.

Answers
5. $12 + 24t - 15t^2$
6. $12x^2 - \frac{1}{12}$
7. $a^8 + 3a^4 - 10$
8. $21xy + 56x - 6y - 16$

9. Expand and multiply $(3x - 2)^2$.

EXAMPLE 9   Find $(4x - 6)^2$.

Solution   Applying the definition of exponents and then the FOIL method, we have

$$(4x - 6)^2 = (4x - 6)(4x - 6)$$
$$= 16x^2 - 24x - 24x + 36$$
$$ \text{F} \quad \text{O} \quad \text{I} \quad \text{L}$$
$$= 16x^2 - 48x + 36$$

10. Expand and simplify.

  a. $(x - y)^2$

  b. $(4t + 5)^2$

  c. $(5x - 3y)^2$

  d. $(6 - a^4)^2$

EXAMPLE 10   Use the formulas in the rule for expanding binomial squares to expand the following.

  a. $(x + 7)^2 = x^2 + 2(x)(7) + 7^2 = x^2 + 14x + 49$

  b. $(3t - 5)^2 = (3t)^2 - 2(3t)(5) + 5^2 = 9t^2 - 30t + 25$

  c. $(4x + 2y)^2 = (4x)^2 + 2(4x)(2y) + (2y)^2 = 16x^2 + 16xy + 4y^2$

  d. $(5 - a^3)^2 = 5^2 - 2(5)(a^3) + (a^3)^2 = 25 - 10a^3 + a^6$

D Products Resulting in the Difference of Two Squares

Another frequently occurring kind of product is found when multiplying two binomials that differ only in the sign between their terms.

11. Let $p(x) = 4x - 3$ and $q(x) = 4x + 3$. Find $p(x) \cdot q(x)$.

EXAMPLE 11   Let $p(x) = 3x - 5$ and $q(x) = 3x + 5$. Find $p(x) \cdot q(x)$.

Solution   Applying the FOIL method, we have

$$p(x) \cdot q(x) = (3x - 5)(3x + 5) = 9x^2 + 15x - 15x - 25 \quad \text{Two middle terms add to 0.}$$
$$ \text{F} \quad \text{O} \quad \text{I} \quad \text{L}$$
$$= 9x^2 - 25$$

The outside and inside products in Example 11 are opposites and therefore add to 0. Here it is in general:

$$(a - b)(a + b) = a^2 + ab - ab - b^2 \quad \text{Two middle terms add to 0.}$$
$$= a^2 - b^2$$

> **RULE**   Difference of Two Squares
>
> To multiply two binomials that differ only in the sign between their two terms, simply subtract the square of the second term from the square of the first term:
> $$(a + b)(a - b) = a^2 - b^2$$

12. Multiply.

  a. $(x + 2)(x - 2)$

  b. $(5a + 7)(5a - 7)$

  c. $(x^3 + 4)(x^3 - 4)$

  d. $(x^4 - 5a)(x^4 + 5a)$

The expression $a^2 - b^2$ is called the *difference of two squares*. Remember, as shown on the previous page, $a^2 - b^2 = (a - b)^2$

EXAMPLE 12   Find the following products:

  a. $(x - 5)(x + 5) = x^2 - 25$

  b. $(2a - 3)(2a + 3) = 4a^2 - 9$

  c. $(x^2 + 4)(x^2 - 4) = x^4 - 16$

  d. $(x^3 - 2a)(x^3 + 2a) = x^6 - 4a^2$

Answers

9. $9x^2 - 12x + 4$

10. a. $x^2 - 2xy + y^2$

  b. $16t^2 + 40t + 25$

  c. $25x^2 - 30xy + 9y^2$

  d. $36 - 12a^4 + a^8$

11. $16x^2 - 9$

12. a. $x^2 - 4$ **b.** $25a^2 - 49$

  c. $x^6 - 16$ **d.** $x^8 - 25a^2$

More About Function Notation

From the introduction to this chapter, we know that the revenue obtained from selling x items at p dollars per item is

$$R = \text{Revenue} = xp \qquad \text{(The number of items} \times \text{price per item)}$$

For example, if a store sells 100 items at \$4.50 per item, the revenue is $100(4.50) = \$450$. If we have an equation that gives the relationship between x and p, then we can write the revenue in terms of x or in terms of p. With function notation, we would write the revenue as either $R(x)$ or $R(p)$, where

> $R(x)$ is the revenue function that gives the revenue R in terms of the number of items x that are sold.
>
> $R(p)$ is the revenue function that gives the revenue R in terms of the price per item p.

With function notation we can see exactly which variables we want our formulas written in terms of.

In the next two examples, we will use function notation to combine a number of problems we have worked previously.

EXAMPLE 13   A company manufactures and sells DVDs. They find that they can sell x DVDs each day at p dollars per disc, according to the equation $x = 230 - 20p$. Find $R(x)$ and $R(p)$.

Solution   The notation $R(p)$ tells us we are to write the revenue equation in terms of the variable p. To do so, we use the formula $R(p) = xp$ and substitute $230 - 20p$ for x to obtain

$$R(p) = xp = (230 - 20p)p = 230p - 20p^2$$

The notation $R(x)$ indicates that we are to write the revenue in terms of the variable x. We need to solve the equation $x = 230 - 20p$ for p. Let's begin by interchanging the two sides of the equation.

$$230 - 20p = x$$

$$-20p = -230 + x \qquad \text{Add } -230 \text{ to each side.}$$

$$p = \frac{-230 + x}{-20} \qquad \text{Divide each side by } -20.$$

$$p = 11.5 - 0.05x \qquad \tfrac{230}{20} = 11.5 \text{ and } \tfrac{1}{20} = 0.05$$

Now we can find $R(x)$ by substituting $11.5 - 0.05x$ for p in the formula $R(x) = xp$.

$$R(x) = xp = x(11.5 - 0.05x) = 11.5x - 0.05x^2$$

Our two revenue functions are actually equivalent. To offer some justification for this, suppose that the company decides to sell each disc for \$5. The equation $x = 230 - 20p$ indicates that, at \$5 per disc, they will sell $x = 230 - 20(5) = 230 - 100 = 130$ discs per day. To find the revenue from selling the discs for \$5 each, we use $R(p)$ with $p = 5$.

$$\text{If} \qquad p = 5$$
$$\text{then} \qquad R(p) = R(5)$$
$$= 230(5) - 20(5)^2$$
$$= 1{,}150 - 500$$
$$= \$650$$

However, to find the revenue from selling 130 discs, we use $R(x)$ with $x = 130$.

$$\text{If} \qquad x = 130$$
$$\text{then} \qquad R(x) = R(130)$$
$$= 11.5(130) - 0.05(130)^2$$
$$= 1{,}495 - 845$$
$$= \$650$$

13. Repeat Example 13 using the equation

$$x = 250 - 25p$$

14. Find $P(x)$ and $P(130)$ if

$$x = 250 - 25p$$
and
$$C(x) = 210 + 1.5x$$

EXAMPLE 14 Suppose the daily cost function for the DVDs in Example 13 is $C(x) = 200 + 2x$. Find the profit function $P(x)$ and then find $P(130)$.

Solution Since profit is equal to the difference of the revenue and the cost, we have

$$P(x) = R(x) - C(x)$$
$$= 11.5x - 0.05x^2 - (200 + 2x)$$
$$= -0.05x^2 + 9.5x - 200$$

Notice that we used the formula for $R(x)$ from Example 13 instead of the formula for $R(p)$. We did so because we were asked to find $P(x)$, meaning we want the profit P only in terms of the variable x.

Next, we use the formula we just obtained to find $P(130)$.

$$P(130) = -0.05(130)^2 + 9.5(130) - 200$$
$$= -0.05(16,900) + 9.5(130) - 200$$
$$= -845 + 1,235 - 200$$
$$= \$190$$

Because $P(130) = \$190$, the company will make a profit of $190 per day by selling 130 discs per day.

USING TECHNOLOGY Graphing Calculators: More about Example 14

We can visualize the three functions given in Example 14 if we set up the functions list and graphing window on our calculator this way:

$Y_1 = 11.5X - 0.05X^2$ *This gives the graph of R(x).*

$Y_2 = 200 + 2X$ *This gives the graph of C(x).*

$Y_3 = Y_1 - Y_2$ *This gives the graph of P(x).*

Window: X from 0 to 250, Y from 0 to 750

The graphs in Figure 2 are similar to what you will obtain using the functions list and window shown here. Next, find the value of $P(x)$ when $R(x)$ and $C(x)$ intersect.

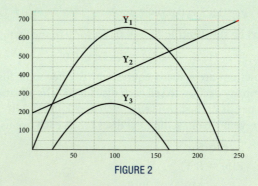

FIGURE 2

GETTING READY FOR CLASS

After reading through the preceding section, respond in your own words and in complete sentences.

A. In words, state the rule for multiplying polynomials.

B. How would you use the FOIL method to multiply two binomials?

C. Write a problem that demonstrates how to expand a binomial square.

D. Discuss the rule for a difference of two squares.

Answers

14. $P(x) = -0.04x^2 + 8.5x - 210$
$P(130) = \$219$

Vocabulary Review

Choose the correct words to fill in the blanks below.

first   second   distributive   difference   binomial

1. To multiply a polynomial by a monomial, you must use the _____ property.

2. To multiply two polynomials, multiply each term in the first polynomial by each term in the _____ polynomial.

3. FOIL is an acronym that stands for the _____, outside, inside, and last terms of a polynomial product.

4. To square a _____, find the sum of the square of the first term, find twice the product of the two inside terms, and find the square of the last term.

5. The _____ of two squares is given by $(a + b)(a - b) = a^2 - b^2$.

Problems

A Multiply the following by applying the distributive property.

1. $2x(6x^2 - 5x + 4)$

2. $-3x(5x^2 - 6x - 4)$

3. $-3a^2(a^3 - 6a^2 + 7)$

4. $4a^3(3a^2 - a + 1)$

5. $2a^2b(a^3 - ab + b^3)$

6. $5a^2b^2(8a^2 - 2ab + b^2)$

Multiply the following vertically.

7. $(x - 5)(x + 3)$

8. $(x + 4)(x + 6)$

9. $(2x^2 - 3)(3x^2 - 5)$

10. $(3x^2 + 4)(2x^2 - 5)$

11. $(x + 3)(x^2 + 6x + 5)$

12. $(x - 2)(x^2 - 5x + 7)$

13. $(a - b)(a^2 + ab + b^2)$

14. $(a + b)(a^2 - ab + b^2)$

15. $(2x + y)(4x^2 - 2xy + y^2)$

16. $(x - 3y)(x^2 + 3xy + 9y^2)$

17. $(2a - 3b)(a^2 + ab + b^2)$

18. $(5a - 2b)(a^2 - ab - b^2)$

B Multiply the following using the FOIL method.

19. $(x - 2)(x + 3)$

20. $(x + 2)(x - 3)$

21. $(2a + 3)(3a + 2)$

22. $(5a - 4)(2a + 1)$

23. $(5 - 3t)(4 + 2t)$

24. $(7 - t)(6 - 3t)$

25. $(x^3 + 3)(x^3 - 5)$

26. $(x^3 + 4)(x^3 - 7)$

27. $(5x - 6y)(4x + 3y)$

28. $(6x - 5y)(2x - 3y)$

29. $\left(3t + \dfrac{1}{3}\right)\left(6t - \dfrac{2}{3}\right)$

30. $\left(5t - \dfrac{1}{5}\right)\left(10t + \dfrac{3}{5}\right)$

31. Let $p(x) = 4x - 3$ and $q(x) = 2x + 1$. Find

  a. $p(x) - q(x)$

  b. $p(x) + q(x)$

  c. $p(x) \cdot q(x)$

32. Let $p(x) = 5x + 2$ and $q(x) = 3x - 7$. Find

  a. $p(x) - q(x)$

  b. $p(x) + q(x)$

  c. $p(x) \cdot q(x)$

C D Find the following special products.

33. $(5x + 2y)^2$

34. $(3x - 4y)^2$

35. $(5 - 3t^3)^2$

36. $(7 - 2t^4)^2$

37. $(2a + 3b)(2a - 3b)$

38. $(6a - 1)(6a + 1)$

39. $(3r^2 + 7s)(3r^2 - 7s)$

40. $(5r^2 - 2s)(5r^2 + 2s)$

41. $\left(y + \dfrac{3}{2}\right)^2$

42. $\left(y - \dfrac{7}{2}\right)^2$

43. $\left(a - \dfrac{1}{2}\right)^2$

44. $\left(a - \dfrac{5}{2}\right)^2$

45. $\left(x + \dfrac{1}{4}\right)^2$

46. $\left(x - \dfrac{3}{8}\right)^2$

47. $\left(t + \dfrac{1}{3}\right)^2$

48. $\left(t - \dfrac{2}{5}\right)^2$

49. $\left(\dfrac{1}{3}x - \dfrac{2}{5}\right)\left(\dfrac{1}{3}x + \dfrac{2}{5}\right)$

50. $\left(\dfrac{3}{4}x - \dfrac{1}{7}\right)\left(\dfrac{3}{4}x + \dfrac{1}{7}\right)$

Find the following products.

51. $(x - 2)^3$

52. $(4x + 1)^3$

53. $\left(x - \dfrac{1}{2}\right)^3$

54. $\left(x + \dfrac{1}{4}\right)^3$

55. $3(x - 1)(x - 2)(x - 3)$

56. $2(x + 1)(x + 2)(x + 3)$

57. $(b^2 + 8)(a^2 + 1)$

58. $(b^2 + 1)(a^4 - 5)$

59. $(x + 1)^2 + (x + 2)^2 + (x + 3)^2$

60. $(x - 1)^2 + (x - 2)^2 + (x - 3)^2$

61. $(2x + 3)^2 - (2x - 3)^2$

62. $(x - 3)^3 - (x + 3)^3$

Simplify.

63. $(x + 3)^2 - 2(x + 3) - 8$

64. $(x - 2)^2 - 3(x - 2) - 10$

65. $(2a - 3)^2 - 9(2a - 3) + 20$

66. $(3a - 2)^2 + 2(3a - 2) - 3$

67. $2(4a + 2)^2 - 3(4a + 2) - 20$

68. $6(2a + 4)^2 - (2a + 4) - 2$

69. Let $a = 2$ and $b = 3$, and evaluate each of the following expressions.

   a. $a^4 - b^4$

   b. $(a - b)^4$

   c. $(a^2 + b^2)(a + b)(a - b)$

70. Let $a = 2$ and $b = 3$, and evaluate each of the following expressions.

   a. $a^3 + b^3$

   b. $(a + b)^3$

   c. $a^3 + 3a^2b + 3ab^2 + b^3$

Applying the Concepts

Text Messages and Voice Minutes The chart shows the average number of text messages sent and voice minutes used by different age groups. Use the chart to answer Problems 71 and 72.

71. If a plan includes 400 text messages for $6, and 5¢ for each additional text, write an equation that describes the cost for text messages for any age group. Then, find how much the average 24–34 year old would owe.

72. A couple in the 35–44 age range have a family plan that includes 700 free text messages and 5¢ for each additional text. Write an expression that describes how much they would owe for text messages and solve.

73. Revenue A store selling art supplies finds that it can sell x sketch pads per week at p dollars each, according to the formula $x = 900 - 300p$. Write formulas for $R(p)$ and $R(x)$. Then find the revenue obtained by selling the pads for $1.60 each.

74. Revenue A company selling CDs finds that it can sell x CDs per day at p dollars per CD, according to the formula $x = 800 - 100p$. Write formulas for $R(p)$ and $R(x)$. Then find the revenue obtained by selling the CDs for $3.80 each.

75. Revenue A company sells an inexpensive accounting program for home computers. If it can sell x programs per week at p dollars per program, according to the formula $x = 350 - 10p$, find formulas for $R(p)$ and $R(x)$. How much will the weekly revenue be if it sells 65 programs?

76. Revenue A company sells boxes of greeting cards through the mail. It finds that it can sell x boxes of cards each week at p dollars per box, according to the formula $x = 1,475 - 250p$. Write formulas for $R(p)$ and $R(x)$. What revenue will it bring in each week if it sells 200 boxes of cards?

77. Profit If the cost to produce the x programs in Problem 75 is $C(x) = 5x + 500$, find $P(x)$ and $P(60)$.

78. Profit If the cost to produce the x CDs in Problem 74 is $C(x) = 2x + 200$, find $P(x)$ and $P(40)$.

79. Interest If you deposit $100 in an account with an interest rate r that is compounded annually, then the amount of money in that account at the end of 4 years is given by the formula $A = 100(1 + r)^4$. Expand the right side of this formula.

80. Interest If you deposit P dollars in an account with an annual interest rate r that is compounded twice a year, then at the end of a year the amount of money in that account is given by the formula

$$A = P\left(1 + \frac{r}{2}\right)^2$$

Expand the right side of this formula.

81. Inflation The future price of an item priced at P dollars is given by

$$F = P(1 + r)^t$$

where r is the inflation rate (expressed as a decimal) and t is the time (in years). If a gallon of gas costs $4.00 today, what will its cost be in 10 years assuming an inflation rate of 8% per year?

82. Gas A politician is quoted as saying, "At this rate, we will be paying $20.00 for a gallon of gas in twenty years." Is the quote from the politician reasonable? Use the formula in the previous problem to answer this question.

83. Tuition Tuition fees for California State University were $5,131 during the 2010 − 2011 academic year. Historically, tuition has increased at an 8% per year rate. Use the formula from problem 81 to find what the tuition was 20 years ago and 30 years ago. Hint: First solve for P, then substitute the given value for F.

84. Tuition Repeat problem 83 for University of California, where the 2010 − 2011 yearly tuition was $11,300.

Getting Ready for the Next Section

85. $\dfrac{8a^3}{a}$

86. $\dfrac{-8a^2}{a}$

87. $\dfrac{-48a}{a}$

88. $\dfrac{-32a}{a}$

89. $\dfrac{16a^5b^4}{8a^2b^3}$

90. $\dfrac{12x^4y^5}{3x^3y^3}$

91. $\dfrac{-24a^5b^5}{8a^5b^3}$

92. $\dfrac{-15x^5y^3}{3x^3y^3}$

93. $\dfrac{x^3y^4}{-x^3}$

94. $\dfrac{x^2y^2}{-x^2}$

Maintaining Your Skills

Solve the following systems of equations.

95. $x + y + z = 6$
$2x - y + z = 3$
$x + 2y - 3z = -4$

96. $x + y + z = 6$
$x - y + 2z = 7$
$2x - y - z = 0$

97. $3x + 4y = 15$
$2x - 5z = -3$
$4y - 3z = 9$

98. $x + 3y = 5$
$6y + z = 12$
$x - 2z = -10$

One Step Further

99. Multiply $(x + y - 4)(x + y + 5)$ by first writing it like this:

$$[(x + y) - 4][(x + y) + 5]$$

and then apply the FOIL method.

100. Multiply $(x - 5 - y)(x - 5 + y)$ by first writing it like this:

$$[(x - 5) - y][(x - 5) + y]$$

and then apply the FOIL method.

Assume n is a positive integer and multiply.

101. $(x^n - 2)(x^n - 3)$

102. $(x^{2n} + 3)(x^{2n} - 3)$

103. $(2x^n + 3)(5x^n - 1)$

104. $(4x^n - 3)(7x^n + 2)$

105. $(x^n + 5)^2$

106. $(x^n - 2)^2$

107. $(x^n + 1)(x^{2n} - x^n + 1)$

108. $(x^{3n} - 3)(x^{6n} + 3x^{3n} + 9)$

Find the Mistake

Each sentence below contains a mistake. Circle the mistake and write the correct word or phrase on the line provided.

1. To multiply binomials using the FOIL method, find the product of the first, the outside, the inside, and the last terms.

_____

2. To square the binomial $(a + b)$, find the sum of the square of the first term, the product of the two inside terms, and the square of the last term. _____

3. The product found by expanding and multiplying $(2x + 4)^2$ is $4x^2 + 8x + 8$. _____

4. Multiplying two binomials that differ only in the sign between their two terms will result in the sum of two squares.

_____

4.3 Greatest Common Factor and Factoring by Grouping

OBJECTIVES

A Factor by factoring out the greatest common factor.

B Factor by grouping.

KEY WORDS

greatest common factor

factor by grouping

The extreme sport of powerbocking enables a person to run, leap, and flip while standing on spring-loaded stilts. Picture a powerbocker slipping his feet into his boots, securing the bindings, and then leaping into the air. The curved path of his leap from the ground, into the air, and back to the ground again creates the shape of a parabola, which we will discuss later in the book. For now, let's say the leap's path can be represented by the polynomial $-16x^2 + 62x + 8$. To begin factoring this polynomial, we need to find the greatest common factor, which we will discuss in this section.

In general, factoring is the reverse of multiplication. The diagram below illustrates the relationship between factoring and multiplication. Reading from left to right, we say the product of 3 and 7 is 21. Reading in the other direction, from right to left, we say 21 factors into 3 times 7. Or 3 and 7 are factors of 21.

$$\text{Multiplication}$$

$$\text{Factors} \rightarrow \quad 3 \cdot 7 = 21 \quad \leftarrow \text{Product}$$

$$\text{Factoring}$$

A Greatest Common Factor

> **DEFINITION** greatest common factor
>
> The greatest common factor for a polynomial is the largest monomial that divides (is a factor of) each term of the polynomial.

Note The term *largest monomial*, as used here, refers to the monomial with the largest integer exponents whose coefficient has the greatest absolute value.

The greatest common factor for the polynomial $25x^5 + 20x^4 - 30x^3$ is $5x^3$ since it is the largest monomial that is a factor of each term. We can apply the distributive property and write

$$25x^5 + 20x^4 - 30x^3 = 5x^3(5x^2) + 5x^3(4x) - 5x^3(6)$$
$$= 5x^3(5x^2 + 4x - 6)$$

The last line is written in factored form.

EXAMPLE 1 Factor the greatest common factor from $8a^3 - 8a^2 - 48a$.

Solution The greatest common factor is $8a$. It is the largest monomial that divides each term of our polynomial. We can write each term in our polynomial as the product of $8a$ and another monomial. Then, we apply the distributive property to factor $8a$ from each term.

$$8a^3 - 8a^2 - 48a = 8a(a^2) - 8a(a) - 8a(6)$$
$$= 8a(a^2 - a - 6)$$

Practice Problems

1. Factor the greatest common factor from $15a^7 - 25a^5 + 30a^3$.

Answer

1. $5a^3(3a^4 - 5a^2 + 6)$

2. Factor $12x^4y^5 - 9x^3y^4 - 15x^5y^3$.

3. Factor.

$4(a + b)^4 - 6(a + b)^3 + 16(a + b)^2$

4. Factor $ab^3 + b^3 + 6a + 6$.

5. Factor $15 - 3y^2 - 5x^2 + x^2y^2$.

6. Factor by grouping.

$x^3 + 5x^2 + 3x + 15$

EXAMPLE 2   Factor the greatest common factor from $16a^5b^4 - 24a^2b^5 - 8a^3b^3$.

Solution   The largest monomial that divides each term is $8a^2b^3$. We write each term of the original polynomial in terms of $8a^2b^3$ and apply the distributive property to write the polynomial in factored form.

$$16a^5b^4 - 24a^2b^5 - 8a^3b^3 = 8a^2b^3(2a^3b) - 8a^2b^3(3b^2) - 8a^2b^3(a)$$
$$= 8a^2b^3(2a^3b - 3b^2 - a)$$

EXAMPLE 3   Factor the greatest common factor from $5x^2(a + b) - 6x(a + b) - 7(a + b)$.

Solution   The greatest common factor is $a + b$. Factoring it from each term, we have

$$5x^2(a + b) - 6x(a + b) - 7(a + b) = (a + b)(5x^2 - 6x - 7)$$

B Factoring by Grouping

The polynomial $5x + 5y + x^2 + xy$ can be factored by noticing that the first two terms have a 5 in common, whereas the last two have an x in common. Applying the distributive property, we have

$$5x + 5y + x^2 + xy = 5(x + y) + x(x + y)$$

This last expression can be thought of as having two terms, $5(x + y)$ and $x(x + y)$, each of which has a common factor $(x + y)$. We apply the distributive property again to factor $(x + y)$ from each term.

$$5(x + y) + x(x + y) = (x + y)(5 + x)$$

EXAMPLE 4   Factor $a^2b^2 + b^2 + 8a^2 + 8$.

Solution   The first two terms have b^2 in common; the last two have 8 in common.

$$a^2b^2 + b^2 + 8a^2 + 8 = b^2(a^2 + 1) + 8(a^2 + 1)$$
$$= (a^2 + 1)(b^2 + 8)$$

EXAMPLE 5   Factor $15 - 5y^4 - 3x^3 + x^3y^4$.

Solution   Let's try factoring a 5 from the first two terms and an $-x^3$ from the last two terms.

$$15 - 5y^4 - 3x^3 + x^3y^4 = 5(3 - y^4) - x^3(3 - y^4)$$
$$= (3 - y^4)(5 - x^3)$$

EXAMPLE 6   Factor by grouping $x^3 + 2x^2 + 9x + 18$.

Solution   We begin by factoring x^2 from the first two terms and 9 from the second two terms.

$$x^3 + 2x^2 + 9x + 18 = x^2(x + 2) + 9(x + 2)$$
$$= (x + 2)(x^2 + 9)$$

GETTING READY FOR CLASS

After reading through the preceding section, respond in your own words and in complete sentences.

A. What is the relationship between multiplication and factoring?

B. What is the greatest common factor for a polynomial?

C. After factoring a polynomial, how can you check your result?

D. When would you try to factor by grouping?

Answers

2. $3x^3y^3(4xy^2 - 3y - 5x^2)$
3. $2(a + b)^2[2(a + b)^2 - 3(a + b) + 8]$
4. $(a + 1)(b^3 + 6)$
5. $(5 - y^2)(3 - x^2)$
6. $(x + 5)(x^2 + 3)$

Problems

A Factor the greatest common factor from each of the following.

1. $10x^3 - 15x^2$

2. $12x^5 + 18x^7$

3. $9y^6 + 18y^3$

4. $24y^4 - 8y^2$

5. $9a^2b - 6ab^2$

6. $30a^3b^4 + 20a^4b^3$

7. $21xy^4 + 7x^2y^2$

8. $14x^6y^3 - 6x^2y^4$

9. $3a^2 - 21a + 33$

10. $3a^2 - 3a + 6$

11. $4x^3 - 16x^2 + 20x$

12. $2x^3 - 14x^2 + 28x$

13. $10x^4y^2 + 20x^3y^3 + 30x^2y^4$

14. $6x^4y^2 + 18x^3y^3 + 24x^2y^4$

15. $-x^2y + xy^2 - x^2y^2$

16. $-x^3y^2 - x^2y^3 - x^2y^2$

17. $4x^3y^2z - 8x^2y^2z^2 + 6xy^2z^3$

SCAN TO ACCESS

18. $7x^4y^3z^2 - 21x^2y^2z^2 - 14x^2y^3z^4$

19. $20a^2b^2c^2 - 30ab^2c + 25a^2bc^2$

20. $8a^3bc^5 - 48a^2b^4c + 16ab^3c^5$

21. $5x(a - 2b) - 3y(a - 2b)$

22. $3a(x - y) - 7b(x - y)$

23. $3x^2(x + y)^2 - 6y^2(x + y)^2$

24. $10x^3(2x - 3y) - 15x^2(2x - 3y)$

25. $2x^2(x + 5) + 7x(x + 5) + 8(x + 5)$

26. $2x^2(x + 2) + 13x(x + 2) + 17(x + 2)$

B Factor each of the following by grouping.

27. $3xy + 3y + 2ax + 2a$

28. $5xy^2 + 5y^2 + 3ax + 3a$

29. $x^2y + x + 3xy + 3$

30. $x^3y^3 + 2x^3 + 5x^2y^3 + 10x^2$

31. $3xy^2 - 6y^2 + 4x - 8$

32. $8x^2y - 4x^2 + 6y - 3$

33. $x^2 - ax - bx + ab$

34. $ax - x^2 - bx + ab$

35. $ab + 5a - b - 5$

36. $x^2 - xy - ax + ay$

37. $a^4b^2 + a^4 - 5b^2 - 5$

38. $2a^2 - a^2b - bc^2 + 2c^2$

39. $x^3 + 3x^2 + 4x + 12$

40. $x^3 + 5x^2 + 4x + 20$

41. $x^3 + 2x^2 + 25x + 50$

42. $x^3 + 4x^2 + 9x + 36$

43. $2x^3 + 3x^2 + 8x + 12$

44. $3x^3 + 2x^2 + 27x + 18$

45. $4x^3 + 12x^2 + 9x + 27$

46. $9x^3 + 18x^2 + 4x + 8$

47. The greatest common factor of the binomial $3x - 9$ is 3. The greatest common factor of the binomial $6x - 2$ is 2. What is the greatest common factor of their product, $(3x - 9)(6x - 2)$, when it has been multiplied out?

48. The greatest common factors of the binomials $5x - 10$ and $2x + 4$ are 5 and 2, respectively. What is the greatest common factor of their product, $(5x - 10)(2x + 4)$, when it has been multiplied out?

Applying the Concepts

49. Investing If P dollars are placed in a savings account in which the rate of interest r is compounded yearly, then at the end of one year the amount of money in the account can be written as $P + Pr$. At the end of two years the amount of money in the account is

$$P + Pr + (P + Pr)r$$

Use factoring by grouping to show that this last expression can be written as $P(1 + r)^2$.

50. Investing At the end of 3 years, the amount of money in the savings account in Problem 49 will be

$$P(1 + r)^2 + P(1 + r)^2 r$$

Use factoring to show that this last expression can be written as $P(1 + r)^3$.

51. Price The weekly revenue equation for a company selling an inexpensive accounting program for home computers is given by the equation

$$R(x) = 35x - 0.1x^2$$

where x is the number of programs they sell per week. What price p should they charge if they want to sell 65 programs per week?

52. Price The weekly revenue equation for a small mail-order company selling boxes of greeting cards is

$$R(x) = 5.9x - 0.004x^2$$

where x is the number of boxes they sell per week. What price p should they charge if they want to sell 200 boxes each week?

Maintaining Your Skills

53. The chart shows attendance at Major League Baseball games from 1999 to 2009. Use the chart to find the equation of the line segment from 2007 to 2009. Then, use the equation to predict the attendance of baseball games for 2010, assuming the trend continues.

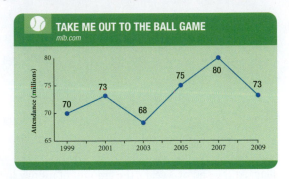

54. The chart shows the close competitions at the 2008 Summer Olympics. If Laszlo Cseh's finishing time was 1 minute, 52.70 seconds, use the chart to find Michael Phelp's finishing time.

Getting Ready for the Next Section

Factor out the greatest common factor.

55. $3x^4 - 9x^3y - 18x^2y^2$

56. $5x^2 + 10x + 30$

57. $2x^2(x - 3) - 4x(x - 3) - 3(x - 3)$

58. $3x^2(x - 2) - 8x(x - 2) + 2(x - 2)$

Multiply.

59. $(x + 2)(3x - 1)$

60. $(x - 2)(3x + 1)$

61. $(x - 1)(3x - 2)$

62. $(x + 1)(3x + 2)$

63. $(x + 2)(x + 3)$

64. $(x - 2)(x - 3)$

65. $(2y + 5)(3y - 7)$

66. $(2y - 5)(3y + 7)$

67. $(4 - 3a)(5 - a)$

68. $(4 - 3a)(5 + a)$

69. $(5 + 2x)(5 - 2x)$

70. $(3 + 2x)^2$

71. Complete the following table.

Two Numbers a and b	Their Product ab	Their Sum a + b
1, −24		
−1, 24		
2, −12		
−2, 12		
3, −8		
−3, 8		
4, −6		
−4, 6		

72. Complete the following table.

Two Numbers a and b	Their Product ab	Their Sum a + b
1, −54		
−1, 54		
2, −27		
−2, 27		
3, −18		
−3, 18		
6, −9		
−6, 9		

73. Find two numbers whose product is 36 and sum is 13.

74. Find two numbers whose product is 40 and sum is -13.

75. Find two numbers whose product is -40 and sum is 3.

76. Find two numbers whose product is -36 and sum is -9.

77. Find two numbers whose product is -48 and sum is -8.

78. Find two numbers whose product is -48 and sum is 13.

Find the Mistake

Each sentence below contains a mistake. Circle the mistake and write the correct word or phrase on the line provided.

1. The greatest common factor for a polynomial is the smallest monomial that is a factor of each term of the polynomial.

2. The greatest common factor for the polynomial $9x^3 + 18x^2 + 36x$ is 9.

3. The greatest common factor for the polynomial $x^5y^3z - x^3y^2 + 6y^4z^2$ is x^3z^2.

4. To begin to factor the polynomial $12 - 4y^2 - 5x^3 + x^3y^5$ by grouping, we can factor 4 from the first two terms and $5x$ from the last two terms.

Landmark Review: Checking Your Progress

Identify each of the following as a monomial, a binomial, or a trinomial. Give the degree of each and name the leading coefficient.

1. $5x - 3$

2. 14

3. $9a^2 + 2x - 3$

4. $3x - 4 + 14x^2$

Simplify each of the following by combining similar terms.

5. $(3x - 4) - (5x + 3)$

6. $5x^2 - 3 + 14x - 3x + 4$

7. $4x + 13y^2 - 3x^2 - 5y^2 + 4x^2 - 3$

8. $11x^3 + 10y - 3x^2 - 3y + 6x^2 + 4x^3 - 13$

Multiply the following expressions.

9. $2a(2a^2 + 4a - 5)$

10. $8xy(5x + 3y + 2xy)$

11. $(x - 1)(x - 8)$

12. $(x - 2)(y + 6)$

13. $(x^2 - 4)(x - 7)$

14. $(x^2 + 6)(x - 8)$

15. $(3a + 5)(a + 2)$

16. $(2y - 1)(3y - 2)$

Factor the greatest common factor from each of the following.

17. $15x^3 + 5x$

18. $24x^2y^2 + 18xy$

19. $14x^3y^2 + 21x^2y^2 + 35xy$

20. $9x^5y^3 + 6x^2y^3 + 18x^3y^2$

Factor each of the following by grouping.

21. $4xy^2 + 12y^2 + 5ax + 15a$

22. $6ab + 18a - 2b - 6$

23. $4a^2 - 2a^2b - 4bc^2 + 8c^2$

24. $6x^3 + 18x^2 + 6x + 18$

4.4 Factoring Trinomials

OBJECTIVES

A Factor trinomials in which the leading coefficient is 1.

B Factor trinomials in which the leading coefficient is a number other than 1.

KEY WORDS

leading coefficient

trinomial

prime polynomial

trial and error

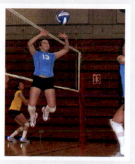

Suppose a volleyball coach had coached a college team for x years. He plans to spend two more years coaching the team before retiring. Considering the cost of uniforms, the polynomial that represents how much the coach will have spent at the end of his coaching career can be given by the trinomial $30x^2 + 70x + 20$. After reading through this section, you will be able to factor this trinomial which has a leading coefficient of 30. However, let's start by practicing factoring trinomials with a leading coefficient of 1.

A Factoring Trinomials with a Leading Coefficient of 1

Earlier in this chapter, we multiplied the following binomials:

$$(x - 2)(x + 3) = x^2 + x - 6$$
$$(x + 5)(x + 2) = x^2 + 7x + 10$$

In each case the product of two binomials is a trinomial. The first term in the resulting trinomial is obtained by multiplying the first term in each binomial. The middle term comes from adding the product of the two inside terms with the product of the two outside terms. The last term is the product of the last terms in each binomial.

In general,

$$(x + a)(x + b) = x^2 + ax + bx + ab$$
$$= x^2 + (a + b)x + ab$$

Writing this as a factoring problem, we have

$$x^2 + (a + b)x + ab = (x + a)(x + b)$$

To factor a trinomial with a leading coefficient of 1, we simply find the two numbers a and b whose sum is the coefficient of the middle term and whose product is the constant term.

EXAMPLE 1 Factor $x^2 + 2x - 15$.

Solution Since the leading coefficient is 1, we need two integers whose product is -15 and whose sum is 2. The integers are 5 and -3.

$$x^2 + 2x - 15 = (x + 5)(x - 3)$$

In the preceding example, we found factors of $x + 5$ and $x - 3$. These are the only two such factors for $x^2 + 2x - 15$. There is no other pair of binomials $x + a$ and $x + b$ whose product is $x^2 + 2x - 15$.

Practice Problems

1. Factor $x^2 - x - 12$.

Answer

1. $(x - 4)(x + 3)$

2. Factor $x^2 + 2xy - 15y^2$.

EXAMPLE 2 Factor $x^2 - xy - 12y^2$.

Solution We need two expressions whose product is $-12y^2$ and whose sum is $-y$. The expressions are $-4y$ and $3y$:

$$x^2 - xy - 12y^2 = (x - 4y)(x + 3y)$$

Checking this result gives

$$(x - 4y)(x + 3y) = x^2 + 3xy - 4xy - 12y^2$$
$$= x^2 - xy - 12y^2$$

3. Factor $x^2 + x + 1$.

EXAMPLE 3 Factor $x^2 - 8x + 6$.

Solution Since there is no pair of integers whose product is 6 and whose sum is -8, the trinomial $x^2 - 8x + 6$ is not factorable. We say it is a *prime polynomial*.

B Factoring When the Lead Coefficient is not 1

4. Factor $5x^2 + 25x + 30$.

EXAMPLE 4 Factor $3x^4 - 15x^3y - 18x^2y^2$.

Solution The leading coefficient is not 1. Each term is divisible by $3x^2$, however. Factoring this out to begin with we have

$$3x^4 - 15x^3y - 18x^2y^2 = 3x^2(x^2 - 5xy - 6y^2)$$

Factoring the resulting trinomial as in the previous examples gives

$$3x^2(x^2 - 5xy - 6y^2) = 3x^2(x - 6y)(x + y)$$

> **Note** As a general rule, it is best to factor out the greatest common factor first.

Factoring Other Trinomials by Trial and Error

We want to turn our attention now to trinomials with leading coefficients other than 1 and with no greatest common factor other than 1.

Suppose we want to factor $3x^2 - x - 2$. The factors will be a pair of binomials. The product of the first terms will be $3x^2$, and the product of the last terms will be -2. We can list all the possible factors along with their products as follows:

Possible Factors	First Term	Middle Term	Last Term
$(x + 2)(3x - 1)$	$3x^2$	$+5x$	-2
$(x - 2)(3x + 1)$	$3x^2$	$-5x$	-2
$(x + 1)(3x - 2)$	$3x^2$	$+x$	-2
$(x - 1)(3x + 2)$	$3x^2$	$-x$	-2

From the last line we see that the factors of $3x^2 - x - 2$ are $(x - 1)(3x + 2)$. That is,

$$3x^2 - x - 2 = (x - 1)(3x + 2)$$

To factor trinomials with leading coefficients other than 1, when the greatest common factor is 1, we must use trial and error or list all the possible factors. In either case the idea is this: look only at pairs of binomials whose products give the correct first and last terms, then look for the combination that will give the correct middle term.

Answers

2. $(x + 5y)(x - 3y)$
3. Does not factor.
4. $5(x + 3)(x + 2)$

EXAMPLE 5   Factor $2x^2 + 13xy + 15y^2$.

5. Factor $3x^2 - x - 2$.

Solution   Listing all possible factors the product of whose first terms is $2x^2$ and the product of whose last terms is $15y^2$ yields

Possible Factors	Middle Term of Product
$(2x - 5y)(x - 3y)$	$-11xy$
$(2x - 3y)(x - 5y)$	$-13xy$
$(2x + 5y)(x + 3y)$	$+11xy$
$(2x + 3y)(x + 5y)$	$+13xy$
$(2x + 15y)(x + y)$	$+17xy$
$(2x - 15y)(x - y)$	$-17xy$
$(2x + y)(x + 15y)$	$+31xy$
$(2x - y)(x - 15y)$	$-31xy$

The fourth line has the correct middle term:

$$2x^2 + 13xy + 15y^2 = (2x + 3y)(x + 5y)$$

Actually, we did not need to check the first two pairs of possible factors in the preceding list. Because all the signs in the trinomial $2x^2 + 13xy + 15y^2$ are positive, the binomial factors must be of the form $(ax + b)(cx + d)$, where a, b, c, and d are all positive.

There are other ways to reduce the number of possible factors to consider. For example, if we were to factor the trinomial $2x^2 - 11x + 12$, we would not have to consider the pair of possible factors $(2x - 4)(x - 3)$. If the original trinomial has no greatest common factor other than 1, then neither of its binomial factors will either. The trinomial $2x^2 - 11x + 12$ has a greatest common factor of 1, but the possible factor $2x - 4$ has a greatest common factor of 2: $2x - 4 = 2(x - 2)$. Therefore, we do not need to consider $2x - 4$ as a possible factor.

EXAMPLE 6   Factor $12x^4 + 17x^2 + 6$.

6. Factor $15x^4 + x^2 - 2$.

Solution   This is a trinomial in x^2:

$$12x^4 + 17x^2 + 6 = (4x^2 + 3)(3x^2 + 2)$$

EXAMPLE 7   Factor $2x^2(x - 3) - 5x(x - 3) - 3(x - 3)$.

7. Factor.

$3x^2(x - 2) - 7x(x - 2) + 2(x - 2)$

Solution   We begin by factoring out the greatest common factor $(x - 3)$. Then we factor the trinomial that remains.

$$2x^2(x - 3) - 5x(x - 3) - 3(x - 3) = (x - 3)(2x^2 - 5x - 3)$$

$$= (x - 3)(2x + 1)(x - 3)$$

$$= (x - 3)^2(2x + 1)$$

Answers
5. $(3x + 2)(x - 1)$
6. $(5x^2 + 2)(3x^2 - 1)$
7. $(x - 2)^2(3x - 1)$

Factoring Trinomials By Grouping

As an alternative to the trial-and-error method of factoring trinomials, we present the following method. The new method does not require as much trial and error. To use this new method, we must rewrite our original trinomial in such a way that the factoring by grouping method can be applied.

> **HOW TO** Factor $ax^2 + bx + c$
>
> **Step 1:** Form the product ac.
>
> **Step 2:** Find a pair of numbers whose product is ac and whose sum is b.
>
> **Step 3:** Rewrite the polynomial to be factored so that the middle term bx is written as the sum of two terms whose coefficients are the two numbers found in step 2.
>
> **Step 4:** Factor by grouping.

8. Factor $8x^2 - 5x - 3$.

EXAMPLE 8 Factor $3x^2 - 10x - 8$ using these steps.

Solution The trinomial $3x^2 - 10x - 8$ has the form $ax^2 + bx + c$, where $a = 3$, $b = -10$, and $c = -8$.

Step 1: The product ac is $3(-8) = -24$.

Step 2: We need to find two numbers whose product is -24 and whose sum is -10. Let's list all the pairs of numbers whose product is -24 to find the pair whose sum is -10.

Product	Sum
$1(-24) = -24$	$1 + (-24) = -23$
$-1(24) = -24$	$-1 + 24 = 23$
$2(-12) = -24$	$2 + (-12) = -10$
$-2(12) = -24$	$-2 + 12 = 10$
$3(-8) = -24$	$3 + (-8) = -5$
$-3(8) = -24$	$-3 + 8 = 5$
$4(-6) = -24$	$4 + (-6) = -2$
$-4(6) = -24$	$-4 + 6 = 2$

As you can see, of all the pairs of numbers whose product is -24, only 2 and -12 have a sum of -10.

Step 3: We now rewrite our original trinomial so the middle term $-10x$ is written as the sum of $-12x$ and $2x$:

$$3x^2 - 10x - 8 = 3x^2 - 12x + 2x - 8$$

Step 4: Factoring by grouping, we have

$$3x^2 - 12x + 2x - 8 = 3x(x - 4) + 2(x - 4)$$
$$= (x - 4)(3x + 2)$$

You can see that this method works by multiplying $x - 4$ and $3x + 2$ to get

$$3x^2 - 10x - 8$$

Answer

8. $(x - 1)(8x + 3)$

EXAMPLE 9 Factor $9x^2 + 15x + 4$.

Solution In this case, $a = 9$, $b = 15$, and $c = 4$. The product ac is $9 \cdot 4 = 36$. Listing all the pairs of numbers whose product is 36 with their corresponding sums, we have

Product	Sum
$1(36) = 36$	$1 + 36 = 37$
$2(18) = 36$	$2 + 18 = 20$
$3(12) = 36$	$3 + 12 = 15$
$4(9) = 36$	$4 + 9 = 13$
$6(6) = 36$	$6 + 6 = 12$

Notice we list only positive numbers since both the product and sum we are looking for are positive. The numbers 3 and 12 are the numbers we are looking for. Their product is 36, and their sum is 15. We now rewrite the original polynomial $9x^2 + 15x + 4$ with the middle term written as $3x + 12x$. We then factor by grouping.

$$9x^2 + 15x + 4 = 9x^2 + 3x + 12x + 4$$
$$= 3x(3x + 1) + 4(3x + 1)$$
$$= (3x + 1)(3x + 4)$$

The polynomial $9x^2 + 15x + 4$ factors into the product

$$(3x + 1)(3x + 4)$$

EXAMPLE 10 Factor $8x^2 - 2x - 15$.

Solution The product ac is $8(-15) = -120$. There are many pairs of numbers whose product is -120. We are looking for the pair whose sum is also -2. The numbers are -12 and 10. Writing $-2x$ as $-12x + 10x$ and then factoring by grouping, we have

$$8x^2 - 2x - 15 = 8x^2 - 12x + 10x - 15$$
$$= 4x(2x - 3) + 5(2x - 3)$$
$$= (2x - 3)(4x + 5)$$

GETTING READY FOR CLASS

After reading through the preceding section, respond in your own words and in complete sentences.

A. What is a prime polynomial?

B. When factoring polynomials, what should you look for first?

C. How can you check to see that you have factored a trinomial correctly?

D. Describe how to determine the binomial factors of $6x^2 + 5x - 25$.

---

9. Factor $4x^2 + 11x + 6$.

10. Factor $4x^2 - 4x - 15$.

Answers

9. $(x + 2)(4x + 3)$

10. $(2x - 5)(2x + 3)$

Vocabulary Review

Choose the correct words to fill in the blanks below.

| trial and error | grouping | middle | common |
| coefficient | product | factor | sum |

1. To factor a trinomial with a leading coefficient of 1, find two numbers whose sum is the _____ of the middle term and whose product is the constant term.

2. To factor a trinomial with a leading coefficient other than 1, make sure to factor out the greatest _____ factor first.

3. To factor a trinomial with a leading coefficient other than 1 by _____, you must list all the possible factors to find the correct terms.

4. To factor the trinomial $ax^2 + bx + c$ by _____, you must follow these steps:

   Step 1: Form the _____ ac.

   Step 2: Find a pair of numbers whose product is ac and whose _____ is b.

   Step 3: Rewrite the polynomial to be factored so that the _____ term bx is written as the sum of two terms whose coefficients are the two numbers found in Step 2.

   Step 4: _____ by grouping.

Problems

A Factor each of the following trinomials.

1. $x^2 + 7x + 12$

2. $x^2 - 7x + 12$

3. $x^2 - x - 12$

4. $x^2 + x - 12$

5. $y^2 + y - 6$

6. $y^2 - y - 6$

7. $16 - 6x - x^2$

8. $3 + 2x - x^2$

9. $12 + 8x + x^2$

10. $15 - 2x - x^2$

11. $16 - x^2$

12. $30 - x - x^2$

13. $x^2 + 3xy + 2y^2$

14. $x^2 - 5xy - 24y^2$

15. $a^2 + 3ab - 18b^2$

16. $a^2 - 8ab - 9b^2$

SCAN TO ACCESS

17. $x^2 - 2xa - 48a^2$

18. $x^2 + 14xa + 48a^2$

19. $x^2 - 12xb + 36b^2$

20. $x^2 + 10xb + 25b^2$

B Factor completely by first factoring out the greatest common factor and then factoring the trinomial that remains.

21. $3a^2 - 21a + 30$

22. $3a^2 - 3a - 6$

23. $4x^3 - 16x^2 - 20x$

24. $2x^3 - 14x^2 + 20x$

Factor completely. Be sure to factor out the greatest common factor first if it is other than 1.

25. $3x^2 - 6xy - 9y^2$

26. $5x^2 + 25xy + 20y^2$

27. $2a^5 + 4a^4b + 4a^3b^2$

28. $3a^4 - 18a^3b + 27a^2b^2$

29. $10x^4y^2 + 20x^3y^3 - 30x^2y^4$

30. $6x^4y^2 + 18x^3y^3 - 24x^2y^4$

31. $2x^2 + 7x - 15$

32. $2x^2 - 7x - 15$

33. $2x^2 + x - 15$

34. $2x^2 - x - 15$

35. $2x^2 - 13x + 15$

36. $2x^2 + 13x + 15$

37. $2x^2 - 11x + 15$

38. $2x^2 + 11x + 15$

39. $2x^2 + 7x + 15$

40. $2x^2 + x + 15$

41. $2 + 7a + 6a^2$

42. $2 - 7a + 6a^2$

43. $60y^2 - 15y - 45$

44. $72y^2 + 60y - 72$

45. $6x^4 - x^3 - 2x^2$

46. $3x^4 + 2x^3 - 5x^2$

47. $40r^3 - 120r^2 + 90r$

48. $40r^3 + 200r^2 + 250r$

49. $4x^2 - 11xy - 3y^2$

50. $3x^2 + 19xy - 14y^2$

51. $10x^2 - 3xa - 18a^2$

52. $9x^2 + 9xa - 10a^2$

53. $18a^2 + 3ab - 28b^2$

54. $6a^2 - 7ab - 5b^2$

55. $600 + 800t - 800t^2$

56. $200 - 600t - 350t^2$

57. $9y^4 + 9y^3 - 10y^2$

58. $4y^5 + 7y^4 - 2y^3$

59. $24a^2 - 2a^3 - 12a^4$

60. $60a^2 + 65a^3 - 20a^4$

61. $8x^4y^2 - 2x^3y^3 - 6x^2y^4$

62. $8x^4y^2 - 47x^3y^3 - 6x^2y^4$

63. $300x^4 + 1{,}000x^2 + 300$

64. $600x^4 - 100x^2 - 200$

65. $20a^4 + 37a^2 + 15$

66. $20a^4 + 13a^2 - 15$

67. $9 + 21r^2 + 12r^4$

68. $2 - 4r^2 - 30r^4$

B Factor each of the following by first factoring out the greatest common factor and then factoring the trinomial that remains.

69. $2x^2(x + 5) + 7x(x + 5) + 6(x + 5)$

70. $2x^2(x + 2) + 13x(x + 2) + 15(x + 2)$

71. $x^2(2x + 3) + 7x(2x + 3) + 10(2x + 3)$

72. $2x^2(x + 1) + 7x(x + 1) + 6(x + 1)$

73. $3x^2(x - 3) + 7x(x - 3) - 20(x - 3)$

74. $4x^2(x + 6) + 23x(x + 6) + 15(x + 6)$

75. $6x^2(x - 2) - 17x(x - 2) + 12(x - 2)$

76. $10x^2(x + 4) - 33x(x + 4) - 7(x + 4)$

77. $12x^2(x + 3) + 7x(x + 3) - 45(x + 3)$

78. $24x^2(x - 6) + 38x(x - 6) + 15(x - 6)$

79. $6x^2(5x - 2) - 11x(5x - 2) - 10(5x - 2)$

80. $14x^2(3x + 4) - 39x(3x + 4) + 10(3x + 4)$

81. $20x^2(2x + 3) + 47x(2x + 3) + 21(2x + 3)$

82. $15x^2(4x - 5) - 2x(4x - 5) - 24(4x - 5)$

83. What polynomial, when factored, gives $(3x + 5y)(3x - 5y)$?

84. What polynomial, when factored, gives $(7x + 2y)(7x - 2y)$?

85. One factor of the trinomial $a^2 + 260a + 2{,}500$ is $a + 10$. What is the other factor?

86. One factor of the trinomial $a^2 - 75a - 2{,}500$ is $a + 25$. What is the other factor?

87. One factor of the trinomial $12x^2 - 107x + 210$ is $x - 6$. What is the other factor?

88. One factor of the trinomial $36x^2 + 134x - 40$ is $2x + 8$. What is the other factor?

89. One factor of the trinomial $54x^2 + 111x + 56$ is $6x + 7$. What is the other factor?

90. One factor of the trinomial $63x^2 + 110x + 48$ is $7x + 6$. What is the other factor?

91. One factor of the trinomial $35x^2 + 19x - 24$ is $5x - 3$. What is the other factor?

92. One factor of the trinomial $36x^2 + 43x - 35$ is $4x + 7$. What is the other factor?

93. Factor the right side of the equation $y = 4x^2 + 18x - 10$, and then use the result to find y when x is $\frac{1}{2}$, when x is -5, and when x is 2.

94. Factor the right side of the equation $y = 9x^2 + 33x - 12$, and use the result to find y when x is $\frac{1}{3}$, when x is -4, and when x is 3.

Maintaining Your Skills

95. $(2x - 3)(2x + 3)$

96. $(4 - 5x)(4 + 5x)$

97. $(2x - 3)^2$

98. $(4 - 5x)^2$

99. $(2x - 3)(4x^2 + 6x + 9)$

100. $(2x + 3)(4x^2 - 6x + 9)$

Getting Ready for the Next Section

For each problem below, place a number or expression inside the parentheses so that the resulting statement is true.

101. $\dfrac{25}{64} = (\quad)^2$

102. $\dfrac{4}{9} = (\quad)^2$

103. $x^6 = (\quad)^2$

104. $x^8 = (\quad)^2$

105. $16x^4 = (\quad)^2$

106. $81y^4 = (\quad)^2$

Write as a perfect cube.

107. $\dfrac{1}{8} = (\quad)^3$

108. $\dfrac{1}{27} = (\quad)^3$

109. $x^6 = (\quad)^3$

110. $x^{12} = (\quad)^3$

111. $27x^3 = (\quad)^3$

112. $125y^3 = (\quad)^3$

113. $8y^3 = (\quad)^3$

114. $1000x^3 = (\quad)^3$

One Step Further

Factor completely.

115. $8x^6 + 26x^3y^2 + 15y^4$

116. $24x^4 + 6x^2y^3 - 45y^6$

117. $3x^2 + 295x - 500$

118. $3x^2 + 594x - 1{,}200$

119. $\dfrac{1}{8}x^2 + x + 2$

120. $\dfrac{1}{9}x^2 + x + 2$

121. $2x^2 + 1.5x + 0.25$

122. $6x^2 + 2x + 0.16$

Find the Mistake

Write true or false for each statement below. If false, circle what is false and write the correct answer on the line provided.

1. True or False: The polynomial $x^2 - 5xy - 24y^2$ has a leading coefficient of 1. _____

2. True or False: A polynomial that cannot be factored is called a prime polynomial. _____

3. True or False: The leading coefficient of the polynomial $4x^4 - 2x^3y^2 - 12x^2y^3$ is 12. _____

4. True or False: The product ac for the polynomial $6a^4 - a^3 - 2x^2$ is -12. _____

KEY WORDS

perfect square trinomial

difference of two squares

sum of two cubes

difference of two cubes

4.5 Factoring Special Products

Image © Patrick Nugent, 2010

In Northern California, a dog named Mutley and a cat named Hawkeye enjoy scuba diving together. That's right, scuba diving! The animals' owner made miniature scuba diving suits for the dog and cat, and the unlikely friends take regular underwater strolls in the owner's backyard pool. They have even frequented shallow waters in the Caribbean. In order to build the suits, the owner had to order special equipment.

Suppose for each suit, the man ordered a small oxygen tank, plastic tubing, an inverted Plexiglas bubble for a helmet, and a small neoprene wetsuit. The items could fit into a box with four sections similar to Figure 1 below which you may recall from earlier in the chapter. In order to find the base area of the box, we can square the length of its side, giving us $(a + b)^2$. However, we can add the areas of the four smaller figures to arrive at the same result.

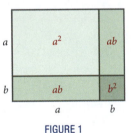

FIGURE 1

Since the area of the large square is the same whether we find it by squaring a side or by adding the four smaller areas, we can write the following relationship:

$$(a + b)^2 = a^2 + 2ab + b^2$$

This is the formula for the square of a binomial. The figure gives us a geometric interpretation for one of the special multiplication formulas. We begin this section by looking at the special multiplication formulas from a factoring perspective.

A Perfect Square Trinomials

We previously listed some special products found in multiplying polynomials. Two of the formulas looked like this:

$$(a + b)^2 = a^2 + 2ab + b^2$$

$$(a - b)^2 = a^2 - 2ab + b^2$$

If we exchange the left and right sides of each formula, we have two special formulas for factoring:

$$a^2 + 2ab + b^2 = (a + b)^2$$

$$a^2 - 2ab + b^2 = (a - b)^2$$

The left side of each formula is called a *perfect square trinomial*. The right sides are binomial squares. Perfect square trinomials can always be factored using the usual methods for factoring trinomials. However, if we notice that the first and last terms of a trinomial are perfect squares, it is wise to see whether the trinomial factors as a binomial square before attempting to factor by the usual method.

EXAMPLE 1   Factor $x^2 - 6x + 9$.

Solution   Since the first and last terms are perfect squares, we attempt to factor according to the preceding formulas.

$$x^2 - 6x + 9 = (x - 3)^2$$

If we expand $(x - 3)^2$, we have $x^2 - 6x + 9$, indicating we have factored correctly.

EXAMPLE 2   Factor each of the following perfect square trinomials.

Solution

a. $16a^2 + 40ab + 25b^2 = (4a + 5b)^2$

b. $49 - 14t + t^2 = (7 - t)^2$

c. $9x^4 - 12x^2 + 4 = (3x^2 - 2)^2$

d. $(y + 3)^2 + 10(y + 3) + 25 = [(y + 3) + 5]^2 = (y + 8)^2$

EXAMPLE 3   Factor $8x^2 - 24xy + 18y^2$.

Solution   We begin by factoring the greatest common factor 2 from each term.

$$8x^2 - 24xy + 18y^2 = 2(4x^2 - 12xy + 9y^2)$$
$$= 2(2x - 3y)^2$$

B The Difference of Two Squares

Recall the formula that results in the difference of two squares:

$$(a + b)(a - b) = a^2 - b^2$$

Writing this as a factoring formula, we have

$$a^2 - b^2 = (a + b)(a - b)$$

EXAMPLE 4   Each of the following is the difference of two squares. Use the formula $a^2 - b^2 = (a + b)(a - b)$ to factor each one.

Solution

a. $x^2 - 25 = x^2 - 5^2 = (x + 5)(x - 5)$

b. $49 - t^2 = 7^2 - t^2 = (7 + t)(7 - t)$

c. $81a^2 - 25b^2 = (9a)^2 - (5b)^2 = (9a + 5b)(9a - 5b)$

d. $4x^6 - 1 = (2x^3)^2 - 1^2 = (2x^3 + 1)(2x^3 - 1)$

e. $x^2 - \dfrac{4}{9} = x^2 - \left(\dfrac{2}{3}\right)^2 = \left(x + \dfrac{2}{3}\right)\left(x - \dfrac{2}{3}\right)$

As our next example shows, the difference of two fourth powers can be factored as the difference of two squares.

Practice Problems

1. Factor $x^2 - 10x + 25$.

2. Factor.

   a. $9a^2 + 42ab + 49b^2$

   b. $25 - 10t + t^2$

   c. $16x^4 - 24x^2 + 9$

   d. $(y + 2)^2 + 8(y + 2) + 16$

3. Factor $27x^2 - 36x + 12$.

4. Factor.

   a. $x^2 - 16$

   b. $64 - t^2$

   c. $25x^2 - 36y^2$

   d. $9x^6 - 1$

   e. $x^2 - \dfrac{25}{64}$

Answers

1. $(x - 5)^2$

2. a. $(3a + 7b)^2$ **b.** $(5 - t)^2$

   c. $(4x^2 - 3)^2$ **d.** $(y + 6)^2$

3. $3(3x - 2)^2$

4. a. $(x + 4)(x - 4)$

   b. $(8 + t)(8 - t)$

   c. $(5x + 6y)(5x - 6y)$

   d. $(3x^3 + 1)(3x^3 - 1)$

   e. $\left(x + \dfrac{5}{8}\right)\left(x - \dfrac{5}{8}\right)$

5. Factor $x^4 - 81$.

Note The sum of two squares never factors into the product of two binomials; that is, if we were to attempt to factor $(4x^2 + 9y^2)$ in the last example, we would be unable to find two binomials (or any other polynomials) whose product is $4x^2 + 9y^2$. The factors do not exist as polynomials.

6. Factor $(x - 4)^2 - 9$.

7. Factor $x^2 - 6x + 9 - y^2$.

8. Factor completely.

$x^3 + 5x^2 - 4x - 20$

Answers

5. $(x^2 + 9)(x + 3)(x - 3)$
6. $(x - 7)(x - 1)$
7. $(x - 3 + y)(x - 3 - y)$
8. $(x + 5)(x + 2)(x - 2)$

EXAMPLE 5   Factor $16x^4 - 81y^4$.

Solution The first and last terms are perfect squares. We factor according to the preceding formula.

$$16x^4 - 81y^4 = (4x^2)^2 - (9y^2)^2$$
$$= (4x^2 + 9y^2)(4x^2 - 9y^2)$$

Notice that the second factor is also the difference of two squares. Factoring completely, we have

$$16x^4 - 81y^4 = (4x^2 + 9y^2)(2x + 3y)(2x - 3y)$$

Here is another example of the difference of two squares.

EXAMPLE 6   Factor $(x - 3)^2 - 25$.

Solution This example has the form $a^2 - b^2$, where a is $x - 3$ and b is 5. We factor it according to the formula for the difference of two squares:

$$(x - 3)^2 - 25 = (x - 3)^2 - 5^2 \qquad \text{Write 25 as } 5^2.$$
$$= [(x - 3) + 5][(x - 3) - 5] \qquad \text{Factor.}$$
$$= (x + 2)(x - 8) \qquad \text{Simplify.}$$

Notice in this example we could have expanded $(x - 3)^2$, subtracted 25, and then factored to obtain the same result.

$$(x - 3)^2 - 25 = x^2 - 6x + 9 - 25 \qquad \text{Expand } (x - 3)^2.$$
$$= x^2 - 6x - 16 \qquad \text{Simplify.}$$
$$= (x - 8)(x + 2) \qquad \text{Factor.}$$

EXAMPLE 7   Factor $x^2 - 10x + 25 - y^2$.

Solution Notice the first three items form a perfect square trinomial; that is, $x^2 - 10x + 25 = (x - 5)^2$. If we replace the first three terms by $(x - 5)^2$, the expression that results has the form $a^2 - b^2$. We can factor as we did in Example 6.

$$x^2 - 10x + 25 - y^2 = (x^2 - 10x + 25) - y^2 \qquad \text{Group first three terms together.}$$
$$= (x - 5)^2 - y^2 \qquad \text{This has the form } a^2 - b^2.$$
$$= [(x - 5) + y][(x - 5) - y] \qquad \text{Factor according to the formula } a^2 - b^2 = (a + b)(a - b).$$
$$= (x - 5 + y)(x - 5 - y) \qquad \text{Simplify.}$$

We could check this result by multiplying the two factors together. (You may want to do that to convince yourself that we have the correct result.)

EXAMPLE 8   Factor completely $x^3 + 2x^2 - 9x - 18$.

Solution We use factoring by grouping to begin and then factor the difference of two squares.

$$x^3 + 2x^2 - 9x - 18 = x^2(x + 2) - 9(x + 2)$$
$$= (x + 2)(x^2 - 9)$$
$$= (x + 2)(x + 3)(x - 3)$$

C The Sum and Difference of Two Cubes

Here are the formulas for factoring the sum and difference of two cubes:

$$a^3 + b^3 = (a + b)(a^2 - ab + b^2)$$
$$a^3 - b^3 = (a - b)(a^2 + ab + b^2)$$

Since these formulas are unfamiliar, it is important that we verify them.

EXAMPLE 9 Verify the two formulas.

Solution We verify the formulas by multiplying the right sides and comparing the results with the left sides:

$$
\begin{array}{r}
a^2 - ab + b^2 \\
a + b \\
\hline
a^3 - a^2b + ab^2 \\
a^2b - ab^2 + b^3 \\
\hline
a^3 \qquad\qquad + b^3
\end{array}
\qquad\qquad
\begin{array}{r}
a^2 + ab + b^2 \\
a - b \\
\hline
a^3 + a^2b + ab^2 \\
- a^2b - ab^2 - b^3 \\
\hline
a^3 \qquad\qquad - b^3
\end{array}
$$

The first formula is correct.   The second formula is correct.

Here are some examples using the formulas for factoring the sum and difference of two cubes:

EXAMPLE 10 Factor $64 + t^3$.

Solution The first term is the cube of 4 and the second term is the cube of t. Therefore,

$$64 + t^3 = 4^3 + t^3$$
$$= (4 + t)(16 - 4t + t^2)$$

EXAMPLE 11 Factor $27x^3 + 125y^3$.

Solution Writing both terms as perfect cubes, we have

$$27x^3 + 125y^3 = (3x)^3 + (5y)^3$$
$$= (3x + 5y)(9x^2 - 15xy + 25y^2)$$

EXAMPLE 12 Factor $a^3 - \dfrac{1}{8}$.

Solution The first term is the cube of a, whereas the second term is the cube of $\frac{1}{2}$.

$$a^3 - \frac{1}{8} = a^3 - \left(\frac{1}{2}\right)^3$$
$$= \left(a - \frac{1}{2}\right)\left(a^2 + \frac{1}{2}a + \frac{1}{4}\right)$$

GETTING READY FOR CLASS

After reading through the preceding section, respond in your own words and in complete sentences.

A. What is a perfect square trinomial?

B. Is it possible to factor the sum of two squares?

C. Write the formula you use to factor the sum of two cubes.

D. Write a problem that uses the formula for the difference of two cubes.

---

9. Multiply $(x - 3)(x^2 + 3x + 9)$.

10. Factor $27 + x^3$.

11. Factor $8x^3 + y^3$.

12. Factor $a^3 - \dfrac{1}{27}$.

Answers

9. $x^3 - 27$

10. $(3 + x)(9 - 3x + x^2)$

11. $(2x + y)(4x^2 - 2xy + y^2)$

12. $\left(a - \frac{1}{3}\right)\left(a^2 + \frac{1}{3}a + \frac{1}{9}\right)$.

EXERCISE SET 4.5

Vocabulary Review

Match the following definitions of special products with the correct names.

1. $a^2 + 2ab + b^2 = (a + b)^2$ **a.** Difference of two squares

2. $a^2 - b^2 = (a - b)(a + b)$ **b.** Perfect square trinomial (−)

3. $a^2 - 2ab + b^2 = (a - b)^2$ **c.** Perfect square trinomial (+)

4. $a^3 - b^3 = (a - b)(a^2 + ab + b^2)$ **d.** Sum of two cubes

5. $a^3 + b^3 = (a + b)(a^2 - ab + b^2)$ **e.** Difference of two cubes

Problems

A Factor each perfect square trinomial.

1. $x^2 - 6x + 9$

2. $x^2 + 10x + 25$

3. $a^2 - 12a + 36$

4. $36 - 12a + a^2$

5. $25 - 10t + t^2$

6. $64 + 16t + t^2$

7. $\dfrac{1}{9}x^2 + 2x + 9$

8. $\dfrac{1}{4}x^2 - 2x + 4$

9. $4y^4 - 12y^2 + 9$

10. $9y^4 + 12y^2 + 4$

11. $16a^2 + 40ab + 25b^2$

12. $25a^2 - 40ab + 16b^2$

13. $\dfrac{1}{25} + \dfrac{1}{10}t^2 + \dfrac{1}{16}t^4$

14. $\dfrac{1}{9} - \dfrac{1}{3}t^3 + \dfrac{1}{4}t^6$

15. $y^2 + 3y + \dfrac{9}{4}$

16. $y^2 - 7y + \dfrac{49}{4}$

17. $a^2 - a + \dfrac{1}{4}$

18. $a^2 - 5a + \dfrac{25}{4}$

19. $x^2 - \dfrac{1}{2}x + \dfrac{1}{16}$

20. $x^2 - \dfrac{3}{4}x + \dfrac{9}{64}$

SCAN TO ACCESS

21. $t^2 + \dfrac{2}{3}t + \dfrac{1}{9}$

22. $t^2 - \dfrac{4}{5}t + \dfrac{4}{25}$

23. $16x^2 - 48x + 36$

24. $36x^2 + 48x + 16$

25. $75a^3 + 30a^2 + 3a$

26. $45a^4 - 30a^3 + 5a^2$

27. $(x + 2)^2 + 6(x + 2) + 9$

28. $(x + 5)^2 + 4(x + 5) + 4$

B Factor each as the difference of two squares. Be sure to factor completely.

29. $x^2 - 9$

30. $x^2 - 16$

31. $49x^2 - 64y^2$

32. $81x^2 - 49y^2$

33. $4a^2 - \dfrac{1}{4}$

34. $25a^2 - \dfrac{1}{25}$

35. $x^2 - \dfrac{9}{25}$

36. $x^2 - \dfrac{25}{36}$

37. $9x^2 - 16y^2$

38. $25x^2 - 49y^2$

39. $250 - 10t^2$

40. $640 - 10t^2$

Factor each as the difference of two squares. Be sure to factor completely.

41. $x^4 - 81$

42. $x^4 - 16$

43. $9x^6 - 1$

44. $25x^6 - 1$

45. $16a^4 - 81$

46. $81a^4 - 16b^4$

47. $\dfrac{1}{81} - \dfrac{y^4}{16}$

48. $\dfrac{1}{25} - \dfrac{y^4}{64}$

49. $\dfrac{x^4}{16} - \dfrac{16}{81}$

50. $81a^4 - 25b$

51. $a^4 - \dfrac{81}{256}$

52. $16a^4 - 625b^4$

Factor completely.

53. $x^6 - y^6$

54. $x^6 - 1$

55. $2a^7 - 128a$

56. $128a^8 - 2a^2$

57. $(x - 2)^2 - 9$

58. $(x + 2)^2 - 9$

59. $(y + 4)^2 - 16$

60. $(y - 4)^2 - 16$

61. $x^2 - 10x + 25 - y^2$

62. $x^2 - 6x + 9 - y^2$

63. $a^2 + 8a + 16 - b^2$

64. $a^2 + 12a + 36 - b^2$

65. $x^2 + 2xy + y^2 - a^2$

66. $a^2 + 2ab + b^2 - y^2$

67. $x^3 + 3x^2 - 4x - 12$

68. $x^3 + 5x^2 - 4x - 20$

69. $x^3 + 2x^2 - 25x - 50$

70. $x^3 + 4x^2 - 9x - 36$

71. $2x^3 + 3x^2 - 8x - 12$

72. $3x^3 + 2x^2 - 27x - 18$

73. $4x^3 + 12x^2 - 9x - 27$

74. $9x^3 + 18x^2 - 4x - 8$

75. $(2x - 5)^2 - 100$

76. $(7a + 5)^2 - 64$

77. $(a - 3)^2 - (4b)^2$

78. $(2x - 5)^2 - (6y)^2$

79. $a^2 - 6a + 9 - 16b^2$

80. $x^2 - 10x + 25 - 9y^2$

81. $x^2(x + 4) - 6x(x + 4) + 9(x + 4)$

82. $x^2(x - 6) + 8x(x - 6) + 16(x - 6)$

C Factor each of the following as the sum or difference of two cubes.

83. $x^3 - y^3$

84. $x^3 + y^3$

85. $a^3 + 8$

86. $a^3 - 8$

87. $27 + x^3$

88. $27 - x^3$

89. $y^3 - 1$

90. $y^3 + 1$

91. $10r^3 - 1{,}250$

92. $10r^3 + 1{,}250$

93. $64 + 27a^3$

94. $27 - 64a^3$

95. $8x^3 - 27y^3$

96. $27x^3 - 8y^3$

97. $t^3 + \dfrac{1}{27}$

98. $t^3 - \dfrac{1}{27}$

99. $27x^3 - \dfrac{1}{27}$

100. $8x^3 + \dfrac{1}{8}$

101. $64a^3 + 125b^3$

102. $125a^3 - 27b^3$

103. Find two values of b that will make $9x^2 + bx + 25$ a perfect square trinomial.

104. Find a value of c that will make $49x^2 - 42x + c$ a perfect square trinomial.

105. Find a value of c that will make $25x^2 - 90x + c$ a perfect square trinomial.

106. Find two values of b that will make $16x^2 + bx + 25$ a perfect square trinomial.

Getting Ready for the Next Section

Factor out the greatest common factor.

107. $y^3 + 25y$

108. $y^4 + 36y^2$

109. $2ab^5 + 8ab^4 + 2ab^3$

110. $3a^2b^3 + 6a^2b^2 - 3a^2b$

Factor by grouping.

111. $4x^2 - 6x + 2ax - 3a$

112. $6x^2 - 4x + 3ax - 2a$

113. $15ax^2 - 10a + 12x - 8$

114. $15bx + 3b - 25x - 5$

Factor the difference of squares.

115. $x^2 - 4$

116. $x^2 - 9$

117. $A^2 - 25$

118. $25x^2 - 36y^2$

Factor the perfect square trinomial.

119. $x^2 - 6x + 9$

120. $x^2 - 10x + 25$

121. $x^2 + 8xy + 16y^2$

122. $a^2 - 12ab + 36b^2$

Factor.

123. $6a^2 - 11a + 4$

124. $6x^2 - x - 15$

125. $12x^2 - 32x - 35$

126. $12a^2 - 7a - 10$

Factor the sum or difference of cubes.

127. $x^3 + 8$

128. $x^3 - 27$

129. $8x^3 - 27$

130. $27a^3 - 8b^3$

Maintaining Your Skills

Solve each system by using the elimination method.

131. $4x - 7y = 3$
$5x + 2y = -3$

132. $9x - 8y = 4$
$2x + 3y = 6$

133. $3x + 4y = 15$
$2x - 5z = -3$
$4y - 3z = 9$

134. $x + 3y = 5$
$6y + z = 12$
$x - 2z = -10$

One Step Further

Factor completely.

135. $a^2 - b^2 + 6b - 9$

136. $a^2 - b^2 - 18b - 81$

137. $(x - 3)^2 - (y + 5)^2$

138. $(a + 7)^2 - (b - 9)^2$

Find k such that each trinomial becomes a perfect square trinomial.

139. $kx^2 - 168xy + 49y^2$

140. $kx^2 + 110xy + 121y^2$

141. $49x^2 + kx + 81$

142. $64x^2 + kx + 169$

Find the Mistake

Each sentence below contains a mistake. Circle the mistake and write the correct word or phrase on the line provided.

1. The left side of the formula $a^2 + 2ab + b^2 = (a + b)^2$ is called a perfect square binomial. _____

2. The difference of two squares $(a^2 - b^2)$ factors into $(a + b)(a + b)$. _____

3. The difference of two cubes $(a^3 - b^3)$ factors into $a^2 + ab + b^2$. _____

4. The sum of two cubes $(27x^3 + y^8)$ factors into $(3x + y^2)(6x^2 - xy^2 + y^4)$. _____

4.6 Factoring: A General Review

©iStockphoto.com/pixdeluxe

A company that sells mountain biking apparel has just received an order for a new helmet and gloves. The helmet is packaged in a box with a volume of a^3 and the gloves are packaged in a box with a volume of b^3. In order to figure out the minimum volume needed for a shipping box to contain both items, we need to set up the following expression:

$$a^3 + b^3$$

Based on what we have learned so far, what type of special product will this expression factor into? Remember from the previous section that $a^3 + b^3$ is called the *sum of two cubes*.

A Factoring Review

In this section, we will review the different methods of factoring that we have presented in the previous sections of this chapter. This section is important because it will give you an opportunity to factor a variety of polynomials.

We begin this section by listing the steps that can be used to factor polynomials of any type.

> **HOW TO** Factor a Polynomial
>
> **Step 1:** If the polynomial has a greatest common factor other than 1, then factor out the greatest common factor.
>
> **Step 2:** If the polynomial has two terms (it is a binomial), then see if it is the difference of two squares or the sum or difference of two cubes, and then factor accordingly. Remember, if it is the sum of two squares it will not factor.
>
> **Step 3:** If the polynomial has three terms (a trinomial), then it is either a perfect square trinomial, which will factor into the square of a binomial, or it is not a perfect square trinomial, in which case we try to write it as the product of two binomials using the methods developed in this chapter.
>
> **Step 4:** If the polynomial has more than three terms, then try to factor it by grouping.
>
> **Step 5:** As a final check, see if any of the factors you have written can be factored further. If you have overlooked a common factor, you can catch it here.

Here are some examples illustrating how we use the steps in our list. There are no new factoring problems in this section. The problems here are all similar to the problems you have seen before. What is different is that they are not all of the same type.

EXAMPLE 1 Factor $2x^5 - 8x^3$.

Solution First we check to see if the greatest common factor is other than 1. Since the greatest common factor is $2x^3$, we begin by factoring it out. Once we have done so, we notice that the binomial that remains is the difference of two squares, which we factor according to the formula $a^2 - b^2 = (a + b)(a - b)$.

$$2x^5 - 8x^3 = 2x^3(x^2 - 4) \qquad \text{\color{green}Factor out the greatest common factor, } 2x^3.$$
$$= 2x^3(x + 2)(x - 2) \quad \text{\color{green}Factor the difference of two squares.}$$

Practice Problems

1. Factor $3x^8 - 27x^6$.

EXAMPLE 2 Factor $3x^4 - 18x^3 + 27x^2$.

Solution Step 1 is to factor out the greatest common factor $3x^2$. After we have done so, we notice that the trinomial that remains is a perfect square trinomial, which will factor as the square of a binomial.

$$3x^4 - 18x^3 + 27x^2 = 3x^2(x^2 - 6x + 9) \quad \text{\color{green}Factor out } 3x^2.$$
$$= 3x^2(x - 3)^2 \qquad \text{\color{green}}x^2 - 6x + 9 \text{ is the square of } x - 3.$$

2. Factor $4x^4 + 40x^3 + 100x^2$.

EXAMPLE 3 Factor $y^3 + 25y$.

Solution We begin by factoring out the y that is common to both terms. The binomial that remains after we have done so is the sum of two squares, which does not factor, so after the first step, we are finished.

$$y^3 + 25y = y(y^2 + 25)$$

3. Factor $y^4 + 36y^2$.

EXAMPLE 4 Factor $6a^2 - 11a + 4$.

Solution Here we have a trinomial that does not have a greatest common factor other than 1. Since it is not a perfect square trinomial, we factor it by trial and error. Without showing all the different possibilities, here is the answer.

$$6a^2 - 11a + 4 = (3a - 4)(2a - 1)$$

4. Factor $6x^2 - x - 15$.

EXAMPLE 5 Factor $2x^4 + 16x$.

Solution This binomial has a greatest common factor of $2x$. The binomial that remains after the $2x$ has been factored from each term is the sum of two cubes, which we factor according to the formula $a^3 + b^3 = (a + b)(a^2 - ab + b^2)$.

$$2x^4 + 16x = 2x(x^3 + 8) \qquad \text{\color{green}Factor } 2x \text{ from each term.}$$
$$= 2x(x + 2)(x^2 - 2x + 4) \quad \text{\color{green}The sum of two cubes}$$

5. Factor $3x^5 - 81x^2$.

EXAMPLE 6 Factor $2ab^5 + 8ab^4 + 2ab^3$.

Solution The greatest common factor is $2ab^3$. We begin by factoring it from each term. After that we find that the trinomial that remains cannot be factored further.

$$2ab^5 + 8ab^4 + 2ab^3 = 2ab^3(b^2 + 4b + 1)$$

6. Factor $3a^2b^3 + 6a^2b^2 - 3a^2b$.

Answers
1. $3x^6(x + 3)(x - 3)$
2. $4x^2(x + 5)^2$
3. $y^2(y^2 + 36)$
4. $(3x - 5)(2x + 3)$
5. $3x^2(x - 3)(x^2 + 3x + 9)$
6. $3a^2b(b^2 + 2b - 1)$

7. Factor $x^2 - 10x + 25 - b^2$.
(*Hint*: Group the first three terms together.)

EXAMPLE 7   Factor $4x^2 - 6x + 2ax - 3a$.

Solution   Our polynomial has four terms, so we factor by grouping.

$$4x^2 - 6x + 2ax - 3a = 2x(2x - 3) + a(2x - 3)$$
$$= (2x - 3)(2x + a)$$

8. Factor $x^6 - 1$.

EXAMPLE 8   Factor $x^6 - y^6$.

Solution   We have a choice of how we want to write the two terms to begin. We can write the expression as the difference of two squares, $(x^3)^2 - (y^3)^2$, or as the difference of two cubes, $(x^2)^3 - (y^2)^3$. It is better to use the difference of two squares if we have a choice.

$$x^6 - y^6 = (x^3)^2 - (y^3)^2$$

$$= (x^3 - y^3)(x^3 + y^3)$$

$$= (x - y)(x^2 + xy + y^2)(x + y)(x^2 - xy + y^2)$$

Try this example again writing the first line as the difference of two cubes instead of the difference of two squares. It will become apparent why it is better to use the difference of two squares.

GETTING READY FOR CLASS

After reading through the preceding section, respond in your own words and in complete sentences.

A. How do you know when you've factored completely?

B. What is the first step in factoring a polynomial?

C. If a polynomial has four terms, what method of factoring should you try?

D. What do we call a polynomial that does not factor?

Answers
7. $(x - 5 + b)(x - 5 - b)$
8. $(x + 1)(x^2 - x + 1)(x - 1)$
$(x^2 + x + 1)$

Vocabulary Review

Choose the correct words to fill in the blanks below.

| common | binomial | trinomial | grouping |
| greatest | polynomial | squares | cubes |

To factor a _____, follow these steps:

Step 1: If the polynomial has a _____ common factor other than 1, then factor it out.

Step 2: If the polynomial is a binomial, then see if it is the difference of two _____ or the sum or difference of two _____, and then factor accordingly.

Step 3: If the polynomial is a _____, then it is either a perfect square trinomial, which will factor into the square of a _____ or it is not a perfect square trinomial, in which case we try to write it as the product of two binomials.

Step 4: If the polynomial has more than three terms, then try to factor it by _____.

Step 5: As a final check, see if any of the factors you have written can be factored further. If you have overlooked a _____ factor, you can catch it here.

Problems

A Factor each of the following polynomials completely. Once you are finished factoring, none of the factors you obtain should be factorable. Also, note that the even-numbered problems are not necessarily similar to the odd-numbered problems that precede them in this problem set.

1. $x^2 - 81$

2. $x^2 - 18x + 81$

3. $x^2 + 2x - 15$

4. $15x^2 + 13x - 6$

5. $x^2(x + 2) + 6x(x + 2) + 9(x + 2)$

6. $12x^2 - 11x + 2$

7. $x^2y^2 + 2y^2 + x^2 + 2$

8. $21y^2 - 25y - 4$

9. $2a^3b + 6a^2b + 2ab$

10. $6a^2 - ab - 15b^2$

11. $x^2 + x + 1$

12. $x^2y + 3y + 2x^2 + 6$

13. $12a^2 - 75$

14. $18a^2 - 50$

15. $9x^2 - 12xy + 4y^2$

16. $x^3 - x^2$

17. $25 - 10t + t^2$

18. $t^2 + 4t + 4 - y^2$

19. $4x^3 + 16xy^2$

20. $16x^2 + 49y^2$

21. $2y^3 + 20y^2 + 50y$

22. $x^2 + 5bx - 2ax - 10ab$

23. $a^7 + 8a^4b^3$

24. $5a^2 - 45b^2$

25. $t^2 + 6t + 9 - x^2$

26. $36 + 12t + t^2$

27. $x^3 + 5x^2 - 9x - 45$

28. $x^3 + 5x^2 - 16x - 80$

29. $5a^2 + 10ab + 5b^2$

30. $3a^3b^2 + 15a^2b^2 + 3ab^2$

Factor completely.

31. $x^2 + 49$

32. $16 - x^4$

33. $3x^2 + 15xy + 18y^2$

34. $3x^2 + 27xy + 54y^2$

35. $9a^2 + 2a + \dfrac{1}{9}$

36. $18 - 2a^2$

37. $x^2(x-3) - 14x(x-3) + 49(x-3)$ **38.** $x^2 + 3ax - 2bx - 6ab$ **39.** $x^2 - 64$

40. $9x^2 - 4$

41. $8 - 14x - 15x^2$

42. $5x^4 + 14x^2 - 3$

43. $49a^7 - 9a^5$

44. $a^6 - b^6$

45. $r^2 - \dfrac{1}{25}$

46. $27 - r^3$

47. $49x^2 + 9y^2$

48. $12x^4 - 62x^3 + 70x^2$

49. $100x^2 - 100x - 600$

50. $100x^2 - 100x - 1{,}200$

51. $25a^3 + 20a^2 + 3a$

52. $16a^5 - 54a^2$

53. $3x^4 - 14x^2 - 5$

54. $8 - 2x - 15x^2$

55. $24a^5b - 3a^2b$

56. $18a^4b^2 - 24a^3b^3 + 8a^2b^4$

57. $64 - r^3$

58. $r^2 - \dfrac{1}{9}$

59. $20x^4 - 45x^2$

60. $16x^3 + 16x^2 + 3x$

61. $400t^2 - 900$

62. $900 - 400t^2$

63. $16x^5 - 44x^4 + 30x^3$

64. $16x^2 + 16x - 1$

65. $y^6 - 1$

66. $25y^7 - 16y^5$

67. $50 - 2a^2$

68. $4a^2 + 2a + \dfrac{1}{4}$

69. $12x^4y^2 + 36x^3y^3 + 27x^2y^4$

70. $16x^3y^2 - 4xy^2$

71. $x^2 - 4x + 4 - y^2$

72. $x^2 - 12x + 36 - b^2$

73. $a^2 - \dfrac{4}{3}ab + \dfrac{4}{9}b^2$

74. $a^2 + \dfrac{3}{2}ab + \dfrac{9}{16}b^2$

75. $x^2 - \dfrac{4}{5}xy + \dfrac{4}{25}y^2$

76. $x^2 - \dfrac{8}{7}xy + \dfrac{16}{49}y^2$

77. $a^2 - \dfrac{5}{3}ab + \dfrac{25}{36}b^2$

78. $a^2 + \dfrac{5}{4}ab + \dfrac{25}{64}b^2$

79. $x^2 - \dfrac{8}{5}xy + \dfrac{16}{25}y^2$

80. $a^2 + \dfrac{3}{5}ab + \dfrac{9}{100}b^2$

81. $2x^2(x + 2) - 13x(x + 2) + 15(x + 2)$

82. $5x^2(x - 4) - 14x(x - 4) - 3(x - 4)$ **83.** $(x - 4)^3 + (x - 4)^4$

84. $(2x - 7)^5 + (2x - 7)^6$

85. $2y^3 - 54$

86. $81 + 3y^3$

87. $2a^3 - 128b^3$

88. $128a^3 + 2b^3$

89. $2x^3 + 432y^3$

90. $432x^3 - 2y^3$

Maintaining Your Skills

The following problems are taken from the book *Algebra for the Practical Man*, written by J. E. Thompson and published by D. Van Nostrand Company in 1931.

91. A man spent $112.80 for 108 geese and ducks, each goose costing 14 dimes and each duck 6 dimes. How many of each did he buy?

92. If 15 pounds of tea and 10 pounds of coffee together cost $15.50, while 25 pounds of tea and 13 pounds of coffee at the same prices cost $24.55, find the price per pound of each.

93. A number of oranges at the rate of three for $0.10 and apples at $0.15 a dozen cost, together, $6.80. Five times as many oranges and one-fourth as many apples at the same rates would have cost $25.45. How many of each were bought?

94. An estate is divided among three persons: *A*, *B*, and *C*. *A*'s share is three times that of *B* and *B*'s share is twice that of *C*. If *A* receives $9,000 more than *C*, how much does each receive?

Getting Ready for the Next Section

Simplify.

95. $x^2 + (x + 1)^2$

96. $x^2 + (x + 3)^2$

97. $\dfrac{16t^2 - 64t + 48}{16}$

98. $\dfrac{100p^2 - 1,300p + 4,000}{100}$

Factor each of the following.

99. $x^2 - 2x - 24$

100. $x^2 - x - 6$

101. $2x^3 - 5x^2 - 3x$

102. $3x^3 - 5x^2 - 2x$

103. $x^3 + 2x^2 - 9x - 18$

104. $x^3 + 5x^2 - 4x - 20$

105. $x^3 + 2x^2 - 5x - 10$

106. $2x^3 + 3x^2 - 18x - 27$

Solve.

107. $x - 6 = 0$

108. $x + 4 = 0$

109. $2x + 1 = 0$

110. $3x + 1 = 0$

Find the Mistake

Each sentence below contains a mistake. Circle the mistake and write the correct word or phrase on the line provided.

1. The first step to factoring a polynomial is to see if it is the difference of two squares.

_____

2. If a polynomial is a perfect square trinomial, we can factor it into the square of a trinomial.

_____

3. If the polynomial has more than three terms, first try to factor it by trial and error. _____

4. A sum of two squares will factor into a perfect square trinomial. _____

4.7 Solving Equations by Factoring

OBJECTIVES

A Solve quadratic equations by factoring.

B Solve polynomial equations by factoring.

C Solve applications of polynomial equations.

KEY WORDS

quadratic equation

quadratic term

linear term

constant term

zero-factor property

Pythagorean theorem

Each year, the International Cherry Pit Spitting Championship is held in southwest Michigan to mark the start of the cherry harvest. The following is an excerpt from the Official Cherry Pit-Spitting Handbook written by the championship host Tree-Mendus Fruit Farm:

> Contestants must select three cherries from the regulation variety (Montmorency) supplied by the tournament committee. Cherries must be washed and chilled to 55-60°F pit temperature. Each cherry must be inserted in the mouth whole, all soluble solids eaten prior to spitting of the pit...Each contestant must spit his/her pit within 60 seconds of the time he/she is called to the line by the tournament judge. Three spits are allowed. The longest of three is recorded. If a pit is swallowed, that spit is forfeited.

Pit-spitting is serious business! Currently, the world record for a cherry pit spit is 100 feet and 4 inches. Believe it or not, we can apply math to a cherry pit spit. If given the initial velocity of the projected cherry pit and its height, we can set up a quadratic equation using its time of flight t. We will learn how to do so in this section. But first, here is the definition of a quadratic equation..

> **DEFINITION** quadratic equation
>
> Any equation that can be written in the form
>
> $$ax^2 + bx + c = 0$$
>
> where a, b, and c are constants and a is not 0 ($a \neq 0$) is called a *quadratic equation*. The form $ax^2 + bx + c = 0$ is called *standard form* for quadratic equations.

Each of the following is a quadratic equation:

$$2x^2 = 5x + 3 \qquad 5x^2 = 75 \qquad 4x^2 - 3x + 2 = 0$$

Notation For a quadratic equation written in standard form, the first term ax^2 is called the *quadratic term*; the second term bx is the *linear term*; and the last term c is called the *constant term*.

In the past we have noticed that the number 0 is a special number. There is another property of 0 that is the key to solving quadratic equations. It is called the *zero-factor property*.

> **PROPERTY** Zero-Factor Property
>
> For all real numbers r and s,
>
> $$r \cdot s = 0 \qquad \text{if and only if} \qquad r = 0 \quad \text{or} \quad s = 0 \quad \text{(or both)}$$

Note The third equation is clearly a quadratic equation since it is in standard form. (Notice that a is 4, b is -3, and c is 2.) The first two equations are also quadratic because they could be put in the form $ax^2 + bx + c = 0$ by using the addition property of equality.

Note What the zero-factor property says in words is that we can't multiply and get 0 without multiplying by 0; that is, if we multiply two numbers and get 0, then one or both of the original two numbers we multiplied must have been 0.

1. Solve $x^2 - x - 6 = 0$.

A Solving Quadratic Equations by Factoring

EXAMPLE 1 Solve $x^2 - 2x - 24 = 0$.

Solution We begin by factoring the left side as $(x - 6)(x + 4)$ and get

$$(x - 6)(x + 4) = 0$$

Now both $(x - 6)$ and $(x + 4)$ represent real numbers. We notice that their product is 0. By the zero-factor property, one or both of them must be 0.

$$x - 6 = 0 \quad \text{or} \quad x + 4 = 0$$

We have used factoring and the zero-factor property to rewrite our original second-degree equation as two first-degree equations connected by the word *or*. Completing the solution, we solve the two first-degree equations.

$$x - 6 = 0 \quad \text{or} \quad x + 4 = 0$$
$$x = 6 \qquad\qquad x = -4$$

We check our solutions in the original equation as follows:

$$\text{Check } x = 6 \qquad\qquad \text{Check } x = -4$$
$$6^2 - 2(6) - 24 \overset{?}{=} 0 \qquad (-4)^2 - 2(-4) - 24 \overset{?}{=} 0$$
$$36 - 12 - 24 \overset{?}{=} 0 \qquad\qquad 16 + 8 - 24 \overset{?}{=} 0$$
$$0 = 0 \qquad\qquad\qquad 0 = 0$$

In both cases the result is a true statement, which means that both 6 and -4 are solutions to the original equation. •

Although the next equation is not quadratic, the method we use is similar.

EXAMPLE 2 Solve $\dfrac{1}{3}x^3 = \dfrac{5}{6}x^2 + \dfrac{1}{2}x$.

Solution We can simplify our work if we clear the equation of fractions. Multiplying both sides by the LCD, 6, we have

$$6 \cdot \frac{1}{3}x^3 = 6 \cdot \frac{5}{6}x^2 + 6 \cdot \frac{1}{2}x$$
$$2x^3 = 5x^2 + 3x$$

2. Solve $\dfrac{1}{2}x^3 = \dfrac{5}{6}x^2 + \dfrac{1}{3}x$.

Next we add $-5x^2$ and $-3x$ to each side so that the right side will become 0.

$$2x^3 - 5x^2 - 3x = 0 \qquad\qquad \textit{Standard form}$$

We factor the left side and then use the zero-factor property to set each factor to 0.

$$x(2x^2 - 5x - 3) = 0 \qquad\qquad \textit{Factor out the greatest common factor.}$$
$$x(2x + 1)(x - 3) = 0 \qquad\qquad \textit{Continue factoring.}$$

$$x = 0 \text{ or } 2x + 1 = 0 \text{ or } x - 3 = 0 \qquad \textit{Zero-factor property}$$

Solving each of the resulting equations, we have

$$x = 0 \quad \text{or} \quad x = -\frac{1}{2} \quad \text{or} \quad x = 3 \qquad •$$

To generalize the preceding example, here are the steps used in solving a quadratic equation by factoring.

HOW TO Solve an Equation by Factoring

Step 1: Write the equation in standard form.

Step 2: Factor the left side.

Step 3: Use the zero-factor property to set each factor equal to 0.

Step 4: Solve the resulting linear equations.

Answers

1. $-2, 3$

2. $0, -\dfrac{1}{3}, 2$

EXAMPLE 3   Solve $100x^2 = 300x$.

Solution   We begin by writing the equation in standard form and factoring.

$$100x^2 = 300x$$

$100x^2 - 300x = 0$	Standard form
$100x(x - 3) = 0$	Factor.

Using the zero-factor property to set each factor to 0, we have

$$100x = 0 \quad \text{or} \quad x - 3 = 0$$
$$x = 0 \qquad\qquad x = 3$$

The two solutions are 0 and 3.

3. Solve $100x^2 = 500x$.

EXAMPLE 4   Solve $(x - 2)(x + 1) = 4$.

Solution   We begin by multiplying the two factors on the left side. (Notice that it would be incorrect to set each of the factors on the left side equal to 4. The fact that the product is 4 does not imply that either of the factors must be 4.)

$(x - 2)(x + 1) = 4$	
$x^2 - x - 2 = 4$	Multiply the left side.
$x^2 - x - 6 = 0$	Standard form
$(x - 3)(x + 2) = 0$	Factor.
$x - 3 = 0 \quad \text{or} \quad x + 2 = 0$	Zero-factor property
$x = 3 \qquad\qquad x = -2$	

4. Solve $(x + 1)(x + 2) = 12$.

EXAMPLE 5   Solve $x^3 + 2x^2 - 9x - 18 = 0$.

Solution   We start with factoring by grouping.

$x^3 + 2x^2 - 9x - 18 = 0$	
$x^2(x + 2) - 9(x + 2) = 0$	
$(x + 2)(x^2 - 9) = 0$	
$(x + 2)(x - 3)(x + 3) = 0$	The difference of two squares
$x + 2 = 0 \quad \text{or} \quad x - 3 = 0 \quad \text{or} \quad x + 3 = 0$	Zero-facor property
$x = -2 \qquad\qquad x = 3 \qquad\qquad x = -3$	

We have three solutions: -2, 3, and -3.

5. Solve $x^3 + 5x^2 - 4x - 20 = 0$.

EXAMPLE 6   Solve $(x + 2)(3x - 1) = (x + 2)(x + 6)$.

Solution   We begin by multiplying the factors on each side.

$(x + 2)(3x - 1) = (x + 2)(x + 6)$	
$3x^2 + 5x - 2 = x^2 + 8x + 12$	FOIL both sides
$2x^2 - 3x - 14 = 0$	Standard form
$(2x - 7)(x + 2) = 0$	Factor
$2x - 7 = 0 \quad \text{or} \quad x + 2 = 0$	Zero-factor Property
$2x = 7$	
$x = \frac{2}{7} \qquad\qquad x = -2$	

We have two solutions: $\frac{2}{7}$ and -2.

6. Solve
$(3x + 1)(x - 4) = (x - 3)(x + 3)$.

Answers

3. $0, 5$

4. $-5, 2$

5. $-5, -2, 2$

6. $\frac{1}{2}, 5$

7. One integer is 3 more than another. The sum of their squares is 29. Find the two integers.

B Solve Polynomial Equations by Factoring

EXAMPLE 7 The sum of the squares of two consecutive integers is 25. Find the two integers.

Solution We apply the Blueprint for Problem Solving to solve this application problem. Remember, step 1 in the blueprint is done mentally.

Step 1: *Read and list.*

Known items: Two consecutive integers. If we add their squares, the result is 25.
Unknown items: The two integers

Step 2: *Assign a variable and translate information.*

Let x = the first integer; then $x + 1$ = the next consecutive integer.

Step 3: *Reread and write an equation.*

Since the sum of the squares of the two integers is 25, the equation that describes the situation is

$$x^2 + (x + 1)^2 = 25$$

Step 4: *Solve the equation.*

$$x^2 + (x + 1)^2 = 25$$
$$x^2 + (x^2 + 2x + 1) = 25$$
$$2x^2 + 2x - 24 = 0$$
$$x^2 + x - 12 = 0$$
$$(x + 4)(x - 3) = 0$$
$$x = -4 \quad \text{or} \quad x = 3$$

Step 5: *Write the answer.*

If $x = -4$, then $x + 1 = -3$. If $x = 3$, then $x + 1 = 4$. The two integers are -4 and -3, or the two integers are 3 and 4.

Step 6: *Reread and check.*

The two integers in each pair are consecutive integers, and the sum of the squares of either pair is 25.

Another application of quadratic equations involves the Pythagorean theorem, an important theorem from geometry. The theorem gives the relationship between the sides of any right triangle (a triangle with a 90-degree angle). We state it here without proof.

FACTS FROM GEOMETRY The Pythagorean Theorem

In any right triangle, the square of the length of the longest side (hypotenuse) is equal to the sum of the squares of the length of the other two sides (legs).

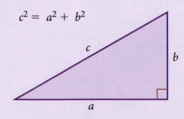

$$c^2 = a^2 + b^2$$

EXAMPLE 8   The lengths of the three sides of a right triangle are given by three consecutive integers. Find the lengths of the three sides.

Solution

Step 1: *Read and list.*

    Known items: A right triangle. The three sides are three consecutive integers.

    Unknown items: The three sides

Step 2: *Assign a variable and translate information.*

    Let x = first integer (shortest side).

    Then $x + 1$ = next consecutive integer

           $x + 2$ = last consecutive integer (longest side)

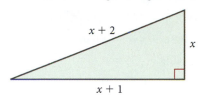

Step 3: *Reread and write an equation.*

    By the Pythagorean theorem, we have

$$(x + 2)^2 = (x + 1)^2 + x^2$$

Step 4: *Solve the equation.*

$$x^2 + 4x + 4 = x^2 + 2x + 1 + x^2$$
$$x^2 - 2x - 3 = 0$$
$$(x - 3)(x + 1) = 0$$
$$x = 3 \quad \text{or} \quad x = -1$$

Step 5: *Write the answer.*

    Since x is the length of a side in a triangle, it must be a positive number. Therefore, $x = -1$ cannot be used.

    The shortest side is 3. The other two sides are 4 and 5.

Step 6: *Reread and check.*

    The three sides are given by consecutive integers. The square of the longest side is equal to the sum of the squares of the two shorter sides.

EXAMPLE 9   A rectangular garden measures 40 meters by 30 meters. A lawn of uniform width surrounds the garden and has an area of 456 square meters. Find the width of the lawn that surrounds the garden.

Solution

Step 1: *Read and list.*

    Known items: The length and width of the garden

                  The area of the lawn

    Unknown items: The width of the lawn

Step 2: *Assign a variable and translate information.*

    Let w = the width of the lawn that surrounds the garden.

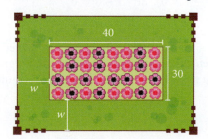

8. The longest side of a right triangle is 4 more than the shortest side. The third side is 2 more than the shortest side. Find the length of each side.

9. A rectangular garden measures 45 meters by 25 meters. A lawn of uniform width surrounds the garden and has an area of 144 square meters. Find the width of the lawn that surrounds the garden.

Answers

8. 6, 8, 10

9. 1 meter

Step 3: *Reread and write an equation.*

The area of the garden is $30(40) = 1,200$ m^2.

The area of the lawn is 456 m^2.

The length of the garden plus the lawn is $40 + 2w$.

The width of the garden plus the lawn is $30 + 2w$.

The product of these expressions is equal to the sum of their areas.

$$(40 + 2w)(30 + 2w) = 1,200 + 456$$

Step 4: *Solve the equation.*

$$(40 + 2w)(30 + 2w) = 1,200 + 456$$
$$1200 + 140w + 4w^2 = 1,656$$
$$4w^2 + 140w - 456 = 0$$
$$4(w^2 + 35w - 114) = 0$$
$$4(w - 3)(w + 38) = 0$$
$$w = 3 \quad \text{or} \quad w = -38$$

Step 5: *Write the answer.*

Because w is measuring length, it must be a positive number. Therefore, $w = -38$ cannot be used. The width of the lawn is 3 meters.

Step 6: *Reread and check.*

$$(40 + 2(3))(30 + 2(3)) \overset{?}{=} 1,200 + 456$$
$$46 \cdot 36 \overset{?}{=} 1,656$$
$$1,656 = 1,656 \qquad \text{A true statement}$$

C Applications of Polynomial Equations

Our next example involves formulas that are quadratic.

EXAMPLE 10   If an object is projected into the air with an initial vertical velocity of v feet/second, its height h, in feet, above the ground after t seconds will be given by

$$h = vt - 16t^2$$

Find t if $v = 64$ feet/second and $h = 48$ feet.

Solution   Substituting $v = 64$ and $h = 48$ into the preceding formula, we have

$$48 = 64t - 16t^2$$

which is a quadratic equation. We write it in standard form and solve by factoring.

$$16t^2 - 64t + 48 = 0$$
$$t^2 - 4t + 3 = 0 \qquad \text{Divide each side by 16.}$$
$$(t - 1)(t - 3) = 0$$
$$t - 1 = 0 \quad \text{or} \quad t - 3 = 0$$
$$t = 1 \qquad\qquad t = 3$$

Here is how we interpret our results: If an object is projected upward with an initial vertical velocity of 64 feet/second, it will be 48 feet above the ground after 1 second and after 3 seconds; that is, it passes 48 feet going up and also coming down.

---

10. Use the formula in Example 10 to find t if $v = 64$ feet/second and $h = 64$ feet.

Answer

10. 2 seconds

EXAMPLE 11 A manufacturer of headphones knows that the number of headphones she can sell each week is related to the price of the headphones by the equation $x = 1{,}300 - 100p$, where x is the number of headphones and p is the price per set. What price should she charge for each set of headphones if she wants the weekly revenue to be $4,000?

Solution The formula for total revenue is $R = xp$. Since we want R in terms of p, we substitute $1{,}300 - 100p$ for x in the equation $R = xp$.

If $\quad R = xp$

and $\quad x = 1{,}300 - 100p$

then $\quad R = (1{,}300 - 100p)p$

We want to find p when R is 4,000. Substituting 4,000 for R in the formula gives us

$$4{,}000 = (1{,}300 - 100p)p$$

$$4{,}000 = 1{,}300p - 100p^2$$

which is a quadratic equation. To write it in standard form, we add $100p^2$ and $-1{,}300p$ to each side, giving us

$$100p^2 - 1{,}300p + 4{,}000 = 0$$

$$p^2 - 13p + 40 = 0 \qquad \textcolor{green}{\textit{Divide each side by 100.}}$$

$$(p - 5)(p - 8) = 0$$

$$p - 5 = 0 \quad \text{or} \quad p - 8 = 0$$

$$p = 5 \qquad\qquad p = 8$$

If she sells the headphones for $5 each or for $8 each she will have a weekly revenue of $4,000.

11. Use the information in Example 10 to find the price that should be charged if the weekly revenue is to be $3,600.

GETTING READY FOR CLASS

After reading through the preceding section, respond in your own words and in complete sentences.

A. What is standard form for a quadratic equation?

B. Describe the zero-factor property in your own words.

C. What is the first step in solving an equation by factoring?

D. Explain the Pythagorean theorem in words.

Answer

11. $4 or $9

Vocabulary Review

Choose the correct words to fill in the blanks below.

standard   zero-factor   quadratic   constants   right   linear

1. An equation written in the form $ax^2 + bx + c = 0$ is called a _____ equation.

2. In a quadratic equation, a, b, and c are _____ and a is not 0.

3. In the quadratic equation $ax^2 + bx + c = 0$, ax^2 is called the quadratic term, bx is called the _____ term, and c is called the constant term.

4. The _____ property states that $r \cdot s = 0$ if and only if $r = 0$ or $s = 0$, or both.

5. To solve a quadratic equation by factoring, we first write the equation in _____ form.

6. For a _____ triangle, the Pythagorean theorem says that $a^2 + b^2 = c^2$.

Problems

A Solve each equation.

1. $x^2 - 5x - 6 = 0$

2. $x^2 + 5x - 6 = 0$

3. $x^3 - 5x^2 + 6x = 0$

4. $x^3 + 5x^2 + 6x = 0$

5. $3y^2 + 11y - 4 = 0$

6. $3y^2 - y - 4 = 0$

7. $60x^2 - 130x + 60 = 0$

8. $90x^2 + 60x - 80 = 0$

9. $\dfrac{1}{10}t^2 - \dfrac{5}{2} = 0$

10. $\dfrac{2}{7}t^2 - \dfrac{7}{2} = 0$

11. $100x^4 = 400x^3 + 2{,}100x^2$

12. $100x^4 = -400x^3 + 2{,}100x^2$

13. $\dfrac{1}{5}y^2 - 2 = -\dfrac{3}{10}y$

14. $\dfrac{1}{2}y^2 + \dfrac{5}{3} = \dfrac{17}{6}y$

15. $9x^2 - 12x = 0$

16. $4x^2 + 4x = 0$

17. $0.02r + 0.01 = 0.15r^2$

18. $0.02r - 0.01 = -0.08r^2$

SCAN TO ACCESS

19. $-100x = 10x^2$

20. $800x = 100x^2$

21. $(x + 6)(x - 2) = -7$

22. $(x - 7)(x + 5) = -20$

23. $(y - 4)(y + 1) = -6$

24. $(y - 6)(y + 1) = -12$

25. $(x + 1)^2 = 3x + 7$

26. $(x + 2)^2 = 9x$

27. $(2r + 3)(2r - 1) = -(3r + 1)$

28. $(3r + 2)(r - 1) = -(7r - 7)$

29. $3x^2 + x = 10$

30. $y^2 + y - 20 = 2y$

31. $12(x + 3) + 12(x - 3) = 3(x^2 - 9)$

32. $8(x + 2) + 8(x - 2) = 3(x^2 - 4)$

33. $(y + 3)^2 + y^2 = 9$

34. $(2y + 4)^2 + y^2 = 4$

35. $(x + 3)^2 + 1 = 2$

36. $(x - 3)^2 + (-1)^2 = 10$

37. $(3x + 1)(x - 4) = (x - 3)(x + 3)$

38. $(x + 3)(x + 6) = (3x + 1)(x - 4)$

39. $(3x - 2)(x + 1) = (x - 4)^2$

40. $(x - 3)(x - 2) = (3x - 2)(x + 1)$

41. $(2x - 3)(x - 5) = (x + 1)(x - 3)$

42. $(3x + 5)(x + 1) = (x - 5)^2$

B Use factoring to solve each polynomial equation.

43. $x^3 + 3x^2 - 4x - 12 = 0$

44. $x^3 + 5x^2 - 4x - 20 = 0$

45. $x^3 + 2x^2 - 25x - 50 = 0$

46. $x^3 + 4x^2 - 9x - 36 = 0$

47. $9a^3 = 16a$

48. $16a^3 = 25a$

49. $2x^3 + 3x^2 - 8x - 12 = 0$

50. $3x^3 + 2x^2 - 27x - 18 = 0$

51. $4x^3 + 12x^2 - 9x - 27 = 0$

52. $9x^3 + 18x^2 - 4x - 8 = 0$

53. Let $f(x) = \left(x + \frac{3}{2}\right)^2$. Find all values for the variable x, for which $f(x) = 0$.

54. Let $f(x) = \left(x - \frac{5}{2}\right)^2$. Find all values for the variable x, for which $f(x) = 0$.

55. Let $f(x) = (x - 3)^2 - 25$. Find all values for the variable x, for which $f(x) = 0$.

56. Let $f(x) = 9x^3 + 18x^2 - 4x - 8$. Find all values for the variable x, for which $f(x) = 0$.

Let $f(x) = x^2 + 6x + 3$. Find all values for the variable x, for which $f(x) = g(x)$.

57. $g(x) = -6$

58. $g(x) = 19$

59. $g(x) = 10$

60. $g(x) = -2$

Let $h(x) = x^2 - 5x$. Find all values for the variable x, for which $h(x) = f(x)$.

61. $f(x) = 0$

62. $f(x) = -6$

63. $f(x) = 2x + 8$

64. $f(x) = -2x + 10$

Find all values for the variable x such that $f(x) = x$.

65. $f(x) = x^2$

66. $f(x) = 4x - 7$

67. $f(x) = \frac{1}{3}x + 1$

68. $f(x) = x^2 - 2$

Applying the Concepts

69. The product of two consecutive odd integers is 99. Find the two integers.

70. The product of two consecutive integers is 132. Find the two integers.

71. The sum of two numbers is 14. Their product is 48. Find the numbers.

72. The sum of two numbers is 12. Their product is 32. Find the numbers.

73. The dimensions of a rectangular photograph are consecutive even integers. The product of the integers is 10 less than 5 times their sum. Find the two integers.

74. The dimensions of a rectangular photograph are consecutive even integers. The sum of the integers squared is 100. Find the integers.

75. In May 2012, fiction authors John Grisham and Toni Morrison had books on the New York Times Hardcover Fiction Best Sellers list. The books, *Calico Joe* and *Home* respectively were ranked as two consecutive odd integers. The product of the integers is 1 less than 4 times their sum. Find the two integers.

76. On Billboard's 2011 year-end Hot 100 Songs list, Katy Perry had songs with rankings that were consecutive integers. The square of the sum of the integers is 49. Find the two integers.

77. **Area Problem** The area of the rectangle below is 126 cm². Find the length and width using the given measurements.

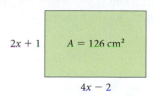

78. **Area Problem** The area of the rectangle is 120 ft². Find the length and width.

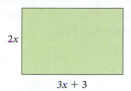

79. **Area Problem** The area of the triangle is 80 m². Find the base and height.

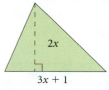

80. **Area Problem** The area of the triangle is 132 cm². Find the base and height.

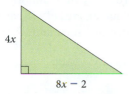

81. **Area Problem** The area of the triangle is 63 in². Find the base and height.

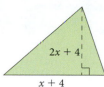

82. **Area Problem** The area of the triangle is 35 cm². Find the base and height.

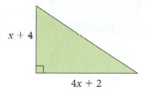

83. **Pythagorean Theorem** Find the value of x.

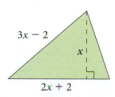

84. **Pythagorean Theorem** Find the value of x.

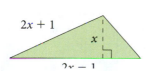

85. **Right Triangle** The lengths of the three sides of a right triangle are given by three consecutive even integers. Find the lengths of the three sides.

86. **Right Triangle** The longest side of a right triangle is 3 less than twice the shortest side. The third side measures 12 inches. Find the length of the shortest side.

87. **Geometry** The length of a rectangle is 2 meters more than 3 times the width. If the area is 16 square meters, find the width and the length.

88. **Geometry** The length of a rectangle is 4 yards more than twice the width. If the area is 70 square yards, find the width and the length.

89. **Geometry** The base of a triangle is 2 cm more than 4 times the height. If the area is 36 cm², find the base and the height.

90. **Geometry** The height of a triangle is 4 feet less than twice the base. If the area is 48 square feet, find the base and the height.

91. Projectile Motion If an object is thrown straight up into the air with an initial velocity of 32 feet per second, then its height above the ground at any time t is given by the formula $h = 32t - 16t^2$. Find the times at which the object is on the ground by letting $h = 0$ in the equation and solving for t.

92. Projectile Motion An object is projected into the air with an initial velocity of 64 feet per second. Its height at any time t is given by the formula $h = 64t - 16t^2$. Find the times at which the object is on the ground.

93. Surrounding Area A rectangular garden measures 40 yards by 35 yards. A lawn of uniform width surrounds the garden and has an area of 316 square yards. Find the width of the lawn that surrounds the garden.

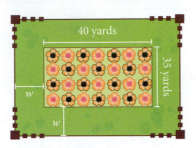

94. Surrounding Area A rectangular garden measures 100 feet by 50 feet. A lawn of uniform width surrounds the garden and has an area of 1,600 square feet. Find the width of the lawn that surrounds the garden.

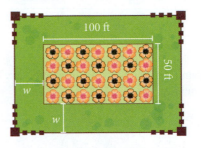

95. Surrounding Area A pool that is 75 feet long and 35 feet wide is surrounded by a cement path. The area of the path and the pool combined is 2,975 square feet. What is the area of the path?

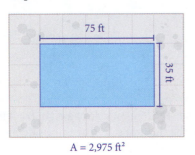

96. Surrounding Area A pool that is 100 feet long and 50 feet wide is surrounded by a cement path. The area of the path and the pool combined is 8,400 square feet. How wide is the path?

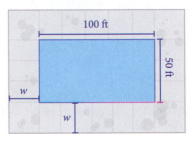

97. Surrounding Area A standard tennis court is 78 feet by 36 feet. There is a path that surrounds the court as shown in the diagram. If the total area of the court and path combined is 7,560 square feet, how wide are the sides of the path? (Solve for x.)

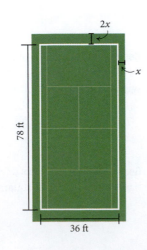

98. Surrounding Area Two standard tennis courts, 78 feet by 36 feet, are side by side with paths surrounding each court as shown. The total area of the courts and paths combined is 15,120 square feet. How wide are the side paths? (Solve for x.)

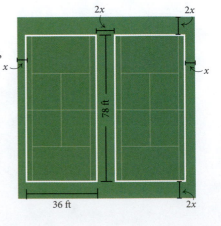

99. Surrounding Area When you open a book, you'll notice the type doesn't touch the edges of the page. The text block is surrounded by margins on all four sides. In the book pictured below, the pages are each 10 inches by 7 inches and the margins are as shown. If there is 23.35 square inches of total margin, how wide is the left margin? (Solve for x.)

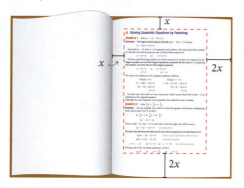

100. Surrounding Area Henry bought a 16 in. by 20 in. flat screen TV. If the front of the TV including the frame has an area of 396 square inches, and the frame is the same width all the way around, how wide is the frame?

C The formula $h = vt - 16t^2$ gives the height h, in feet, of an object projected into the air with an initial vertical velocity v, in feet per second, after t seconds.

101. Projectile Motion If an object is projected upward with an initial velocity of 48 feet per second, at what times will it reach a height of 32 feet above the ground?

102. Projectile Motion If an object is projected upward into the air with an initial velocity of 80 feet per second, at what times will it reach a height of 64 feet above the ground?

103. Projectile Motion An object is projected into the air with a vertical velocity of 24 feet per second. At what times will the object be on the ground? (It is on the ground when h is 0.)

104. Projectile Motion An object is projected into the air with a vertical velocity of 20 feet per second. At what times will the object be on the ground?

105. Height of a Bullet A bullet is fired into the air with an initial upward velocity of 80 feet per second from the top of a building 96 feet high. The equation that gives the height of the bullet at any time t is $h = 96 + 80t - 16t^2$. At what times will the bullet be 192 feet in the air?

106. Height of an Arrow An arrow is shot into the air with an upward velocity of 48 feet per second from a hill 32 feet high. The equation that gives the height of the arrow at any time t is $h = 32 + 48t - 16t^2$. Find the times at which the arrow will be 64 feet above the ground.

Maintaining Your Skills

Solve each system.

107. $2x - 5y = -8$
$3x + y = 5$

108. $4x - 7y = -2$
$-5x + 6y = -3$

109. $\frac{1}{3}x - \frac{1}{6}y = 3$
$-\frac{1}{5}x + \frac{1}{4}y = 0$

110. $2x - 5y = 16$
$y = 3x + 8$

Graph the solution set for each system.

111. $3x + 2y < 6$
$-2x + 3y < 6$

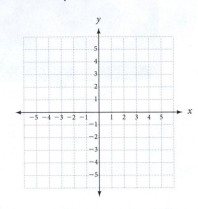

112. $y \le x + 3$
$y > x - 4$

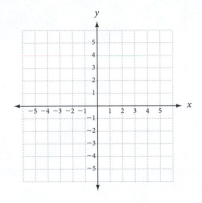

113. $x \le 4$
$y < 2$

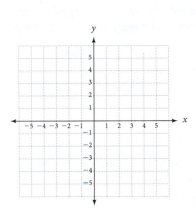

114. $2x + y < 4$
$x \ge 0$
$y \ge 0$

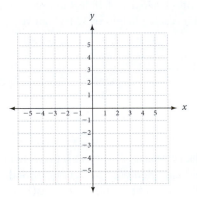

Find the Mistake

Each sentence below contains a mistake. Circle the mistake and write the correct word or phrase on the line provided.

1. The quadratic term for the equation $4x^2 + 4x - 3$ is 4. _____

2. The quadratic equation $x^2 - 5x = 6$ written in standard form is $x^2 = 5x + 6$. _____

3. The constant term for the equation $\frac{1}{10}y^2 - \frac{5}{2} = 0$ is 0. _____

4. The sum of two cubes $(27x^3 + y^6)$ factors into $(3x + y^2)(6x^2 - xy^2 + y^4)$. _____

A Access to the internet or a library for research

B Computer with a spreadsheet program

C Piece of paper and pencil

Blaze Your Own Trail: Starting a Business

Starting a business is a tricky road to navigate successfully as there are numerous costs to consider. For this project, work in groups to put together a plan for a new business.

1. Research the definitions of the following start-up cost categories. How much money would you need to fulfill these costs? Include an itemized list of any subcategories (e.g., administrative costs include rent, utilities, office supplies, etc.) and a total cost for each category. Create a spreadsheet to clearly present this information.

   a. Cost of sales

   b. Professional fees

   c. Technology costs

   d. Administrative costs

   e. Sales and marketing costs

   f. Wages and benefits

2. Would any of the above costs change once you have established your business?

3. Using the above information, calculate how much it would cost per month to get your business off the ground? This would include money spent setting up your business before you are able to sell your product or service.

4. Are there any hidden costs associated with starting your specific type of business?

5. Some analysts recommend considering any miscellaneous expenses to equal 10% of your total budget. How much money would this be for your business? Why would it be important to include this amount in your budget?

6. What is profit and how would you calculate it for your business? From where is profit coming for your business? How much money would you need to make from your business to turn a profit each month?

7. How would you use rational expressions when running your business?

8. What mathematical functions might you use when running your business?

Chapter **4** Summary

Addition of Polynomials [4.1]

1. $(3x^2 + 2x - 5) + (4x^2 - 7x + 2)$

$= 7x^2 - 5x - 3$

To add two polynomials, simply combine the coefficients of similar terms.

Negative Signs Preceding Parentheses [4.1]

2. $-(2x^2 - 8x - 9)$

$= -2x^2 + 8x + 9$

If there is a negative sign directly preceding the parentheses surrounding a polynomial, we may remove the parentheses and preceding negative sign by changing the sign of each term within the parentheses. (This procedure is actually just another application of the distributive property.)

Multiplication of Polynomials [4.2]

3. $(3x - 5)(x + 2)$

$= 3x^2 + 6x - 5x - 10$

$= 3x^2 + x - 10$

To multiply two polynomials, multiply each term in the first by each term in the second.

Special Products [4.2]

4. The following are examples of the three special products.

$(x + 3)^2 = x^2 + 6x + 9$

$(5 - x)^2 = 25 - 10x + x^2$

$(x + 7)(x - 7) = x^2 - 49$

$$(a + b)^2 = a^2 + 2ab + b^2 \qquad \text{Binomial squares}$$

$$(a - b)^2 = a^2 - 2ab + b^2$$

$$(a + b)(a - b) = a^2 - b^2 \qquad \text{Difference of two squares}$$

Business Applications [4.2, 4.3, 4.4]

5. A company makes x items each week and sells them for p dollars each, according to the equation $p = 35 - 0.1x$. Then, the revenue is $R = x(35 - 0.1x) = 35x - 0.1x^2$ If the total cost to make all x items is $C = 8x + 500$, then the profit gained by selling the x items is

$P(x) = 35x - 0.1x^2 - (8x + 500)$

$= -500 + 27x - 0.1x^2$

If a company manufacturers and sells x items at p dollars per item, then the revenue R is given by the formula

$$R(x) = xp$$

If the total cost to manufacture all x items is $C(x)$, then the profit, $P(x)$, obtained from selling all x items is

$$P(x) = R(x) - C(x)$$

Greatest Common Factor [4.3]

6. The greatest common factor of $10x^5 - 15x^4 + 30x^3$ is $5x^3$. Factoring it out of each term, we have

$5x^3(2x^2 - 3x + 6)$

The greatest common factor of a polynomial is the largest monomial (the monomial with the largest coefficient and highest exponent) that divides each term of the polynomial. The first step in factoring a polynomial is to factor the greatest common factor (if it is other than 1) out of each term.

Factoring Trinomials [4.4]

We factor a trinomial by writing it as the product of two binomials. (This refers to trinomials whose greatest common factor is 1.) Each factorable trinomial has a unique set of factors. Finding the factors is sometimes a matter of trial and error.

7. $x^2 + 5x + 6 = (x + 2)(x + 3)$

$x^2 - 5x + 6 = (x - 2)(x - 3)$

$x^2 + x - 6 = (x - 2)(x + 3)$

$x^2 - x - 6 = (x + 2)(x - 3)$

Special Factoring [4.5]

$a^2 + 2ab + b^2 = (a + b)^2$ Perfect square trinomials

$a^2 - 2ab + b^2 = (a - b)^2$

$a^2 - b^2 = (a - b)(a + b)$ Difference of two squares

$a^3 - b^3 = (a - b)(a^2 + ab + b^2)$ Difference of two cubes

$a^3 + b^3 = (a + b)(a^2 - ab + b^2)$ Sum of two cubes

A sum of two squares does not result n the product of two binomials; that is, there is no binomial factorization for a sum of two squares.

8. Here are some binomials that have been factored this way:

$x^2 + 6x + 9 = (x + 3)^2$

$x^2 - 6x + 9 = (x - 3)^2$

$x^2 - 9 = (x + 3)(x - 3)$

$x^3 - 27 = (x - 3)(x^2 + 3x + 9)$

$x^3 + 27 = (x + 3)(x^2 - 3x + 9)$

To Factor Polynomials in General [4.6]

Step 1: If the polynomial has a greatest common factor other than 1, then factor out the greatest common factor.

Step 2: If the polynomial has two terms (it is a binomial), then see if it is the difference of two squares, or the sum or difference of two cubes, and then factor accordingly. Remember, if it is the sum of two squares it will not factor.

Step 3: If the polynomial has three terms (a trinomial), then it is either a perfect square trinomial, which will factor into the square of a binomial, or it is not a perfect square trinomial, in which case you use one of the methods developed in Section 4.5.

Step 4: If the polynomial has more than three terms, then try to factor it by grouping.

Step 5: As a final check, see if any of the factors you have written can be factored further. If you have overlooked a common factor, you can catch it here.

9. Factor completely.

a. $3x^3 - 6x^2 = 3x^2(x - 2)$

b. $x^2 - 9 = (x + 3)(x - 3)$
$x^3 - 8 = (x - 2(x^2 + 2x + 4)$
$x^3 + 27 = (x + 3)$
$(x^2 - 3x + 9)$

c. $x^2 - 6x + 9 = (x - 3)^2$
$6x^2 - 7x - 5 = (2x + 1)(3x - 5)$

d. $x^2 + ax + bx + ab$
$= x(x + a) + b(x + a)$
$= (x + a)(x + b)$

To Solve an Equation by Factoring [4.7]

Step 1: Write the equation in standard form.

Step 2: Factor the left side.

Step 3: Use the zero-factor property to set each factor equal to zero.

Step 4: Solve the resulting linear equations.

10. Solve $x^2 - 5x = -6$.

$x^2 - 5x + 6 = 0$

$(x - 3)(x - 2) = 0$

$x - 3 = 0$ or $x - 2 = 0$

$x = 3$ $x = 2$

COMMON MISTAKE

When we subtract one polynomial from another, it is common to forget to add the opposite of each term in the second polynomial. For example

$(6x - 5) - (3x + 4) = 6x - 5 - 3x + 4$ Mistake

This mistake occurs if the negative sign outside the second set of parentheses is not distributed over all terms inside the parentheses. To avoid this mistake, remember: the opposite of a sum is the sum of the opposites, or,

$-(3x + 4) = -3x + (-4)$

CHAPTER 4 Review

Simplify the following expressions. [4.1]

1. $\left(\frac{2}{3}x^3 - 3x - \frac{4}{3}\right) - \left(\frac{1}{3}x^2 - \frac{2}{3}x + \frac{4}{3}\right)$

2. $6 - 3[4(5x + 2) - 13x]$

Profit, Revenue, and Cost A company making stuffed animal toys finds that it can sell x toys per week at p dollars each, according to the formula $p(x) = 24 - 0.2x$. If the total cost to produce and sell x toys is $C(x) = 3x + 25$, find the following. [4.1, 4.2]

3. An equation for the revenue that gives the revenue in terms of x

4. The profit equation

5. The revenue brought in by selling 100 toys

6. The cost of producing 100 toys

7. The profit obtained by making and selling 100 toys

Multiply. [4.2]

8. $(x + 3)(-3x + 5)$

9. $(2x + 3)(2x^2 - 4x - 1)$

10. $(4a - 3)^2$

11. $(2x - 5)(2x + 5)$

12. $x(x + 4)(2x - 5)$

13. $\left(3x - \frac{1}{5}\right)\left(5x + \frac{1}{3}\right)$

Factor the following expressions. [4.3, 4.4, 4.5, 4.6]

14. $x^2 - 3x - 10$

15. $6x^4 + 14x^2 - 40$

16. $16x^4 - 81y^4$

17. $6ab + 3ay - 8bx - 4xy$

18. $y^3 - \frac{1}{64}$

19. $2x^4y^3 + 6x^3y^4 - 20x^2y^5$

20. $4a^2 - 4ab + b^2 - 16$

21. $81 - x^4$

Solve each equation [4.7]

22. $\frac{1}{2}x^2 = \frac{23}{6}x - \frac{7}{3}$

23. $36x^4 = 9x^3$

24. $(x - 3)(x + 4) = 8$

25. $x^3 + 3x^2 - 16x - 48 = 0$

Let $f(x) = x^2 + 4x - 28$. Find all values for the variable x for which $f(x) = g(x)$. [4.7]

26. $g(x) = 7$

27. $g(x) = -2x - 12$

28. Find the value of the variable x in the following figure. [4.7]

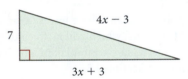

29. The area of the figure below is 35 square inches. Find the value of x. [4.7]

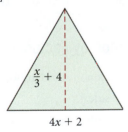

30. The area of the figure below is 18 square centimeters. Find the value of x. [4.7]

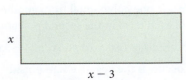

Simplify each of the following.

1. $(-3)^3$

2. $(-3)^{-3}$

3. $|-14 - 61| - 17$

4. $5^3 + 3(4^2 - 2^2)$

5. $84 \div 7 \cdot 4$

6. $77 \div 7 \cdot 11$

7. $\dfrac{7 - 2^3}{-4(6^2 - 5^2)}$

8. $\dfrac{8^2 - 32}{2^3(14 - 3 \cdot 6)}$

Let $P(x) = 12.25x - 0.01x^2$. Find the following.

9. $P(1)$

10. $P(-1)$

Let $Q(x) = \dfrac{5}{3}x^2 + \dfrac{2}{3}x - 7$. Find the following.

11. $Q(3)$

12. $Q(-3)$

Simplify each of the following expressions.

13. $17\left(x - \dfrac{1}{17}\right)$

14. $6\left(\dfrac{5}{3}x - \dfrac{7}{2}y\right)$

15. $(4x + 1) - (5x - 2)$

16. $5a - 3[4 + (a - 2)]$

17. $(4x^3 + 8) - (5x^2 - 8) + (3x^3 - 2) + (6x^2 + 1)$

18. $\left(\dfrac{4}{5}x^2 - \dfrac{1}{3}\right) - \left(\dfrac{2}{15}x^2 + \dfrac{8}{5}x\right) + \left(\dfrac{5}{3}x^2 - \dfrac{1}{15}\right) + \left(\dfrac{2}{5}x^2 + \dfrac{2}{5}\right)$

Multiply or divide as indicated.

19. $(2x^2y)^5$

20. $(-5ab^3)(4a^2b^2)$

21. $\dfrac{48x^2y}{16xy^9}$

22. $\dfrac{(5a^2b)^3}{(a^4b)^2}$

23. $(4x + 5)(2x - 1)$

24. $2x(9x^2 + 4x + 7)$

25. $\left(x - \dfrac{3}{4}\right)^2$

26. $\left(x + \dfrac{3}{4}\right)^3$

27. $(8 \times 10^4)(2 \times 10^3)$

28. $\dfrac{9 \times 10^8}{3 \times 10^2}$

Factor each of the following expressions.

29. $x^2 + ax - bx - ab$

30. $8a^2 + 10a - 7$

31. $(x - 1)^2 - 16$

32. $\dfrac{1}{27} + t^3$

Solve the following equations.

33. $5x - 3 = 12$

34. $|x + 7| - 4 = -2$

35. $\dfrac{2}{3}x + 8 = -12$

36. $\dfrac{2}{3}x = 2 - \dfrac{4}{3}x$

37. $16a^4 = 49a^2$

38. $(x + 5)^2 = 2x + 9$

Solve each system.

39. $-7x - 7y = -28$
$-7x + \ \ y = -4$

40. $y = -\dfrac{1}{5}x - 2$
$y = -3x - 2$

Find the slope of the line through the given points.

41. $(-3, 2)$ and $(4, 1)$

42. $(-16, 2)$ and $(3, 4)$

Let $f(x) = \dfrac{4x - 2}{3}$ and $g(x) = \dfrac{3x + 2}{4}$. Find the following.

43. $f(-4)$

44. $g(8)$

45. $(g \circ f)(11)$

46. $(f \circ g)(2)$

47. Geometry The height of a triangle is 4 feet less than 4 times the base. If the area is 12 square feet, find the base and height.

48. Geometry Find all three angles in a triangle if the smallest angle is one-third the largest angle and the remaining angle is 30 degrees more than the smallest angle.

The results of a survey of 80 internet users is displayed here as a Venn diagram. Use this information to answer Questions 49 and 50.

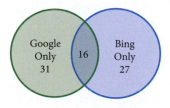

49. How many of the internet users surveyed use either Bing or Google?

50. How many of the internet users surveyed use neither Bing nor Google?

CHAPTER 4 Test

Simplify the following expressions. [4.1]

1. $\left(\frac{6}{5}x^3 - 2x - \frac{3}{5}\right) - \left(\frac{6}{5}x^2 - \frac{2}{5}x + \frac{3}{5}\right)$

2. $5 - 7[9(2x + 1) - 16x]$

Profit, Revenue, and Cost A company making ceramic coffee cups finds that it can sell x cups per week at p dollars each, according to the formula $p(x) = 36 - 0.3x$. If the total cost to produce and sell x coffee cups is $C(x) = 4x + 50$, find the following. [4.1, 4.2]

3. An equation for the revenue that gives the revenue in terms of x

4. The profit equation

5. The revenue brought in by selling 100 coffee cups

6. The cost of producing 100 coffee cups

7. The profit obtained by making and selling 100 coffee cups

Multiply. [4.2]

8. $(x + 7)(-5x + 4)$ **9.** $(3x - 2)(2x^2 + 6x - 5)$

10. $(3a^4 - 7)^2$ **11.** $(2x + 3)(2x - 3)$

12. $x(x - 7)(3x + 4)$ **13.** $\left(2x - \frac{1}{7}\right)\left(7x + \frac{1}{2}\right)$

Factor the following expressions. [4.3, 4.4, 4.5, 4.6]

14. $x^2 - 6x + 5$ **15.** $15x^4 + 33x^2 - 36$

16. $81x^4 - 16y^4$ **17.** $6ax - ay + 18b^2x - 3b^2y$

18. $y^3 + \frac{1}{27}$ **19.** $3x^4y^4 + 15x^3y^5 - 72x^2y^6$

20. $a^2 - 2ab - 36 + b^2$ **21.** $16 - x^4$

Solve each equation [4.7]

22. $\frac{1}{4}x^2 = -\frac{21}{8}x - \frac{5}{4}$ **23.** $243x^3 = 81x^4$

24. $(x + 5)(x - 2) = 8$ **25.** $x^3 + 5x^2 - 9x - 45 = 0$

Let $f(x) = x^2 - 2x - 15$. Find all values for the variable x for which $f(x) = g(x)$. [4.7]

26. $g(x) = 0$ **27.** $g(x) = 5 - 3x$

28. Find the value of the variable x in the following figure. [4.7]

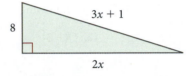

29. The area of the figure below is 12 square inches. Find the value of x. [4.7]

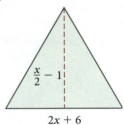

30. The area of the figure below is 12 square centimeters. Find the value of x. [4.7]

Rational Expressions, Equations, and Functions

5

Image © 2010 DigitalGlobe
Perrine Memorial Bridge
Twin Falls, Idaho

B ASE jumping is an extreme sport where participants jump off fixed objects and have mere seconds to deploy their parachute. The hazardous sport is named for the acronym representing the objects enthusiasts jump from: **B**uildings, **A**ntennae, **S**pans, and **E**arth. Despite the danger associated with this sport, there are many popular jumping destinations. Perrine Memorial Bridge in Twin Falls, Idaho is one such location. BASE jumping is allowed here year round without a permit, making it the only man-made structure in the United States where this is true. Standing at 486 feet, more than 10,000 jumps are made from Perrine each year.

Suppose a BASE jumper jumped from Perrine Memorial Bridge and fell 64 feet in the first 2 seconds of her jump. The table here summarizes the distances she will fall.

Input (time in seconds)	Output (distance in feet)
t	$d(t)$
0	0
1	16
2	64
3	144
4	256
5	400

Later in the chapter, you will learn how we arrived at these distances and how to use that formula, which uses a rational expression to determine how long the jumper will fall before she deploys her parachute.

OBJECTIVES

A Reduce rational expressions to lowest terms.

B Find function values for rational functions.

C Evaluate difference quotients.

KEY WORDS

rational expression

reduce

lowest terms

rational function

difference quotient

Archaeologists recently discovered the first use of pressurized plumbing in the New World. They unearthed a stone aqueduct in the ancient Mayan city of Palenque, Mexico. Water from a nearby stream would have spilled down the 60-meter-long aqueduct. The downward force of the water would have powered it through a narrow buried conduit for the last 2 meters, thereby building pressure and propelling it up out of the ground. Whether to control flooding or to feed a fountain, the exact purpose for the aqueduct is still unknown.

In order to mathematically understand the concept of pressure, like that used in the aqueduct, we need to explore rational expressions. For pressure P, we have to observe the rational relationship between force F and area A using the following equation:

$$P = \frac{F}{A}$$

Recall from a previous math class that a *rational number* is any number that can be expressed as the ratio of two integers:

$$\text{Rational numbers} = \left\{ \frac{a}{b} \,\middle|\, a \text{ and } b \text{ are integers, } b \neq 0 \right\}$$

A *rational expression* is defined similarly as any expression that can be written as the ratio of two polynomials:

$$\text{Rational expressions} = \left\{ \frac{P}{Q} \,\middle|\, P \text{ and } Q \text{ are polynomials, } Q \neq 0 \right\}$$

Some examples of rational expressions are

$$\frac{2x - 3}{x + 5} \qquad \frac{x^2 - 5x - 6}{x^2 - 1} \qquad \frac{a - b}{b - a}$$

Basic Properties

For rational expressions, multiplying the numerator and denominator by the same nonzero expression may change the form of the rational expression, but it will always produce an expression equivalent to the original one. The same is true when dividing the numerator and denominator by the same nonzero quantity.

> **PROPERTY** Properties of Rational Expressions
>
> If P, Q, and K are polynomials with $Q \neq 0$ and $K \neq 0$, then
>
> $$\frac{P}{Q} = \frac{PK}{QK} \qquad \text{and} \qquad \frac{P}{Q} = \frac{\frac{P}{K}}{\frac{Q}{K}}$$

A Reducing to Lowest Terms

The fraction $\frac{6}{8}$ can be written in lowest terms as $\frac{3}{4}$. The process is shown here:

$$\frac{6}{8} = \frac{3 \cdot 2}{4 \cdot 2} = \frac{3}{4}$$

Reducing $\frac{6}{8}$ to $\frac{3}{4}$ involves dividing the numerator and denominator by 2, the factor they have in common. Before dividing out the common factor 2, we must notice that the common factor *is* 2. (This may not be as obvious with other numbers. We are very familiar with the numbers 6 and 8 and therefore do not have to put much thought into finding what number divides both of them.)

We reduce rational expressions to lowest terms by first factoring the numerator and denominator and then dividing both numerator and denominator by any factors they have in common.

Practice Problems

EXAMPLE 1   Reduce $\dfrac{x^2 - 9}{x - 3}$ to lowest terms.

Solution   Factoring, we have

$$\frac{x^2 - 9}{x - 3} = \frac{(x + 3)(x - 3)}{x - 3}$$

1. Reduce $\dfrac{x^2 - 4}{x^2 - 2x - 8}$ to lowest terms.

The numerator and denominator have the factor $x - 3$ in common. Dividing the numerator and denominator by $x - 3$, we have

$$\frac{(x + 3)(x - 3)}{x - 3} = \frac{x + 3}{1} = x + 3$$

> **Note**   The lines drawn through the $(x - 3)$ in the numerator and denominator indicate that we have divided through by $(x - 3)$. As the problems become more involved, these lines will help keep track of which factors have been divided out and which have not.

For the problem in Example 1, there is an implied restriction on the variable x: It cannot be 3. If x were 3, the expression $\frac{(x^2 - 9)}{(x - 3)}$ would become $\frac{0}{0}$, an expression that we cannot associate with a real number. For all problems involving rational expressions, we restrict the variable to only those values that result in a nonzero denominator. When we state the relationship

$$\frac{x^2 - 9}{x - 3} = x + 3$$

we are assuming that it is true for all values of x except $x = 3$.

It is also a common mistake, when working a problem like Example 1, to begin by dividing common terms.

$$\frac{x^2 - 9}{x - 3} \neq \frac{x^2}{2} - \frac{9}{3}$$

Factoring the numerator is essential since we can only divide out common factors, not common terms.

Here are some other examples of reducing rational expressions to lowest terms:

EXAMPLE 2   Reduce $\dfrac{y^2 - 5y - 6}{y^2 - 1}$ to lowest terms.

Solution

$$\frac{y^2 - 5y - 6}{y^2 - 1} = \frac{(y - 6)(y + 1)}{(y - 1)(y + 1)}$$

$$= \frac{y - 6}{y - 1}$$

2. Reduce $\dfrac{x^2 + 2x - 8}{x^2 - 4}$ to lowest terms.

EXAMPLE 3   Reduce $\dfrac{2a^3 - 16}{4a^2 - 12a + 8}$ to lowest terms.

Solution

$$\frac{2a^3 - 16}{4a^2 - 12a + 8} = \frac{2(a^3 - 8)}{4(a^2 - 3a + 2)}$$

$$= \frac{2(a - 2)(a^2 + 2a + 4)}{4(a - 2)(a - 1)}$$

$$= \frac{a^2 + 2a + 4}{2(a - 1)}$$

3. Reduce $\dfrac{x^3 - 27}{3x^2 + 3x - 36}$ to lowest terms.

Answers
1. $\dfrac{x - 2}{x - 4}$
2. $\dfrac{x + 4}{x + 2}$
3. $\dfrac{x^2 + 3x + 9}{3(x + 4)}$

4. Reduce $\dfrac{xa + 3x - 4a - 12}{xa - 2a + 3x - 6}$ to lowest terms.

EXAMPLE 4   Reduce $\dfrac{x^2 - 3x + ax - 3a}{x^2 - ax - 3x + 3a}$ to lowest terms.

Solution

$$f(x) = \frac{x^2 - 3x + ax - 3a}{x^2 - ax - 3x + 3a}$$

$$= \frac{x(x - 3) + a(x - 3)}{x(x - a) - 3(x - a)}$$

$$= \frac{(x - 3)(x + a)}{(x - a)(x - 3)}$$

$$= \frac{x + a}{x - a}$$

The answer to Example 4 cannot be reduced further. It is a fairly common mistake to attempt to divide out an x or an a in this last expression. Remember, again we can divide out only the factors common to the numerator and denominator of a rational expression.

The next example involves what we call a trick. The trick is to reverse the order of the terms in a difference by factoring -1 from each term in either the numerator or the denominator. Here is how this is done.

5. Reduce $\dfrac{3 - a}{a - 3}$ to lowest terms.

EXAMPLE 5   Reduce $\dfrac{a - b}{b - a}$ to lowest terms.

Solution   The relationship between $a - b$ and $b - a$ is that they are opposites. We can show this fact by factoring -1 from each term in the numerator.

$$\frac{a - b}{b - a} = \frac{-1(-a + b)}{b - a} \qquad \text{Factor } -1 \text{ from each term in the numerator.}$$

$$= \frac{-1(b - a)}{b - a} \qquad \text{Reverse the order of the terms in the numerator.}$$

$$= -1 \qquad \text{Divide out common factor } b - a.$$

6. Reduce $\dfrac{2x^2 - 8}{2 - x}$ to lowest terms.

EXAMPLE 6   Reduce $\dfrac{x^2 - 25}{5 - x}$ to lowest terms.

Solution   We begin by factoring the numerator.

$$\frac{x^2 - 25}{5 - x} = \frac{(x - 5)(x + 5)}{5 - x}$$

The factors $x - 5$ and $5 - x$ are similar but are not exactly the same. We can reverse the order of either by factoring -1 from it. That is: $5 - x = -1(-5 + x) = -1(x - 5)$.

$$\frac{(x - 5)(x + 5)}{5 - x} = \frac{(x - 5)(x + 5)}{-1(x - 5)}$$

$$= \frac{x + 5}{-1}$$

$$= -(x + 5)$$

B Rational Functions

We can extend our knowledge of rational expressions to functions with the following definition:

> **DEFINITION**   rational function
>
> A *rational function* is any function that can be written in the form
>
> $$f(x) = \frac{P(x)}{Q(x)}$$
>
> where $P(x)$ and $Q(x)$ are polynomials and $Q(x) \neq 0$.

Answers

4. $\dfrac{x - 4}{x - 2}$

5. -1

6. $-2(x + 2)$

EXAMPLE 7   For the rational function $f(x) = \dfrac{x-4}{x-2}$, find $f(0)$, $f(-4)$, $f(4)$, $f(-2)$, and $f(2)$.

7. If $f(x) = \dfrac{x+5}{x-3}$ find $f(5)$.

Solution   To find these function values, we substitute the given value of x into the rational expression, and then simplify if possible.

$$f(0) = \frac{0-4}{0-2} = \frac{-4}{-2} = 2 \qquad\qquad f(-2) = \frac{-2-4}{-2-2} = \frac{-6}{-4} = \frac{3}{2}$$

$$f(-4) = \frac{-4-4}{-4-2} = \frac{-8}{-6} = \frac{4}{3} \qquad f(2) = \frac{2-4}{2-2} = \frac{-2}{0} \quad \text{Undefined}$$

$$f(4) = \frac{4-4}{4-2} = \frac{0}{2} = 0$$

Because the rational function in Example 7 is not defined when x is 2, the domain of that function does not include 2. We have more to say about the domain of a rational function next.

The Domain of a Rational Function

If the domain of a rational function is not specified, it is assumed to be all real numbers for which the function is defined. That is, the domain of the rational function

$$f(x) = \frac{P(x)}{Q(x)}$$

is all x for which $Q(x)$ is nonzero.

EXAMPLE 8   Find the domain for each function.

8. Find the domain of $q(x) = \dfrac{x}{2x^2 - 8}$.

a. $f(x) = \dfrac{x-4}{x-2}$ **b.** $g(x) = \dfrac{x^2+5}{x+1}$ **c.** $h(x) = \dfrac{x}{x^2-9}$

Solution

a. The domain for $f(x) = \dfrac{x-4}{x-2}$ is $\{x \mid x \neq 2\}$.

b. The domain for $g(x) = \dfrac{x^2+5}{x+1}$ is $\{x \mid x \neq -1\}$.

c. The domain for $h(x) = \dfrac{x}{x^2-9}$ is $\{x \mid x \neq -3, x \neq 3\}$.

Notice that, for these functions, $f(2)$, $g(-1)$, $h(-3)$, and $h(3)$ are all undefined, and that is why the domains are written as shown.

C Difference Quotients

Figure 1 is an important diagram from calculus. Although it may look complicated, the point of it is simple: The slope of the line passing through the points P and Q is given by the following formula.

$$\text{Slope of line through } PQ = m = \frac{f(x) - f(a)}{x - a}$$

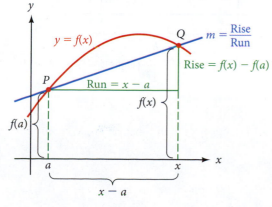

FIGURE 1

Answers

7. 5

8. $\{x \mid x \neq -2, x \neq 2\}$

The expression $\frac{f(x) - f(a)}{x - a}$ is called a *difference quotient*. When $f(x)$ is a polynomial, it will be a rational expression. It is also important to know that a is a constant and x is a variable.

9. If $f(x) = 4x + 2$, find $\frac{f(x) - f(a)}{x - a}$.

EXAMPLE 9   If $f(x) = 3x - 5$, find $\frac{f(x) - f(a)}{x - a}$.

Solution

$$
\begin{aligned}
\frac{f(x) - f(a)}{x - a} &= \frac{(3x - 5) - (3a - 5)}{x - a} \\
&= \frac{3x - 3a}{x - a} \\
&= \frac{3(x - a)}{x - a} \\
&= 3
\end{aligned}
$$

10. If $f(x) = 2x^2 - 5$, find $\frac{f(x) - f(a)}{x - a}$.

EXAMPLE 10   If $f(x) = x^2 - 4$, find $\frac{f(x) - f(a)}{x - a}$ and simplify.

Solution   Because $f(x) = x^2 - 4$ and $f(a) = a^2 - 4$, we have

$$
\begin{aligned}
\frac{f(x) - f(a)}{x - a} &= \frac{(x^2 - 4) - (a^2 - 4)}{x - a} \\
&= \frac{x^2 - 4 - a^2 + 4}{x - a} \\
&= \frac{x^2 - a^2}{x - a} \\
&= \frac{(x + a)(x - a)}{x - a} \qquad \text{Factor and divide out common factor.} \\
&= x + a
\end{aligned}
$$

Figure 2 is similar to the one in Figure 1. The main difference is in how we label the points. From Figure 2, we can see another difference quotient that gives us the slope of the line through the points P and Q. Similar to the formula used in Figure 1, h is a constant and x is a variable.

$$
\text{Slope of line through } PQ = m = \frac{f(x + h) - f(x)}{h}
$$

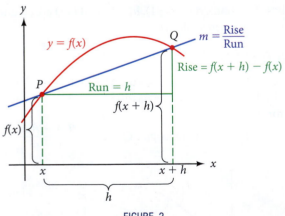

FIGURE 2

Examples 11 and 12 use the same functions used in Examples 9 and 10, but this time the new difference quotient is used.

EXAMPLE 11   If $f(x) = 3x - 5$, find $\dfrac{f(x + h) - f(x)}{h}$.

Solution   The expression $f(x + h)$ is given by

$$f(x + h) = 3(x + h) - 5$$
$$= 3x + 3h - 5$$

Using this result gives us

$$\frac{f(x + h) - f(x)}{h} = \frac{(3x + 3h - 5) - (3x - 5)}{h}$$
$$= \frac{3h}{h}$$
$$= 3$$

EXAMPLE 12   If $f(x) = x^2 - 4$, find $\dfrac{f(x + h) - f(x)}{h}$.

Solution   The expression $f(x + h)$ is given by

$$f(x + h) = (x + h)^2 - 4$$
$$= x^2 + 2xh + h^2 - 4$$

Using this result gives us

$$\frac{f(x + h) - f(x)}{h} = \frac{(x^2 + 2xh + h^2 - 4) - (x^2 - 4)}{h}$$
$$= \frac{2xh + h^2}{h}$$
$$= \frac{h(2x + h)}{h}$$
$$= 2x + h$$

GETTING READY FOR CLASS

After reading through the preceding section, respond in your own words and in complete sentences.

A. What is a rational expression?

B. Explain how to determine if a rational expression is in "lowest terms."

C. What is a rational function?

D. What is a difference quotient?

11. If $f(x) = 4x + 2$, find $\dfrac{f(x + h) - f(x)}{h}$.

12. If $f(x) = 2x^2 - 5$, find $\dfrac{f(x + h) - f(x)}{h}$.

Answers

11. 4
12. $2(2x + h)$

Vocabulary Review

Choose the correct words to fill in the blanks below.

nonzero   rational   domain   quotient   function

1. For _____ expressions, $\dfrac{P}{Q} = \dfrac{PK}{QK}$ and $\dfrac{P}{Q} = \dfrac{\frac{P}{K}}{\frac{Q}{K}}$.

2. Problems involving rational expressions contain a variable with a value that results in a _____ denominator.

3. $P(x)$ and $Q(x)$ in the rational _____ $f(x) = \dfrac{P(x)}{Q(x)}$ are polynomials.

4. The _____ of the rational function $f(x) = \dfrac{P(x)}{Q(x)}$ is all x for which $Q(x)$ is nonzero.

5. The expression $\dfrac{f(x) - f(a)}{(x - a)}$ is called a difference _____.

Problems

A Reduce each rational expression to lowest terms.

1. $\dfrac{x^2 - 16}{6x + 24}$

2. $\dfrac{12x - 9y}{3x^2 + 3xy}$

3. $\dfrac{a^4 - 81}{a - 3}$

4. $\dfrac{a^2 - 4a - 12}{a^2 + 8a + 12}$

5. $\dfrac{20y^2 - 45}{10y^2 - 5y - 15}$

6. $\dfrac{20x^2 - 93x + 34}{4x^2 - 9x - 34}$

7. $\dfrac{12y - 2xy - 2x^2y}{6y - 4xy - 2x^2y}$

8. $\dfrac{250a + 100ax + 10ax^2}{50a - 2ax^2}$

9. $\dfrac{(x - 3)^2(x + 2)}{(x + 2)^2(x - 3)}$

10. $\dfrac{(x - 4)^3(x + 3)}{(x + 3)^2(x - 4)}$

11. $\dfrac{x^3 + 1}{x^2 - 1}$

12. $\dfrac{x^3 - 1}{x^2 - 1}$

SCAN TO ACCESS

13. $\dfrac{4am - 4an}{3n - 3m}$

14. $\dfrac{ad - ad^2}{d - 1}$

15. $\dfrac{ab - a + b - 1}{ab + a + b + 1}$

16. $\dfrac{6cd - 4c - 9d + 6}{6d^2 - 13d + 6}$

17. $\dfrac{21x^2 - 23x + 6}{21x^2 + x - 10}$

18. $\dfrac{36x^2 - 11x - 12}{20x^2 - 39x + 18}$

19. $\dfrac{8x^2 - 6x - 9}{8x^2 - 18x + 9}$

20. $\dfrac{42x^2 + 23x - 10}{14x^2 + 45x - 14}$

21. $\dfrac{4x^2 + 29x + 45}{8x^2 - 10x - 63}$

22. $\dfrac{30x^2 - 61x + 30}{60x^2 + 22x - 60}$

23. $\dfrac{3a^2 + 10ax - 25x^2}{25x^2 - 9a^2}$

24. $\dfrac{6x^2 + xy - 12y^2}{16y^2 - 9x^2}$

25. $\dfrac{a^3 + b^3}{a^2 - b^2}$

26. $\dfrac{a^2 - b^2}{a^3 - b^3}$

27. $\dfrac{8x^4 - 8x}{4x^4 + 4x^3 + 4x^2}$

28. $\dfrac{6x^5 - 48x^3}{12x^3 + 24x^2 + 48x}$

29. $\dfrac{ax + 2x + 3a + 6}{ay + 2y - 4a - 8}$

30. $\dfrac{x^2 - 3ax - 2x + 6a}{x^2 - 3ax + 2x - 6a}$

31. $\dfrac{x^3 + 3x^2 - 4x - 12}{x^2 + x - 6}$

32. $\dfrac{6a^2 - 19ab + 15b^2}{12b^2 + ab - 6a^2}$

33. $\dfrac{9x^2 - 4y^2}{2y^2 + 3xy - 9x^2}$

34. $\dfrac{x^3 + 5x^2 - 4x - 20}{x^2 + 7x + 10}$

35. $\dfrac{x^3 - 8}{x^2 - 4}$

36. $\dfrac{y^2 - 9}{y^3 + 27}$

37. $\dfrac{8x^2 - 27}{4x^2 - 9}$

38. $\dfrac{25y^2 - 4}{125y^3 + 8}$

39. $\dfrac{x + 2}{x^3 + 8}$

40. $\dfrac{x + 3}{x^3 + 27}$

Refer to Examples 5 and 6 in this section, and reduce the following to lowest terms.

41. $\dfrac{x - 4}{4 - x}$

42. $\dfrac{6 - x}{x - 6}$

43. $\dfrac{y^2 - 36}{6 - y}$

44. $\dfrac{1 - y}{y^2 - 1}$

45. $\dfrac{1 - 9a^2}{9a^2 - 6a + 1}$

46. $\dfrac{1 - a^2}{a^2 - 2a + 1}$

Simplify each expression.

47. $\dfrac{(3x - 5) - (3a - 5)}{x - a}$

48. $\dfrac{(2x + 3) - (2a + 3)}{x - a}$

49. $\dfrac{(x^2 - 4) - (a^2 - 4)}{x - a}$

50. $\dfrac{(x^2 - 1) - (a^2 - 1)}{x - a}$

51. $\dfrac{(x^2 + 3x) - (a^2 + 3a)}{x - a}$

52. $\dfrac{(x^2 - 2x) - (a^2 - 2a)}{x - a}$

B State the domain for each rational function.

53. $f(x) = \dfrac{x - 3}{x - 1}$

54. $f(x) = \dfrac{x + 4}{x - 2}$

55. $g(x) = \dfrac{x^2 - 4}{x - 2}$

56. $g(x) = \dfrac{x^2 - 9}{x - 3}$

57. $h(t) = \dfrac{t - 4}{t^2 - 16}$

58. $h(t) = \dfrac{t - 5}{t^2 - 25}$

59. If $g(x) = \dfrac{x + 3}{x - 1}$, find $g(0), g(-3), g(3), g(-1)$, and $g(1)$, if possible.

60. If $g(x) = \dfrac{x - 2}{x - 1}$, find $g(0), g(-2), g(2), g(-1)$, and $g(1)$, if possible.

61. If $h(t) = \dfrac{t - 3}{t + 1}$, find $h(0), h(-3), h(3), h(-1)$, and $h(1)$, if possible.

62. If $h(t) = \dfrac{t - 2}{t + 1}$, find $h(0), h(-2), h(2), h(-1)$, and $h(1)$, if possible.

C For the functions below, evaluate each difference quotient of the form:

a. $\dfrac{f(x) - f(a)}{x - a}$

b. $\dfrac{f(x + h) - f(x)}{h}$

63. $f(x) = 4x$

64. $f(x) = -3x$

65. $f(x) = 5x + 3$

66. $f(x) = 6x - 5$

67. $f(x) = x^2$

68. $f(x) = 3x^2$

69. $f(x) = x^2 + 1$

70. $f(x) = x^2 - 3$

71. $f(x) = x^2 - 3x + 4$

72. $f(x) = x^2 + 4x - 7$

73. $f(x) = 2x^2 + 3x - 4$

74. $f(x) = 5x^2 + 3x - 7$

Applying the Concepts

75. Diet The following rational function is a mathematical model of a person's weekly progress on a weight-loss diet. The quantity $W(x)$ is the weight (in pounds) of the person after x weeks of dieting. Use the function to fill in the table.

$$W(x) = \frac{80(2x + 15)}{x + 6}$$

Weeks	Weight (lb)
x	$W(x)$
0	
1	
4	
12	
24	

76. Drag Racing The following rational function gives the speed $V(x)$, in miles per hour, of a dragster at each second x during a quarter-mile race. Use the function to fill in the table.

$$V(x) = \frac{340x}{x + 3}$$

Time (sec)	Speed (mph)
x	$V(x)$
0	
1	
2	
3	
4	
5	
6	

77. A new refrigerator cost $1,200 and costs approximately $540 per year to operate. Let n represent the total years you own the refrigerator.

a. Express the total cost C as a function of the number of years n. (Assume there are no additional costs).

b. Express the average yearly cost $\overline{C}$ as a function of the number of years n.

c. If you operate the refrigerator for 10 years, what is your average yearly cost?

d. If you operate the refrigerator for 15 years, what is your average yearly cost?

e. Is the average yearly cost increasing or decreasing as n increases? Explain your answer.

78. A new washer and dryer cost $2,200 and cost approximately $140 per year to operate. Let n represent the total years you own the washer and dryer.

a. Express the total cost C as a function of the number of years n. (Assume there are no additional costs).

b. Express the average yearly cost $\overline{C}$ as a function of the number of years n.

c. If you operate the washer and dryer for 10 years, what is your average yearly cost?

d. If you operate the washer and dryer for 15 years, what is your average yearly cost?

e. Is the average yearly cost increasing or decreasing as n increases? Explain your answer.

Getting Ready for the Next Section

Multiply or divide, as indicated.

79. $\dfrac{6}{7} \cdot \dfrac{14}{18}$

80. $\dfrac{6}{8} \div \dfrac{3}{5}$

81. $5y^2 \cdot 4x^2$

82. $6x^4 \cdot 12y^5$

Factor.

83. $x^2 - 4$

84. $x^2 - 6x + 9$

85. $x^3 - x^2y$

86. $2y^2 - 2$

One Step Further

The graphs of two rational functions are given in Figures 3 and 4. Use the graphs to find the following.

87. a. $f(2)$

   b. $f(-1)$

   c. $f(0)$

   d. $g(3)$

   e. $g(6)$

   f. $g(-1)$

   g. $f(g(6))$

88. a. $f(4)$

   b. $f(-2)$

   c. $g(-3)$

   d. $g(-6)$

   e. $g(f(4))$

   f. $f(g(-6))$

   g. $f(-2) + g(3)$

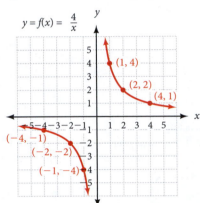

$y = f(x) = \dfrac{4}{x}$

FIGURE 3

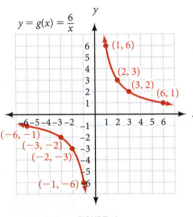

$y = g(x) = \dfrac{6}{x}$

FIGURE 4

Find the Mistake

Each sentence below contains a mistake. Circle the mistake and write the correct word or expression on the line provided.

1. The rational expression $\dfrac{8x^2 + 2x - 1}{16x^2 - 6x - 7}$ reduced to lowest terms is $\dfrac{4x - 1}{4x - 7}$. _____

2. To reduce the rational expression $\dfrac{15x^2 - 16x + 4}{5x^2 - 17x + 6}$ to lowest terms, we must divide the numerator and denominator by the factor $(3x - 2)$. _____

3. The rational function $f(x) = \dfrac{x^3 - 4x^2 + 4x + 16}{x^2 - 6x + 8}$ is undefined when $f(-2)$ and $f(1)$. _____

4. To find the difference quotient for $f(x) = x^2 + x - 7$, we use $f(a) = x + a + 1$. _____

Navigation Skills: Prepare, Study, Achieve

Think about your current study routine. Has it been successful? There are many things that you must consider when creating a routine. One important aspect is the environment in which you choose to study. Think about the location you typically study. Are you able to focus there without distraction? Consider what things may distract you (e.g., cell phone, television, noise, friends who socialize rather than study) and find a place to study where these things are absent. Other important aspects of a productive study routine are time of day you choose to study and sights and sounds around you during your study time. A study environment that is not distracting will help you focus and foster further success in this course.

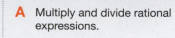

OBJECTIVES

A Multiply and divide rational expressions.

To have a home videotape transferred to DVD, the amount you pay for the service depends on the number of DVDs you have made. In other words, The more DVDs you have made, the lower the charge per copy. The following demand function gives the price (in dollars) per disk $p(x)$ a company charges for making x DVDs. As you can see, it is a rational function.

$$p(x) = \frac{2(x + 60)}{x + 5}$$

The graph in Figure 1 shows this function from $x = 0$ to $x = 100$. As you can see, the more DVDs that are made, the lower the price per copy.

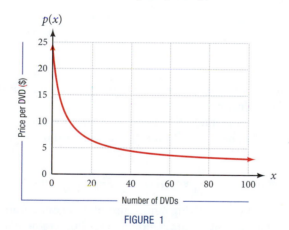

FIGURE 1

If we were interested in finding the revenue function for this situation, we would multiply the number of copies made x by the price per copy $p(x)$. This involves multiplication with a rational expression, which is one of the topics we cover in this section.

In the last section, we found the process of reducing rational expressions to lowest terms to be the same process used in reducing fractions to lowest terms. The similarity also holds for the process of multiplication or division of rational expressions.

Multiplication with fractions is the simplest of the four basic operations. To multiply two fractions, we simply multiply numerators and multiply denominators. That is, if a, b, c, and d are real numbers, with $b \neq 0$ and $d \neq 0$, then

$$\frac{a}{b} \cdot \frac{c}{d} = \frac{ac}{bd}$$

Practice Problems

1. Multiply $\frac{3}{4} \cdot \frac{10}{21}$.

A Multiplying and Dividing Rational Expressions

EXAMPLE 1   Multiply $\frac{6}{7} \cdot \frac{14}{18}$.

Solution

$$\frac{6}{7} \cdot \frac{14}{18} = \frac{6(14)}{7(18)} \qquad \text{Multiply numerators and denominators.}$$

$$= \frac{2 \cdot 3(2 \cdot 7)}{7(2 \cdot 3 \cdot 3)} \qquad \text{Factor.}$$

$$= \frac{2}{3} \qquad \text{Divide out common factors.}$$

Our next example is similar to some of the problems you may have worked previously. We multiply fractions whose numerators and denominators are monomials by multiplying numerators and multiplying denominators and then reducing to lowest terms. Here is how it looks:

2. Multiply $\frac{6x^4}{10y^5} \cdot \frac{15y^3}{3x^2}$.

EXAMPLE 2   Multiply $\frac{8x^3}{27y^8} \cdot \frac{9y^3}{12x^2}$.

Solution   We multiply numerators and denominators and then divide out common factors.

$$\frac{8x^3}{27y^8} \cdot \frac{9y^3}{12x^2} = \frac{8 \cdot 9x^3y^3}{27 \cdot 12x^2y^8} \qquad \begin{array}{l} \text{Multiply numerators.} \\ \text{Multiply denominators.} \end{array}$$

$$= \frac{4 \cdot 2 \cdot 9x^3y^3}{9 \cdot 3 \cdot 4 \cdot 3x^2y^8} \qquad \text{Factor coefficients.}$$

$$= \frac{2x}{9y^5} \qquad \text{Divide out common factors.}$$

The product of two rational expressions is the product of their numerators over the product of their denominators.

Once again, we should mention that the little slashes we have drawn through the factors are simply used to denote the factors we have divided out of the numerator and denominator.

3. Let $f(x) = \frac{x - 5}{x + 2}$ and

$g(x) = \frac{x^2 - 4}{3x - 15}$.

Find $f(x) \cdot g(x)$.

EXAMPLE 3   Let $f(x) = \frac{x - 3}{x^2 - 4}$ and $g(x) = \frac{x + 2}{x^2 - 6x + 9}$. Find $f(x) \cdot g(x)$.

Solution   We begin by multiplying numerators and denominators. We then factor all polynomials and divide out factors common to the numerator and denominator.

$$f(x) \cdot g(x) = \frac{x - 3}{x^2 - 4} \cdot \frac{x + 2}{x^2 - 6x + 9}$$

$$= \frac{(x - 3)(x + 2)}{(x^2 - 4)(x^2 - 6x + 9)} \qquad \text{Multiply.}$$

$$= \frac{(x - 3)(x + 2)}{(x + 2)(x - 2)(x - 3)(x - 3)} \qquad \text{Factor.}$$

$$= \frac{1}{(x - 2)(x - 3)}$$

The first two steps can be combined to save time. In the next example, we'll show how to perform the multiplication and factoring steps together.

Answers

1. $\frac{5}{14}$

2. $\frac{3x^2}{y^2}$

3. $f(x) \cdot g(x) = \frac{x - 2}{3}$

EXAMPLE 4 Multiply $\dfrac{2y^2 - 4y}{2y^2 - 2} \cdot \dfrac{y^2 - 2y - 3}{y^2 - 5y + 6}$.

Solution

$$\frac{2y^2 - 4y}{2y^2 - 2} \cdot \frac{y^2 - 2y - 3}{y^2 - 5y + 6} = \frac{2y(y-2)(y-3)(y+1)}{2(y+1)(y-1)(y-3)(y-2)}$$

$$= \frac{y}{y-1}$$

Notice in both of the preceding examples that we did not actually multiply the polynomials. It would be senseless to do that because we would then have to factor each of the resulting products to reduce them to lowest terms.

The quotient of two rational expressions is the product of the first and the reciprocal of the second. That is, we find the quotient of two rational expressions the same way we find the quotient of two fractions. Here is an example that reviews division with fractions:

EXAMPLE 5 Divide $\dfrac{6}{8} \div \dfrac{3}{5}$.

Solution

$$\frac{6}{8} \div \frac{3}{5} = \frac{6}{8} \cdot \frac{5}{3} \qquad \text{Write division in terms of multiplication.}$$

$$= \frac{6(5)}{8(3)} \qquad \text{Multiply numerators and denominators.}$$

$$= \frac{2 \cdot 3(5)}{2 \cdot 2 \cdot 2(3)} \qquad \text{Factor.}$$

$$= \frac{5}{4} \qquad \text{Divide out common factors.}$$

To divide one rational expression by another, we use the definition of division to multiply by the reciprocal of the expression that follows the division symbol.

EXAMPLE 6 Divide $\dfrac{8x^3}{5y^2} \div \dfrac{4x^2}{10y^6}$.

Solution First, we rewrite the problem in terms of multiplication. Then we multiply.

$$\frac{8x^3}{5y^2} \div \frac{4x^2}{10y^6} = \frac{8x^3}{5y^2} \cdot \frac{10y^6}{4x^2}$$

$$= \frac{\overset{2}{8} \cdot \overset{2}{10}x^3y^6}{\underset{}{4} \cdot 5x^2y^2}$$

$$= 4xy^4$$

EXAMPLE 7 Divide $\dfrac{x^2 - y^2}{x^2 - 2xy + y^2} \div \dfrac{x^3 + y^3}{x^3 - x^2y}$.

Solution We begin by writing the problem as the product of the first and the reciprocal of the second and then proceed as in the previous two examples.

$$\frac{x^2 - y^2}{x^2 - 2xy + y^2} \div \frac{x^3 + y^3}{x^3 - x^2y}$$

$$= \frac{x^2 - y^2}{x^2 - 2xy + y^2} \cdot \frac{x^3 - x^2y}{x^3 + y^3} \qquad \begin{array}{l}\text{Multiply by the reciprocal}\\\text{of the divison}\end{array}$$

$$= \frac{(x-y)(x+y)(x^2)(x-y)}{(x-y)(x-y)(x+y)(x^2 - xy + y^2)} \qquad \text{Factor and multiply.}$$

$$= \frac{x^2}{x^2 - xy + y^2} \qquad \text{Divide out common factors.}$$

4. Multiply
$$\frac{4x^2 - 16x}{2x^2 - 14x + 24} \cdot \frac{x^2 - 5x + 6}{x^2 + x - 6}.$$

5. Divide $\dfrac{3}{4} \div \dfrac{5}{8}$.

6. Divide $\dfrac{9x^2y}{10x^3y^2} \div \dfrac{6xy^3}{5x^5y^4}$.

> **Note** A common mistake is that students divide the 8 and the 10 in the original problem by 2 before multiplying. It is important to multiply by the reciprocal before dividing out common factors.

7. Divide $\dfrac{4x + 8}{x^2 - x - 6} \div \dfrac{x^2 + 7x + 12}{x^2 - 9}$.

Answers

4. $\dfrac{2x}{x + 3}$

5. $\dfrac{6}{5}$

6. $\dfrac{3x^3}{4}$

7. $\dfrac{4}{x + 4}$

Here are some more examples of multiplication and division with rational expressions.

EXAMPLE 8   Perform the indicated operations.

$$\frac{a^2 - 8a + 15}{a + 4} \cdot \frac{a + 2}{a^2 - 5a + 6} \div \frac{a^2 - 3a - 10}{a^2 + 2a - 8}$$

Solution   First, we rewrite the division as multiplication by the reciprocal. Then we proceed as usual.

$$\frac{a^2 - 8a + 15}{a + 4} \cdot \frac{a + 2}{a^2 - 5a + 6} \div \frac{a^2 - 3a - 10}{a^2 + 2a - 8}$$ *Change division to multiplication by the reciprocal.*

$$= \frac{(a^2 - 8a + 15)(a + 2)(a^2 + 2a - 8)}{(a + 4)(a^2 - 5a + 6)(a^2 - 3a - 10)}$$ *Factor.*

$$= \frac{(a - 5)(a - 3)(a + 2)(a + 4)(a - 2)}{(a + 4)(a - 3)(a - 2)(a - 5)(a + 2)}$$ *Divide out common factors.*

$$= 1$$

Our next example involves factoring by grouping. As you may have noticed, working the problems in this chapter gives you a very detailed review of factoring. You also may have noticed that when multiplying or dividing rational expressions we do not find a common denominator. This process would only complicate our work, and is only necessary when adding or subtracting rational expressions.

EXAMPLE 9   Multiply $\dfrac{xa + xb + ya + yb}{xa - xb - ya + yb} \cdot \dfrac{xa + xb - ya - yb}{xa - xb + ya - yb}$.

Solution   We will factor each polynomial by grouping, which takes two steps.

$$\frac{xa + xb + ya + yb}{xa - xb - ya + yb} \cdot \frac{xa + xb - ya - yb}{xa - xb + ya - yb}$$

$$= \frac{x(a + b) + y(a + b)}{x(a - b) - y(a - b)} \cdot \frac{x(a + b) - y(a + b)}{x(a - b) + y(a - b)}$$ *Factor by grouping.*

$$= \frac{(a + b)(x + y)(a + b)(x - y)}{(a - b)(x - y)(a - b)(x + y)}$$

$$= \frac{(a + b)^2}{(a - b)^2}$$

EXAMPLE 10   Multiply $(4x^2 - 36) \cdot \dfrac{12}{4x + 12}$.

Solution   We can think of $4x^2 - 36$ as having a denominator of 1. Thinking of it in this way allows us to proceed as we did in the previous examples.

$$(4x^2 - 36) \cdot \frac{12}{4x + 12}$$

$$= \frac{4x^2 - 36}{1} \cdot \frac{12}{4x + 12}$$ *Write $4x^2 - 36$ with denominator 1.*

$$= \frac{4(x - 3)(x + 3)12}{4(x + 3)}$$ *Factor.*

$$= 12(x - 3)$$ *Divide out common factors.*

---

8. Perform the indicated operations.

$$\frac{x^2 - 9x + 20}{x^2 + 5x + 6} \cdot \frac{x + 2}{x^2 - 4x - 5} \div \frac{x - 4}{x^2 + 2x - 3}$$

9. Multiply.

$$\frac{xy - 2x - 4y + 8}{xy + 3x + 2y + 6} \cdot \frac{xy + 3x - 4y - 12}{xy + 2y - 2x - 4}$$

10. Multiply $(x^2 - 9) \cdot \dfrac{x + 2}{x + 3}$.

Answers

8. $\dfrac{x - 1}{x + 1}$

9. $\dfrac{(x - 4)^2}{(x + 2)^2}$

10. $(x - 3)(x + 2)$

EXAMPLE 11 Multiply $3(x-2)(x-1) \cdot \dfrac{5}{x^2 - 3x + 2}$.

Solution This problem is very similar to the problem in Example 10. Writing the first rational expression with a denominator of 1, we have

$$\dfrac{3(x-2)(x-1)}{1} \cdot \dfrac{5}{x^2 - 3x + 2} = \dfrac{3(x-2)(x-1)5}{(x-2)(x-1)}$$

$$= 3 \cdot 5$$

$$= 15$$

GETTING READY FOR CLASS

After reading through the preceding section, respond in your own words and in complete sentences.

A. Summarize the steps used to multiply fractions.

B. What is the first step in multiplying two rational expressions?

C. Why is factoring important when multiplying and dividing rational expressions?

D. How is division with rational expressions different than multiplication of rational expressions?

11. Multiply $a(a+2)(a-4)\dfrac{a+5}{a^2 - 4a}$.

Vocabulary Review

Choose the correct words to fill in the blanks below.

numerators   quotient   reciprocal   product

1. The _____ of two rational expressions is the product of their _____

over the product of their denominators.

2. The _____ of two rational expressions is the product of the first and the _____

of the second.

Problems

A Perform the indicated operations.

1. $\dfrac{2}{9} \cdot \dfrac{3}{4}$

2. $\dfrac{5}{6} \cdot \dfrac{7}{8}$

3. $\dfrac{3}{4} \div \dfrac{1}{3}$

4. $\dfrac{3}{8} \div \dfrac{5}{4}$

5. $\dfrac{3}{7} \cdot \dfrac{14}{24} \div \dfrac{1}{2}$

6. $\dfrac{6}{5} \cdot \dfrac{10}{36} \div \dfrac{3}{4}$

7. $\dfrac{10x^2}{5y^2} \cdot \dfrac{15y^3}{2x^4}$

8. $\dfrac{8x^3}{7y^4} \cdot \dfrac{14y^6}{16x^2}$

9. $\dfrac{11a^2b}{5ab^2} \div \dfrac{22a^3b^2}{10ab^4}$

10. $\dfrac{8ab^3}{9a^2b} \div \dfrac{16a^2b^2}{18ab^3}$

11. $\dfrac{6x^2}{5y^3} \cdot \dfrac{11z^2}{2x^2} \div \dfrac{33z^5}{10y^8}$

12. $\dfrac{4x^3}{7y^2} \cdot \dfrac{6z^5}{5x^6} \div \dfrac{24z^2}{35x^6}$

Perform the indicated operations. Be sure to write all answers in lowest terms.

13. $\dfrac{x^2-9}{x^2-4} \cdot \dfrac{x-2}{x-3}$

14. $\dfrac{x^2-16}{x^2-25} \cdot \dfrac{x-5}{x-4}$

15. $\dfrac{y^2-1}{y+2} \cdot \dfrac{y^2+5y+6}{y^2+2y-3}$

16. $\dfrac{y-1}{y^2-y-6} \cdot \dfrac{y^2+5y+6}{y^2-1}$

17. $\dfrac{3x-12}{x^2-4} \cdot \dfrac{x^2+6x+8}{x-4}$

18. $\dfrac{x^2+5x+1}{4x-4} \cdot \dfrac{x-1}{x^2+5x+1}$

19. $\dfrac{xy}{xy+1} \div \dfrac{x}{y}$

20. $\dfrac{y}{x} \div \dfrac{xy}{xy-1}$

21. $\dfrac{1}{x^2-9} \div \dfrac{1}{x^2+9}$

22. $\dfrac{1}{x^2-9} \div \dfrac{1}{(x-3)^2}$

23. $\dfrac{5x - 10}{6x - 9} \div \dfrac{6 - 3x}{8x - 12}$

24. $\dfrac{10b - 15}{4b + 2} \div \dfrac{6 - 4b}{6b + 3}$

25. $\dfrac{5x + 2y}{25x^2 - 5xy - 6y^2} \cdot \dfrac{20x^2 - 7xy - 3y^2}{4x + y}$

26. $\dfrac{7x + 3y}{42x^2 - 17xy - 15y^2} \cdot \dfrac{12x^2 - 4xy - 5y^2}{2x + y}$

27. $\dfrac{a^2 - 5a + 6}{a^2 - 2a - 3} \div \dfrac{a - 5}{a^2 + 3a + 2}$

28. $\dfrac{a^2 + 7a + 12}{a - 5} \div \dfrac{a^2 + 9a + 18}{a^2 - 7a + 10}$

29. $\dfrac{4t^2 - 1}{6t^2 + t - 2} \div \dfrac{8t^3 + 1}{27t^3 + 8}$

30. $\dfrac{9t^2 - 1}{6t^2 + 7t - 3} \div \dfrac{27t^3 + 1}{8t^3 + 27}$

31. $\dfrac{2x^2 - 5x - 12}{4x^2 + 8x + 3} \div \dfrac{x^2 - 16}{2x^2 + 7x + 3}$

32. $\dfrac{x^2 - 2x + 1}{3x^2 + 7x - 20} \div \dfrac{x^2 + 3x - 4}{3x^2 - 2x - 5}$

33. $\dfrac{2a^2 - 21ab - 36b^2}{a^2 - 11ab - 12b^2} \div \dfrac{10a + 15b}{a^2 - b^2}$

34. $\dfrac{3a^2 + 7ab - 20b^2}{a^2 + 5ab + 4b^2} \div \dfrac{3a^2 - 17ab + 20b^2}{3a - 12b}$

35. $\dfrac{6c^2 - c - 15}{9c^2 - 25} \cdot \dfrac{15c^2 + 22c - 5}{6c^2 + 5c - 6}$

36. $\dfrac{m^2 + 4m - 21}{m^2 - 12m + 27} \cdot \dfrac{m^2 - 7m + 12}{m^2 + 3m - 28}$

37. $\dfrac{6x^2 - 11xy + 3y^2}{2x^2 - 7xy + 6y^2} \cdot \dfrac{3x^2 + 8xy + 4y^2}{2y^2 - 3xy - 9x^2}$

38. $\dfrac{a^2 + ab - 12b^2}{3a^2 + 8ab - 16b^2} \cdot \dfrac{2a^2 + 7ab + 6b^2}{9b^2 + 3ab - 2a^2}$

39. $\dfrac{360x^3 - 490x}{36x^2 + 84x + 49} \cdot \dfrac{30x^2 + 83x + 56}{150x^3 + 65x^2 - 280x}$

40. $\dfrac{490x^2 - 640}{49x^2 - 112x + 64} \cdot \dfrac{28x^2 - 95x + 72}{56x^3 - 62x^2 - 144x}$

41. $\dfrac{x^5 - x^2}{5x^2 - 5x} \cdot \dfrac{10x^4 - 10x^2}{2x^4 + 2x^3 + 2x^2}$

42. $\dfrac{2x^4 - 16x}{3x^6 - 48x^2} \cdot \dfrac{6x^5 + 24x^3}{4x^4 + 8x^3 + 16x^2}$

43. $\dfrac{a^2 - 16b^2}{a^2 - 8ab + 16b^2} \cdot \dfrac{a^2 - 9ab + 20b^2}{a^2 - 7ab + 12b^2} \div \dfrac{a^2 - 25b^2}{a^2 - 6ab + 9b^2}$

44. $\dfrac{a^2 - 6ab + 9b^2}{a^2 - 4b^2} \cdot \dfrac{a^2 - 5ab + 6b^2}{(a - 3b)^2} \div \dfrac{a^2 - 9b^2}{a^2 - ab - 6b^2}$

45. $\dfrac{2y^2 - 7y - 15}{42y^2 - 29y - 5} \cdot \dfrac{12y^2 - 16y + 5}{7y^2 - 36y + 5} \div \dfrac{4y^2 - 9}{49y^2 - 1}$

46. $\dfrac{8y^2 + 18y - 5}{21y^2 - 16y + 3} \cdot \dfrac{35y^2 - 22y + 3}{6y^2 + 17y + 5} \div \dfrac{16y^2 - 1}{9y^2 - 1}$

47. $\dfrac{xy - 2x + 3y - 6}{xy + 2x - 4y - 8} \cdot \dfrac{xy + x - 4y - 4}{xy - x + 3y - 3}$

48. $\dfrac{ax + bx + 2a + 2b}{ax - 3a + bx - 3b} \cdot \dfrac{ax - bx - 3a + 3b}{ax - bx - 2a + 2b}$

49. $\dfrac{xy^2 - y^2 + 4xy - 4y}{xy - 3y + 4x - 12} \div \dfrac{xy^3 + 2xy^2 + y^3 + 2y^2}{xy^2 - 3y^2 + 2xy - 6y}$

50. $\dfrac{4xb - 8b + 12x - 24}{xb^2 + 3b^2 + 3xb + 9b} \div \dfrac{4xb - 8b - 8x + 16}{xb^2 + 3b^2 - 2xb - 6b}$

51. $\dfrac{2x^3 + 10x^2 - 8x - 40}{x^3 + 4x^2 - 9x - 36} \cdot \dfrac{x^2 + x - 12}{2x^2 + 14x + 20}$

52. $\dfrac{x^3 + 2x^2 - 9x - 18}{x^4 + 3x^3 - 4x^2 - 12x} \cdot \dfrac{x^3 + 5x^2 + 6x}{x^2 - x - 6}$

53. $\dfrac{w^3 - w^2x}{wy - w} \div \left(\dfrac{w - x}{y - 1}\right)^2$

54. $\dfrac{a^3 - a^2b}{ac - a} \div \left(\dfrac{a - b}{c - 1}\right)^2$

55. $\dfrac{mx + my + 2x + 2y}{6x^2 - 5xy - 4y^2} \div \dfrac{2mx - 4x + my - 2y}{3mx - 6x - 4my + 8y}$

56. $\dfrac{ax - 2a + 2xy - 4y}{ax + 2a - 2xy - 4y} \div \dfrac{ax + 2a + 2xy + 4y}{ax - 2a - 2xy + 4y}$

57. $\dfrac{1 - 4d^2}{(d - c)^2} \cdot \dfrac{d^2 - c^2}{1 + 2d}$

58. $\dfrac{k^2 - 1}{k^3 + 1} \div \dfrac{k^3 - 1}{k - 1} \cdot (k^2 + k + 1)$

59. $\dfrac{r^2 - s^2}{r^2 + rs + s^2} \cdot \dfrac{r^3 - s^3}{r^2 + s^2} \div \dfrac{r^4 - s^4}{r^2 - s^2}$

60. $\dfrac{r^3 - s^3}{s - r} \cdot \dfrac{(r + s)^2}{r^2 + s^2}$

Use the method shown in Examples 10 and 11 to find the following products.

61. $(3x - 6) \cdot \dfrac{x}{x - 2}$

62. $(4x + 8) \cdot \dfrac{x}{x + 2}$

63. $(x^2 - 25) \cdot \dfrac{2}{x - 5}$

64. $(x^2 - 49) \cdot \dfrac{5}{x + 7}$

65. $(x^2 - 3x + 2) \cdot \dfrac{3}{3x - 3}$

66. $(x^2 - 3x + 2) \cdot \dfrac{-1}{x - 2}$

67. $(y - 3)(y - 4)(y + 3) \cdot \dfrac{-1}{y^2 - 9}$

68. $(y + 1)(y + 4)(y - 1) \cdot \dfrac{3}{y^2 - 1}$

69. $a(a + 5)(a - 5) \cdot \dfrac{a + 1}{a^2 + 5a}$

70. $a(a + 3)(a - 3) \cdot \dfrac{a - 1}{a^2 - 3a}$

The next two problems are intended to give you practice reading, and paying attention to, the instructions that accompany the problems you are working. Working these problems is an excellent way to get ready for a test or a quiz.

71. Work each problem according to the instructions.

    a. Simplify: $\dfrac{16 - 1}{64 - 1}$.

    b. Reduce: $\dfrac{25x^2 - 9}{125x^3 - 27}$.

    c. Multiply: $\dfrac{25x^2 - 9}{125x^3 - 27} \cdot \dfrac{5x - 3}{5x + 3}$.

    d. Divide: $\dfrac{25x^2 - 9}{125x^3 - 27} \div \dfrac{5x - 3}{25x^2 + 15x + 9}$.

72. Work each problem according to the instructions.

    a. Simplify: $\dfrac{64 - 49}{64 + 112 + 49}$.

    b. Reduce: $\dfrac{9x^2 - 49}{9x^2 + 42x + 49}$.

    c. Multiply: $\dfrac{9x^2 - 49}{9x^2 + 42x + 49} \cdot \dfrac{3x + 7}{3x - 7}$.

    d. Divide: $\dfrac{9x^2 - 49}{9x^2 + 42x + 49} \div \dfrac{3x + 7}{3x - 7}$.

The work you did with algebra of functions will help with these.

73. Let $f(x) = \dfrac{x^2 - x - 6}{x - 1}$ and $g(x) = \dfrac{x + 2}{x^2 - 4x + 3}$, find

    a. $f(x) \cdot g(x)$

    b. $f(x) \div g(x)$

    c. $g(x) \div f(x)$

74. Let $f(x) = \dfrac{x^2 - x - 12}{x^2 - 4x + 3}$ and $g(x) = \dfrac{x^2 - x - 12}{x^2 - 5x + 4}$, find

    a. $f(x) \cdot g(x)$

    b. $f(x) \div g(x)$

    c. $g(x) \div f(x)$

Applying the Concepts

75. Demand Equation At the beginning of this section, we introduced the demand equation shown here.

$$p(x) = \frac{2(x + 60)}{x + 5}$$

a. Use the demand equation to fill in the table. Then compare your results with the graph shown in Figure 1 of this section.

Number of Copies	Price per Copy ($)
x	p(x)
1	
10	
20	
50	
100	

b. To find the revenue for selling 50 copies of a tape, we multiply the price per tape by 50. Find the revenue for selling 50 tapes.

c. Find the revenue for selling 100 tapes.

d. Find the revenue equation $R(x)$.

76. Rate Equation The drug concentration t hours after an injection of a certain drug is given by the rational function

$$c(t) = \frac{0.04t + 5}{0.02t^2}$$

a. Complete the following table using the rational function given above.

Time	Concentration
t	c(t)
5	
10	
15	
20	

b. Find the concentration after 1 day and 2 days.

Maintaining Your Skills

77. The chart shows the ratio of a planet's diameter to Earth's diameter. If Earth has a diameter of 7,900 miles, what is the diameter of Venus? Of Saturn?

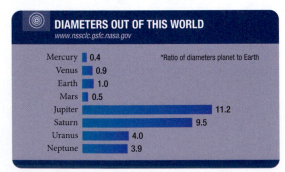

78. The following chart shows the number of dogs registered in the most popular breeds. Use the chart to write the numbers in scientific notation.

a. Labrador Retriever

b. German Shepherd

c. Beagle

Getting Ready for the Next Section

Combine.

79. $\dfrac{4}{9} + \dfrac{2}{9}$

80. $\dfrac{3}{8} + \dfrac{1}{8}$

81. $\dfrac{3}{14} + \dfrac{7}{30}$

82. $\dfrac{3}{10} + \dfrac{11}{42}$

Multiply.

83. $-1(7 - x)$

84. $-1(3 - x)$

85. $-2(2x - 5)$

86. $-3(4 - 7x)$

Factor.

87. $x^2 - 1$

88. $x^2 - 2x - 3$

89. $2x + 10$

90. $x^2 + 4x + 3$

91. $a^3 - b^3$

92. $8y^3 - 27$

93. $4x^2 - 9$

94. $5x^2 - 125$

One Step Further

Perform the following operations. Be sure to write all answers in lowest terms.

95. $\dfrac{x^6 + y^6}{x^4 + 4x^2y^2 + 3y^4} \div \dfrac{x^4 + 3x^2y^2 + 2y^4}{x^4 + 5x^2y^2 + 6y^4}$

96. $\dfrac{x^2 + 9xy + 8y^2}{x^2 + 7xy - 8y^2} \div \dfrac{x^2 - y^2}{x^2 + 5xy - 6y^2}$

97. $\dfrac{a^2(2a + b) + 6a(2a + b) + 5(2a + b)}{3a^2(2a + b) - 2a(2a + b) + (2a + b)} \div \dfrac{a + 1}{a - 1}$

98. $\dfrac{2x^2(x - 3z) - 5x(x - 3z) + 2(x - 3z)}{4x^2(x - 3z) - 11x(x - 3z) + 6(x - 3z)} \div \dfrac{4x - 3}{4x + 1}$

99. $\dfrac{p^3 + q^3}{q - p} \div \dfrac{(p + q)^2}{p^2 - q^2}$

100. $\dfrac{x^3 - y^3}{x^3 + y^3} \cdot \dfrac{x + y}{x - y}$

Find the Mistake

Each sentence below contains a mistake. Circle the mistake and write the correct word(s) or expression on the line provided.

1. Multiplying $\dfrac{x^2 - 4}{x - 2} \cdot \dfrac{x^2 + 2x - 8}{x^2 + 6x + 8}$ we get $x + 2$. _____

2. Multiplying $\dfrac{2x - 6}{4x^2 - 4x - 24} \cdot \dfrac{6x^2 + 2x - 20}{3x - 5}$ we get $\dfrac{2x - 6}{x - 3}$. _____

3. To divide the rational expression $\dfrac{5x^2 + 21x - 20}{15x^2 - 17x + 4}$ by $\dfrac{x^2 - 25}{3x^2 - 4x + 1}$ we multiply the numerators and the denominators.

_____

4. Dividing $\dfrac{4x^2 + 27xy + 18y^2}{x^2 + 7xy + 6y^2} \div \dfrac{16x + 12y}{x^2 - y^2}$, we get $x - y$. _____

A Add and subtract rational expressions with the same denominator.

B Add and subtract rational expressions with different denominators.

KEY WORDS

least common denominator

LCD

5.3 Addition and Subtraction of Rational Expressions

©iStockphoto.com/ArthurBraunstein

Have you ever heard of the competitive sport of joggling? Imagine a track runner competing in the 100-meter dash while juggling five balls! Previously, we have used the rate equation $d = rt$ to help us solve some application problems. If we solve this equation for rate r, we get the rational equation $r = \frac{d}{t}$. Suppose a runner can complete the 100-meter dash without juggling in 12 seconds, but he takes 31 seconds to joggle the same distance. After reading this section, we can create two rational expressions for rate r to determine the juggler's difference in speed.

This section is concerned with addition and subtraction of rational expressions. In the first part of this section, we will look at addition of expressions that have the same denominator. In the second part of this section, we will look at addition of expressions that have different denominators.

A Addition and Subtraction with the Same Denominator

To add two expressions that have the same denominator, we simply add numerators and put the sum over the common denominator. Because the process we use to add and subtract rational expressions is the same process used to add and subtract fractions, we will begin with an example involving fractions.

Practice Problems

1. Add $\frac{2}{x} + \frac{7}{x}$.

EXAMPLE 1   Add $\frac{4}{9} + \frac{2}{9}$.

Solution   We add fractions with the same denominator by using the distributive property. Here is a detailed look at the steps involved:

$$\frac{4}{9} + \frac{2}{9} = 4\left(\frac{1}{9}\right) + 2\left(\frac{1}{9}\right)$$

$$= (4 + 2)\left(\frac{1}{9}\right) \qquad \text{Distributive property}$$

$$= 6\left(\frac{1}{9}\right)$$

$$= \frac{6}{9}$$

$$= \frac{2}{3} \qquad \text{Divide the numerator and denominator by common factor 3.}$$

Note that the important thing about the fractions in this example is that they each have a denominator of 9. If they did not have the same denominator, we could not have written them as two terms with a factor of $\frac{1}{9}$ in common. Without the $\frac{1}{9}$ common to each term, we couldn't apply the distributive property. Without the distributive property, we would not have been able to add the two fractions in this form.

Answer

1. $\frac{9}{x}$

Chapter 5   Rational Expressions, Equations, and Functions

In the following examples, we will not show all the steps we showed in Example 1. The steps are shown in Example 1 so you will see why both fractions must have the same denominator before we can add them. In practice, we simply add numerators and place the result over the common denominator.

We add and subtract rational expressions with the same denominator by combining numerators and writing the result over the common denominator. Then we reduce the result to lowest terms, if possible. Example 2 shows this process in detail. If you see the similarities between operations on rational numbers and operations on rational expressions, this chapter will look like an extension of rational numbers rather than a completely new set of topics.

EXAMPLE 2   Add $\dfrac{x}{x^2 - 1} + \dfrac{1}{x^2 - 1}$.

Solution   Because the denominators are the same, we simply add numerators.

$$\frac{x}{x^2 - 1} + \frac{1}{x^2 - 1} = \frac{x + 1}{x^2 - 1} \qquad \text{Add numerators.}$$

$$= \frac{x + 1}{(x - 1)(x + 1)} \qquad \text{Factor denominator.}$$

$$= \frac{1}{x - 1} \qquad \text{Divide out common factor } x + 1.$$

Our next example involves subtraction of rational expressions. Pay careful attention to what happens to the signs of the terms in the numerator of the second expression when we subtract it from the first expression.

EXAMPLE 3   Let $f(x) = \dfrac{2x - 5}{x - 2}$ and $g(x) = \dfrac{x - 3}{x - 2}$. Find and simplify $f(x) - g(x)$.

Solution   Because each expression has the same denominator, we simply subtract the numerator in the second expression from the numerator in the first expression and write the difference over the common denominator $x - 2$. We must be careful, however, that we subtract both terms in the second numerator. To ensure that we do, we will enclose that numerator in parentheses.

$$f(x) - g(x) = \frac{2x - 5}{x - 2} - \frac{x - 3}{x - 2}$$

$$= \frac{2x - 5 - (x - 3)}{x - 2} \qquad \text{Subtract numerators.}$$

$$= \frac{2x - 5 - x + 3}{x - 2} \qquad \text{Remove parentheses.}$$

$$= \frac{x - 2}{x - 2} \qquad \text{Combine similar terms in the numerator.}$$

$$= 1 \qquad \text{Reduce (or divide).}$$

Note the 3 in the numerator of the second step. It is a common mistake to write this as -3, by forgetting to subtract both terms in the numerator of the second expression. Whenever the expression we are subtracting has two or more terms in its numerator, we have to watch for this mistake.

Next we consider addition and subtraction of fractions and rational expressions that have different denominators.

2. Add $\dfrac{2x}{x^2 - 1} + \dfrac{2}{x^2 - 1}$.

3. Let $f(x) = \dfrac{4x - 7}{x - 4}$ and $g(x) = \dfrac{3x - 3}{x - 4}$.

Find and simplify $f(x) - g(x)$.

Answers

2. $\dfrac{2}{x - 1}$

3. 1

B Addition and Subtraction With Different Denominators

Before we look at an example of addition of fractions with different denominators, we need to review the definition for the *least common denominator (LCD).*

> **DEFINITION** least common denominator
>
> The *least common denominator* for a set of denominators is the smallest expression that is divisible by each of the denominators. It is also considered the least common multiple of the given set of denominators.

The first step in combining two fractions is to find the LCD. Once we have the common denominator, we rewrite each fraction as an equivalent fraction with the common denominator. After that, we simply add or subtract as we did in our first three examples.

Example 4 is a review of the step-by-step procedure used to add two fractions with different denominators.

4. Add $\dfrac{1}{21} + \dfrac{4}{15}$.

EXAMPLE 4   Add $\dfrac{3}{14} + \dfrac{7}{30}$.

Solution

Step 1: *Find the LCD.*

To do this, we first factor both denominators into prime factors.

Factor 14: $14 = 2 \cdot 7$

Factor 30: $30 = 2 \cdot 3 \cdot 5$

Because the LCD must be divisible by 14, it must have factors of $2 \div 7$. It must also be divisible by 30 and, therefore, have factors of $2 \div 3 \div 5$. We do not need to repeat the 2 that appears in both the factors of 14 and those of 30. Therefore,

$$LCD = 2 \cdot 3 \cdot 5 \cdot 7 = 210$$

Step 2: *Change to equivalent fractions.*

Because we want each fraction to have a denominator of 210 and at the same time keep its original value, we multiply each by 1 in the appropriate form.

Change $\frac{3}{14}$ to a fraction with denominator 210.

$$\frac{3}{14} \cdot \frac{15}{15} = \frac{45}{210}$$

Change $\frac{7}{30}$ to a fraction with denominator 210.

$$\frac{7}{30} \cdot \frac{7}{7} = \frac{49}{210}$$

Step 3: *Add numerators of equivalent fractions found in step 2.*

$$\frac{45}{210} + \frac{49}{210} = \frac{94}{210}$$

Step 4: *Reduce to lowest terms, if necessary.*

$$\frac{94}{210} = \frac{47}{105}$$

The main idea in adding fractions is to write each fraction again with the LCD for a denominator. In doing so, we must be sure not to change the value of either of the original fractions.

Answer

4. $\dfrac{11}{35}$

EXAMPLE 5   Add $\dfrac{-2}{x^2 - 2x - 3} + \dfrac{3}{x^2 - 9}$.

Solution

Step 1: *Find the LCD.*

$$x^2 - 2x - 3 = (x - 3)(x + 1)$$
$$x^2 - 9 = (x - 3)(x + 3)$$

$$LCD = (x - 3)(x + 3)(x + 1)$$

Step 2: *Change to equivalent rational expressions.*

$$\frac{-2}{x^2 - 2x - 3} = \frac{-2}{(x - 3)(x + 1)} \cdot \frac{(x + 3)}{(x + 3)} = \frac{-2x - 6}{(x - 3)(x + 3)(x + 1)}$$

$$\frac{3}{x^2 - 9} = \frac{3}{(x - 3)(x + 3)} \cdot \frac{(x + 1)}{(x + 1)} = \frac{3x + 3}{(x - 3)(x + 3)(x + 1)}$$

Step 3: *Add numerators.*

$$\frac{-2x - 6}{(x - 3)(x + 3)(x + 1)} + \frac{3x + 3}{(x - 3)(x + 3)(x + 1)} = \frac{x - 3}{(x - 3)(x + 3)(x + 1)}$$

Step 4: *Reduce to lowest terms.*

$$\frac{\cancel{x - 3}}{\cancel{(x - 3)}(x + 3)(x + 1)} = \frac{1}{(x + 3)(x + 1)}$$

EXAMPLE 6   Subtract $\dfrac{x + 4}{2x + 10} - \dfrac{5}{x^2 - 25}$.

Solution   We begin by factoring each denominator.

$$\frac{x + 4}{2x + 10} - \frac{5}{x^2 - 25} = \frac{x + 4}{2(x + 5)} - \frac{5}{(x + 5)(x - 5)}$$

The LCD is $2(x + 5)(x - 5)$. Completing the problem, we have

$$= \frac{x + 4}{2(x + 5)} \cdot \frac{(x - 5)}{(x - 5)} - \frac{5}{(x + 5)(x - 5)} \cdot \frac{2}{2}$$

$$= \frac{x^2 - x - 20}{2(x + 5)(x - 5)} - \frac{10}{2(x + 5)(x - 5)}$$

$$= \frac{x^2 - x - 30}{2(x + 5)(x - 5)}$$

To see if this expression will reduce, we factor the numerator into $(x - 6)(x + 5)$.

$$= \frac{(x - 6)\cancel{(x + 5)}}{2\cancel{(x + 5)}(x - 5)}$$

$$= \frac{x - 6}{2(x - 5)}$$

5. Add $\dfrac{1}{x^2 + 8x + 15} + \dfrac{3}{x^2 - 9}$.

6. Subtract $\dfrac{x + 3}{2x + 8} - \dfrac{4}{x^2 - 16}$.

Answers

5. $\dfrac{4}{(x + 5)(x - 3)}$

6. $\dfrac{x - 5}{2(x - 4)}$

7. Subtract $\dfrac{x-1}{x^2-x-2} - \dfrac{x}{x^2+x-6}$.

EXAMPLE 7 Subtract $\dfrac{2x-2}{x^2+4x+3} - \dfrac{x-1}{x^2+5x+6}$.

Solution We factor each denominator and build the LCD from those factors.

$$\frac{2x-2}{x^2+4x+3} - \frac{x-1}{x^2+5x+6}$$

$$= \frac{2x-2}{(x+3)(x+1)} - \frac{x-1}{(x+3)(x+2)}$$

$$= \frac{2x-2}{(x+3)(x+1)} \cdot \frac{(x+2)}{(x+2)} - \frac{x-1}{(x+3)(x+2)} \cdot \frac{(x+1)}{(x+1)} \qquad \text{The LCD is } (x+1)(x+2)(x+3).$$

$$= \frac{2x^2+2x-4}{(x+1)(x+2)(x+3)} - \frac{x^2-1}{(x+1)(x+2)(x+3)} \qquad \text{Multiply out each numerator.}$$

$$= \frac{(2x^2+2x-4)-(x^2-1)}{(x+1)(x+2)(x+3)} \qquad \text{Subtract numerators.}$$

$$= \frac{x^2+2x-3}{(x+1)(x+2)(x+3)} \qquad \text{Factor numerator to see if we can reduce.}$$

$$= \frac{(x+3)(x-1)}{(x+1)(x+2)(x+3)} \qquad \text{Reduce.}$$

$$= \frac{x-1}{(x+1)(x+2)}$$

---

8. Let $f(x) = \dfrac{x^2}{x-3}$ and $g(x) = \dfrac{12-x}{-x+3}$. Find and simplify $f(x)+g(x)$.

EXAMPLE 8 Let $f(x) = \dfrac{x^2}{x-7}$ and $g(x) = \dfrac{6x+7}{7-x}$. Find and simplify $f(x)+g(x)$.

Solution In Section 5.1, we were able to reverse the terms in a factor such as $7-x$ by factoring -1 from each term. In a problem like this, the same result can be obtained by multiplying the numerator and denominator by -1.

$$f(x)+g(x) = \frac{x^2}{x-7} + \frac{6x+7}{7-x}\cdot\frac{-1}{-1} = \frac{x2}{x-7} + \frac{-6x-7}{x-7}$$

$$= \frac{x^2-6x-7}{x-7} \qquad \text{Add numerators.}$$

$$= \frac{(x-7)(x+1)}{(x-7)} \qquad \text{Factor numerator.}$$

$$= x+1 \qquad \text{Divide out } x-7.$$

For our next example, we will look at a problem in which we combine a whole number and a rational expression.

Answers

7. $\dfrac{x-3}{(x+1)(x-2)(x+3)}$

8. $x+4$

EXAMPLE 9 Subtract $2 - \dfrac{9}{3x + 1}$.

Solution To subtract these two expressions, we think of 2 as a rational expression with a denominator of 1.

$$2 - \frac{9}{3x + 1} = \frac{2}{1} - \frac{9}{3x + 1}$$

The LCD is $3x + 1$. Multiplying the numerator and denominator of the first expression by $3x + 1$ gives us a rational expression equivalent to 2, but with a denominator of $3x + 1$.

$$\frac{2}{1} \cdot \frac{(3x + 1)}{(3x + 1)} - \frac{9}{3x + 1} = \frac{6x + 2 - 9}{3x + 1}$$

$$= \frac{6x - 7}{3x + 1}$$

The numerator and denominator of this last expression do not have any factors in common other than 1, so the expression is in lowest terms.

9. Subtract $3 - \dfrac{7}{4x - 7}$.

EXAMPLE 10 Write an expression for the sum of a number and twice its reciprocal. Then, simplify that expression.

Solution If x is the number, then its reciprocal is $\frac{1}{x}$. Twice its reciprocal is $\frac{2}{x}$. The sum of the number and twice its reciprocal is

$$x + \frac{2}{x}$$

To combine these two expressions, we think of the first term x as a rational expression with a denominator of 1. The LCD is x.

$$x + \frac{2}{x} = \frac{x}{1} + \frac{2}{x}$$

$$= \frac{x}{1} \cdot \frac{x}{x} + \frac{2}{x}$$

$$= \frac{x^2 + 2}{x}$$

10. Write an expression for the difference of twice a number and its reciprocal. Then simplify.

GETTING READY FOR CLASS

After reading through the preceding section, respond in your own words and in complete sentences.

A. How would you use the distributive property to add two rational expressions that have the same denominator?

B. What is the definition of the least common denominator?

C. Why is factoring important in finding a least common denominator?

D. What is the last step in adding or subtracting two rational expressions?

Answers

9. $\dfrac{4(3x - 7)}{4x - 7}$

10. $\dfrac{2x^2 - 1}{x}$

Vocabulary Review

Choose the correct words to fill in the blanks below.

common	equivalent	numerators
lowest	different	denominator

1. To add rational expressions with the same denominator, add numerators and put the sum over the _____ denominator.

2. LCD is an acronym for least common _____.

3. To add two fractions with _____ denominators, follow these steps:

   Step 1: Find the lowest common denominator.

   Step 2: Change to _____ fractions.

   Step 3: Add _____ of equivalent fractions found in Step 2.

   Step 4: Reduce to _____ terms, if necessary.

Problems

A Combine the following rational expressions. Reduce all answers to lowest terms.

1. $\dfrac{x}{x+3} + \dfrac{3}{x+3}$

2. $\dfrac{2}{5x-2} - \dfrac{5x}{5x-2}$

3. $\dfrac{4}{y-4} - \dfrac{y}{y-4}$

4. $\dfrac{8}{y-8} - \dfrac{y}{y-8}$

5. $\dfrac{x}{x^2-y^2} - \dfrac{y}{x^2-y^2}$

6. $\dfrac{x}{x^2-y^2} + \dfrac{y}{x^2-y^2}$

7. $\dfrac{2x-3}{x-2} - \dfrac{x-1}{x-2}$

8. $\dfrac{2x-4}{x+2} - \dfrac{x-6}{x+2}$

9. $\dfrac{1}{a} + \dfrac{2}{a^2} - \dfrac{3}{a^3}$

10. $\dfrac{3}{a} + \dfrac{2}{a^2} - \dfrac{1}{a^3}$

11. $\dfrac{7x-2}{2x+1} - \dfrac{5x-3}{2x+1}$

12. $\dfrac{7x-1}{3x+2} - \dfrac{4x-3}{3x+2}$

B Combine the following fractions.

13. $\dfrac{3}{4} + \dfrac{1}{2}$

14. $\dfrac{5}{6} + \dfrac{1}{3}$

15. $\dfrac{2}{5} - \dfrac{1}{15}$

16. $\dfrac{5}{8} - \dfrac{1}{4}$

17. $\dfrac{5}{6} + \dfrac{7}{8}$

18. $\dfrac{3}{4} + \dfrac{2}{3}$

19. $\dfrac{9}{48} - \dfrac{3}{54}$

SCAN TO ACCESS

20. $\dfrac{6}{28} - \dfrac{5}{42}$

21. $\dfrac{3}{4} - \dfrac{1}{8} + \dfrac{2}{3}$

22. $\dfrac{1}{3} - \dfrac{5}{6} + \dfrac{5}{12}$

23. $\dfrac{1}{3} - \dfrac{1}{2} + \dfrac{5}{6}$

24. $\dfrac{1}{4} - \dfrac{5}{6} - \dfrac{1}{8}$

25. Work each problem according to the instructions.

    a. Multiply: $\dfrac{3}{8} \cdot \dfrac{1}{6}$.

    b. Divide: $\dfrac{3}{8} \div \dfrac{1}{6}$.

    c. Add: $\dfrac{3}{8} + \dfrac{1}{6}$.

    d. Multiply: $\dfrac{x + 3}{x - 3} \cdot \dfrac{5x + 15}{x^2 - 9}$.

    e. Divide: $\dfrac{x + 3}{x - 3} \div \dfrac{5x + 15}{x^2 - 9}$.

    f. Subtract: $\dfrac{x + 3}{x - 3} - \dfrac{5x + 15}{x^2 - 9}$.

26. Work each problem according to the instructions.

    a. Multiply: $\dfrac{16}{49} \cdot \dfrac{1}{28}$.

    b. Divide: $\dfrac{16}{49} \div \dfrac{1}{28}$.

    c. Subtract: $\dfrac{16}{49} - \dfrac{1}{28}$.

    d. Multiply: $\dfrac{3x - 2}{3x + 2} \cdot \dfrac{15x + 6}{9x^2 - 4}$.

    e. Divide: $\dfrac{3x - 2}{3x + 2} \div \dfrac{15x + 6}{9x^2 - 4}$.

    f. Subtract: $\dfrac{3x + 2}{3x - 2} - \dfrac{15x + 6}{9x^2 - 4}$.

B Find and simplify $f(x) - g(x)$ for the following functions.

27. $f(x) = \dfrac{3x + 1}{2x - 6}, g(x) = \dfrac{x + 2}{x - 3}$

28. $f(x) = \dfrac{x + 1}{x - 2}, g(x) = \dfrac{4x + 7}{5x - 10}$

29. $f(x) = \dfrac{6x + 5}{5x - 25}, g(x) = \dfrac{x + 2}{x - 5}$

30. $f(x) = \dfrac{4x + 2}{3x + 12}, g(x) = \dfrac{x - 2}{x + 4}$

Combine the following rational expressions. Reduce all answers to lowest terms.

31. $\dfrac{x + 1}{2x - 2} - \dfrac{2}{x^2 - 1}$

32. $\dfrac{x + 7}{2x + 12} + \dfrac{6}{x^2 - 36}$

33. $\dfrac{1}{a - b} - \dfrac{3ab}{a^3 - b^3}$

34. $\dfrac{1}{a + b} + \dfrac{3ab}{a^3 + b^3}$

35. $\dfrac{x - 2}{2x + 3} - \dfrac{2x - 11}{3 - 4x - 4x^2}$

36. $\dfrac{5}{2x - 1} + \dfrac{6x + 7}{1 - 4x^2}$

37. $\dfrac{1}{2y - 3} - \dfrac{18y}{8y^3 - 27}$

38. $\dfrac{1}{3y - 2} - \dfrac{18y}{27y^3 - 8}$

39. $\dfrac{x}{x^2 - 5x + 6} - \dfrac{3}{3 - x}$

40. $\dfrac{x}{x^2 + 4x + 4} - \dfrac{2}{2 + x}$

41. $\dfrac{2}{4t - 5} + \dfrac{9}{8t^2 - 38t + 35}$

42. $\dfrac{3}{2t - 5} + \dfrac{21}{8t^2 - 14t - 15}$

43. $\dfrac{1}{a^2 - 5a + 6} + \dfrac{3}{a^2 - a - 2}$

44. $\dfrac{-3}{a^2 + a - 2} + \dfrac{5}{a^2 - a - 6}$

45. $\dfrac{1}{8x^3 - 1} - \dfrac{1}{4x^2 - 1}$

46. $\dfrac{1}{27x^3 - 1} - \dfrac{1}{9x^2 - 1}$

47. $\dfrac{4}{4x^2 - 9} - \dfrac{6}{8x^2 - 6x - 9}$

48. $\dfrac{9}{9x^2 + 6x - 8} - \dfrac{6}{9x^2 - 4}$

49. $\dfrac{4a}{a^2 + 6a + 5} - \dfrac{3a}{a^2 + 5a + 4}$

50. $\dfrac{3a}{a^2 + 7a + 10} - \dfrac{2a}{a^2 + 6a + 8}$

51. $\dfrac{2x - 1}{x^2 + x - 6} - \dfrac{x + 2}{x^2 + 5x + 6}$

52. $\dfrac{4x + 1}{x^2 + 5x + 4} - \dfrac{x + 3}{x^2 + 4x + 3}$

53. $\dfrac{2x - 8}{3x^2 + 8x + 4} + \dfrac{x + 3}{3x^2 + 5x + 2}$

54. $\dfrac{5x + 3}{2x^2 + 5x + 3} - \dfrac{3x + 9}{2x^2 + 7x + 6}$

55. $\dfrac{2}{x^2 + 5x + 6} - \dfrac{4}{x^2 + 4x + 3} + \dfrac{3}{x^2 + 3x + 2}$

56. $\dfrac{-5}{x^2 + 3x - 4} + \dfrac{5}{x^2 + 2x - 3} + \dfrac{1}{x^2 + 7x + 12}$

57. $\dfrac{2x + 8}{x^2 + 5x + 6} - \dfrac{x + 5}{x^2 + 4x + 3} - \dfrac{x - 1}{x^2 + 3x + 2}$

58. $\dfrac{2x + 11}{x^2 + 9x + 20} - \dfrac{x + 1}{x^2 + 7x + 12} - \dfrac{x + 6}{x^2 + 8x + 15}$

59. $2 + \dfrac{3}{2x + 1}$

60. $3 - \dfrac{2}{2x + 3}$

61. $5 + \dfrac{2}{4 - t}$

62. $7 + \dfrac{3}{5 - t}$

63. $x - \dfrac{4}{2x + 3}$

64. $x - \dfrac{5}{3x + 4} + 1$

65. $\dfrac{x}{x + 2} + \dfrac{1}{2x + 4} - \dfrac{3}{x^2 + 2x}$

66. $\dfrac{x}{x + 3} + \dfrac{7}{3x + 9} - \dfrac{2}{x^2 + 3x}$

67. $\dfrac{1}{x} + \dfrac{x}{2x + 4} - \dfrac{2}{x^2 + 2x}$

68. $\dfrac{1}{x} + \dfrac{x}{3x + 9} - \dfrac{3}{x^2 + 3x}$

69. Let $f(x) = \dfrac{2}{x + 4}$ and $g(x) = \dfrac{x - 1}{x^2 + 3x - 4}$.
Find $f(x) + g(x)$.

70. Let $f(t) = \dfrac{5}{3t - 2}$ and $g(t) = \dfrac{t - 3}{3t^2 + 7t - 6}$.
Find $f(t) - g(t)$.

71. Let $f(x) = \dfrac{2x}{x^2 - x - 2}$ and $g(x) = \dfrac{5}{x^2 + x - 6}$.
Find $f(x) + g(x)$.

72. Let $f(x) = \dfrac{7}{x^2 - x - 12}$ and $g(x) = \dfrac{5}{x^2 + x - 6}$.
Find $f(x) - g(x)$.

Applying the Concepts

73. Optometry The formula

$$P = \frac{1}{a} + \frac{1}{b}$$

is used by optometrists to help determine how strong to make the lenses for a pair of eyeglasses. If a is 10 and b is 0.2, find the corresponding value of P.

74. Quadratic Formula Later in the book we will work with the quadratic formula. The derivation of the formula requires that you can add the fractions below. Add the fractions.

$$\frac{-c}{a} + \frac{b^2}{(2a)^2}$$

75. Painting The formula

$$T = \frac{ab}{a+b}$$

gives the time for two people to paint a room together if their individual times are a and b. If two people take 2 hours and 4 hours to paint a room alone, how long will it take them to paint the room together?

76. Electronics The formula

$$R = \frac{R_1 R_2}{R_1 + R_2}$$

gives the overall resistance of a circuit with individual resistors R_1 and R_2. If two resistors are 4 ohms and 3 ohms, what is the overall resistance of the circuit?

77. Number Problem Write an expression for the sum of a number and 4 times its reciprocal. Then simplify that expression.

78. Number Problem Write an expression for the sum of a number and 3 times its reciprocal. Then simplify that expression.

79. Number Problem Write an expression for the sum of the reciprocals of two consecutive integers. Then simplify that expression.

80. Number Problem Write an expression for the sum of the reciprocals of two consecutive even integers. Then simplify that expression.

Getting Ready for the Next Section

Divide.

81. $\dfrac{3}{4} \div \dfrac{5}{8}$

82. $\dfrac{2}{3} \div \dfrac{5}{6}$

83. $\dfrac{1}{15} \div \dfrac{3}{5}$

84. $\dfrac{4}{9} \div \dfrac{2}{3}$

Multiply.

85. $x\left(1 + \dfrac{2}{x}\right)$

86. $3\left(x + \dfrac{1}{3}\right)$

87. $3x\left(\dfrac{1}{x} - \dfrac{1}{3}\right)$

88. $3x\left(\dfrac{1}{x} + \dfrac{1}{3}\right)$

Factor.

89. $x^2 - 4$

90. $x^2 - x - 6$

91. $x^3 - 27$

92. $x^4 - 16$

One Step Further

Simplify.

93. $\left(1 - \dfrac{1}{x}\right)\left(1 - \dfrac{1}{x+1}\right)\left(1 - \dfrac{1}{x+2}\right)\left(1 - \dfrac{1}{x+3}\right)$

94. $\left(1 + \dfrac{1}{x}\right)\left(1 + \dfrac{1}{x+1}\right)\left(1 + \dfrac{1}{x+2}\right)\left(1 + \dfrac{1}{x+3}\right)$

95. $\left(\dfrac{a^2 - b^2}{u^2 - v^2}\right)\left(\dfrac{av - au}{b - a}\right) + \left(\dfrac{a^2 - av}{u + v}\right)\left(\dfrac{1}{a}\right)$

96. $\left(\dfrac{6r^2}{r^2 - 1}\right)\left(\dfrac{r + 1}{3}\right) - \left(\dfrac{2r^2}{r - 1}\right)$

97. $\dfrac{18x - 19}{4x^2 + 27x - 7} - \dfrac{12x - 41}{3x^2 + 17x - 28}$

98. $\dfrac{42 - 22y}{3y^2 - 13y - 10} - \dfrac{21 - 13y}{2y^2 - 9y - 5}$

99. $\left(\dfrac{1}{y^2 - 1} \div \dfrac{1}{y^2 + 1}\right)\left(\dfrac{y^3 + 1}{y^4 - 1}\right) + \dfrac{1}{(y + 1)^2\,(y - 1)}$

100. $\left(\dfrac{a^3 - 64}{a^2 - 16} \div \dfrac{a^2 - 4a + 16}{a^2 - 4} \div \dfrac{a^2 + 4a + 16}{a^3 + 64}\right) + 4 - a^2$

Find the Mistake

Each sentence below contains a mistake. Circle the mistake and write the correct word or expression on the line provided.

1. If we add the rational expressions $\dfrac{x + 5}{3x + 9}$ and $\dfrac{4}{x^2 - 9}$ and reduce to the lowest terms, we get $x - 1$. _____

2. The least common denominator for the rational expressions $\dfrac{2x + 3}{3x^2 + 6x - 45}$ and $\dfrac{x + 7}{3x^2 + 14x - 5}$ is $3x - 1$. _____

3. If we subtract $\dfrac{5}{2x^2 + x - 3}$ from $\dfrac{2}{x^2 - 1}$, we get $\dfrac{1}{(x + 1)(2x + 3)}$. _____

4. To subtract $\dfrac{x - 4}{x^2 - x - 20}$ from $\dfrac{2x + 5}{x^2 - 4x - 5}$, make the expressions equivalent by first using LCD $(x + 4)(2x + 5)$.

_____

KEY WORDS

complex fraction

complex rational expression

5.4 Complex Rational Expressions

Image © Katherine Heistand Shields, 2010

Every week in Portland, Oregon, bicyclists known as zoobombers convene on a street corner near the local zoo. Bicycles are decorated in streamers, strings of lights, and other ornate details to coincide with the often-costumed riders. Even the bikes themselves are known to cause a stir, falling into a variety of categories, such as swing bikes, mini bikes, tall bikes, and cruisers. Amidst laughter and screams of joy, the bicyclists spend the evening speeding together down the steep hills of the neighborhood.

Imagine you are a spectator on the sidewalk of a street the zoobombers are speeding down. You hear the sound of a single screaming bicyclist zooming by because of the Doppler effect, where the pitch of the sound changes as the bicyclist approaches and then eventually passes you. The formula for the sound frequency h you hear is

$$h = \frac{f}{1 + \dfrac{v}{s}}$$

where f is the actual frequency of the sound being produced, s is the speed of the sound (about 740 miles per hour), and v is the velocity of the bicyclist. This formula uses a *complex rational expression*, a quotient of two rational expressions, which is the focus of this section. You will encounter this exact formula later in the problem set.

A Simplifying Complex Rational Expressions

EXAMPLE 1 Simplify $\dfrac{\frac{3}{4}}{\frac{5}{8}}$.

Solution There are generally two methods that can be used to simplify complex fractions.

Method 1 We can multiply the numerator and denominator of the complex fractions by the LCD for both of the fractions, which in this case is 8.

$$\frac{\frac{3}{4}}{\frac{5}{8}} = \frac{\frac{3}{4} \cdot 8}{\frac{5}{8} \cdot 8} = \frac{6}{5}$$

Method 2 Instead of dividing by $\frac{5}{8}$ we can multiply by $\frac{8}{5}$.

$$\frac{\frac{3}{4}}{\frac{5}{8}} = \frac{3}{4} \cdot \frac{8}{5} = \frac{24}{20} = \frac{6}{5}$$

Practice Problems

1. Simplify $\dfrac{\frac{1}{3}}{\frac{3}{4}}$.

Answer

1. $\frac{4}{9}$

Here are some examples of complex fractions involving rational expressions. Most can be solved using either of the two methods shown in Example 1.

EXAMPLE 2   Simplify $\dfrac{\dfrac{1}{x} + \dfrac{1}{y}}{\dfrac{1}{x} - \dfrac{1}{y}}$.

Solution   This problem is most easily solved using Method 1. We begin by multiplying both the numerator and denominator by the quantity xy, which is the LCD for all the fractions.

$$\frac{\dfrac{1}{x} + \dfrac{1}{y}}{\dfrac{1}{x} - \dfrac{1}{y}} = \frac{\left(\dfrac{1}{x} + \dfrac{1}{y}\right) \cdot xy}{\left(\dfrac{1}{x} - \dfrac{1}{y}\right) \cdot xy}$$

$$= \frac{\dfrac{1}{x}(xy) + \dfrac{1}{y}(xy)}{\dfrac{1}{x}(xy) - \dfrac{1}{y}(xy)}$$

Apply the distributive property to distribute xy over both terms in the numerator and denominator.

$$= \frac{y + x}{y - x}$$

EXAMPLE 3   Let $f(x) = \dfrac{x-2}{x^2-9}$ and $g(x) = \dfrac{x^2-4}{x+3}$. Simplify $\dfrac{f(x)}{g(x)}$.

Solution   Applying Method 2, we have

$$\frac{f(x)}{g(x)} = \frac{\dfrac{x-2}{x^2-9}}{\dfrac{x^2-4}{x+3}} = \frac{x-2}{x^2-9} \cdot \frac{x+3}{x^2-4}$$

$$= \frac{(x-2)(x+3)}{(x+3)(x-3)(x+2)(x-2)}$$

$$= \frac{1}{(x-3)(x+2)}$$

EXAMPLE 4   Simplify $\dfrac{1 - \dfrac{4}{x^2}}{1 - \dfrac{1}{x} - \dfrac{6}{x^2}}$.

Solution   The easiest way to simplify this complex fraction is to multiply the numerator and denominator by the LCD, x^2.

$$\frac{1 - \dfrac{4}{x^2}}{1 - \dfrac{1}{x} - \dfrac{6}{x^2}} = \frac{x^2\left(1 - \dfrac{4}{x^2}\right)}{x^2\left(1 - \dfrac{1}{x} - \dfrac{6}{x^2}\right)}$$

Multiply numerator and denominator by x^2.

$$= \frac{x^2 \cdot 1 - x^2 \cdot \dfrac{4}{x^2}}{x^2 \cdot 1 - x^2 \cdot \dfrac{1}{x} - x^2 \cdot \dfrac{6}{x^2}}$$

Distributive property

$$= \frac{x^2 - 4}{x^2 - x - 6}$$

Simplify.

$$= \frac{(x-2)(x+2)}{(x-3)(x+2)}$$

Factor.

$$= \frac{x-2}{x-3}$$

Reduce.

2. Simplify $\dfrac{\dfrac{1}{x} - \dfrac{1}{y}}{\dfrac{1}{y} + \dfrac{1}{x}}$.

3. Let $f(x) = \dfrac{x-1}{x^2-16}$ and $g(x) = \dfrac{x^2-1}{x+4}$.

Simplify $\dfrac{f(x)}{g(x)}$.

4. Simplify $\dfrac{1 - \dfrac{9}{x^2}}{1 - \dfrac{5}{x} + \dfrac{6}{x^2}}$.

Answers

2. $\frac{y-x}{x+y}$

3. $\frac{f(x)}{g(x)} = \frac{1}{(x+4)(x+1)}$

4. $\frac{x+3}{x-2}$

5. Simplify $4 - \dfrac{2}{x + \dfrac{1}{2}}$.

EXAMPLE 5 Simplify $2 - \dfrac{3}{x + \dfrac{1}{3}}$.

Solution First, we simplify the expression that follows the subtraction sign.

$$2 - \frac{3}{x + \dfrac{1}{3}} = 2 - \frac{3 \cdot 3}{3\left(x + \dfrac{1}{3}\right)} = 2 - \frac{9}{3x + 1}$$

Now we subtract by rewriting the first term, 2, with the LCD, $3x + 1$.

$$2 - \frac{9}{3x + 1} = \frac{2}{1} \cdot \frac{3x + 1}{3x + 1} - \frac{9}{3x + 1}$$

$$= \frac{6x + 2 - 9}{3x + 1} = \frac{6x - 7}{3x + 1}$$

6. Simplify $\dfrac{1 - 4x^{-2}}{1 + 2x^{-1} - 8x^{-2}}$.

EXAMPLE 6 Simplify $\dfrac{1 - 9x^{-2}}{1 - 5x^{-1} + 6x^{-2}}$.

Solution Let's begin by writing this expression with only positive exponents.

$$\frac{1 - \dfrac{9}{x^2}}{1 - \dfrac{5}{x} + \dfrac{6}{x^2}} \qquad \text{Negative exponent property}$$

Now we simplify this complex fraction by multiplying the numerator and denominator by x^2.

$$\frac{1 - \dfrac{9}{x^2}}{1 - \dfrac{5}{x} + \dfrac{6}{x^2}} = \frac{x^2\left(1 - \dfrac{9}{x^2}\right)}{x^2\left(1 - \dfrac{5}{x} + \dfrac{6}{x^2}\right)} \qquad \text{Multiply numerator and denominator by } x^2.$$

$$= \frac{x^2 - 9}{x^2 - 5x + 6} \qquad \text{Simplify.}$$

$$= \frac{(x - 3)(x + 3)}{(x - 3)(x - 2)} \qquad \text{Factor.}$$

$$= \frac{x + 3}{x - 2} \qquad \text{Reduce.}$$

GETTING READY FOR CLASS

After reading through the preceding section, respond in your own words and in complete sentences.

A. What is a complex rational expression?

B. Explain how to simplify a complex fraction using a least common denominator.

C. When is it more efficient to convert a complex fraction to a division problem of rational expressions?

D. Which method of simplifying complex fractions do you prefer? Why?

Answers

5. $\dfrac{8x}{2x + 1}$

6. $\dfrac{x + 2}{x + 4}$

Vocabulary Review

Choose the correct words to fill in the blanks in the paragraph.

numerator   reciprocal   LCD   complex   quotient   denominator

A _____ rational expression is a _____ of two rational expressions. Method 1 for simplifying a complex rational expression requires multiplying the _____ and denominator of the complex fraction by the _____ for both fractions. Method 2 for simplifying a complex rational expression requires multiplying the numerator of the complex fraction by the _____ of the fraction in the _____.

Problems

A Simplify each of the following as much as possible.

1. $\dfrac{\frac{3}{4}}{\frac{2}{3}}$

2. $\dfrac{\frac{5}{9}}{\frac{7}{12}}$

3. $\dfrac{\frac{1}{3} - \frac{1}{4}}{\frac{1}{2} + \frac{1}{8}}$

4. $\dfrac{\frac{1}{6} - \frac{1}{3}}{\frac{1}{4} - \frac{1}{8}}$

5. $\dfrac{3 + \frac{2}{5}}{1 - \frac{3}{7}}$

6. $\dfrac{2 + \frac{5}{6}}{1 - \frac{7}{8}}$

7. $\dfrac{\frac{1}{x}}{1 + \frac{1}{x}}$

8. $\dfrac{1 - \frac{1}{x}}{\frac{1}{x}}$

9. $\dfrac{1 + \frac{1}{a}}{1 - \frac{1}{a}}$

10. $\dfrac{1 - \frac{2}{a}}{1 - \frac{3}{a}}$

11. $\dfrac{\frac{1}{x} - \frac{1}{y}}{\frac{1}{x} + \frac{1}{y}}$

12. $\dfrac{\frac{1}{x} + \frac{2}{y}}{\frac{2}{x} + \frac{1}{y}}$

13. $\dfrac{\frac{x - 5}{x^2 - 4}}{\frac{x^2 - 25}{x + 2}}$

14. $\dfrac{\frac{3x + 1}{x^2 - 49}}{\frac{9x^2 - 1}{x - 7}}$

15. $\dfrac{\frac{4a}{2a^3 + 2}}{\frac{8a}{4a + 4}}$

16. $\dfrac{\frac{2a}{3a^3 - 3}}{\frac{4a}{6a - 6}}$

17. $\dfrac{1 - 9x^{-2}}{1 - x^{-1} - 6x^{-2}}$

18. $\dfrac{4 - x^{-2}}{4 + 4x^{-1} + x^{-2}}$

19. $\dfrac{2 + \dfrac{5}{a} - \dfrac{3}{a^2}}{2 - \dfrac{5}{a} + \dfrac{2}{a^2}}$

20. $\dfrac{3 + \dfrac{5}{a} - \dfrac{2}{a^2}}{3 - \dfrac{10}{a} + \dfrac{3}{a^2}}$

21. $\dfrac{1 + \dfrac{1}{x+3}}{1 + \dfrac{7}{x-3}}$

22. $\dfrac{1 + \dfrac{1}{x-2}}{1 - \dfrac{3}{x+2}}$

23. $\dfrac{1 - (a+1)^{-1}}{1 + (a-1)^{-1}}$

24. $\dfrac{(a-1)^{-1} + 1}{(a+1)^{-1} - 1}$

25. $\dfrac{(x+3)^{-1} + (x-3)^{-1}}{(x+3)^{-1} - (x-3)^{-1}}$

26. $\dfrac{(x+a)^{-1} + (x-a)^{-1}}{(x+a)^{-1} - (x-a)^{-1}}$

27. $\dfrac{\dfrac{y+1}{y-1} + \dfrac{y-1}{y+1}}{\dfrac{y+1}{y-1} - \dfrac{y-1}{y+1}}$

28. $\dfrac{\dfrac{y-1}{y+1} - \dfrac{y+1}{y-1}}{\dfrac{y-1}{y+1} + \dfrac{y+1}{y-1}}$

29. $1 - \dfrac{x}{1 - \dfrac{1}{x}}$

30. $x - \dfrac{1}{x - \dfrac{1}{2}}$

31. $1 + \dfrac{1}{1 + \dfrac{1}{1+1}}$

32. $1 - \dfrac{1}{1 - \dfrac{1}{1 - \dfrac{1}{2}}}$

33. $\dfrac{1 - \dfrac{1}{x + \dfrac{1}{2}}}{1 + \dfrac{1}{x + \dfrac{1}{2}}}$

34. $\dfrac{2 + \dfrac{1}{x - \dfrac{1}{3}}}{2 - \dfrac{1}{x - \dfrac{1}{3}}}$

35. $\dfrac{\dfrac{1}{x+h} - \dfrac{1}{x}}{h}$

36. $\dfrac{\dfrac{1}{(x+h)^2} - \dfrac{1}{x^2}}{h}$

37. $\dfrac{\dfrac{1}{x+8} - \dfrac{1}{x}}{8}$

38. $\dfrac{\dfrac{1}{x+7} - \dfrac{1}{x}}{7}$

39. $\dfrac{1}{(x+6)^2} - \dfrac{1}{x^2}$

40. $\dfrac{1}{(x+7)^2} - \dfrac{1}{x^2}$

41. $\dfrac{\dfrac{3}{ab} + \dfrac{4}{bc} - \dfrac{2}{ac}}{\dfrac{5}{abc}}$

42. $\dfrac{\dfrac{x}{yz} - \dfrac{y}{xz} + \dfrac{z}{xy}}{\dfrac{1}{x^2 y^2} - \dfrac{1}{x^2 z^2} + \dfrac{1}{y^2 z^2}}$

43. $\dfrac{\dfrac{t^2 - 2t - 8}{t^2 + 7t + 6}}{\dfrac{t^2 - t - 6}{t^2 + 2t + 1}}$

44. $\dfrac{\dfrac{y^2 - 5y - 14}{y^2 + 3y - 10}}{\dfrac{y^2 - 8y + 7}{y^2 + 6y + 5}}$

45. $\dfrac{5 + \left(\frac{b-1}{4}\right)^{-1}}{\left(\frac{b+5}{7}\right)^{-1} - \left(\frac{b-1}{3}\right)^{-1}}$

46. $\dfrac{(k - 7k + 12)^{-1}}{(k - 3)^{-1} + (k - 4)^{-1}}$

47. $\dfrac{\dfrac{3}{x^2 - x - 6}}{\dfrac{2}{x + 2} - \dfrac{4}{x - 3}}$

48. $\dfrac{\dfrac{9}{a - 7} + \dfrac{8}{2a + 3}}{\dfrac{10}{2a^2 - 11a - 21}}$

49. $\dfrac{\dfrac{1}{m - 4} + \dfrac{1}{m - 5}}{\dfrac{1}{m^2 - 9m + 20}}$

50. $\dfrac{\dfrac{1}{k^2 - 7k + 12}}{\dfrac{1}{k - 3} + \dfrac{1}{k - 4}}$

51. $\dfrac{\dfrac{1}{x + 2} - \dfrac{1}{x - 2}}{\dfrac{1}{x + 2} + \dfrac{1}{x - 2}}$

52. $1 + \dfrac{1}{1 + \dfrac{1}{1 + \dfrac{1}{x}}}$

For the following functions, find $\dfrac{f(x)}{g(x)}$, and simplify.

53. $f(x) = 2 + \dfrac{3}{x} - \dfrac{18}{x^2} - \dfrac{27}{x^3}, g(x) = 2 + \dfrac{9}{x} + \dfrac{9}{x^2}$

54. $f(x) = 3 + \dfrac{5}{x} - \dfrac{12}{x^2} - \dfrac{20}{x^3}, g(x) = 3 + \dfrac{11}{x} + \dfrac{10}{x^2}$

55. $f(x) = 1 + \dfrac{1}{x + 3}, g(x) = 1 - \dfrac{1}{x + 3}$

56. $f(x) = 1 + \dfrac{1}{x - 2}, g(x) = 1 - \dfrac{1}{x - 2}$

Applying the Concepts

To solve the following problems, recall our discussion of difference quotients at the beginning of this chapter.

57. Difference Quotient For each rational function below, find the difference quotient

$$\dfrac{f(x) - f(a)}{x - a}$$

a. $f(x) = \dfrac{4}{x}$

b. $f(x) = \dfrac{1}{x + 1}$

c. $f(x) = \dfrac{1}{x^2}$

58. Difference Quotient For each rational function below, find the difference quotient

$$\dfrac{f(x + h) - f(x)}{h}$$

a. $f(x) = \dfrac{4}{x}$

b. $f(x) = \dfrac{1}{x + 1}$

c. $f(x) = \dfrac{1}{x^2}$

59. Doppler Effect As mentioned in the opening to this section, the change in the pitch of a sound as an object passes is called the Doppler effect, named after C. J. Doppler (1803–1853). A person will *hear* a sound with a frequency, *h*, according to the formula

$$h = \frac{f}{1 + \dfrac{v}{s}}$$

where *f* is the actual frequency of the sound being produced, *s* is the speed of sound (about 740 miles per hour), and *v* is the velocity of the moving object.

a. Examine this fraction, and then explain why *h* and *f* approach the same value as *v* becomes smaller and smaller.

b. Solve this formula for *v*.

60. Work Problem A water storage tank has two drains. It can be shown that the time it takes to empty the tank if both drains are open is given by the formula

$$\frac{1}{\dfrac{1}{a} + \dfrac{1}{b}}$$

where *a* = time it takes for the first drain to empty the tank, and *b* = time for the second drain to empty the tank.

a. Simplify this complex fraction.

b. Find the amount of time needed to empty the tank using both drains if, used alone, the first drain empties the tank in 4 hours and the second drain can empty the tank in 3 hours.

Getting Ready for the Next Section

Multiply.

61. $x(y - 2)$

62. $x(y - 1)$

63. $6\left(\dfrac{x}{2} - 3\right)$

64. $6\left(\dfrac{x}{3} + 1\right)$

65. $xab \cdot \dfrac{1}{x}$

66. $xab\left(\dfrac{1}{b} + \dfrac{1}{a}\right)$

67. $st^2\left(\dfrac{2}{s} - \dfrac{3}{t^2}\right)$

68. $4x^2y\left(\dfrac{1}{2xy} - \dfrac{3}{x^2}\right)$

Factor.

69. $y^2 - 25$

70. $x^2 - 3x + 2$

71. $xa + xb$

72. $xy - y$

Solve.

73. $5x - 4 = 6$

74. $y^2 + y - 20 = 2y$

75. $x^2 - 2x - 12 = 2x$

76. $3x - 4 = 9$

Find the Mistake

Each sentence below contains a mistake. Circle the mistake and write the correct words or expression on the line provided.

1. Simplifying $\dfrac{\frac{2}{m} + \frac{4}{n}}{\frac{2}{m} - \frac{6}{n}}$ gives us $-\dfrac{2}{3}$. _____

2. If $f(x) = \dfrac{x+2}{x^2-16}$ and $g(x) = \dfrac{x^2-x-6}{x+4}$, then $\dfrac{f(x)}{g(x)} = \dfrac{(x-2)^2(x+3)}{(x-4)(x+4)^2}$. _____

3. Simplifying $5 - \dfrac{2}{x+\frac{1}{4}}$ gives us $-\dfrac{3}{4x+1}$. _____

4. The first step to simplify $\dfrac{1-36x^{-2}}{1-9x^{-1}+18x^{-2}}$ is to multiply the numerator and denominator by x. _____

_____

Landmark Review: Checking Your Progress

Simplify the following expressions.

1. $\dfrac{3a^2 - 27}{6a - 18}$

2. $\dfrac{y-3}{y^2 - 6y + 9}$

3. $\dfrac{3x^3 - 6x^2 - 45x}{x^2 + 5x + 6}$

4. $\dfrac{5x^2 + 22x + 8}{25x^2 - 4}$

Perform the indicated operations.

5. $\dfrac{3a^4}{(6a+18)} \cdot \dfrac{(a+3)}{a^3}$

6. $\dfrac{(4x^2 - 12x)}{(x^2 - 1)} \cdot \dfrac{(x^2 + x)}{4x}$

7. $\dfrac{(y^2 - 3y)}{(2y^2 - 8)} \div \dfrac{(y^2 - 25)}{(y^2 + 3y - 10)}$

8. $(x^2 + 8x + 15) \cdot \dfrac{2x}{x+5}$

9. $\dfrac{8}{(4x+16)} \div \dfrac{2}{(x+4)}$

10. $\dfrac{4}{y+3} + \dfrac{24}{y^2 - 9}$

11. $\dfrac{a+5}{a+4} + \dfrac{8}{a^2 - 16}$

12. $\dfrac{x^2}{x+4} - \dfrac{16}{x+4}$

13. $\dfrac{x+2}{x-1} - \dfrac{6}{x^2 - 1}$

14. $\dfrac{x}{(x+2)(x+3)} - \dfrac{2}{(x+2)(x+3)}$

15. $\dfrac{1 - \frac{3}{a}}{1 - \frac{4}{a}}$

16. $\dfrac{\frac{2}{x} + \frac{1}{y}}{\frac{3}{x} + \frac{1}{y}}$

17. $\dfrac{\frac{4x+1}{x^2-9}}{\frac{16x^2-1}{x-3}}$

18. $\dfrac{\frac{y-1}{y+1} - \frac{y+1}{y-1}}{\frac{y-1}{y+1} + \frac{y+1}{y-1}}$

©iStockphoto.com/AndreyPS

For more than fifty years, the Japanese city of Ito has hosted a wooden wash tub race down the Matsukawa River. Each contestant must sit in a 1-meter-diameter tub and use large wooden rice paddles to steer a 400-meter-long course. We can use the following rational equation to calculate how long it would take one tub racer to complete the course:

$$t = \frac{400}{v + c}$$

where the total time t (in seconds) depends on the speed of the water c (in meters per second) and the speed of the tub v. In this section, we will work more with *rational equations*, such as this one. When working with rational equations, our goal is to find a solution for the variable that makes the equation a true statement. Until now, we have only worked with rational *expressions* where our goal was only to simplify.

A Solving Equations with Rational Expressions

The first step in solving an equation that contains one or more rational expressions is to find the LCD for all denominators in the equation. Then, we multiply both sides of the equation by the LCD to clear the equation of all fractions. That is, after we have multiplied through by the LCD, each term in the resulting equation will have a denominator of 1.

EXAMPLE 1 Solve $\frac{x}{2} - 3 = \frac{2}{3}$.

Solution The LCD for 2 and 3 is 6. Multiplying both sides by 6, we have

$$6\left(\frac{x}{2} - 3\right) = 6\left(\frac{2}{3}\right)$$

$$6\left(\frac{x}{2}\right) - 6(3) = 6\left(\frac{2}{3}\right)$$

$$3x - 18 = 4$$

$$3x = 22$$

$$x = \frac{22}{3}$$

The solution is $\frac{22}{3}$.

Multiplying both sides of an equation by the LCD clears the equation of fractions because the LCD has the property that all the denominators divide it evenly.

EXAMPLE 2 Solve $\dfrac{6}{a-4} = \dfrac{3}{8}$.

Solution The LCD for $a-4$ and 8 is $8(a-4)$. Multiplying both sides by this quantity yields

$$8(a-4) \cdot \frac{6}{a-4} = 8(a-4) \cdot \frac{3}{8}$$

$$48 = (a-4) \cdot 3$$

$$48 = 3a - 12$$

$$60 = 3a$$

$$20 = a$$

The solution is 20, which checks in the original equation.

When we multiply both sides of an equation by an expression containing the variable, we must be sure to check our solutions. The multiplication property of equality does not allow multiplication by 0. If the expression we multiply by contains the variable, then it has the possibility of being 0. In the last example, we multiplied both sides by $8(a-4)$. This gives a restriction $a \neq 4$ for any solution we come up with.

EXAMPLE 3 Solve $\dfrac{x}{x-2} + \dfrac{2}{3} = \dfrac{2}{x-2}$.

Solution The LCD is $3(x-2)$. We are assuming $x \neq 2$ when we multiply both sides of the equation by $3(x-2)$.

$$3(x-2) \cdot \left(\frac{x}{x-2} + \frac{2}{3} \right) = 3(x-2) \cdot \frac{2}{x-2}$$

$$3x + (x-2) \cdot 2 = 3 \cdot 2$$

$$3x + 2x - 4 = 6$$

$$5x - 4 = 6$$

$$5x = 10$$

$$x = 2$$

The only possible solution is $x = 2$. Let's check this value in the original equation.

$$\frac{2}{2-2} + \frac{2}{3} \overset{?}{=} \frac{2}{2-2}$$

$$\frac{2}{0} + \frac{2}{3} \overset{?}{=} \frac{2}{0}$$

The first and last terms are undefined. The proposed solution, $x = 2$, does not check in the original equation. There is no solution to the original equation.

When the proposed solution to an equation is not actually a solution, it is called an *extraneous solution*. In the last example, $x = 2$ is an extraneous solution.

---

2. Solve $\dfrac{4}{x-5} = \dfrac{4}{3}$.

3. Solve $\dfrac{x}{x-5} + \dfrac{5}{2} = \dfrac{5}{x-5}$.

Note In the process of solving the equation, we multiplied both sides by $3(x-2)$, solved for x, and got $x = 2$ for our solution. But when x is 2, the quantity $3(x-2) = 3(2-2) = 3(0) = 0$, which means we multiplied both sides of our equation by 0, which is not allowed under the multiplication property of equality.

Answers

2. 8

3. No solution

4. Solve $\dfrac{x}{x^2 - 4} - \dfrac{2}{x - 2} = \dfrac{1}{x + 2}$.

EXAMPLE 4   Solve $\dfrac{5}{x^2 - 3x + 2} - \dfrac{1}{x - 2} = \dfrac{1}{3x - 3}$.

Solution   Writing the equation again with the denominators in factored form, we have

$$\frac{5}{(x - 2)(x - 1)} - \frac{1}{x - 2} = \frac{1}{3(x - 1)}$$

The LCD is $3(x - 2)(x - 1)$. Multiplying through by the LCD, we have

$$3(x - 2)(x - 1)\,\frac{5}{(x - 2)(x - 1)} - 3(x - 2)(x - 1) \cdot \frac{1}{(x - 2)}$$

$$= 3(x - 2)(x - 1) \cdot \frac{1}{3(x - 1)}$$

$$3 \cdot 5 - 3(x - 1) \cdot 1 = (x - 2) \cdot 1$$

$$15 - 3x + 3 = x - 2$$

$$-3x + 18 = x - 2$$

$$-4x + 18 = -2$$

$$-4x = -20$$

$$x = 5$$

Checking the proposed solution $x = 5$ in the original equation yields a true statement. Try it and see.

> **Note**   We can check the proposed solution in any of the equations obtained before multiplying through by the LCD. We cannot check the proposed solution in an equation obtained after multiplying through by the LCD because, if we have multiplied by 0, the resulting equations will not be equivalent to the original one.

5. Let $f(x) = 2 + \dfrac{9}{x}$ and $g(x) = -\dfrac{10}{x^2}$.
Find all values of x where $f(x) = g(x)$.

EXAMPLE 5   Let $f(x) = 3 + \dfrac{1}{x}$ and $g(x) = \dfrac{10}{x^2}$. Find all values of x where $f(x) = g(x)$.

Solution   To clear the equation of denominators, we multiply both sides by x^2.

$$x^2\left(3 + \frac{1}{x}\right) = x^2\left(\frac{10}{x^2}\right)$$

$$3(x^2) + \left(\frac{1}{x}\right)(x^2) = \left(\frac{10}{x^2}\right)(x^2)$$

$$3x^2 + x = 10$$

Rewrite in standard form, and solve.

$$3x^2 + x - 10 = 0$$

$$(3x - 5)(x + 2) = 0$$

$$3x - 5 = 0 \quad \text{or} \quad x + 2 = 0$$

$$x = \frac{5}{3} \qquad\qquad x = -2$$

Both solutions check in the original equation. So $f(x) = g(x)$ when $x = \frac{5}{3}$ or $x = -2$. Remember, we have to check all solutions any time we multiply both sides of the equation by an expression that contains the variable, just to be sure we haven't multiplied by 0.

6. Solve $\dfrac{a + 2}{a^2 + 3a} = \dfrac{-2}{a^2 - 9}$.

EXAMPLE 6   Solve $\dfrac{y - 4}{y^2 - 5y} = \dfrac{2}{y^2 - 25}$.

Solution   Factoring each denominator, we find the LCD is $y(y - 5)(y + 5)$. Multiplying each side of the equation by the LCD clears the equation of denominators and leads us to our possible solutions.

$$y(y - 5)(y + 5) \cdot \frac{y - 4}{y(y - 5)} = \frac{2}{(y - 5)(y + 5)} \cdot y(y - 5)(y + 5)$$

$$(y + 5)(y - 4) = 2y$$

$$y^2 + y - 20 = 2y \qquad \text{Multiply out the left side.}$$

$$y^2 - y - 20 = 0 \qquad \text{Add } -2y \text{ to each side.}$$

$$(y - 5)(y + 4) = 0$$

$$y - 5 = 0 \quad \text{or} \quad y + 4 = 0$$

$$y = 5 \qquad\qquad y = -4$$

Answers

4. -1

5. $x = -\dfrac{5}{2}, -2$

6. 2

The two possible solutions are 5 and -4. If we substitute -4 for y in the original equation, we find that it leads to a true statement. It is therefore a solution. On the other hand, if we substitute 5 for y in the original equation, we find that both sides of the equation are undefined. The only solution to our original equation is $y = -4$. The other possible solution $y = 5$ is extraneous.

B Formulas

EXAMPLE 7   Solve $x = \dfrac{y - 4}{y - 2}$ for y.

7. Solve $x = \dfrac{y - 3}{y + 3}$ for y.

Solution   To solve for y, we first multiply each side by $y - 2$ to obtain

$$x(y - 2) = y - 4$$

$$xy - 2x = y - 4 \qquad \text{Distributive property}$$

$$xy - y = 2x - 4 \qquad \text{Collect all terms containing } y \text{ on the left side.}$$

$$y(x - 1) = 2x - 4 \qquad \text{Factor } y \text{ from each term on the left side.}$$

$$y = \frac{2x - 4}{x - 1} \qquad \text{Divide each side by } x - 1.$$

EXAMPLE 8   Solve the formula $\dfrac{1}{x} = \dfrac{1}{b} + \dfrac{1}{a}$ for x.

8. Solve $\dfrac{2}{x} = \dfrac{3}{b} + \dfrac{1}{a}$ for x.

Solution   We begin by multiplying both sides by the least common denominator xab. As you can see from our previous examples, multiplying both sides of an equation by the LCD is equivalent to multiplying each term of both sides by the LCD.

$$xab \cdot \frac{1}{x} = \frac{1}{b} \cdot xab + \frac{1}{a} \cdot xab$$

$$ab = xa + xb$$

$$ab = (a + b)x \qquad \text{Factor } x \text{ from the right side.}$$

$$\frac{ab}{a + b} = x$$

We know we are finished because the variable we were solving for is alone on one side of the equation and does not appear on the other side.

C Graphing Rational Functions

> **DEFINITION**   rational function
>
> A *rational function* is any function that is a polynomial quotient; that is, a polynomial is divided by another polynomial.

9. Graph $f(x) = \dfrac{6}{x - 4}$.

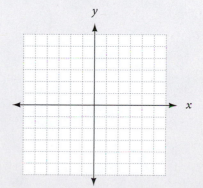

Our next example begins our investigation of the graphs of rational functions.

EXAMPLE 9   Graph the rational function $f(x) = \dfrac{6}{x - 2}$.

Solution   To find the y-intercept, we let x equal 0.

When $x = 0$: $y = \dfrac{6}{0 - 2} = \dfrac{6}{-2} = -3$ y-intercept

The graph will not cross the x-axis. If it did, we would have a solution to the equation

$$0 = \frac{6}{x - 2}$$

which has no solution because there is no number to divide 6 by to obtain 0.

Answers

7. $y = \dfrac{-3(x + 1)}{x - 1}$

8. $x = \dfrac{2ab}{3a + b}$

Note   The graph of $f(x) = \frac{6}{x-2}$ also has a horizontal asymptote of $y = 0$. We will discuss horizontal asymptotes further in the next section.

The graph of our equation is shown in Figure 1 along with a table giving values of x and y that satisfy the equation. Notice that y is undefined when x is 2. This means that the graph will not cross the vertical line $x = 2$. (If it did, there would be a value of y for $x = 2$.) The line $x = 2$ is called a *vertical asymptote* of the graph. The graph will get very close to the vertical asymptote, but will never touch or cross it.

x	y
-4	-1
-1	-2
0	-3
1	-6
2	Undefined
3	6
4	3
5	2

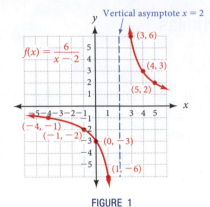

FIGURE 1

If you were to graph $y = \frac{6}{x}$ on the coordinate system in Figure 1, you would see that the graph of $y = \frac{6}{x-2}$ is the graph of $y = \frac{6}{x}$ with all points shifted 2 units to the right.

USING TECHNOLOGY   More About Example 9

We know the graph of $f(x) = \frac{6}{x-2}$ will not cross the vertical asymptote $x = 2$ because replacing x with 2 in the equation gives us an undefined expression, meaning there is no value of y to associate with $x = 2$. We can use a graphing calculator to explore the behavior of this function when x gets closer and closer to 2 by using the table function on the calculator. We want to put our own values for X into the table, so we set the independent variable to Ask. (On a TI-83/84, use the TBLSET key to set up the table.) To see how the function behaves as x gets close to 2, we let X take on values of 1.9, 1.99, and 1.999. Then we move to the other side of 2 and let X become 2.1, 2.01, and 2.001.

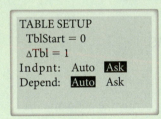

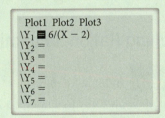

The table will look like this:

X	Y_1
1.9	-60
1.99	-600
1.999	-6000
2.1	60
2.01	600
2.001	6000

Again, the calculator asks us for a table increment. Because we are inputting the x values ourselves, the increment value does not matter.

As you can see, the values in the table support the shape of the curve in Figure 1 around the vertical asymptote $x = 2$.

EXAMPLE 10 Graph $g(x) = \dfrac{6}{x+2}$.

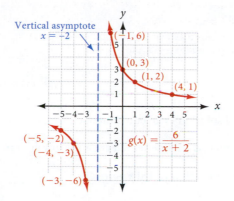

FIGURE 2

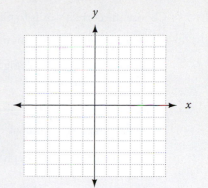

Solution The only difference between this equation and the equation in Example 9 is in the denominator. This graph will have the same shape as the graph in Example 9, but the vertical asymptote will be $x = -2$ instead of $x = 2$. Figure 2 shows the graph.

Notice that the graphs shown in Figures 1 and 2 are both graphs of functions because no vertical line will cross either graph in more than one place. Notice the similarities and differences in our two functions,

$$f(x) = \frac{6}{x-2} \qquad \text{and} \qquad g(x) = \frac{6}{x+2}$$

and their graphs. The vertical asymptotes shown in Figures 1 and 2 correspond to the fact that both $f(2)$ and $g(-2)$ are undefined. The domain for the function $f(x)$ is all real numbers except $x = 2$, while the domain for $g(x)$ is all real numbers except $x = -2$. We would write the domain for $f(x)$ as $\{x \mid x \neq 2\}$ and the domain for $g(x)$ as $\{x \mid x \neq -2\}$

GETTING READY FOR CLASS

After reading through the preceding section, respond in your own words and in complete sentences.

A. Explain how a least common denominator can be used to simplify an equation.

B. What is an extraneous solution?

C. How does the location of the vertical asymptote in the graph of a rational function relate to the equation of the function?

D. What is the last step in solving an equation that contains rational expressions?

Vocabulary Review

Choose the correct words to fill in the blanks below.

asymptote   rational   extraneous   function

1. The first step when solving an equation that contains _____ expressions is to find the least common denominator.

2. An _____ solution is a proposed solution to an equation but is not an actual solution.

3. When graphing a rational _____ no vertical line should cross the graph in more than one place.

4. A vertical line that a graph may get very close to but never touch is called a vertical _____.

Problems

A Solve each of the following equations.

1. $\dfrac{x}{5} + 4 = \dfrac{5}{3}$

2. $\dfrac{x}{5} = \dfrac{x}{2} - 9$

3. $\dfrac{a}{3} + 2 = \dfrac{4}{5}$

4. $\dfrac{a}{4} + \dfrac{1}{2} = \dfrac{2}{3}$

5. $\dfrac{y}{2} + \dfrac{y}{4} + \dfrac{y}{6} = 3$

6. $\dfrac{y}{3} - \dfrac{y}{6} + \dfrac{y}{2} = 1$

7. $\dfrac{5}{2x} = \dfrac{1}{x} + \dfrac{3}{4}$

8. $\dfrac{1}{2a} = \dfrac{2}{a} - \dfrac{3}{8}$

9. $\dfrac{1}{x} = \dfrac{1}{3} - \dfrac{2}{3x}$

10. $\dfrac{5}{2x} = \dfrac{2}{x} - \dfrac{1}{12}$

11. $\dfrac{2x}{x-3} + 2 = \dfrac{2}{x-3}$

12. $\dfrac{2}{x+5} = \dfrac{2}{5} - \dfrac{x}{x+5}$

13. $1 - \dfrac{1}{x} = \dfrac{12}{x^2}$

14. $2 + \dfrac{5}{x} = \dfrac{3}{x^2}$

15. $y - \dfrac{4}{3y} = -\dfrac{1}{3}$

16. $\dfrac{y}{2} - \dfrac{4}{y} = -\dfrac{7}{2}$

SCAN TO ACCESS

17. $\dfrac{x+2}{x+1} = \dfrac{1}{x+1} + 2$

18. $\dfrac{x+6}{x+3} = \dfrac{3}{x+3} + 2$

19. $\dfrac{x+2}{3x-1} = \dfrac{x-12}{4-9x-9x^2}$

20. $\dfrac{2x+3}{2x+4} = \dfrac{5x+4}{16+4x-2x^2}$

21. $6 - \dfrac{5}{x^2} = \dfrac{7}{x}$

22. $10 - \dfrac{3}{x^2} = -\dfrac{1}{x}$

23. $\dfrac{1}{x-1} - \dfrac{1}{x+1} = \dfrac{3x}{x^2-1}$

24. $\dfrac{5}{x-1} + \dfrac{2}{x-1} = \dfrac{4}{x+1}$

25. $\dfrac{t-4}{t^2-3t} = \dfrac{-2}{t^2-9}$

26. $\dfrac{t+3}{t^2-2t} = \dfrac{10}{t^2-4}$

27. $\dfrac{3}{y-4} - \dfrac{2}{y+1} = \dfrac{5}{y^2-3y-4}$

28. $\dfrac{1}{y+2} - \dfrac{2}{y-3} = \dfrac{-2y}{y^2-y-6}$

29. $\dfrac{2}{1+a} = \dfrac{3}{1-a} + \dfrac{5}{a}$

30. $\dfrac{1}{a+3} - \dfrac{a}{a^2-9} = \dfrac{2}{3-a}$

31. $\dfrac{3}{2x-6} - \dfrac{x+1}{4x-12} = 4$

32. $\dfrac{2x-3}{5x+10} + \dfrac{3x-2}{4x+8} = 1$

33. $\dfrac{y+2}{y^2-y} - \dfrac{6}{y^2-1} = 0$

34. $\dfrac{y+3}{y^2-y} - \dfrac{8}{y^2-1} = 0$

35. $\dfrac{4}{2x-6} - \dfrac{12}{4x+12} = \dfrac{12}{x^2-9}$

36. $\dfrac{1}{x+2} + \dfrac{1}{x-2} = \dfrac{4}{x^2-4}$

37. $\dfrac{2}{y^2-7y+12} - \dfrac{1}{y^2-9} = \dfrac{4}{y^2-y-12}$

38. $\dfrac{1}{y^2+5y+4} + \dfrac{3}{y^2-1} = \dfrac{-1}{y^2+3y-4}$

Given the functions $f(x)$ and $g(x)$, find where $f(x) = g(x)$.

39. $f(x) = \dfrac{2}{x-3} + \dfrac{x}{x^2-9}, g(x) = \dfrac{4}{x+3}$

40. $f(x) = \dfrac{2}{x+5} + \dfrac{3}{x+4}, g(x) = \dfrac{2x}{x^2+9x+20}$

41. $f(x) = \dfrac{3}{2} - \dfrac{1}{x-4}, g(x) = \dfrac{-2}{2x-8}$

42. $f(x) = \dfrac{2}{x} - \dfrac{1}{x+1}, g(x) = \dfrac{-2}{5x+5}$

43. Let $f(x) = \dfrac{1}{x-3}$ and $g(x) = \dfrac{1}{x+3}$. Find x if

    a. $f(x) + g(x) = \dfrac{5}{8}$

    b. $\dfrac{f(x)}{g(x)} = 5$

    c. $f(x) = g(x)$

44. Let $f(x) = \dfrac{4}{x+2}$ and $g(x) = \dfrac{4}{x-2}$. Find x if

    a. $f(x) - g(x) = -\dfrac{4}{3}$

    b. $\dfrac{g(x)}{f(x)} = -7$

    c. $f(x) = -g(x)$

45. Solve each equation.

    a. $6x - 2 = 0$

    b. $\dfrac{6}{x} - 2 = 0$

    c. $\dfrac{x}{6} - 2 = -\dfrac{1}{2}$

    d. $\dfrac{6}{x} - 2 = -\dfrac{1}{2}$

    e. $\dfrac{6}{x^2} + 6 = \dfrac{20}{x}$

46. Solve each equation.

    a. $5x - 2 = 0$

    b. $5 - \dfrac{2}{x} = 0$

    c. $\dfrac{x}{2} - 5 = -\dfrac{3}{4}$

    d. $\dfrac{2}{x} - 5 = -\dfrac{3}{4}$

    e. $-\dfrac{3}{x} + \dfrac{2}{x^2} = 5$

47. Work each problem according to the instructions.

    a. Divide: $\dfrac{6}{x^2-2x-8} \div \dfrac{x+3}{x+2}$.

    b. Add: $\dfrac{6}{x^2-2x-8} + \dfrac{x+3}{x+2}$.

    c. Solve: $\dfrac{6}{x^2-2x-8} + \dfrac{x+3}{x+2} = 2$.

48. Work each problem according to the instructions.

    a. Divide: $\dfrac{-10}{x^2-25} \div \dfrac{x-4}{x-5}$.

    b. Add: $\dfrac{-10}{x^2-25} + \dfrac{x-4}{x-5}$.

    c. Solve: $\dfrac{-10}{x^2-25} + \dfrac{x-4}{x-5} = \dfrac{4}{5}$.

B Solve for the given variable

49. Solve $\dfrac{1}{x} = \dfrac{1}{b} - \dfrac{1}{a}$ for x.

50. Solve $\dfrac{1}{x} = \dfrac{1}{a} - \dfrac{1}{b}$ for x.

Solve for *y*.

51. $x = \dfrac{y-3}{y-1}$

52. $x = \dfrac{y-2}{y-3}$

53. $x = \dfrac{2y+1}{3y+1}$

54. $x = \dfrac{3y+2}{5y+1}$

C Graph each function. Show the vertical asymptote.

55. $f(x) = \dfrac{1}{x-3}$

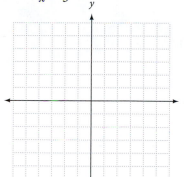

56. $f(x) = \dfrac{1}{x+3}$

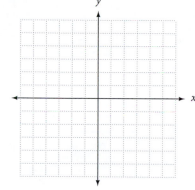

57. $f(x) = \dfrac{4}{x+2}$

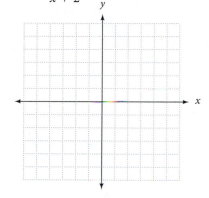

58. $f(x) = \dfrac{4}{x-2}$

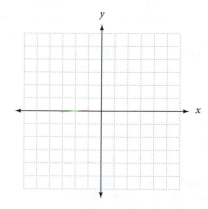

59. $g(x) = \dfrac{2}{x-4}$

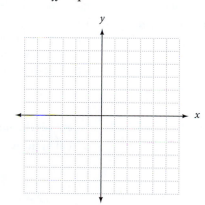

60. $g(x) = \dfrac{2}{x+4}$

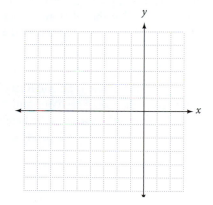

61. $g(x) = \dfrac{6}{x+1}$

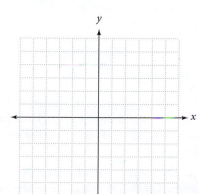

62. $g(x) = \dfrac{6}{x-1}$

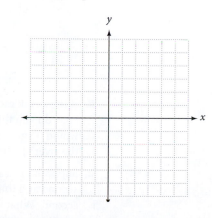

For each function given, find the location of any vertical asymptotes. Also give the domain of the function. (Do not graph.)

63. $f(x) = \dfrac{x}{3x - 5}$

64. $f(x) = \dfrac{2x}{4x + 7}$

65. $\dfrac{3x + 1}{2x - 3}$

66. $g(x) = \dfrac{x^2}{x^2 - 4}$

67. $g(x) = \dfrac{5x}{x^2 - 9}$

68. $g(x) = \dfrac{x + 2}{5x^2 + x}$

69. $h(x) = \dfrac{4 - 3x}{2x^2 + 6x}$

70. $h(x) = \dfrac{x^3}{x^2 + x - 2}$

71. $h(x) = \dfrac{x^2}{x^2 - 3x + 2}$

72. $y(x) = \dfrac{x + 3}{x^2 - 5x + 4}$

73. $y(x) = \dfrac{5}{2x^2 - 5x - 3}$

74. $y(x) = \dfrac{7}{3x^2 + 5x - 2}$

Applying the Concepts

75. Geometry From plane geometry and the principle of similar triangles, the relationship between y_1, y_2, and h shown in Figure 3 can be expressed as

$$\frac{1}{h} = \frac{1}{y_1} + \frac{1}{y_2}$$

Two poles are 12 feet high and 8 feet high. If a cable is attached to the top of each one and stretched to the bottom of the other, what is the height above the ground at which the two wires will meet?

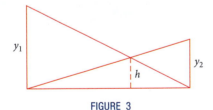

FIGURE 3

76. Kayak Race In a kayak race, the participants must paddle a kayak 450 meters down a river and then return 450 meters up the river to the starting point. Susan has correctly deduced that the total time t (in seconds) depends on the speed of the current c (in meters per second) according to the following expression:

$$t = \frac{450}{v + c} + \frac{450}{v - c}$$

where v is the speed of the kayak (in still water).

a. Fill in the following table.

Time	Speed of Kayak in Still Water	Speed of the Current
t(sec)	v(m/sec)	c(m/sec)
240		1
300		2
	4	3
	3	1
540	3	
	3	3

b. If the kayak race were conducted in the still waters of a lake, do you think that the total time of a given participant would be greater than, equal to, or smaller than the time in the river? Justify your answer.

c. Suppose Peter can paddle his kayak at 4.1 meters per second and that the speed of the current is 4.1 meters per second. What will happen when Peter makes the turn and tries to come back up the river? How does this situation show up in the equation for total time?

Getting Ready for the Next Section

Multiply.

77. $39.3 \cdot 60$

78. $1{,}100 \cdot 60 \cdot 60$

79. $2x\left(\dfrac{1}{x} + \dfrac{1}{2x}\right)$

80. $3x\left(\dfrac{1}{x} + \dfrac{1}{3x}\right)$

Divide. Round to the nearest tenth, if necessary.

81. $65{,}000 \div 5{,}280$

82. $3{,}960{,}000 \div 5{,}280$

Solve.

83. $12(x + 3) + 12(x - 3) = 3(x^2 - 9)$

84. $40 + 2x = 60 - 3x$

85. $\dfrac{1}{10} - \dfrac{1}{12} = \dfrac{1}{x}$

86. $\dfrac{1}{x} + \dfrac{1}{2x} = 2$

One Step Further

Solve each equation.

87. $\dfrac{12}{x} + \dfrac{8}{x^2} - \dfrac{75}{x^3} - \dfrac{50}{x^4} = 0$

88. $\dfrac{45}{x} + \dfrac{18}{x^2} - \dfrac{80}{x^3} - \dfrac{32}{x^4} = 0$

89. $\dfrac{1}{x^3} - \dfrac{1}{3x^2} - \dfrac{1}{4x} + \dfrac{1}{12} = 0$

90. $\dfrac{1}{x^3} - \dfrac{1}{2x^2} - \dfrac{1}{9x} + \dfrac{1}{18} = 0$

91. Solve $\dfrac{2}{x} + \dfrac{4}{x + a} = \dfrac{-6}{a - x}$ for x.

92. Solve $\dfrac{1}{b - x} - \dfrac{1}{x} = \dfrac{-2}{b + x}$ for x.

93. Solve $\dfrac{s - vt}{t^2} = -16$ for v.

94. Solve $A = P\left(1 + \dfrac{r}{n}\right)$ for r.

95. Solve $\dfrac{1}{p} = \dfrac{1}{f} + \dfrac{1}{g}$ for f.

96. Solve $h = \dfrac{v^2}{2g} + \dfrac{p}{c}$ for p.

Find the Mistake

Each sentence below contains a mistake. Circle the mistake and write the correct word or expression on the line provided.

1. When we solve the rational equation $\dfrac{1}{x - 3} - \dfrac{5}{x + 3} = \dfrac{x}{x^2 - 9}$, we get $x = 3$. _____

2. To solve the rational equation $\dfrac{x + 5}{x^2 - 5x} = \dfrac{20}{x^2 - 25}$, we must use the LCD $x - 5$. _____

3. The graph for $f(x) = \dfrac{3}{x - 1}$ will come very close but not touch the line $x = 3$. _____

4. The graph for $f(x) = \dfrac{2}{x + 2}$ has a horizontal asymptote of $x = -2$. _____

5.6 Applications

©iStockphoto.com/FotoW

Late one night in 2009, an octopus at the Santa Monica Aquarium used her strong arm to open a valve at the top of her tank. Two hundred gallons of water spilled onto the floor and flooded nearby exhibits. Aquarium employees discovered the octopus alive and still in her tank the next morning. Let us imagine for a moment that an employee was tasked with refilling the octopus's tank, but he forgot to close the valve. If water from an inlet pipe could fill the tank in 2 hours, and it took 3 hours to empty the tank through the valve, how long would it take to fill the tank with the valve still open? In this section, we will work with similar application problems, the solutions to which involve equations that contain rational expressions.

As you will see, the solutions to the examples show only the essential steps from our Blueprint for Problem Solving. Recall that Step 1 was done mentally; we read the problem and mentally list the items that are known and the items that are unknown. This is an essential part of problem solving. Now that you have had experience with application problems, however, you should be doing Step 1 automatically.

A Applications

EXAMPLE 1 One number is twice another. The sum of their reciprocals is 2. Find the numbers.

Solution Let x = the smaller number. The larger number is $2x$. Their reciprocals are $\frac{1}{x}$ and $\frac{1}{2x}$. The equation that describes the situation is

$$\frac{1}{x} + \frac{1}{2x} = 2$$

Multiplying both sides by the LCD $2x$, we have

$$2x \cdot \frac{1}{x} + 2x \cdot \frac{1}{2x} = 2x(2)$$

$$2 + 1 = 4x$$

$$3 = 4x$$

$$x = \frac{3}{4}$$

The smaller number is $\frac{3}{4}$. The larger is $\left(2 \cdot \frac{3}{4}\right) = \frac{6}{4} = \frac{3}{2}$. Adding their reciprocals, we have

$$\frac{4}{3} + \frac{2}{3} = \frac{6}{3} = 2$$

The sum of the reciprocals of $\frac{3}{4}$ and $\frac{3}{2}$ is 2.

EXAMPLE 2   Two families from the same neighborhood plan a ski trip together. The first family is driving a newer vehicle and makes the 455-mile trip at a speed 5 miles per hour faster than the second family who is traveling in an older vehicle. The second family takes a half-hour longer to make the trip. What are the speeds of the two families?

Solution   The following table will be helpful in finding the equation necessary to solve this problem.

	Distance d (miles)	Rate r (mph)	Time t (hours)
First family			
Second family			

If we let x be the speed of the second family, then the speed of the first family will be $x + 5$. Both families travel the same distance of 455 miles.

Putting this information into the table we have

	d	r	t
First family	455	$x + 5$	
Second family	455	x	

To fill in the last two spaces in the table, we use the relationship $d = r \cdot t$. Since the last column of the table is the time, we solve the equation $d = r \cdot t$ for t and get

$$t = \frac{d}{r}$$

Taking the distance and dividing by the rate (speed) for each family, we complete the table.

	d	r	t
First family	455	$x + 5$	$\frac{455}{x + 5}$
Second family	455	x	$\frac{455}{x}$

Reading the problem again, we find that the time for the second family is longer than the time for the first family by one-half hour. In other words, the time for the second family can be found by adding one-half hour to the time for the first family, or

$$\frac{455}{x + 5} + \frac{1}{2} = \frac{455}{x}$$

Multiplying both sides by the LCD of $2x(x + 5)$ gives

$$2x \cdot (455) + x(x + 5) \cdot 1 = 455 \cdot 2(x + 5)$$

$$910x + x^2 + 5x = 910x + 4550$$

$$x^2 + 5x - 4550 = 0$$

$$(x + 70)(x - 65) = 0$$

$$x = -70 \quad \text{or} \quad x = 65$$

Since we cannot have a negative speed, the only solution is $x = 65$. Then

$$x + 5 = 65 + 5 = 70$$

The speed of the first family is 70 miles per hour, and the speed of the second family is 65 miles per hour.

2. If the two families in Example 2 drive 330 miles instead of 455, what are the speeds of the two families?

Answers

2. 55 mph, 60 mph

3. A boat travels 26 miles up a river in the same amount of time it takes to travel 34 miles down the same river. The current is 2 miles per hour, what is the speed of the boat in still water?

EXAMPLE 3   The current of a river is 3 miles per hour. It takes a motorboat a total of 3 hours to travel 12 miles upstream and return 12 miles downstream. What is the speed of the boat in still water?

Solution   This time we let x = the speed of the boat in still water. Then, we fill in as much of the table as possible using the information given in the problem. For instance, because we let x = the speed of the boat in still water, the rate upstream (against the current) must be $x - 3$. The rate downstream (with the current) is $x + 3$.

	Distance d (miles)	Rate r (mph)	Time t (hours)
Upstream	12	$x - 3$	
Downstream	12	$x + 3$	

The last two boxes can be filled in using the relationship $t = \dfrac{d}{r}$.

	d	r	t
Upstream	12	$x - 3$	$\dfrac{12}{x - 3}$
Downstream	12	$x + 3$	$\dfrac{12}{x + 3}$

The total time for the trip up and back is 3 hours:

$$\text{Time upstream} + \text{Time downstream} = \text{Total time}$$

$$\frac{12}{x - 3} \quad + \quad \frac{12}{x + 3} \quad = \quad 3$$

Multiplying both sides by $(x - 3)(x + 3)$, we have

$$12(x + 3) + 12(x - 3) = 3(x^2 - 9)$$

$$12x + 36 + 12x - 36 = 3x^2 - 27$$

$$3x^2 - 24x - 27 = 0$$

$$x^2 - 8x - 9 = 0 \qquad \text{\color{green}{Divide both sides by 3.}}$$

$$(x - 9)(x + 1) = 0$$

$$x = 9 \quad \text{or} \quad x = -1$$

The speed of the motorboat in still water is 9 miles per hour. (We don't use $x = -1$ because the speed of the motorboat cannot be a negative number.)

4. A tub can be filled by the cold water faucet in 8 minutes. The drain empties the tub in 12 minutes. How long will it take to fill the tub if both the faucet and the drain are open?

EXAMPLE 4   An inlet pipe can fill a pool in 10 hours, while the drain can empty it in 12 hours. If the pool is empty and both the inlet pipe and drain are open, how long will it take to fill the pool?

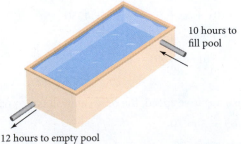

10 hours to fill pool

12 hours to empty pool

Answers

3. 15 miles per hour

4. 24 minutes

Solution   It is helpful to think in terms of how much work is done by each pipe in 1 hour.

Let $x =$ the time it takes to fill the pool with both pipes open.

If the inlet pipe can fill the pool in 10 hours, then in 1 hour it is $\frac{1}{10}$ full. If the outlet pipe empties the pool in 12 hours, then in 1 hour it is $\frac{1}{12}$ empty. If the pool can be filled in x hours with both the inlet pipe and the drain open, then in 1 hour it is $\frac{1}{x}$ full when both pipes are open.

Here is the equation:

$$\left[\begin{array}{c}\text{Amount filled by} \\ \text{inlet pipe in 1 hour}\end{array}\right] - \left[\begin{array}{c}\text{Amount emptied by} \\ \text{the drain in 1 hour}\end{array}\right] = \left[\begin{array}{c}\text{Fraction of pool filled} \\ \text{with both pipes in 1 hour}\end{array}\right]$$

$$\frac{1}{10} \qquad - \qquad \frac{1}{12} \qquad = \qquad \frac{1}{x}$$

Multiplying through by $60x$, we have

$$60x \cdot \frac{1}{10} - 60x \cdot \frac{1}{12} = 60x \cdot \frac{1}{x}$$

$$6x - 5x = 60$$

$$x = 60$$

It takes 60 hours to fill the pool if both the inlet pipe and the drain are open.

B More About Graphing Rational Functions

We continue our investigation of the graphs of rational functions by considering the graph of a rational function with binomials in the numerator and denominator.

EXAMPLE 5   Graph the rational function $y = \dfrac{x-4}{x-2}$.

Solution   In addition to making a table to find some points on the graph, we can analyze the graph as follows:

1. The graph will have a y-intercept of 2, because when $x = 0$, $y = \frac{-4}{-2} = 2$.

2. To find the x-intercept, we let $y = 0$ to get

$$0 = \frac{x-4}{x-2}$$

    The only way this expression can be 0 is if the numerator is 0, which happens when $x = 4$. (If you want to solve this equation, multiply both sides by $x - 2$. You will get the same solution, $x = 4$.)

3. The graph will have a vertical asymptote at $x = 2$, because $x = 2$ will make the denominator of the function 0, meaning y is undefined when x is 2.

4. The graph will have a *horizontal asymptote* at $y = 1$ because for very large values of x, $\frac{x-4}{x-2}$ is very close to 1 but never crosses it. The larger x is, the closer $\frac{x-4}{x-2}$ is to 1. The same is true for very small values of x, such as $-1{,}000$ and $-10{,}000$.

Putting this information together with the ordered pairs in the table next to the figure, we have the graph shown in Figure 1.

5. Graph the rational function $y = \dfrac{x-4}{x+2}$.

x	y
-1	$\frac{5}{3}$
0	2
1	3
2	Undefined
3	-1
4	0
5	$\frac{1}{3}$

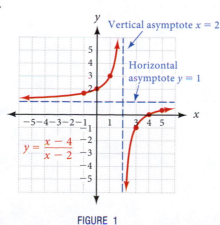

FIGURE 1

USING TECHNOLOGY   More About Example 5

In the previous section, we used technology to explore the graph of a rational function around a vertical asymptote. This time, we are going to explore the graph near the horizontal asymptote. In Figure 1, the horizontal asymptote is at $y = 1$. To show that the graph approaches this line as x becomes very large, we use the table function on our graphing calculator, with X taking values of 100, 1,000, and 10,000. To show that the graph approaches the line $y = 1$ on the left side of the coordinate system, we let X become -100, $-1,000$ and $-10,000$.

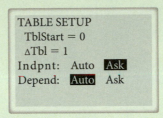

```
TABLE SETUP
  TblStart = 0
  ΔTbl = 1
Indpnt:  Auto  Ask
Depend:  Auto  Ask
```

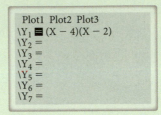

```
Plot1  Plot2  Plot3
\Y₁ ▇ (X − 4)(X − 2)
\Y₂ =
\Y₃ =
\Y₄ =
\Y₅ =
\Y₆ =
\Y₇ =
```

The table will look like this:

X	Y_1	
100	.97959	
1000	.998	
10000	.9998	
−100	1.0196	
−1000	1.002	
−100000	1.0002	

As you can see, as x becomes very large in the positive direction, the graph approaches the line $y = 1$ from below. As x becomes very small in the negative direction, the graph approaches the line $y = 1$ from above.

GETTING READY FOR CLASS

After reading through the preceding section, respond in your own words and in complete sentences.

A. Briefly list the steps in the Blueprint for Problem Solving that you have used previously to solve application problems.

B. Write an application problem for which the solution depends on solving the equation $\frac{1}{2} + \frac{1}{3} = \frac{1}{x}$.

C. How may a table be helpful when solving application problems similar to those in this section?

D. What is a horizontal asymptote?

Problems

Solve the following word problems. Use techniques illustrated in the examples of this section to set up an equation, then solve it to answer the question.

A Number Problems

1. One number is 3 times another. The sum of their reciprocals is $\frac{20}{3}$. Find the numbers.

2. One number is 3 times another. The sum of their reciprocals is $\frac{4}{5}$. Find the numbers.

3. The sum of a number and its reciprocal is $\frac{10}{3}$. Find the number.

4. The sum of a number and twice its reciprocal is $\frac{27}{5}$. Find the number.

5. The sum of the reciprocals of two consecutive integers is $\frac{7}{12}$. Find the two integers.

6. Find two consecutive even integers, the sum of whose reciprocals is $\frac{3}{4}$.

7. If a certain number is added to the numerator and denominator of $\frac{7}{9}$, the result is $\frac{5}{6}$. Find the number.

8. Find the number you would add to both the numerator and denominator of $\frac{8}{11}$ so that the result would be $\frac{6}{7}$.

SCAN TO ACCESS

Rate Problems

9. The speed of a boat in still water is 5 miles per hour. If the boat travels 3 miles downstream in the same amount of time it takes to travel 1.5 miles upstream, what is the speed of the current?

   a. Let x be the speed of the current. Complete the distance d and rate r columns in the table.

	d (miles)	r (mph)	t (hours)
Upstream			
Downstream			

   b. Now use the distance and rate information to complete the time t column.

   c. What does the problem tell us about the two times? Use this fact to write an equation involving the two expressions for time.

   d. Solve the equation. Write your answer as a complete sentence.

10. A boat, which moves at 18 miles per hour in still water, travels 14 miles downstream in the same amount of time it takes to travel 10 miles upstream. Find the speed of the current.

   a. Let x be the speed of the current. Complete the distance d and rate r columns in the table.

	d (miles)	r (mph)	t (hours)
Upstream			
Downstream			

   b. Now use the distance and rate information to complete the time t column.

   c. What does the problem tell us about the two times? Use this fact to write an equation involving the two expressions for time.

   d. Solve the equation. Write your answer as a complete sentence.

11. The current of a river is 2 miles per hour. A boat travels to a point 8 miles upstream and back again in 3 hours. What is the speed of the boat in still water?

12. A motorboat travels at 4 miles per hour in still water. It goes 12 miles upstream and 12 miles back again in a total of 8 hours. Find the speed of the current of the river.

13. Train A has a speed 15 miles per hour greater than that of train B. If train A travels 150 miles in the same time train B travels 120 miles, what are the speeds of the two trains?

   a. Let x be the speed of the train B. Complete the distance d and rate r columns in the table.

	d (miles)	r (mph)	t (hours)
Train A			
Train B			

   b. Now use the distance and rate information to complete the time column.

   c. What does the problem tell us about the two times? Use this fact to write an equation involving the two expressions for time.

   d. Solve the equation. Write your answer as a complete sentence.

14. A train travels 30 miles per hour faster than a car. If the train covers 120 miles in the same time the car covers 80 miles, what are the speeds of each of them?

   a. Let x be the speed of the car. Complete the distance d and rate r columns in the table.

	d (miles)	r (mph)	t (hours)
Car			
Train			

   b. Now use the distance and rate information to complete the time column.

   c. What does the problem tell us about the two times? Use this fact to write an equation involving the two expressions for time.

   d. Solve the equation. Write your answer as a complete sentence.

15. A small airplane flies 810 miles from Los Angeles to Portland, OR, with an average speed of 270 miles per hour. An hour and a half after the plane leaves, a Boeing 747 leaves Los Angeles for Portland. Both planes arrive in Portland at the same time. What was the average speed of the 747?

16. Lou leaves for a cross-country excursion on a bicycle traveling at 20 miles per hour. His friends are driving the trip and will meet him at several rest stops along the way. The first stop is scheduled 30 miles from the original starting point. If the people driving leave 15 minutes after Lou from the same place, how fast will they have to drive to reach the first rest stop at the same time as Lou?

17. A tour bus leaves Sacramento every Friday evening at 5:00 p.m. for a 270-mile trip to Las Vegas. This week, however, the bus leaves at 5:30 p.m. To arrive in Las Vegas on time, the driver drives 6 miles per hour faster than usual. What is the bus' usual speed?

18. A bakery delivery truck leaves the bakery at 5:00 a.m. each morning on its 140-mile route. One day the driver gets a late start and does not leave the bakery until 5:30 a.m. To finish her route on time the driver drives 5 miles per hour faster than usual. At what speed does she usually drive?

Work Problems

19. A water tank can be filled by an inlet pipe in 8 hours. It takes twice that long for the outlet pipe to empty the tank. How long will it take to fill the tank if both pipes are open?

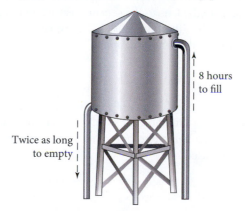

8 hours to fill

Twice as long to empty

20. A sink can be filled from the faucet in 5 minutes. It takes only 3 minutes to empty the sink when the drain is open. If the sink is full and both the faucet and the drain are open, how long will it take to empty the sink?

5 min to fill

3 min to empty

21. It takes 10 hours to fill a pool with the inlet pipe. It can be emptied in 15 hours with the outlet pipe. If the pool is half full to begin with, how long will it take to fill it from there if both pipes are open?

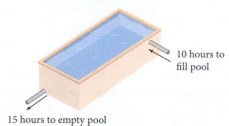

10 hours to fill pool

15 hours to empty pool

22. A sink is one-quarter full when both the faucet and the drain are opened. The faucet alone can fill the sink in 6 minutes, while it takes 8 minutes to empty it with the drain. How long will it take to fill the remaining three quarters of the sink?

23. A sink has two faucets: one for hot water and one for cold water. The sink can be filled by a cold water faucet in 3.5 minutes. If both faucets are open, the sink is filled in 2.1 minutes. How long does it take to fill the sink with just the hot water faucet open?

24. A water tank is being filled by two inlet pipes. Pipe A can fill the tank in $4\frac{1}{2}$ hours, but both pipes together can fill the tank in 2 hours. How long does it take to fill the tank using only pipe B?

25. A hot water faucet can fill a sink in 3.5 minutes. The cold water faucet can fill the sink in 2.7 minutes. How long does it take to fill the sink if both faucets are turned on?

26. If Amelie can paint a room in 2.3 hours and her roommate can paint it in 4.2 hours, how long will it take to paint the room if they work together?

Miscellaneous Problems

27. Rhind Papyrus Nearly 4,000 years ago, Egyptians worked mathematical exercises involving reciprocals. The *Rhind Papyrus* contains a wealth of such problems, and one of them is as follows:

"A quantity and its two thirds are added together, one third of this is added, then one third of the sum is taken, and the result is 10."

Write an equation and solve this exercise.

28. Photography For clear photographs, a camera must be properly focused. Professional photographers use a mathematical relationship relating the distance from the camera lens to the object being photographed, a; the distance from the lens to the sensor creating an image, b; and the focal length of the lens, f. These quantities, a, b, and f, are related by the equation

$$\frac{1}{a} + \frac{1}{b} = \frac{1}{f}$$

A camera has a focal length of 3 inches. If the lens is 5 inches from the film, how far should the lens be placed from the object being photographed for the camera to be perfectly focused?

The Periodic Table If you take a chemistry class, you will work with the Periodic Table of Elements. Figure 2 shows three of the elements listed in the periodic table. As you can see, the bottom number in each figure is the molecular weight of the element. In chemistry, a mole is the amount of a substance that will give the weight in grams equal to the molecular weight. For example, 1 mole of lead is 207.2 grams.

FIGURE 2

29. Chemistry For the element carbon, 1 mole = 12.01 grams.

   a. To the nearest gram, how many grams of carbon are in 2.5 moles of carbon?

   b. How many moles of carbon are in 39 grams of carbon? Round to the nearest hundredth.

30. Chemistry For the element sulfur, 1 mole = 32.07 grams.

   a. How many grams of sulfur are in 3 moles of sulfur?

   b. How many moles of sulfur are found in 80.2 grams of sulfur?

B Graph each rational function. In each case, show the vertical asymptote, the horizontal asymptote, and any intercepts that exist.

31. $f(x) = \dfrac{x-3}{x-1}$

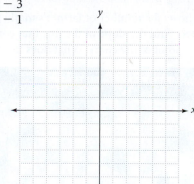

32. $f(x) = \dfrac{x+4}{x-2}$

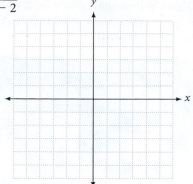

33. $f(x) = \dfrac{x+3}{x-1}$

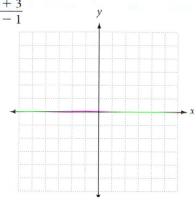

34. $f(x) = \dfrac{x-2}{x-1}$

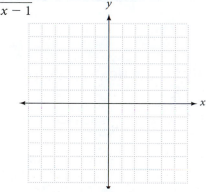

35. $g(x) = \dfrac{x-3}{x+1}$

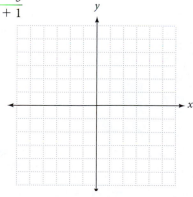

36. $g(x) = \dfrac{x-2}{x+1}$

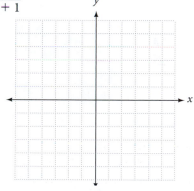

37. $g(x) = \dfrac{x+2}{x+4}$

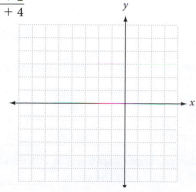

38. $g(x) = \dfrac{x}{x+2}$

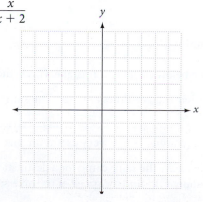

Maintaining Your Skills

39. The chart shows the number of Facebook fans for several popular fan pages. How many more fans does Homer Simpson have than Nutella?

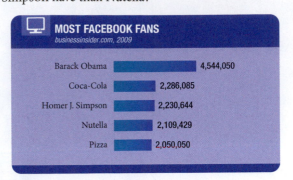

MOST FACEBOOK FANS
businessinsider.com, 2009

Barack Obama	4,544,050
Coca-Cola	2,286,085
Homer J. Simpson	2,230,644
Nutella	2,109,429
Pizza	2,050,050

40. The chart shows the number of farmer's markets in the United States. Find the slope of the line that connects the first and last point of the chart. What does this mean?

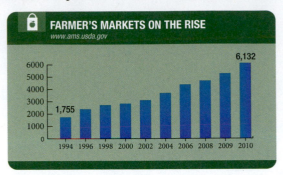

FARMER'S MARKETS ON THE RISE
www.ams.usda.gov

Getting Ready for the Next Section

Divide.

41. $\dfrac{10x^2}{5x^2}$

42. $\dfrac{-15x^4}{5x^2}$

43. $\dfrac{4x^4y^3}{-2x^2y}$

44. $\dfrac{10a^4b^2}{4a^2b^2}$

45. $4{,}628 \div 25$

46. $7{,}546 \div 35$

47. $\dfrac{(x-3)^2}{x-3}$

48. $\dfrac{(x+4)^2}{(x+4)^4}$

Multiply.

49. $2x^2(2x-4)$

50. $3x^2(x-2)$

51. $(2x-4)(2x^2+4x+5)$

52. $(x-2)(3x^2+6x+15)$

Subtract.

53. $(2x^2-7x+9)-(2x^2-4x)$

54. $(x^2-6xy-7y^2)-(x^2+xy)$

Factor.

55. x^2-a^2

56. x^2-1

57. $x^2-6xy-7y^2$

58. $2x^2-5xy+3y^2$

Find the Mistake

Each sentence below contains a mistake. Circle the mistake and write the correct word or expression on the line provided.

1. One number is three times another. The sum of their reciprocals is 8. The equation to find the numbers is $\dfrac{1}{x} + \dfrac{3}{x} = 8$.

2. The rate equation $t = \dfrac{r}{d}$ is very useful for many applications of rational equations.

3. Two hikers leave the same camp for a destination 10 miles away. One is 2 miles per hour faster than the other and arrives $\frac{1}{2}$ an hour sooner. To find the speed of the hikers, you must recognize that their travel time is the same.

4. One worker can mow a park in 3 hours and another can finish in 4 hours. To find how long it takes to finish together, you would use the equation $\dfrac{1}{4} - \dfrac{1}{3} = x$.

Suppose a bank charges \$2.00 per month for a regular checking account plus \$0.15 per check written. If you write x checks in one month, the total monthly cost of the checking account will be $C(x) = 2.00 + 0.15x$. From this formula, we see that the more checks we write in a month, the more we pay for the account. But it is also true that the more checks we write in a month, the lower the average cost per check. To find the average cost per check, we use the *average cost* function. To find the average cost function, we divide the total cost by the number of checks written.

$$\text{Average Cost} = \overline{C}(x) = \frac{C(x)}{x} = \frac{2.00 + 0.15x}{x}$$

This last expression gives us the average cost per check for each of the x checks written. To work with this last expression, we need to know something about division with polynomials, and that is what we will cover in this section.

We begin this section by considering division of a polynomial by a monomial. This is the simplest kind of polynomial division. The rest of the section is devoted to division of a polynomial by a polynomial. This kind of division is similar to long division with whole numbers.

A Dividing a Polynomial by a Monomial

To divide a polynomial by a monomial, we use the definition of division and apply the distributive property. The following example illustrates the procedure:

EXAMPLE 1   Divide $\dfrac{10x^5 - 15x^4 + 20x^3}{5x^2}$.

Solution

$$= (10x^5 - 15x^4 + 20x^3) \cdot \frac{1}{5x^2} \qquad \text{Dividing by } 5x^2 \text{ is the same as}$$
$$\text{multiplying by } \tfrac{1}{5x^2}.$$

$$= 10x^5 \cdot \frac{1}{5x^2} - 15x^4 \cdot \frac{1}{5x^2} + 20x^3 \cdot \frac{1}{5x^2} \qquad \text{Distributive property}$$

$$= \frac{10x^5}{5x^2} - \frac{15x^4}{5x^2} + \frac{20x^3}{5x^2} \qquad \text{Multiplying by } \tfrac{1}{5x^2} \text{ is the same}$$
$$\text{as multiplying by } 5x^2.$$

$$= 2x^3 - 3x^2 + 4x$$

Notice that division of a polynomial by a monomial is accomplished by dividing each term of the polynomial by the monomial. The first two steps are usually not shown in a problem like this. They are part of Example 1 to justify distributing $5x^2$ under all three terms of the polynomial $10x^5 - 15x^4 + 20x^3$.

Here are some more examples of this kind of division:

EXAMPLE 2 Divide $\dfrac{8x^3y^5 - 16x^2y^2 + 4x^4y^3}{-2x^2y}$. Write the result with positive exponents.

Solution

$$\frac{8x^3y^5 - 16x^2y^2 + 4x^4y^3}{-2x^2y} = \frac{8x^3y^5}{-2x^2y} + \frac{-16x^2y^2}{-2x^2y} + \frac{4x^4y^3}{-2x^2y}$$

$$= -4xy^4 + 8y - 2x^2y^2$$

EXAMPLE 3 Divide $\dfrac{10a^4b^2 + 8ab^3 - 12a^3b + 6ab}{4a^2b^2}$. Write the result with positive exponents.

Solution

$$\frac{10a^4b^2 + 8ab^3 - 12a^3b + 6ab}{4a^2b^2} = \frac{10a^4b^2}{4a^2b^2} + \frac{8ab^3}{4a^2b^2} - \frac{12a^3b}{4a^2b^2} + \frac{6ab}{4a^2b^2}$$

$$= \frac{5a^2}{2} + \frac{2b}{a} - \frac{3a}{b} + \frac{3}{2ab}$$

Notice in Example 3 that the result is not a polynomial because of the last three terms. If we were to write each as a product, some of the variables would have negative exponents. For example, the second term would be

$$\frac{2b}{a} = 2a^{-1}b$$

The divisor in each of the preceding examples was a monomial. We now want to turn our attention to division of polynomials in which the divisor has two or more terms.

B Dividing a Polynomial by Factoring

EXAMPLE 4 Divide: $\dfrac{x^2 - 6xy - 7y^2}{x + y}$

Solution In this case, we can factor the numerator and perform division by simply dividing out common factors, just like we did in previous sections.

$$\frac{x^2 - 6xy - 7y^2}{x + y} = \frac{(x + y)(x - 7y)}{x + y}$$

$$= x - 7y$$

C Dividing a Polynomial by Long Division

For the type of division shown in Example 4, the denominator must be a factor of the numerator. When the denominator is not a factor of the numerator, or in the case where we can't factor the numerator, the method used in Example 4 won't work. We need to develop a new method for these cases. Because this new method is very similar to *long division* with whole numbers, we will review the method of long division here.

---

2. Divide $\dfrac{21x^3y^2 + 14x^2y^2 - 7x^2y^3}{7x^2y}$.

3. Divide.

$$\frac{12a^2b - 24a^3b^4 - 10a^5b + 4ab^3}{8a^3b^2}$$

4. Divide $\dfrac{x^2 + x - 12}{x + 4}$.

Note It is important to remember that when dividing a polynomial by factoring, only divide the common factors, not common terms.

Answers

2. $3xy + 2y - y^2$

3. $\dfrac{3}{2ab} - 3b^2 - \dfrac{5a^2}{4b} + \dfrac{b}{2a^2}$

4. $x - 3$

EXAMPLE 5 Divide $25\overline{)4{,}628}$.

Solution

$$
\begin{array}{r}
1 \\
25\overline{)4628} \\
\underline{25} \\
21
\end{array}
$$

Estimate 25 into 46.

Multiply $1 \times 25 = 25$.

Subtract $46 - 25 = 21$.

$$
\begin{array}{r}
1 \\
25\overline{)4628} \\
\underline{25\downarrow} \\
212
\end{array}
$$

Bring down the 2.

These are the four basic steps in long division: estimate, multiply, subtract, and bring down the next term. To complete the problem, we simply perform the same four steps.

$$
\begin{array}{r}
18 \\
25\overline{)4628} \\
\underline{25} \\
212 \\
\underline{200\downarrow} \\
128
\end{array}
$$

8 is the estimate.

Multiply to get 200.

Subtract to get 12, then bring down the 8.

One more time:

$$
\begin{array}{r}
185 \\
25\overline{)4628} \\
\underline{25} \\
212 \\
\underline{200\downarrow} \\
128 \\
\underline{125} \\
3
\end{array}
$$

5 is the estimate.

Multiply to get 125.

Subtract to get 3.

Because 3 is less than 25 and we have no more terms to bring down, we have our answer:

$$\frac{4{,}628}{25} = 185 + \frac{3}{25} \text{ or } 185\,\frac{3}{25}$$

To check our answer, we multiply 185 by 25 and then add 3 to the result.

$$25(185) + 3 = 4{,}625 + 3 = 4{,}628$$

Long division with polynomials is similar to long division with whole numbers. Both use the same four basic steps: estimate, multiply, subtract, and bring down the next term. We use long division with polynomials when the denominator has two or more terms and is not a factor of the numerator. Here is an example:

EXAMPLE 6 Divide $\dfrac{2x^2 - 7x + 9}{x - 2}$.

Solution

$$
\begin{array}{r}
2x \phantom{{}- 7x + 9} \\
x - 2\overline{)2x^2 - 7x + 9} \\
\underline{2x^2 - 4x} \\
-3x
\end{array}
$$

Estimate $2x^2 \div x = 2x$.

Multiply $2x(x - 2) = 2x^2 - 4x$.

Subtract $(2x^2 - 7x) - (2x^2 - 4x) = -3x$.

$$
\begin{array}{r}
2x \phantom{{}- 7x + 9} \\
x - 2\overline{)2x^2 - 7x + 9} \\
\underline{2x^2 - 4x} \downarrow \\
-3x + 9
\end{array}
$$

Bring down the 9.

5. Divide $35\overline{)4{,}281}$.

6. Divide $\dfrac{x^2 - 5x + 8}{x - 2}$.

Answers

5. $122 + \frac{11}{35}$ or $122\frac{11}{35}$

6. $x - 3 + \frac{2}{x - 2}$

Notice we change the signs on $2x^2 - 4x$ and add in the subtraction step. Subtracting a polynomial is equivalent to adding its opposite.

We repeat the four steps.

$$
\begin{array}{r}
2x - 3 \\
x - 2 \overline{)\ 2x^2 - 7x + 9} \\
\end{array}
$$

-3 is the estimate: $-3x \div x = -3$.

$$
\begin{array}{r}
-\quad + \\
2x^2 - 4x \\
\hline
-3x + 9 \\
+ \quad - \\
-3x + 6 \\
\hline
3
\end{array}
$$

Multiply $-3(x - 2) = -3x + 6$.
Subtract $(-3x + 9) - (-3x + 6) = 3$.

Because we have no other term to bring down, we have our answer:

$$\frac{2x^2 - 7x + 9}{x - 2} = 2x - 3 + \frac{3}{x - 2}$$

To check, we multiply $(2x - 3)(x - 2)$ to get $2x^2 - 7x + 6$; then, adding the remainder 3 to this result, we have $2x^2 - 7x + 9$.

In setting up a division problem involving two polynomials, we must remember two things: (1) both polynomials should be in decreasing powers of the variable, and (2) neither should skip any powers from the highest power down to the constant term. If there are any missing terms, they can be filled in using a coefficient of 0.

EXAMPLE 7   Divide $p(x) = 4x^3 - 6x - 11$ by $q(x) = 2x - 4$.

Solution   Because the trinomial is missing a term in x^2, we can fill it in with $0x^2$.

$$p(x) = 4x^3 - 6x - 11 = 4x^3 + 0x^2 - 6x - 11$$

Adding $0x^2$ does not change our original problem. Using $q(x)$ as the divisor and $p(x)$ as the dividend in our long division we have

$$
\begin{array}{r}
2x^2 + 4x + 5 \\
2x - 4 \overline{)\ 4x^3 + 0x^2 - 6x - 11} \\
-\quad + \\
4x^3 - 8x^2 \\
\hline
+ 8x^2 - 6x \\
- \quad + \\
8x^2 - 16x \\
\hline
+ 10x - 11 \\
- \quad + \\
10x - 20 \\
\hline
+ 9
\end{array}
$$

$$\frac{p(x)}{q(x)} = \frac{4x^3 - 6x - 11}{2x - 4} = 2x^2 + 4x + 5 + \frac{9}{2x - 4}$$

To check this result, we multiply $2x - 4$ and $2x^2 + 4x + 5$.

$$
\begin{array}{r}
2x^2 + 4x + 5 \\
\times \quad 2x - 4 \\
\hline
4x^3 + 8x^2 + 10x \\
+ \quad -8x^2 - 16x - 20 \\
\hline
4x^3 \qquad\ - 6x - 20
\end{array}
$$

Adding 9 (the remainder) to this result gives us the polynomial $4x^3 - 6x - 11$. Our answer checks.

---

7. Divide $p(x) = 3x^3 - 2x + 1$ by $q(x) = x - 3$.

Note   Adding the $0x^2$ term gives us a column in which to write $-8x^2$.

Answer

7. $3x^2 + 9x + 25 + \dfrac{76}{x - 3}$

For our next example, let's do Example 4 again, but this time use long division.

EXAMPLE 8   Divide $\dfrac{x^2 - 6xy - 7y^2}{x + y}$.

Solution

$$
\begin{array}{r}
x \phantom{{}- 7y} \\
\end{array}
$$

$$
\begin{array}{r}
x\ -\ 7y \\
x + y\overline{)x^2\ -\ 6xy\ -\ 7y^2} \\
\underline{+\ x^2\ +\ xy} \\
-\ 7xy\ -\ 7y^2 \\
\underline{+\ +} \\
-\ 7xy\ -\ 7y^2 \\
0
\end{array}
$$

In this case, the remainder is 0, and we have

$$\frac{x^2 - 6xy - 7y^2}{x + y} = x - 7y$$

which is easy to check because

$$(x + y)(x - 7y) = x^2 - 6xy - 7y^2$$

EXAMPLE 9   Factor $x^3 + 9x^2 + 26x + 24$ completely if $x + 2$ is one of its factors.

Solution   Because $x + 2$ is one of the factors of the polynomial we are trying to factor, it must divide that polynomial evenly — that is, without a remainder. Therefore, we begin by dividing the polynomial by $x + 2$.

$$
\begin{array}{r}
x^2\ +\ 7x\ +\ 12 \\
x + 2\overline{)x^3\ +\ 9x^2\ +\ 26x\ +\ 24} \\
\underline{+\ x^3\ +\ 2x^2} \\
+\ 7x^2\ +\ 26x \\
\underline{+\ 7x^2\ +\ 14x} \\
+\ 12x\ +\ 24 \\
\underline{+\ 12x\ +\ 24} \\
0
\end{array}
$$

Now we know that the polynomial we are trying to factor is equal to the product of $x + 2$ *and* $x^2 + 7x + 12$. To factor completely, we simply factor $x^2 + 7x + 12$.

$$x^3 + 9x^2 + 26x + 24 = (x + 2)(x^2 + 7x + 12)$$
$$= (x + 2)(x + 3)(x + 4)$$

GETTING READY FOR CLASS

After reading through the preceding section, respond in your own words and in complete sentences.

A. What property of real numbers is the key to dividing a polynomial by a monomial?

B. What are the four steps used in long division with polynomials?

C. What does it mean to have a remainder of 0?

D. When must long division be performed, and when can factoring be used to divide polynomials?

8. Divide $2x + 3\overline{)8x^2 + 10x - 3}$.

9. Is $x - 3$ a factor of $x^3 + 3x^2 - 4x - 12$?

Answers

8. $4x - 1$

9. No

Vocabulary Review

Choose the correct words to fill in the blanks below.

| factor | monomial | average | numerator | long division | estimate |

1. The _____ cost of an item is the total cost divided by the number of items in question.
2. To divide a polynomial by a _____ , use the definition of division and apply the distributive property.
3. If the numerator and the denominator of a polynomial have a common _____ , divide it out.
4. Dividing a polynomial by_____ is very similar to dividing whole numbers with long division.
5. Long division uses four steps: _____ , multiply, subtract, and bring down the next term.
6. Use long division to divide a polynomial when the denominator has two or more terms and is not an obvious factor of the _____ .

Problems

A Find the following quotients.

1. $\dfrac{4x^3 - 8x^2 + 6x}{2x}$

2. $\dfrac{6x^3 + 12x^2 - 9x}{3x}$

3. $\dfrac{10x^4 + 15x^3 - 20x^2}{-5x^2}$

4. $\dfrac{12x^5 - 18x^4 - 6x^3}{6x^3}$

5. $\dfrac{8y^5 + 10y^3 - 6y}{4y^3}$

6. $\dfrac{6y^4 - 3y^3 + 18y^2}{9y^2}$

7. $\dfrac{5x^3 - 8x^2 - 6x}{-2x^2}$

8. $\dfrac{-9x^5 + 10x^3 - 12x}{-6x^4}$

9. $\dfrac{28a^3b^5 + 42a^4b^3}{7a^2b^2}$

10. $\dfrac{a^2b + ab^2}{ab}$

11. $\dfrac{10x^3y^2 - 20x^2y^3 - 30x^3y^3}{-10x^2y}$

12. $\dfrac{9x^4y^4 + 18x^3y^4 - 27x^2y^4}{-9xy^3}$

SCAN TO ACCESS

B Divide by factoring numerators and then dividing out common factors.

13. $\dfrac{x^2 - x - 6}{x - 3}$

14. $\dfrac{x^2 - x - 6}{x + 2}$

15. $\dfrac{2a^2 - 3a - 9}{2a + 3}$

16. $\dfrac{2a^2 + 3a - 9}{2a - 3}$

17. $\dfrac{5x^2 - 14xy - 24y^2}{x - 4y}$

18. $\dfrac{5x^2 - 26xy - 24y^2}{5x + 4y}$

19. $\dfrac{x^3 - y^3}{x - y}$

20. $\dfrac{x^3 + 8}{x + 2}$

21. $\dfrac{y^4 - 16}{y - 2}$

22. $\dfrac{y^4 - 81}{y - 3}$

23. $\dfrac{x^3 + 2x^2 - 25x - 50}{x - 5}$

24. $\dfrac{x^3 + 2x^2 - 25x - 50}{x + 5}$

25. $\dfrac{4x^3 + 12x^2 - 9x - 27}{x + 3}$

26. $\dfrac{9x^3 + 18x^2 - 4x - 8}{x + 2}$

C Divide using the long division method.

27. $\dfrac{x^2 - 5x - 7}{x + 2}$

28. $\dfrac{x^2 + 4x - 8}{x - 3}$

29. $\dfrac{6x^2 + 7x - 18}{3x - 4}$

30. $\dfrac{8x^2 - 26x - 9}{2x - 7}$

31. $\dfrac{2y^3 - 9y^2 - 17y + 39}{2y - 3}$

32. $\dfrac{3y^3 - 19y^2 + 17y + 4}{3y - 4}$

33. $\dfrac{2x^3 - 9x^2 + 11x - 6}{2x^2 - 3x + 2}$

34. $\dfrac{6x^3 + 7x^2 - x + 3}{3x^2 - x + 1}$

35. $\dfrac{6y^3 - 8y + 5}{2y - 4}$

36. $\dfrac{9y^3 - 6y^2 + 8}{3y - 3}$

37. $\dfrac{a^4 - 2a + 5}{a - 2}$

38. $\dfrac{a^4 + a^3 - 1}{a + 2}$

39. $\dfrac{y^4 - 16}{y - 2}$

40. $\dfrac{y^4 - 81}{y - 3}$

Given the functions $p(x)$ and $q(x)$, use long division to find $\dfrac{p(x)}{q(x)}$.

41. $p(x) = 2x^3 - 3x^2 - 4x + 5, q(x) = x + 1$

42. $p(x) = 3x^3 - 5x^2 + 2x - 1, q(x) = x - 2$

43. $p(x) = x^4 + x^3 - 3x^2 - x + 2, q(x) = x^2 + 3x + 2$

44. $p(x) = 2x^4 + x^3 + 4x - 3, q(x) = 2x^2 - x + 3$

45. Factor $x^3 + 6x^2 + 11x + 6$ completely if one of its factors is $x + 3$.

46. Factor $x^3 + 10x^2 + 29x + 20$ completely if one of its factors is $x + 4$.

47. Factor $x^3 + 5x^2 - 2x - 24$ completely if one of its factors is $x + 3$.

48. Factor $x^3 + 3x^2 - 10x - 24$ completely if one of its factors is $x + 2$.

49. Problems 21 and 39 are the same problem. Are the two answers you obtained equivalent?

50. Problems 22 and 40 are the same problem. Are the two answers you obtained equivalent?

51. Find $P(-2)$ if $P(x) = x^2 - 5x - 7$. Compare it with the remainder in Problem 27.

52. Find $P(3)$ if $P(x) = x^2 + 4x - 8$. Compare it with the remainder in Problem 28.

Applying the Concepts

53. The Factor Theorem The factor theorem of algebra states that if $x - a$ is a factor of a polynomial, $P(x)$, then $P(a) = 0$. Verify the following.

a. That $x - 2$ is a factor of $P(x) = x^3 - 3x^2 + 5x - 6$, and that $P(2) = 0$.

b. That $x - 5$ is a factor of $P(x) = x^4 - 5x^3 - x^2 + 6x - 5$, and that $P(5) = 0$.

54. The Remainder Theorem The remainder theorem of algebra states that if a polynomial, $P(x)$, is divided by $x - a$, then the remainder is $P(a)$. Verify the remainder theorem by showing that when $P(x) = x^2 - x + 3$ is divided by $x - 2$ the remainder is 5, and that $P(2) = 5$.

55. Let $p(x) = x^3 + x^2 - 14x - 24$.

a. Verify that $p(4) = 0$, and thus $x - 4$ is a factor of p(x).

b. Use long division to factor $p(x)$ completely.

56. Let $p(x) = x^3 - 4x^2 + x + 6$.

a. Verify that p(2) = 0, and thus $x - 2$ is a factor for p(x).

b. Use long division to factor p(x) completely.

57. Checking Account A bank charges $2.00 per month and $0.15 per check for a regular checking account. As we mentioned in the introduction to this section, the total monthly cost of this account is $C(x) = 2.00 + 0.15x$. To find the average cost of each of the x checks, we divide the total cost by the number of checks written. That is,

$$\overline{C}(x) = \frac{C(x)}{x}$$

a. Use the total cost function to fill in the following table.

x	1	5	10	15	20
$C(x)$					

b. Find the formula for the average cost function, $\overline{C}(x)$.

c. Use the average cost function to fill in the following table.

x	1	5	10	15	20
$\overline{C}(x)$					

d. What happens to the average cost as more checks are written?

e. Assume that you write at least 1 check a month, but never more than 20 checks per month, and graph both $y = C(x)$ and $y = \overline{C}(x)$ on the same set of axes.

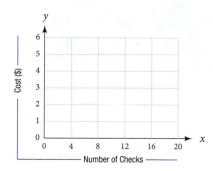

f. Give the domain and range of each of the functions you graphed in part e.

58. Average Cost A company that manufactures flash drives uses the function $C(x) = 200 + 2x$ to represent the daily cost of producing x drives.

a. Find the average cost function, $\overline{C}(x)$.

b. Use the average cost function to fill in the following table:

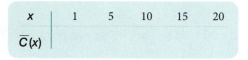

x	1	5	10	15	20
$\overline{C}(x)$					

c. What happens to the average cost as more items are produced?

d. Graph the function $y = \overline{C}(x)$ for $x > 0$.

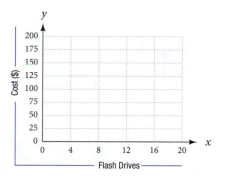

e. What is the domain of this function?

f. What is the range of this function?

59. Average Cost For long distance service, a particular phone company charges a monthly fee of $4.95 plus $0.07 per minute of calling time used. The relationship between the number of minutes of calling time used, m, and the amount of the monthly phone bill $T(m)$ is given by the function $T(m) = 4.95 + 0.07m$.

a. Find the total cost when 100, 400, and 500 minutes of calling time is used in 1 month.

b. Find a formula for the average cost per minute function $\overline{T}(m)$.

c. Find the average cost per minute of calling time used when 100, 400, and 500 minutes are used in 1 month.

60. Average Cost A company manufactures electric pencil sharpeners. Each month they have fixed costs of $40,000 and variable costs of $8.50 per sharpener. Therefore, the total monthly cost to manufacture x sharpeners is given by the function $C(x) = 40{,}000 + 8.5x$.

a. Find the total cost to manufacture 1,000, 5,000, and 10,000 sharpeners a month.

b. Write an expression for the average cost per sharpener function $\overline{C}(x)$.

c. Find the average cost per sharpener to manufacture 1,000, 5,000, and 10,000 sharpeners per month.

Maintaining Your Skills

61. The chart shows the percentage of Google Chrome users. If there are 239,893,600 internet users in America, how many are using Google Chrome? Round to the nearest person.

62. The chart shows the percentage of free and paid applications for the iPad and iPhone. If there were 133,979 apps available for the iPhone, how many were free? Round to the nearest app.

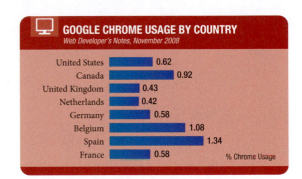

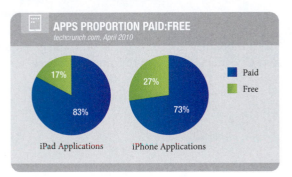

Perform the indicated operations.

63. $\dfrac{2a + 10}{a^3} \cdot \dfrac{a^2}{3a + 15}$

64. $\dfrac{4a + 8}{a^2 - a - 6} \div \dfrac{a^2 + 7a + 12}{a^2 - 9}$

65. $(x^2 - 9)\left(\dfrac{x + 2}{x + 3}\right)$

66. $\dfrac{1}{x + 4} + \dfrac{8}{x^2 - 16}$

67. $\dfrac{2x - 7}{x - 2} - \dfrac{x - 5}{x - 2}$

68. $2 + \dfrac{25}{5x - 1}$

Simplify each expression.

69. $\dfrac{\dfrac{1}{x} - \dfrac{1}{3}}{\dfrac{1}{x} + \dfrac{1}{3}}$

70. $\dfrac{1 - \dfrac{9}{x^2}}{1 - \dfrac{1}{x} - \dfrac{6}{x^2}}$

Solve each equation.

71. $\dfrac{x}{x-3} + \dfrac{3}{2} = \dfrac{3}{x-3}$

72. $1 - \dfrac{3}{x} = \dfrac{-2}{x^2}$

Getting Ready for the Next Section

Simplify.

73. $16(3.5)^2$

74. $\dfrac{2,400}{100}$

75. $\dfrac{180}{45}$

76. $4(2)(4)^2$

77. $\dfrac{0.0005(200)}{(0.25)^2}$

78. $\dfrac{0.2(0.5)^2}{100}$

79. If $y = kx$, find k if $x = 5$ and $y = 15$.

80. If $d = kt^2$, find k if $t = 2$ and $d = 64$.

81. If $P = \dfrac{k}{V}$, find k if $P = 48$ and $V = 50$.

82. If $y = kxz^2$, find k if $x = 5$, $z = 3$, and $y = 180$.

Find the Mistake

Each sentence below contains a mistake. Circle the mistake and write the correct word or expression on the line provided.

1. Dividing the polynomial $15x^3 + 6x^2 + 9x$ by the monomial $3x$ is the same as multiplying by $\dfrac{1}{15x^3 + 6x^2 + 9x}$.

_____

2. Dividing $\dfrac{4x^3 + 12x^2 - 20x}{4x}$, we get $16x^4 + 48x^3 - 80x^2$. _____

3. When using long division to divide $\dfrac{2x^3 - 9x^2 + 19x - 15}{2x - 3}$, we first estimate $2x^3 \div 2x = 3$. _____

4. Dividing $3x^3 - 10x^2 + 7$ by $x - 3$ leaves us with a remainder of 3. _____

OBJECTIVES

A Set up and solve problems with direct variation.

B Set up and solve problems with inverse variation.

C Set up and solve problems with joint variation.

KEY WORDS

direct variation

constant of variation

directly proportional

inverse variation

inversely proportional

joint variation

5.8 Variation

If you are a runner and you average t minutes for every mile you run during one of your workouts, then your speed s in miles per hour is given by the equation and graph shown here. The graph (Figure 1) is shown in the first quadrant only because both t and s are positive.

$$s = \frac{60}{t}$$

Input	Output
t	s
4	15
6	10
8	7.5
10	6
12	5
14	4.3

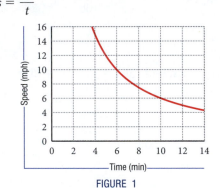

FIGURE 1

You know intuitively that as your average time per mile t increases, your speed s decreases. Likewise, lowering your time per mile will increase your speed. The equation and Figure 1 also show this to be true: Increasing t decreases s, and decreasing t increases s. Quantities that are connected in this way are said to *vary inversely* with each other. Inverse variation is one of the topics we will study in this section.

There are two main types of variation: *direct variation* and *inverse variation*. Variation problems are most common in the sciences, particularly in chemistry and physics.

A Direct Variation

When we say the variable y *varies directly* with the variable x, we mean that the relationship can be written in symbols as $y = kx$, where k is a nonzero constant called the *constant of variation* (or *constant of proportionality*). Another way of saying y varies directly with x is to say y is *directly proportional* to x.

Study the following list. It gives the mathematical equivalent of some direct variation statements.

Verbal phrase	Algebraic equation
y varies directly with x.	$y = kx$
s varies directly with the square of t.	$s = kt^2$
y is directly proportional to the cube of z.	$y = kz^3$
u is directly proportional to the square root of v.	$u = k\sqrt{v}$

EXAMPLE 1   Suppose y varies directly with x. If y is 15 when x is 5, find y when x is 7.

Solution   The first sentence gives us the general relationship between x and y. The equation equivalent to the statement "y varies directly with x" is

$$y = kx$$

The first part of the second sentence in our example gives us the information necessary to evaluate the constant k.

When → $\quad y = 15$

and → $\quad x = 5$

the equation → $\quad y = kx$

becomes → $\quad 15 = k \cdot 5$

$\qquad\qquad k = 3$

The equation can now be written specifically as

$$y = 3x$$

Letting $x = 7$, we have

$$y = 3 \cdot 7$$

$$y = 21$$

EXAMPLE 2   Let's return to the BASE jumper from the beginning of the chapter. Like any object that falls toward earth, the distance the jumper falls is directly proportional to the square of the time she has been falling, until she reaches her terminal velocity. If the BASE jumper falls 64 feet in the first 2 seconds of the jump, then

a. How far will she have fallen after 3.5 seconds?
b. Graph the relationship between distance and time.
c. How long will it take her to fall 256 feet?

Solution   We let t represent the time the jumper has been falling, then we can use function notation and let $d(t)$ represent the distance she has fallen.

a. Since $d(t)$ is directly proportional to the square of t, we have the general function that describes this situation:

$$d(t) = kt^2$$

Next, we use the fact that $d(2) = 64$ to find k.

$$64 = k \cdot 4$$

$$k = 16$$

The specific equation that describes this situation is

$$d(t) = 16t^2$$

To find how far a BASE jumper will fall after 3.5 seconds, we find $d(3.5)$,

$$d(3.5) = 16(3.5)^2$$

$$d(3.5) = 196$$

A BASE jumper will fall 196 feet after 3.5 seconds.

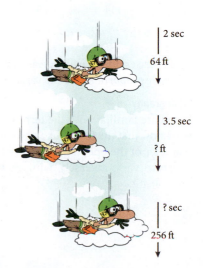

2 sec
64 ft

3.5 sec
? ft

? sec
256 ft

1. Suppose y varies directly with x. If y is 24 when x is 8, find y when x is 2.

2. Use the information in Example 2 to find how far the BASE jumper will fall in 2.5 seconds.

Note   When an object falls and there is no further external force on the object, the vertical acceleration of the object is zero. With no acceleration, the object will abide by Isaac Newton's first law of motion, remaining at a constant velocity. This constant is called terminal velocity.

Answers
1. 6
2. 100 feet

b. To graph this equation, we use a table:

Input	Output
t	$d(t)$
0	0
1	16
2	64
3	144
4	256
5	400

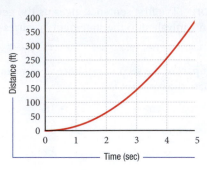

FIGURE 2

c. From the table or the graph (Figure 2), we see that it will take 4 seconds for the BASE jumper to fall 256 feet.

B Inverse Variation

Running

From the introduction to this section, we know that the relationship between the number of minutes t it takes a person to run a mile and his average speed in miles per hour s can be described with the following equation and table, and with Figure 3.

$$s = \frac{60}{t}$$

Input	Output
t	s
4	15
6	10
8	7.5
10	6
12	5
14	4.3

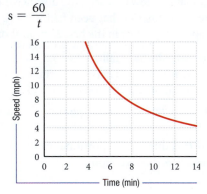

FIGURE 3

If t decreases, then s will increase, and if t increases, then s will decrease. The variable s is *inversely proportional* to the variable t. In this case, the *constant of proportionality* is 60.

Photography

If you are familiar with the terminology and mechanics associated with photography, you know that the *f*-stop for a particular lens will increase as the aperture (the maximum diameter of the opening of the lens) decreases. In mathematics, we say that *f*-stop and aperture vary inversely with each other. The following diagram illustrates this relationship.

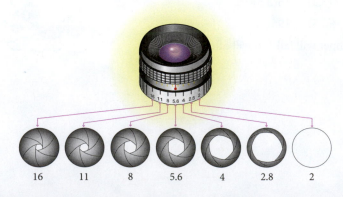

16   11   8   5.6   4   2.8   2

If f is the f-stop and d is the aperture, then their relationship can be written

$$f = \frac{k}{d}$$

In this case, k is the constant of proportionality. (Those of you familiar with photography know that k is also the focal length of the camera lens.)

We generalize this discussion of inverse variation as follows: If y varies inversely with x, then

$$y = k\frac{1}{x} \qquad \text{or} \qquad y = \frac{k}{x}$$

We can also say y is inversely proportional to x. The constant k is again called the constant of variation or proportionality constant.

Verbal phrase	Algebraic equation
y is inversely proportional to x.	$y = \dfrac{k}{x}$
s varies inversely with the square of t.	$s = \dfrac{k}{t^2}$
y is inversely proportional to x^4.	$y = \dfrac{k}{x^4}$
z varies inversely with the cube root of t.	$z = \dfrac{k}{\sqrt[3]{t}}$

EXAMPLE 3   The volume of a gas is inversely proportional to the pressure of the gas on its container. If a pressure of 48 pounds per square inch corresponds to a volume of 50 cubic feet, what pressure is needed to produce a volume of 100 cubic feet?

Solution   We can represent volume with V and pressure with P.

$$V = \frac{k}{P}$$

Using $P = 48$ and $V = 50$, we have

$$50 = \frac{k}{48}$$

$$k = 50(48)$$

$$k = 2,400$$

The equation that describes the relationship between P and V is

$$V = \frac{2,400}{P}$$

Here is a graph of this relationship:

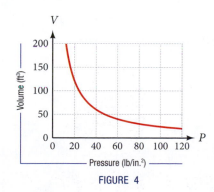

FIGURE 4

3. In Example 3, what pressure is needed to produce a volume of 150 cubic feet?

Note   The relationship between pressure and volume as given in this example is known as Boyle's law and applies to situations such as those encountered in a piston-cylinder arrangement. It was Robert Boyle (1627–1691) who, in 1662, published the results of some of his experiments that showed, among other things, that the volume of a gas decreases as the pressure increases. This is an example of inverse variation.

Answer

3. 16 pounds per square inch

Substituting $V = 100$ into our last equation, we get

$$100 = \frac{2{,}400}{P}$$

$$100P = 2{,}400$$

$$P = \frac{2{,}400}{100}$$

$$P = 24$$

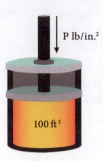
P lb/in.²
100 ft³

A volume of 100 cubic feet is produced by a pressure of 24 pounds per square inch.

C Joint Variation and Other Variation Combinations

Many times relationships among different quantities are described in terms of more than two variables. If the variable y varies directly with *two* other variables, say x and z, then we say y varies *jointly* with x and z. In addition to *joint variation*, there are many other combinations of direct and inverse variation involving more than two variables. The following table is a list of some variation statements and their equivalent mathematical forms:

Verbal phrase	Algebraic equation
y varies jointly with x and z.	$y = kxz$
z varies jointly with r and the square of s.	$z = krs^2$
V is directly proportional to T and inversely proportional to P.	$V = \dfrac{kT}{P}$
F varies jointly with m_1 and m_2 and inversely with the square of r.	$F = \dfrac{km_1m_2}{r^2}$

EXAMPLE 4   Suppose y varies jointly with x and the square of z. When x is 5 and z is 3, y is 180. Find y when x is 2 and z is 4.

Solution   The general equation is given by

$$y = kxz^2$$

Substituting $x = 5$, $z = 3$, and $y = 180$, we have

$$180 = k(5)(3)^2$$

$$180 = 45k$$

$$k = 4$$

The specific equation is

$$y = 4xz^2$$

When $x = 2$ and $z = 4$, the last equation becomes

$$y = 4(2)(4)^2$$

$$y = 128$$

EXAMPLE 5   In electricity, the resistance of a cable is directly proportional to its length and inversely proportional to the square of the diameter. If a 100-foot cable 0.5 inch in diameter has a resistance of 0.2 ohm, what will be the resistance of a cable made from the same material if it is 200 feet long with a diameter of 0.25 inch?

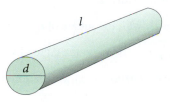

Solution   Let R = resistance, l = length, and d = diameter. The equation is

$$R = \frac{kl}{d^2}$$

When $R = 0.2$, $l = 100$, and $d = 0.5$, the equation becomes

$$0.2 = \frac{k(100)}{(0.5)^2}$$

or

$$k = 0.0005$$

Using this value of k in our original equation, the result is

$$R = \frac{0.0005l}{d^2}$$

When $l = 200$ and $d = 0.25$, the equation becomes

$$R = \frac{0.0005(200)}{(0.25)^2}$$

$$R = 1.6 \text{ ohms}$$

5. Use the information in Example 5 to find the resistance of a 300-foot cable made of the same material that is 0.25 inch in diameter.

GETTING READY FOR CLASS

After reading through the preceding section, respond in your own words and in complete sentences.

A. Give an example of a direct variation statement, and then translate it into symbols.

B. Translate the equation $y = \frac{k}{x}$ into words.

C. For the inverse variation equation $y = \frac{3}{x}$, what happens to the values of y as x gets larger?

D. How are direct variation statements and linear equations in two variables related?

Answer

5. 2.4 ohms

Vocabulary Review

Match the following algebraic equations with the sentence that describes the correct variation.

1. $u = k\sqrt{v}$

2. $s = kt^2$

3. $y = kx^3$

4. $z = krs^2$

5. $s = \dfrac{k}{t^2}$

6. $F = \dfrac{km_1m_2}{r^2}$

7. $y = \dfrac{k}{x^4}$

8. $z = \dfrac{k}{\sqrt[3]{t}}$

9. $y = kxz$

10. $y = kx$

11. $V = \dfrac{kT}{P}$

12. $y = \dfrac{k}{x}$

a. y varies directly with x.

b. s varies directly with the square of t.

c. y is directly proportional to the cube of x.

d. u is directly proportional to the square root of v.

e. y is inversely proportional to x.

f. s varies inversely with the square of t.

g. y is inversely proportional to x^4.

h. z varies inversely with the cube root of t.

i. y varies jointly with x and z.

j. z varies jointly with r and square of s.

k. V is directly proportional to T and inversely proportional to P.

l. F varies jointly with m_1 and m_2 and inversely with the square of r.

Problems

A B C For the following problems, assume y varies directly with x.

1. If y is 10 when x is 2, find y when x is 6.

2. If y is -32 when x is 4, find x when y is -40.

For the following problems, assume r is inversely proportional to s.

3. If r is -3 when s is 4, find r when s is 2.

4. If r is 8 when s is 3, find s when r is 48.

For the following problems, assume d varies directly with the square of r.

5. If $d = 10$ when $r = 5$, find d when $r = 10$.

6. If $d = 12$ when $r = 6$, find d when $r = 9$.

SCAN TO ACCESS

For the following problems, assume y varies inversely with the square of x.

7. If $y = 45$ when $x = 3$, find y when x is 5.

8. If $y = 12$ when $x = 2$, find y when x is 6.

For the following problems, assume z varies jointly with x and the square of y.

9. If z is 54 when x and y are 3, find z when $x = 2$ and $y = 4$. **10.** If z is 27 when $x = 6$ and $y = 3$, find x when $z = 50$ and $y = 4$.

For the following problems, assume I varies inversely with the cube of w.

11. If $I = 32$ when $w = \dfrac{1}{2}$, find I when $w = \dfrac{1}{3}$.

12. If $I = \dfrac{1}{25}$ when $w = 5$, find I when $w = 10$.

For the following problems, assume z varies jointly with y and the square of x.

13. If $z = 72$ when $x = 3$ and $y = 2$, find z when $x = 5$ and $y = 3$.

14. If $z = 240$ when $x = 4$ and $y = 5$, find z when $x = 6$ and $y = 3$.

15. If $x = 1$ when $z = 25$ and $y = 5$, find x when $z = 160$ and $y = 8$.

16. If $x = 4$ when $z = 96$ and $y = 2$, find x when $z = 108$ and $y = 1$.

For the following problems, assume F varies directly with m and inversely with the square of d.

17. If $F = 150$ when $m = 240$ and $d = 8$, find F when $m = 360$ and $d = 3$.

18. If $F = 72$ when $m = 50$ and $d = 5$, find F when $m = 80$ and $d = 6$.

19. If $d = 5$ when $F = 24$ and $m = 20$, find d when $F = 18.75$ and $m = 40$.

20. If $d = 4$ when $F = 75$ and $m = 20$, find d when $F = 200$ and $m = 120$.

Applying the Concepts

21. Length of a Spring The length a spring stretches is directly proportional to the force applied. If a force of 5 pounds stretches a spring 3 inches, how much force is necessary to stretch the same spring 10 inches?

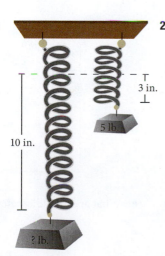

10 in.

3 in.

5 lb.

? lb.

22. Weight and Surface Area The weight of a certain material varies directly with the surface area of that material. If 8 square feet weighs half a pound, how much will 10 square feet weigh?

23. **Pressure and Temperature** The temperature of a gas varies directly with its pressure. A temperature of 200 Kelvin (K) produces a pressure of 50 pounds per square inch.

    a. Find the equation that relates pressure and temperature.

    b. Graph the equation from part a in the first quadrant only.

    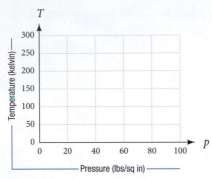

    c. What pressure will the gas have at 280 K?

24. **Circumference and Diameter** The circumference of a wheel is directly proportional to its diameter. A wheel has a circumference of 8.5 feet and a diameter of 2.7 feet.

    a. Find the equation that relates circumference and diameter.

    b. Graph the equation from part a in the first quadrant only.

    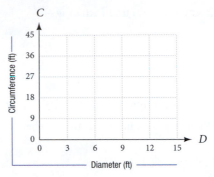

    c. What is the circumference of a wheel that has a diameter of 11.3 feet?

25. **Volume and Pressure** The volume of a gas is inversely proportional to the pressure applied. If a pressure of 36 pounds per square inch corresponds to a volume of 25 cubic feet, what pressure is needed to produce a volume of 75 cubic feet?

26. **Wave Frequency** The frequency of an electromagnetic wave varies inversely with the wavelength. If a wavelength of 200 meters has a frequency of 800 kilocycles per second, what frequency will be associated with a wavelength of 500 meters?

27. **f-Stop and Aperture Diameter** The relative aperture, or f-stop, for a camera lens is inversely proportional to the diameter of the aperture. An f-stop of 2 corresponds to an aperture diameter of 40 millimeters for the lens on an automatic camera.

    a. Find the equation that relates f-stop and diameter.

    b. Graph the equation from part a in the first quadrant only.

    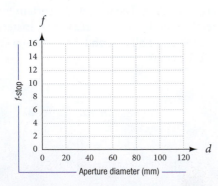

    c. What is the f-stop of this camera when the aperture diameter is 10 millimeters?

28. **f-Stop and Aperture Diameter** The relative aperture, or f-stop, for a camera lens is inversely proportional to the diameter of the aperture. An f-stop of 2.8 corresponds to an aperture diameter of 75 millimeters for a certain telephoto lens.

    a. Find the equation that relates f-stop and diameter.

    b. Graph the equation from part a in the first quadrant only.

    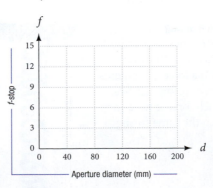

    c. What aperture diameter corresponds to an f-stop of 5.6?

29. Surface Area of a Cylinder The surface area of a hollow cylinder varies jointly with the height and radius of the cylinder. If a cylinder with radius 3 inches and height 5 inches has a surface area of 94 square inches, what is the surface area of a cylinder with radius 2 inches and height 8 inches?

30. Capacity of a Cylinder The capacity of a cylinder varies jointly with its height and the square of its radius. If a cylinder with a radius of 3 centimeters and a height of 6 centimeters has a capacity of 169.56 cubic centimeters, what will be the capacity of a cylinder with radius 4 centimeters and height 9 centimeters?

31. Electrical Resistance The resistance of a wire varies directly with its length and inversely with the square of its diameter. If 100 feet of wire with diameter 0.01 inch has a resistance of 10 ohms, what is the resistance of 60 feet of the same type of wire if its diameter is 0.02 inch?

32. Volume and Temperature The volume of a gas varies directly with its temperature and inversely with the pressure. If the volume of a certain gas is 30 cubic feet at a temperature of 300 Kelvin (K) and a pressure of 20 pounds per square inch, what is the volume of the same gas at 340 K when the pressure is 30 pounds per square inch?

33. Period of a Pendulum The time it takes for a pendulum to complete one period varies directly with the square root of the length of the pendulum. A 100-centimeter pendulum takes 2.1 seconds to complete one period.

   a. Find the equation that relates period and pendulum length.

   b. Graph the equation from part a in quadrant I only.

   c. How long does it take to complete one period if the pendulum hangs 225 centimeters?

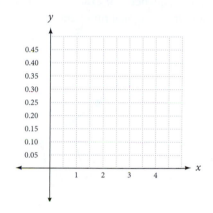

Maintaining Your Skills

Multiply.

34. $x(35 - 0.1x)$

35. $0.6(M - 70)$

36. $(4x - 3)(x - 1)$

37. $(4x - 3)(4x^2 - 7x + 3)$

Simplify.

38. $(35x - 0.1x^2) - (8x + 500)$

39. $(4x - 3) + (4x^2 - 7x + 3)$

40. $(4x^2 + 3x + 2) - (2x^2 - 5x - 6)$

41. $(4x^2 + 3x + 2) + (2x^2 - 5x - 6)$

42. $4(2)^2 - 3(2)$

43. $4(-1)^2 - 7(-1)$

Find the Mistake

Each sentence below contains a mistake. Circle the mistake and write the correct word or expression on the line provided.

1. If y varies directly with x and $y = 9$ when $x = 27$, then $x = \frac{1}{3}$ when $y = 3$. _____

2. If s is directly proportional to the square root of t, and $t = 100$ when $s = 40$, then $s = 8\sqrt{2}$ when $t = 8$. _____

3. If z varies inversely with the square of t and $z = 16$ when $t = \frac{1}{4}$, then $z = \frac{256}{9}$ when $t = \frac{1}{3}$. _____

4. If w is directly proportional to l and inversely proportional to p when $w = 18$, $l = 9$ and $p = 3$, then $w = 26.67$ when $l = 4$ and $p = 10$. _____

A Access to the internet for research

B Pencil and paper

Figurate Numbers

Research the following categories of numbers. Explain three unique characteristics of each category.

1. Triangular numbers: 1, 3, 6, 10, 15,...$T_n = \dfrac{n(n+1)}{2}$

2. Pentagonal numbers: 1, 5, 12, 22, 35,...$P_n = \dfrac{3n^2 - n}{2}$

3. Hexagonal numbers: 1, 6, 15, 28, 45,...$H_n = \dfrac{2n(2n-1)}{2}$

Note that we have provided the first five numbers in each sequence. Use the given formulas for the nth terms in the sequences (all of which include a rational expression) to generate the next five numbers in each sequence. Then draw a visual representation of the first ten numbers in each sequence. For example, the figure below is a visual representation of the first four pentagonal numbers.

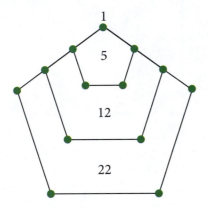

Chapter 5 Summary

Rational Numbers and Expressions [5.1]

A *rational number* is any number that can be expressed as the ratio of two integers.

$$\text{Rational numbers} = \left\{ \frac{a}{b} \,\middle|\, a \text{ and } b \text{ are integers, } b \neq 0 \right\}$$

A *rational expression* is any quantity that can be expressed as the ratio of two polynomials.

$$\text{Rational expressions} = \left\{ \frac{P}{Q} \,\middle|\, P \text{ and } Q \text{ are polynomials, } Q \neq 0 \right\}$$

Properties of Rational Expressions [5.1]

If P, Q, and K are polynomials with $Q \neq 0$ and $K \neq 0$, then

$$\frac{P}{Q} = \frac{PK}{QK} \qquad \text{and} \qquad \frac{P}{Q} = \frac{\dfrac{P}{K}}{\dfrac{Q}{K}}$$

which is to say that multiplying or dividing the numerator and denominator of a rational expression by the same nonzero quantity always produces an equivalent rational expression.

Reducing to Lowest Terms [5.1]

To reduce a rational expression to lowest terms, we first factor the numerator and denominator and then divide the numerator and denominator by any factors they have in common.

Multiplication of Rational Expressions [5.2]

To multiply two rational numbers or rational expressions, multiply numerators and multiply denominators. In symbols,

$$\frac{P}{Q} \cdot \frac{R}{S} = \frac{PR}{QS} \qquad (Q \neq 0 \text{ and } S \neq 0)$$

In practice, we don't really multiply, but rather, we factor and then divide out common factors.

Division of Rational Expressions [5.2]

To divide one rational expression by another, we use the definition of division to rewrite our division problem as an equivalent multiplication problem. To divide by a rational expression we multiply by its reciprocal. In symbols,

$$\frac{P}{Q} \div \frac{R}{S} = \frac{P}{Q} \cdot \frac{S}{R} = \frac{PS}{QR} \qquad (Q \neq 0, S \neq 0, R \neq 0)$$

EXAMPLES

1. $\frac{3}{4}$ is a rational number. $\frac{x-3}{x^2-9}$ is a rational expression.

2. $\dfrac{x-3}{x^2-9} = \dfrac{x-3}{(x-3)(x+3)}$

   $= \dfrac{1}{x+3}$

3. $\dfrac{x+1}{x^2-4} \cdot \dfrac{x+2}{3x+3}$

   $= \dfrac{(x+1)(x+2)}{(x-2)(x+2)(3)(x+1)}$

   $= \dfrac{1}{3(x-2)}$

4. $\dfrac{x^2-y^2}{x^3+y^3} \div \dfrac{x-y}{x^2-xy+y^2}$

   $= \dfrac{x^2-y^2}{x^3+y^3} \cdot \dfrac{x^2-xy+y^2}{x-y}$

   $= \dfrac{(x+y)(x-y)(x^2-xy+y^2)}{(x+y)(x^2-xy+y^2)(x-y)}$

   $= 1$

Least Common Denominator [5.3]

The *least common denominator,* LCD, for a set of denominators is the smallest quantity divisible by each of the denominators.

5. The LCD for $\frac{2}{x-3}$ and $\frac{3}{5}$ is $5(x-3)$.

Addition and Subtraction of Rational Expressions [5.3]

If P, Q, and R represent polynomials, $R \neq 0$, then

$$\frac{P}{R} + \frac{Q}{R} = \frac{P+Q}{R} \quad \text{and} \quad \frac{P}{R} - \frac{Q}{R} = \frac{P-Q}{R}$$

When adding or subtracting rational expressions with different denominators, we must find the LCD for all denominators and change each rational expression to an equivalent expression that has the LCD.

6. $\dfrac{2}{x-3} + \dfrac{3}{5}$

$= \dfrac{2}{x-3} \cdot \dfrac{5}{5} + \dfrac{3}{5} \cdot \dfrac{x-3}{x-3}$

$= \dfrac{3x+1}{5(x-3)}$

Complex Rational Expressions [5.4]

A rational expression that contains, in its numerator or denominator, other rational expressions is called a *complex fraction*, or complex rational expression. One method of simplifying a complex fraction is to multiply the numerator and denominator by the LCD for all denominators.

7. $\dfrac{\frac{1}{x} + \frac{1}{y}}{\frac{1}{x} - \frac{1}{y}} = \dfrac{xy\left(\frac{1}{x} + \frac{1}{y}\right)}{xy\left(\frac{1}{x} - \frac{1}{y}\right)}$

$= \dfrac{y+x}{y-x}$

Rational Equations [5.5]

To solve an equation involving rational expressions, we first find the LCD for all denominators appearing on either side of the equation. We then multiply both sides by the LCD to clear the equation of all fractions and solve as usual.

8. Solve $\dfrac{x}{2} + 3 = \dfrac{1}{3}$.

$6\left(\dfrac{x}{2}\right) + 6 \cdot 3 = 6 \cdot \dfrac{1}{3}$

$3x + 18 = 2$

$x = -\dfrac{16}{3}$

Dividing a Polynomial by a Monomial [5.7]

To divide a polynomial by a monomial, divide each term of the polynomial by the monomial.

9. $\dfrac{15x^3 - 20x^2 + 10x}{5x}$

$= 3x^2 - 4x + 2$

Long Division with Polynomials [5.7]

If division with polynomials cannot be accomplished by dividing out factors common to the numerator and denominator, then we use a process similar to long division with whole numbers. The steps in the process are estimate, multiply, subtract, and bring down the next term.

10.

$$
\begin{array}{r}
x - 2 \\
x - 3\overline{\smash{\big)}\,x^2 - 5x + 8} \\
\underline{+x^2 - 3x} \downarrow \\
-2x + 8 \\
\underline{+2x + 6} \\
2
\end{array}
$$

Variation [5.8]

If *y varies directly* with x (y is directly proportional to x), then

$$y = kx$$

If *y varies inversely* with x (y is inversely proportional to x), then

$$y = \frac{k}{x}$$

If *z varies jointly* with x and y (z is directly proportional to both x and y), then

$$z = kxy$$

In each case, k is called the *constant of variation.*

> ### COMMON MISTAKES
>
> 1. Attempting to divide the numerator and denominator of a rational expression by a quantity that is not a factor of both. Like this:
>
> $$\frac{x^2 - \overset{3}{9x} + \overset{2}{20}}{x^2 - 3x - \underset{1}{10}} \quad \text{Mistake}$$
>
> This makes no sense at all. The numerator and denominator must be completely before any factors they have in common can be recognized.
>
> $$\frac{x^2 - 9x + 20}{x^2 - 3x - 10} = \frac{(x - 5)(x - 4)}{(x - 5)(x + 2)}$$
> $$= \frac{x - 4}{x + 2}$$
>
> 2. Forgetting to check solutions to equations involving rational expressions. When we multiply both sides of an equation by a quantity containing the variable, we must be sure to check for extraneous solutions.

11. If y varies directly with x, then

$$y = kx$$

If y is 18 when x is 6,

$$18 = k \cdot 6$$

Then

$$k = 3$$

So the equation can be written more specifically as

$$y = 3x$$

If we want to know what y is when x is 4, we simply substitute.

$$y = 3 \cdot 4$$
$$y = 12$$

Reduce to lowest terms. [5.1]

1. $\dfrac{2x^2 - 3xy + y^2}{2x - y}$ **2.** $\dfrac{3x^2 - 4x - 4}{3x^2 - x - 2}$

3. Average Speed A person riding a Ferris wheel with a diameter of 75 feet travels once around the wheel in 70 seconds. What is the average speed of the rider in feet per minute? Use 3.14 as an approximation for π and round your answer to the nearest tenth. [5.1]

For the functions in Problems 4 and 5, evaluate the given difference quotients. [5.1]

   a. $\dfrac{f(x + h) - f(x)}{h}$ **b.** $\dfrac{f(x) - f(a)}{x - a}$

4. $f(x) = 3x + 5$ **5.** $f(x) = x^2 + 2x - 8$

Multiply and divide as indicated. [5.2]

6. $\dfrac{x^2 - 16}{3x + 12} \cdot \dfrac{3(x + 3)^2}{x^2 - x - 12}$

7. $\dfrac{x^4 - 81}{2x^2 - 9x + 9} \div \dfrac{x^2 + 9}{10x - 15}$

8. $\dfrac{4x^2 + 6x - 4}{2x^2 - 11x + 5} \div \dfrac{x^3 - 8}{x^2 - 7x + 10}$

Add and subtract as indicated. [5.3]

9. $\dfrac{3}{14} + \dfrac{9}{20}$ **10.** $\dfrac{1}{2} - \dfrac{7}{24} + \dfrac{5}{6}$

11. $\dfrac{a + 7}{a^2 - 81} + \dfrac{2}{a^2 - 81}$

12. $\dfrac{3}{x - 1} + \dfrac{6}{2x - 1}$

13. $\dfrac{4x}{x^2 + 2x - 3} - \dfrac{2x}{x^2 - 1}$

14. $\dfrac{2x - 1}{x^2 - 1} - \dfrac{x + 4}{x^2 + 4x + 3}$

Simplify each complex rational expression. [5.4]

15. $\dfrac{4 - \dfrac{1}{a + 2}}{4 + \dfrac{1}{a + 2}}$ **16.** $\dfrac{1 - \dfrac{16}{x^2}}{1 - \dfrac{1}{x} - \dfrac{12}{x^2}}$

Solve each of the following equations. [5.5]

17. $5 - \dfrac{2}{x} = \dfrac{9}{5}$ **18.** $\dfrac{3x}{x + 2} + \dfrac{3}{2} = \dfrac{9}{x + 1}$

19. $\dfrac{4}{x + 5} + \dfrac{x + 6}{8x} = \dfrac{7}{8}$ **20.** $1 = \dfrac{11}{x} - \dfrac{30}{x^2}$

Solve the following applications. Be sure to show the equation in each case. [5.6]

21. Number Problem What number must be added to the denominator of $\frac{23}{63}$ to make the result $\frac{1}{3}$?

22. Speed of a Boat The current of a river is 7 miles per hour. It takes a motorboat a total of 11 hours to travel 36 miles upstream and return 36 miles downstream. What is the speed of the boat in still water?

23. Filling a Pool An inlet pipe can fill a tub in 5 minutes, and the drain can empty it in 6 minutes. If both the inlet pipe and the drain are left open, how long will it take to fill the tub?

Divide. [5.7]

24. $\dfrac{8x^4y^2 - 20x^5y + 28x^2y^3}{4x^2y}$ **25.** $\dfrac{9x^3 - 9x^2 + 2x - 12}{3x - 5}$

26. Inverse Variation y varies inversely with the square of x. If $y = 6$ when $x = 3$, find y when $x = 6$. [5.8]

27. Joint Variation z varies jointly with x and the cube of y. If $z = 216$ when $x = 4$ and $y = 3$, find z when $x = 7$ and $y = 2$. [5.8]

Simplify.

1. $\left(-\dfrac{5}{3}\right)^3$

2. $\left(\dfrac{8}{7}\right)^{-2}$

3. $|-18 - 32| - 90$

4. $3^3 - 6(5^2 - 6^2)$

5. $81 \div 18 \cdot 2$

6. $36 \div 12 \cdot 4$

7. $\dfrac{2^4 - 5}{11(6^2 - 5^2)}$

8. $\dfrac{-5^2(4 - 4 \cdot 2)}{19 + 3^4}$

9. Subtract $\dfrac{4}{5}$ from the product of -3 and $\dfrac{7}{15}$.

10. Write in symbols: The difference of $2a$ and $9b$ is greater than their sum.

Let $P(x) = 14x - 0.01x^2$ and find the following.

11. $P(10)$

12. $P(-10)$

Let $Q(x) = \dfrac{5}{4}x^2 + \dfrac{1}{5}x - 33$ and find the following.

13. $Q(10)$

14. $Q(-10)$

Simplify each of the following expressions.

15. $-4(4x + 5) - 11x$

16. $(x + 5)^2 - (x - 5)^2$

Simplify.

17. $\dfrac{5}{y^2 + y - 6} - \dfrac{4}{y^2 - 4}$

18. $\dfrac{x}{y^2 - x^2} + \dfrac{y}{y^2 - x^2}$

Multiply.

19. $\left(5t^2 + \dfrac{1}{4}\right)\left(4t^2 - \dfrac{1}{5}\right)$

20. $\dfrac{x^2 - 81}{x^2 + 5x + 6} \cdot \dfrac{x + 2}{x + 9}$

Divide.

21. $\dfrac{9x^{7n} - 6x^{4n}}{3x^{2n}}$

22. $\dfrac{2a^5 + a^2 - 8}{a - 1}$

Reduce to lowest terms. Write the answer with positive exponents only.

23. $\dfrac{x^{-6}}{x^{-11}}$

24. $\left(\dfrac{x^{-5}y^{-8}}{x^2 y^{-6}}\right)^{-1}$

25. $\dfrac{x^3 + 3x^2 - 25x - 75}{x^2 - 2x - 15}$

26. $\dfrac{\dfrac{7a}{4a^3 - 4}}{\dfrac{14a}{8a - 8}}$

Find the equation of the line through the two points.

27. $(5, 4)$ and $(10, 1)$

28. $(3, -1)$ and $(11, -1)$

Factor each of the following expressions.

29. 396

30. $x^2 + 3x - 40$

31. $x^2 + 2xy - 36 + y^2$

32. $y^3 - \dfrac{64}{125}x^3$

Solve the following equations.

33. $\dfrac{9}{2}x - 6 = 57$

34. $9y + 27 = 2y - 8$

35. $\dfrac{2}{3}(9x - 1) - \dfrac{1}{3} = -7$

36. $\dfrac{3}{7}(14x - 4) + \dfrac{5}{7} = 8$

37. $\dfrac{4}{y - 1} = \dfrac{7}{y + 6}$

38. $3 - \dfrac{16}{x} = \dfrac{12}{x^2}$

Solve each system.

39. $2x + 5y = 17$
$\quad\ 9x + 3y = 2x + 16$

40. $3x - 2y = 5$
$\quad\ 2x + 9y = 6$

Solve each inequality, and graph the solution.

41. $-3(x - 2) \le -2(x - 6)$

42. $|x - 4| + 3 < 5$

Graph on a rectangular coordinate system.

43. $-6x + 5y = -15$

44. $3x + y < -4$

Use the graph to work Problems 45–48.

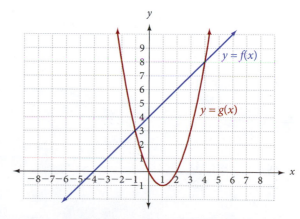

45. Find $(f - g)(3)$.

46. Find $(f + g)(-2)$.

47. Find $(f \circ g)(-1)$.

48. Find $(g \circ f)(-1)$.

Reduce to lowest terms. [5.1]

1. $\dfrac{x^2 + 2xy + y^2}{x + y}$

2. $\dfrac{4x^2 - 4x - 3}{2x^2 - x - 1}$

3. Average Speed A person riding a Ferris wheel with a diameter of 65 feet travels once around the wheel in 50 seconds. What is the average speed of the rider in feet per second? Use 3.14 as an approximation for π and round your answer to the nearest tenth. [5.1]

For the functions in Problems 4 and 5, evaluate the given difference quotients. [5.1]

a. $\dfrac{f(x + h) - f(x)}{h}$

b. $\dfrac{f(x) - f(a)}{x - a}$

4. $f(x) = 8x - 5$

5. $f(x) = x^2 + x - 6$

Multiply and divide as indicated. [5.2]

6. $\dfrac{a^2 - 9}{4a + 12} \cdot \dfrac{12(a + 2)^2}{a^2 - a - 6}$

7. $\dfrac{a^4 - 16}{2a^2 + a - 6} \div \dfrac{a^2 + 4}{8a - 12}$

8. $\dfrac{2x^2 + 3x + 1}{2x^2 + 6x + 18} \div \dfrac{2x^2 - 5x - 3}{x^3 - 27}$

Add and subtract as indicated. [5.3]

9. $\dfrac{3}{10} + \dfrac{6}{25}$

10. $\dfrac{1}{2} - \dfrac{5}{32} + \dfrac{3}{16}$

11. $\dfrac{a + 2}{a^2 - 25} - \dfrac{7}{a^2 - 25}$

12. $\dfrac{2}{x + 1} + \dfrac{4}{2x - 1}$

13. $\dfrac{5x}{x^2 + 3x - 4} - \dfrac{4x}{x^2 + 2x - 3}$

14. $\dfrac{2x + 4}{x^2 - 10x - 39} - \dfrac{x + 2}{x^2 - 18x + 65}$

Simplify each complex rational expression. [5.4]

15. $\dfrac{2 - \dfrac{1}{a + 6}}{2 + \dfrac{1}{a + 6}}$

16. $\dfrac{1 - \dfrac{25}{x^2}}{1 + \dfrac{7}{x} + \dfrac{10}{x^2}}$

Solve each of the following equations. [5.5]

17. $5 + \dfrac{2}{x} = \dfrac{6}{5}$

18. $\dfrac{2x}{x + 4} - 3 = \dfrac{-8}{x + 4}$

19. $\dfrac{9}{y + 3} + \dfrac{y + 4}{7y} = \dfrac{1}{7}$

20. $1 - \dfrac{10}{x} = \dfrac{-21}{x^2}$

Solve the following applications. Be sure to show the equation in each case. [5.6]

21. Number Problem What number must be added to the denominator of $\frac{13}{54}$ to make the result $\frac{1}{4}$?

22. Speed of a Boat The current of a river is 5 miles per hour. It takes a motorboat a total of 14 hours to travel 24 miles upstream and return 24 miles downstream. What is the speed of the boat in still water?

23. Filling a Pool An inlet pipe can fill a pool in 6 hours, and the drain can empty it in 10 hours. If the pool is one-third full and both the inlet pipe and the drain are left open, how long will it take to fill the pool the rest of the way?

Divide. [5.7]

24. $\dfrac{6x^4y + 33x^3y^2 - 6x^2y^3}{3x^2y}$

25. $\dfrac{4x^3 - 9x^2 - x + 8}{4x + 3}$

26. Inverse Variation y varies inversely with the square of x. If $y = 7$ when $x = 2$, find y when $x = 14$. [5.8]

27. Joint Variation z varies jointly with x and the square of y. If z is 27 when $x = 6$ and $y = 3$, find x when $z = 50$ and $y = 4$. [5.8]

Rational Exponents and Radicals

Image © 2010 GeoEye
Chichen Itza, Mexico

On Mexico's Yucatan Peninsula lies one of the world's great mysteries: the ruins of the Mayan city of Chichen Itza. The Temple of Kukulkan is the largest and most recognizable structure at Chichen Itza. This 90-foot-tall pyramid is a 4-sided structure, each side containing a stairway with 91 steps. One of the most puzzling facts about these stairs can be observed from inside of the Temple. If you were to clap your hands inside the pyramid, the echoes from each of the stairs in the staircases sound similar to the primary call of the Mayan sacred bird, the Quetzal. Experts believe that the Mayan people intentionally coded the echoes within their temple to sound like their sacred bird. If this is true, this ancient people's knowledge of sound and acoustics remains astounding.

The time delay of an echo is the time it takes for the sound to travel to the reflection surface and back to the origin of the sound. In other words, to begin calculating the echo's time delay, you must first double the distance to the reflection surface. Therefore, if you were to clap your hands inside the Temple of Kukulkan at a distance of $15\sqrt{5}$ feet from the base of the stairs, you would have to know how to double $15\sqrt{5}$. The formula to calculate the time delay of an echo, as well as the techniques needed to double expressions involving square roots, will be covered as you work through this chapter.

KEY WORDS

positive square root

root

radical expression

index

radical sign

radicand

rational exponent

golden rectangle

golden ratio

6.1 Rational Exponents

Kite fighting is a sport where two competitors each fly a kite and attempt to cut down the other's kite string with their own. In some contests, a competitor uses a small nylon kite tethered by a string coated in a paste made from glue and powdered glass. This paste helps to strengthen the line and gives it the sharpness it needs to slice through the opponent's string.

Suppose we are competing in a kite fighting competition. Our opponent's kite is already flying in the air at 100 feet high. Using the Pythagorean theorem, can you set up an equation that would give us the length of the string that is tethering the kite? In this section, we will work with properties of exponents, such as those needed to solve problems that include the Pythagorean theorem. Then we can return to this question to find out how much line we will need to let out for our kite to reach the same height as our opponent's.

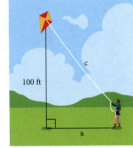

FIGURE 1

Figure 2 shows a square in which each of the four sides is 1 inch long. To find the square of the length of the diagonal c, we apply the Pythagorean theorem:

$$c^2 = 1^2 + 1^2$$
$$c^2 = 2$$

1 inch

1 inch

FIGURE 2

Because we know that c is positive and that its square is 2, we call c the *positive square root* of 2, and we write $c = \sqrt{2}$. Associating numbers, such as $\sqrt{2}$, with the diagonal of a square or rectangle allows us to analyze some interesting items from geometry. One particularly interesting geometric object that we will study in this section is called the *golden rectangle*, shown in Figure 3. It is constructed from a right triangle, and the length of the diagonal is found from the Pythagorean theorem. We will come back to this figure at the end of this section.

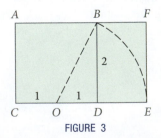

FIGURE 3

A Radical Expressions

In Chapter 0, we developed notation (exponents) to give us the square, cube, or any other power of a number. For instance, if we wanted the square of 3, we wrote $3^2 = 9$. If we wanted the cube of 3, we wrote $3^3 = 27$. In this section, we will develop notation that will take us in the reverse direction; that is, from the square of a number, say 25, back to the original number, 5.

> **DEFINITION** positive square root
>
> If x is a nonnegative real number, then the expression $\sqrt{x}$ is called the ***positive square root*** of x and is such that
>
> $$(\sqrt{x})^2 = x$$
>
> *In words:* $\sqrt{x}$ is the positive number we square to get x.

The negative square root of x, $-\sqrt{x}$, is defined in a similar manner.

EXAMPLE 1   The positive square root of 64 is 8 because 8 is the positive number with the property $8^2 = 64$. The negative square root of 64 is -8 because -8 is the negative number whose square is 64. We can summarize both these facts by saying

$$\sqrt{64} = 8 \qquad \text{and} \qquad -\sqrt{64} = -8$$

The higher roots, cube roots, fourth roots, and so on, are defined by definitions similar to that of square roots.

> **DEFINITION** roots
>
> If x is a real number and n is a positive integer, then
>
> Positive square root of x, $\sqrt{x}$, is such that $(\sqrt{x})^2 = x$ $x \geq 0$
>
> Cube root of x, $\sqrt[3]{x}$, is such that $(\sqrt[3]{x})^3 = x$
>
> Positive fourth root of x, $\sqrt[4]{x}$, is such that $(\sqrt[4]{x})^4 = x$ $x \geq 0$
>
> Fifth root of x, $\sqrt[5]{x}$, is such that $(\sqrt[5]{x})^5 = x$
>
> $$\vdots \qquad \vdots$$
>
> The nth root of x, $\sqrt[n]{x}$, is such that $(\sqrt[n]{x})^n = x$ $x \geq 0$ if n is even.

The following is a table of the most common roots used in this book. Any of the roots that are unfamiliar should be memorized.

Square Roots		Cube Roots	Fourth Roots
$\sqrt{0} = 0$	$\sqrt{49} = 7$	$\sqrt[3]{0} = 0$	$\sqrt[4]{0} = 0$
$\sqrt{1} = 1$	$\sqrt{64} = 8$	$\sqrt[3]{1} = 1$	$\sqrt[4]{1} = 1$
$\sqrt{4} = 2$	$\sqrt{81} = 9$	$\sqrt[3]{8} = 2$	$\sqrt[4]{16} = 2$
$\sqrt{9} = 3$	$\sqrt{100} = 10$	$\sqrt[3]{27} = 3$	$\sqrt[4]{81} = 3$
$\sqrt{16} = 4$	$\sqrt{121} = 11$	$\sqrt[3]{64} = 4$	
$\sqrt{25} = 5$	$\sqrt{144} = 12$	$\sqrt[3]{125} = 5$	
$\sqrt{36} = 6$	$\sqrt{169} = 13$		

Note It is a common mistake to assume that an expression like $\sqrt{25}$ indicates both square roots, 5 and -5. The expression $\sqrt{25}$ indicates only the positive square root of 25, which is 5. If we want the negative square root, we must use a negative sign: $-\sqrt{25} = -5$.

Practice Problems

1. Give the two square roots of 36.

Answers

1. 6, -6

Notation An expression like $\sqrt[3]{8}$ that involves a root is called a *radical expression*. In the expression $\sqrt[3]{8}$, the 3 is called the *index*, the $\sqrt{}$ is the *radical sign*, and 8 is called the *radicand*. The index of a radical must be a positive integer greater than 1. If no index is written, it is assumed to be 2.

Roots and Negative Numbers

When dealing with negative numbers and radicals, the only restriction concerns negative numbers under even roots. We can have negative signs in front of radicals and negative numbers under odd roots and still obtain real numbers. Here are some examples to help clarify this. In the last section of this chapter, we will see how to deal with even roots of negative numbers.

2. Simplify, if possible.

   a. $\sqrt[3]{-64}$

   b. $\sqrt{-25}$

   c. $-\sqrt{4}$

   d. $\sqrt[5]{-1}$

   e. $\sqrt[4]{-16}$

EXAMPLE 2 Simplify each expression, if possible.

   a. $\sqrt[3]{-8} = -2$ because $(-2)^3 = -8$.

   b. $\sqrt{-4}$ is not a real number because there is no real number whose square is -4.

   c. $-\sqrt{25} = -5$, because -5 is the negative square root of 25.

   d. $\sqrt[5]{-32} = -2$ because $(-2)^5 = -32$.

   e. $\sqrt[4]{-81}$ is not a real number because there is no real number we can raise to the fourth power and obtain -81.

Variables Under a Radical

From the preceding examples, it is clear that we must be careful that we do not try to take an even root of a negative number. For the purpose of this section, we will assume that all variables appearing under a radical sign represent nonnegative numbers. We will address even roots of negative numbers later in this book.

3. Simplify.

   a. $\sqrt{81a^4b^8}$

   b. $\sqrt[3]{8x^3y^9}$

   c. $\sqrt[4]{81a^4b^8}$

EXAMPLE 3 Assume all variables represent nonnegative numbers, and simplify each expression as much as possible.

   a. $\sqrt{25a^4b^6} = 5a^2b^3$ because $(5a^2b^3)^2 = 25a^4b^6$.

   b. $\sqrt[3]{x^6y^{12}} = x^2y^4$ because $(x^2y^4)^3 = x^6y^{12}$.

   c. $\sqrt[4]{81r^8s^{20}} = 3r^2s^5$ because $(3r^2s^5)^4 = 81r^8s^{20}$.

B Rational Numbers as Exponents

We will now develop a second kind of notation involving exponents that will allow us to designate square roots, cube roots, and so on in another way.

Consider the equation $x = 8^{1/3}$. The exponent $\frac{1}{3}$ is called a rational exponent. Although we have not encountered fractions as exponents before, let's assume that all the properties of exponents hold in this case. Cubing both sides of the equation, we have

$$x^3 = (8^{1/3})^3$$

$$x^3 = 8^{(1/3)(3)}$$

$$x^3 = 8^1$$

$$x^3 = 8$$

The last line tells us that x is the number whose cube is 8. It must be true, then, that x is the cube root of 8, $x = \sqrt[3]{8}$. Because we started with $x = 8^{1/3}$, it follows that

$$8^{1/3} = \sqrt[3]{8}$$

It seems reasonable, then, to define rational exponents as indicating roots. Here is the formal property:

> **PROPERTY** Rational Exponent Property
>
> If x is a real number and n is a positive integer greater than 1, then
>
> $$x^{1/n} = \sqrt[n]{x} \qquad (x \geq 0 \text{ when } n \text{ is even})$$
>
> *In words:* The quantity $x^{1/n}$ is the nth root of x.

With this property, we have a way of representing roots with exponents. Here are some examples:

EXAMPLE 4 Write each expression as a root and then simplify, if possible.

a. $8^{1/3} = \sqrt[3]{8} = 2$

b. $36^{1/2} = \sqrt{36} = 6$

c. $-25^{1/2} = -\sqrt{25} = -5$

d. $(-25)^{1/2} = \sqrt{-25}$, which is not a real number

e. $\left(\dfrac{4}{9}\right)^{1/2} = \sqrt{\dfrac{4}{9}} = \dfrac{2}{3}$

The properties of exponents developed in Chapter 0 were applied to integer exponents only. We will now extend these properties to include rational exponents also. We do so without proof.

> **PROPERTY** Properties of Exponents
>
> If a and b are real numbers and r and s are rational numbers, and a and b are nonnegative whenever r and s indicate even roots, then
>
> | | | |
> |---|---|---|
> | **1.** $a^r \cdot a^s = a^{r+s}$ | | Product Property for Exponents |
> | **2.** $(a^r)^s = a^{rs}$ | | Power Property for Exponents |
> | **3.** $(ab)^r = a^r b^r$ | | Distributive Property for Exponents |
> | **4.** $a^{-r} = \dfrac{1}{a^r}$ | $(a \neq 0)$ | Negative Exponent Property |
> | **5.** $\left(\dfrac{a}{b}\right)^r = \dfrac{a^r}{b^r}$ | $(b \neq 0)$ | Expanded Distributive Property for Exponents |
> | **6.** $\dfrac{a^r}{a^s} = a^{r-s}$ | $(a \neq 0)$ | Quotient Property for Exponents |
> | **7.** $a^{1/r} = \sqrt[r]{a}$ | $(a \neq 0)$ | Rational Exponent Property |

Sometimes rational exponents can simplify our work with radicals. Here are Examples 3b and 3c again, but this time we will work them using rational exponents.

EXAMPLE 5 Write each radical with a rational exponent, then simplify.

a.
$$\sqrt[3]{x^6 y^{12}} = (x^6 y^{12})^{1/3}$$
$$= (x^6)^{1/3}(y^{12})^{1/3}$$
$$= x^2 y^4$$

b.
$$\sqrt[4]{81 r^8 s^{20}} = (81 r^8 s^{20})^{1/4}$$
$$= 81^{1/4}(r^8)^{1/4}(s^{20})^{1/4}$$
$$= 3 r^2 s^5$$

---

4. Simplify.

a. $9^{1/2}$

b. $27^{1/3}$

c. $-49^{1/2}$

d. $(-49)^{1/2}$

e. $\left(\dfrac{16}{25}\right)^{1/2}$

Note In part d of Example 4, we use parentheses to notate that we are asking for the square root of -25. If the parentheses are absent, as is part c, we consider the negative sign as a -1 multiplier and work the exponent first.

$$-25^{1/2} = -(25^{1/2})$$
$$= -\sqrt{25}$$
$$= -5$$

5. Simplify.

a. $\sqrt[3]{8 x^3 y^9}$

b. $\sqrt[4]{81 a^4 b^8}$

Answers

4. a. 3 **b.** 3 **c.** -7
d. Not a real number **e.** $\dfrac{4}{5}$
5. a. $2xy^3$ **b.** $3ab^2$

So far, the numerators of all the rational exponents we have encountered have been 1. The next theorem extends the work we can do with rational exponents to rational exponents with numerators other than 1.

We can extend our properties of exponents with the following theorem.

> **THEOREM**   Rational Exponent Theorem
>
> If a is a nonnegative real number, m is an integer, and n is a positive integer, then
>
> $$a^{m/n} = (a^{1/n})^m = (a^m)^{1/n}$$

Proof   We can prove the rational exponent theorem using the properties of exponents. Because $m/n = m(1/n)$, we have

$$a^{m/n} = a^{m(1/n)} \qquad\qquad a^{m/n} = a^{(1/n)(m)}$$
$$= (a^m)^{1/n} \qquad\qquad\quad = (a^{1/n})^m$$

Here are some examples that illustrate how we use this theorem:

EXAMPLE 6   Simplify as much as possible.

a.

$8^{2/3}$	$= (8^{1/3})^2$	Rational exponent theorem
	$= 2^2$	Rational exponent property
	$= 4$	The square of 2 is 4.

b.

$25^{3/2}$	$= (25^{1/2})^3$	Rational exponent theorem
	$= 5^3$	Rational exponent property
	$= 125$	The cube of 5 is 125.

c.

$9^{-3/2}$	$= (9^{1/2})^{-3}$	Rational exponent theorem
	$= 3^{-3}$	Rational exponent property
	$= \dfrac{1}{3^3}$	Negative exponent property
	$= \dfrac{1}{27}$	The cube of 3 is 27.

d.

$\left(\dfrac{27}{8}\right)^{-4/3}$	$= \left[\left(\dfrac{27}{8}\right)^{1/3}\right]^{-4}$	Rational exponent theorem
	$= \left(\dfrac{3}{2}\right)^{-4}$	Rational exponent property
	$= \left(\dfrac{2}{3}\right)^4$	Negative exponent property
	$= \dfrac{16}{81}$	The fourth power of $\frac{2}{3}$ is $\frac{16}{81}$.

In each part of Example 6, you may have noticed that we first worked the $a^{1/n}$ part of each expression before raising to the power m. This approach typically results in easier calculations than raising a to the power m and then working the rational exponent. In the end, either approach will yield the same correct answer.

The following examples show the application of the properties of exponents to rational exponents.

Note   It is important to note that the rational exponent theorem also extends to negative values for a but we have chosen to focus on nonnegative real numbers for this section.

6. Simplify.

a. $9^{3/2}$

b. $16^{3/4}$

c. $8^{-2/3}$

d. $\left(\dfrac{16}{81}\right)^{-3/4}$

Note   On a scientific calculator, Example 6a would look like this:

$8\ \boxed{y^x}\ \boxed{(}\ 2\ \boxed{\div}\ 3\ \boxed{)}\ \boxed{=}$

Answers

6. a. 27   b. 8   c. $\frac{1}{4}$ **d.** $\frac{27}{8}$

EXAMPLE 7 Assume all variables represent positive quantities, and simplify as much as possible.

a. $x^{1/3} \cdot x^{5/6} = x^{1/3 + 5/6}$ Product property for exponents

$\quad = x^{2/6 + 5/6}$ LCD is 6.

$\quad = x^{7/6}$ Add fractions.

b. $(y^{2/3})^{3/4} = y^{(2/3)(3/4)}$ Power property or exponents

$\quad = y^{1/2}$ Multiply fractions: $\frac{2}{3} \cdot \frac{3}{4} = \frac{6}{12} = \frac{1}{2}$.

c. $\dfrac{z^{1/3}}{z^{1/4}} = z^{1/3 - 1/4}$ Quotient property for exponents

$\quad = z^{4/12 - 3/12}$ LCD is 12.

$\quad = z^{1/12}$ Subtract fractions.

d. $\left(\dfrac{a^{-1/3}}{b^{1/2}} \right)^6 = \dfrac{(a^{-1/3})^6}{(b^{1/2})^6}$ Expanded distributive property for exponents

$\quad = \dfrac{a^{-2}}{b^3}$ Power property for exponents

$\quad = \dfrac{1}{a^2 b^3}$ Negative exponent property

e. $\dfrac{(x^{-3}y^{1/2})^4}{x^{10}y^{3/2}} = \dfrac{(x^{-3})^4(y^{1/2})^4}{x^{10}y^{3/2}}$ Distributive property for exponents

$\quad = \dfrac{x^{-12}y^2}{x^{10}y^{3/2}}$ Power property for exponents

$\quad = x^{-22}y^{1/2}$ Quotient property for exponents

$\quad = \dfrac{y^{1/2}}{x^{22}}$ Negative exponent property

7. Simplify.

a. $x^{1/2} \cdot x^{1/4}$

b. $(y^{3/5})^{5/6}$

c. $\dfrac{z^{3/4}}{z^{2/3}}$

d. $\left(\dfrac{a^{3/4}}{b^{-1/2}} \right)^4$

e. $\dfrac{(x^{1/3}y^{-3})^6}{x^4 y^{10}}$

FACTS FROM GEOMETRY   More about the Pythagorean Theorem

Now that we have had some experience working with square roots, we can rewrite the Pythagorean theorem using a square root. If triangle ABC is a right triangle with $C = 90°$, then the length of the longest side is the **positive square root** of the sum of the squares of the other two sides (see Figure 4).

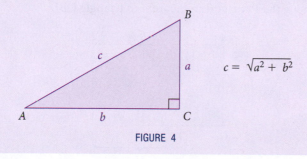

$$c = \sqrt{a^2 + b^2}$$

FIGURE 4

Constructing a Golden Rectangle From a Square of Side 2

Recall the golden rectangle. Its origins can be traced back over 2,000 years to the Greek civilization that produced Pythagoras, Socrates, Plato, Aristotle, and Euclid. The most important mathematical work to come from that Greek civilization was Euclid's *Elements*, an elegantly written summary of all that was known about geometry at that time in history. Euclid's *Elements*, according to Howard Eves, an authority on the history of mathematics, exercised a greater influence on scientific thinking than any other work. Here is how we construct a golden rectangle from a square of side 2, using the same method that Euclid used in his *Elements*.

Answers

7. a. $x^{3/4}$ **b.** $y^{1/2}$ **c.** $z^{1/12}$

d. $a^3 b^2$ **e.** $\dfrac{1}{x^2 y^{28}}$

Step 1: Draw a square with a side of length 2. Connect the midpoint of side *CD* to corner *B*. (Note that we have labeled the midpoint of segment *CD* with the letter *O*.)

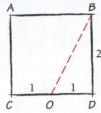

Step 2: Drop the diagonal from step 1 down so it aligns with side *CD*.

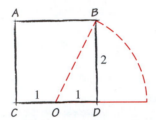

Step 3: Form rectangle *ACEF*. This is a golden rectangle.

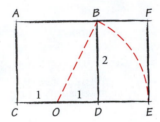

All golden rectangles are constructed from squares. Every golden rectangle, no matter how large or small it is, will have the same shape. To associate a number with the shape of the golden rectangle, we use the ratio of its length to its width. This ratio is called the *golden ratio*. To calculate the golden ratio, we must first find the length of the diagonal we used to construct the golden rectangle. Figure 5 shows the golden rectangle we constructed from a square of side 2. The length of the diagonal *OB* is found by applying the Pythagorean theorem to triangle *OBD*.

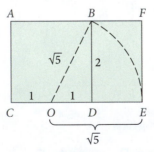

FIGURE 5

The length of segment *OE* is equal to the length of diagonal *OB*; both are $\sqrt{5}$. Because the distance from *C* to *O* is 1, the length *CE* of the golden rectangle is $1 + \sqrt{5}$. Now we can find the golden ratio:

$$\text{Golden ratio} = \frac{\text{Length}}{\text{Width}} = \frac{CE}{EF} = \frac{1 + \sqrt{5}}{2}$$

USING TECHNOLOGY Graphing Calculators—A Word of Caution

Some graphing calculators give surprising results when evaluating expressions such as $(-8)^{2/3}$. As you know from reading this section, the expression $(-8)^{2/3}$ simplifies to 4, either by taking the cube root first and then squaring the result, or by squaring the base first and then taking the cube root of the result. Here are three different ways to evaluate this expression on your calculator:

1. $(-8)^\wedge(2/3)$ To evaluate $(-8)^{2/3}$
2. $((-8)^\wedge 2)^\wedge(1/3)$ To evaluate $((-8)^2)^{1/3}$
3. $((-8)^\wedge(1/3))^\wedge 2$ To evaluate $((-8)^{1/3})^2$

Note any differences in the results.
   Next, graph each of the following functions, one at a time.

1. $Y_1 = X^{2/3}$ **2.** $Y_2 = (X^2)^{1/3}$ **3.** $Y_3 = (X^{1/3})^2$

The correct graph is shown in Figure 6. Note which of your graphs match the correct graph.

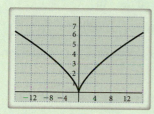

FIGURE 6

   Different calculators evaluate exponential expressions in different ways. You should use the method (or methods) that gave you the correct graph.

GETTING READY FOR CLASS

After reading through the preceding section, respond in your own words and in complete sentences.

A. Define positive square root.

B. Explain why a square root of -4 is not a real number.

C. For the expression $a^{m/n}$, explain the significance of the numerator m and the significance of the denominator n in the exponent.

D. Briefly explain the concept behind the golden rectangle.

Vocabulary Review

Choose the correct words to fill in the blanks below.

ratio	cube	positive	index	root
rational	radical	radicand	exponent	

1. If x is a nonnegative real number, then the expression $\sqrt{x}$ is called the _____ square root of x where x is greater than zero.

2. The _____ root of x, $\sqrt[3]{x}$ is such that $(\sqrt[3]{x})^3 = x$.

3. An expression like $\sqrt[3]{64}$ that involves a root is called a _____ expression.

4. In the expression $\sqrt[3]{64}$, the ^{3} is called the index, the $\sqrt{}$ is called the radical sign, and 64 is called the _____.

5. If no _____ of a radical is written, it is assumed to be 2.

6. The quantity x with a fractional exponent $\frac{1}{n}$, such that $x^{1/n}$, is considered the nth _____ of x.

7. The properties of exponents can be extended from integer exponents to also include _____ exponents.

8. The rational _____ theorem states that if a is a nonnegative real number, m is an integer, and n is a positive integer, then $a^{m/n} = (a^{1/n})^m = (a^m)^{1/n}$.

9. The ratio of length to width, known as the golden _____, is used to associate a number with the shape of a golden rectangle.

Problems

A Find each of the following roots, if possible.

1. $\sqrt{144}$

2. $-\sqrt{144}$

3. $\sqrt{-144}$

4. $\sqrt{-49}$

5. $-\sqrt{49}$

6. $\sqrt{49}$

7. $\sqrt[3]{-27}$

8. $-\sqrt[3]{27}$

9. $\sqrt[4]{16}$

10. $-\sqrt[4]{16}$

11. $\sqrt[4]{-16}$

12. $-\sqrt[4]{-16}$

SCAN TO ACCESS

13. $\sqrt{0.04}$

14. $\sqrt{0.81}$

15. $\sqrt[3]{0.008}$

16. $\sqrt[3]{0.125}$

Assume all variables represent nonnegative numbers.

18. $\sqrt{49a^{10}}$

19. $\sqrt[3]{27a^{12}}$

20. $\sqrt[3]{8a^{15}}$

22. $\sqrt[3]{x^6 y^3}$

23. $\sqrt[5]{32x^{10}y^5}$

24. $\sqrt[5]{32x^5 y^{10}}$

25. $\sqrt[4]{16a^{12}b^{20}}$

26. $\sqrt[4]{81a^{24}b^8}$

27. $\sqrt[3]{-27x^9 y12}$

28. $\sqrt[3]{-8x^{15}y^{21}}$

B Convert the following radicals to expressions with rational exponents and simplify where possible. Assume all variables represent positive numbers.

29. $\sqrt[5]{x^4}$

30. $\sqrt[8]{y^4}$

31. $\sqrt[3]{n^2}$

32. $\sqrt{a}$

33. $\sqrt[6]{a^3}$

34. $\sqrt[10]{y^2}$

35. $\sqrt[8]{m^5}$

36. $\sqrt[12]{b^3}$

Use the rational exponent property to write each of the following with the appropriate root. Then simplify.

37. $36^{1/2}$

38. $49^{1/2}$

39. $-9^{1/2}$

40. $-16^{1/2}$

41. $8^{1/3}$

42. $-8^{1/3}$

43. $(-8)^{1/3}$

44. $-27^{1/3}$

45. $32^{1/5}$

46. $81^{1/4}$

47. $\left(\dfrac{81}{25}\right)^{1/2}$

48. $\left(\dfrac{9}{16}\right)^{1/2}$

49. $\left(\dfrac{64}{125}\right)^{1/3}$

50. $\left(\dfrac{8}{27}\right)^{1/3}$

51. $\left(\dfrac{16}{81}\right)^{1/4}$

52. $\left(-\dfrac{27}{8}\right)^{1/3}$

Use the rational exponent theorem to simplify each of the following as much as possible.

53. $27^{2/3}$

54. $8^{4/3}$

55. $25^{3/2}$

56. $9^{3/2}$

57. $16^{3/4}$

58. $81^{3/4}$

59. $(-27)^{2/3}$

60. $(-32)^{3/5}$

Simplify each expression. Remember, negative exponents give reciprocals.

61. $27^{-1/3}$

62. $9^{-1/2}$

63. $81^{-3/4}$

64. $4^{-3/2}$

65. $\left(\dfrac{25}{36}\right)^{-1/2}$

66. $\left(\dfrac{16}{49}\right)^{-1/2}$

67. $\left(\dfrac{81}{16}\right)^{-3/4}$

68. $\left(\dfrac{27}{8}\right)^{-2/3}$

69. $16^{1/2} + 27^{1/3}$

70. $25^{1/2} + 100^{1/2}$

71. $8^{-2/3} + 4^{-1/2}$

72. $49^{-1/2} + 25^{-1/2}$

Use the properties of exponents to simplify each of the following as much as possible. Assume all bases are positive.

73. $x^{3/5} \cdot x^{1/5}$

74. $x^{3/4} \cdot x^{5/4}$

75. $(a^{3/4})^{4/3}$

76. $(a^{2/3})^{3/4}$

77. $\dfrac{x^{1/5}}{x^{3/5}}$

78. $\dfrac{x^{2/7}}{x^{5/7}}$

79. $\dfrac{x^{5/6}}{x^{2/3}}$

80. $\dfrac{x^{7/8}}{x^{8/7}}$

81. $(a^{2/3}b^{1/2}c^{1/4})^{1/3}$

82. $(a^{5/8}b^{1/2}c^{3/4})^{2/5}$

83. $(x^{3/5}y^{5/6}z^{1/3})^{3/5}$

84. $(x^{3/4}y^{1/8}z^{5/6})^{4/5}$

85. $\dfrac{a^{3/4}b^2}{a^{7/8}b^{1/4}}$

86. $\dfrac{a^{1/3}b^4}{a^{3/5}b^{1/3}}$

87. $\dfrac{(y^{2/3})^{3/4}}{(y^{1/3})^{3/5}}$

88. $\dfrac{(y^{5/4})^{2/5}}{(y^{1/4})^{4/3}}$

89. $\left(\dfrac{x^{-1/2}}{y^{1/4}}\right)^4$

90. $\left(\dfrac{a^{-3/4}}{b^{2/3}}\right)^{12}$

91. $\left(\dfrac{a^{-1/4}}{b^{1/2}}\right)^8$

92. $\left(\dfrac{a^{-1/5}}{b^{1/3}}\right)^{15}$

93. $\dfrac{(r^{-2}s^{1/3})^6}{r^8s^{3/2}}$

94. $\dfrac{(r^{-5}s^{1/2})^4}{r^{12}s^{5/2}}$

95. $\dfrac{(25a^6b^4)^{1/2}}{(8a^{-9}b^3)^{-1/3}}$

96. $\dfrac{(27a^3b^6)^{1/3}}{(81a^8b^{-4})^{1/4}}$

97. Show that the expression $(a^{1/2} + b^{1/2})^2$ is not equal to $a + b$ by replacing a with 9 and b with 4 in both expressions and then simplifying each.

98. Show that the statement $(a^2 + b^2)^{1/2} = a + b$ is not, in general, true by replacing a with 3 and b with 4 and then simplifying both sides.

99. You may have noticed, if you have been using a calculator to find roots, that you can find the fourth root of a number by pressing the square root button twice. Written in symbols, this fact looks like this:

$$\sqrt{\sqrt{a}} = \sqrt[4]{a} \qquad (a \geq 0)$$

Show that this statement is true by rewriting each side with exponents instead of radical notation and then simplifying the left side.

100. Show that the statement is true by rewriting each side with exponents instead of radical notation and then simplifying the left side.

$$\sqrt[3]{\sqrt{a}} = \sqrt[6]{a} \qquad (a \geq 0)$$

Applying the Concepts

101. Maximum Speed The maximum speed v that an automobile can travel around a curve of radius r without skidding is given by the equation

$$v = \left(\frac{5r}{2} \right)^{1/2}$$

where v is in miles per hour and r is measured in feet. What is the maximum speed a car can travel around a curve with a radius of 250 feet without skidding?

102. Relativity The equation

$$L = \left(1 - \frac{v^2}{c^2} \right)^{1/2}$$

gives the relativistic length of a 1-foot ruler traveling with velocity v. Find L if

$$\frac{v}{c} = \frac{3}{5}$$

103. Golden Ratio The golden ratio is the ratio of the length to the width in any golden rectangle. The exact value of this number is $\frac{1 + \sqrt{5}}{2}$. Use a calculator to find a decimal approximation of this number and round it to the nearest thousandth.

104. Golden Ratio The reciprocal of the golden ratio is $\frac{2}{1 + \sqrt{5}}$. Find a decimal approximation to this number that is accurate of the nearest thousandth.

105. Chemistry Figure 7 shows part of a model of a magnesium oxide (MgO) crystal. Each corner of the square is at the center of one oxygen ion (O^{2-}), and the center of the middle ion is at the center of the square. The radius for each oxygen ion is 60 picometers (pm), and the radius for each magnesium ion (Mg^{2+}) is 150 picometers.

a. Find the length of the side of the square. Write your answer in picometers.

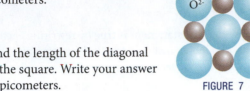

b. Find the length of the diagonal of the square. Write your answer in picometers.

FIGURE 7

c. If 1 meter is 10^{12} picometers, give the length of the diagonal of the square in meters.

106. Geometry The length of each side of the cube shown in Figure 8 is 1 inch.

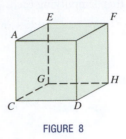

FIGURE 8

a. Find the length of the diagonal CH.

b. Find the length of the diagonal CF.

Below is a list of some well known pyramids in the world.

Pyramid Name	Location Culture	b	h	Slant height (ft)	Volume (ft³)
Temple of Kukulkan	Mexico/Maya	181 ft	98 ft		
Temple IV at Tecal	Guatemala/Maya	118 ft	187 ft		
Pyramid of the Sun	Mexico/Aztec	733 ft	246 ft		
Great Pyramid of Khufu	Gizal/Egypt	755 ft	455 ft		
Luxor Hotel	Las Vegas/Modern	600 ft	350 ft		
Transamerica Pyramid	San Francisco/Modern	175 ft	850 ft		

Source: skyscrapercity.com

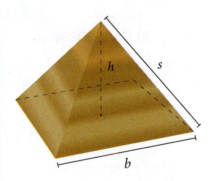

107. The slant height s of a pyramid with a square base with side length b and height h can be found using the formula

$$s = \sqrt{\frac{b^2}{4} + h^2}.$$

Use this formula to find the slant height of each pyramid in the table.

108. The volume v of a pyramid with a square base with side length b and height h is given by the formula

$$v = \frac{1}{3}b^2h.$$

Find the volume of each pyramid. (Round to the nearest thousand cubic feet.)

109. If you were to climb to the top of each pyramid along its slant height, which pyramid is the largest?

110. Which pyramid has the largest volume?

Getting Ready for the Next Section

Simplify. Assume all variable are positive real numbers.

111. $\sqrt{25}$

112. $\sqrt{4}$

113. $\sqrt{6^2}$

114. $\sqrt{3^2}$

115. $\sqrt{16x^4y^2}$

116. $\sqrt{4x^6y^8}$

117. $\sqrt{(5y)^2}$

118. $\sqrt{(8x^3)^2}$

119. $\sqrt[3]{27}$

120. $\sqrt[3]{-8}$

121. $\sqrt[3]{2^3}$

122. $\sqrt[3]{(-5)^3}$

123. $\sqrt[3]{8a^3b^3}$

124. $\sqrt[3]{64a^6b^3}$

125. $\sqrt[4]{625x^8y^4}$

126. $\sqrt[5]{(-x)^5}$

Fill in the blank.

127. $50 = \underline{\quad} \cdot 2$

128. $12 = \underline{\quad} \cdot 3$

129. $48x^4y^3 = \underline{\quad} \cdot y$

130. $40a^5b^4 = \underline{\quad} \cdot 5a^2b$

131. $12x^7y^6 = \underline{\quad} \cdot 3x$

132. $54a^6b^2c^4 = \underline{\quad} \cdot 2b^2c$

133. $10x^2y^5 = \underline{\quad} \cdot 5xy$

134. $-40a^4b^7 = \underline{\quad} \cdot 5a^3b^2$

Find the Mistake

Each sentence below contains a mistake. Circle the mistake and write the correct word or expression on the line provided.

1. $\sqrt{x}$ is the positive square number we add to get x. _____

2. If we raise the fifth root of x to the fifth power, we get x^2. _____

3. For the radical expression $\sqrt[4]{16}$, the 4 is the index and the 16 is the radical. _____

4. Using the rational exponent theorem, we can say that $16^{-3/2} = (16^{1/2})^{-3} = 64$. _____

Navigation Skills: Prepare, Study, Achieve

Your academic self-image is how you see yourself as a student and the level of success you see yourself achieving. Do you believe you are capable of learning any subject and succeeding in any class you take? If you believe in yourself and work hard by applying the appropriate study methods, you will succeed. If you have a poor outlook for a class, most likely your performance in that class will match that outlook. Self-doubt or questioning the purpose of this course will negatively affect your focus.

Furthermore, an inner dialogue of negative statements, such as "I'll never be able to learn this material" or "I'm never going to use this stuff," distract you from achieving success. Consider replacing those thoughts with three positive statements you can say when you notice your mind participating in a negative inner dialogue. Make a commitment to change your attitude for the better. Begin by thinking positively, having confidence in your abilities, and utilizing your resources if you are having difficulty. Asking for help is a sign of a successful student.

product property for radicals

quotient property for radicals

simplified form

rationalize the denominator

perfect square

absolute value property for radicals

spiral of roots

recursive

6.2 Simplifying Radicals

Nikola Tesla was an inventor and engineer in the late 19th and early 20th centuries. Tesla acquired the first patent for a remote-controlled device. He demonstrated this device, a small boat with an iron hull, on an indoor pond. Onlookers were in awe of this new robotic contraption, having never seen anything like it before. Tesla prompted the audience to ask the boat a question, such as "What is the cube root of 64?" Then, as the boat continued to move over the surface of the pond, Tesla used the remote control to make the lights on the boat flash four times. In this section, we will continue our work with radicals, including cube roots.

Earlier in this chapter, we showed how the Pythagorean theorem can be used to construct a golden rectangle. In a similar manner, the Pythagorean theorem can be used to construct the attractive spiral shown here.

This spiral is called the Spiral of Roots because each of the diagonals is the positive square root of one of the positive integers. At the end of this section, we will use the Pythagorean theorem and some of the material in this section to construct this spiral.

In this section, we will use radical notation instead of rational exponents. We will begin by stating the first two properties of radicals. Following this, we will give a definition for simplified form for radical expressions. The examples in this section show how we use the properties of radicals to write radical expressions in simplified form.

The Spiral of Roots

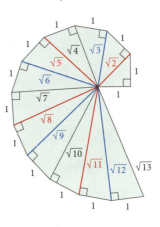

A Simplified Form

Here are the first two properties of radicals. For these two properties, we will assume a and b are nonnegative real numbers whenever n is an even number.

Note There is not a property for radicals that says the nth root of a sum is the sum of the nth roots. That is,

$$\sqrt[n]{a + b} \neq \sqrt[n]{a} + \sqrt[n]{b}$$

PROPERTY Product Property for Radicals

$$\sqrt[n]{ab} = \sqrt[n]{a} \cdot \sqrt[n]{b}$$

In words: The nth root of a product is the product of the nth roots.

Proof of the Product Property for Radicals

$$\sqrt[n]{ab} = (ab)^{1/n} \qquad \text{Definition of fractional exponents}$$
$$= a^{1/n}b^{1/n} \qquad \text{Exponents distribute over products.}$$
$$= \sqrt[n]{a} \cdot \sqrt[n]{b} \qquad \text{Definition of fractional exponents}$$

> **PROPERTY** Quotient Property for Radicals
>
> $$\sqrt[n]{\frac{a}{b}} = \frac{\sqrt[n]{a}}{\sqrt[n]{b}} = \qquad (b \neq 0)$$
>
> *In words:* The *n*th root of a quotient is the quotient of the *n*th roots.

The proof of the quotient property for radicals is similar to the proof of the product property for radicals.

These two properties of radicals allow us to change the form of and simplify radical expressions without changing their value.

> **RULE** Simplified Form for Radical Expressions
>
> A radical expression is in *simplified form* if
>
> 1. None of the factors of the radicand (the quantity under the radical sign) can be written as powers greater than or equal to the index—that is, no perfect squares can be factors of the quantity under a square root sign, no perfect cubes can be factors of what is under a cube root sign, and so forth.
>
> 2. There are no fractions under the radical sign.
>
> 3. There are no radicals in the denominator.

Satisfying the first condition for simplified form actually amounts to taking as much out from under the radical sign as possible. The following examples illustrate the first condition for simplified form.

EXAMPLE 1 Write $\sqrt{50}$ in simplified form.

Solution The largest perfect square that divides 50 is 25. We write 50 as $25 \cdot 2$ and apply the product property for radicals.

$$\sqrt{50} = \sqrt{25 \cdot 2} \qquad \textcolor{green}{50 = 25 \cdot 2}$$

$$= \sqrt{25}\sqrt{2} \qquad \textcolor{green}{\text{Product property for radicals}}$$

$$= 5\sqrt{2} \qquad \textcolor{green}{\sqrt{25} = 5}$$

We have taken as much as possible out from under the radical sign — in this case, factoring 25 from 50 and then writing $\sqrt{25}$ as 5.

EXAMPLE 2 Write $\sqrt{48x^4y^3}$, where $x, y \geq 0$ in simplified form.

Solution The largest perfect square that is a factor of the radicand is $16x^4y^2$. Applying the product property for radicals again, we have

$$\sqrt{48x^4y^3} = \sqrt{16x^4y^2 \cdot 3y}$$

$$= \sqrt{16x^4y^2}\sqrt{3y}$$

$$= 4x^2y\sqrt{3y}$$

Practice Problems

1. Write $\sqrt{18}$ in simplified form.

2. Write $\sqrt{50x^2y^3}$ in simplified form. Assume $x, y \geq 0$.

Answers

1. $3\sqrt{2}$

2. $5xy\sqrt{2y}$

3. Write $\sqrt[3]{54a^4b^3}$ in simplified form.

EXAMPLE 3   Write $\sqrt[3]{40a^5b^4}$ in simplified form.

Solution   We now want to factor the largest perfect cube from the radicand. We write $40a^5b^4$ as $8a^3b^3 \cdot 5a^2b$ and proceed as we did in Examples 1 and 2.

$$\sqrt[3]{40a^5b^4} = \sqrt[3]{8a^3b^3 \cdot 5a^2b}$$

$$= \sqrt[3]{8a^3b^3}\sqrt[3]{5a^2b}$$

$$= 2ab\sqrt[3]{5a^2b}$$

Here are some further examples concerning the first condition for simplified form.

4. Write each expression in simplified form.

  a. $\sqrt{75x^5y^8}$

  b. $\sqrt[4]{48a^8b^5c^4}$

EXAMPLE 4   Write each expression in simplified form:

a. $\sqrt{12x^7y^6} = \sqrt{4x^6y^6 \cdot 3x}$

$$= \sqrt{4x^6y^6}\sqrt{3x}$$

$$= 2x^3y^3\sqrt{3x}$$

b. $\sqrt[3]{54a^6b^2c^4} = \sqrt[3]{27a^6c^3 \cdot 2b^2c}$

$$= \sqrt[3]{27a^6c^3}\sqrt[3]{2b^2c}$$

$$= 3a^2c\sqrt[3]{2b^2c}$$

B Rationalizing the Denominator

The quotient property of radicals is used to simplify a radical that contains a fraction.

5. Simplify $\sqrt{\dfrac{5}{9}}$.

EXAMPLE 5   Simplify $\sqrt{\dfrac{3}{4}}$.

Solution   Applying quotient property for radicals, we have

$$\sqrt{\frac{3}{4}} = \frac{\sqrt{3}}{\sqrt{4}} \qquad \textcolor{green}{\text{Quotient property for radicals}}$$

$$= \frac{\sqrt{3}}{2} \qquad \textcolor{green}{\sqrt{4} = 2}$$

The last expression is in simplified form because it satisfies all three conditions for simplified form.

6. Write $\sqrt{\dfrac{2}{3}}$ in simplified form.

EXAMPLE 6   Write $\sqrt{\dfrac{5}{6}}$ in simplified form.

Solution   Proceeding as in Example 5, we have

$$\sqrt{\frac{5}{6}} = \frac{\sqrt{5}}{\sqrt{6}}$$

The resulting expression satisfies the second condition for simplified form because neither radical contains a fraction. It does, however, violate condition 3 because it has a radical in the denominator. Getting rid of the radical in the denominator is called *rationalizing the denominator* and is accomplished, in this case, by multiplying the numerator and denominator by $\sqrt{6}$.

$$\frac{\sqrt{5}}{\sqrt{6}} = \frac{\sqrt{5}}{\sqrt{6}} \cdot \frac{\sqrt{6}}{\sqrt{6}}$$

$$= \frac{\sqrt{30}}{(\sqrt{6})^2}$$

$$= \frac{\sqrt{30}}{6}$$

Answers

3. $3ab\sqrt[3]{2a}$

4. a. $5x^2y^4\sqrt{3x}$ **b.** $2a^2bc\sqrt[4]{3b}$

5. $\dfrac{\sqrt{5}}{3}$

6. $\dfrac{\sqrt{6}}{3}$

EXAMPLE 7 Rationalize the denominator.

a. $\dfrac{4}{\sqrt{3}} = \dfrac{4}{\sqrt{3}} \cdot \dfrac{\sqrt{3}}{\sqrt{3}}$

$\quad = \dfrac{4\sqrt{3}}{(\sqrt{3})^2}$

$\quad = \dfrac{4\sqrt{3}}{3}$

b. $\dfrac{2\sqrt{3x}}{\sqrt{5y}} = \dfrac{2\sqrt{3x}}{\sqrt{5y}} \cdot \dfrac{\sqrt{5y}}{\sqrt{5y}}$

$\quad = \dfrac{2\sqrt{15xy}}{(\sqrt{5y})^2}$

$\quad = \dfrac{2\sqrt{15xy}}{5y}$

7. Rationalize the denominator.

  a. $\dfrac{5}{\sqrt{2}}$

  b. $\dfrac{3\sqrt{5x}}{\sqrt{2y}}$

When the denominator involves a cube root, we must multiply by a radical that will produce a perfect cube under the cube root sign in the denominator, as Example 8 illustrates.

EXAMPLE 8 Rationalize the denominator in $\dfrac{7}{\sqrt[3]{4}}$.

Solution Because $4 = 2^2$, we can multiply both numerator and denominator by $\sqrt[3]{2}$ and obtain $\sqrt[3]{2^3}$ in the denominator.

$$\dfrac{7}{\sqrt[3]{4}} = \dfrac{7}{\sqrt[3]{2^2}}$$

$$\phantom{\dfrac{7}{\sqrt[3]{4}}} = \dfrac{7}{\sqrt[3]{2^2}} \cdot \dfrac{\sqrt[3]{2}}{\sqrt[3]{2}}$$

$$\phantom{\dfrac{7}{\sqrt[3]{4}}} = \dfrac{7\sqrt[3]{2}}{\sqrt[3]{2^3}}$$

$$\phantom{\dfrac{7}{\sqrt[3]{4}}} = \dfrac{7\sqrt[3]{2}}{2}$$

8. Rationalize the denominator.

  $\dfrac{5}{\sqrt[3]{9}}$

EXAMPLE 9 Simplify $\sqrt{\dfrac{12x^5 y^3}{5z}}$.

Solution We use the quotient property for radicals to write the numerator and denominator as two separate radicals.

$$\sqrt{\dfrac{12x^5 y^3}{5z}} = \dfrac{\sqrt{12x^5 y^3}}{\sqrt{5z}}$$

Simplifying the numerator, we have

$$\dfrac{\sqrt{12x^5 y^3}}{\sqrt{5z}} = \dfrac{\sqrt{4x^4 y^2}\sqrt{3xy}}{\sqrt{5z}}$$

$$\phantom{\dfrac{\sqrt{12x^5 y^3}}{\sqrt{5z}}} = \dfrac{2x^2 y\sqrt{3xy}}{\sqrt{5z}}$$

To rationalize the denominator, we multiply the numerator and denominator by $\sqrt{5z}$.

$$\dfrac{2x^2 y\sqrt{3xy}}{\sqrt{5z}} \cdot \dfrac{\sqrt{5z}}{\sqrt{5z}} = \dfrac{2x^2 y\sqrt{15xyz}}{(\sqrt{5z})^2}$$

$$\phantom{\dfrac{2x^2 y\sqrt{3xy}}{\sqrt{5z}} \cdot \dfrac{\sqrt{5z}}{\sqrt{5z}}} = \dfrac{2x^2 y\sqrt{15xyz}}{5z}$$

9. Simplify $\sqrt{\dfrac{48x^3 y^4}{7z}}$.

Answers

7. **a.** $\dfrac{5\sqrt{2}}{2}$ **b.** $\dfrac{3\sqrt{10xy}}{2y}$

8. $\dfrac{5\sqrt[3]{3}}{3}$

9. $\dfrac{4xy^2\sqrt{21xz}}{7z}$

Square Root of a Perfect Square

So far in this chapter, we have assumed that all our variables are nonnegative when they appear under a square root symbol. There are times, however, when this is not the case.

Consider the following two statements:

$$\sqrt{3^2} = \sqrt{9} = 3 \qquad \text{and} \qquad \sqrt{(-3)^2} = \sqrt{9} = 3$$

Whether we operate on 3 or -3, the result is the same: both expressions simplify to 3. The other operation we have worked with in the past that produces the same result is absolute value. That is,

$$|3| = 3 \qquad \text{and} \qquad |-3| = 3$$

This leads us to the next property of radicals.

> **PROPERTY** Absolute Value Property for Radicals
>
> If a is a real number, then $\sqrt{a^2} = |a|$.

The result of this discussion and the absolute value property for radicals is simply this:

If we know a is positive, then $\sqrt{a^2} = a$.

If we know a is negative, then $\sqrt{a^2} = |a|$.

If we don't know if a is positive or negative, then $\sqrt{a^2} = |a|$.

10. Simplify each expression. Do not assume the variables represent nonnegative numbers.

a. $\sqrt{16x^2}$

b. $\sqrt{25x^3}$

c. $\sqrt{x^2 + 10x + 25}$

d. $\sqrt{2x^3 + 7x^2}$

EXAMPLE 10 Simplify each expression. Do *not* assume the variables represent positive numbers.

a. $\sqrt{9x^2} = 3|x|$

b. $\sqrt{x^3} = |x|\sqrt{x}$

c. $\sqrt{x^2 - 6x + 9} = \sqrt{(x-3)^2} = |x-3|$

d. $\sqrt{x^3 - 5x^2} = \sqrt{x^2(x-5)} = |x|\sqrt{x-5}$

As you can see, the square root of a sum or difference does not equal the sum or difference of the square roots. This is a common mistake. We must use absolute value symbols when we take a square root of a perfect square, unless we know the base of the perfect square is a positive number. The same idea holds for higher even roots, but not for odd roots. With odd roots, no absolute value symbols are necessary.

11. Simplify each expression.

a. $\sqrt[3]{(-3)^3}$

b. $\sqrt[3]{(-1)^3}$

EXAMPLE 11 Simplify each expression.

a. $\sqrt[3]{(-2)^3} = \sqrt[3]{-8} = -2$

b. $\sqrt[3]{(-5)^3} = \sqrt[3]{-125} = -5$

We can extend this discussion to all roots as follows:

> **PROPERTY** Extending the Absolute Value Property for Radicals
>
> If a is a real number, then
>
> $$\sqrt[n]{a^n} = |a| \qquad \text{if} \qquad n \text{ is even}$$
>
> $$\sqrt[n]{a^n} = a \qquad \text{if} \qquad n \text{ is odd}$$

Answers

10. a. $4|x|$ **b.** $5|x|\sqrt{x}$ **c.** $|x+5|$

d. $|x|\sqrt{2x+7}$

11. a. -3 **b.** -1

The Spiral of Roots

To visualize the square roots of the positive integers, we can construct a spiral of roots. To begin, we draw two line segments, each of length 1, at right angles to each other. Then we use the Pythagorean theorem to find the length of the diagonal. Figure 1 illustrates this procedure.

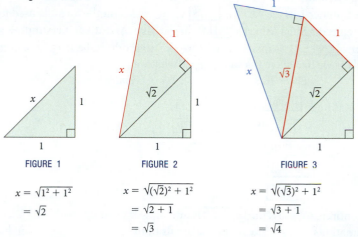

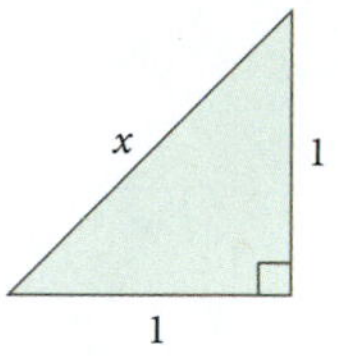

FIGURE 1

$x = \sqrt{1^2 + 1^2}$
$ = \sqrt{2}$

FIGURE 2

$x = \sqrt{(\sqrt{2})^2 + 1^2}$
$ = \sqrt{2 + 1}$
$ = \sqrt{3}$

FIGURE 3

$x = \sqrt{(\sqrt{3})^2 + 1^2}$
$ = \sqrt{3 + 1}$
$ = \sqrt{4}$

Next, we construct a second triangle by connecting a line segment of length 1 to the end of the first diagonal so that the angle formed is a right angle. We find the length of the second diagonal using the Pythagorean theorem. Figure 2 illustrates this procedure. Continuing to draw new triangles by connecting line segments of length 1 to the end of each new diagonal, so that the angle formed is a right angle, the spiral of roots begins to appear (Figure 3).

The Spiral of Roots and Function Notation

Looking over the diagrams and calculations in the preceding discussion, we see that each diagonal in the spiral of roots is found by using the length of the previous diagonal.

First diagonal: $\quad \sqrt{1^2 + 1^2} = \sqrt{2}$

Second diagonal: $\quad \sqrt{(\sqrt{2})^2 + 1^2} = \sqrt{3}$

Third diagonal: $\quad \sqrt{(\sqrt{3})^2 + 1^2} = \sqrt{4}$

Fourth diagonal: $\quad \sqrt{(\sqrt{4})^2 + 1^2} = \sqrt{5}$

A process like this one, in which the answer to one calculation is used to find the answer to the next calculation, is called a *recursive* process. In this particular case, we can use function notation to model the process. If we let x represent the length of any diagonal, then the length of the next diagonal is given by

$$f(x) = \sqrt{x^2 + 1}$$

To begin the process of finding the diagonals, we let $x = 1$.

$$f(1) = \sqrt{1^2 + 1} = \sqrt{2}$$

To find the next diagonal, we substitute $\sqrt{2}$ for x to obtain

$$f(f(1)) = f(\sqrt{2}) = \sqrt{(\sqrt{2})^2 + 1} = \sqrt{3}$$

$$f(f(f(1))) = f(\sqrt{3}) = \sqrt{(\sqrt{3})^2 + 1} = \sqrt{4}$$

We can concisely describe the process of finding the diagonals of the spiral of roots this way:

$$f(1), f(f(1)), f(f(f(1))), \ldots \qquad \text{where } f(x) = \sqrt{x^2 + 1}$$

First input
x

function machine

Output $f(x)$

USING TECHNOLOGY

As our preceding discussion indicates, the length of each diagonal in the spiral of roots is used to calculate the length of the next diagonal. The ANS key on a graphing calculator can be used effectively in a situation like this. To begin, we store the number 1 in the variable ANS. Next, we key in the formula used to produce each diagonal using ANS for the variable. After that, it is simply a matter of pressing ENTER, as many times as we like, to produce the lengths of as many diagonals as we like. Here is a summary of what we do:

Enter This	Display Shows
1 ENTER	1.000
$\sqrt{\ }$ (ANS^2 + 1) ENTER	1.414
ENTER	1.732
ENTER	2.000
ENTER	2.236

If you continue to press the ENTER key, you will produce decimal approximations for as many of the diagonals in the spiral of roots as you like.

GETTING READY FOR CLASS

After reading through the preceding section, respond in your own words and in complete sentences.

A. Explain why this statement is false: "The square root of a sum is the sum of the square roots."

B. What is simplified form for an expression that contains a square root?

C. Why is it not necessarily true that $\sqrt{a^2} = a$?

D. What does it mean to rationalize the denominator in an expression?

Vocabulary Review

Choose the correct words to fill in the blanks below.

product   absolute value   spiral of roots   quotient   simplified form

1. The _____ property for radicals states that the nth root of a product is the product of the nth roots.

2. The _____ property for radicals is used to rationalize the denominator of a radical that contains a fraction.

3. A radical expression is in _____ if none of the factors of the radicand can be written as powers greater than or equal to the index, there are no fractions under the radical sign, and there are no radicals in the denominator.

4. The _____ property for radicals states that if a is a real number, then $\sqrt{a^2} = |a|$.

5. We use the Pythagorean theorem to calculate the hypotenuse of each triangle that makes up a _____.

Problems

A Use the product property for radicals to write each of the following expressions in simplified form. (Assume all variables are nonnegative through Problem 74.)

1. $\sqrt{8}$

2. $\sqrt{32}$

3. $\sqrt{98}$

4. $\sqrt{75}$

5. $\sqrt{288}$

6. $\sqrt{128}$

7. $\sqrt{80}$

8. $\sqrt{200}$

9. $\sqrt{48}$

10. $\sqrt{27}$

11. $\sqrt{675}$

12. $\sqrt{972}$

13. $\sqrt[3]{54}$

14. $\sqrt[3]{24}$

15. $\sqrt[3]{128}$

16. $\sqrt[3]{162}$

17. $\sqrt[3]{432}$

18. $\sqrt[3]{1,536}$

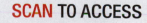

19. $\sqrt[5]{64}$

20. $\sqrt[4]{48}$

21. $\sqrt{18x^3}$

22. $\sqrt{27x^5}$

23. $\sqrt[4]{32y^7}$

24. $\sqrt[5]{32y^7}$

25. $\sqrt[3]{40x^4y^7}$

26. $\sqrt[3]{128x^6y^2}$

27. $\sqrt{48a^2b^3c^4}$

28. $\sqrt{72a^4b^3c^2}$

29. $\sqrt[3]{48a^2b^3c^4}$

30. $\sqrt[3]{72a^4b^3c^2}$

31. $\sqrt[5]{64x^8y^{12}}$

32. $\sqrt[4]{32x^9y^{10}}$

33. $\sqrt[5]{243x^7y^{10}z^5}$

34. $\sqrt[5]{64x^8y^4z^{11}}$

35. $\sqrt[3]{250x^5y^{10}}$

36. $\sqrt[3]{-128x^{15}y^{20}}$

37. $\sqrt[3]{-27x^{10}y^{13}}$

38. $\sqrt[4]{48x^{13}y^{23}}$

Substitute the given numbers into the expression $\sqrt{b^2 - 4ac}$, and then simplify.

39. $a = 2, b = -6, c = 3$

40. $a = 6, b = 7, c = -5$

41. $a = 1, b = 2, c = 6$

42. $a = 2, b = 5, c = 3$

43. $a = \dfrac{1}{2}, b = -\dfrac{1}{2}, c = -\dfrac{5}{4}$

44. $a = \dfrac{7}{4}, b = -\dfrac{3}{4}, c = -2$

B Rationalize the denominator in each of the following expressions.

45. $\dfrac{2}{\sqrt{3}}$

46. $\dfrac{3}{\sqrt{2}}$

47. $\dfrac{5}{\sqrt{6}}$

48. $\dfrac{7}{\sqrt{5}}$

49. $\sqrt{\dfrac{1}{2}}$

50. $\sqrt{\dfrac{1}{3}}$

51. $\sqrt{\dfrac{1}{5}}$

52. $\sqrt{\dfrac{1}{6}}$

53. $\dfrac{4}{\sqrt[3]{2}}$ **54.** $\dfrac{5}{\sqrt[3]{3}}$ **55.** $\dfrac{2}{\sqrt[3]{9}}$ **56.** $\dfrac{3}{\sqrt[3]{4}}$

57. $\sqrt[4]{\dfrac{3}{2x^2}}$ **58.** $\sqrt[4]{\dfrac{5}{3x^2}}$ **59.** $\sqrt[4]{\dfrac{8}{y}}$ **60.** $\sqrt[4]{\dfrac{27}{y}}$

61. $\sqrt[3]{\dfrac{4x}{3y}}$ **62.** $\sqrt[3]{\dfrac{7x}{6y}}$ **63.** $\sqrt[3]{\dfrac{2x}{9y}}$ **64.** $\sqrt[3]{\dfrac{5x}{4y}}$

65. $\sqrt[4]{\dfrac{1}{8x^3}}$ **66.** $\sqrt[4]{\dfrac{8}{9x^3}}$ **67.** $\sqrt[4]{\dfrac{9}{8x^2}}$ **68.** $\sqrt[4]{\dfrac{5}{27y}}$

Write each of the following in simplified form.

69. $\sqrt{\dfrac{27x^3}{5y}}$ **70.** $\sqrt{\dfrac{12x^5}{7y}}$ **71.** $\sqrt{\dfrac{75x^3y^2}{2z}}$ **72.** $\sqrt{\dfrac{50x^2y^3}{3z}}$

73. $\sqrt[3]{\dfrac{16a^4b^3}{9c}}$ **74.** $\sqrt[3]{\dfrac{54a^5b^4}{25c^2}}$ **75.** $\sqrt[3]{\dfrac{8x^3y^6}{9z}}$ **76.** $\sqrt[3]{\dfrac{27x^6y^3}{2z^2}}$

Simplify each expression. Do *not* assume the variables represent positive numbers.

77. $\sqrt{25x^2}$ **78.** $\sqrt{49x^2}$ **79.** $\sqrt{27x^3y^2}$ **80.** $\sqrt{40x^3y^2}$

81. $\sqrt{x^2 - 10x + 25}$ **82.** $\sqrt{x^2 - 16x + 64}$ **83.** $\sqrt{4x^2 + 12x + 9}$ **84.** $\sqrt{16x^2 + 40x + 25}$

85. $\sqrt{4a^4 + 16a^3 + 16a^2}$ **86.** $\sqrt{9a^4 + 18a^3 + 9a^2}$ **87.** $\sqrt{4x^3 - 8x^2}$ **88.** $\sqrt{18x^3 - 9x^2}$

89. Show that the statement $\sqrt{a + b} = \sqrt{a} + \sqrt{b}$ is not true by replacing a with 9 and b with 16 and simplifying both sides.

90. Find a pair of values for a and b that will make the statement $\sqrt{a + b} = \sqrt{a} + \sqrt{b}$ true.

Applying the Concepts

91. Diagonal Distance The distance d between opposite corners of a rectangular room with length l and width w is given by
$$d = \sqrt{l^2 + w^2}$$
How far is it between opposite corners of a living room that measures 10 by 15 feet?

92. Radius of a Sphere The radius r of a sphere with volume V can be found by using the formula
$$r = \sqrt[3]{\frac{3V}{4\pi}}$$
Find the radius of a sphere with volume 9 cubic feet. Write your answer in simplified form.
(Use $\frac{22}{7}$ for π.)

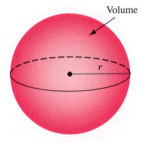

Volume

r

93. Distance to the Horizon If you are at a point k miles above the surface of the earth, the distance you can see towards the horizon, in miles, is approximated by the equation
$$d = \sqrt{8000k + k^2}$$

a. How far towards the horizon can you see from a point that is 1 mile above the surface of the earth?

b. How far can you see from a point that is 2 miles above the surface of the earth?

c. How far can you see from a point that is 3 miles above the surface of the earth?

94. Investing If you invest P dollars and you want the investment to grow to A dollars in t years, the interest rate that must be earned if interest is compounded annually is given by the formula
$$r = \sqrt[t]{\frac{A}{P}} - 1$$
If you invest \$4,000 and want to have \$7,000 in 8 years, what interest rate must be earned?

95. Spiral of Roots Construct your own spiral of roots by using a ruler. Draw the first triangle by using two 1-inch lines. The first diagonal will have a length of $\sqrt{2}$ inches. Each new triangle will be formed by drawing a 1-inch line segment at the end of the previous diagonal so the angle formed is 90°.

96. Spiral of Roots Construct a spiral of roots by using line segments of length 2 cm. The length of the first diagonal will be $2\sqrt{2}$ cm. The length of the second diagonal will be $2\sqrt{3}$ cm.

97. Spiral of Roots If $f(x) = \sqrt{x^2 + 1}$, find the first six terms in the following sequence. Use your results to predict the value of the 10th term and the 100th term.

$$f(1), f(f(1)), f(f(f(1))), \ldots$$

98. Spiral of Roots If $f(x) = \sqrt{x^2 + 4}$, find the first six terms in the following sequence. Use your results to predict the value of the 10th term and the 100th term. (The numbers in this sequence are the lengths of the diagonals of the spiral you drew in Problem 94.)

$$f(2), f(f(2)), f(f(f(2))), \ldots$$

Getting Ready for the Next Section

Simplify the following.

99. $5x - 4x + 6x$

100. $12x + 8x - 7x$

101. $35xy^2 - 8xy^2$

102. $20a^2b + 33a^2b$

103. $\dfrac{1}{2}x + \dfrac{1}{3}x$

104. $\dfrac{2}{3}x + \dfrac{5}{8}x$

105. $5(x + 1)^2 - 12(x + 1)^2$

106. $-7(a + 2)^3 + 4(a + 2)^3$

Write in simplified form for radicals.

107. $\sqrt{18}$

108. $\sqrt{8}$

109. $\sqrt{75xy^3}$

110. $\sqrt{12xy}$

111. $\sqrt[3]{8a^4b^2}$

112. $\sqrt[3]{27ab^2}$

113. $\sqrt[4]{32x^8y^5}$

114. $\sqrt[4]{81a^6b^{11}}$

Find the Mistake

Write true or false for each statement below. If false, circle what is false and write the correct answer on the line provided.

1. True or False: To write the radical expression $\sqrt{4x^7y^2}$ in simplified form, we apply the quotient property for radicals. _____

2. True or False: The radical expression in statement 1 above written in simplified form is $2x^3y\sqrt{xy}$.

_____

3. True or False: To rationalize the denominator of the radical expression $\sqrt{\dfrac{8x^{4y}}{3x}}$, we use the quotient property for radicals and then multiply the numerator and denominator by $\sqrt{8x^{4y}}$. _____

4. True of False: For the spiral of roots, each new line segment drawn to the end of a diagonal is at a complementary angle to the previous line segment. _____

6.3 Addition and Subtraction of Radical Expressions

Image © Luca Giarelli

As the sun sets on a high mountain valley called Val Camonica in Northern Italy, shadows bring out ancient rock carvings on the surrounding cliffs. Archaeologists believe the Camuni tribe may have lit the carvings with candlelight, thus making them "move" as the candle flame flickered. The location of the nearby rock walls would also make a storyteller's voice echo. This recurring echo may have provided a cinematic experience for the tribe.

The echo's time delay is the time it takes for the sound of the storyteller's voice to travel to the nearby rock wall, bounce off it, and be audible to the storyteller and audience. To determine this time delay, we need to know the distance of the storyteller to the nearby wall. Suppose the storyteller was standing $12\sqrt{3}$ meters from the opposing rock wall. We can use the following formula for the echo's time delay t:

$$t = \frac{d}{s}$$

where d is the doubled distance (distance of the storyteller to the wall and back), and s is the speed of sound. Since the formula calls for us to double the distance, we need to know how to add $12\sqrt{3}$ and $12\sqrt{3}$ together. In this section, we will learn how to apply our knowledge of similar terms to the addition and subtraction of radical expressions, such as these.

A Combining Similar Terms

> **DEFINITION** similar, or like, radicals
>
> Two radicals are said to be *similar*, or *like*, *radicals* if they have the same index and the same radicand.

The expressions $5\sqrt[3]{7}$ and $-8\sqrt[3]{7}$ are similar since the index is ^{3} in both cases and the radicands are 7. The expressions $3\sqrt[4]{5}$ and $7\sqrt[3]{5}$ are not similar because they have different indices, and the expressions $2\sqrt[5]{8}$ and $3\sqrt[5]{9}$ are not similar because the radicands are not the same.

We add and subtract radical expressions in the same way we add and subtract polynomials—by combining similar terms under the distributive property.

Practice Problems

1. Combine $3\sqrt{5} - 2\sqrt{5} + 4\sqrt{5}$.

EXAMPLE 1 Combine $5\sqrt{3} - 4\sqrt{3} + 6\sqrt{3}$.

Solution All three radicals are similar. We apply the distributive property to get

$$5\sqrt{3} - 4\sqrt{3} + 6\sqrt{3} = (5 - 4 + 6)\sqrt{3}$$
$$= 7\sqrt{3}$$

Answer

1. $5\sqrt{5}$

EXAMPLE 2   Combine $3\sqrt{8} + 5\sqrt{18}$.

Solution   The two radicals do not seem to be similar. We must write each in simplified form before applying the distributive property.

$$
\begin{aligned}
3\sqrt{8} + 5\sqrt{18} &= 3\sqrt{4 \cdot 2} + 5\sqrt{9 \cdot 2} \\
&= 3\sqrt{4} \cdot \sqrt{2} + 5\sqrt{9} \cdot \sqrt{2} \\
&= 3 \cdot 2\sqrt{2} + 5 \cdot 3\sqrt{2} \\
&= 6\sqrt{2} + 15\sqrt{2} \\
&= (6 + 15)\sqrt{2} \\
&= 21\sqrt{2}
\end{aligned}
$$

The result of Example 2 can be generalized to the following rule for sums and differences of radical expressions.

> **RULE**   Adding or Subtracting Radical Expressions
>
> To add or subtract radical expressions, put each in simplified form and apply the distributive property, if possible. We can add only similar radicals. We must write each expression in simplified form for radicals before we can tell if the radicals are similar.

EXAMPLE 3   Combine $7\sqrt{75xy^3} - 4y\sqrt{12xy}$, where $x, y \geq 0$.

Solution   We write each expression in simplified form and combine similar radicals.

$$
\begin{aligned}
7\sqrt{75xy^3} - 4y\sqrt{12xy} &= 7\sqrt{25y^2} \cdot \sqrt{3xy} - 4y\sqrt{4} \cdot \sqrt{3xy} \\
&= 35y\sqrt{3xy} - 8y\sqrt{3xy} \\
&= (35y - 8y)\sqrt{3xy} \\
&= 27y\sqrt{3xy}
\end{aligned}
$$

EXAMPLE 4   Combine $10\sqrt[3]{8a^4b^2} + 11a\sqrt[3]{27ab^2}$.

Solution   Writing each radical in simplified form and combining similar terms, we have

$$
\begin{aligned}
10\sqrt[3]{8a^4b^2} + 11a\sqrt[3]{27ab^2} &= 10\sqrt[3]{8a^3} \cdot \sqrt[3]{ab^2} + 11a\sqrt[3]{27} \cdot \sqrt[3]{ab^2} \\
&= 20a\sqrt[3]{ab^2} + 33a\sqrt[3]{ab^2} \\
&= 53a\sqrt[3]{ab^2}
\end{aligned}
$$

EXAMPLE 5   Combine $\dfrac{\sqrt{3}}{2} + \dfrac{1}{\sqrt{3}}$.

Solution   We begin by writing the second term in simplified form.

$$
\begin{aligned}
\frac{\sqrt{3}}{2} + \frac{1}{\sqrt{3}} &= \frac{\sqrt{3}}{2} + \frac{1}{\sqrt{3}} \cdot \frac{\sqrt{3}}{\sqrt{3}} \\
&= \frac{\sqrt{3}}{2} + \frac{\sqrt{3}}{3} \\
&= \frac{1}{2}\sqrt{3} + \frac{1}{3}\sqrt{3} \\
&= \left(\frac{1}{2} + \frac{1}{3}\right)\sqrt{3} \\
&= \frac{5}{6}\sqrt{3} \\
&= \frac{5\sqrt{3}}{6}
\end{aligned}
$$

2. Combine $4\sqrt{50} + 3\sqrt{8}$.

3. Assume $x, y \geq 0$ and combine.
$$4\sqrt{18x^2y} - 3x\sqrt{50y}$$

4. Combine $2\sqrt[3]{27a^2b^4} + 3b\sqrt[3]{125a^2b}$.

5. Combine $\dfrac{\sqrt{5}}{3} + \dfrac{1}{\sqrt{5}}$.

Answers
2. $26\sqrt{2}$
3. $-3x\sqrt{2y}$
4. $21b\sqrt[3]{a^2b}$
5. $\frac{8\sqrt{5}}{15}$

B The Golden Rectangle

Recall our discussion of the golden rectangle from the first section of this chapter. Now that we have practiced adding and subtracting radical expressions, let's put our new skills to use and find a golden rectangle of a different size.

6. Construct a golden rectangle from a square of side 6. Then show that the ratio of the length to the width is the golden ratio.

EXAMPLE 6   Construct a golden rectangle from a square of side 4. Then show that the ratio of the length to the width is the golden ratio $\frac{1+\sqrt{5}}{2}$.

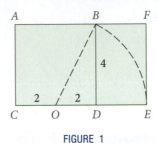

FIGURE 1

Solution   Figure 1 shows the golden rectangle constructed from a square of side 4. The length of the diagonal OB is found from the Pythagorean theorem.

$$OB = \sqrt{2^2 + 4^2} = \sqrt{4 + 16} = \sqrt{20} = 2\sqrt{5}$$

The ratio of the length to the width for the rectangle is the golden ratio.

$$\text{Golden ratio} = \frac{CE}{EF} = \frac{2 + 2\sqrt{5}}{4} = \frac{2(1 + \sqrt{5})}{2 \cdot 2} = \frac{1 + \sqrt{5}}{2}$$

As you can see, showing that the ratio of length to width in this rectangle is the golden ratio depends on our ability to write $\sqrt{20}$ as $2\sqrt{5}$ and our ability to reduce to lowest terms by factoring and then dividing out the common factor 2 from the numerator and denominator.

GETTING READY FOR CLASS

After reading through the preceding section, respond in your own words and in complete sentences.

A. What are similar radicals?

B. When can we add two radical expressions?

C. What is the first step when adding or subtracting expressions containing radicals?

D. How would you construct a golden rectangle from a square of side 8?

Problems

A Combine the following expressions. (Assume any variables under an even root are nonnegative.)

1. $3\sqrt{5} + 4\sqrt{5}$

2. $6\sqrt{3} - 5\sqrt{3}$

3. $3x\sqrt{7} - 4x\sqrt{7}$

4. $6y\sqrt{a} + 7y\sqrt{a}$

5. $5\sqrt[3]{10} - 4\sqrt[3]{10}$

6. $6\sqrt[4]{2} + 9\sqrt[4]{2}$

7. $8\sqrt[5]{6} - 2\sqrt[5]{6} + 3\sqrt[5]{6}$

8. $7\sqrt[6]{7} - \sqrt[6]{7} + 4\sqrt[6]{7}$

9. $3x\sqrt{2} - 4x\sqrt{2} + x\sqrt{2}$

10. $5x\sqrt{6} - 3x\sqrt{6} - 2x\sqrt{6}$

11. $\sqrt{20} - \sqrt{80} + \sqrt{45}$

12. $\sqrt{8} - \sqrt{32} - \sqrt{18}$

13. $4\sqrt{8} - 2\sqrt{50} - 5\sqrt{72}$

14. $\sqrt{48} - 3\sqrt{27} + 2\sqrt{75}$

15. $5x\sqrt{8} + 3\sqrt{32x^2} - 5\sqrt{50x^2}$

16. $2\sqrt{50x^2} - 8x\sqrt{18} - 3\sqrt{72x^2}$

17. $5\sqrt{5y} + \sqrt{18y}$

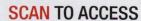

SCAN TO ACCESS

18. $\sqrt{250x} - 2\sqrt{40x} + 5\sqrt{10x^3}$

19. $5\sqrt[3]{16} - 4\sqrt[3]{54}$

20. $\sqrt[3]{81} + 3\sqrt[3]{24}$

21. $\sqrt[3]{x^4y^2} + 7x\sqrt[3]{xy^2}$

22. $2\sqrt[3]{x^8y^6} - 3y^2\sqrt[3]{8x^8}$

23. $5a^2\sqrt{27ab^3} - 6b\sqrt{12a^5b}$

24. $9a\sqrt{20a^3b^2} + 7b\sqrt{45a^5}$

25. $\sqrt[3]{24a^5b^4} + \sqrt[3]{81a^8b^4}$

26. $\sqrt[4]{32x^7y^5} - \sqrt[4]{162x^7y^4}$

27. $b\sqrt[3]{24a^5b} + 3a\sqrt[3]{81a^2b^4}$

28. $7\sqrt[3]{a^4b^3c^2} - 6ab\sqrt[3]{ac^2}$

29. $5x\sqrt[4]{3y^5} + y\sqrt[4]{243x^4y} + \sqrt[4]{48x^4y^5}$

30. $x\sqrt[4]{5xy^8} + y\sqrt[4]{405x^5y^4} + y^2\sqrt[4]{80x^5}$

31. $\dfrac{\sqrt{2}}{2} + \dfrac{1}{\sqrt{2}}$

32. $\dfrac{\sqrt{3}}{3} + \dfrac{1}{\sqrt{3}}$

33. $\dfrac{\sqrt{5}}{3} + \dfrac{1}{\sqrt{5}}$

34. $\dfrac{\sqrt{6}}{2} + \dfrac{1}{\sqrt{6}}$

35. $\sqrt{x} - \dfrac{1}{\sqrt{x}}$

36. $\sqrt{x} + \dfrac{1}{\sqrt{x}}$

37. $\dfrac{\sqrt{18}}{6} + \sqrt{\dfrac{1}{2}} + \dfrac{\sqrt{2}}{2}$

38. $\dfrac{\sqrt{12}}{6} + \sqrt{\dfrac{1}{3}} + \dfrac{\sqrt{3}}{3}$

39. $\sqrt{6} - \sqrt{\dfrac{2}{3}} + \sqrt{\dfrac{1}{6}}$

40. $\sqrt{15} - \sqrt{\dfrac{3}{5}} + \sqrt{\dfrac{5}{3}}$

41. $\sqrt[3]{25} + \dfrac{3}{\sqrt[3]{5}}$

42. $\sqrt[4]{8} + \dfrac{1}{\sqrt[4]{2}}$

43. $\dfrac{\sqrt{45}}{10} + \sqrt{\dfrac{1}{5}} + \dfrac{\sqrt{5}}{5}$

44. $\sqrt[3]{49} + \dfrac{4}{\sqrt[3]{7}}$

45. Use a calculator to find a decimal approximation for $\sqrt{12}$ and for $2\sqrt{3}$.

46. Use a calculator to find decimal approximations for $\sqrt{50}$ and $5\sqrt{2}$.

47. Use a calculator to find a decimal approximation for $\sqrt{8} + \sqrt{18}$. Is it equal to the decimal approximation for $\sqrt{26}$ or $\sqrt{50}$?

48. Use a calculator to find a decimal approximation for $\sqrt{3} + \sqrt{12}$. Is it equal to the decimal approximation for $\sqrt{15}$ or $\sqrt{27}$?

Each of the following statements is false. Correct the right side of each one to make the statement true.

49. $3\sqrt{2x} + 5\sqrt{2x} = 8\sqrt{4x}$

50. $5\sqrt{3} - 7\sqrt{3} = -2\sqrt{9}$

51. $\sqrt{9 + 16} = 3 + 4$

52. $\sqrt{36 + 64} = 6 + 8$

Applying the Concepts

53. Golden Rectangle Construct a golden rectangle from a square of side 8. Then show that the ratio of the length to the width is the golden ratio $\frac{1 + \sqrt{5}}{2}$.

54. Golden Rectangle Construct a golden rectangle from a square of side 10. Then show that the ratio of the length to the width is the golden ratio $\frac{1 + \sqrt{5}}{2}$.

55. Golden Rectangle Use a ruler to construct a golden rectangle from a square of side 1 inch. Then show that the ratio of the length to the width is the golden ratio.

56. Golden Rectangle Use a ruler to construct a golden rectangle from a square of side $\frac{2}{3}$ inch. Then show that the ratio of the length to the width is the golden ratio.

57. Golden Rectangle To show that all golden rectangles have the same ratio of length to width, construct a golden rectangle from a square of side $2x$. Then show that the ratio of the length to the width is the golden ratio.

58. Golden Rectangle To show that all golden rectangles have the same ratio of length to width, construct a golden rectangle from a square of side x. Then show that the ratio of the length to the width is the golden ratio.

59. Isosceles Right Triangles A triangle is isosceles if it has two equal sides, and a triangle is a right triangle if it has a right angle in it. Sketch an isosceles right triangle, and find the ratio of the hypotenuse to a leg.

60. Equilateral Triangles A triangle is equilateral if it has three equal sides. The triangle in the figure is equilateral with each side of length $2x$. Find the ratio of the height to a side.

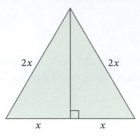

61. Pyramids The following solid is called a regular square pyramid because its base is a square and all eight edges are the same length, 5. It is also true that the vertex, V, is directly above the center of the base.

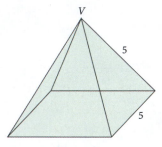

a. Find the ratio of a diagonal of the base to the length of a side.

b. Find the ratio of the area of the base to the diagonal of the base.

c. Find the ratio of the area of the base to the perimeter of the base.

62. Pyramids Refer to the diagram of a square pyramid below. Find the ratio of the height h of the pyramid to the altitude a.

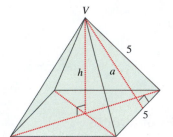

Getting Ready for the Next Section

Simplify the following.

63. $3 \cdot 2$

64. $5 \cdot 7$

65. $(x + y)(4x - y)$

66. $(2x + y)(x - y)$

67. $(x + 3)^2$

68. $(3x - 2y)^2$

69. $(x - 2)(x + 2)$

70. $(2x + 5)(2x - 5)$

Simplify the following expressions.

71. $2\sqrt{18}$

72. $5\sqrt{36}$

73. $(\sqrt{6})^2$

74. $(\sqrt{2})^2$

75. $(3\sqrt{x})^2$

76. $(2\sqrt{y})^2$

77. $(4\sqrt{3x})^2$

78. $(-5\sqrt{2x})^2$

Rationalize the denominator.

79. $\dfrac{\sqrt{3}}{\sqrt{2}}$

80. $\dfrac{\sqrt{5}}{\sqrt{6}}$

81. $\dfrac{\sqrt{6}}{\sqrt{8}}$

82. $\dfrac{\sqrt{10}}{\sqrt{12}}$

Find the Mistake

Write true or false for each statement below. If false, circle what is false and write the correct answer on the line provided.

1. True or False: When adding or subtracting radical expressions, put each in simplified form and apply the product property for radicals. _____

2. True or False: We can only tell if radicals are similar if they are in simplified form. _____

3. True or False: When combining $3\sqrt{5} + 5\sqrt{3} - 2\sqrt{5}$, all three radicals are similar and can be combined to get $5\sqrt{5} - \sqrt{3}$. _____

4. True or False: Combining $4\sqrt{8} + 3\sqrt{32}$, we get $13\sqrt{2}$. _____

6.4 Multiplication and Division of Radical Expressions

Image © sxc.hu, cempey, 2006

The insects of a London park now have a new hotel in which to stay. The city held a unique competition that requested participants to build a structure that could house multiple varieties of bugs at once. A 1.5 meter tall box with a façade mimicking the multi-polygon design of a dragonfly's wing, also called voronoi, won first prize. The designers stacked 25 layers of 20mm thick birch plywood front to back, with each layer having the voronoi design. Each polygonal hole caters to a specific type of insect, and is loosely stuffed with recycled or organic materials. Beetles, bees, butterflies, spiders, and more are all encouraged to make their homes in this new Bug Hotel.

A pentagon is one of the shapes present in the Bug Hotel's façade. We can find the area A of a regular pentagon with a side length t by using the formula $A = \frac{t^2\sqrt{25 + 10\sqrt{5}}}{4}$. This formula uses a radical expression to find the pentagon's area.

In this section, we will look at multiplication and division of expressions like this one that contain radicals. As you will see, multiplication of expressions that contain radicals is very similar to multiplication of polynomials. The division problems in this section are just an extension of the work we did previously when we rationalized denominators.

A Multiplication of Radical Expressions

EXAMPLE 1   Multiply $(3\sqrt{5})(2\sqrt{7})$.

Solution   We can rearrange the order and grouping of the numbers in this product by applying the commutative and associative properties. Following this, we apply the product property for radicals and multiply.

$$(3\sqrt{5})(2\sqrt{7}) = (3 \cdot 2)(\sqrt{5}\sqrt{7}) \quad \text{Communicative and associative properties}$$
$$= (3 \cdot 2)(\sqrt{5 \cdot 7}) \quad \text{Product property for radicals}$$
$$= 6\sqrt{35} \quad \text{Multiply.}$$

In practice, it is not necessary to show the first two steps.

EXAMPLE 2   Multiply $\sqrt{3}(2\sqrt{6} - 5\sqrt{12})$.

Solution   Applying the distributive property, we have
$$\sqrt{3}(2\sqrt{6} - 5\sqrt{12}) = \sqrt{3} \cdot 2\sqrt{6} - \sqrt{3} \cdot 5\sqrt{12}$$
$$= 2\sqrt{18} - 5\sqrt{36}$$

Writing each radical in simplified form gives
$$2\sqrt{18} - 5\sqrt{36} = 2\sqrt{9}\sqrt{2} - 5\sqrt{36}$$
$$= 6\sqrt{2} - 30$$

Practice Problems

1. Multiply $(7\sqrt{3})(5\sqrt{11})$.

2. Multiply $\sqrt{2}(3\sqrt{5} - 4\sqrt{2})$.

Answers

1. $35\sqrt{33}$

2. $3\sqrt{10} - 8$

EXAMPLE 3   Multiply $(\sqrt{3} + \sqrt{5})(4\sqrt{3} - \sqrt{5})$.

Solution   The same principle that applies when multiplying two binomials applies to this product. We must multiply each term in the first expression by each term in the second one. Any convenient method can be used. Let's use the FOIL method.

$$(\sqrt{3} + \sqrt{5})(4\sqrt{3} - \sqrt{5}) = \overset{F}{\sqrt{3} \cdot 4\sqrt{3}} - \overset{O}{\sqrt{3} \cdot \sqrt{5}} + \overset{I}{\sqrt{5} \cdot 4\sqrt{3}} - \overset{L}{\sqrt{5} \cdot \sqrt{5}}$$

$$= 4 \cdot 3 - \sqrt{15} + 4\sqrt{15} - 5$$

$$= 12 + 3\sqrt{15} - 5$$

$$= 7 + 3\sqrt{15}$$

EXAMPLE 4   Expand and simplify $(\sqrt{x} + 3)^2$.

Solution 1   We can write this problem as a multiplication problem and proceed as we did in Example 3.

$$(\sqrt{x} + 3)^2 = (\sqrt{x} + 3)(\sqrt{x} + 3)$$

$$= \overset{F}{\sqrt{x} \cdot \sqrt{x}} + \overset{O}{3\sqrt{x}} + \overset{I}{3\sqrt{x}} + \overset{L}{3 \cdot 3}$$

$$= x + 3\sqrt{x} + 3\sqrt{x} + 9$$

$$= x + 6\sqrt{x} + 9$$

Solution 2   We can obtain the same result by applying the formula for the square of a sum: $(a + b)^2 = a^2 + 2ab + b^2$.

$$(\sqrt{x} + 3)^2 = (\sqrt{x})^2 + 2(\sqrt{x})(3) + 3^2$$

$$= x + 6\sqrt{x} + 9$$

EXAMPLE 5   Expand $(3\sqrt{x} - 2\sqrt{y})^2$ and simplify the result.

Solution   Let's apply the formula for the square of a difference, $(a - b)^2 = a^2 - 2ab + b^2$.

$$(3\sqrt{x} - 2\sqrt{y})^2 = (3\sqrt{x})^2 - 2(3\sqrt{x})(2\sqrt{y}) + (2\sqrt{y})^2$$

$$= 9x - 12\sqrt{xy} + 4y$$

EXAMPLE 6   Expand and simplify $(\sqrt{x + 2} - 1)^2$.

Solution   Applying the formula $(a - b)^2 = a^2 - 2ab + b^2$, we have

$$(\sqrt{x + 2} - 1)^2 = (\sqrt{x + 2})^2 - 2\sqrt{x + 2}(1) + 1^2$$

$$= x + 2 - 2\sqrt{x + 2} + 1$$

$$= x + 3 - 2\sqrt{x + 2}$$

EXAMPLE 7   Multiply $(\sqrt{6} + \sqrt{2})(\sqrt{6} - \sqrt{2})$.

Solution   We notice the product is of the form $(a + b)(a - b)$, which always gives the difference of two squares, $a^2 - b^2$.

$$(\sqrt{6} + \sqrt{2})(\sqrt{6} - \sqrt{2}) = (\sqrt{6})^2 - (\sqrt{2})^2$$

$$= 6 - 2$$

$$= 4$$

---

3. Multiply $(\sqrt{2} + \sqrt{7})(\sqrt{2} - 3\sqrt{7})$.

4. Expand and simplify $(\sqrt{x} + 5)^2$.

5. Expand $(5\sqrt{a} - 3\sqrt{b})^2$ and simplify the result.

6. Expand and simplify $(\sqrt{x + 3} - 1)^2$.

7. Multiply $(\sqrt{5} + \sqrt{3})(\sqrt{5} - \sqrt{3})$.

Answers

3. $-19 - 2\sqrt{14}$
4. $x + 10\sqrt{x} + 25$
5. $25a - 30\sqrt{ab} + 9b$
6. $x + 4 - 2\sqrt{x + 3}$
7. 2

In Example 7, the two expressions $(\sqrt{6} + \sqrt{2})$ and $(\sqrt{6} - \sqrt{2})$ are called *conjugates*. In general, the conjugate of $\sqrt{a} + \sqrt{b}$ is $\sqrt{a} - \sqrt{b}$. If a and b are integers, multiplying conjugates of this form always produces a rational number. That is, if a and b are positive integers, then

$$(\sqrt{a} + \sqrt{b})(\sqrt{a} - \sqrt{b}) = \sqrt{a}\sqrt{a} - \sqrt{a}\sqrt{b} + \sqrt{a}\sqrt{b} - \sqrt{b}\sqrt{b}$$
$$= a - \sqrt{ab} + \sqrt{ab} - b$$
$$= a - b$$

which is rational if a and b are rational.

B Division of Radical Expressions

Division with radical expressions is the same as rationalizing the denominator. In Section 6.2, we were able to divide $\sqrt{3}$ by $\sqrt{2}$ by rationalizing the denominator:

$$\frac{\sqrt{3}}{\sqrt{2}} = \frac{\sqrt{3}}{\sqrt{2}} \cdot \frac{\sqrt{2}}{\sqrt{2}} = \frac{\sqrt{6}}{2}$$

We can accomplish the same result with expressions such as

$$\frac{6}{\sqrt{5} - \sqrt{3}}$$

by multiplying the numerator and denominator by the conjugate of the denominator.

8. Divide $\dfrac{3}{\sqrt{7} - \sqrt{3}}$.

EXAMPLE 8 Divide $\dfrac{6}{\sqrt{5} - \sqrt{3}}$. (Hint: rationalize the denominator.)

Solution Because the product of two conjugates is a rational number, we multiply the numerator and denominator by the conjugate of the denominator.

$$\frac{6}{\sqrt{5} - \sqrt{3}} = \frac{6}{\sqrt{5} - \sqrt{3}} \cdot \frac{(\sqrt{5} + \sqrt{3})}{(\sqrt{5} + \sqrt{3})}$$

$$= \frac{6\sqrt{5} + 6\sqrt{3}}{(\sqrt{5})^2 - (\sqrt{3})^2}$$

$$= \frac{6\sqrt{5} + 6\sqrt{3}}{5 - 3}$$

$$= \frac{6\sqrt{5} + 6\sqrt{3}}{2}$$

$$= \frac{2(3\sqrt{5} + 3\sqrt{3})}{2} \qquad \text{Divide out the common factor, 2.}$$

$$= 3\sqrt{5} + 3\sqrt{3}$$

9. Rationalize the denominator.

$$\frac{\sqrt{10} - 3}{\sqrt{10} + 3}$$

EXAMPLE 9 Rationalize the denominator $\dfrac{\sqrt{5} - 2}{\sqrt{5} + 2}$.

Solution To rationalize the denominator, we multiply the numerator and denominator by the conjugate of the denominator.

$$\frac{\sqrt{5} - 2}{\sqrt{5} + 2} = \frac{\sqrt{5} - 2}{\sqrt{5} + 2} \cdot \frac{(\sqrt{5} - 2)}{(\sqrt{5} - 2)}$$

$$= \frac{5 - 2\sqrt{5} - 2\sqrt{5} + 4}{(\sqrt{5})^2 - 2^2}$$

$$= \frac{9 - 4\sqrt{5}}{5 - 4}$$

$$= \frac{9 - 4\sqrt{5}}{1}$$

$$= 9 - 4\sqrt{5}$$

Answers

8. $\dfrac{3\sqrt{7} + 3\sqrt{3}}{4}$

9. $19 - 6\sqrt{10}$

We have worked with the golden rectangle more than once in this chapter. Remember the following diagram, the general representation of a golden rectangle:
By now you know that, in any golden rectangle constructed from a square (of any size), the ratio of the length to the width will be

$$\frac{1 + \sqrt{5}}{2}$$

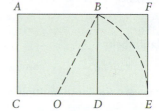

which we call the golden ratio. What is interesting is that the smaller rectangle on the right, *BFED*, is also a golden rectangle. We will now use the mathematics developed in this section to confirm this fact.

EXAMPLE 10   A golden rectangle constructed from a square of side 2 is shown in Figure 1. Show that the smaller rectangle *BDEF* is also a golden rectangle by finding the ratio of its length to its width.

Solution   First, find expressions for the length and width of the smaller rectangle.

$$\text{Length} = EF = 2$$

$$\text{Width} = DE = \sqrt{5} - 1$$

Next, we find the ratio of length to width.

$$\text{Ratio of length to width} = \frac{EF}{DE} = \frac{2}{\sqrt{5} - 1}$$

To show that the small rectangle is a golden rectangle, we must show that the ratio of length to width is the golden ratio. We do so by rationalizing the denominator.

$$\frac{2}{\sqrt{5} - 1} = \frac{2}{\sqrt{5} - 1} \cdot \frac{\sqrt{5} + 1}{\sqrt{5} + 1}$$

$$= \frac{2(\sqrt{5} + 1)}{5 - 1}$$

$$= \frac{2(\sqrt{5} + 1)}{4}$$

$$= \frac{\sqrt{5} + 1}{2} \qquad \text{Divide out common factor 2.}$$

Because addition is commutative, this last expression is the golden ratio. Therefore, the small rectangle in Figure 1 is a golden rectangle.

FIGURE 1

GETTING READY FOR CLASS

After reading through the preceding section, respond in your own words and in complete sentences.

A. Explain why $(\sqrt{5} + \sqrt{2})^2 \neq 5 + 2$.

B. Explain in words how you would rationalize the denominator in the expression $\dfrac{\sqrt{3}}{\sqrt{5} - \sqrt{2}}$.

C. What are conjugates?

D. What result is guaranteed when multiplying radical expressions that are conjugates?

10. If side *AC* in Figure 1 were 4 instead of 2, show that the smaller rectangle *BDEF* is also a golden rectangle by finding the ratio of its length to its width.

Answer

10. $\dfrac{4}{2\sqrt{5} - 2} = \dfrac{4}{2\sqrt{5} - 2} \cdot \dfrac{2\sqrt{5} + 2}{2\sqrt{5} + 2}$

$= \dfrac{8\sqrt{5} + 8}{4(5) - 4}$

$= \dfrac{8(\sqrt{5} + 1)}{16}$

$= \dfrac{\sqrt{5} + 1}{2}$

Vocabulary Review

Choose the correct words to fill in the blanks below.

denominator   polynomials   radical   conjugate

1. Multiplication of radical expressions is very similar to multiplication of _____.

2. The expression $(\sqrt{a} - \sqrt{b})$ is the _____ of the expression $(\sqrt{a} + \sqrt{b})$.

3. Division with _____ expressions is the same as rationalizing the denominator.

4. We can rationalize a denominator that contains a radical expression by multiplying the numerator and denominator by the conjugate of the _____.

Problems

A Multiply. (Assume all expressions appearing under a square root symbol represent nonnegative numbers throughout this exercise set.)

1. $\sqrt{6} \cdot \sqrt{3}$

2. $\sqrt{6} \cdot \sqrt{2}$

3. $(2\sqrt{3})(5\sqrt{7})$

4. $(3\sqrt{5})(2\sqrt{7})$

5. $(4\sqrt{6})(2\sqrt{15})(3\sqrt{10})$

6. $(4\sqrt{35})(2\sqrt{21})(5\sqrt{15})$

7. $(3\sqrt[3]{3})(6\sqrt[3]{9})$

8. $(2\sqrt[3]{2})(6\sqrt[3]{4})$

9. $\sqrt{3}(\sqrt{2} - 3\sqrt{3})$

10. $\sqrt{2}(5\sqrt{3} + 4\sqrt{2})$

11. $6\sqrt[3]{4}(2\sqrt[3]{2} + 1)$

12. $7\sqrt[3]{5}(3\sqrt[3]{25} - 2)$

13. $(\sqrt{3} + \sqrt{2})(3\sqrt{3} - \sqrt{2})$

14. $(\sqrt{5} - \sqrt{2})(3\sqrt{5} + 2\sqrt{2})$

SCAN TO ACCESS

15. $(\sqrt{x} + 5)(\sqrt{x} - 3)$

16. $(\sqrt{x} + 4)(\sqrt{x} + 2)$

17. $(3\sqrt{6} + 4\sqrt{2})(\sqrt{6} + 2\sqrt{2})$

18. $(\sqrt{7} - 3\sqrt{3})(2\sqrt{7} - 4\sqrt{3})$

19. $(\sqrt{3} + 4)^2$

20. $(\sqrt{5} - 2)^2$

21. $(\sqrt{x} - 3)^2$

22. $(\sqrt{x} + 4)^2$

23. $(2\sqrt{a} - 3\sqrt{b})^2$

24. $(5\sqrt{a} - 2\sqrt{b})^2$

25. $(\sqrt{x - 4} + 2)^2$

26. $(\sqrt{x - 3} + 2)^2$

27. $(\sqrt{x - 5} - 3)^2$

28. $(\sqrt{x - 3} - 4)^2$

29. $(\sqrt{3} - \sqrt{2})(\sqrt{3} + \sqrt{2})$

30. $(\sqrt{5} - \sqrt{2})(\sqrt{5} + \sqrt{2})$

31. $(\sqrt{a} + 7)(\sqrt{a} - 7)$

32. $(\sqrt{a} + 5)(\sqrt{a} - 5)$

33. $(5 - \sqrt{x})(5 + \sqrt{x})$

34. $(3 - \sqrt{x})(3 + \sqrt{x})$

35. $(\sqrt{x - 4} + 2)(\sqrt{x - 4} - 2)$

36. $(\sqrt{x + 3} + 5)(\sqrt{x + 3} - 5)$

37. $(\sqrt{3} + 1)^3$

38. $(\sqrt{5} - 2)^3$

39. $(\sqrt{3} - 3)^3$

40. $(\sqrt{2} - 3)^3$

B Rationalize the denominator in each of the following.

41. $\dfrac{\sqrt{2}}{\sqrt{6} - \sqrt{2}}$

42. $\dfrac{\sqrt{5}}{\sqrt{5} + \sqrt{3}}$

43. $\dfrac{\sqrt{5}}{\sqrt{5} + 1}$

44. $\dfrac{\sqrt{7}}{\sqrt{7} - 1}$

45. $\dfrac{\sqrt{x}}{\sqrt{x} - 3}$

46. $\dfrac{\sqrt{x}}{\sqrt{x} + 2}$

47. $\dfrac{\sqrt{5}}{2\sqrt{5} - 3}$

48. $\dfrac{\sqrt{7}}{3\sqrt{7} - 2}$

49. $\dfrac{3}{\sqrt{x} - \sqrt{y}}$

50. $\dfrac{2}{\sqrt{x} + \sqrt{y}}$

51. $\dfrac{\sqrt{6} + \sqrt{2}}{\sqrt{6} - \sqrt{2}}$

52. $\dfrac{\sqrt{5} - \sqrt{3}}{\sqrt{5} + \sqrt{3}}$

53. $\dfrac{\sqrt{7} - 2}{\sqrt{7} + 2}$

54. $\dfrac{\sqrt{11} + 3}{\sqrt{11} - 3}$

55. $\dfrac{\sqrt{a} + \sqrt{b}}{\sqrt{a} - \sqrt{b}}$

56. $\dfrac{\sqrt{a} - \sqrt{b}}{\sqrt{a} + \sqrt{b}}$

57. $\dfrac{\sqrt{x} + 2}{\sqrt{x} - 2}$

58. $\dfrac{\sqrt{x} - 3}{\sqrt{x} + 3}$

59. $\dfrac{2\sqrt{3} - \sqrt{7}}{3\sqrt{3} + \sqrt{7}}$

60. $\dfrac{5\sqrt{6} + 2\sqrt{2}}{\sqrt{6} - \sqrt{2}}$

61. $\dfrac{3\sqrt{x} + 2}{1 + \sqrt{x}}$

62. $\dfrac{5\sqrt{x} - 1}{2 + \sqrt{x}}$

63. $\dfrac{2\sqrt{x} - 1}{1 - 3\sqrt{x}}$

64. $\dfrac{4\sqrt{x} - 3}{2\sqrt{x} + 3}$

65. Show that the product below is 5.

$$(\sqrt[3]{2} + \sqrt[3]{3})(\sqrt[3]{4} - \sqrt[3]{6} + \sqrt[3]{9})$$

66. Show that the product below is $x + 8$.

$$(\sqrt[3]{x} + 2)(\sqrt[3]{x^2} - 2\sqrt[3]{x} + 4)$$

Each of the following statements below is false. Correct the right side of each one to make it true.

67. $5(2\sqrt{3}) = 10\sqrt{15}$

68. $3(2\sqrt{x}) = 6\sqrt{3x}$

69. $(\sqrt{x} + 3)^2 = x + 9$

70. $(\sqrt{x} - 7)^2 = x - 49$

71. $(5\sqrt{3})^2 = 15$

72. $(3\sqrt{5})^2 = 15$

73. $(x + \sqrt{3})^2 = x^2 + 3$

74. $(\sqrt{x} - \sqrt{3})^2 = x - 3$

Applying the Concepts

75. Gravity If an object is dropped from the top of a 100-foot building, the amount of time t (in seconds) that it takes for the object to be h feet from the ground is given by the formula

$$t = \frac{\sqrt{100 - h}}{4}$$

How long does it take before the object is 50 feet from the ground? How long does it take to reach the ground? (When it is on the ground, h is 0.)

76. Gravity Use the formula given in Problem 73 to determine h if t is 1.25 seconds.

77. Golden Rectangle Rectangle *ACEF* in Figure 2 is a golden rectangle. If side *AC* is 6 inches, show that the smaller rectangle *BDEF* is also a golden rectangle.

78. Golden Rectangle Rectangle *ACEF* in Figure 2 is a golden rectangle. If side *AC* is 1 inch, show that the smaller rectangle *BDEF* is also a golden rectangle.

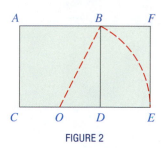

FIGURE 2

79. Golden Rectangle If side *AC* in Figure 2 is $2x$, show that rectangle *BDEF* is a golden rectangle.

80. Golden Rectangle If side *AC* in Figure 2 is x, show that rectangle *BDEF* is a golden rectangle.

Getting Ready for the Next Section

Simplify.

81. $(t + 5)^2$

82. $(x - 4)^2$

83. $\sqrt{x} \cdot \sqrt{x}$

84. $\sqrt{3x} \cdot \sqrt{3x}$

Solve.

85. $3x + 4 = 5^2$

86. $4x - 7 = 3^2$

87. $t^2 + 7t + 12 = 0$

88. $x^2 - 3x - 10 = 0$

89. $t^2 + 10t + 25 = t + 7$

90. $x^2 - 4x + 4 = x - 2$

91. $(x + 4)^2 = x + 6$

92. $(x - 6)^2 = x - 4$

93. Is 7 a solution to $\sqrt{3x + 4} = 5$?

94. Is 4 a solution to $\sqrt{4x - 7} = -3$?

95. Is -6 a solution to $t + 5 = \sqrt{t + 7}$?

96. Is -3 a solution to $t + 5 = \sqrt{t + 7}$?

Maintaining Your Skills

97. The chart shows where the most iPhones are being sold. If Apple has sold 50 million iPhones, how many iPhones are in North America? In Asia? In Africa?

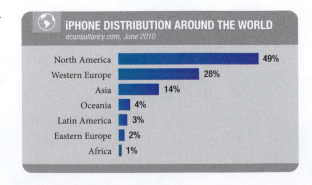

iPHONE DISTRIBUTION AROUND THE WORLD
econsultancy.com, June 2010

North America 49%
Western Europe 28%
Asia 14%
Oceania 4%
Latin America 3%
Eastern Europe 2%
Africa 1%

Find the Mistake

Each sentence below contains a mistake. Circle the mistake and write the correct word or expression on the line provided.

1. To multiply $\sqrt{5}(\sqrt{7} + 6\sqrt{5})$, we use the communicative property to get our answer $30 + \sqrt{35}$.

_____

2. To multiply $(2\sqrt{2} + \sqrt{7})(4\sqrt{2} - 3\sqrt{7})$, we apply the associative property and regroup the terms to get our answer $(6\sqrt{2}) - (4\sqrt{7})$. _____

3. To divide 4 by $(2\sqrt{5} - 2)$, we rationalize the denominator by multiplying the numerator and the denominator by 4.

_____

4. Dividing $\frac{(2\sqrt{3} - \sqrt{2})}{(4\sqrt{3} + \sqrt{2})}$ we get $\frac{(24 - 3\sqrt{6})}{44}$. _____

Landmark Review: Checking Your Progress

Find each of the following roots if possible.

1. $\sqrt{81}$ **2.** $-\sqrt{81}$ **3.** $\sqrt{-81}$ **4.** $\sqrt[3]{-27}$

Simplify each expression. Assume all variables represent nonnegative numbers.

5. $\sqrt[3]{x^6 y^{12}}$ **6.** $64^{1/2}$ **7.** $\left(\frac{16}{25}\right)^{-1/2}$ **8.** $81^{1/2} + 100^{1/2}$

Write each of the following expressions in simplified form.

9. $\sqrt{164}$ **10.** $\sqrt{18x^5 y^6}$ **11.** $\sqrt[5]{486 x^{12} y^{10} z^{18}}$ **12.** $\frac{4}{\sqrt{5}}$

Perform the indicated operations.

13. $6\sqrt{150} - 4\sqrt{150} + 3\sqrt{216}$ **14.** $\frac{\sqrt{2}}{3} + \frac{1}{\sqrt{2}}$ **15.** $\sqrt{6} \cdot \sqrt{2}$

16. $(3\sqrt{5})(2\sqrt{10})(4\sqrt{8})$ **17.** $(2\sqrt{3} + 3\sqrt{2})(\sqrt{3} + 4\sqrt{2})$ **18.** $(\sqrt{5} - 1)^2$

19. $\frac{\sqrt{2}}{\sqrt{6} - \sqrt{2}}$ **20.** $\frac{\sqrt{5} + 3}{\sqrt{5} - 3}$

A Solve equations containing radicals.

B Graph simple square root and cube root functions in two variables.

6.5 Radical Equations and Functions

Image © bewinca, 2007

The Three Gorges Dam stretches across the Yangtze River in China. Every year, the Chinese government spends more than a million dollars to remove tons of domestic garbage from the river. The garbage has the potential to block and jam the shipping locks in the dam. It may also damage boats and pollute the water. Approximately 150 million people live upstream from the dam. Unfortunately, it is common practice to allow household waste and agricultural pollutants to run off into the river. In this section, we will work more with radicals and solving equations that contain them. For instance, if a pollutant enters the river at a constant rate, its plume disperses down the river and may be represented by the equation $y = \sqrt{x}$. We will learn how to solve for the variables in this equation. The first step is to eliminate the radical. To do so, we need an additional property.

A Solving Equations with Radicals

> **PROPERTY** Squaring Property of Equality
>
> If both sides of an equation are squared, the solutions to the original equation are solutions to the resulting equation.

We will never lose solutions to our equations by squaring both sides. We may, however, introduce *extraneous solutions*. Extraneous solutions satisfy the equation obtained by squaring both sides of the original equation, but do not satisfy the original equation.

We know that if two real numbers a and b are equal, then so are their squares:

$$\text{If} \qquad a = b$$

$$\text{then} \qquad a^2 = b^2$$

On the other hand, extraneous solutions are introduced when we square opposites. That is, even though opposites are not equal, their squares are. For example,

$$5 = -5 \qquad \textit{A false statement}$$

$$(5)^2 = (-5)^2 \qquad \textit{Square both sides.}$$

$$25 = 25 \qquad \textit{A true statement}$$

We are free to square both sides of an equation any time it is convenient. We must be aware, however, that doing so may introduce extraneous solutions. We must, therefore, check all our solutions in the original equation if at any time we square both sides of the original equation.

EXAMPLE 1 Solve $\sqrt{3x + 4} = 5$.

Solution We square both sides and proceed as usual.

$$\sqrt{3x + 4} = 5$$

$$(\sqrt{3x + 4})^2 = 5^2$$

$$3x + 4 = 25$$

$$3x = 21$$

$$x = 7$$

Checking $x = 7$ in the original equation, we have

$$\sqrt{3(7) + 4} \stackrel{?}{=} 5$$

$$\sqrt{21 + 4} \stackrel{?}{=} 5$$

$$\sqrt{25} \stackrel{?}{=} 5$$

$$5 = 5$$

The solution 7 satisfies the original equation.

EXAMPLE 2 Solve $\sqrt{4x - 7} = -3$.

Solution Squaring both sides, we have

$$\sqrt{4x - 7} = -3$$

$$(\sqrt{4x - 7})^2 = (-3)^2$$

$$4x - 7 = 9$$

$$4x = 16$$

$$x = 4$$

Checking $x = 4$ in the original equation gives

$$\sqrt{4(4) - 7} \stackrel{?}{=} -3$$

$$\sqrt{16 - 7} \stackrel{?}{=} -3$$

$$\sqrt{9} \stackrel{?}{=} -3$$

$$3 = -3$$

The value 4 produces a false statement when checked in the original equation. Because $x = 4$ was the only possible solution, there is no solution to the original equation. The possible solution $x = 4$ is an extraneous solution. It satisfies the equation obtained by squaring both sides of the original equation, but does not satisfy the original equation.

EXAMPLE 3 Solve $\sqrt{5x - 1} + 3 = 7$.

Solution We must isolate the radical on the left side of the equation. If we attempt to square both sides without doing so, the resulting equation will also contain a radical. Adding -3 to both sides, we have

$$\sqrt{5x - 1} + 3 = 7$$

$$\sqrt{5x - 1} = 4$$

We can now square both sides and proceed as usual.

$$(\sqrt{5x - 1})^2 = 4^2$$

$$5x - 1 = 16$$

$$5x = 17$$

$$x = \frac{17}{5}$$

---

1. Solve $\sqrt{2x + 4} = 4$.

2. Solve $\sqrt{7x - 3} = -5$.

Note The fact that there is no solution to the equation in Example 2 was obvious to begin with. Notice that the left side of the equation is the positive square root of $4x - 7$, which must be a positive number or 0. The right side of the equation is -3. Because we cannot have a number that is either positive or zero equal to a negative number, there is no solution to the equation. It is always important to check our work just in case any extraneous solutions exist.

3. Solve $\sqrt{4x + 5} + 2 = 7$.

Answers

1. 6
2. No solution
3. 5

Checking $x = \dfrac{17}{5}$, we have

$$\sqrt{5\left(\dfrac{17}{5}\right) - 1} + 3 \stackrel{?}{=} 7$$

$$\sqrt{17 - 1} + 3 \stackrel{?}{=} 7$$

$$\sqrt{16} + 3 \stackrel{?}{=} 7$$

$$4 + 3 \stackrel{?}{=} 7$$

$$7 = 7$$

The solution is $\dfrac{17}{5}$.

4. Solve $t - 6 = \sqrt{t - 4}$.

EXAMPLE 4   Solve $t + 5 = \sqrt{t + 7}$.

Solution   This time, squaring both sides of the equation results in a quadratic equation.

$$(t + 5)^2 = (\sqrt{t + 7})^2 \qquad \text{Square both sides.}$$

$$t^2 + 10t + 25 = t + 7$$

$$t^2 + 9t + 18 = 0 \qquad \text{Standard form}$$

$$(t + 3)(t + 6) = 0 \qquad \text{Factor the left side.}$$

$$t + 3 = 0 \quad \text{or} \quad t + 6 = 0 \qquad \text{Set factors equal to 0.}$$

$$t = -3 \qquad\qquad t = -6$$

We must check each solution in the original equation.

Check $t = -3$ Check $t = -6$

$-3 + 5 \stackrel{?}{=} \sqrt{-3 + 7}$ $-6 + 5 \stackrel{?}{=} \sqrt{-6 + 7}$

$2 \stackrel{?}{=} \sqrt{4}$ $-1 \stackrel{?}{=} \sqrt{1}$

$2 = 2$ A true statement $-1 = 1$ A false statement

Because $t = -6$ does not check, our only solution is -3.

5. Solve $\sqrt{x - 9} = \sqrt{x} - 3$.

EXAMPLE 5   Solve $\sqrt{x - 3} = \sqrt{x} - 3$.

Solution   We begin by squaring both sides. Note carefully what happens when we square the right side of the equation, and compare the square of the right side with the square of the left side. You must convince yourself that these results are correct. (The note in the margin will help if you are having trouble convincing yourself that what is written below is true.)

$$(\sqrt{x - 3})^2 = (\sqrt{x} - 3)^2$$

$$x - 3 = x - 6\sqrt{x} + 9$$

Now we still have a radical in our equation, so we will have to square both sides again. Before we do, though, let's isolate the remaining radical.

$$x - 3 = x - 6\sqrt{x} + 9$$

$$-3 = -6\sqrt{x} + 9 \qquad \text{Add } -x \text{ to each side.}$$

$$-12 = -6\sqrt{x} \qquad \text{Add } -9 \text{ to each side.}$$

$$2 = \sqrt{x} \qquad \text{Divide each side by } -6.$$

$$4 = x \qquad \text{Square each side.}$$

Note   It is very important that you realize that the square of $(\sqrt{x} - 3)$ is not $x + 9$. Remember, when we square a difference with two terms, we use the formula

$$(a - b)^2 = a^2 - 2ab + b^2$$

Applying this formula to $(\sqrt{x} - 3)^2$, we have

$$(\sqrt{x} - 3)^2 =$$
$$(\sqrt{x})^2 - 2(\sqrt{x})(3) + 3^2$$
$$= x - 6\sqrt{x} + 9$$

Answers

4. 8

5. 9

Our only possible solution is $x = 4$, which we check in our original equation as follows:

$$\sqrt{4-3} \stackrel{?}{=} \sqrt{4} - 3$$

$$\sqrt{1} \stackrel{?}{=} 2 - 3$$

$$1 = -1 \qquad \text{A false statement}$$

Substituting 4 for x in the original equation yields a false statement. Because 4 was our only possible solution, there is no solution to our equation.

Here is another example of an equation for which we must apply our squaring property twice before all radicals are eliminated.

EXAMPLE 6 Solve $\sqrt{x+1} = 1 - \sqrt{2x}$.

Solution This equation has two separate terms involving radical signs.
Squaring both sides gives

$$x + 1 = 1 - 2\sqrt{2x} + 2x$$

$$-x = -2\sqrt{2x} \qquad \text{Add } -2x \text{ and } -1 \text{ to both sides.}$$

$$x^2 = 4(2x) \qquad \text{Square both sides.}$$

$$x^2 - 8x = 0 \qquad \text{Standard form}$$

Our equation is a quadratic equation in standard form. To solve for x, we factor the left side and set each factor equal to 0.

$$x(x-8) = 0 \qquad \text{Factor left side.}$$

$$x = 0 \quad \text{or} \quad x - 8 = 0 \qquad \text{Set factors equal to 0.}$$

$$x = 8$$

Because we squared both sides of our equation, we have the possibility that one or both of the solutions are extraneous. We must check each one in the original equation.

Check $x = 8$ Check $x = 0$

$$\sqrt{8+1} \stackrel{?}{=} 1 - \sqrt{2 \cdot 8} \qquad\qquad \sqrt{0+1} \stackrel{?}{=} 1 - \sqrt{2 \cdot 0}$$

$$\sqrt{9} \stackrel{?}{=} 1 - \sqrt{16} \qquad\qquad\quad \sqrt{1} \stackrel{?}{=} 1 - \sqrt{0}$$

$$3 \stackrel{?}{=} 1 - 4 \qquad\qquad\qquad\quad 1 \stackrel{?}{=} 1 - 0$$

$$3 = -3 \quad \text{A false statement} \qquad 1 = 1 \quad \text{A true statement}$$

Because 8 does not check, it is an extraneous solution. Our only solution is 0.

EXAMPLE 7 Solve $\sqrt{x+1} = \sqrt{x+2} - 1$.

Solution Squaring both sides we have

$$(\sqrt{x+1})^2 = (\sqrt{x+2} - 1)^2$$

$$x + 1 = x + 2 - 2\sqrt{x+2} + 1$$

Once again, we are left with a radical in our equation. Before we square each side again, we must isolate the radical on the right side of the equation.

$$x + 1 = x + 3 - 2\sqrt{x+2} \qquad \text{Simplify the right side.}$$

$$1 = 3 - 2\sqrt{x+2} \qquad \text{Add } -x \text{ to each side.}$$

$$-2 = -2\sqrt{x+2} \qquad \text{Add } -3 \text{ to each side.}$$

$$1 = \sqrt{x+2} \qquad \text{Divide each side by } -2.$$

$$1 = x + 2 \qquad \text{Square both sides.}$$

$$-1 = x \qquad \text{Add } -2 \text{ to each side.}$$

6. Solve $\sqrt{x+4} = 2 - \sqrt{3x}$.

7. Solve $\sqrt{x+2} = \sqrt{x+3} - 1$.

Answers

6. 0

7. -2

Checking our only possible solution, -1, in our original equation, we have

$$\sqrt{-1+1} \stackrel{?}{=} \sqrt{-1+2} - 1$$

$$\sqrt{0} \stackrel{?}{=} \sqrt{1} - 1$$

$$0 \stackrel{?}{=} 1 - 1$$

$$0 = 0 \qquad \text{A true statement}$$

Our solution checks.

8. Solve
$\sqrt{5x-4} - \sqrt{x+5} = \sqrt{x-3}$.

EXAMPLE 8   Solve $\sqrt{2x-10} = \sqrt{x-4} + \sqrt{x-12}$.

Solution   We start by squaring both sides of the equation.

$$(\sqrt{2x-10})^2 = (\sqrt{x-4} + \sqrt{x-12})^2$$

$$2x - 10 = x - 4 + 2\sqrt{x^2-16x+48} + x - 12 \quad (a+b)^2 = a^2 + 2ab + b^2$$

$$2x - 10 = 2x - 16 + 2\sqrt{x^2-16x+48} \qquad \text{Combine like terms.}$$

Then we need to square both sides again and put the equation in standard form.

$$6 = 2\sqrt{x^2-16x+48} \qquad \text{Add } -2x \text{ and } 16 \text{ to both sides.}$$

$$3 = \sqrt{x^2-16x+48} \qquad \text{Divide both sides by 2.}$$

$$3^2 = (\sqrt{x^2-16x+48})^2 \qquad \text{Square both sides.}$$

$$9 = x^2 - 16x + 48$$

$$x^2 - 16x + 39 = 0 \qquad \text{Put the equation in standard form.}$$

$$(x-3)(x-13) = 0 \qquad \text{Factor.}$$

$$x - 3 = 0 \quad \text{or} \quad x - 13 = 0 \qquad \text{Zero factor property}$$

$$x = 3 \quad \text{or} \quad x = 13$$

Now we must check our answers in the original equation.

$$\sqrt{2(3)-10} \stackrel{?}{=} \sqrt{3-4} + \sqrt{3-12} \qquad \text{Check } x = 3.$$

$$\sqrt{6-10} \stackrel{?}{=} \sqrt{-1} + \sqrt{-9}$$

Since the square root of a negative number is not a real number, $x = 3$ is not a solution.

$$\sqrt{2(13)-10} \stackrel{?}{=} \sqrt{13-4} + \sqrt{13-12} \quad \text{Check } x = 13.$$

$$\sqrt{16} \stackrel{?}{=} \sqrt{9} + \sqrt{1}$$

$$4 = 3 + 1$$

So $x = 13$ is the only real solution.

It is also possible to raise both sides of an equation to powers greater than 2. We only need to check for extraneous solutions when we raise both sides of an equation to an even power. Raising both sides of an equation to an odd power will not produce extraneous solutions.

9. Solve $\sqrt[3]{3x-7} = 2$.

EXAMPLE 9   Solve $\sqrt[3]{4x+5} = 3$.

Solution   Cubing both sides, we have

$$(\sqrt[3]{4x+5})^3 = 3^3$$

$$4x + 5 = 27$$

$$4x = 22$$

$$x = \frac{22}{4}$$

$$x = \frac{11}{2}$$

Answers

8. 4

9. 5

We do not need to check $x = \frac{11}{2}$ because we raised both sides to an odd power.

B Graphing Radical Functions

We end this section by looking at graphs of some functions that contain radicals.

EXAMPLE 10 Graph $y = \sqrt{x}$ and $y = \sqrt[3]{x}$.

Solution The graphs are shown in Figures 1 and 2. Notice that aside from beginning at the origin, the graph of $y = \sqrt{x}$ appears in the first quadrant because in the equation $y = \sqrt{x}$, x and y cannot be negative.

The graph of $y = \sqrt[3]{x}$ appears in Quadrants 1 and 3 because the cube root of a positive number is also a positive number, and the cube root of a negative number is a negative number. That is, when x is positive, y will be positive, and when x is negative, y will be negative.

The graphs of both equations will contain the origin, because $y = 0$ when $x = 0$ in both equations.

10. Graph $y = \sqrt{x} + 3$ and $y = \sqrt{x + 3}$.

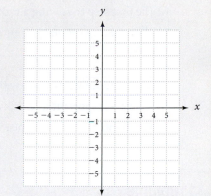

x	y
-4	Undefined
-1	Undefined
0	0
1	1
4	2
9	3
16	4

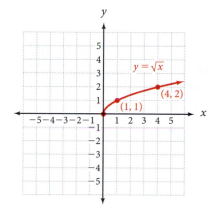

FIGURE 1

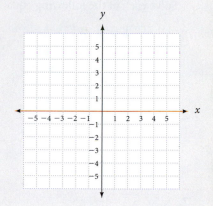

x	y
-27	-3
-8	-2
-1	-1
0	0
1	1
8	2
27	3

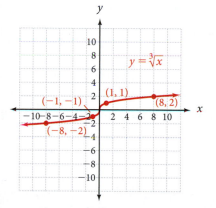

FIGURE 2

Vocabulary Review

Choose the correct words to fill in the blanks below.

negative   solutions   original   extraneous   radical

1. The first step to solving a radical equation is to eliminate the _____.

2. The squaring property of equality states that if both sides of an equation are squared, the solutions to the original equation are _____ to the resulting equation.

3. _____ solutions satisfy the equation obtained by squaring both sides of the original equation, but do not satisfy the original equation.

4. It is very important to check any solutions in the _____ equation to avoid extraneous solutions.

5. The graph of $y = \sqrt{x}$ appears in the first quadrant only because x and y cannot be _____.

Problems

A Solve each of the following equations.

1. $\sqrt{2x + 1} = 3$

2. $\sqrt{3x + 1} = 4$

3. $\sqrt{4x + 1} = -5$

4. $\sqrt{6x + 1} = -5$

5. $\sqrt{2y - 1} = 3$

6. $\sqrt{3y - 1} = 2$

7. $\sqrt{5x - 7} = -1$

8. $\sqrt{8x + 3} = -6$

9. $\sqrt{2x - 3} - 2 = 4$

10. $\sqrt{3x + 1} - 4 = 1$

11. $\sqrt{4a + 1} + 3 = 2$

12. $\sqrt{5a - 3} + 6 = 2$

13. $\sqrt[4]{3x + 1} = 2$

14. $\sqrt[4]{4x + 1} = 3$

15. $\sqrt[3]{2x - 5} = 1$

SCAN TO ACCESS

16. $\sqrt[3]{5x + 7} = 2$

17. $\sqrt[3]{3a + 5} = -3$

18. $\sqrt[3]{2a + 7} = -2$

19. $\sqrt{y-3} = y-3$

20. $\sqrt{y+3} = y-3$

21. $\sqrt{a+2} = a+2$

22. $\sqrt{a+10} = a-2$

23. $\sqrt{2x+4} = \sqrt{1-x}$

24. $\sqrt{3x+4} = -\sqrt{2x+3}$

25. $\sqrt{4a+7} = -\sqrt{a+2}$

26. $\sqrt{7a-1} = \sqrt{2a+4}$

27. $\sqrt[4]{5x-8} = \sqrt[4]{4x-1}$

28. $\sqrt[4]{6x+7} = \sqrt[4]{x+2}$

29. $x+1 = \sqrt{5x+1}$

30. $x-1 = \sqrt{6x+1}$

31. $t+5 = \sqrt{2t+9}$

32. $t+7 = \sqrt{2t+13}$

33. $\sqrt{y-8} = \sqrt{8-y}$

34. $\sqrt{2y+5} = \sqrt{5y+2}$

35. $\sqrt[3]{3x+5} = \sqrt[3]{5-2x}$

36. $\sqrt[3]{4x+9} = \sqrt[3]{3-2x}$

The following equations will require that you square both sides twice before all the radicals are eliminated. Solve each equation using the methods shown in Examples 5, 6, 7, and 8.

37. $\sqrt{x-8} = \sqrt{x}-2$

38. $\sqrt{x+3} = \sqrt{x}-3$

39. $\sqrt{x+1} = \sqrt{x}+1$

40. $\sqrt{x-1} = \sqrt{x}-1$

41. $\sqrt{x+8} = \sqrt{x-4}+2$

42. $\sqrt{x+5} = \sqrt{x-3}+2$

43. $\sqrt{x-5} - 3 = \sqrt{x-8}$

44. $\sqrt{x-3} - 4 = \sqrt{x-3}$

45. $\sqrt{x+4} = 2 - \sqrt{2x}$

46. $\sqrt{5x+1} = 1 + \sqrt{5x}$

47. $\sqrt{2x+4} = \sqrt{x+3} + 1$

48. $\sqrt{2x-1} = \sqrt{x-4} + 2$

49. $\sqrt{2x-5} + \sqrt{3x-5} = \sqrt{2x+3}$

50. $\sqrt{3x+5} + \sqrt{3x-3} = \sqrt{4x+4}$

51. $\sqrt{2a-2} + \sqrt{4a+3} = \sqrt{2a+5}$

52. $\sqrt{2a-3} + \sqrt{a+4} = \sqrt{4a+5}$

53. $\sqrt{4x-5} - \sqrt{x-5} = \sqrt{3x-4}$

54. $\sqrt{4x+1} - \sqrt{2x+4} = \sqrt{x-5}$

Applying the Concepts

55. Solving a Formula Solve the following formula for h:

$$t = \frac{\sqrt{100-h}}{4}$$

56. Solving a Formula Solve the following formula for h:

$$t = \sqrt{\frac{2h-40t}{g}}$$

57. Pendulum Clock The length of time T in seconds it takes the pendulum of a clock to swing through one complete cycle is given by the formula

$$T = 2\pi\sqrt{\frac{L}{32}}$$

where L is the length, in feet, of the pendulum, and π is approximately $\frac{22}{7}$. How long must the pendulum be if one complete cycle takes 2 seconds?

58. Pendulum Clock Solve the formula in Problem 51 for L.

59. Similar Rectangles Two rectangles are similar if their vertices lie along the same diagonal, as shown in the following diagram.

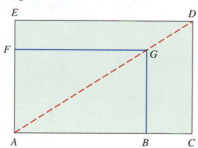

If two rectangles are similar, then corresponding sides are in proportion, which means that $\frac{ED}{DC} = \frac{FG}{GB}$. Use these facts in the following diagram to express the length of the larger rectangle l in terms of x.

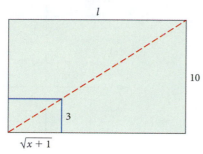

60. Volume Recall that the volume of a box V can be found from the formula $V = (\text{length})(\text{width})(\text{height})$. Find the volume of the following box in terms of x.

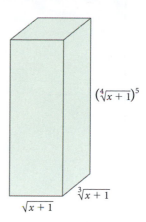

Pollution A long straight river, 100 meters wide, is flowing at 1 meter per second. A pollutant is entering the river at a constant rate from one of its banks. As the pollutant disperses in the water, it forms a plume that is modeled by the equation $y = \sqrt{x}$. Use this information to answer the following questions.

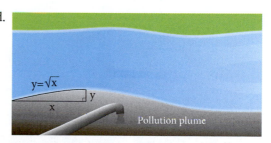

61. How wide is the plume 25 meters down river from the source of the pollution?

62. How wide is the plume 100 meters down river from the source of the pollution?

63. How far down river from the source of the pollution does the plume reach halfway across the river?

64. How far down the river from the source of the pollution does the plume reach the other side of the river?

65. For the situation described in the instructions and modeled by the equation $y = \sqrt{x}$, what is the range of values that y can assume?

66. If the river was moving at 2 meters per second, would the plume be larger or smaller 100 meters downstream from the source?

B Graph each equation.

67. $y = 2\sqrt{x}$

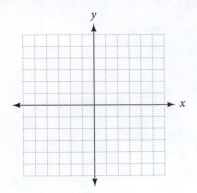

68. $y = -2\sqrt{x}$

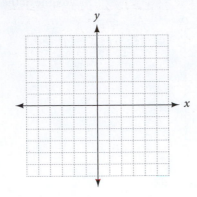

69. $y = \sqrt{x} - 2$

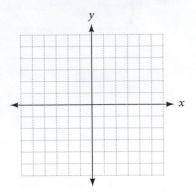

70. $y = \sqrt{x} + 2$

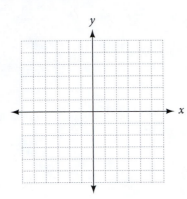

71. $y = \sqrt{x - 2}$

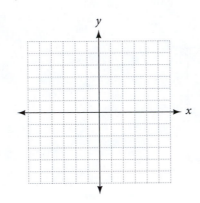

72. $y = \sqrt{x + 2}$

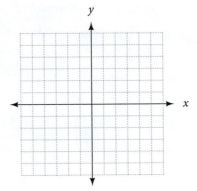

73. $y = 3\sqrt[3]{x}$

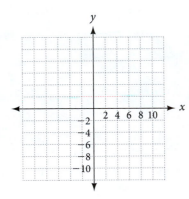

74. $y = -3\sqrt[3]{x}$

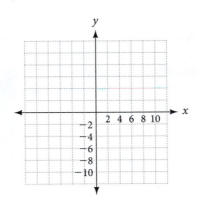

75. $y = \sqrt[3]{x} + 3$

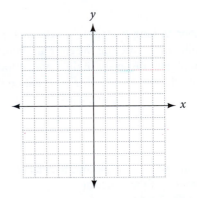

76. $y = \sqrt[3]{x} - 3$

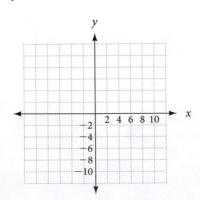

77. $y = \sqrt[3]{x + 3}$

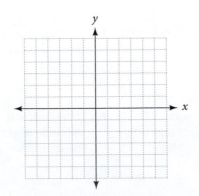

78. $y = \sqrt[3]{x - 3}$

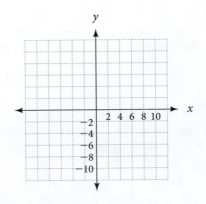

Getting Ready for the Next Section

Simplify.

79. $\sqrt{25}$

80. $\sqrt{49}$

81. $\sqrt{12}$

82. $\sqrt{50}$

83. $(-1)^{15}$

84. $(-1)^{20}$

85. $(-1)^{50}$

86. $(-1)^5$

Solve each equation.

87. $3x = 12$

88. $4 = 8y$

89. $4x - 3 = 5$

90. $7 = 2y - 1$

Perform the indicated operation.

91. $(3 + 4x) + (7 - 6x)$

92. $(2 - 5x) + (-1 + 7x)$

93. $(7 + 3x) - (5 + 6x)$

94. $(5 - 2x) - (9 - 4x)$

95. $(3 - 4x)(2 + 5x)$

96. $(8 + x)(7 - 3x)$

97. $2x(4 - 6x)$

98. $3x(7 + 2x)$

99. $(2 + 3x)^2$

100. $(3 + 5x)^2$

101. $(2 - 3x)(2 + 3x)$

102. $(4 - 5x)(4 + 5x)$

Find the Mistake

Each sentence below contains a mistake. Circle the mistake and write the correct word or phrase on the line provided.

1. Solving $\sqrt{6x + 1} = -5$ gives us a solution of $x = 4$. _____

2. Solving $\sqrt[4]{3x - 8} = 2$ gives us an extraneous solution. _____

3. Solving $\sqrt{x + 8} = x + 6$ gives a solution of $x = -7$. _____

4. The graphs of $y = \sqrt{x - 1}$ and $y = \sqrt{2x - 6}$ will intersect at $(-5, 2)$. _____

KEY WORDS

the number *i*

imaginary unit

complex number

a + *bi* standard form

imaginary number

complex conjugates

©iStockphoto.com/phuxy

In 2010, a magnitude 7.2 earthquake struck Baja California. Residents in the capital city of Mexicali reportedly felt intense shaking that lasted for nearly a minute and a half. Widespread soil liquefaction damaged much of the city's buildings and infrastructure. Earthquake liquefaction occurs when shaking from a seismic event turns the once solid ground to a consistency similar to quicksand. Buildings and bridges can literally sink into the ground! Some civil engineers study soil liquefaction using a special group of numbers called complex numbers, which is the focus of this section.

The equation $x^2 = -9$ has no real number solutions because the square of a real number is always positive. We have been unable to work with square roots of negative numbers like $\sqrt{-25}$ and $\sqrt{-16}$ for the same reason. Complex numbers allow us to expand our work with radicals to include square roots of negative numbers and to solve equations like $x^2 = -9$ and $x^2 = -64$. Our work with complex numbers is based on the following definition.

A Square Roots of Negative Numbers

> **DEFINITION** the number *i*
>
> The **number *i*,** called the *imaginary unit*, is such that $i = \sqrt{-1}$ (which is the same as saying $i^2 = -1$).

The number *i*, as we have defined it here, is not a real number. Because of the way we have defined *i*, we can use it to simplify square roots of negative numbers.

> **PROPERTY** Square Roots of Negative Numbers
>
> If a is a positive number, then $\sqrt{-a}$ can always be written as $i\sqrt{a}$. That is,
>
> $$\sqrt{-a} = i\sqrt{a} \qquad \text{if } a \text{ is a positive number}$$

To justify our rule, we simply square the quantity $i\sqrt{a}$ to obtain $-a$. Here is what it looks like when we do so:

$$(i\sqrt{a})^2 = i^2 \cdot (\sqrt{a})^2$$
$$= -1 \cdot a$$
$$= -a$$

Here are some examples that illustrate the use of our new rule:

EXAMPLE 1 Write each square root in terms of the number i.

a. $\sqrt{-25} = i\sqrt{25} = i \cdot 5 = 5i$

b. $\sqrt{-49} = i\sqrt{49} = i \cdot 7 = 7i$

c. $\sqrt{-12} = i\sqrt{12} = i \cdot 2\sqrt{3} = 2i\sqrt{3}$

d. $\sqrt{-17} = i\sqrt{17}$

B Powers of i

If we assume all the properties of exponents hold when the base is i, we can write any power of i as i, -1, $-i$, or 1. Using the fact that $i^2 = -1$, we have

$$i^1 = i$$
$$i^2 = -1$$
$$i^3 = i^2 \cdot i = -1(i) = -i$$
$$i^4 = i^2 \cdot i^2 = -1(-1) = 1$$

Because $i^4 = 1$, i^5 will simplify to i, and we will begin repeating the sequence i, -1, $-i$, 1 as we simplify higher powers of i: Any power of i simplifies to i, -1, $-i$, or 1. The easiest way to simplify higher powers of i is to write them in terms of i^2. For instance, to simplify i^{21}, we would write it as

$$(i^2)^{10} \cdot i \qquad \text{because} \qquad 2 \cdot 10 + 1 = 21$$

Then, because $i^2 = -1$, we have

$$(-1)^{10} \cdot i = 1 \cdot i = i$$

EXAMPLE 2 Simplify as much as possible.

a. $i^{30} = (i^2)^{15} = (-1)^{15} = -1$

b. $i^{11} = (i^2)^5 \cdot i = (-1)^5 \cdot i = (-1)i = -i$

c. $i^{40} = (i^2)^{20} = (-1)^{20} = 1$

DEFINITION complex number

A *complex number* is any number that can be put in the form

$$a + bi$$

where a and b are real numbers and $i = \sqrt{-1}$. The form $a + bi$ is called *standard form* for complex numbers. The number a is called the *real part* of the complex number. The number b is called the *imaginary part* of the complex number.

Every real number is a complex number. For example, 8 can be written as $8 + 0i$. Likewise, $-\frac{1}{2}$, π, $\sqrt{3}$, and 29 are complex numbers because they can all be written in the form $a + bi$.

$$-\frac{1}{2} = -\frac{1}{2} + 0i \qquad \pi = \pi + 0i \qquad \sqrt{3} = \sqrt{3} + 0i \qquad -9 = -9 + 0i$$

The real numbers occur when $b = 0$. When $b \neq 0$, we have complex numbers that contain i, such as $2 + 5i$, $6 - i$, $4i$, and $\frac{1}{2}i$. These numbers are called *imaginary numbers*. The following diagram explains this further.

Practice Problems

1. Write each square root in terms of the number i.
   a. $\sqrt{-36}$
   b. $-\sqrt{-64}$
   c. $\sqrt{-18}$
   d. $-\sqrt{-19}$

Note In Example 1 parts c and d, we wrote i before the radical simply to avoid confusion. If we were to write the answer to 3 as $2\sqrt{3}i$, some people would think the i was under the radical sign, but it is not.

2. Simplify.
   a. i^{20}
   b. i^{23}
   c. i^{50}

Answers
1. a. $6i$ b. $-8i$ c. $3i\sqrt{2}$ d. $i\sqrt{19}$
2. a. 1 b. $-i$ c. -1

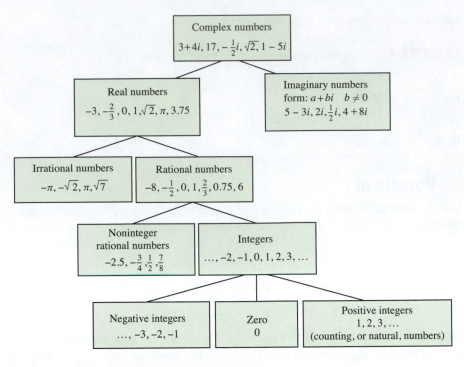

C Equality for Complex Numbers

Two complex numbers are equal if and only if their real parts are equal and their imaginary parts are equal. That is, for real numbers a, b, c, and d,

$$a + bi = c + di \quad \text{if and only if} \quad a = c \quad \text{and} \quad b = d$$

3. Find x and y if

$$4x + 7i = 8 - 14yi.$$

EXAMPLE 3   Find x and y if $3x + 4i = 12 - 8yi$.

Solution   Because the two complex numbers are equal, their real parts are equal and their imaginary parts are equal.

$$3x = 12 \quad \text{and} \quad 4 = -8y$$
$$x = 4 \qquad\qquad y = -\frac{1}{2}$$

4. Find x and y if

$$(2x - 1) + 9i = 5 + (4y + 1)i.$$

EXAMPLE 4   Find x and y if $(4x - 3) + 7i = 5 + (2y - 1)i$.

Solution   The real parts are $4x - 3$ and 5. The imaginary parts are 7 and $2y - 1$.

$$4x - 3 = 5 \quad \text{and} \quad 7 = 2y - 1$$
$$4x = 8 \qquad\qquad 8 = 2y$$
$$x = 2 \qquad\qquad y = 4$$

D Addition and Subtraction of Complex Numbers

To add two complex numbers, add their real parts and their imaginary parts. That is, if a, b, c, and d are real numbers, then

$$(a + bi) + (c + di) = (a + c) + (b + d)i$$

If we assume that the commutative, associative, and distributive properties hold for the number i, then the definition of addition is simply an extension of these properties.

We define subtraction in a similar manner. If a, b, c, and d are real numbers, then

$$(a + bi) - (c + di) = (a - c) + (b - d)i$$

Answers

3. $x = 2, y = -\frac{1}{2}$

4. $x = 3, y = 2$

EXAMPLE 5   Add or subtract as indicated.

a. $(3 + 4i) + (7 - 6i) = (3 + 7) + (4 - 6)i = 10 - 2i$

b. $(7 + 3i) - (5 + 6i) = (7 - 5) + (3 - 6)i = 2 - 3i$

c. $(5 - 2i) - (9 - 4i) = (5 - 9) + (-2 + 4)i = -4 + 2i$

E Multiplication of Complex Numbers

Because complex numbers have the same form as binomials, we find the product of two complex numbers the same way we find the product of two binomials.

EXAMPLE 6   Multiply $(3 - 4i)(2 + 5i)$.

Solution   Multiplying each term in the second complex number by each term in the first, we have

$$\overset{F\quad O\quad I\quad L}{(3 - 4i)(2 + 5i) = 3 \cdot 2 + 3 \cdot 5i - 2 \cdot 4i - 4i(5i)}$$

$$= 6 + 15i - 8i - 20i^2$$

Combining similar terms and using the fact that $i^2 = -1$, we can simplify as follows:

$$6 + 15i - 8i - 20i^2 = 6 + 7i - 20(-1)$$

$$= 6 + 7i + 20$$

$$= 26 + 7i$$

The product of the complex numbers $3 - 4i$ and $2 + 5i$ is the complex number $26 + 7i$.

EXAMPLE 7   Multiply $2i(4 - 6i)$.

Solution   Applying the distributive property gives us

$$2i(4 - 6i) = 2i \cdot 4 - 2i \cdot 6i$$

$$= 8i - 12i^2$$

$$= 12 + 8i$$

EXAMPLE 8   Expand $(3 + 5i)^2$.

Solution   We treat this like the square of a binomial. Remember, $(a + b)^2 = a^2 + 2ab + b^2$.

$$(3 + 5i)^2 = 3^2 + 2(3)(5i) + (5i)^2$$

$$= 9 + 30i + 25i^2$$

$$= 9 + 30i - 25$$

$$= -16 + 30i$$

EXAMPLE 9   Multiply $(2 - 3i)(2 + 3i)$.

Solution   This product has the form $(a - b)(a + b)$, which we know results in the difference of two squares, $a^2 - b^2$.

$$(2 - 3i)(2 + 3i) = 2^2 - (3i)^2$$

$$= 4 - 9i^2$$

$$= 4 + 9$$

$$= 13$$

The product of the two complex numbers $2 - 3i$ and $2 + 3i$ is the real number 13. The two complex numbers $2 - 3i$ and $2 + 3i$ are called complex conjugates. The fact that their product is a real number is very useful.

5. Add or subtract as indicated.
   a. $(2 + 6i) + (3 - 4i)$
   b. $(6 + 5i) - (4 + 3i)$
   c. $(7 - i) - (8 - 2i)$

6. Multiply $(2 + 3i)(1 - 4i)$.

7. Multiply $-3i(2 + 3i)$.

8. Expand $(2 + 4i)^2$.

9. Multiply $(3 - 5i)(3 + 5i)$.

Answers
5. **a.** $5 + 2i$ **b.** $2 + 2i$ **c.** $-1 + i$
6. $14 - 5i$
7. $9 - 6i$
8. $-12 + 16i$
9. 34

> **DEFINITION**   complex conjugates
>
> The complex numbers $a + bi$ and $a - bi$ are called *complex conjugates*. One important property they have is that their product is the real number $a^2 + b^2$. Here's why:
>
> $$(a + bi)(a - bi) = a^2 - (bi)^2$$
> $$= a^2 - b^2 i^2$$
> $$= a^2 - b^2(-1)$$
> $$= a^2 + b^2$$

F Division With Complex Numbers

The fact that the product of two complex conjugates is a real number is the key to division with complex numbers.

10. Divide $\dfrac{3 + 2i}{2 - 5i}$.

EXAMPLE 10   Divide $\dfrac{2 + i}{3 - 2i}$.

Solution   We want a complex number in standard form that is equivalent to the quotient $\frac{2+i}{3-2i}$. We need to eliminate i from the denominator. Multiplying the numerator and denominator by $3 + 2i$ will give us what we want.

$$\frac{2 + i}{3 - 2i} = \frac{2 + i}{3 - 2i} \cdot \frac{(3 + 2i)}{(3 + 2i)}$$

$$= \frac{6 + 4i + 3i + 2i^2}{9 - 4i^2}$$

$$= \frac{6 + 7i - 2}{9 + 4}$$

$$= \frac{4 + 7i}{13}$$

$$= \frac{4}{13} + \frac{7}{13}i$$

Dividing the complex number $2 + i$ by $3 - 2i$ gives the complex number $\frac{4}{13} + \frac{7}{13}i$.

11. Divide $\dfrac{3 + 2i}{i}$.

EXAMPLE 11   Divide $\dfrac{7 - 4i}{i}$.

Solution   The conjugate of the denominator is $-i$. Multiplying numerator and denominator by this number, we have

$$\frac{7 - 4i}{i} = \frac{7 - 4i}{i} \cdot \frac{-i}{-i}$$

$$= \frac{-7i + 4i^2}{-i^2}$$

$$= \frac{-7i + 4(-1)}{-(-1)}$$

$$= -4 - 7i$$

> **GETTING READY FOR CLASS**
>
> *After reading through the preceding section, respond in your own words and in complete sentences.*
>
> **A.** What is the number i?
>
> **B.** Explain why every real number is a complex number.
>
> **C.** What kind of number will always result when we multiply complex conjugates?
>
> **D.** Explain how to divide complex numbers.

Answers

10. $-\dfrac{4}{29} + \dfrac{19}{29}i$

11. $2 - 3i$

Vocabulary Review

Choose the correct words to fill in the blanks below.

zero　　　equal　　　complex number　　　number i　　　conjugates　　　imaginary

1. The _____ is such that $i = \sqrt{-1}$, which is not a real number.

2. A _____ is any number than can be put in the form $a + bi$.

3. Every real number is a complex number because the variable b in the standard form for complex numbers is _____ .

4. Two complex numbers are equal if and only if their real parts are _____ and their imaginary parts are equal.

5. To add two complex numbers, add their real parts and their _____ parts, and assume that the commutative, associative, and distributive properties hold true for the number i.

6. The complex numbers $a + bi$ and $a - bi$ are complex _____ , and have a product that is the real number $a^2 + b^2$.

Problems

A Write the following in terms of i, and simplify as much as possible.

1. $\sqrt{-36}$

2. $\sqrt{-49}$

3. $-\sqrt{-25}$

4. $-\sqrt{-81}$

5. $\sqrt{-72}$

6. $\sqrt{-48}$

7. $-\sqrt{-12}$

8. $-\sqrt{-75}$

B Write each of the following as i, -1, $-i$, or 1.

9. i^{28}

10. i^{31}

11. i^{26}

12. i^{37}

13. i^{75}

14. i^{42}

15. $(-i)^{12}$

16. $(-i)^{22}$

C Find x and y so each of the following equations is true.

17. $2x + 3yi = 6 - 3i$

18. $4x - 2yi = 4 + 8i$

19. $2 - 5i = -x + 10yi$

20. $4 + 7i = 6x - 14yi$

21. $2x + 10i = -16 - 2yi$

22. $4x - 5i = -2 + 3yi$

23. $(2x - 4) - 3i = 10 - 6yi$

24. $(4x - 3) - 2i = 8 + yi$

25. $(7x - 1) + 4i = 2 + (5y + 2)i$

26. $(5x + 2) - 7i = 4 + (2y + 1)i$

D Combine the following complex numbers.

27. $(2 + 3i) + (3 + 6i)$

28. $(4 + i) + (3 + 2i)$

29. $(3 - 5i) + (2 + 4i)$

30. $(7 + 2i) + (3 - 4i)$

31. $(5 + 2i) - (3 + 6i)$

32. $(6 + 7i) - (4 + i)$

33. $(3 - 5i) - (2 + i)$

34. $(7 - 3i) - (4 + 10i)$

35. $[(3 + 2i) - (6 + i)] + (5 + i)$

36. $[(4 - 5i) - (2 + i)] + (2 + 5i)$

37. $[(7 - i) - (2 + 4i)] - (6 + 2i)$

38. $[(3 - i) - (4 + 7i)] - (3 - 4i)$

39. $(3 + 2i) - [(3 - 4i) - (6 + 2i)]$

40. $(7 - 4i) - [(-2 + i) - (3 + 7i)]$

41. $(4 - 9i) + [(2 - 7i) - (4 + 8i)]$

42. $(10 - 2i) - [(2 + i) - (3 - i)]$

E Find the following products.

43. $3i(4 + 5i)$

44. $2i(3 + 4i)$

45. $6i(4 - 3i)$

46. $11i(2 - i)$

47. $(3 + 2i)(4 + i)$

48. $(2 - 4i)(3 + i)$

49. $(4 + 9i)(3 - i)$

50. $(5 - 2i)(1 + i)$

51. $(1 + i)^3$

52. $(1 - i)^3$

53. $(2 - i)^3$

54. $(2 + i)^3$

55. $(2 + 5i)^2$

56. $(3 + 2i)^2$

57. $(1 - i)^2$

58. $(1 + i)^2$

59. $(3 - 4i)^2$

60. $(6 - 5i)^2$

61. $(2 + i)(2 - i)$

62. $(3 + i)(3 - i)$

63. $(6 - 2i)(6 + 2i)$

64. $(5 + 4i)(5 - 4i)$

65. $(2 + 3i)(2 - 3i)$

66. $(2 - 7i)(2 + 7i)$

67. $(10 + 8i)(10 - 8i)$

68. $(11 - 7i)(11 + 7i)$

69. $(\sqrt{3} + 2i)(\sqrt{3} - 2i)$

70. $(\sqrt{7} + \sqrt{5}i)(\sqrt{7} - \sqrt{5}i)$

F Find the following quotients. Write all answers in standard form for complex numbers.

71. $\dfrac{2 - 3i}{i}$

72. $\dfrac{3 + 4i}{i}$

73. $\dfrac{5 + 2i}{-i}$

74. $\dfrac{4 - 3i}{-i}$

75. $\dfrac{4}{2 - 3i}$

76. $\dfrac{3}{4 - 5i}$

77. $\dfrac{6}{-3 + 2i}$

78. $\dfrac{-1}{-2 - 5i}$

79. $\dfrac{2 + 3i}{2 - 3i}$

80. $\dfrac{4 - 7i}{4 + 7i}$

81. $\dfrac{5 + 4i}{3 + 6i}$

82. $\dfrac{2 + i}{5 - 6i}$

Applying the Concepts

83. Electric Circuits Complex numbers may be applied to electrical circuits. Electrical engineers use the fact that resistance R, electrical current I and voltage V are related by the formula $V = RI$. (Voltage is measured in volts, resistance in ohms, and current in amperes.) Find the resistance to electrical flow in a circuit that has a voltage $V = (80 + 20i)$ volts and current $I = (-6 + 2i)$ amps.

84. Electric Circuits Refer to the information about electrical circuits in Problem 83, and find the current in a circuit that has a resistance of $(4 + 10i)$ ohms and a voltage of $(5 - 7i)$ volts.

Maintaining Your Skills.

Solve each equation.

85. $\dfrac{t}{3} - \dfrac{1}{2} = -1$

86. $\dfrac{x}{x-2} + \dfrac{2}{3} = \dfrac{2}{x-2}$

87. $2 + \dfrac{5}{y} = \dfrac{3}{y^2}$

88. $1 - \dfrac{1}{y} = \dfrac{12}{y^2}$

Solve each application problem.

89. The sum of a number and its reciprocal is $\dfrac{41}{20}$. Find the number.

90. It takes an inlet pipe 8 hours to fill a tank. The drain can empty the tank in 6 hours. If the tank is full and both the inlet pipe and drain are open, how long will it take to drain the tank?

91. The chart shows the average movie ticket price from 2000 to 2009. Find the slope of the line that connects the first and last point. Then, use it to predict the cost of a movie ticket in 2015. Round to the nearest cent.

Find the Mistake

Each sentence below contains a mistake. Circle the mistake and write the correct word or phrase on the line provided.

1. The number i is such that $i = \sqrt{-1}$ is the same as saying $i^2 = \dfrac{1}{i}$. _____

2. The form $a + bi$ is called standard form for complex numbers and the i is the imaginary part of the number.

_____

3. The product of the complex numbers $(5 - 3i)$ and $(2 + 4i)$ is $48i$. _____

4. The product of two complex conjugates $6 + 4i$ and $6 - 4i$ is $6^2 - 4^2$. _____

Imaginary Numbers

Knowing that complex numbers are created from imaginary numbers can make it difficult to picture their use in real life. However, the use of complex numbers in science is vast, and the history of this imaginary concept that led to our modern advancements dates as far back as the 16th century. The Italian mathematician Gerolamo Cardano (1501–1576) was the first to acknowledge in print the existence of imaginary numbers; however, his colleague Rafael Bombelli (1526–1572) was the first to describe and solve equations with them in detail. Today, fields such as applied mathematics, engineering, and quantum physics often use the concepts derived from these Italian mathematicians. Research the history of imaginary and complex numbers, as well as the history behind the use of imaginary numbers in today's modern sciences. Create a timeline that documents the discoveries and contributions by important people related to this concept.

Chapter 6 Summary

The numbers in brackets refer to the section(s) in which the topic can be found.

EXAMPLES

1. The number 49 has two square roots, 7 and −7. They are written like this:

$$\sqrt{49} = 7 \qquad -\sqrt{49} = -7$$

Square Roots [6.1]

Every positive real number x has two square roots. The *positive square root* of x is written $\sqrt{x}$, and the *negative square root* of x is written $-\sqrt{x}$. Both the positive and the negative square roots of x are numbers we square to get x; that is,

$$\left. \begin{array}{c} (\sqrt{x})^2 = x \\ \text{and} \quad (-\sqrt{x})^2 = x \end{array} \right\} \text{ for } x \geq 0$$

General Roots [6.1]

2. $\sqrt[3]{8} = 2$

$\sqrt[3]{-27} = -3$

In the expression $\sqrt[n]{a}$, n is the *index*, a is the *radicand*, and $\sqrt{}$ is the *radical sign*. The expression $\sqrt[n]{a}$ is such that

$$(\sqrt[n]{a})^n = a \qquad a \geq 0 \text{ when } n \text{ is even}$$

Rational Exponents [6.1]

3. $25^{1/2} = \sqrt{25} = 5$

$8^{2/3} = (\sqrt[3]{8})^2 = 2^2 = 4$

$9^{3/2} = (\sqrt{9})^3 = 3^3 = 27$

Rational exponents are used to indicate roots. The relationship between rational exponents and roots is as follows:

$$a^{1/n} = \sqrt[n]{a} \qquad \text{and} \qquad a^{m/n} = (a^{1/n})^m = (a^m)^{1/n}$$

$$a \geq 0 \text{ when } n \text{ is even}$$

Properties of Radicals [6.2]

4. $\sqrt{4 \cdot 5} = \sqrt{4} \cdot \sqrt{5} = 2\sqrt{5}$

$\sqrt{\dfrac{7}{9}} = \dfrac{\sqrt{7}}{\sqrt{9}} = \dfrac{\sqrt{7}}{3}$

If a and b are nonnegative real numbers whenever n is even, then

1. $\sqrt[n]{ab} = \sqrt[n]{a} \, \sqrt[n]{b}$ Product property for radicals

2. $\sqrt[n]{\dfrac{a}{b}} = \dfrac{\sqrt[n]{a}}{\sqrt[n]{b}} \qquad (b \neq 0)$ Quotient property for radicals

Simplifying Radicals [6.2]

5. $\sqrt{\dfrac{4}{5}} = \dfrac{\sqrt{4}}{\sqrt{5}}$

$= \dfrac{2}{\sqrt{5}} \cdot \dfrac{\sqrt{5}}{\sqrt{5}}$

$= \dfrac{2\sqrt{5}}{5}$

A radical expression is said to be in *simplified form* if

1. there is no factor of the radicand that can be written as a power greater than or equal to the index;

2. there are no fractions under the radical sign; and

3. there are no radicals in the denominator.

Addition and Subtraction of Radical Expressions [6.3]

We add and subtract radical expressions by using the distributive property to combine similar radicals. Similar radicals are radicals with the same index and the same radicand.

6. $5\sqrt{3} - 7\sqrt{3} = (5 - 7)\sqrt{3}$
$= -2\sqrt{3}$

$\sqrt{20} + \sqrt{45} = 2\sqrt{5} + 3\sqrt{5}$
$= (2 + 3)\sqrt{5}$
$= 5\sqrt{5}$

Multiplication of Radical Expressions [6.4]

We multiply radical expressions in the same way that we multiply polynomials. We can use the distributive property and the FOIL method.

7. $(\sqrt{x} + 2)(\sqrt{x} + 3)$
$= \sqrt{x} \cdot \sqrt{x} + 3\sqrt{x} + 2\sqrt{x} + 2 \cdot 3$
$= x + 5\sqrt{x} + 6$

Rationalizing the Denominator [6.2, 6.4]

When a fraction contains a square root in the denominator, we rationalize the denominator by multiplying the numerator and denominator by

1. The square root itself if there is only one term in the denominator, or

2. The conjugate of the denominator if there are two terms in the denominator.

Rationalizing the denominator is also called division of radical expressions.

8. $\dfrac{3}{\sqrt{2}} = \dfrac{3}{\sqrt{2}} \cdot \dfrac{\sqrt{2}}{\sqrt{2}} = \dfrac{3\sqrt{2}}{2}$

$\dfrac{3}{\sqrt{5} - \sqrt{3}} = \dfrac{3}{\sqrt{5} - \sqrt{3}} \cdot \dfrac{\sqrt{5} + \sqrt{3}}{\sqrt{5} + \sqrt{3}}$

$= \dfrac{3\sqrt{5} + 3\sqrt{3}}{5 - 3}$

$= \dfrac{3\sqrt{5} + 3\sqrt{3}}{2}$

Squaring Property of Equality [6.5]

We may square both sides of an equation any time it is convenient to do so, as long as we check all resulting solutions in the original equation.

9. $\sqrt{2x + 1} = 3$
$(\sqrt{2x + 1})^2 = 3^2$
$2x + 1 = 9$
$x = 4$

Complex Numbers [6.6]

A *complex number* is any number that can be put in the form

$$a + bi$$

where a and b are real numbers and $i = \sqrt{-1}$. The *real part* of the complex number is a, and b is the *imaginary part*.

If a, b, c, and d are real numbers, then we have the following definitions associated with complex numbers:

1. *Equality*

$$a + bi = c + di \quad \text{if and only if} \quad a = c \text{ and } b = d$$

2. *Addition and subtraction*

$$(a + bi) + (c + di) = (a + c) + (b + d)i$$
$$(a + bi) - (c + di) = (a - c) + (b - d)i$$

3. *Multiplication*

$$(a + bi)(c + di) = (ac - bd) + (ad + bc)i$$

4. *Division is similar to rationalizing the denominator.*

10. $3 + 4i$ is a complex number.

Addition
$(3 + 4i) + (2 - 5i) = 5 - i$

Multiplication
$(3 + 4i)(2 - 5i)$
$= 6 - 15i + 8i - 20i^2$
$= 6 - 7i + 20$
$= 26 - 7i$

Division
$\dfrac{2}{3 + 4i} = \dfrac{2}{3 + 4i} \cdot \dfrac{3 - 4i}{3 - 4i}$

$= \dfrac{6 - 8i}{9 + 16}$

$= \dfrac{6}{25} - \dfrac{8}{25}i$

Simplify each of the following. (Assume all variable bases are positive integers and all variable exponents are positive real numbers throughout this test.) [6.1]

1. $27^{-2/3}$

2. $\left(\dfrac{144}{49}\right)^{-1/2}$

3. $x^{2/3} \cdot x^{1/5}$

4. $\dfrac{(a^{1/4}b)^{-2}}{(a^{3/2}b^6)^{-1}}$

5. $\sqrt{49x^8y^{12}}$

6. $\sqrt[3]{27x^6y^{12}}$

7. $\dfrac{(49a^8b^{10})^{1/2}}{(27a^{15}b^3)^{1/3}}$

8. $\dfrac{(x^{1/n}y^{2/n})^n}{(x^{1/n}y^n)^{n_2}}$

Write in simplified form. [6.2]

9. $\sqrt{50x^3y^7}$

10. $\sqrt[3]{135x^5y^7}$

11. $\sqrt{\dfrac{5}{7}}$

12. $\sqrt{\dfrac{12a^3b^5}{5c^2}}$

Rationalize the denominator. [6.2]

13. $\dfrac{3}{\sqrt{5}-2}$

14. $\dfrac{\sqrt{x}-\sqrt{5}}{\sqrt{x}+\sqrt{5}}$

Combine. [6.3]

15. $\dfrac{3}{x^{1/3}} + x^{2/3}$

16. $\dfrac{x^4}{(x^4-6)^{1/2}} - (x^4-6)^{1/2}$

17. $2\sqrt{12} - 2\sqrt{48}$

18. $3\sqrt[3]{40a^3b^5} - a\sqrt[3]{5b^5}$

Multiply. [6.4]

19. $3a^{5/2}(2a^{3/2} - a^{7/2})$

20. $(3a^{4/5} - 2)^2$

21. $(\sqrt{x} - 3)(\sqrt{x} + 5)$

22. $(4\sqrt{3} - \sqrt{5})^2$

Solve for x. [6.5]

23. $\sqrt{x+7} = x - 5$

24. $\sqrt[3]{5x+4} = 4$

25. $\sqrt{x+8} = \sqrt{x-16} + 4$

Graph. [6.5]

26. $y = \sqrt{x} - 3$

27. $y = \sqrt[3]{x} - 1$

Match each equation with its graph. [6.5]

28. $y = \sqrt[3]{x} - 2$

29. $y = \sqrt{x} - 2$

30. $y = \sqrt{x}$

31. $y = \sqrt[3]{x} - 2$

32. $y = \sqrt{x} + 2$

33. $y = -\sqrt{x} - 2$

A.

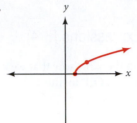

B.

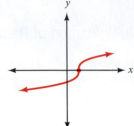

C.

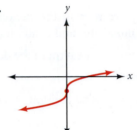

D.

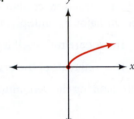

E.

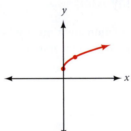

F.

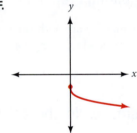

34. Find x and y so that the following equation is true. [6.6]
$$9 - (2x - 3)i = (4y - 3) + 5i$$

Perform the indicated operations. [6.6]

35. $(5 + 3i) - [(4 - 2i) - (3 + 7i)]$

36. $(3 - 2i)(4 + 3i)$

37. $(5 - 3i)^2$

38. $\dfrac{4i - 1}{4i + 1}$

39. Show that i^{39} can be written as $-i$. [6.6]

Simplify.

1. $2 - [1 + 4(8 - 9)]$

2. $3 + 21 \div 7 + 4 \cdot 2$

3. $5 + (7 - 3)^2 - (4 + 1)^2$

4. $\sqrt[4]{128}$

5. $27^{2/3} + 25^{1/2}$

6. $\dfrac{1 - \dfrac{4}{5}}{1 + \dfrac{1}{5}}$

7. $\left(-\dfrac{8}{125}\right)^{-1/3}$

8. $32 + [4 - 9 \div 3(6 - 9)]$

Reduce.

9. $\dfrac{104}{117}$

10. $\dfrac{18x^2 - 21xy - 15y^2}{6x + 3y}$

11. $\dfrac{x^2 + x - 2}{x^2 + 5x + 6}$

12. $\dfrac{4a^8 b^{-2}}{16a^3 b^{-8}}$

Multiply or divide.

13. $(3\sqrt{x} - 5)(\sqrt{x} + 1)$

14. $(8 + 2i)(6 - i)$

15. $\dfrac{3 - i}{3 - 2i}$

16. $\dfrac{\sqrt{5}}{2 - \sqrt{5}}$

17. $(4x + 2y)(7x - y)$

18. $(16x^2 - 9) \cdot \dfrac{x - 7}{4x - 3}$

19. $\dfrac{24x^7 y^{-3}}{5x^2 y^4} \div \dfrac{6x^2 y^2}{25x^3 y}$

20. $\dfrac{\sqrt{x} - \sqrt{y}}{\sqrt{x} + \sqrt{y}}$

Solve.

21. $\dfrac{1}{2} + \dfrac{2}{a + 3} = \dfrac{1}{a + 3}$

22. $\sqrt{5 - x} = x - 5$

23. $(3x + 4)^2 = 7$

24. $0.02x^2 + 0.07x = 0.04$

25. $\sqrt{x - 7} = 7 - \sqrt{x}$

26. $3x^3 - 4x = -4x^2$

27. $\dfrac{1}{10}x^2 + \dfrac{2}{5}x = \dfrac{1}{2}$

28. $\dfrac{1}{8}(16x - 2) + \dfrac{1}{4} = 6$

29. $|8x - 9| - 4 = 3$

30. $\sqrt[3]{15 - 4x} = -1$

Solve each inequality, and graph the solution.

31. $-1 \le \dfrac{x}{5} - 5 \le 1$

32. $|7x + 2| > 3$

Solve each system.

33. $-9x + 6y = 30$
$\quad\ 4x - 5y = -25$

34. $2x + 5y = -23$
$\quad\ 7x - 3y = 22$

Graph on a rectangular coordinate system.

35. $2x + 3y = 3$

36. $-x + 2y < -2$

Find the slope of the line passing through the points.

37. $(-5, 7)$ and $(-8, -5)$

38. $(6, 13)$ and $(5, 2)$

Rationalize the denominator.

39. $\dfrac{6}{\sqrt[4]{27}}$

40. $\dfrac{5}{\sqrt{7} - 3}$

Factor completely.

41. $81a^4 - 256b^4$

42. $24a^4 - 44a^2 - 28$

43. Inverse Variation y varies inversely with the square root of x. If $y = -1$ when $x = 9$, find y when $x = 16$.

44. Joint Variation z varies jointly with x and the cube of y. If $z = -32$ when $x = 4$, and $y = 2$, find z when $x = 2$ and $y = 3$.

Let $f(x) = 2x + 2$ and $g(x) = x^2 + x$. Find the following.

45. $f(b)$

46. $(g - f)(3)$

47. $(g \circ f)(1)$

48. $(f \circ g)(-1)$

The chart shows the yearly sales for the top frozen pizza retailers. Use the information to answer the following questions.

TOP FROZEN PIZZA SELLERS
www.aibonline.org, 2009

DiGiorno	$591,262,700
Tombstone	$270,412,700
Red Baron	$256,308,000
California Pizza Kitchen	$175,750,800
Totino's Party Pizza	$152,630,700

49. What is the difference in sales between California Pizza Kitchen and Red Baron?

50. What is the difference in sales between DiGiorno and Totinos?

Simplify each of the following. (Assume all variable bases are positive integers and all variable exponents are positive real numbers throughout this test.) [6.1]

1. $16^{-3/4}$

2. $\left(\dfrac{81}{64}\right)^{-1/2}$

3. $x^{1/7} \cdot x^{-2/3}$

4. $\dfrac{(a^{1/3}b^2)^{-1}}{(a^{2/5}b^{-5})^{-2}}$

5. $\sqrt{25x^6 y^{20}}$

6. $\sqrt[4]{81x^4 y^{20}}$

7. $\dfrac{(8a^{12}b^3)^{1/3}}{(49a^{12}b^8)^{1/2}}$

8. $\dfrac{(x^{1/n}y^n)^{n_2}}{(x^n y^{1/n})^{n_2}}$

Write in simplified form. [6.2]

9. $\sqrt{27x^5 y^3}$

10. $\sqrt[3]{128x^2 y^7}$

11. $\sqrt{\dfrac{3}{5}}$

12. $\sqrt{\dfrac{8a^2 b^7}{7c}}$

Rationalize the denominator. [6.2]

13. $\dfrac{7}{\sqrt{2}+1}$

14. $\dfrac{\sqrt{x}+\sqrt{3}}{\sqrt{x}-\sqrt{3}}$

Combine. [6.3]

15. $\dfrac{5}{x^{1/2}} - x^{1/2}$

16. $\dfrac{x^3}{(x^3-17)^{1/2}} - (x^3-17)^{1/2}$

17. $3\sqrt{8} - 2\sqrt{18}$

18. $4\sqrt[3]{32a^6 b^3} - 5a^2\sqrt[3]{4b^3}$

Multiply. [6.4]

19. $4a^{3/2}(a^{5/2} - 3a^{1/2})$

20. $(2a^{5/2} - 3)^2$

21. $(\sqrt{x} - 1)(\sqrt{x} + 5)$

22. $(2\sqrt{3} - \sqrt{2})^2$

Solve for x. [6.5]

23. $\sqrt{2x-3} = x - 3$

24. $\sqrt[3]{4x-2} = -2$

25. $\sqrt{x+6} = \sqrt{x-9} + 3$

Graph. [6.5]

26. $y = \sqrt{x+1}$

27. $y = \sqrt[3]{x} - 2$

Match each equation with its graph. [6.5]

28. $y = \sqrt{x}$

29. $y = \sqrt{x-1}$

30. $y = \sqrt{x} + 1$

31. $y = \sqrt[3]{x} + 1$

32. $y = \sqrt[3]{x-1}$

33. $y = -\sqrt[3]{x+1}$

A.

B.

C.

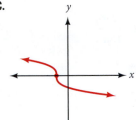

D.

E.

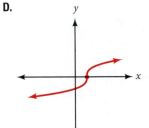

F.

34. Find x and y so that the following equation is true. [6.6]

$$8 - (x-2)i = (7y+1) + 4i$$

Perform the indicated operations. [6.6]

35. $(2 + i) - [(5 - 2i) - (3 + 4i)]$

36. $(5 + 3i)(2 - i)$

37. $(3 - 6i)^2$

38. $\dfrac{3i+2}{3i-2}$

39. Show that i^{36} can be written as 1. [6.6]

Quadratic Equations and Functions

Gray Buildings © 2008 Sanborn
St. Louis, Missouri

Known as the "Gateway to the West," the St. Louis Gateway Arch is one of the most recognizable structures in the United States. As the tallest monument in the country, it stands 630 feet tall and spans 630 feet across its base. The cross-sections of its legs are isosceles triangles and are hollow to accommodate a unique tram system. Visitors ride these trams up the inside of the arch's legs to an enclosed observation deck at the top. Nearly a million visitors ride the trams each year. The trams have been in operation for over 30 years, traveling a total of 250,000 miles and carrying over 25 million passengers.

While the architect used what is called a catenary curve in the arch's design, the shape of the arch can be closely modeled by the quadratic equation $y = -\frac{1}{150}x^2 + \frac{21}{5}x$. The equation can be graphed on a rectangular coordinate system, and the result is called a parabola, shown below.

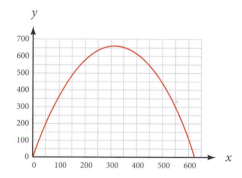

The parabola is one of the topics covered in this chapter. You will also be asked to sketch this graph in one of the problem sets.

A Solve quadratic equations by taking the square root of both sides.

B Solve quadratic equations by completing the square.

C Use quadratic equations to solve for missing parts of right triangles.

7.1 Completing the Square

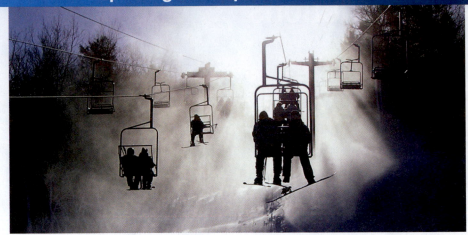

Image © sxc.hu, abejo, 2009

Table 1 is taken from the trail map given to skiers at the Northstar at Tahoe Ski Resort in Lake Tahoe, California. The table gives the length of each chair lift at Northstar, along with the change in elevation from the beginning of the lift to the end of the lift.

Table 1	Lift Information from the Trail Map for the Northstar at Tahoe Ski Resort	
Lift	Vertical Rise (feet)	Length (feet)
Big Springs Gondola	480	4,100
Bear Paw Double	120	790
Echo Triple	710	4,890
Aspen Express Quad	900	5,100
Forest Double	1,170	5,750
Lookout Double	960	4,330
Comstock Express Quad	1,250	5,900
Rendezvous Triple	650	2,900
Schaffer Camp Triple	1,860	6,150
Chipmunk Tow Lift	28	280
Bear Cub Tow Lift	120	750

Right triangles are good mathematical models for chair lifts. Can you picture why? In this section, we will use our knowledge of right triangles, along with the new material developed in the section, to solve problems involving chair lifts and a variety of other examples.

A The Square Root Method

In this section, we will develop the first of our new methods of solving quadratic equations. Remember, a quadratic equation is an equation in the form $ax^2 + bx + c = 0$, where a, b, c are constants and a is not 0. The new method is called *completing the square*. Completing the square on a quadratic equation allows us to obtain solutions, regardless of whether the equation can be factored. Before we solve equations by completing the square, we need to learn how to solve equations by taking square roots of both sides.

Consider the following equation:

$$x^2 = 16$$

We could solve the equation $x^2 = 16$ by writing it in standard form and factoring the left side. We can shorten our work considerably, however, if we simply notice that x must be either the positive square root of 16 or the negative square root of 16. That is,

$$\text{If} \qquad x^2 = 16$$
$$\text{then} \qquad x = \sqrt{16} \qquad \text{or} \qquad x = -\sqrt{16}$$
$$x = 4 \qquad \qquad \qquad x = -4$$

We can generalize this result as follows:

> **PROPERTY** **Square Root Property for Equations**
>
> If $a^2 = b$, where b is a real number, then $a = \sqrt{b}$ or $a = -\sqrt{b}$.

Notation The expression $a = b$ or $a = -b$ can be written in shorthand form as $a = \pm b$. The symbol $\pm$ is read "plus or minus."

We can apply the square root property for equations to some fairly complicated quadratic equations.

EXAMPLE 1 Solve $(2x - 3)^2 = 25$.

Solution

$$(2x - 3)^2 = 25$$

$$\begin{align} 2x - 3 &= \pm\sqrt{25} & &\text{Square root property for equations} \\ 2x - 3 &= \pm 5 & &\sqrt{25} = 5 \\ 2x &= 3 \pm 5 & &\text{Add 3 to both sides.} \\ x &= \frac{3 \pm 5}{2} & &\text{Divide both sides by 2.} \end{align}$$

The last equation can be written as two separate statements:

$$x = \frac{3 + 5}{2} \qquad \text{or} \qquad x = \frac{3 - 5}{2}$$
$$= \frac{8}{2} \qquad \qquad \qquad = \frac{-2}{2}$$
$$= 4 \qquad \qquad \qquad \quad = -1$$

The solutions are $4, -1$.

Notice that we could have solved the equation in Example 1 by expanding the left side, writing the resulting equation in standard form, and then factoring. The problem would look like this:

$$\begin{align} (2x - 3)^2 &= 25 & &\text{Original equation} \\ 4x^2 - 12x + 9 &= 25 & &\text{Expand the left side.} \\ 4x^2 - 12x - 16 &= 0 & &\text{Add } -25 \text{ to each side.} \\ 4(x^2 - 3x - 4) &= 0 & &\text{Begin factoring.} \\ 4(x - 4)(x + 1) &= 0 & &\text{Factor completely.} \\ x - 4 = 0 \quad \text{or} \quad x + 1 &= 0 & &\text{Set variable factors equal to 0.} \\ x = 4 \qquad \qquad x &= -1 \end{align}$$

Notice that solving the equation by factoring leads to the same two solutions.

Practice Problems

1. Solve $(3x + 2)^2 = 16$.

Answers

1. $\frac{2}{3}, -2$

2. Solve $(4x - 3)^2 = -50$.

> **Note** We cannot solve the equation in Example 2 by factoring. If we expand the left side and write the resulting equation in standard form, we are left with a quadratic equation that does not factor.
>
> $$(3x - 1)^2 = -12$$
>
> Equation from Example 2
>
> $$9x^2 - 6x + 1 = -12$$
>
> Expand the left side.
>
> $$9x^2 - 6x + 13 = 0$$
>
> Standard form, but not factorable.

EXAMPLE 2   Solve $(3x - 1)^2 = -12$ for x.

Solution

$$(3x - 1)^2 = -12$$

$$3x - 1 = \pm\sqrt{-12} \qquad \text{Square root property}$$

$$3x - 1 = \pm 2i\sqrt{3} \qquad \sqrt{-12} = 2i\sqrt{3}$$

$$3x = 1 \pm 2i\sqrt{3} \qquad \text{Add 1 to both sides.}$$

$$x = \frac{1 \pm 2i\sqrt{3}}{3} \qquad \text{Divide both sides by 3.}$$

The solutions are $\dfrac{1 + 2i\sqrt{3}}{3}, \dfrac{1 - 2i\sqrt{3}}{3}$. We can combine these as $\dfrac{1 \pm 2i\sqrt{3}}{3}$.

Both solutions are complex. Here is a check of the first solution:

$$\text{When} \rightarrow \qquad x = \frac{1 + 2i\sqrt{3}}{3}$$

$$\text{the equation} \rightarrow \qquad (3x - 1)^2 = -12$$

$$\text{becomes} \rightarrow \qquad \left(3 \cdot \frac{1 + 2i\sqrt{3}}{3} - 1\right)^2 \overset{?}{=} -12$$

$$(1 + 2i\sqrt{3} - 1)^2 \overset{?}{=} -12$$

$$(2i\sqrt{3})^2 \overset{?}{=} -12$$

$$4 \cdot i^2 \cdot 3 \overset{?}{=} -12$$

$$12(-1) \overset{?}{=} -12$$

$$-12 = -12$$

3. Solve $x^2 + 10x + 25 = 20$.

EXAMPLE 3   Solve $x^2 + 6x + 9 = 12$.

Solution   We can solve this equation as we have the equations in Examples 1 and 2 if we first write the left side as $(x + 3)^2$.

$$x^2 + 6x + 9 = 12 \qquad \text{Original equation}$$

$$(x + 3)^2 = 12 \qquad \text{Write } x^2 + 6x + 9 \text{ as } (x + 3)^2.$$

$$x + 3 = \pm 2\sqrt{3} \qquad \text{Square root property}$$

$$x = -3 \pm 2\sqrt{3} \qquad \text{Add } -3 \text{ to each side.}$$

We have two irrational solutions: $-3 + 2\sqrt{3}$ and $-3 - 2\sqrt{3}$. What is important about this problem, however, is the fact that the equation was easy to solve because the left side was a perfect square trinomial.

B Completing the Square

The method of completing the square is simply a way of transforming any quadratic equation into an equation of the form found in the preceding three examples. The key to understanding the method of completing the square lies in recognizing the relationship between the last two terms of any perfect square trinomial whose leading coefficient is 1.

Consider the following list of perfect square trinomials and their corresponding binomial squares:

$$x^2 - 6x + 9 = (x - 3)^2$$

$$x^2 + 8x + 16 = (x + 4)^2$$

$$x^2 - 10x + 25 = (x - 5)^2$$

$$x^2 + 12x + 36 = (x + 6)^2$$

Answers

2. $\dfrac{3 \pm 5i\sqrt{2}}{4}$

3. $-5 \pm 2\sqrt{5}$

In each case, the leading coefficient is 1. A more important observation comes from noticing the relationship between the linear and constant terms (middle and last terms) in each trinomial. Observe that the constant term in each case is the square of half the coefficient of x in the middle term. For example, in the last expression, the constant term 36 is the square of half of 12, where 12 is the coefficient of x in the middle term. (Notice also that the second terms in all the binomials on the right side are half the coefficients of the middle terms of the trinomials on the left side.) We can use these observations to build our own perfect square trinomials and, in doing so, solve some quadratic equations.

EXAMPLE 4   Solve $x^2 - 6x + 5 = 0$ by completing the square.

Solution   We begin by adding -5 to both sides of the equation. We want just $x^2 - 6x$ on the left side so that we can add on our own final term to get a perfect square trinomial.

$$x^2 - 6x + 5 = 0$$

$$x^2 - 6x \quad\; = -5 \qquad \text{Add } -5 \text{ to both sides.}$$

Now we can add 9 to both sides and the left side will be a perfect square.

$$x^2 - 6x + 9 = -5 + 9$$

$$(x - 3)^2 = 4$$

The final line is in the form of the equations we solved previously.

$$x - 3 = \pm 2$$

$$x = 3 \pm 2 \qquad \text{Add 3 to both sides.}$$

$$x = 3 + 2 \quad \text{or} \quad x = 3 - 2$$

$$x = 5 \qquad\qquad\; x = 1$$

The two solutions are 5 and 1.

- - -

EXAMPLE 5   Solve $x^2 + 5x - 2 = 0$ by completing the square.

Solution   We must begin by adding 2 to both sides. (The left side of the equation, as it is, is not a perfect square, because it does not have the correct constant term. We will simply "move" that term to the other side and use our own constant term.)

$$x^2 + 5x = 2 \qquad \text{Add 2 to each side.}$$

We complete the square by adding the square of half the coefficient of the linear term to both sides.

$$x^2 + 5x + \frac{25}{4} = 2 + \frac{25}{4} \qquad \text{Half of 5 is } \tfrac{5}{2}, \text{ the square of which}$$
$$\text{is } \left(\tfrac{5}{2}\right)^2 = \tfrac{25}{4}.$$

$$\left(x + \frac{5}{2}\right)^2 = \frac{33}{4} \qquad 2 + \frac{25}{4} = \frac{8}{4} + \frac{25}{4} = \frac{33}{4}$$

$$x + \frac{5}{2} = \pm\sqrt{\frac{33}{4}} \qquad \text{Square root property}$$

$$x + \frac{5}{2} = \pm\frac{\sqrt{33}}{2} \qquad \text{Simplify the radical.}$$

$$x = -\frac{5}{2} \pm \frac{\sqrt{33}}{2} \qquad \text{Add } -\tfrac{5}{2} \text{ to both sides.}$$

$$x = \frac{-5 \pm \sqrt{33}}{2}$$

The solutions are $\dfrac{-5 \pm \sqrt{33}}{2}$.

4. Solve $x^2 + 3x - 4 = 0$ by completing the square.

Note   The equation in Example 4 can be solved quickly by factoring.

$$x^2 - 6x + 5 = 0$$
$$(x - 5)(x - 1) = 0$$
$$x - 5 = 0 \quad \text{or} \quad x - 1 = 0$$
$$x = 5 \qquad\qquad x = 1$$

The reason we didn't solve it by factoring is we want to practice completing the square on some simple equations.

5. Solve $x^2 - 5x + 2 = 0$ by completing the square.

Answers

4. $1, -4$

5. $\dfrac{5 \pm \sqrt{17}}{2}$

We can use a calculator to get decimal approximations to these solutions. If $\sqrt{33} \approx 5.74$, then

$$\frac{-5 + 5.74}{2} = 0.37$$

$$\frac{-5 - 5.74}{2} = -5.37$$

6. Solve $5x^2 - 3x + 2 = 0$.

EXAMPLE 6   Solve $3x^2 - 8x + 7 = 0$.

Solution

$$3x^2 - 8x + 7 = 0$$

$$3x^2 - 8x = -7 \qquad \text{Add } -7 \text{ to both sides.}$$

We cannot complete the square on the left side because the leading coefficient is not 1. We take an extra step and divide both sides by 3.

$$\frac{3x^2}{3} - \frac{8x}{3} = -\frac{7}{3}$$

$$x^2 - \frac{8}{3}x = -\frac{7}{3}$$

Half of $\frac{8}{3}$ is $\frac{4}{3}$, the square of which is $\left(\frac{4}{3}\right)^2 = \frac{16}{9}$.

$$x^2 - \frac{8}{3}x + \frac{16}{9} = -\frac{7}{3} + \frac{16}{9} \qquad \text{Add } -\frac{6}{9} \text{ to both sides.}$$

$$\left(x - \frac{4}{3}\right)^2 = -\frac{5}{9} \qquad \text{Simplify right side.}$$

$$x - \frac{4}{3} = \pm\sqrt{-\frac{5}{9}} \qquad \text{Square root property}$$

$$x - \frac{4}{3} = \pm\frac{i\sqrt{5}}{3} \qquad \sqrt{-\frac{5}{9}} = \frac{\sqrt{-5}}{3} = \frac{i\sqrt{5}}{3}$$

$$x = \frac{4}{3} \pm \frac{i\sqrt{5}}{3} \qquad \text{Add } \frac{4}{3} \text{ to both sides.}$$

$$x = \frac{4 \pm i\sqrt{5}}{3}$$

The solutions are $\dfrac{4 \pm i\sqrt{5}}{3}$.

7. Find the x-intercepts of the function $f(x) = x^2 - 4x + 1$.

EXAMPLE 7   Find the x-intercepts of the function $f(x) = x^2 + 6x + 7$.

Solution   Recall that the x-intercepts occur where $f(x) = 0$.

$$x^2 + 6x + 7 = 0$$

$$x^2 + 6x = -7 \qquad \text{Add } -7 \text{ to both sides.}$$

Half of 6 is 3, the square of which is $3^2 = 9$.

$$x^2 + 6x + 9 = -7 + 9 \qquad \text{Add 9 to both sides.}$$

$$(x + 3)^2 = 2 \qquad \text{Simplify right side.}$$

$$x + 3 = \pm\sqrt{2} \qquad \text{Square root property}$$

$$x = -3 \pm \sqrt{2} \qquad \text{Add } -3 \text{ to both sides.}$$

The x-intercepts are $-3 + \sqrt{2}$ and $-3 - \sqrt{2}$.

Note   The x-intercepts cannot be complex, since they represent points on the x-axis.

Answers

6. $\dfrac{3 \pm i\sqrt{31}}{10}$

7. $2 \pm \sqrt{3}$

HOW TO Solve a Quadratic Equation by Completing the Square

To summarize the method used in the preceding two examples, we list the following steps:

Step 1: Write the equation in the form $ax^2 + bx = c$.

Step 2: If the leading coefficient is not 1, divide both sides by the coefficient so that the resulting equation has a leading coefficient of 1. That is, if $a \neq 1$, then divide both sides by a.

Step 3: Add the square of half the coefficient of the linear term to both sides of the equation.

Step 4: Write the left side of the equation as the square of a binomial, and simplify the right side if possible.

Step 5: Apply the square root property for equations, and solve as usual.

C Quadratic Equations and Right Triangles

FACTS FROM GEOMETRY More Special Triangles

The triangles shown in Figures 1 and 2 occur frequently in mathematics.

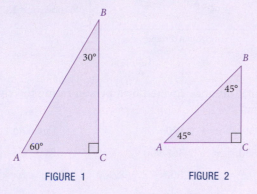

FIGURE 1   FIGURE 2

Note that both of the triangles are right triangles. We refer to the triangle in Figure 1 as a 30°–60°–90° triangle, and the triangle in Figure 2 as a 45°–45°–90° triangle.

EXAMPLE 8   If the shortest side in a 30°–60°–90° triangle is 1 inch, find the lengths of the other two sides.

Solution   In Figure 3, triangle ABC is a 30°–60°–90° triangle in which the shortest side AC is 1 inch long. Triangle DBC is also a 30°–60°–90° triangle in which the shortest side DC is 1 inch long.

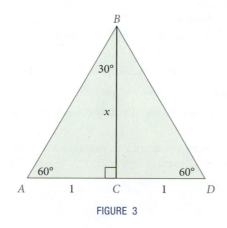

FIGURE 3

8. If the shortest side in a 30°–60°–90° triangle is 2 inches long, find the lengths of the other two sides.

Notice that the large triangle ABD is an equilateral triangle because each of its interior angles is 60°. Each side of triangle ABD is 2 inches long. Side AB in triangle ABC is therefore 2 inches. To find the length of side BC, we use the Pythagorean theorem.

$$BC^2 + AC^2 = AB^2$$
$$x^2 + 1^2 = 2^2$$
$$x^2 + 1 = 4$$
$$x^2 = 3$$
$$x = \sqrt{3} \text{ inches}$$

Note that we write only the positive square root because x is the length of a side in a triangle and is therefore a positive number.

9. Table 1 in the introduction to this section gives the vertical rise of the Lookout Double chair lift as 960 feet and the length of the chair lift as 4,330 feet. To the nearest foot, find the horizontal distance covered by a person riding the lift.

EXAMPLE 9   Table 1 in the introduction to this section gives the vertical rise of the Forest Double chair lift as 1,170 feet and the length of the chair lift as 5,750 feet. To the nearest foot, find the horizontal distance covered by a person riding this lift.

Solution   Figure 4 is a model of the Forest Double chair lift. A rider gets on the lift at point A and exits at point B. The length of the lift is AB.

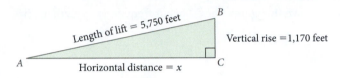

FIGURE 4

To find the horizontal distance covered by a person riding the chair lift, we use the Pythagorean theorem.

$$5,750^2 = x^2 + 1,170^2 \qquad \text{Pythagorean theorem}$$
$$33,062,500 = x^2 + 1,368,900 \qquad \text{Simplify squares.}$$
$$x^2 = 33,062,500 - 1,368,900 \qquad \text{Solve for } x^2.$$
$$x^2 = 31,693,600 \qquad \text{Simplify the right side.}$$
$$x = \sqrt{31,693,600} \qquad \text{Square root property}$$
$$\approx 5,630 \text{ feet} \qquad \text{Round to the nearest foot.}$$

A rider getting on the lift at point A and riding to point B will cover a horizontal distance of approximately 5,630 feet.

GETTING READY FOR CLASS

After reading through the preceding section, respond in your own words and in complete sentences.

A. What kind of equation do we solve using the method of completing the square?

B. Explain in words how you would complete the square on $x^2 - 16x = 4$.

C. What is the relationship between the shortest side and the longest side in a 30°–60°–90° triangle?

D. What two expressions together are equivalent to $x = \pm 4$?

Vocabulary Review

The following is a list of steps for solving a quadratic equation by completing the square. Write the correct step number in the blanks.

_____ If the leading coefficient is not 1, divide both sides by the coefficient so that the resulting equation has a leading coefficient of 1.

_____ Write the equation in the form $ax^2 + bx = c$.

_____ Write the left side of the equation as the square of a binomial, and simplify the right side if possible.

_____ Add the square of half the coefficient of the linear term to both sides of the equation.

_____ Apply the square root property for equations, and solve as usual.

Problems

A Solve the following equations.

1. $x^2 = 25$ **2.** $x^2 = 16$ **3.** $a^2 = -9$ **4.** $a^2 = -49$

5. $y^2 = \dfrac{3}{4}$ **6.** $y^2 = \dfrac{5}{9}$ **7.** $x^2 + 12 = 0$ **8.** $x^2 + 8 = 0$

9. $4a^2 - 45 = 0$ **10.** $9a^2 - 20 = 0$ **11.** $(2y - 1)^2 = 25$ **12.** $(3y + 7)^2 = 1$

13. $(2a + 3)^2 = -9$ **14.** $(3a - 5)^2 = -49$ **15.** $(5x + 2)^2 = -8$ **16.** $(6x - 7)^2 = -75$

17. $x^2 + 8x + 16 = -27$ **18.** $x^2 - 12x + 36 = -8$ **19.** $4a^2 - 12a + 9 = -4$ **20.** $9a^2 - 12a + 4 = -9$

B Copy each of the following, and fill in the blanks so the left side of each is a perfect square trinomial. That is, complete the square.

21. $x^2 + 12x + \underline{\ \ \ } = (x + \underline{\ \ \ })^2$ **22.** $x^2 + 6x + \underline{\ \ \ } = (x + \underline{\ \ \ })^2$

23. $x^2 - 4x + \underline{\ \ \ } = (x - \underline{\ \ \ })^2$ **24.** $x^2 - 2x + \underline{\ \ \ } = (x - \underline{\ \ \ })^2$

25. $a^2 - 10a + \underline{\ \ \ } = (a - \underline{\ \ \ })^2$ **26.** $a^2 - 8a + \underline{\ \ \ } = (a - \underline{\ \ \ })^2$

SCAN TO ACCESS

27. $x^2 + 5x +$ ___ $= (x +$ ___$)^2$

28. $x^2 + 3x +$ ___ $= (x +$ ___$)^2$

29. $y^2 - 7y +$ ___ $= (y -$ ___$)^2$

30. $y^2 - y +$ ___ $= (y -$ ___$)^2$

31. $x^2 + \dfrac{1}{2}x +$ ___ $= (x +$ ___$)^2$

32. $x^2 - \dfrac{3}{4}x +$ ___ $= (x -$ ___$)^2$

33. $x^2 + \dfrac{2}{3}x +$ ___ $= (x +$ ___$)^2$

34. $x^2 - \dfrac{4}{5}x +$ ___ $= (x -$ ___$)^2$

Solve each of the following quadratic equations by completing the square.

35. $x^2 + 4x = 12$

36. $x^2 - 2x = 8$

37. $x^2 + 12x = -27$

38. $x^2 - 6x = 16$

39. $a^2 - 2a + 5 = 0$

40. $a^2 + 10a + 22 = 0$

41. $y^2 - 8y + 1 = 0$

42. $y^2 + 6y - 1 = 0$

43. $x^2 - 5x - 3 = 0$

44. $x^2 - 5x - 2 = 0$

45. $2x^2 - 4x - 8 = 0$

46. $3x^2 - 9x - 12 = 0$

47. $3t^2 - 8t + 1 = 0$

48. $5t^2 + 12t - 1 = 0$

49. $4x^2 - 3x + 5 = 0$

50. $7x^2 - 5x + 2 = 0$

51. $3x^2 + 4x - 1 = 0$

52. $2x^2 + 6x - 1 = 0$

53. $2x^2 - 10x = 11$

54. $25x^2 - 20x = 1$

55. $4x^2 - 10x + 11 = 0$

56. $4x^2 - 6x + 1 = 0$

57. $27x^2 - 90x + 71 = 0$

58. $18x^2 + 12x - 23 = 0$

59. Consider the equation $x^2 = -9$.

    a. Can it be solved by factoring?

    b. Solve it.

61. Solve the equation $x^2 - 6x = 0$

    a. by factoring.

    b. by completing the square.

63. Solve the equation $x^2 + 2x = 35$

    a. by factoring.

    b. by completing the square.

65. Is $x = -3 + \sqrt{2}$ a solution to $x^2 - 6x = 7$?

67. Solve each equation.

    a. $5x - 7 = 0$

    b. $5x - 7 = 8$

    c. $(5x - 7)^2 = 8$

    d. $\sqrt{5x - 7} = 8$

    e. $\dfrac{5}{2} - \dfrac{7}{2x} = \dfrac{4}{x}$

60. Consider the equation $x^2 - 10x + 18 = 0$.

    a. Can it be solved by factoring?

    b. Solve it.

62. Solve the equation $x^2 + ax = 0$

    a. by factoring.

    b. by completing the square.

64. Solve the equation $8x^2 - 10x - 25 = 0$

    a. by factoring.

    b. by completing the square.

66. Is $x = 2 - \sqrt{5}$ a solution to $x^2 - 4x = 1$?

68. Solve each equation.

    a. $5x + 11 = 0$

    b. $5x + 11 = 9$

    c. $(5x + 11)^2 = 9$

    d. $\sqrt{5x + 11} = 9$

    e. $\dfrac{5}{3} - \dfrac{11}{3x} = \dfrac{3}{x}$

Find the x-intercepts of the given function.

69. $f(x) = x^2 - 4x - 5$ **70.** $f(x) = x^2 + 5x - 6$ **71.** $f(x) = x^2 + 3x - 5$ **72.** $f(x) = x^2 - 5x + 1$

73. $f(x) = x^2 - 7x - 9$ **74.** $f(x) = x^2 + 3x + 1$ **75.** $f(x) = x^2 + 4x + 9$ **76.** $f(x) = x^2 - 6x + 12$

77. $f(x) = 2x^2 + 3x - 4$ **78.** $f(x) = 3x^2 - 2x - 4$ **79.** $f(x) = 4x^2 + 4x - 11$ **80.** $f(x) = 9x^2 - 12x - 28$

C Applying the Concepts

81. Geometry If the shortest side in a 30°–60°–90° triangle is a $\frac{1}{2}$ meter long, find the lengths of the other two sides.

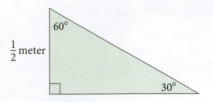

82. Geometry If the length of the longest side of a 30°–60°–90° triangle is x, find the lengths of the other two sides in terms of x.

83. Geometry If the length of the shorter sides of a 45°–45°–90° triangle is 1 inch, find the length of the hypotenuse.

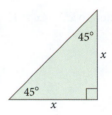

84. Geometry If the length of the shorter sides of a 45°–45°–90° triangle is x, find the length of the hypotenuse, in terms of x.

85. Chair Lift Use Table 1 from the introduction to this section to find the horizontal distance covered by a person riding the Bear Paw Double chair lift. Round your answer to the nearest foot.

86. Fermat's Last Theorem Fermat's last theorem states that if n is an integer greater than 2, then there are no positive integers x, y, and z that will make the formula $x^n + y^n = z^n$ true. Use the formula $x^n + y^n = z^n$ to find the following.

   a. Find z if $n = 2$, $x = 6$, and $y = 8$.

   b. Find y if $n = 2$, $x = 5$, and $z = 13$.

87. Interest Rate Suppose a loan of $3,000 that charges an annual interest rate r (compounded yearly) increases to $3,456 after 2 years. Using the formula $A = P(1 + r)^t$, we have

$$3{,}456 = 3{,}000(1 + r)^2$$

Solve for r to find the annual interest rate.

88. Special Triangles In Figure 5, triangle ABC has angles 45° and 30°, and height x. Find the lengths of sides AB, BC, and AC, in terms of x.

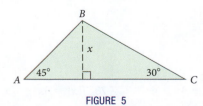

FIGURE 5

89. Length of an Escalator An escalator in a department store is made to carry people a vertical distance of 20 feet between floors. How long is the escalator if it makes an angle of 45° with the ground? (See Figure 6.)

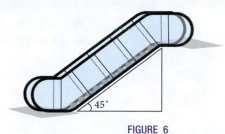

FIGURE 6

90. Dimensions of a Tent A two-person tent is to be made so the height at the center is 4 feet. If the sides of the tent are to meet the ground at an angle of 60° and the tent is to be 6 feet in length, how many square feet of material will be needed to make the tent? (Figure 7; assume that the tent has a floor and is closed at both ends.) Give your answer to the nearest tenth of a square foot.

FIGURE 7

Getting Ready for the Next Section

Simplify.

91. $49 - 4(6)(-5)$

92. $49 - 4(6)(2)$

93. $(-27)^2 - 4(0.1)(1,700)$

94. $25 - 4(4)(-10)$

95. $-7 + \dfrac{169}{12}$

96. $-7 - \dfrac{169}{12}$

97. $\dfrac{-4 + \sqrt{36}}{2}$

98. $\dfrac{-5 - \sqrt{81}}{4}$

Factor.

99. $27t^3 - 8$

100. $125t^3 + 1$

101. $2x^3 + 54$

102. $8y^3 - 27$

Maintaining Your Skills

Volunteering The chart shows how many hours per week volunteers spend volunteering. Use the chart to answer the following questions.

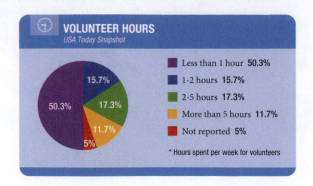

103. If 2,000 volunteers were surveyed, how many say

    a. they volunteer less than 1 hour per week?

    b. they volunteer between 2 and 5 hours per week?

104. If 785 people said they volunteer between 1 and 2 hours per week, how many volunteers were surveyed?

Find the Mistake

Each sentence below contains a mistake. Circle the mistake and write the correct word or expression on the line provided.

1. The square root property for equations says that if $a^2 = b$, then $a = \sqrt{b}$. _____

2. Solving $(4x + 1)^2 = 16$ gives us only one solution. _____

3. The first step to solving a quadratic equation by completing the square is to put the equation in standard form. _____

4. The method of completing the square says that adding the square of half the coefficient of the constant term to both sides will give a perfect square trinomial. _____

Navigation Skills: Prepare, Study, Achieve

It is very common for students to feel burned out or feel a decrease in motivation for continuing to put in the hard work this course requires. Many factors may contribute to this burn out and vary depending on the student. It is important to recognize these factors and implement ways to combat them. If you notice yourself feeling burned out, the following suggestions may help:

- Join a study group for emotional support during tough times in this class.
- Set short-term and long-term goals.
- Take breaks during your study sessions.
- Begin each study session with the most difficult topics, then work through the easier topics toward the end of your session.
- Revisit your study calendar to see if you have overbooked your time.
- Change your time and location for studying.
- Examine your diet and introduce healthier and more balanced meals.
- Get adequate restful sleep between studying and going to class.
- Schedule regular physical activity during your day.
- Make use of your on-campus health services for either physical or psychological needs.

Image © sxc.hu, andrewb, 2006

On April 22nd of each year, the people of Springfield, Missouri celebrate Secretaries Day by hurling typewriters off a fifty-foot-high platform. Each participating secretary takes her turn tossing a typewriter off the platform toward a large bull's-eye chalked on the ground below. The machine shatters as it hits the pavement. Then a representative from the sponsoring radio station measures the distance of the largest remnant from the center of the bull's eye. The participant with the closest measurement is the winner.

Let's suppose one of the typewriters is thrown downward with an initial velocity of 15 feet per second. The distance $s(t)$, in feet, the typewriter travels in t seconds is given by the function $s(t) = 15t + 16t^2$. After working similar problems in this section, you will be able to return to this one to determine how long it takes for the typewriter to reach the ground.

In this section, we will use the method of completing the square from the preceding section to derive the quadratic formula. The *quadratic formula* is a very useful tool in mathematics. It allows us to solve all types of quadratic equations.

A The Quadratic Formula

> **THEOREM** The Quadratic Formula
>
> For any quadratic equation in the form $ax^2 + bx + c = 0$, $a \neq 0$, the two solutions are
>
> $$x = \frac{-b + \sqrt{b^2 - 4ac}}{2a} \quad \text{and} \quad x = \frac{-b - \sqrt{b^2 - 4ac}}{2a}$$

Proof We will prove the quadratic formula by completing the square on $ax^2 + bx + c = 0$.

$$ax^2 + bx + c = 0$$

$$ax^2 + bx = -c \qquad \text{Add } -c \text{ to both sides.}$$

$$x^2 + \frac{b}{a}x = -\frac{c}{a} \qquad \text{Divide both sides by } a.$$

To complete the square on the left side, we add the square of $\frac{1}{2}$ of $\frac{b}{a}$ to both sides $\left(\frac{1}{2} \text{ of } \frac{b}{a} \text{ is } \frac{b}{2a} \right)$.

$$x^2 + \frac{b}{a}x + \left(\frac{b}{2a} \right)^2 = -\frac{c}{a} + \left(\frac{b}{2a} \right)^2$$

We now simplify the right side as a separate step. We combine the two terms by writing each with the least common denominator $4a^2$.

$$-\frac{c}{a} + \left(\frac{b}{2a}\right)^2 = -\frac{c}{a} + \frac{b^2}{4a^2} = \frac{4a}{4a}\left(\frac{-c}{a}\right) + \frac{b^2}{4a^2} = \frac{-4ac + b^2}{4a^2}$$

It is convenient to write this last expression as

$$\frac{b^2 - 4ac}{4a^2}$$

Continuing with the proof, we have

$$x^2 + \frac{b}{a}x + \left(\frac{b}{2a}\right)^2 = \frac{b^2 - 4ac}{4a^2}$$

$$\left(x + \frac{b}{2a}\right)^2 = \frac{b^2 - 4ac}{4a^2} \qquad \text{Write left side as a binomial square.}$$

$$x + \frac{b}{2a} = \pm\frac{\sqrt{b^2 - 4ac}}{2a} \qquad \text{Square root property for equations}$$

$$x = -\frac{b}{2a} \pm \frac{\sqrt{b^2 - 4ac}}{2a} \qquad \text{Add } -\frac{b}{2a} \text{ to both sides.}$$

$$= \frac{-b \pm \sqrt{b^2 - 4ac}}{2a}$$

Our proof is now complete. What we have is this: If our equation is in the form $ax^2 + bx + c = 0$ (standard form), where $a \neq 0$, the two solutions are always given by the formula

$$x = \frac{-b \pm \sqrt{b^2 - 4ac}}{2a}$$

This formula is known as the *quadratic formula*. If we substitute the coefficients a, b, and c of any quadratic equation in standard form into the formula, we need only perform some basic arithmetic to arrive at the solution set.

EXAMPLE 1   Solve $x^2 - 5x - 6 = 0$ using the quadratic formula.

Solution   To use the quadratic formula, we must make sure the equation is in standard form; identify a, b, and c; substitute them into the formula; and work out the arithmetic.
For the equation $x^2 - 5x - 6 = 0$, $a = 1$, $b = -5$, and $c = -6$.

$$x = \frac{-b \pm \sqrt{b^2 - 4ac}}{2a}$$

$$= \frac{-(-5) \pm \sqrt{(-5)^2 - 4(1)(-6)}}{2(1)}$$

$$= \frac{5 \pm \sqrt{49}}{2}$$

$$= \frac{5 \pm 7}{2}$$

$$x = \frac{5 + 7}{2} \quad \text{or} \quad x = \frac{5 - 7}{2}$$

$$x = \frac{12}{2} \qquad\qquad x = -\frac{2}{2}$$

$$x = 6 \qquad\qquad x = -1$$

The two solutions are 6 and -1.

EXAMPLE 2   Solve $2x^2 = -4x + 3$ using the quadratic formula.

Solution   Before we can identify a, b, and c, we must write the equation in standard form. To do so, we add $4x$ and -3 to each side of the equation.

$$2x^2 = -4x + 3$$

$$2x^2 + 4x - 3 = 0 \qquad \text{Add } 4x \text{ and } -3 \text{ to each side.}$$

Practice Problems

1. Solve $x^2 - 9x + 18 = 0$ using the quadratic formula.

Note   Whenever the solutions to our quadratic equations turn out to be rational numbers, as in Example 1, it means the original equation could have been solved by factoring. (We didn't solve the equation in Example 1 by factoring because we were trying to get some practice with the quadratic formula.)

2. Solve $6x^2 + 2 = -7x$ using the quadratic formula.

Answers

1. 3, 6

2. $-\frac{1}{2}, -\frac{2}{3}$

Now that the equation is in standard form, we see that $a = 2$, $b = 4$, and $c = -3$. Using the quadratic formula we have

$$x = \frac{-b \pm \sqrt{b^2 - 4ac}}{2a}$$

$$= \frac{-4 \pm \sqrt{4^2 - 4(2)(-3)}}{2(2)}$$

$$= \frac{-4 \pm \sqrt{40}}{4}$$

$$= \frac{-4 \pm 2\sqrt{10}}{4}$$

We can reduce the final expression in the preceding equation to lowest terms by factoring 2 from the numerator and denominator and then dividing it out.

$$x = \frac{2(-2 \pm \sqrt{10})}{2 \cdot 2}$$

$$= \frac{-2 \pm \sqrt{10}}{2}$$

Our two solutions are $\dfrac{-2 + \sqrt{10}}{2}$ and $\dfrac{-2 - \sqrt{10}}{2}$.

EXAMPLE 3   Solve $x^2 - 6x = -7$ using the quadratic formula.

Solution   We begin by writing the equation in standard form.

$$x^2 - 6x = -7$$

$$x^2 - 6x + 7 = 0 \qquad \text{Add 7 to each side.}$$

Using $a = 1$, $b = -6$, and $c = 7$ in the quadratic formula

$$x = \frac{-b \pm \sqrt{b^2 - 4ac}}{2a}$$

we have

$$x = \frac{-(-6) \pm \sqrt{(-6)^2 - 4(1)(7)}}{2(1)}$$

$$= \frac{6 \pm \sqrt{36 - 28}}{2}$$

$$= \frac{6 \pm \sqrt{8}}{2}$$

$$= \frac{6 \pm 2\sqrt{2}}{2}$$

The two terms in the numerator have a 2 in common. We reduce to lowest terms by factoring the 2 from the numerator and then dividing numerator and denominator by 2.

$$= \frac{2(3 \pm \sqrt{2})}{2}$$

$$= 3 \pm \sqrt{2}$$

The two solutions are $3 + \sqrt{2}$ and $3 - \sqrt{2}$. For practice, let's check our solutions in the original equation $x^2 - 6x = -7$.

Checking $x = 3 + \sqrt{2}$, we have

$$(3 + \sqrt{2})^2 - 6(3 + \sqrt{2}) \overset{?}{=} -7$$

$$9 + 6\sqrt{2} + 2 - 18 - 6\sqrt{2} \overset{?}{=} -7 \qquad \text{Multiply.}$$

$$11 - 18 + 6\sqrt{2} - 6\sqrt{2} \overset{?}{=} -7 \qquad \text{Add 9 and 2.}$$

$$-7 + 0 \overset{?}{=} -7 \qquad \text{Subtract.}$$

$$-7 = -7 \qquad \text{A true statement}$$

3. Solve $x^2 - 5x = 4$ using the quadratic formula.

Checking $x = 3 - \sqrt{2}$, we have

$$(3 - \sqrt{2})^2 - 6(3 - \sqrt{2}) \stackrel{?}{=} -7$$

$$9 - 6\sqrt{2} + 2 - 18 + 6\sqrt{2} \stackrel{?}{=} -7 \qquad \text{Multiply.}$$

$$11 - 18 - 6\sqrt{2} + 6\sqrt{2} \stackrel{?}{=} -7 \qquad \text{Add 9 and 2.}$$

$$-7 + 0 \stackrel{?}{=} -7 \qquad \text{Subtract.}$$

$$-7 = -7 \qquad \text{A true statement}$$

As you can see, both solutions yield true statements when used in place of the variable in the original equation.

4. Solve $\dfrac{x^2}{2} + x = \dfrac{1}{3}$. using the quadratic formula.

EXAMPLE 4   Solve $\dfrac{1}{10}x^2 - \dfrac{1}{5}x = -\dfrac{1}{2}$ using the quadratic formula.

Solution   It will be easier to apply the quadratic formula if we clear the equation of fractions. Multiplying both sides of the equation by the LCD 10 gives us

$$x^2 - 2x = -5$$

Next, we add 5 to both sides to put the equation into standard form.

$$x^2 - 2x + 5 = 0 \qquad \text{Add 5 to both sides.}$$

Applying the quadratic formula with $a = 1$, $b = -2$, and $c = 5$, we have

$$x = \frac{-(-2) \pm \sqrt{(-2)^2 - 4(1)(5)}}{2(1)} = \frac{2 \pm \sqrt{-16}}{2} = \frac{2 \pm 4i}{2}$$

Dividing the numerator and denominator by 2, we have the two solutions.

$$x = 1 \pm 2i$$

The two solutions are $1 + 2i$ and $1 - 2i$.

5. Solve $(3x - 2)(x - 1) = -3$ using the quadratic formula.

EXAMPLE 5   Solve $(2x - 3)(2x - 1) = -4$ using the quadratic formula.

Solution   We multiply the binomials on the left side and then add 4 to each side to write the equation in standard form. From there we identify a, b, and c and apply the quadratic formula.

$$(2x - 3)(2x - 1) = -4$$

$$4x^2 - 8x + 3 = -4 \qquad \text{Multiply binomials on left side.}$$

$$4x^2 - 8x + 7 = 0 \qquad \text{Add 4 to each side.}$$

Placing $a = 4$, $b = -8$, and $c = 7$ in the quadratic formula we have

$$x = \frac{-(-8) \pm \sqrt{(-8)^2 - 4(4)(7)}}{2(4)}$$

$$= \frac{8 \pm \sqrt{64 - 112}}{8}$$

$$= \frac{8 \pm \sqrt{-48}}{8}$$

$$= \frac{8 \pm 4i\sqrt{3}}{8} \qquad \qquad \sqrt{-48} = i\sqrt{48} = i\sqrt{16}\sqrt{3} = 4i\sqrt{3}$$

Note   It would be a mistake to try to reduce this final expression further. Sometimes students will try to divide the 2 in the denominator into the 2 in the numerator, which is a mistake. Remember, when we reduce to lowest terms, we do so by dividing the numerator and denominator by any factors they have in common. In this case 2 is not a factor of the numerator. This expression is in lowest terms.

To reduce this final expression to lowest terms, we factor a 4 from the numerator and then divide the numerator and denominator by 4.

$$= \frac{4(2 \pm i\sqrt{3})}{4 \cdot 2}$$

$$= \frac{2 \pm i\sqrt{3}}{2}$$

Answers

4. $\dfrac{-3 \pm \sqrt{15}}{3}$

5. $\dfrac{5 \pm i\sqrt{35}}{6}$

Although the equation in our next example is not a quadratic equation, we solve it by using both factoring and the quadratic formula.

EXAMPLE 6 Solve $27t^3 - 8 = 0$.

Solution It would be a mistake to add 8 to each side of this equation and then take the cube root of each side because we would lose two of our solutions. Instead, we factor the left side, and then set the factors equal to 0.

$$27t^3 - 8 = 0 \qquad \text{Equation in standard form}$$

$$(3t - 2)(9t^2 + 6t + 4) = 0 \qquad \text{Factor as the difference of two cubes.}$$

$$3t - 2 = 0 \quad \text{or} \quad 9t^2 + 6t + 4 = 0 \quad \text{Set each factor equal to 0.}$$

The first equation leads to a solution of $t = \frac{2}{3}$. The second equation does not factor, so we use the quadratic formula with $a = 9$, $b = 6$, and $c = 4$.

$$t = \frac{-6 \pm \sqrt{36 - 4(9)(4)}}{2(9)}$$

$$= \frac{-6 \pm \sqrt{36 - 144}}{18}$$

$$= \frac{-6 \pm \sqrt{-108}}{18}$$

$$= \frac{-6 \pm 6i\sqrt{3}}{18} \qquad \sqrt{-108} = i\sqrt{36 \cdot 3} = 6i\sqrt{3}$$

$$= \frac{6(-1 \pm i\sqrt{3})}{6 \cdot 3} \qquad \text{Factor 6 from the numerator and denominator.}$$

$$= \frac{-1 \pm i\sqrt{3}}{3} \qquad \text{Divide out common factor 6.}$$

The three solutions to our original equation are

$$\frac{2}{3}, \quad \frac{-1 + i\sqrt{3}}{3}, \text{ and } \frac{-1 - i\sqrt{3}}{3}$$

B Applications

EXAMPLE 7 If an object is thrown downward with an initial velocity of 20 feet per second, the distance $s(t)$, in feet, it travels in t seconds is given by the function $s(t) = 20t + 16t^2$. How long does it take the object to fall 40 feet?

Solution We let $s(t) = 40$, and solve for t.

$$\text{When} \rightarrow \qquad s(t) = 40$$

$$\text{the function} \rightarrow \qquad s(t) = 20t + 16t^2$$

$$\text{becomes} \rightarrow \qquad 40 = 20t + 16t^2$$

$$16t^2 + 20t - 40 = 0$$

$$4t^2 + 5t - 10 = 0 \qquad \text{Divide by 4.}$$

Using the quadratic formula, we have

$$t = \frac{-5 \pm \sqrt{25 - 4(4)(-10)}}{2(4)}$$

$$= \frac{-5 \pm \sqrt{185}}{8}$$

$$t = \frac{-5 + \sqrt{185}}{8} \quad \text{or} \quad t = \frac{-5 - \sqrt{185}}{8}$$

The second solution is impossible because it is a negative number and time t must be positive. It takes

$$t = \frac{-5 + \sqrt{185}}{8} \quad \text{or approximately} \quad \frac{-5 + 13.60}{8} \approx 1.08 \text{ seconds}$$

for the object to fall 40 feet.

6. Solve $8t^3 - 27 = 0$.

7. An object thrown upward with an initial velocity of 32 feet per second rises and falls according to the equation $s = 32t - 16t^2$ where s is the height of the object above the ground at any time t. At what times t will the object be 12 feet above the ground?

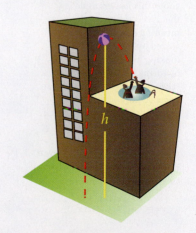

Recall that the relationship between profit, revenue, and cost is given by the formula

$$P(x) = R(x) - C(x)$$

where $P(x)$ is the profit, $R(x)$ is the total revenue, and $C(x)$ is the total cost of producing and selling x items.

8. Use the information in Example 8 to find the number of programs the company must sell each week for its weekly profit to be $1,320.

EXAMPLE 8   A company produces and sells copies of an accounting program for home computers. The total weekly cost (in dollars) to produce x copies of the program is $C(x) = 8x + 500$, and the weekly revenue for selling all x copies of the program is $R(x) = 35x - 0.1x^2$. How many programs must be sold each week for the weekly profit to be $1,200?

Solution   Substituting the given expressions for $R(x)$ and $C(x)$ in the equation $P(x) = R(x) - C(x)$, we have a polynomial in x that represents the weekly profit $P(x)$.

$$P(x) = R(x) - C(x)$$

$$= 35x - 0.1x^2 - (8x + 500)$$

$$= 35x - 0.1x^2 - 8x - 500$$

$$= -500 + 27x - 0.1x^2$$

By setting this expression equal to 1,200, we have a quadratic equation to solve that gives us the number of programs x that need to be sold each week to bring in a profit of $1,200.

$$1,200 = -500 + 27x - 0.1x^2$$

We can write this equation in standard form by adding the opposite of each term on the right side of the equation to both sides of the equation. Doing so produces the following equation:

$$0.1x^2 - 27x + 1,700 = 0$$

Applying the quadratic formula to this equation with $a = 0.1$, $b = -27$, and $c = 1,700$, we have

$$x = \frac{27 \pm \sqrt{(-27)^2 - 4(0.1)(1,700)}}{2(0.1)}$$

$$= \frac{27 \pm \sqrt{729 - 680}}{0.2}$$

$$= \frac{27 \pm \sqrt{49}}{0.2}$$

$$= \frac{27 \pm 7}{0.2}$$

> **Note**   What is interesting about this last example is that it has rational solutions, meaning it could have been solved by factoring. But looking back at the equation, factoring does not seem like a reasonable method of solution because the coefficients are either very large or very small. So there are times when using the quadratic formula is a faster method of solution, even though the equation you are solving is factorable.

Writing this last expression as two separate expressions, we have our two solutions:

$$x = \frac{27 + 7}{0.2} \quad \text{or} \quad x = \frac{27 - 7}{0.2}$$

$$= \frac{34}{0.2} \qquad\qquad = \frac{20}{0.2}$$

$$= 170 \qquad\qquad\quad = 100$$

The weekly profit will be $1,200 if the company produces and sells 100 programs or 170 programs.

Answer

8. 130 or 140 programs

USING TECHNOLOGY Graphing Calculators

More About Example 7

We can solve the problem discussed in Example 7 by graphing the function $Y_1 = 20X + 16X^2$ in a window with X from 0 to 2 (because X is taking the place of t and we know t is a positive quantity) and Y from 0 to 50 (because we are looking for X when Y_1 is 40). Graphing Y_1 gives a graph similar to the graph in Figure 1. Using the Zoom and Trace features at $Y_1 = 40$ gives us X = 1.08 to the nearest hundredth, matching the results we obtained by solving the original equation algebraically.

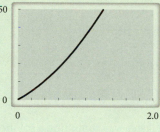

FIGURE 1

More About Example 8

To visualize the functions in Example 8, we set up our calculator this way:

$$Y_1 = 35X - .1X^2 \qquad \text{Revenue function}$$

$$Y_2 = 8X + 500 \qquad \text{Cost function}$$

$$Y_3 = Y_1 - Y_2 \qquad \text{Profit function}$$

Window: X from 0 to 350, Y from 0 to 3,500

Graphing these functions produces graphs similar to the ones shown in Figure 2. The lowest graph is the graph of the profit function. Using the Zoom and Trace features on the lowest graph at $Y_3 = 1,200$ produces two corresponding values of X, 170 and 100, which match the results in Example 8.

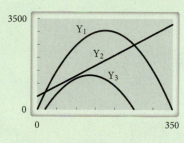

FIGURE 2

We will continue this discussion of the relationship between graphs of functions and solutions to equations in the Using Technology material in the next section.

GETTING READY FOR CLASS

After reading through the preceding section, respond in your own words and in complete sentences.

A. What is the quadratic formula?

B. Under what circumstances should the quadratic formula be applied?

C. When would the quadratic formula result in complex solutions?

D. When will the quadratic formula result in only one solution?

Problems

A Solve each equation. Use factoring or the quadratic formula, whichever is appropriate. (Try factoring first. If you have any difficulty factoring, then try the quadratic formula.)

1. $x^2 + 5x + 6 = 0$

2. $x^2 + 5x - 6 = 0$

3. $a^2 - 4a + 1 = 0$

4. $a^2 + 4a + 1 = 0$

5. $\frac{1}{6}x^2 - \frac{1}{2}x + \frac{1}{3} = 0$

6. $\frac{1}{6}x^2 + \frac{1}{2}x + \frac{1}{3} = 0$

7. $\frac{x^2}{2} + 1 = \frac{2x}{3}$

8. $\frac{x^2}{2} + \frac{2}{3} = -\frac{2x}{3}$

9. $y^2 - 5y = 0$

10. $2y^2 + 10y = 0$

11. $30x^2 + 40x = 0$

12. $50x^2 - 20x = 0$

13. $\frac{2t^2}{3} - t = -\frac{1}{6}$

14. $\frac{t^2}{3} - \frac{t}{2} = -\frac{3}{2}$

15. $0.01x^2 + 0.06x - 0.08 = 0$

16. $0.02x^2 - 0.03x + 0.05 = 0$

17. $2x + 3 = -2x^2$

18. $2x - 3 = 3x^2$

19. $100x^2 - 200x + 100 = 0$

20. $100x^2 - 600x + 900 = 0$

21. $\frac{1}{2}r^2 = \frac{1}{6}r - \frac{2}{3}$

SCAN TO ACCESS

22. $\frac{1}{4}r^2 = \frac{2}{5}r + \frac{1}{10}$

23. $(x - 3)(x - 5) = 1$

24. $(x - 3)(x + 1) = -6$

25. $(x + 3)^2 + (x - 8)(x - 1) = 16$

26. $(x - 4)^2 + (x + 2)(x + 1) = 9$

27. $\frac{x^2}{3} - \frac{5x}{6} = \frac{1}{2}$

28. $\frac{x^2}{6} + \frac{5}{6} = -\frac{x}{3}$

29. $\sqrt{x} = x - 1$

30. $\sqrt{x} = x - 2$

Multiply both sides of each equation by its LCD. Then solve the resulting equation.

31. $\frac{1}{x + 1} - \frac{1}{x} = \frac{1}{2}$

32. $\frac{1}{x + 1} + \frac{1}{x} = \frac{1}{3}$

33. $\frac{1}{y - 1} + \frac{1}{y + 1} = 1$

34. $\frac{2}{y + 2} + \frac{3}{y - 2} = 1$

35. $\frac{1}{x + 2} + \frac{1}{x + 3} = 1$

36. $\frac{1}{x + 3} + \frac{1}{x + 4} = 1$

37. $\frac{6}{r^2 - 1} - \frac{1}{2} = \frac{1}{r + 1}$

38. $2 + \frac{5}{r - 1} = \frac{12}{(r - 1)^2}$

39. $\frac{1}{x^2 - 4} + \frac{x}{x - 2} = 2$

40. $\frac{1}{x^2} - \frac{1}{x} = 2$

41. $2 - \frac{3}{x} = \frac{1}{x^2}$

42. $4 - \frac{1}{x} = \frac{2}{x^2}$

Solve each equation. In each case you will have three solutions.

43. $x^3 - 8 = 0$

44. $x^3 - 27 = 0$

45. $8a^3 + 27 = 0$

46. $27a^3 + 8 = 0$

47. $125t^3 - 1 = 0$

48. $64t^3 + 1 = 0$

Each of the following equations has three solutions. Look for the greatest common factor; then use the quadratic formula to find all solutions.

49. $2x^3 + 2x^2 + 3x = 0$

50. $6x^3 - 4x^2 + 6x = 0$

51. $3y^4 = 6y^3 - 6y^2$

52. $4y^4 = 16y^3 - 20y^2$

53. $6t^5 + 4t^4 = -2t^3$

54. $8t^5 + 2t^4 = -10t^3$

55. Which two of the expressions below are equivalent?

a. $\dfrac{6 + 2\sqrt{3}}{4}$

b. $\dfrac{3 + \sqrt{3}}{2}$

c. $6 + \dfrac{\sqrt{3}}{2}$

56. Which two of the expressions below are equivalent?

a. $\dfrac{8 - 4\sqrt{2}}{4}$

b. $2 - 4\sqrt{3}$

c. $2 - \sqrt{2}$

57. Solve $3x^2 - 5x = 0$

    a. by factoring.

    b. using the quadratic formula.

58. Solve $3x^2 + 23x - 70 = 0$

    a. by factoring.

    b. using the quadratic formula.

59. Can the equation $x^2 - 4x + 7 = 0$ be solved by factoring? Solve it.

60. Can the equation $x^2 = 5$ be solved by factoring? Solve it.

61. Is $x = -1 + i$ a solution to $x^2 + 2x = -2$?

62. Is $x = 2 + 2i$ a solution to $(x - 2)^2 = -4$?

Recall from section 2.4 that fixed points for a function are solutions to the equation $f(x) = x$. For each of the following functions, find the fixed points.

63. $f(x) = x^2 - 3x$

64. $f(x) = x^2 + 5x$

65. $f(x) = x(1.4 - x)$

66. $f(x) = x(1.6 - x)$

67. $f(x) = x^2 + 5x + 2$

68. $f(x) = x^2 - 4x + 3$

69. $f(x) = \dfrac{x + 3}{x}$

70. $f(x) = \dfrac{5 - x}{x}$

71. $f(x) = \dfrac{2x^2 - 6}{x}$

72. $f(x) = \dfrac{x^2 + 5}{3x}$

B Applying the Concepts

73. Falling Object   An object is thrown downward with an initial velocity of 5 feet per second. The relationship between the distance s it travels and time t is given by $s = 5t + 16t^2$. How long does it take the object to fall 74 feet?

74. Coin Toss   A coin is tossed upward with an initial velocity of 32 feet per second from a height of 16 feet above the ground. The equation giving the object's height h at any time t is $h = 16 + 32t - 16t^2$. Does the object ever reach a height of 32 feet?

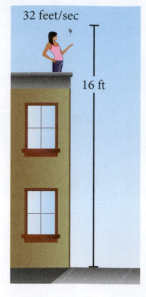

75. Profit   The total cost (in dollars) for a company to manufacture and sell x items per week is $C = 60x + 300$, whereas the revenue brought in by selling all x items is $R = 100x - 0.5x^2$. How many items must be sold to obtain a weekly profit of $300?

76. Profit   Suppose a company manufactures and sells x picture frames each month with a total cost of $C = 1{,}200 + 3.5x$ dollars. If the revenue obtained by selling x frames is $R = 9x - 0.002x^2$, find the number of frames it must sell each month if its monthly profit is to be $2,300.

77. Photograph Cropping   The following figure shows a photographic image on a 10.5-centimeter by 8.2-centimeter background. The overall area of the background is to be reduced to 80% of its original area by cutting off (cropping) equal strips on all four sides. What is the width of the strip that is cut from each side?

Image © Staci Truelson, 2011

78. Area of a Garden   A garden measures 20.3 meters by 16.4 meters. To double the area of the garden, strips of equal width are added to all four sides.

a. Draw a diagram that illustrates these conditions.

b. What are the new overall dimensions of the garden?

79. Area and Perimeter A rectangle has a perimeter of 20 yards and an area of 15 square yards.

a. Write two equations that state these facts in terms of the rectangle's length, l, and its width, w.

b. Solve the two equations from part a to determine the actual length and width of the rectangle.

c. Explain why two answers are possible to part b.

80. Population Size Writing in 1829, former President James Madison made some predictions about the growth of the population of the United States. The populations he predicted fit the equation

$$P(x) = 0.029x^2 - 1.39x + 42$$

where $P(x)$ is the population in millions of people x years from 1829.

a. Use the equation to determine the approximate year President Madison would have predicted that the U.S. population would reach 100,000,000.

b. If the U.S. population in 2010 was approximately 308,000,000, were President Madison's predictions accurate in the long term? Explain why or why not.

Getting Ready for the Next Section

Find the value of $b^2 - 4ac$ when

81. $a = 1, b = -3, c = -40$

82. $a = 2, b = 3, c = 4$

83. $a = 4, b = 12, c = 9$

84. $a = -3, b = 8, c = -1$

Solve.

85. $k^2 - 144 = 0$

86. $36 - 20k = 0$

Multiply.

87. $(x - 3)(x + 2)$

88. $(t - 5)(t + 5)$

89. $(x - 3)(x - 3)(x + 2)$

90. $(t - 5)(t + 5)(t - 3)$

Find the Mistake

Each sentence below contains a mistake. Circle the mistake and write the correct word or expression on the line provided.

1. Using the quadratic formula to solve a quadratic equation gives the solutions $x = -b \pm 4ac$. _____

2. Using the quadratic formula to solve $x^2 + 10x = 11$, $a = 1$, $b = 10$, and $c = 11$. _____

3. To solve $x^2 - 10x = -24$ using the quadratic formula, begin by finding the greatest common factor on both sides of the equals sign. _____

4. Solving $3x^2 - 2x - 4 = 0$ using the quadratic formula, we get $\frac{(1 \pm \sqrt{7})}{3}$ for our answer. _____

7.3 The Discriminant and Multiplicity

OBJECTIVES

A Find the number and types of solutions to a quadratic equation by using the discriminant.

B Find an unknown constant in a quadratic equation so that there is exactly one solution.

C Find an equation given its solutions and multiplicities.

KEY WORDS

discriminant

unknown constant

multiplicity

©iStockphoto.com/fredrocko

A shortstop can jump straight up with an initial velocity of 2 feet per second. If he wants to catch a line drive hit over his head, he needs to be able to jump 5 feet off the ground. The equation that models the jump is $h = -16t^2 + 2t$. Can he make the out? Problems like this can be answered without having to name the solutions. We can use an expression called the *discriminant* to determine if there are any real solutions to this equation. In this section, we will define the discriminant and use it to find the number and types of solutions an equation may have. We will also use the zero-factor property to build equations from their solutions, which may help us decide the initial velocity the shortstop will need to catch the line drive.

A The Discriminant

The quadratic formula

$$x = \frac{-b \pm \sqrt{b^2 - 4ac}}{2a}$$

gives the solutions to any quadratic equation in standard form. There are times when working with quadratic equations that it is important only to know what kind of solutions the equation has.

> **DEFINITION** discriminant
>
> The expression under the radical in the quadratic formula is called the *discriminant*:
>
> $$\text{discriminant} = D = b^2 - 4ac$$

The discriminant indicates the number and type of solutions to a quadratic equation, when the original equation has integer coefficients. For example, if we were to use the quadratic formula to solve the equation $2x^2 + 2x + 3 = 0$, we would find the discriminant to be

$$b^2 - 4ac = 2^2 - 4(2)(3) = -20$$

Because the discriminant appears under a square root symbol, we have the square root of a negative number in the quadratic formula. Our solutions would therefore be complex numbers. Similarly, if the discriminant were 0, the quadratic formula would yield

$$x = \frac{-b \pm \sqrt{0}}{2a} = \frac{-b \pm 0}{2a} = \frac{-b}{2a}$$

and the equation would have one rational solution, the number $\frac{-b}{2a}$.

The following table gives the relationship between the discriminant and the type of solutions to the equation.

For the equation $ax^2 + bx + c = 0$ where a, b, and c are integers and $a \neq 0$:

If the discriminant $b^2 - 4ac$ is	Then the equation will have
Negative	Two complex solutions containing i
Zero	One rational solution
A positive number that is also a perfect square	Two rational solutions
A positive number that is not a perfect square	Two irrational solutions

In the second and third cases, when the discriminant is 0 or a positive perfect square, the solutions are rational numbers. The quadratic equations in these two cases are the ones that can be factored.

EXAMPLE 1   For each equation, give the number and kind of solutions.

a. $x^2 - 3x - 40 = 0$

Solution   Using $a = 1$, $b = -3$, and $c = -40$ in $b^2 - 4ac$, we have

$$(-3)^2 - 4(1)(-40) = 9 + 160 = 169.$$

The discriminant is a perfect square. The equation therefore has two rational solutions.

b. $2x^2 - 3x + 4 = 0$

Solution   Using $a = 2$, $b = -3$, and $c = 4$, we have

$$b^2 - 4ac = (-3)^2 - 4(2)(4) = 9 - 32 = -23$$

The discriminant is negative, implying the equation has two complex solutions that contain i.

c. $4x^2 - 12x + 9 = 0$

Solution   Using $a = 4$, $b = -12$, and $c = 9$, the discriminant is

$$b^2 - 4ac = (-12)^2 - 4(4)(9) = 144 - 144 = 0$$

Because the discriminant is 0, the equation will have one rational solution.

d. $x^2 + 6x = 8$

Solution   We must first put the equation in standard form by adding -8 to each side. If we do so, the resulting equation is

$$x^2 + 6x - 8 = 0$$

Now we identify a, b, and c as 1, 6, and -8, respectively.

$$b^2 - 4ac = 6^2 - 4(1)(-8) = 36 + 32 = 68$$

The discriminant is a positive number, but not a perfect square. The equation will therefore have two irrational solutions.

●

B Finding an Unknown Constant

EXAMPLE 2   Find an appropriate k so that the equation $4x^2 - kx = -9$ has exactly one rational solution.

Solution   We begin by writing the equation in standard form.

$$4x^2 - kx + 9 = 0$$

Using $a = 4$, $b = -k$, and $c = 9$, we have

$$b^2 - 4ac = (-k)^2 - 4(4)(9)$$
$$= k^2 - 144$$

1. Give the number and kind of solutions to each equation.

a. $x^2 - 3x - 28 = 0$

b. $x^2 - 6x + 9 = 0$

c. $3x^2 - 2x + 4 = 0$

d. $x^2 + 1 = 4x$

2. Find k so that the equation $9x^2 + kx = -4$ has exactly one rational solution.

Answers

1. a. Two, rational   **b.** One, rational
c. Two, complex   **d.** Two, irrational

2. ± 12

An equation has exactly one rational solution when the discriminant is 0. We set the discriminant equal to 0 and solve.

$$k^2 - 144 = 0$$

$$k^2 = 144$$

$$k = \pm 12$$

Choosing k to be 12 or -12 will result in an equation with one rational solution.

C Building Equations From Their Solutions

Suppose we know that the solutions to an equation are $x = 3$ and $x = -2$. We can find equations with these solutions by using the zero-factor property. First, let's write our solutions as equations with 0 on the right side.

If	$x = 3$	First solution
then	$x - 3 = 0$	Add -3 to each side.
and if	$x = -2$	Second solution
then	$x + 2 = 0$	Add 2 to each side.

Now, because both $x - 3$ and $x + 2$ are 0, their product must be 0 also. Therefore, we can write

$$(x - 3)(x + 2) = 0 \qquad \text{Zero-factor property}$$

$$x^2 - x - 6 = 0 \qquad \text{Multiply out the left side.}$$

Many other equations have 3 and -2 as solutions. For example, any constant multiple of $x^2 - x - 6 = 0$, such as $5x^2 - 5x - 30 = 0$, also has 3 and -2 as solutions. Similarly, any equation built from positive integer powers of the factors $x - 3$ and $x + 2$ will also have 3 and -2 as solutions. One such equation is

$$(x - 3)^2(x + 2) = 0$$

$$(x^2 - 6x + 9)(x + 2) = 0$$

$$x^3 - 4x^2 - 3x + 18 = 0$$

In mathematics, we distinguish between the solutions to this last equation and those to the equation $x^2 - x - 6 = 0$ by saying $x = 3$ is a solution of *multiplicity 2* in the equation $x^3 - 4x^2 - 3x + 18 = 0$ because it appears as a value of zero twice when substituted for x in the equation, and a solution of *multiplicity 1* in the equation $x^2 - x - 6 = 0$ because it only appears once as a value of zero. Here is a formal definition for multiplicity:

> **DEFINITION** multiplicity
>
> The number of times a solution to a polynomial equation appears as zero is its *multiplicity*.

EXAMPLE 3 Find an equation that has solutions $t = 5$, $t = -5$, and $t = 3$.

Solution First, we use the given solutions to write equations that have 0 on their right sides.

If	$t = 5$	$t = -5$	$t = 3$
then	$t - 5 = 0$	$t + 5 = 0$	$t - 3 = 0$

Since $t - 5$, $t + 5$, and $t - 3$ are all 0, their product is also 0 by the zero-factor property. An equation with solutions of 5, -5, and 3 is

$$(t - 5)(t + 5)(t - 3) = 0 \qquad \text{Zero-factor property}$$

$$(t^2 - 25)(t - 3) = 0 \qquad \text{Multiply first two binomials.}$$

3. Find an equation that has solutions $t = -2$, $t = 2$, and $t = 3$.

Answer

3. $t^3 - 3t^2 - 4t + 12$

$$t^3 - 3t^2 - 25t + 75 = 0 \qquad \textit{Complete the multiplication.}$$

The last line $t^3 - 3t^2 - 25t + 75 = 0$ gives us an equation with solutions of 5, -5, and 3, all of which are of multiplicty 1 because they are each zero once in this equation. Remember, many other equations have these same solutions.

4. Find an equation with solutions $x = -\frac{3}{4}$ and $x = \frac{1}{5}$.

EXAMPLE 4   Find an equation with solutions $x = -\frac{2}{3}$ and $x = \frac{4}{5}$, both of multiplicty 1.

Solution   The solution $x = -\frac{2}{3}$ can be rewritten as $3x + 2 = 0$ as follows:

$$x = -\frac{2}{3} \qquad \textit{First solution}$$

$$3x = -2 \qquad \textit{Multiply each side by 3.}$$

$$3x + 2 = 0 \qquad \textit{Add 2 to each side.}$$

Similarly, the solution $x = \frac{4}{5}$ can be rewritten as $5x - 4 = 0$.

$$x = \frac{4}{5} \qquad \textit{Second solution}$$

$$5x = 4 \qquad \textit{Multiply each side by 5.}$$

$$5x - 4 = 0 \qquad \textit{Add } -4 \textit{ to each side.}$$

Because both $3x + 2$ and $5x - 4$ are 0, their product is 0 also, giving us the equation we are looking for.

$$(3x + 2)(5x - 4) = 0 \qquad \textit{Zero-factor property}$$

$$15x^2 - 2x - 8 = 0 \qquad \textit{Multiply.}$$

5. Find an equation that has a solution of $x = 5$ of multiplicity 1 and solution of $x = -2$ of multiplicity 2.

EXAMPLE 5   Find an equation that has a solution of $x = -3$ of multiplicity 1 and a solution $x = 4$ of multiplicity 2.

Solution   The solution $x = -3$ can be written as $x + 3 = 0$. Similarily, the solution $x = 4$ can be written as $x - 4 = 0$. The solution $x = 4$ has a multiplicity of 2 so it will appear in the equation twice.
Because both $x + 3$ and $x - 4$ are 0, their product is also zero, giving us the equation:

$$(x + 3)(x - 4)(x - 4) = 0$$

$$(x + 3)(x^2 - 8x + 16) = 0$$

$$x^3 - 5x^2 - 8x + 48 = 0$$

6. Find an equation with solutions $x = \frac{2 - 3i}{3}$ and $x = \frac{2 + 3i}{3}$.

EXAMPLE 6   Find an equation with solutions $x = \frac{-2 + i}{3}$ and $x = \frac{-2 - i}{3}$.

Solution   We will adapt the technique used in Example 4. Write the solutions together as

$$x = \frac{-2 \pm i}{3}$$

$$3x = -2 \pm i \qquad \textit{Multiply each side by 3.}$$

$$3x + 2 = \pm i \qquad \textit{Add 2 to each side.}$$

Now we will square each side of the equation.

$$(3x + 2)^2 = (\pm i)^2 \qquad \textit{Square each side.}$$

$$9x^2 + 12x + 4 = -1 \qquad \textit{Simplify.}$$

$$9x^2 + 12x + 5 = 0 \qquad \textit{Add 1 to each side.}$$

The original equation is $9x^2 + 12x + 5 = 0$.

Answers

4. $20x^2 + 11x - 3 = 0$
5. $x^3 - x^2 - 16x - 20 = 0$
6. $9x^2 - 12x + 13 = 0$

EXAMPLE 7 Find an equation with solutions $x = -4 \pm \sqrt{5}$.

Solution Using the same technique,

$$x = -4 \pm \sqrt{5}$$

$$x + 4 = \pm\sqrt{5} \qquad \text{Add 4 to each side.}$$

$$(x + 4)^2 = (\pm\sqrt{5})^2 \qquad \text{Square.}$$

$$x^2 + 8x + 16 = 5 \qquad \text{Simplify.}$$

$$x^2 + 8x + 11 = 0 \qquad \text{Add } -5 \text{ to each side.}$$

The original equation is $x^2 + 8x + 11 = 0$.

●

USING TECHNOLOGY Graphing Calculators

Solving Equations

Now that we have explored the relationship between equations and their solutions, we can look at how a graphing calculator can be used in the solution process. To begin, let's solve the equation $x^2 = x + 2$ using techniques from algebra: writing it in standard form, factoring, and then setting each factor equal to 0.

$$x^2 - x - 2 = 0 \qquad \text{Standard form}$$

$$(x - 2)(x + 1) = 0 \qquad \text{Factor.}$$

$$x - 2 = 0 \quad \text{or} \quad x + 1 = 0 \qquad \text{Set each factor equal to 0.}$$

$$x = 2 \qquad\qquad x = -1 \qquad \text{Solve.}$$

Our original equation, $x^2 = x + 2$, has two solutions: $x = 2$ and $x = -1$. To solve the equation using a graphing calculator, we need to associate it with an equation (or equations) in two variables. One way to do this is to associate the left side with the equation $y = x^2$ and the right side of the equation with $y = x + 2$. To do so, we set up the functions list in our calculator this way:

$$Y_1 = X^2$$

$$Y_2 = X + 2$$

Window: X from -5 to 5, Y from -5 to 5

Graphing these functions in this window will produce a graph similar to the one shown in Figure 1.

If we use the Trace feature to find the coordinates of the points of intersection, we find that the two curves intersect at $(-1, 1)$ and $(2, 4)$. We note that the x-coordinates of these two points match the solutions to the equation $x^2 = x + 2$, which we found using algebraic techniques. This makes sense because if two graphs intersect at a point (x, y), then the coordinates of that point satisfy both equations. If a point (x, y) satisfies both $y = x^2$ and $y = x + 2$, then for that particular point, $x^2 = x + 2$. From this, we conclude that the x-coordinates of the points of intersection are solutions to our original equation. The next page shows a summary of what we have discovered.

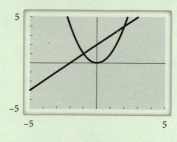

FIGURE 1

7. Find an equation with solutions $x = 5 \pm \sqrt{6}$.

Answer

7. $x^2 - 10x + 19 = 0$

Conclusion 1 If the graphs of two functions $y = f(x)$ and $y = g(x)$ intersect in the coordinate plane, then the x-coordinates of the points of intersection are solutions to the equation $f(x) = g(x)$.

A second method of solving our original equation $x^2 = x + 2$ graphically requires the use of one function instead of two. To begin, we write the equation in standard form as $x^2 - x - 2 = 0$. Next, we graph the function $y = x^2 - x - 2$. The x-intercepts of the graph are the points with y-coordinates of 0. They therefore satisfy the equation $0 = x^2 - x - 2$, which is equivalent to our original equation. The graph in Figure 2 shows $Y_1 = X^2 - X - 2$ in a window with X from -5 to 5 and Y from -5 to 5.

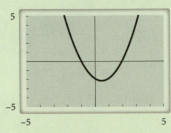

FIGURE 2

Using the Trace feature, we find that the x-intercepts of the graph are $x = -1$ and $x = 2$, which match the solutions to our original equation $x^2 = x + 2$. We can summarize the relationship between solutions to an equation and the intercepts of its associated graph this way:

Conclusion 2 If $y = f(x)$ is a function, then any x-intercept on the graph of $y = f(x)$ is a solution to the equation $f(x) = 0$.

GETTING READY FOR CLASS

After reading through the preceding section, respond in your own words and in complete sentences.

A. What is the discriminant?

B. What kind of solutions do we get to a quadratic equation when the discriminant is negative?

C. What does it mean for a solution to have multiplicity 3?

D. When will a quadratic equation have two rational solutions?

Vocabulary Review

Choose the correct words to fill in the blanks below.

rational	irrational	multiplicity	discriminant
quadratic	zero	negative	

1. The _____ is the expression under the radical in the quadratic formula.
2. The discriminant indicates the number and type of solutions to a _____ equation.
3. If a discriminant is _____, then the equation will have two complex solutions containing the number i.
4. If a discriminant is _____, then the equation will have one rational solution.
5. If a discriminant is a positive number that is also a perfect square, then the equation will have two _____ solutions.
6. If a discriminant is a positive number that is not a perfect square, then the equation will have two _____ solutions.
7. The number of times a solution to a quadratic equation appears as zero is its _____.

Problems

A Use the discriminant to find the number and kind of solutions for each of the following equations.

1. $x^2 - 6x + 5 = 0$ **2.** $x^2 - x - 12 = 0$ **3.** $4x^2 - 4x = -1$ **4.** $9x^2 + 12x = -4$

5. $x^2 + x - 1 = 0$ **6.** $x^2 - 2x + 3 = 0$ **7.** $2y^2 = 3y + 1$ **8.** $3y^2 = 4y - 2$

9. $x^2 - 9 = 0$ **10.** $4x^2 - 81 = 0$ **11.** $5a^2 - 4a = 5$ **12.** $3a = 4a^2 - 5$

B Determine k so that each of the following has exactly one rational solution.

13. $x^2 - kx + 25 = 0$ **14.** $x^2 + kx + 25 = 0$ **15.** $x^2 = kx - 36$

16. $x^2 = kx - 49$ **17.** $4x^2 - 12x + k = 0$ **18.** $9x^2 + 30x + k = 0$

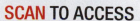

SCAN TO ACCESS

19. $kx^2 - 40x = 25$

20. $kx^2 - 2x = -1$

21. $3x^2 - kx + 2 = 0$

22. $5x^2 + kx + 1 = 0$

23. $3x^2 + kx + k = 0$

24. $kx^2 + 5x + 4k = 0$

C For each of the following problems, find an equation that has the given solutions.

25. $x = 5, x = 2$

26. $x = -5, x = -2$

27. $t = -3, t = 6$

28. $t = -4, t = 2$

29. $y = 2, y = -2, y = 4$

30. $y = 1, y = -1, y = 3$

31. $x = \dfrac{1}{2}, x = 3$

32. $x = \dfrac{1}{3}, x = 5$

33. $t = -\dfrac{3}{4}, t = 3$

34. $t = -\dfrac{4}{5}, t = 2$

35. $x = 3, x = -3, x = \dfrac{5}{6}$

36. $x = 5, x = -5, x = \dfrac{2}{3}$

37. $a = -\dfrac{1}{2}, a = \dfrac{3}{5}$

38. $a = -\dfrac{1}{3}, a = \dfrac{4}{7}$

39. $x = -\dfrac{2}{3}, x = \dfrac{2}{3}, x = 1$

40. $x = -\dfrac{4}{5}, x = \dfrac{4}{5}, x = -1$

41. $x = 2, x = -2, x = 3, x = -3$

42. $x = 1, x = -1, x = 5, x = -5$

43. $x = \sqrt{7}, x = -\sqrt{7}$

44. $x = -\sqrt{3}, x = \sqrt{3}$

45. $x = 5i, x = -5i$

46. $x = -2i, x = 2i$

47. $x = 1 + i, x = 1 - i$

48. $x = 2 + 3i, x = 2 - 3i$

49. $x = -2 - 3i, x = -2 + 3i$

50. $x = -1 + i, x = -1 - i$

51. $x = -4 \pm 3i$

52. $x = 2 \pm i$

53. $x = \dfrac{-2 \pm 5i}{4}$

54. $x = \dfrac{-3 \pm 2i}{6}$

55. $x = -1 \pm \sqrt{6}$

56. $x = 2 \pm \sqrt{5}$

57. $x = \dfrac{-4 \pm \sqrt{2}}{3}$

58. $x = \dfrac{5 \pm \sqrt{3}}{6}$

59. $x = \dfrac{3 \pm i\sqrt{5}}{7}$

60. $x = \dfrac{-4 \pm i\sqrt{3}}{5}$

61. Find an equation that has a solution of $x = 3$ of multiplicity 1 and a solution $x = -5$ of multiplicity 2.

62. Find an equation that has a solution of $x = 5$ of multiplicity 1 and a solution $x = -3$ of multiplicity 2.

63. Find an equation that has solutions $x = 3$ and $x = -3$, both of multiplicity 2.

64. Find an equation that has solutions $x = 4$ and $x = -4$, both of multiplicity 2.

65. Find all solutions to $x^3 + 6x^2 + 11x + 6 = 0$, if $x = -3$ is one of its solutions.

66. Find all solutions to $x^3 + 10x^2 + 29x + 20 = 0$, if $x = -4$ is one of its solutions.

67. One solution to $y^3 + 5y^2 - 2y - 24 = 0$ is $y = -3$. Find all solutions.

68. One solution to $y^3 + 3y^2 - 10y - 24 = 0$ is $y = -2$. Find all solutions.

69. If $x = 3$ is one solution to $x^3 - 5x^2 + 8x = 6$, find the other solutions.

70. If $x = 2$ is one solution to $x^3 - 6x^2 + 13x = 10$, find the other solutions.

71. Find all solutions to $t^3 = 13t^2 - 65t + 125$, if $t = 5$ is one of the solutions.

72. Find all solutions to $t^3 = 8t^2 - 25t + 26$, if $t = 2$ is one of the solutions.

Getting Ready for the Next Section

Simplify.

73. $(x + 3)^2 - 2(x + 3) - 8$

74. $(x - 2)^2 - 3(x - 2) - 10$

75. $(2a - 3)^2 - 9(2a - 3) + 20$

76. $(3a - 2)^2 + 2(3a - 2) - 3$

77. $2(4a + 2)^2 - 3(4a + 2) - 20$

78. $6(2a + 4)^2 - (2a + 4) - 2$

Solve.

79. $x^2 = \dfrac{1}{4}$

80. $x^2 = -2$

81. $\sqrt{x} = -3$

82. $\sqrt{x} = 2$

83. $x + 3 = 4$

84. $x + 3 = -2$

85. $y^2 - 2y - 8 = 0$

86. $y^2 + y - 6 = 0$

87. $4y^2 + 7y - 2 = 0$

88. $6x^2 - 13x - 5 = 0$

89. $12x^2 = x + 1$

90. $15x^2 = x + 2$

Maintaining Your Skills

91. Comics The chart shows how many stars (out of five) the most popular comic strips have. Use the chart to write the following ratings in English.

a. Far Side

b. Peanuts

92. Cars The chart shows the most expensive cars in the world. Use the chart to write the cost of the following cars in scientific notation.

a. Bugatti Veyron

b. Ferrari Enzo

Find the Mistake

Each sentence below contains a mistake. Circle the mistake and write the correct word or phrase on the line provided.

1. The equation $3x^2 + 2x + 5 = 0$ will have one rational solution.

2. The equation $5x^2 - 10x + 5 = 0$ will have two rational solutions.

3. The equation $9x^2 + 8x - 3 = 0$ will have two complex solutions containing i.

4. The equation $6x^2 - 13x + 2 = 0$ will have one irrational solution.

Landmark Review: Checking Your Progress

Solve the following equations.

1. $x^2 = 9$

2. $y^2 = \dfrac{3}{4}$

3. $(x + 3)^2 = 16$

4. $y^2 - 6y + 9 = 25$

5. $x^2 + 8x = 9$

6. $y^2 + 10y = -24$

7. $x^2 + 6x = 16$

8. $3x^2 - 9x - 12 = 0$

9. $x^2 + 5x - 3 = 0$

10. $x^2 + x + 8 = 0$

11. $2y^2 - 4y = 12$

12. $3x^2 + 9x = 9$

Use the discriminant to find the number and types of solutions.

13. $x^2 + 4x - 5 = 0$

14. $x^2 + 2x - 5 = 0$

15. $2x^2 + 3x = -3$

16. $6x^2 + 5x + 1 = 0$

For each of the following problems, find an equation that has the given solutions.

17. $x = -6, x = -2$

18. $x = 3, x = -3, x = 5$

A Solve equations that are reducible to a quadratic equation.

B Solve application problems using equations quadratic in form.

7.4 Equations Quadratic in Form

©iStockphoto.com/Sportlibrary

An Olympic platform diver jumps off a platform of height h with an upward velocity of about 8 feet per second. The time it takes her to reach the water is modeled by the equation $16t^2 - 8t - h = 0$. We can use this equation to calculate how long before she passes the platform. We can also use it to determine how long until she reaches the water. Let's put our knowledge of quadratic equations to work to solve a variety of equations, which can be converted to quadratic equations. The next few examples will illustrate these techniques.

A Equations Quadratic in Form

EXAMPLE 1 Solve $(x + 3)^2 - 2(x + 3) - 8 = 0$.

Solution We can see that this equation is quadratic in form by replacing $x + 3$ with another variable, say, y. Replacing $x + 3$ with y we have

$$y^2 - 2y - 8 = 0$$

We can solve this equation by factoring the left side and then setting each factor equal to 0.

$$\begin{aligned} y^2 - 2y - 8 &= 0 \\ (y - 4)(y + 2) &= 0 \qquad \textcolor{green}{\text{Factor.}} \\ y - 4 = 0 \quad &\text{or} \quad y + 2 = 0 \qquad \textcolor{green}{\text{Set factors equal to 0.}} \\ y = 4 \quad\quad & \qquad\quad y = -2 \end{aligned}$$

Because our original equation was written in terms of the variable x, we want our solutions in terms of x also. Replacing y with $x + 3$ and then solving for x, we have

$$\begin{aligned} x + 3 = 4 \quad &\text{or} \quad x + 3 = -2 \\ x = 1 \quad\quad & \qquad\quad x = -5 \end{aligned}$$

The solutions to our original equation are 1 and -5.

The method we have just shown lends itself well to other types of equations that are quadratic in form, as we will see. In this example, however, there is another method that works just as well. Let's solve our original equation again, but this time, let's begin by expanding $(x + 3)^2$ and $2(x + 3)$.

$$\begin{aligned} (x + 3)^2 - 2(x + 3) - 8 &= 0 \\ x^2 + 6x + 9 - 2x - 6 - 8 &= 0 \qquad \textcolor{green}{\text{Multiply.}} \\ x^2 + 4x - 5 &= 0 \qquad \textcolor{green}{\text{Combine similar terms.}} \\ (x - 1)(x + 5) &= 0 \qquad \textcolor{green}{\text{Factor.}} \\ x - 1 = 0 \quad \text{or} \quad x + 5 &= 0 \qquad \textcolor{green}{\text{Set factors equal to 0.}} \\ x = 1 \quad\quad x &= -5 \end{aligned}$$

As you can see, either method produces the same result.

EXAMPLE 2 Solve $4x^4 + 7x^2 = 2$.

Solution This equation is quadratic in x^2. We can make this easier to see by using the substitution $y = x^2$. (The choice of the letter y is arbitrary. We could just as easily use the substitution $m = x^2$.) Making the substitution $y = x^2$ and then solving the resulting equation, we have

$$4y^2 + 7y = 2$$

$$4y^2 + 7y - 2 = 0 \qquad \textcolor{green}{\text{Standard form}}$$

$$(4y - 1)(y + 2) = 0 \qquad \textcolor{green}{\text{Factor.}}$$

$$4y - 1 = 0 \quad \text{or} \quad y + 2 = 0 \qquad \textcolor{green}{\text{Set factors equal to 0.}}$$

$$y = \frac{1}{4} \qquad\qquad y = -2$$

Now we replace y with x^2 to solve for x.

$$x^2 = \frac{1}{4} \quad \text{or} \quad x^2 = -2$$

$$x = \pm\sqrt{\frac{1}{4}} \qquad x = \pm\sqrt{-2} \qquad \textcolor{green}{\text{Rational exponent theorem}}$$

$$x = \pm\frac{1}{2} \qquad\qquad = \pm i\sqrt{2}$$

The solutions are $\frac{1}{2}, -\frac{1}{2}, i\sqrt{2}, -i\sqrt{2}$.

EXAMPLE 3 Solve $x + \sqrt{x} - 6 = 0$.

Solution To see that this equation is quadratic in form, we have to notice that $(\sqrt{x})^2 = x$. That is, the equation can be rewritten as

$$(\sqrt{x})^2 + \sqrt{x} - 6 = 0$$

Replacing $\sqrt{x}$ with y and solving as usual, we have

$$y^2 + y - 6 = 0$$

$$(y + 3)(y - 2) = 0$$

$$y + 3 = 0 \quad \text{or} \quad y - 2 = 0$$

$$y = -3 \qquad\qquad y = 2$$

Again, to find x, we replace y with x and solve.

$$\sqrt{x} = -3 \quad \text{or} \quad \sqrt{x} = 2$$

$$x = 9 \qquad\qquad x = 4 \qquad \textcolor{green}{\text{Square both sides of each equation.}}$$

Because we squared both sides of each equation, we have the possibility of obtaining extraneous solutions. We have to check both solutions in our original equation.

When $\rightarrow$	$x = 9$	When $\rightarrow$	$x = 4$
the equation $\rightarrow$	$x + \sqrt{x} - 6 = 0$	the equation $\rightarrow$	$x + \sqrt{x} - 6 = 0$
becomes $\rightarrow$	$9 + \sqrt{9} - 6 \overset{?}{=} 0$	becomes $\rightarrow$	$4 + \sqrt{4} - 6 \overset{?}{=} 0$
	$9 + 3 - 6 \overset{?}{=} 0$		$4 + 2 - 6 \overset{?}{=} 0$
	$6 \neq 0$		$0 = 0$

$$\textcolor{green}{\text{This means 9 is extraneous.}} \qquad \textcolor{green}{\text{This means 4 is a solution.}}$$

The only solution to the equation $x + \sqrt{x} - 6 = 0$ is 4.

2. Solve $6x^4 - 13x^2 = 5$.

3. Solve $x - \sqrt{x} - 12 = 0$.

Answers

2. $\pm\frac{i\sqrt{3}}{3}, \pm\frac{\sqrt{10}}{-2}$

3. 16

We should note here that the two possible solutions, 9 and 4, to the equation in Example 3 can be obtained by another method. Instead of substituting for x, we can isolate the radical on one side of the equation and then square both sides to clear the equation of radicals.

$$x + \sqrt{x} - 6 = 0$$

$$\sqrt{x} = -x + 6 \qquad \text{Isolate } \sqrt{x}.$$

$$x = x^2 - 12x + 36 \qquad \text{Square both sides.}$$

$$0 = x^2 - 13x + 36 \qquad \text{Add } -x \text{ to both sides.}$$

$$0 = (x - 4)(x - 9) \qquad \text{Factor.}$$

$$x - 4 = 0 \quad \text{or} \quad x - 9 = 0$$

$$x = 4 \qquad\qquad x = 9$$

We obtain the same two possible solutions. Because we squared both sides of the equation to find them, we would have to check each one in the original equation. As was the case in Example 3, only 4 is a solution; 9 is extraneous.

B Applications

4. Solve $16t^2 - 10t - h = 0$ for t.

EXAMPLE 4   If an object is tossed into the air with an upward velocity of 12 feet per second from the top of a building h feet high, the time it takes for the object to hit the ground below is given by the formula

$$16t^2 - 12t - h = 0$$

Solve this formula for t.

Solution   The formula is in standard form and is quadratic in t. The coefficients a, b, and c that we need to apply to the quadratic formula are $a = 16$, $b = -12$, and $c = -h$. Substituting these quantities into the quadratic formula, we have

$$t = \frac{12 \pm \sqrt{144 - 4(16)(-h)}}{2(16)}$$

$$= \frac{12 \pm \sqrt{144 + 64h}}{32}$$

We can factor the perfect square 16 from the two terms under the radical and simplify our radical somewhat.

$$t = \frac{12 \pm \sqrt{16(9 + 4h)}}{32}$$

$$= \frac{12 \pm 4\sqrt{9 + 4h}}{32}$$

Now we can reduce to lowest terms by factoring a 4 from the numerator and denominator.

$$t = \frac{4(3 \pm \sqrt{9 + 4h})}{4 \cdot 8}$$

$$= \frac{3 \pm \sqrt{9 + 4h}}{8}$$

If we were given a value of h, we would find that one of the solutions to this last formula would be a negative number. Because time is always measured in positive units, we wouldn't use that solution. So the solution is $t = \frac{3 \pm \sqrt{9 + 4h}}{8}$.

USING TECHNOLOGY   Graphing Calculators

More About Example 1

As we mentioned before, algebraic expressions entered into a graphing calculator do not have to be simplified to be evaluated. This fact also applies to equations. We can graph the equation $y = (x + 3)^2 - 2(x + 3) - 8$ to assist us in solving the equation in Example 1. The graph is shown in Figure 1. Using the Zoom and Trace features at the x-intercepts gives us $x = 1$ and $x = -5$ as the solutions to the equation $0 = (x + 3)^2 - 2(x + 3) - 8$.

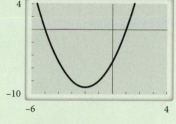

FIGURE 1

More About Example 2

Figure 2 shows the graph of $y = 4x^4 + 7x^2 - 2$. As we expect, the x-intercepts give the real number solutions to the equation $0 = 4x^4 + 7x^2 - 2$. The complex solutions do not appear on the graph.

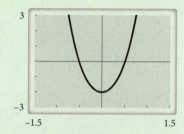

FIGURE 2

More About Example 3

In solving the equation in Example 3, we found that one of the possible solutions was an extraneous solution. If we solve the equation $x + \sqrt{x} - 6 = 0$ by graphing the function $y = x + \sqrt{x} - 6$, we find that the extraneous solution, 9, is not an x-intercept. Figure 3 shows that the only solution to the equation occurs at the x-intercept 4.

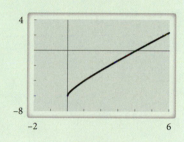

FIGURE 3

GETTING READY FOR CLASS

After reading through the preceding section, respond in your own words and in complete sentences.

A. What does it mean for an equation to be quadratic in form?

B. What are all the circumstances in solving equations (that we have studied) in which it is necessary to check for extraneous solutions?

C. How would you start to solve the equation $x + \sqrt{x} - 6 = 0$?

D. Is 9 a solution to $x + \sqrt{x} - 6 = 0$?

Problems

A Solve each equation.

1. $(x - 3)^2 + 3(x - 3) + 2 = 0$

2. $(x + 4)^2 - (x + 4) - 6 = 0$

3. $2(x + 4)^2 + 5(x + 4) - 12 = 0$

4. $3(x - 5)^2 + 14(x - 5) - 5 = 0$

5. $x^4 - 6x^2 - 27 = 0$

6. $x^4 + 2x^2 - 8 = 0$

7. $x^4 + 9x^2 = -20$

8. $x^4 - 11x^2 = -30$

9. $(2a - 3)^2 - 9(2a - 3) = -20$

10. $(3a - 2)^2 + 2(3a - 2) = 3$

11. $2(4a + 2)^2 = 3(4a + 2) + 20$

12. $6(2a + 4)^2 = (2a + 4) + 2$

13. $6t^4 = -t^2 + 5$

14. $3t^4 = -2t^2 + 8$

SCAN TO ACCESS

15. $9x^4 - 49 = 0$

16. $25x^4 - 9 = 0$

Solve each of the following equations. Remember, if you square both sides of an equation in the process of solving it, you have to check all solutions in the original equation.

17. $x - 7\sqrt{x} + 10 = 0$

18. $x - 6\sqrt{x} + 8 = 0$

19. $t - 2\sqrt{t} - 15 = 0$

20. $t - 3\sqrt{t} - 10 = 0$

21. $6x + 11\sqrt{x} = 35$

22. $2x + \sqrt{x} = 15$

23. $(a - 2) - 11\sqrt{a - 2} + 30 = 0$

24. $(a - 3) - 9\sqrt{a - 3} + 20 = 0$

25. $(2x + 1) - 8\sqrt{2x + 1} + 15 = 0$

26. $(2x - 3) - 7\sqrt{2x - 3} + 12 = 0$

27. Solve the formula $16t^2 - vt - h = 0$ for t.

28. Solve the formula $16t^2 + vt + h = 0$ for t.

29. Solve the formula $kx^2 + 8x + 4 = 0$ for x.

30. Solve the formula $k^2x^2 + kx + 4 = 0$ for x.

31. Solve $x^2 + 2xy + y^2 = 0$ for x by using the quadratic formula with $a = 1$, $b = 2y$, and $c = y^2$.

32. Solve $x^2 - 2xy + y^2 = 0$ for x by using the quadratic formula, with $a = 1$, $b = -2y$, $c = y^2$.

B Applying the Concepts

For Problems 33 and 34, t is in seconds.

33. Falling Object An object is tossed into the air with an upward velocity of 8 feet per second from the top of a building h feet high. The time it takes for the object to hit the ground below is given by the formula $16t^2 - 8t - h = 0$. Solve this formula for t.

34. Falling Object An object is tossed into the air with an upward velocity of 6 feet per second from the top of a building h feet high. The time it takes for the object to hit the ground below is given by the formula $16t^2 - 6t - h = 0$. Solve this formula for t.

35. Falling Object A competitive diving tower has three diving heights: 16.5 feet, 24.7 feet, and 33 feet. Use the equation $16t^2 - 8t - h = 0$, given at the introduction of this section, to find how long it takes the diver to reach the water level. Round to the closest hundredth of a second. Note that h represents the height of the platform.

36. Falling Object A recent Red Bull cliff diving championship featured a dive from 88.6 feet. The Farred La Quebrada Cliffs in Acapulco, Mexico sit 148 feet above the water. Using the equation in problem 35, find the time before each diver would hit the water level from the Acapulco cliffs.

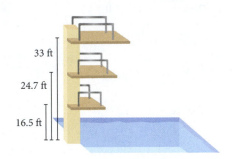

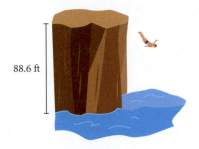

37. Listed below are some of the longest natural arches in the world. The height refers to the height at the center, and the span refers to the x-intercept, using the first x-intercept of $(0, 0)$. Find the equation of each arch.

Arch name	Location	Height (ft)	Span (ft)
a. Fairy Bridge	Guangi River, China	230	400
b. Landscape Arch	Arches National Park, Utah	77	290
c. Aloba Arch	Ennedi Range, Chad	400	250
d. Rainbow Bridge	Rainbow Bridge National Monument, Utah	246	275

38. Saint Louis Arch As we discussed in the beginning of this chapter, the shape of the famous "Gateway to the West" arch in Saint Louis can be modeled by a parabola. The equation for one such parabola is

$$y = -\frac{1}{150}x^2 + \frac{21}{5}x$$

a. Sketch the graph of the arch's equation on a coordinate axis.

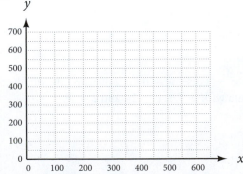

b. Approximately how far do you have to walk to get from one side of the arch to the other?

39. Area In the following diagram, *ABC* is a right triangle. Find its area.

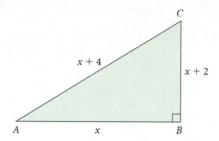

40. Area In the following diagram, *ABCD* is a rectangle with diagonal *AC*. Find its area.

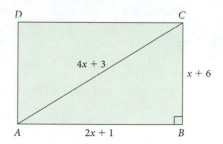

41. Area and Perimeter A total of 160 meters of fencing is to be used to enclose part of a lot that borders on a river. This situation is shown in the following diagram.

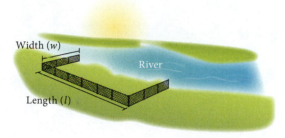
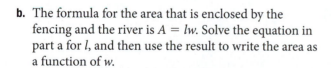

a. Write an equation that gives the relationship between the length and width and the 160 meters of fencing.

b. The formula for the area that is enclosed by the fencing and the river is $A = lw$. Solve the equation in part a for *l*, and then use the result to write the area as a function of *w*.

c. Make a table that gives at least five possible values of *w* and associated area *A*.

w (m)	A (m²)

d. From the pattern in your table shown in part c, what is the largest area that can be enclosed by the 160 meters of fencing? (Try some other table values if necessary.)

42. Area and Perimeter Refer to problem 41. Use the information in the following questions if you need to have an opening 2 meters wide in one of the shorter sides of the fence, as shown in the diagram.

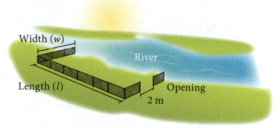

a. Write an equation that gives the relationship between the length and width and the amount of fencing.

b. The formula for the area that is enclosed by the fencing and the river is $A = lw$. Solve the equation in part a for *l*, and then use the result to write the area as a function of *w*.

c. Make a table that gives at least five possible values of *w* and associated area *A*.

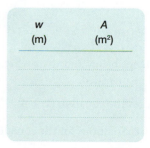

w (m)	A (m²)

d. From the pattern in your table shown in part c, what is the largest area that can be enclosed by the of fencing? (Try some other table values if necessary.)

Getting Ready for the Next Section

43. Evaluate $y = 3x^2 - 6x + 1$ for $x = 1$.

44. Evaluate $y = -2x^2 + 6x - 5$ for $x = \frac{3}{2}$.

45. Let $P(x) = -0.1x^2 + 27x - 500$ and find $P(135)$.

46. Let $P(x) = -0.1x^2 + 12x - 400$ and find $P(600)$.

Solve.

47. $a(80)^2 + 70 = 0$

48. $a(80)^2 + 90 = 0$

49. $x^2 - 6x + 5 = 0$

50. $x^2 - 3x - 4 = 0$

51. $-x^2 - 2x + 3 = 0$

52. $-x^2 + 4x + 12 = 0$

53. $2x^2 - 6x + 5 = 0$

54. $x^2 - 4x + 5 = 0$

Fill in the blanks to complete the square.

55. $x^2 - 6x + \Box = (x - \Box)^2$

56. $x^2 - 10x + \Box = (x - \Box)^2$

57. $y^2 + 2y + \Box = (y + \Box)^2$

58. $y^2 - 12y + \Box = (x - \Box)^2$

Find the Mistake

Read each sentence below and write whether it is true or false. If false, circle the mistake and write the correct expression or words on the line provided.

1. True or False: The equation $(x + 4) - 3(x + 4) - 7 = 0$ is quadratic in form. _____

2. True or False: Solutions for the equation $(y + 2)^2 - 3(y + 2) - 18 = 0$ are $y = 6$ and $y = -3$. _____

3. True or False: Squaring both sides of an equation has the potential to give extraneous solutions. _____

4. True or False: The equation $x + \sqrt{x} - 12 = 0$ has two real solutions. _____

OBJECTIVES

A Graph a quadratic function.

B Solve application problems using information from a graph.

C Find an equation from its graph.

KEY WORDS

parabola

quadratic function

vertex

vertical parabola

concave up

concave down

Participants of parkour, also called freerunning, make their own paths of travel by overcoming whatever physical obstacle stands before them. Their intense, and sometime dangerous training requires them to climb walls, flip off ramps, leap over great distances, and land with body rolls to ease the impact on their bones and joints. The art of parkour has a French origin and has been practiced for nearly a century; however, its American popularity has only begun to grow in the past decade.

Imagine a freerunner as a falling object. He has just leaped over a railing at the top of a ten-foot-high brick wall. If he is falling downward at an initial velocity of 5 feet per second, the relationship between the distance s traveled and the time t is $s = 5t + 16t^2$. Problems like this one can be found by graphing a parabola. In this section, we will continue our work with parabolas and look at several ways to graph them.

A Graphing Quadratic Functions

The solution set to the equation $y = x^2 - 3$ consists of ordered pairs. One method of graphing the solution set is to find a number of ordered pairs that satisfy the equation and to graph them. We can obtain some ordered pairs that are solutions to $y = x^2 - 3$ by use of a table as follows:

x	$y = x^2 - 3$	y	Solutions
-3	$y = (-3)^2 - 3 = 9 - 3 = 6$	6	$(-3, 6)$
-2	$y = (-2)^2 - 3 = 4 - 3 = 1$	1	$(-2, 1)$
-1	$y = (-1)^2 - 3 = 1 - 3 = -2$	-2	$(-1, -2)$
0	$y = 0^2 - 3 = 0 - 3 = -3$	-3	$(0, -3)$
1	$y = 1^2 - 3 = 1 - 3 = -2$	-2	$(1, -2)$
2	$y = 2^2 - 3 = 4 - 3 = 1$	1	$(2, 1)$
3	$y = 3^2 - 3 = 9 - 3 = 6$	6	$(3, 6)$

Graphing these solutions and then connecting them with a smooth curve, we have the graph of $y = x^2 - 3$. (See Figure 1.)

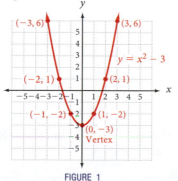

FIGURE 1

This graph is an example of a *parabola*. All functions of the form $f(x) = ax^2 + bx + c$, $a \neq 0$, have parabolas for graphs.

Although it is always possible to graph parabolas by making a table of values of x and y that satisfy the equation, there are other methods that are faster and, in some cases, more accurate.

The important points associated with the graph of a parabola are the highest (or lowest) point on the graph and the x-intercepts. The y-intercepts can also be useful.

Intercepts for Parabolas

The graph of the *quadratic function* $f(x) = ax^2 + bx + c$ crosses the y-axis at $y = c$, because $f(0) = a(0)^2 + b(0) + c = c$.

Because the graph crosses the x-axis when $y = 0$, the x-coordinates of the x-intercepts are solutions to the quadratic equation $0 = ax^2 + bx + c$.

The Vertex of a Parabola

The highest or lowest point on a parabola is called the *vertex*. The vertex for the graph of $f(x) = ax^2 + bx + c$ will always occur when

$$x = \frac{-b}{2a}$$

To see this, we must transform the right side of $f(x) = ax^2 + bx + c$ into an expression that contains x in just one of its terms. This is accomplished by completing the square on the first two terms. Here is what it looks like:

$$f(x) = ax^2 + bx + c$$

$$= a\left(x^2 + \frac{b}{a}x\right) + c$$

$$= a\left[x^2 + \frac{b}{a}x + \left(\frac{b}{2a}\right)^2\right] + c - a\left(\frac{b}{2a}\right)^2$$

$$= a\left(x + \frac{b}{2a}\right)^2 + \frac{4ac - b^2}{4a}$$

It may not look like it, but this last line indicates that the vertex of the graph of $y = ax^2 + bx + c$ has an x-coordinate of $\frac{-b}{2a}$. Because a, b, and c are constants, the only quantity that is varying in the last expression is the x in $\left(x + \frac{b}{2a}\right)^2$. Because the quantity $\left(x + \frac{b}{2a}\right)^2$ is the square of $x + \frac{b}{2a}$, the smallest it will ever be is 0, and that will happen when $x = \frac{-b}{2a}$.

The graph of the simplest quadratic function $f(x) = x^2$ appears as a parabola with its vertex at the origin $(0, 0)$ because the value for a is 1 and the value for b is 0. (See Figure 2.)

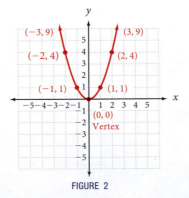

FIGURE 2

We can use the vertex point along with the x- and y-intercepts to sketch the graph of any equation of the form $y = ax^2 + bx + c$.

Here is a summary of the preceding information:

PROPERTY Graphing Parabolas I

The graph of the quadratic function $f(x) = ax^2 + bx + c, a \neq 0$, will be a vertical parabola with

1. y-coordinate of y-intercept at $y = c$

2. x-coordinate of x-intercepts (if they exist) at

$$x = \frac{-b \pm \sqrt{b^2 - 4ac}}{2a}$$

3. x-coordinates of the vertex when $x = \dfrac{-b}{2a}$

EXAMPLE 1   Sketch the graph of $f(x) = x^2 - 6x + 5$.

Solution   To find the x-intercepts, we let $y = 0$ and solve for x.

$$0 = x^2 - 6x + 5$$

$$0 = (x - 5)(x - 1)$$

$$x = 5 \quad \text{or} \quad x = 1$$

To find the coordinates of the vertex, we first find

$$x = \frac{-b}{2a} = \frac{-(-6)}{2(1)} = 3$$

The x-coordinate of the vertex is 3. To find the y-coordinate, we substitute 3 for x in our original equation.

$$f(3) = 3^2 - 6(3) + 5 = 9 - 18 + 5 = -4$$

The graph crosses the x-axis at 1 and 5 and has its vertex at $(3, -4)$. Plotting these points and connecting them with a smooth curve, we have the graph shown in Figure 3.

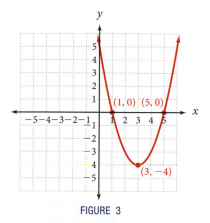

FIGURE 3

The graph is a *vertical parabola* that opens up, so we say the graph is *concave up*. The vertex is the lowest point on the graph. (Note that the graph crosses the y-axis at 5, which is the value of y we obtain when we let $x = 0$.)

Finding the Vertex by Completing the Square

Another way to locate the vertex of the parabola in Example 1 is by completing the square on the first two terms on the right side of the equation $f(x) = x^2 - 6x + 5$. In this case, we would do so by adding 9 to and subtracting 9 from the right side of the equation. This amounts to adding 0 to the equation, so we know we haven't changed its solutions.

Practice Problems

1. Graph $f(x) = x^2 - 2x - 3$.

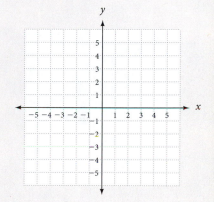

This is what it looks like:

$$f(x) = (x^2 - 6x \quad\quad) + 5$$
$$= (x^2 - 6x + 9) + 5 - 9$$
$$= (x - 3)^2 - 4$$

You may have to look at this last equation awhile to see this, but when $x = 3$, then $f(3) = (3 - 3)^2 - 4 = 0^2 - 4 = -4$ is the smallest y will ever be. That is why the vertex is at $(3, -4)$. As a matter of fact, this is the same kind of reasoning we used when we derived the formula $x = -\frac{b}{2a}$ for the x-coordinate of the vertex.

EXAMPLE 2   Graph $f(x) = -x^2 - 2x + 3$.

Solution   To find the x-intercepts, we let $y = 0$.

$$0 = -x^2 - 2x + 3$$
$$0 = x^2 + 2x - 3 \quad\quad \text{\color{green}Multiply each side by } -1.$$
$$0 = (x + 3)(x - 1)$$
$$x = -3 \quad \text{or} \quad x = 1$$

The x-coordinate of the vertex is given by

$$x = \frac{-b}{2a} = \frac{-(-2)}{2(-1)} = \frac{2}{-2} = -1$$

To find the y-coordinate of the vertex, we substitute -1 for x in our original equation to get

$$f(-1) = -(-1)^2 - 2(-1) + 3 = -1 + 2 + 3 = 4$$

Our parabola has x-intercepts at -3 and 1, and a vertex at $(-1, 4)$. Figure 4 shows the graph.

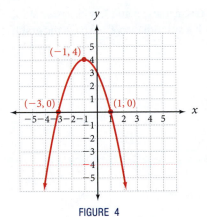

FIGURE 4

We say the graph is *concave down* because it opens downward. Again, we could have obtained the coordinates of the vertex by completing the square on the first two terms on the right side of our equation. To do so, we must first factor -1 from the first two terms. (Remember, the leading coefficient must be 1 to complete the square.) When we complete the square, we add 1 inside the parentheses, which actually decreases the right side of the equation by -1 because everything in the parentheses is multiplied by -1. To make up for it, we add 1 outside the parentheses.

$$f(x) = -1(x^2 + 2x \quad\quad) + 3$$
$$= -1(x^2 + 2x + 1) + 3 + 1$$
$$= -1(x + 1)^2 + 4$$

The last line tells us that the *largest* value of y will be 4, and that will occur when $x = -1$.

---

2. Graph $f(x) = -x^2 + 2x + 8$.

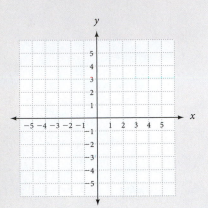

EXAMPLE 3 Graph $f(x) = 3x^2 - 6x + 1$.

Solution To find the x-intercepts, we let $y = 0$ and solve for x.

$$0 = 3x^2 - 6x + 1$$

Because the right side of this equation does not factor, we can look at the discriminant to see what kind of solutions are possible. The discriminant for this equation is

$$b^2 - 4ac = 36 - 4(3)(1) = 24$$

Because the discriminant is a positive number but not a perfect square, the equation will have irrational solutions. This means that the x-intercepts are irrational numbers and will have to be approximated with decimals using the quadratic formula. Rather than use the quadratic formula, we will find some other points on the graph, but first let's find the vertex.

Here are both methods of finding the vertex:

Using the formula that gives us the x-coordinate of the vertex, we have	To complete the square on the right side of the equation, we factor 3 from the first two terms, add 1 inside the parentheses, and add -3 outside the parentheses (this amounts to adding 0 to the right side).
$$x = \frac{-b}{2a} = \frac{-(-6)}{2(3)} = 1$$	
Substituting 1 for x in the equation gives us the y-coordinate of the vertex.	$$f(x) = 3(x^2 - 2x \quad) + 1$$
$$f(1) = 3 \cdot 1^2 - 6 \cdot 1 + 1 = -2$$	$$= 3(x^2 - 2x + 1) + 1 - 3$$
	$$= 3(x - 1)^2 - 2$$

In either case, the vertex is $(1, -2)$.

If we can find two points, one on each side of the vertex, we can sketch the graph. Let's let $x = 0$ and $x = 2$, because each of these numbers is the same distance from $x = 1$, and $x = 0$ will give us the y-intercept.

When $x = 0$	When $x = 2$
$$f(0) = 3(0)^2 - 6(0) + 1$$	$$f(2) = 3(2)^2 - 6(2) + 1$$
$$= 0 - 0 + 1$$	$$= 12 - 12 + 1$$
$$= 1$$	$$= 1$$

The two points just found are $(0, 1)$ and $(2, 1)$. Plotting these two points along with the vertex $(1, -2)$, we have the graph shown in Figure 5.

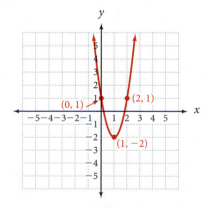

FIGURE 5

3. Graph $f(x) = 2x^2 - 4x + 1$.

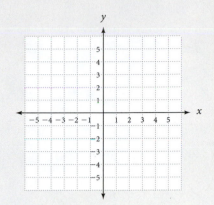

Note In a similar manner used in Example 2, we add -3 outside the parenthesis because of the coefficient we factored from the first two terms of the function and the 1 we added inside the parenthesis. It is important to take some time to understand how this process works. Mastering this will come in handy when we work with two variables later in the book.

4. Graph $f(x) = -x^2 + 4x - 5$.

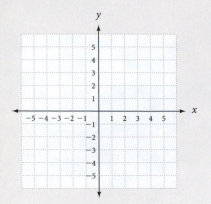

EXAMPLE 4   Graph $f(x) = -2x^2 + 6x - 5$.

Solution   Letting $y = 0$, we have

$$0 = -2x^2 + 6x - 5$$

Again, the right side of this equation does not factor. The discriminant is $b^2 - 4ac = 36 - 4(-2)(-5) = -4$, which indicates that the solutions are complex numbers. This means that our original equation does not have x-intercepts. The graph does not cross the x-axis.

Let's find the vertex.

Using our formula for the x-coordinate of the vertex, we have

$$x = \frac{-b}{2a} = \frac{-6}{2(-2)} = \frac{6}{4} = \frac{3}{2}$$

To find the y-coordinate, we let $x = \frac{3}{2}$.

$$f\left(\frac{3}{2}\right) = -2\left(\frac{3}{2}\right)^2 + 6\left(\frac{3}{2}\right) - 5$$

$$= \frac{-18}{4} + \frac{18}{2} - 5$$

$$= \frac{-18 + 36 - 20}{4}$$

$$= -\frac{1}{2}$$

Finding the vertex by completing the square is a more complicated matter. To make the coefficient of x^2 a 1, we must factor -2 from the first two terms. To complete the square inside the parentheses, we add $\frac{9}{4}$. Since each term inside the parentheses is multiplied by -2, we add $\frac{9}{2}$ outside the parentheses so that the net result is the same as adding 0 to the right side.

$$f(x) = -2(x^2 - 3x \qquad) - 5$$

$$= -2\left(x^2 - 3x + \frac{9}{4}\right) - 5 + \frac{9}{2}$$

$$= -2\left(x - \frac{3}{2}\right)^2 - \frac{1}{2}$$

The vertex is $\left(\frac{3}{2}, -\frac{1}{2}\right)$. Because this is the only point we have so far, we must find two others. Let's let $x = 3$ and $x = 0$, because each point is the same distance from $x = \frac{3}{2}$ and on either side.

When $x = 3$

$$f(3) = -2(3)^2 + 6(3) - 5$$

$$= -18 + 18 - 5$$

$$= -5$$

When $x = 0$

$$f(0) = -2(0)^2 + 6(0) - 5$$

$$= 0 + 0 - 5$$

$$= -5$$

The two additional points on the graph are $(3, -5)$ and $(0, -5)$. Figure 6 shows the graph.

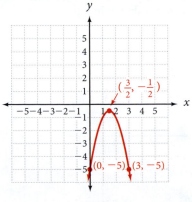

FIGURE 6

The graph is concave down. The vertex is the highest point on the graph.

By looking at the equations and graphs in Examples 1 through 4, we can conclude that the graph of $f(x) = ax^2 + bx + c$ will be concave up when a is positive, and concave down when a is negative. Taking this even further, if $a > 0$, then the vertex is the lowest point on the graph, and if $a < 0$, the vertex is the highest point on the graph. Finally, if we complete the square on x in the equation $f(x) = ax^2 + bx + c, a \neq 0$, we can rewrite the equation of our parabola as $f(x) = a(x - h)^2 + k$. When the equation is in this form, the vertex is at the point (h, k). Here is a summary:

> **PROPERTY** Graphing Parabolas II
>
> The graph of the quadratic function
>
> $$f(x) = a(x - h)^2 + k, a \neq 0$$
>
> will be a vertical parabola with a vertex at (h, k). The vertex will be the highest point on the graph when $a < 0$, and the lowest point on the graph when $a > 0$. When the vertex is the highest point, the parabola will be concave down; whereas, when the vertex is the lowest point, the parabola will be concave up.

B Applications

EXAMPLE 5   A company selling copies of an accounting program for home computers finds that it will make a weekly profit of P dollars from selling x copies of the program, according to the function

$$P(x) = -0.1x^2 + 27x - 500$$

How many copies of the program should it sell to make the largest possible profit, and what is the largest possible profit?

Solution   Because the coefficient of x^2 is negative, we know the graph of this parabola will be concave down, meaning that the vertex is the highest point of the curve. We find the vertex by first finding its x-coordinate.

$$x = \frac{-b}{2a} = \frac{-27}{2(-0.1)} = \frac{27}{0.2} = 135$$

This represents the number of programs the company needs to sell each week to make a maximum profit. To find the maximum profit, we substitute 135 for x in the original equation. (A calculator is helpful for these kinds of calculations.)

$$P(135) = -0.1(135)^2 + 27(135) - 500$$

$$= -0.1(18,225) + 3,645 - 500$$

$$= -1,822.5 + 3,645 - 500$$

$$= 1,322.5$$

The maximum weekly profit is \$1,322.50 and is obtained by selling 135 programs a week.

EXAMPLE 6   An art supply store finds that they can sell x sketch pads each week at p dollars each, according to the equation $x = 900 - 300p$. Graph the revenue function $R = xp$. Then use the graph to find the price p that will bring in the maximum revenue. Finally, find the maximum revenue.

Solution   As it stands, the revenue equation contains three variables. Because we are asked to find the value of p that gives us the maximum value of R, we rewrite the equation using just the variables R and p. Because $x = 900 - 300p$, we have

$$R(p) = xp = (900 - 300p)p$$

Note It is important to recognize the subtraction sign that precedes h in the function. Students often mistake the value for h in (h, k) as a negative value simply because of the subtraction sign. Remember that the value of h is positive if it appears as $f(x) = a(x - h)^2 + k$ and the value of h is negative if $f(x)$ appears as $f(x) = (a + h)^2 + k$.

5. Find the largest value of $p(x)$ if $p(x) = -0.01x^2 + 12x - 400$.

6. Repeat Example 9 if the number of sketch pads they can sell each week is related to the price by the equation $x = 800 - 200p$.

Answers

5. $x = 600, y = 3200$

6. $R(2) = \$800$

The graph of this equation is shown in Figure 7. The graph appears in the first quadrant only, because R and p are both positive quantities.

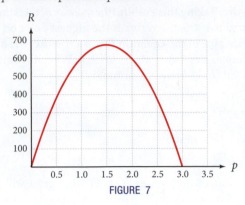

FIGURE 7

From the graph in figure 10, we see that the maximum value of R occurs when $p = \$1.50$. We can calculate the maximum value of R from the equation.

$$\text{When} \rightarrow \qquad\qquad p = 1.5$$
$$\text{the equation} \rightarrow \quad R(p) = (900 - 300p)p$$
$$\text{becomes} \rightarrow \quad\; R(1.5) = (900 - 300 \cdot 1.5)1.5$$
$$= (900 - 450)1.5$$
$$= 450 \cdot 1.5$$
$$= 675$$

The maximum revenue is $675. It is obtained by setting the price of each sketch pad at $p = \$1.50$.

USING TECHNOLOGY   Graphing Calculators

If you have been using a graphing calculator for some of the material in this course, you are well aware that your calculator can draw all the graphs in this section very easily. It is important, however, that you be able to recognize and sketch the graph of any parabola by hand. It is a skill that all successful intermediate algebra students should possess, even if they are proficient in the use of a graphing calculator. My suggestion is that you work the problems in this section and problem set without your calculator. Then use your calculator to check your results.

C Finding the Equation from the Graph

EXAMPLE 7   At the 1997 Washington County Fair in Oregon, David Smith, Jr., The Bullet, was shot from a cannon. As a human cannonball, he reached a height of 70 feet before landing in a net 160 feet from the cannon. Sketch the graph of his path, and then find the equation of the graph.

Solution   We assume that the path taken by the human cannonball is a parabola. If the origin of the coordinate system is at the opening of the cannon, then the net that catches him will be at 160 on the x-axis. Figure 8 shows the graph.

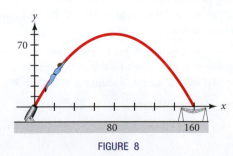

FIGURE 8

Because the curve is a parabola, we know the function will have the form

$$f(x) = a(x - h)^2 + k$$

Because the vertex of the parabola is at $(80, 70)$, we can fill in two of the three constants in our equation, giving us

$$f(x) = a(x - 80)^2 + 70$$

To find a, we note that the landing point will be $(160, 0)$. Substituting the coordinates of this point into the equation, we solve for a.

$$f(160) = a(160 - 80)^2 + 70$$

$$0 = a(80)^2 + 70$$

$$0 = 6,400a + 70$$

$$a = -\frac{70}{6,400} = -\frac{7}{640}$$

The equation that describes the path of the human cannonball is

$$f(x) = -\frac{7}{640}(x - 80)^2 + 70 \quad \text{for} \quad 0 \le x \le 160$$

USING TECHNOLOGY Graphing Calculators

Graph the equation found in Example 10 on a graphing calculator using the window shown here. (We will use this graph later in the book to find the angle between the cannon and the horizontal.)

      Window: X from 0 to 160, increment 20
                     Y from 0 to 80, increment 10

On the TI-83/84, an increment of 20 for X means Xscl = 20.

GETTING READY FOR CLASS

After reading through the preceding section, respond in your own words and in complete sentences.

A. What is a parabola?

B. What part of the equation of a parabola determines whether the graph is concave up or concave down?

C. Suppose $f(x) = ax^2 + bx + c$ is the equation for a parabola. Explain how $f(4) = 1$ relates to the graph of the parabola.

D. A line can be graphed with two points. How many points are necessary to get a reasonable sketch of a parabola? Explain.

Vocabulary Review

Choose the correct words to fill in the blanks below.

up	down	$x = -\frac{b}{2a}$	(h, k)
$y = -\frac{b}{2a}$	vertex	x-intercepts	y-intercept

1. The _____ for the function $f(x) = ax^2 + bx + c$ is $y = c$.

2. The _____ for the function $f(x) = ax^2 + bx + c$ are at $x = \dfrac{-b \pm \sqrt{b^2 - 4ac}}{2a}$.

3. The highest or lowest point on a parabola is called the _____.

4. The vertex for the graph of $f(x) = ax^2 + bx + c$ will always occur when _____.

5. If the graph of a parabola opens up, the graph is said to be concave _____.

6. If the graph of a parabola opens _____, the graph is said to be concave down.

7. The graph of $f(x) = a(x - h)^2 + k$ will be a parabola with a vertex at _____.

8. Parabolas that open left to right will have a vertex when _____.

Problems

A For each of the following functions, give the x-intercepts and the coordinates of the vertex, and sketch the graph.

1. $f(x) = x^2 + 2x - 3$

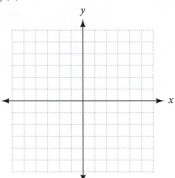

2. $f(x) = x^2 - 2x - 3$

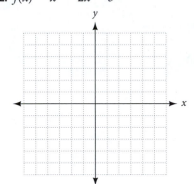

3. $f(x) = -x^2 - 4x + 5$

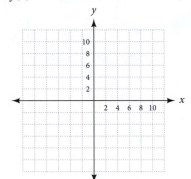

4. $f(x) = x^2 + 4x - 5$

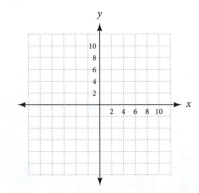

5. $f(x) = x^2 - 1$

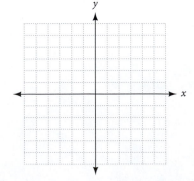

SCAN TO ACCESS

6. $f(x) = x^2 - 4$

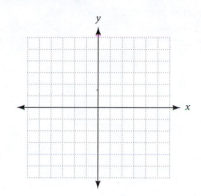

7. $f(x) = -x^2 + 9$

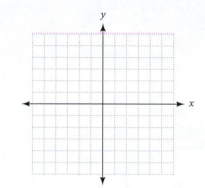

8. $f(x) = -x^2 + 1$

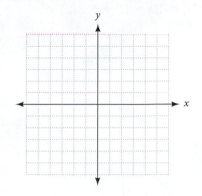

9. $f(x) = 2x^2 - 4x - 6$

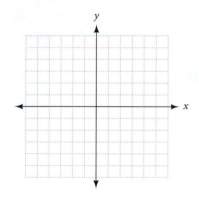

10. $f(x) = 2x^2 + 4x - 6$

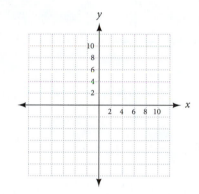

11. $f(x) = x^2 - 2x - 4$

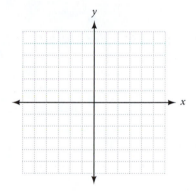

12. $f(x) = x^2 - 2x - 2$

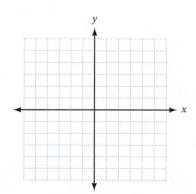

13. $f(x) = -x^2 - 2x + 3$

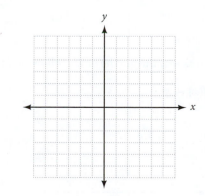

14. $f(x) = 2x^2 - 20x - 48$

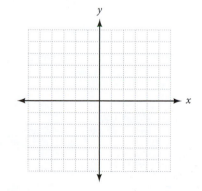

Graph each quadratic function. Label the vertex and any intercepts that exist.

15. $f(x) = 2(x - 1)^2 + 3$

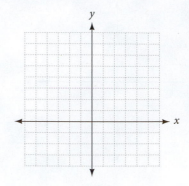

16. $f(x) = 2(x + 1)^2 - 3$

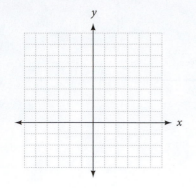

17. $f(x) = -(x + 2)^2 + 4$

18. $f(x) = -(x - 3)^2 + 1$

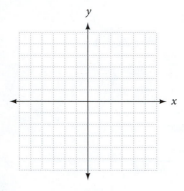

19. $g(x) = \frac{1}{2}(x - 2)^2 - 4$

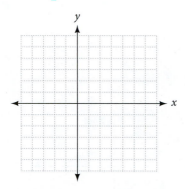

20. $g(x) = \frac{1}{3}(x - 3)^2 - 3$

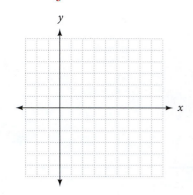

21. $f(x) = -2(x - 4)^2 - 1$

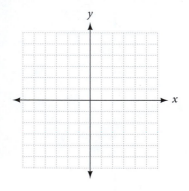

22. $f(x) = -4(x - 1)^2 + 4$

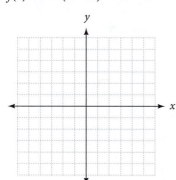

Find the vertex and any two convenient points to sketch the graphs of the following functions.

23. $f(x) = x^2 - 4x - 4$

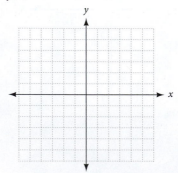

24. $f(x) = x^2 - 2x + 3$

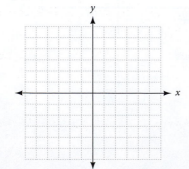

25. $f(x) = -x^2 + 2x - 5$

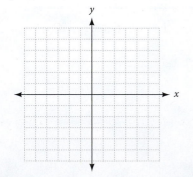

26. $f(x) = -x^2 + 4x - 2$

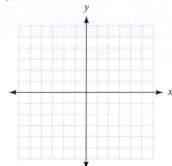

27. $f(x) = x^2 + 1$

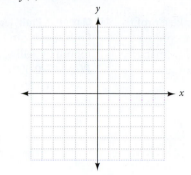

28. $f(x) = x^2 + 4$

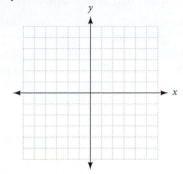

29. $f(x) = -x^2 - 3$

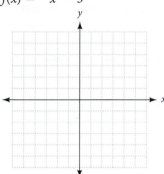

30. $f(x) = -x^2 - 2$

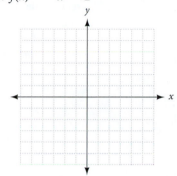

31. $g(x) = 3x^2 + 4x + 1$

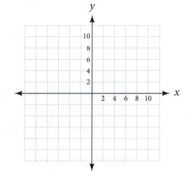

32. $g(x) = 2x^2 + 4x + 3$

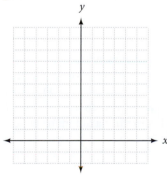

33. $g(x) = -2x^2 + 8x - 3$

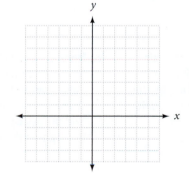

34. $g(x) = -3x^2 + 6x$

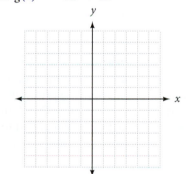

For each of the following equations, find the coordinates of the vertex, and indicate whether the vertex is the highest point on the graph or the lowest point on the graph. (Do not graph.)

35. $f(x) = x^2 - 6x + 5$

36. $f(x) = -x^2 + 6x - 5$

37. $f(x) = -x^2 + 2x + 8$

38. $f(x) = x^2 - 2x - 8$

39. $f(x) = 12 + 4x - x^2$

40. $f(x) = -12 - 4x + x^2$

41. $f(x) = -x^2 - 8x$

42. $f(x) = x^2 + 8x$

B C Applying the Concepts

43. Maximum Profit A company finds that it can make a profit of P dollars each month by selling x patterns, according to the formula $P(x) = -0.002x^2 + 3.5x - 800$. How many patterns must it sell each month to have a maximum profit? What is the maximum profit?

44. Maximum Profit A company selling picture frames finds that it can make a profit of P dollars each month by selling x frames, according to the formula $P(x) = -0.002x^2 + 5.5x - 1,200$. How many frames must it sell each month to have a maximum profit? What is the maximum profit?

45. Maximum Height Chaudra is tossing a softball into the air with an underhand motion. The distance of the ball above her hand at any time is given by the function

$$h(t) = 32t - 16t^2 \quad \text{for}$$
$$0 \le t \le 2$$

where $h(t)$ is the height of the ball (in feet) and t is the time (in seconds). Find the times at which the ball is in her hand, and the maximum height of the ball.

h

46. Maximum Area Justin wants to fence three sides of a rectangular exercise yard for his dog. The fourth side of the exercise yard will be a side of the house. He has 80 feet of fencing available. Find the dimensions of the exercise yard that will enclose the maximum area.

47. Maximum Revenue A company that manufactures hair ribbons knows that the number of ribbons x it can sell each week is related to the price p of each ribbon by the equation $x = 1,200 - 100p$. Graph the revenue equation $R = xp$. Then use the graph to find the price p that will bring in the maximum revenue. Finally, find the maximum revenue.

48. Maximum Revenue A company that manufactures CDs for home computers finds that it can sell x CDs each day at p dollars per CD, according to the equation $x = 800 - 100p$. Graph the revenue equation $R = xp$. Then use the graph to find the price p that will bring in the maximum revenue. Finally, find the maximum revenue.

49. Maximum Revenue The relationship between the number of calculators x a company sells each day and the price p of each calculator is given by the equation $x = 1,700 - 100p$. Graph the revenue equation $R = xp$, and use the graph to find the price p that will bring in the maximum revenue. Then find the maximum revenue.

50. Maximum Revenue The relationship between the number x of pencil sharpeners a company sells each week and the price p of each sharpener is given by the equation $x = 1,800 - 100p$. Graph the revenue equation $R = xp$, and use the graph to find the price p that will bring in the maximum revenue. Then find the maximum revenue.

51. Human Cannonball A human cannonball is shot from a cannon at the county fair. He reaches a height of 60 feet before landing in a net 180 feet from the cannon. Sketch the graph of his path, and then find the equation of the graph.

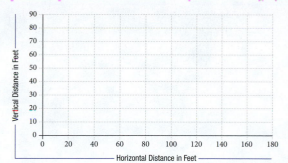

52. Interpreting Graphs The graph below shows the different paths taken by the human cannonball when his velocity out of the cannon is 50 miles/hour, and his cannon is inclined at varying angles.

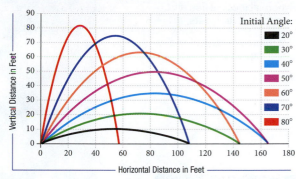

Initial Velocity: 50 miles per hour

a. If his landing net is placed 104 feet from the cannon, at what angle should the cannon be inclined so that he lands in the net.

b. Approximately where do you think he would land if the cannon was inclined at 45°?

c. The fact that every landing point can come from two different paths makes us think that the equations that give us the landing points must be what type of equations?

Getting Ready for the Next Section

Solve each equation.

53. $x^2 - 2x - 8 = 0$

54. $x^2 - x - 12 = 0$

55. $6x^2 - x = 2$

56. $3x^2 - 5x = 2$

57. $x^2 - 6x + 9 = 0$

58. $x^2 + 8x + 16 = 0$

59. $x - \dfrac{4}{x} = 0$

60. $1 - \dfrac{9}{x^2} = 0$

Find the Mistake

Each sentence below contains a mistake. Circle the mistake and write the correct word or phrase on the line provided.

1. The graph of $f(x) = -x^2 - 2x + 3$ is concave up. _____

2. The graph of $f(x) = \frac{1}{4}(x + 1)^2 - 4$ opens to the right. _____

3. The graph of $x = -2(y + 1)^2 - 1$ is concave down. _____

4. The graph of $x = y^2 - 6x + 4$ opens to the left. _____

OBJECTIVES

A Solve quadratic and polynomial inequalities and graph the solution sets.

B Solve inequalities involving rational expressions

KEY WORDS

quadratic inequality

7.6 Polynomial and Rational Inequalities

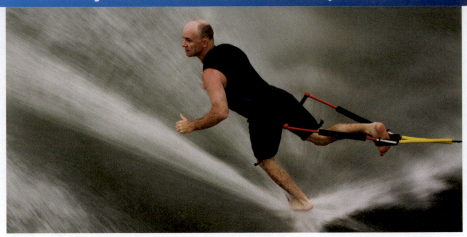

©iStockphoto.com/dlanier

Some talented water skiers ski on their bare feet while being towed behind a motor boat. During training, the barefoot skiers may use foot skis, which are small thin skis just a few inches longer and a little wider than the skier's foot. With foot skis, the skier can learn maneuvers faster and with less wipeouts than without.

Let's say the manufacturer of foot skis knows that the weekly revenue produced by selling x foot skis is given by the equation $R = 13{,}000p - 100p^2$, where p is the price of each pair of foot skis (in dollars). To find the price that needs to be charged for each pair of foot skis if the weekly revenue is to be at least \$400,000, we need to understand quadratic inequalities, which is the focus of this section.

A Quadratic Inequalities

Quadratic inequalities in one variable are inequalities of the form

$$ax^2 + bx + c < 0 \qquad ax^2 + bx + c > 0$$
$$ax^2 + bx + c \leq 0 \qquad ax^2 + bx + c \geq 0$$

where a, b, and c are constants, with $a \neq 0$. The technique we will use to solve inequalities of this type involves graphing. Suppose, for example, we want to find the solution set for the inequality $x^2 - x - 6 > 0$. We begin by factoring the left side to obtain

$$(x - 3)(x + 2) > 0$$

For any chosen value of x, each expression $x - 3$ and $x + 2$ will be a real number whose product $(x - 3)(x + 2)$ is greater than zero. That is, their product is positive. The only way the product can be positive is either if both factors, $(x - 3)$ and $(x + 2)$, are positive or if they are both negative. To help visualize where $x - 3$ is positive and where it is negative, we draw a real number line and label it accordingly.

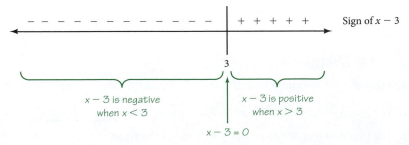

Here is a similar diagram showing where the factor $x + 2$ is positive and where it is negative:

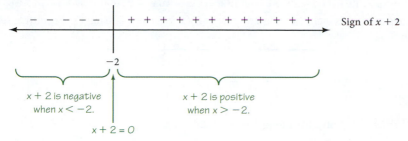

Drawing the two number lines together and eliminating the unnecessary numbers, we have

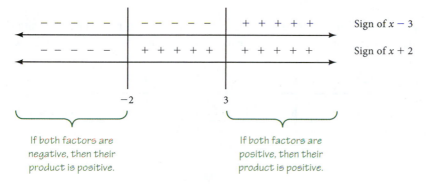

We can see from the preceding diagram that the graph of the solution to the inequality $x^2 - x - 6 > 0$ is

The solution (in interval notation) is $(-\infty, -2) \cup (3, \infty)$. Notice that neither boundary is included in the solution.

USING TECHNOLOGY Graphical Solutions to Quadratic Inequalities

We can solve the preceding problem by using a graphing calculator to visualize where the product $(x - 3)(x + 2)$ is positive. First, we graph the function $y = (x - 3)(x + 2)$ as shown in Figure 1.

Next, we observe where the graph is above the x-axis. As you can see, the graph is above the x-axis to the right of 3 and to the left of -2, as shown in Figure 2.

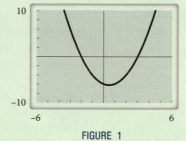

FIGURE 1

Graph is above the x-axis when x is here.

Graph is above the x-axis when x is here.

FIGURE 2

Practice Problems

1. Solve the inequality.
   $x^2 + 2x - 15 \leq 0$

When the graph is above the x-axis, we have points whose y-coordinates are positive. Because these y-coordinates are the same as the expression $(x - 3)(x + 2)$, the values of x for which the graph of $y = (x - 3)(x + 2)$ is above the x-axis are the values of x for which the inequality $(x - 3)(x + 2) > 0$ is true. Our solution set is therefore

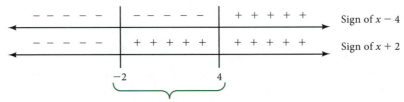

$$x < -2 \quad \text{or} \quad x > 3$$

EXAMPLE 1 Solve the inequality $x^2 - 2x - 8 \leq 0$.

Algebraic Solution We begin by factoring.

$$x^2 - 2x - 8 \leq 0$$
$$(x - 4)(x + 2) \leq 0$$

The product $(x - 4)(x + 2)$ is negative or zero. The factors must have opposite signs. We draw a diagram showing where each factor is positive and where each factor is negative.

From the diagram, we have the graph of the solution set.

The solution (in interval notation) is $[-2, 4]$.

Graphical Solution To solve this inequality with a graphing calculator, we graph the function $y = (x - 4)(x + 2)$ and observe where the graph is below the x-axis. These points have negative y-coordinates, which means that the product $(x - 4)(x + 2)$ is negative for these points. Figure 3 shows the graph of $y = (x - 4)(x + 2)$, along with the region on the x-axis where the graph contains points with negative y-coordinates.

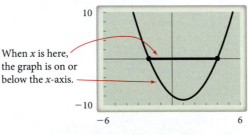

When x is here, the graph is on or below the x-axis.

FIGURE 3

As you can see, the graph is below the x-axis when x is between -2 and 4. Because our original inequality includes the possibility that $(x - 4)(x + 2)$ is 0, we include the endpoints, -2 and 4, with our solution set.

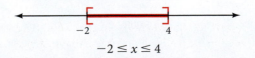

$$-2 \leq x \leq 4$$

Answer

1. $[-5, 3]$

EXAMPLE 2 Solve the inequality $6x^2 - x \geq 2$.

Algebraic Solution

$$6x^2 - x \geq 2$$

$$6x^2 - x - 2 \geq 0 \quad \leftarrow \textit{Standard form}$$

$$(3x - 2)(2x + 1) \geq 0$$

The product is positive or zero, so the factors must agree in sign. Here is the diagram showing where that occurs:

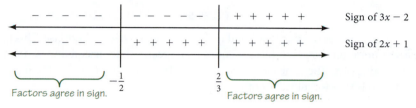

Because the factors agree in sign below $-\frac{1}{2}$ and above $\frac{2}{3}$, the graph of the solution set is

The solution (in interval notation) is $\left(-\infty, -\frac{1}{2} \right] \cup \left[\frac{2}{3}, \infty \right)$.

Graphical Solution To solve this inequality with a graphing calculator, we graph the function $y = (3x - 2)(2x + 1)$ and observe where the graph is above the x-axis. These are the points that have positive y-coordinates, which means that the product $(3x - 2)$ $(2x + 1)$ is positive for these points. Figure 4 shows the graph of $y = (3x - 2)(2x + 1)$, along with the regions on the x-axis where the graph is on or above the x-axis.

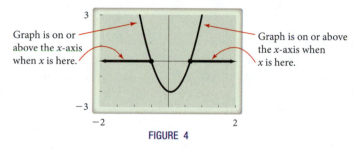

FIGURE 4

To find the points where the graph crosses the x-axis, we need to use either the Trace and Zoom features to zoom in on each point, or the calculator function that finds the intercepts automatically (on the TI-83/84 this is the root/zero function under the CALC key). Whichever method we use, we will obtain the following result:

$$x \leq -0.5 \qquad \text{or} \qquad x \geq 0.67$$

EXAMPLE 3 Solve the inequality $x^2 - 6x + 9 \geq 0$.

Algebraic Solution

$$x^2 - 6x + 9 \geq 0$$

$$(x - 3)^2 \geq 0$$

This is a special case in which both factors are the same. Because $(x - 3)^2$ is always positive or zero, the solution set is all real numbers. That is, any real number that is used in place of x in the original inequality will produce a true statement.

2. Solve the inequality $2x^2 + 3x > 2$.

3. Solve the inequality.
$x^2 + 10x + 25 < 0$

Answers

2. $(-\infty, -2) \cup \left(\frac{1}{2}, \infty \right)$

3. No solution

Graphical Solution The graph of $y = (x - 3)^2$ is shown in Figure 5.

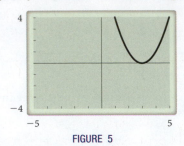

FIGURE 5

Notice that it touches the x-axis at 3 and is above the x-axis everywhere else. This means that every point on the graph has a y-coordinate greater than or equal to 0, no matter what the value of x. The conclusion that we draw from the graph is that the inequality $(x - 3)^2 \geq 0$ is true for all values of x.

The techniques used to solve quadratic inequalities can be applied to higher-order polynomial inequalities, as the next example illustrates.

4. Solve the inequality.
$(x + 4)(x - 1)(x - 5) \geq 0$

EXAMPLE 4   Solve the inequality $(x + 3)(x - 2)(x - 4) < 0$.

Algebraic Solution We construct a sign chart for all three factors.

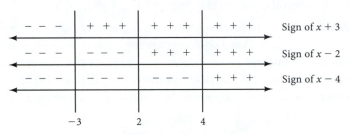

The product of factors is negative (less than 0) when $x < -3$ or $2 < x < 4$. The solution (in interval notation) is $(-\infty, -3) \cup (2, 4)$.

Graphical Solution The graph of $y = (x + 3)(x - 2)(x - 4)$ is shown here.

Notice that $y < 0$ when $x < -3$ or when $2 < x < 4$.

Our next two examples involve inequalities that contain rational expressions.

B Inequalities Involving Rational Expressions

5. Solve the inequality $\dfrac{x + 5}{x - 1} > 0$.

EXAMPLE 5   Solve the inequality $\dfrac{x - 4}{x + 1} \leq 0$.

Solution   The inequality indicates that the quotient of $(x - 4)$ and $(x + 1)$ is negative or 0 (less than or equal to 0). We can use the same reasoning we used to solve the first three examples, because quotients are positive or negative under the same conditions that products are positive or negative. Here is the diagram that shows where each factor is positive and where each factor is negative.

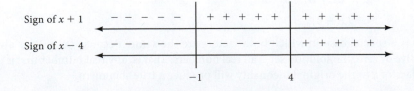

Between -1 and 4 the factors have opposite signs, making the quotient negative. Thus, the region between -1 and 4 is where the solutions lie, because the original inequality indicates the quotient $\frac{x-4}{x+1}$ is negative. The solution set and its graph are shown here:

The solution (in interval notation) is $(-1, 4]$.

Notice that the left endpoint is open—that is, it is not included in the solution set—because $x = -1$ would make the denominator in the original inequality 0. It is important to check all endpoints of solution sets to inequalities that involve rational expressions.

> **Note** This is an important difference from polyomial inequalities. Be aware of restrictions on the variable when considering if a boundary is included in the solution.

EXAMPLE 6 Solve the inequality $\dfrac{3}{x-2} - \dfrac{2}{x-3} > 0$.

Solution We begin by adding the two rational expressions on the left side. The common denominator is $(x-2)(x-3)$.

$$\frac{3}{x-2} \cdot \frac{(x-3)}{(x-3)} - \frac{2}{x-3} \cdot \frac{(x-2)}{(x-2)} > 0$$

$$\frac{3x - 9 - 2x + 4}{(x-2)(x-3)} > 0$$

$$\frac{x-5}{(x-2)(x-3)} > 0$$

This time the quotient involves three factors. Here is the diagram that shows the signs of the three factors:

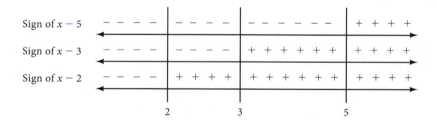

The original inequality indicates that the quotient is positive. For this to happen, either all three factors must be positive, or exactly two factors must be negative. Looking back at the diagram, we see the regions that satisfy these conditions are between 2 and 3 or above 5. Here is our solution set.

The solution (in interval notation) is $(2, 3) \cup (5, \infty)$.

---

6. Solve the inequality.
$$\frac{5}{x+3} + \frac{4}{x+4} > 0$$

Answer

6. $\left(-4, -\frac{32}{9}\right) \cup (-3, \infty)$

7. Solve the inequality $\dfrac{x-1}{2x+6} \le \dfrac{2}{x^2-4x+3}$.

EXAMPLE 7   Solve the inequality $\dfrac{x+1}{3x-6} \le \dfrac{1}{x^2-3x+2}$.

Solution   We begin by putting the rational inequality into standard form.

$$\frac{x+1}{3x-6} - \frac{1}{x^2-3x+2} \le 0$$

Then we factor both denominators so we can find the common denominator.

$$\frac{x+1}{3(x-2)} - \frac{1}{(x-2)(x-1)} \le 0$$

We subtract the two rational expressions, using the LCD $3(x-2)(x-1)$.

$$\frac{(x+1)(x-1) - 3}{3(x-2)(x-1)} \le 0$$

Finally, we simplify and factor the numerator.

$$\frac{x^2-4}{3(x-2)(x-1)} \le 0 \qquad \textcolor{green}{\text{Simplify the numerator.}}$$

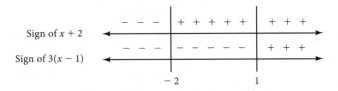

$$\frac{(x+2)(x-2)}{3(x-2)(x-1)} \le 0 \qquad \textcolor{green}{\text{Factor the numerator.}}$$

$$\frac{(x+2)}{3(x-1)} \le 0 \qquad \textcolor{green}{\text{Divide out the common factor } (x-2).}$$

Sign of $x+2$ ⟵ $-\ -\ -$ | $+\ +\ +\ +\ +$ | $+\ +\ +$

Sign of $3(x-1)$ ⟵ $-\ -\ -$ | $-\ -\ -\ -\ -$ | $+\ +\ +$

-2 1

In order for the inequality to be true, the factors must have opposite signs. In addition, $x \ne 1$ since the denominator would be zero. Here is our solution set.

-2 0 1

The solution in interval notation is $[-2, 1)$.

GETTING READY FOR CLASS

After reading through the preceding section, respond in your own words and in complete sentences.

A. What is the first step in solving a quadratic inequality?

B. How do you show that the endpoint of a line segment is not part of the graph of a quadratic inequality?

C. How would you use the graph of $y = ax^2 + bx + c$ to help you find the graph of $ax^2 + bx + c < 0$?

D. Can a quadratic inequality have exactly one solution? Give an example.

Answer

7. $[-1, 1)$

Vocabulary Review

Choose the correct words to fill in the blanks below.

opposite   factors   constants   real number line

same   inequalities

1. Quadratic _____ in one variable are of the forms $ax^2 + bx + c < 0$, $ax^2 + bx + c > 0$, $ax^2 + bx + c \leq 0$, and $ax^2 + bx + c \geq 0$.

2. In a quadratic inequality in one variable, the coefficients a, b, and c are _____.

3. Drawing a _____ that shows the sign of the _____ to a quadratic inequality will help determine the graph of the solutions to the inequality.

4. If the product of the factors of a quadratic inequality is positive, then the factors must have the _____ sign.

5. If the product of the factors of a quadratic inequality is negative or zero, then the factors must have _____ signs.

Problems

A Solve each of the following inequalities, and graph the solution set.

1. $x^2 + x - 6 > 0$

2. $x^2 + x - 6 < 0$

3. $x^2 - x - 12 \leq 0$

4. $x^2 - x - 12 \geq 0$

5. $x^2 + 5x \geq -6$

6. $x^2 - 5x > 6$

7. $6x^2 < 5x - 1$

8. $4x^2 \geq -5x + 6$

9. $x^2 - 9 < 0$

10. $x^2 - 16 \geq 0$

11. $4x^2 - 9 \geq 0$

12. $9x^2 - 4 < 0$

13. $2x^2 - x - 3 < 0$

14. $x^2 + 2x - 12 > 3$

15. $x^2 - 4x - 5 \geq -8$

16. $x^2 - 8x - 16 \leq -3x + 8$

17. $x^2 + 3x - 4 \geq -4 + 2x$

18. $3x^2 + x - 10 \geq 0$

19. $x^2 - 4x + 4 \geq 0$

20. $x^2 - 4x + 4 < 0$

21. $x^2 - 10x + 25 < 0$

22. $x^2 - 10x + 25 > 0$

23. $(x - 2)(x - 3)(x - 4) > 0$

24. $(x - 2)(x - 3)(x - 4) < 0$

25. $(x + 1)(x + 2)(x + 3) \leq 0$

26. $(x + 1)(x + 2)(x + 3) \geq 0$

27. $x(x + 4)(x - 2) \geq 0$

28. $x(x - 1)(x + 2) > 0$

B Solve each of the following inequalities, and graph the solution set.

29. $\dfrac{x - 1}{x + 4} \leq 0$

30. $\dfrac{x + 4}{x - 1} \leq 0$

31. $\dfrac{3x}{x + 6} - \dfrac{8}{x + 6} < 0$

32. $\dfrac{5x}{x + 1} - \dfrac{3}{x + 1} < 0$

33. $\dfrac{4}{x - 6} + 1 > 0$

34. $\dfrac{2}{x - 3} + 1 \geq 0$

35. $\dfrac{x - 2}{(x + 3)(x - 4)} < 0$

36. $\dfrac{x - 1}{(x + 2)(x - 5)} < 0$

37. $\dfrac{2}{x - 4} - \dfrac{1}{x - 3} > 0$

38. $\dfrac{4}{x+3} - \dfrac{3}{x+2} > 0$

39. $\dfrac{x+7}{2x+12} + \dfrac{6}{x^2-36} \leq 0$

40. $\dfrac{x+1}{2x-2} - \dfrac{2}{x^2-1} \leq 0$

41. $\dfrac{x+1}{4x-4} \leq \dfrac{1}{x^2-1}$

42. $\dfrac{x+1}{2x+4} \geq \dfrac{2}{x^2-4}$

43. The graph of $y = x^2 - 4$ is shown in Figure 6. Use the graph to write the solution set for each of the following:

   a. $x^2 - 4 < 0$

   b. $x^2 - 4 > 0$

   c. $x^2 - 4 = 0$

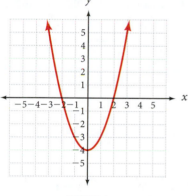

FIGURE 6

44. The graph of $y = 4 - x^2$ is shown in Figure 7. Use the graph to write the solution set for each of the following:

   a. $4 - x^2 < 0$

   b. $4 - x^2 > 0$

   c. $4 - x^2 = 0$

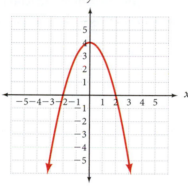

FIGURE 7

45. The graph of $y = x^2 - 3x - 10$ is shown in Figure 8. Use the graph to write the solution set for each of the following:

   a. $x^2 - 3x - 10 < 0$

   b. $x^2 - 3x - 10 > 0$

   c. $x^2 - 3x - 10 = 0$

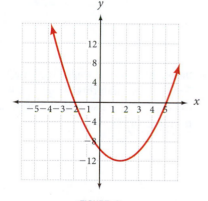

FIGURE 8

46. The graph of $y = x^2 + x - 12$ is shown in Figure 9. Use the graph to write the solution set for each of the following:

   a. $x^2 + x - 12 < 0$

   b. $x^2 + x - 12 > 0$

   c. $x^2 + x - 12 = 0$

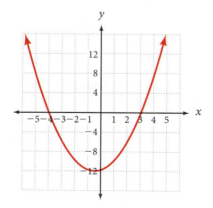

FIGURE 9

47. The graph of $y = x^3 - 3x^2 - x + 3$ is shown in Figure 10. Use the graph to write the solution set for each of the following:

a. $x^3 - 3x^2 - x + 3 < 0$

b. $x^3 - 3x^2 - x + 3 > 0$

c. $x^3 - 3x^2 - x + 3 = 0$

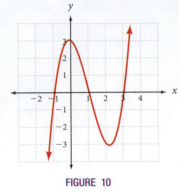

FIGURE 10

48. The graph of $y = x^3 + 4x^2 - 4x - 16$ is shown in Figure 11. Use the graph to write the solution set for each of the following:

a. $x^3 + 4x^2 - 4x - 16 < 0$

b. $x^3 + 4x^2 - 4x - 16 > 0$

c. $x^3 + 4x^2 - 4x - 16 = 0$

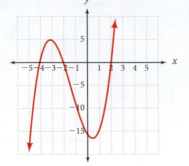

FIGURE 11

Recall that the domain of a function is the set of all possible values that x can have. Use this idea to find the domain of the following functions.

49. $f(x) = \dfrac{1}{x^2 - 3x - 4}$

50. $f(x) = \dfrac{1}{x^2 + 4x - 5}$

51. $g(x) = \sqrt{x^2 - 7x}$

52. $g(x) = \sqrt{2x^2 - 5x}$

53. $h(x) = \sqrt{2x^2 - 5x - 3}$

54. $h(x) = \sqrt{3x^2 + 4x - 4}$

55. $f(x) = \dfrac{1}{\sqrt{5x^2 - 17x + 6}}$

56. $f(x) = \dfrac{1}{\sqrt{6x^2 - 23x + 20}}$

Applying the Concepts

57. Dimensions of a Rectangle The length of a rectangle is 3 inches more than twice the width. If the area is to be at least 44 square inches, what are the possibilities for the width?

58. Dimensions of a Rectangle The length of a rectangle is 5 inches less than three times the width. If the area is to be less than 12 square inches, what are the possibilities for the width?

59. Revenue A manufacturer of portable radios knows that the weekly revenue produced by selling x radios is given by the equation $R = 1{,}300p - 100p^2$, where p is the price of each radio (in dollars). What price should be charged for each radio if the weekly revenue is to be at least $4,000?

60. Revenue A manufacturer of small calculators knows that the weekly revenue produced by selling x calculators is given by the equation $R = 1{,}700p - 100p^2$, where p is the price of each calculator (in dollars). What price should be charged for each calculator if the revenue is to be at least $7,000 each week?

61. Union Dues A labor union has 10,000 members. For every $10 increase in union dues, membership is decreased by 200 people. If the current dues are $100, what should be the new dues (to the nearest multiple of $10) so income from dues is greatest, and what is that income? *Hint:* Because Income = (membership)(dues), we can let $x =$ the number of $10 increases in dues, and then this will give us income of $y = (10{,}000 - 200x)(100 + 10x)$.

62. Bookstore Receipts The owner of a used book store charges $2 for quality paperbacks and usually sells 40 per day. For every 10-cent increase in the price of these paperbacks, he thinks that he will sell two fewer per day. What is the price he should charge (to the nearest 10 cents) for these books to maximize his income, and what would be that income? *Hint:* Let $x =$ the number of 10-cent increases in price.

63. Jiffy-Lube The owner of a quick oil-change business charges $20 per oil change and has 40 customers per day. If each increase of $2 results in 2 fewer daily customers, what price should the owner charge (to the nearest $2) for an oil change if the income from this business is to be as great as possible?

64. Computer Sales A computer manufacturer charges $2,200 for its basic model and sells 1,500 computers per month at this price. For every $200 increase in price, it is believed that 75 fewer computers will be sold. What price should the company place on its basic model of computer (to the nearest $100) to have the greatest income?

Maintaining Your Skills

Use a calculator to evaluate, give answers to 4 decimal places.

65. $\dfrac{50{,}000}{32{,}000}$

66. $\dfrac{2.4362}{1.9758} - 1$

67. $\dfrac{1}{2}\left(\dfrac{4.5926}{1.3876} - 2\right)$

68. $1 + \dfrac{0.06}{12}$

Solve each equation.

69. $2\sqrt{3t - 1} = 2$

70. $\sqrt{4t + 5} + 7 = 3$

71. $\sqrt{x + 3} = x - 3$

72. $\sqrt{x + 3} = \sqrt{x} - 3$

Graph each function.

73. $f(x) = \sqrt{x - 2}$

74. $g(x) = \sqrt{x} - 2$

75. $f(x) = \sqrt[3]{x - 1}$

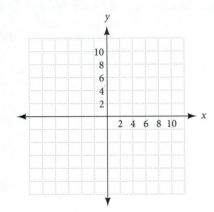

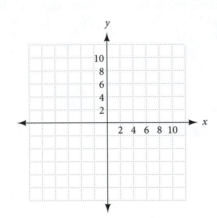

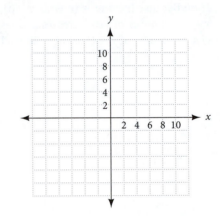

Find the Mistake

Each sentence below contains a mistake. Circle the mistake and write the correct word or phrase on the line provided.

1. The solution to the inequality $4x^2 - 31x > -21$ is $\left(\pm\dfrac{3}{4}, \infty\right)$. _____

2. The solution to the inequality $x(x + 5)(x - 6) \geq 0$ is $(-\infty, -5] \cup [0, \infty)$. _____

3. The solution to the inequality $\dfrac{x + 8}{x - 3} > 0$ is $(-8, 3)$. _____

4. For the inequality $\dfrac{4}{x + 5} - \dfrac{3}{x + 4} > 0$, the solution lies within the interval $(-\infty, 5)$. _____

A Access to the internet for research

B Pencil and paper

Parabolic Focus

A parabolic reflector microphone and a parabolic antennae both use the shape of a parabola to focus frequencies originating from a great distance away. How is this possible? Research these two technologies to answer the following questions:

1. Compare and contrast the shape, size, and use of the microphone and the antennae. What mathematics are required to make each work?

2. What information does each interpret?

3. How does this information change the construction and shape of each technology?

EXAMPLES

1. If $(x - 3)^2 = 25$
then $x - 3 = \pm 5$
$x = 3 \pm 5$
$x = 8$ or $x = -2$

The Square Root Property for Equations [7.1]

If $a^2 = b$, where b is a real number, then

$$a = \sqrt{b} \quad \text{or} \quad a = -\sqrt{b}$$

which can be written as $a = \pm\sqrt{b}$.

To Solve a Quadratic Equation by Completing the Square [7.1]

2. Solve $x^2 - 6x - 6 = 0$
$x^2 - 6x = 6$
$x^2 - 6x + 9 = 6 + 9$
$(x - 3)^2 = 15$
$x - 3 = \pm\sqrt{15}$
$x = 3 \pm \sqrt{15}$

Step 1: Write the equation in the form $ax^2 + bx = c$.
Step 2: If $a \neq 1$, divide through by the constant a so the coefficient of x^2 is 1.
Step 3: Complete the square on the left side by adding the square of $\frac{1}{2}$ the coefficient of x to both sides.
Step 4: Write the left side of the equation as the square of a binomial. Simplify the right side if possible.
Step 5: Apply the square root property for equations, and solve as usual.

The Quadratic Formula [7.2]

3. If $2x^2 + 3x - 4 = 0$, then

$$x = \frac{-3 \pm \sqrt{9 - 4(2)(-4)}}{2(2)}$$

$$= \frac{-3 \pm \sqrt{41}}{4}$$

For any quadratic equation in the form $ax^2 + bx + c = 0$, $a \neq 0$, the two solutions are

$$x = \frac{-b \pm \sqrt{b^2 - 4ac}}{2a}$$

This equation is known as the *quadratic formula*.

The Discriminant [7.3]

4. The discriminant for
$x^2 + 6x + 9 = 0$
is $D = 36 - 4(1)(9) = 0$, which means the equation has one rational solution.

The expression $b^2 - 4ac$ that appears under the radical sign in the quadratic formula is known as the *discriminant*.

We can classify the solutions to $ax^2 + bx + c = 0$:

The solutions are	When the discriminant is
Two complex numbers containing i	Negative
One rational number	Zero
Two rational numbers	A positive perfect square
Two irrational numbers	A positive number, but not a perfect square

Equations Quadratic in Form [7.4]

There are a variety of equations whose form is quadratic. We solve most of them by making a substitution so the equation becomes quadratic, and then solving the equation by factoring or the quadratic formula. For example,

The equation	*is quadratic in*
$(2x - 3)^2 + 5(2x - 3) - 6 = 0$	$2x - 3$
$4x^4 - 7x^2 - 2 = 0$	x^2
$2x - 7\sqrt{x} + 3 = 0$	$\sqrt{x}$

5. The equation $x^4 - x^2 - 12 = 0$ is quadratic in x^2. Letting $y = x^2$ we have

$$y^2 - y - 12 = 0$$
$$(y - 4)(y + 3) = 0$$
$$y = 4 \quad \text{or} \quad y = -3$$

Resubstituting x^2 for y, we have
$$x^2 = 4 \quad \text{or} \quad x^2 = -3$$
$$x = \pm 2 \qquad x = \pm i\sqrt{3}$$

Graphing Quadratic Functions [7.5]

The graph of any function of the form

$$f(x) = ax^2 + bx + c \qquad a \neq 0$$

is a *parabola*. The graph is *concave up* if $a > 0$ and *concave down* if $a < 0$. The highest or lowest point on the graph is called the *vertex* and always has an *x*-coordinate of

$$x = \frac{-b}{2a}.$$

6. The graph of $y = x^2 - 4$ will be a parabola. It will cross the *x*-axis at 2 and -2, and the vertex will be $(0, -4)$.

Quadratic Inequalities [7.6]

We solve quadratic inequalities by manipulating the inequality to get 0 on the right side and then factoring the left side. We then make a diagram that indicates where the factors are positive and where they are negative. From this sign diagram and the original inequality we graph the appropriate solution set.

7. Solve $x^2 - 2x - 8 > 0$. We factor and draw the sign diagram.
$$(x - 4)(x + 2) > 0$$

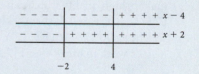

The solution set is $(-\infty, -2) \cup (4, \infty)$.

Rational Inequalities [7.6]

We solve rational inequalities by simplifying the rational expressions on the left side and obtaining 0 on the right side. We then make a diagram that indicates where the factors are positive and where they are negative. From this sign diagram and the original inequality we graph the appropriate solution set, making sure to avoid zero denominators.

8. Solve $\dfrac{3}{x + 2} - \dfrac{2}{x - 1} \leq 0$.
$$\frac{3(x - 1) - 2(x + 2)}{(x + 2)(x - 1)} \leq 0$$
$$\frac{x - 7}{(x + 2)(x - 1)} \leq 0$$

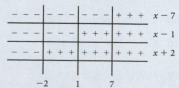

The solution set is $(-\infty, -2) \cup (1, 7]$.

Solve each equation. [7.1, 7.2]

1. $(3x + 4)^2 = 25$

2. $(2x - 3)^2 = -18$

3. $y^2 - 6y + 9 = 36$

4. $(x - 4)(x + 7) = -10$

5. $125x^3 - 64 = 0$

6. $\dfrac{1}{a} - \dfrac{1}{3} = \dfrac{1}{a - 2}$

7. Solve the formula $36(r - 4)^2 = A$ for r. [7.1]

8. Solve $x^2 - 8x = -13$ by completing the square. [7.1]

9. If the length of the longest side of $30° - 60° - 90°$ triangle is 12 cm, find the lengths of the other two sides. [7.1]

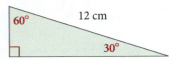

10. Projectile Motion An object projected upward with an initial velocity of 48 feet per second will rise and fall according to the equation $s(t) = 48t - 16t^2$, where s is its distance above the ground at time t. At what times will the object be 20 feet above the ground? [7.2]

11. Revenue The total weekly cost for a company to make x high quality watches is given by the formula $C(x) = 10x + 300$. If the weekly revenue from selling all x watches is $R(x) = 161x - 0.5x^2$, how many watches must it sell a week to break even? [7.2]

12. Find k so that $3x^2 + 18x = -k$ has one rational solution. [7.3]

13. Use the discriminant to identify the number and kind of solutions to $3x^2 = -16x + 35$. [7.3]

Find equations that have the given solutions. [7.3]

14. $x = 3, x = \dfrac{3}{5}$

15. $x = -4, x = 4, x = 3$

Solve each equation. [7.4]

16. $4x^4 - 11x^2 - 45 = 0$

17. $(3x + 2)^2 - 4(3x + 2) - x - 6 = 0$

18. $2t - 5\sqrt{t} - 3 = 0$

19. Projectile Motion An object is tossed into the air with an upward velocity of 12 feet per second from the top of a building h feet high. The time it takes for the object to hit the ground below is given by the formula $16t^2 - 12t - h = 0$. Solve this formula for t. [7.4]

Sketch the graph of each of the following. Give the coordinates of the vertex in each case. [7.5]

20. $y = x^2 - 4x - 6$

21. $y = -x^2 - 2x + 4$

22. Find an equation in the form $y = a(x - h)^2 + k$ that describes the graph of the parabola. [7.5]

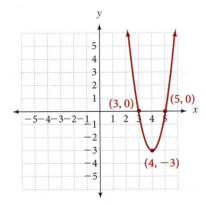

23. Profit Find the maximum weekly profit for a company with weekly costs of $C = 4x + 200$ and weekly revenue of $R = 26x - 0.4x^2$. [7.5]

Solve each inequality and graph the solution. [7.6]

24. $x^2 + 3x - 10 \leq 0$

25. $3x^2 - 8x > 3$

26. The graph of $y = x^3 + 2x^2 - 5x - 6$ is shown below.

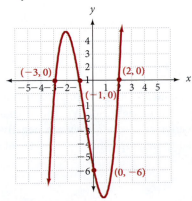

Use the graph to write the solution set for each of the following. [7.6]

a. $x^3 + 2x^2 - 5x - 6 < 0$

b. $x^3 + 2x^2 - 5x - 6 > 0$

c. $x^3 + 2x^2 - 5x - 6 = 0$

Simplify.

1. $6 - 36 \div 6 + 2 \cdot 2$

2. $5(13 - 17)^2 - 2(16 - 17)^2$

3. $\left(-\dfrac{5}{2}\right)^3$

4. $\left(-\dfrac{3}{4}\right)^2$

5. $3 + 2x - 5(3x - 2)$

6. $7 - 2[4x - 3(x + 1)]$

7. $\left(\dfrac{x^{-3}y^{-3}}{x^{-2}y^4}\right)^{-1}$

8. $\left(\dfrac{a^{-2}b^{-1}}{a^5b^{-2}}\right)^{-3}$

9. $\sqrt[3]{54}$

10. $\sqrt{32x^5}$

11. $64^{-2/3} - 16^{-1/2}$

12. $\left(\dfrac{125}{8}\right)^{-2/3}$

13. $\dfrac{\frac{4}{7} - 1}{\frac{4}{7} + 1}$

14. $\dfrac{\frac{1}{8} + 3}{1 - \frac{3}{2}}$

Reduce.

15. $\dfrac{3x^2 - 17xy - 28y^2}{3x + 4y}$

16. $\dfrac{x^2 - x - 20}{x + 4}$

Divide.

17. $\dfrac{-4 + i}{i}$

18. $\dfrac{4 + 2i}{5 - i}$

Solve.

19. $\dfrac{5}{4}x + 3 = 8$

20. $\dfrac{1}{3}a + \dfrac{3}{4} = 5$

21. $|a| + 4 = 7$

22. $|a + 1| = -2$

23. $\dfrac{x}{4} + \dfrac{15}{4x} = -2$

24. $-\dfrac{1}{3} + \dfrac{a}{a + 2} = \dfrac{2}{a + 2}$

25. $(2x - 3)^2 = 8$

26. $4x^3 + x = 9x^2$

27. $\dfrac{1}{10}x^2 - \dfrac{1}{2}x + \dfrac{2}{5} = 0$

28. $0.09a^2 + 0.02a = -0.04$

29. $\sqrt{y + 4} = y + 4$

30. $\sqrt{x - 5} = 5 - \sqrt{x}$

31. $x + 5\sqrt{x} - 24 = 0$

32. $x^{2/3} - x^{1/3} = 30$

Solve each inequality and graph the solution.

33. $-5 \le \dfrac{1}{3}x + 7 \le 1$

34. $|3x + 2| \ge 1$

35. $x^2 - 3x < 18$

36. $\dfrac{x - 4}{x + 2} \ge 0$

Solve each system.

37. $\begin{aligned} 7x - 3y &= 43 \\ -14x + 6y &= -5 \end{aligned}$

38. $\begin{aligned} -2x + 4y &= 8 \\ 3x - 6y &= 8 \end{aligned}$

Graph on a rectangular coordinate system.

39. $2x + 3y = 12$

40. $y = -x^2 + 4x - 3$

41. Find the equation of the line passing through the points $\left(\dfrac{1}{2}, \dfrac{13}{4}\right)$ and $\left(-\dfrac{3}{2}, \dfrac{1}{4}\right)$.

42. Find the equation of the line perpendicular to $2x + 8y = 17$ and containing the point $(1, 2)$.

Factor completely.

43. $x^2 - 6x + 9 - y^2$

44. $(x - 3)^2 + 5(x - 4) - 1$

Rationalize the denominator.

45. $\dfrac{6}{\sqrt[3]{4}}$

46. $\dfrac{\sqrt{3}}{\sqrt{3} - \sqrt{5}}$

If $A = \{1, 4, 9, 16\}$ and $B = \{1, 2, 4, 8\}$, find the following.

47. $\{x \mid x \in A \text{ and } x \in B\}$

48. $\{x \mid x \notin A \text{ and } x \in B\}$

49. Geometry Find all three angles in a triangle if the smallest angle is three-tenths the largest angle and the remaining angle is 20° more than the smallest angle.

50. Investing A total of $6,000 is invested in two accounts. One account earns 3% per year, and the other earns 2% per year. If the total interest for the first year is $145, how much was invested in each account?

51. Direct Variation y varies directly with the square of x. If $y = -4$ when $x = -\dfrac{2}{3}$, find y when $x = \dfrac{5}{9}$.

52. Inverse Variation w varies inversely with the square root of c. If w is 8 when c is 4, find w when c is 16.

The chart shows what other types of phones iPhone users had before purchasing the iPhone. If 426,350 people were surveyed, use the information to answer the following questions. Round to the nearest whole number.

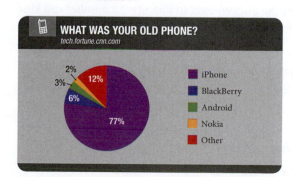

WHAT WAS YOUR OLD PHONE?
tech.fortune.cnn.com

- iPhone — 77%
- BlackBerry — 12%
- Android — 6%
- Nokia — 3%
- Other — 2%

53. How many people formerly owned another iPhone?

54. How many people formerly owned either a Blackberry or an Android?

Solve each equation. [7.1, 7.2]

1. $(2x - 3)^2 = 36$

2. $(4x + 1)^2 = -12$

3. $y^2 + 8y + 16 = -25$

4. $(x + 4)(x + 6) = -3$

5. $64t^3 - 27 = 0$

6. $\dfrac{1}{a} - \dfrac{1}{2} = \dfrac{1}{a - 1}$

7. Solve the formula $49(r - 7)^2 = A$ for r. [7.1]

8. Solve $x^2 - 6x = -11$ by completing the square. [7.1]

9. If the length of the longest side of $30° - 60° - 90°$ triangle is 8 meters, find the lengths of the other two sides. [7.1]

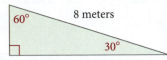

10. Projectile Motion An object projected upward with an initial velocity of 32 feet per second will rise and fall according to the equation $s(t) = 32t - 16t^2$, where s is its distance above the ground at time t. At what times will the object be 7 feet above the ground? [7.2]

11. Revenue The total weekly cost for a company to make x ceramic letter openers is given by the formula $C(x) = 9x + 100$. If the weekly revenue from selling all x letter openers is $R(x) = 30x - 0.2x^2$, how many letter openers must it sell a week to break even? [7.2]

12. Find k so that $kx^2 = -16x - 8$ has one rational solution. [7.3]

13. Use the discriminant to identify the number and kind of solutions to $5x^2 - 3x = -7$. [7.3]

Find equations that have the given solutions. [7.3]

14. $x = -7, x = \dfrac{5}{2}$

15. $x = -3, x = 3, x = 5$

Solve each equation. [7.4]

16. $16x^4 - 39x^2 - 27 = 0$

17. $(2t + 1)^2 - 6(2t + 1) - t + 7 = 0$

18. $3t - 7\sqrt{t} + 2 = 0$

19. Projectile Motion An object is tossed into the air with an upward velocity of 10 feet per second from the top of a building h feet high. The time it takes for the object to hit the ground below is given by the formula $16t^2 - 10t - h = 0$. Solve this formula for t. [7.4]

Sketch the graph of each of the following. Give the coordinates of the vertex in each case. [7.5]

20. $y = x^2 - 2x + 1$

21. $y = -x^2 + 4x - 2$

22. Find an equation in the form $y = a(x - h)^2 + k$ that describes the graph of the parabola. [7.5]

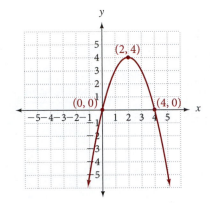

23. Profit Find the maximum weekly profit for a company with weekly costs of $C = 4x + 200$ and weekly revenue of $R = 30x - 0.2x^2$. [7.5]

Solve each inequality and graph the solution. [7.6]

24. $x^2 - 3x - 4 \le 0$

25. $2x^2 + 3x > 2$

26. The graph of $y = -x^3 - 2x^2 + x + 2$ is shown below. Use the graph to write the solution set for each of the following. [7.6]

a. $-x^3 - 2x^2 + x + 2 < 0$

b. $-x^3 - 2x^2 + x + 2 > 0$

c. $-x^3 - 2x^2 + x + 2 = 0$

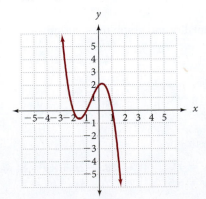

Exponential and Logarithmic Functions

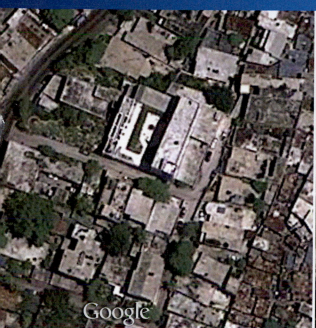

Large earthquakes are often so destructive that their effects are difficult to miss (as the before and after photos above show of the 2010 earthquake in Haiti.) You may be surprised to know that thousands of earthquakes occur every day all over the world. Approximately 80 percent of these quakes occur along the rim of the Pacific Ocean, also known as the "Ring of Fire" for its large amount of volcanic activity. For some earthquakes to occur, large sections of the upper layers of Earth, called tectonic plates, collide and slide against each other. However, much of this sliding is usually gradual and goes unnoticed here on the surface.

The following table shows the number of earthquakes occurring from 2000 through most of 2010 based on their strength.

Magnitude	2000	2002	2004	2006	2008	2010
8.0–9.9	1	0	2	2	0	1
7.0–7.9	14	13	14	9	12	19
6.0–6.9	146	127	141	142	168	131
5.0–5.9	1,344	1,201	1,515	1,712	1,768	1,488
4.0–4.9	8,008	8,541	10,888	12,838	12,291	7,687
3.0–3.9	4,827	7,068	7,932	9,990	11,735	3,463
2.0–2.9	3,765	6,419	6,316	4,207	3,860	3,380
1.0–1.9	1,026	1,137	1,344	18	21	23
0.1–0.9	5	10	103	2	0	0
Total	22,256	27,454	31,194	29,568	31,777	16,217
Estimated Deaths	231	1,685	228,802	6,605	88,011	226,487

Source: earthquake.usgs.gov

In this chapter, we will discuss earthquakes and how they are measured. You will learn that the magnitude of an earthquake is measured on the Richter Scale, which is a logarithmic scale. As you work through this chapter, you will use mathematics to understand what that means and how to compare the strengths of earthquakes using logarithms.

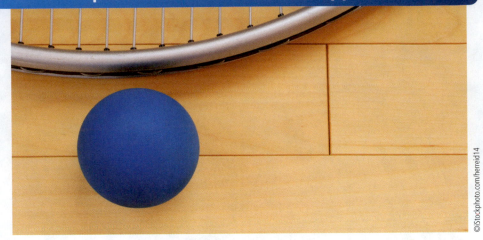

To obtain an intuitive idea of how exponential functions behave, we can consider the heights attained by a bouncing ball. When a ball used in the game of racquetball is dropped from any height, the first bounce will reach a height that is $\frac{2}{3}$ of the original height. The second bounce will reach $\frac{2}{3}$ of the height of the first bounce, and so on, as shown in Figure 1.

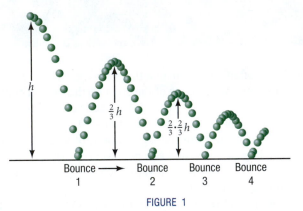

FIGURE 1

If the ball is initially dropped from a height of 1 meter, then during the first bounce it will reach a height of $\frac{2}{3}$ meter. The height of the second bounce will reach $\frac{2}{3}$ of the height reached on the first bounce. The maximum height of any bounce is $\frac{2}{3}$ of the height of the previous bounce.

$$\text{Initial height: } h = 1$$

$$\text{Bounce 1:} \quad h = \frac{2}{3}(1) = \frac{2}{3}$$

$$\text{Bounce 2:} \quad h = \frac{2}{3}\left(\frac{2}{3}\right) = \left(\frac{2}{3}\right)^2$$

$$\text{Bounce 3:} \quad h = \frac{2}{3}\left(\frac{2}{3}\right)^2 = \left(\frac{2}{3}\right)^3$$

$$\text{Bounce 4:} \quad h = \frac{2}{3}\left(\frac{2}{3}\right)^3 = \left(\frac{2}{3}\right)^4$$

$$\vdots \qquad \qquad \vdots$$

$$\text{Bounce } n: \quad h = \frac{2}{3}\left(\frac{2}{3}\right)^{n-1} = \left(\frac{2}{3}\right)^n$$

This last equation is exponential in form. We classify all exponential functions together with the following definition.

> **DEFINITION** exponential function
>
> An *exponential function* is any function that can be written in the form
>
> $$f(x) = b^x$$
>
> where b is a positive real number other than 1.

Each of the following is an exponential function:

$$f(x) = 2^x \qquad g(x) = 3^x \qquad f(x) = \left(\frac{1}{4}\right)^x$$

A Evaluating Exponential Functions

The first step in becoming familiar with exponential functions is to find some values for specific exponential functions.

EXAMPLE 1 If the exponential functions $f(x)$ and $g(x)$ are defined by $f(x) = 2^x$ and $g(x) = 3^x$ then

$$f(0) = 2^0 = 1 \qquad\qquad g(0) = 3^0 = 1$$

$$f(1) = 2^1 = 2 \qquad\qquad g(1) = 3^1 = 3$$

$$f(2) = 2^2 = 4 \qquad\qquad g(2) = 3^2 = 9$$

$$f(3) = 2^3 = 8 \qquad\qquad g(3) = 3^3 = 27$$

$$f(-2) = 2^{-2} = \frac{1}{2^2} = \frac{1}{4} \qquad g(-2) = 3^{-2} = \frac{1}{3^2} = \frac{1}{9}$$

$$f(-3) = 2^{-3} = \frac{1}{2^3} = \frac{1}{8} \qquad g(-3) = 3^{-3} = \frac{1}{3^3} = \frac{1}{27}$$

Half-life is a term used in science and mathematics that quantifies the amount of decay or reduction in an element. The half-life of iodine-131 is 8 days, which means that every 8 days a sample of iodine-131 will decrease by half of its remaining amount. If we start with A_0 micrograms of iodine-131, then the function $A(t)$ represents the number of micrograms of iodine -131 in the sample after t days.

$$A(t) = A_0 \cdot 2^{-t/8}$$

EXAMPLE 2 A patient is administered a 1,200-microgram dose of iodine-131. How much iodine-131 will be in the patient's system after 10 days, and after 16 days?

Solution The initial amount of iodine-131 is $A_0 = 1,200$, so the function that gives the amount left in the patient's system after t days is

$$A(t) = 1,200 \cdot 2^{-t/8}$$

After 10 days, the amount left in the patient's system is

$$A(10) = 1,200 \cdot 2^{-10/8} = 1,200 \cdot 2^{-1.25} \approx 504.5 \text{ micrograms}$$

After 16 days, the amount left in the patient's system is

$$A(16) = 1,200 \cdot 2^{-16/8} = 1,200 \cdot 2^{-2} = 300 \text{ micrograms}$$

It is also important to note that the domain of an exponential function is the set of all real numbers, $(\infty, -\infty)$; whereas, the range is the set of all positive real numbers, $(0, \infty)$. This will become clearer as we start graphing these functions.

Practice Problems

1. If $f(x) = 4^x$, find
   a. $f(0)$
   b. $f(1)$
   c. $f(2)$
   d. $f(3)$
   e. $f(-1)$
   f. $f(-2)$

2. Referring to Example 2, how much iodine-131 will be in the patient's system after 12 days and after 24 days?

Answers

1. a. 1 b. 4 c. 16
   d. 64 e. $\frac{1}{4}$ f. $\frac{1}{16}$
2. $A(12) \approx 424.3$ micrograms
   $A(24) = 150$ micrograms

3. Graph $f(x) = 3^x$.

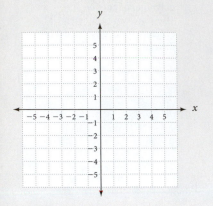

B Graphing Exponential Functions

EXAMPLE 3   Sketch the graph of the exponential function $f(x) = 2^x$.

Solution   Using the results of Example 1, we produce the following table. Graphing the ordered pairs given in the table and connecting them with a smooth curve, we have the graph of $f(x) = 2^x$ shown in Figure 2.

x	f(x)
−3	$\frac{1}{8}$
−2	$\frac{1}{4}$
0	1
1	2
2	4
3	8

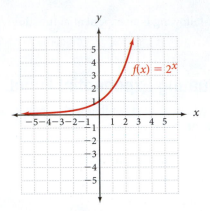

FIGURE 2

Notice that the graph does not cross the x-axis. It *approaches* the x-axis—in fact, we can get it as close to the x-axis as we want without it actually intersecting the x-axis. For the graph of $f(x) = 2^x$ to intersect the x-axis, we would have to find a value of x that would make $2^x = 0$. Because no such value of x exists, the graph of $f(x) = 2^x$ cannot intersect the x-axis.

4. Graph $f(x) = \left(\frac{1}{2}\right)^x$.

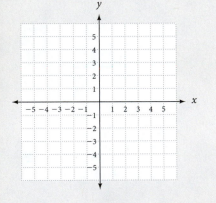

EXAMPLE 4   Sketch the graph of $f(x) = \left(\frac{1}{3}\right)^x$.

Solution   The table beside Figure 3 gives some ordered pairs that satisfy the equation. Using the ordered pairs from the table, we have the graph shown in Figure 3.

x	f(x)
−3	27
−2	9
−1	3
0	1
1	$\frac{1}{3}$
2	$\frac{1}{9}$
3	$\frac{1}{27}$

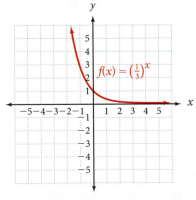

FIGURE 3

The graphs of all exponential functions have two things in common: (1) Each crosses the y-axis at $(0, 1)$ because $b^0 = 1$; and (2) none can cross the x-axis because $b^x = 0$ is impossible due to the restrictions on b.

Figures 4 and 5 show some families of exponential curves to help you become more familiar with them on an intuitive level.

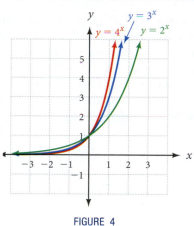

FIGURE 4

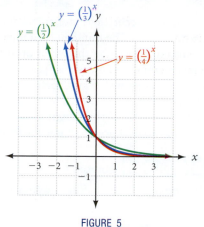

FIGURE 5

C Compound Interest

Among the many applications of exponential functions are the applications having to do with interest-bearing accounts including savings accounts and borrowed money accounts. Here are the details:

If P dollars are deposited in an account with annual interest rate r, compounded n times per year, then the amount of money in the account after t years is given by the formula

$$A(t) = P\left(1 + \frac{r}{n}\right)^{nt}$$

EXAMPLE 5 Suppose you borrow $500 from an account with an annual interest rate of 8% compounded quarterly. Find an equation that gives the amount of money owed after t years. Then find

a. The amount of money owed after 5 years.

b. The number of years it will take for the account balance owed to reach $1,000.

Solution First, we note that $P = 500$ and $r = 0.08$. Interest that is compounded quarterly is compounded four times a year, giving us $n = 4$. Substituting these numbers into the preceding formula, we have our function

$$A(t) = 500\left(1 + \frac{0.08}{4}\right)^{4t} = 500(1.02)^{4t}$$

a. To find the amount after 5 years, we let $t = 5$.

$$A(5) = 500(1.02)^{4 \cdot 5} = 500(1.02)^{20} \approx \$742.97$$

Our answer is found on a calculator, and then rounded to the nearest cent.

b. To see how long it will take for this account to total $1,000, we graph the equation $Y_1 = 500(1.02)^{4X}$ on a graphing calculator, and then look to see where it intersects the line $Y_2 = 1,000$. The two graphs are shown in Figure 6.

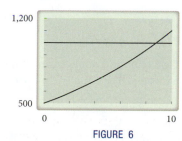

FIGURE 6

Using Zoom and Trace, or the Intersect function on the graphing calculator, we find that the two curves intersect at $X \approx 8.75$ and $Y = 1,000$. This means that our account balance will be $1,000 after the money has been borrowed for 8.75 years.

5. Repeat Example 5 if $600 is borrowed from an account that charges 6% annual interest, compounded monthly.

Answers

5 a. $809.31 **b.** About 8.5 years

The Natural Exponential Function

A commonly occurring exponential function is based on a special number we denote with the letter e. The number e is a number like π. It is irrational and occurs in many formulas that describe the world around us. Like π, it can be approximated with a decimal number. Whereas π is approximately 3.1416, e is approximately 2.7183. (If you have a calculator with a key labeled e^x, you can use it to find e^1 to find a more accurate approximation to e.) We cannot give a more precise definition of the number e without using some of the topics taught in calculus. For the work we are going to do with the number e, we only need to know the following.

> **DEFINITION** number e
>
> The *number e* is an irrational number that is approximately 2.7183.

> **DEFINITION** natural exponential function
>
> The *natural exponential function* is any function that can be written in the form
>
> $$f(x) = e^x$$
>
> where e is the number e with an approximate value 2.7183.

Here are a table and graph (Figure 7) for the natural exponential function $f(x) = e^x$:

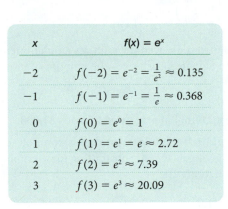

x	$f(x) = e^x$
-2	$f(-2) = e^{-2} = \frac{1}{e^2} \approx 0.135$
-1	$f(-1) = e^{-1} = \frac{1}{e} \approx 0.368$
0	$f(0) = e^0 = 1$
1	$f(1) = e^1 = e \approx 2.72$
2	$f(2) = e^2 \approx 7.39$
3	$f(3) = e^3 \approx 20.09$

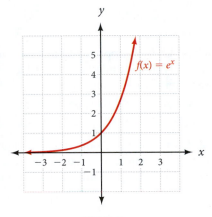

FIGURE 7

As you may have noticed, the domain and range of the natural exponential function is the same as the exponential function.

$$\begin{array}{ccc} & Domain & Range \\ f(x) = b^x & (-\infty, \infty) & (-\infty, \infty) \\ f(x) = e^x & (-\infty, \infty) & (-\infty, \infty) \end{array}$$

One common application of natural exponential functions is with interest-bearing accounts. In Example 5, we worked with the formula

$$A(t) = P\left(1 + \frac{r}{n}\right)^{nt}$$

that gives the amount of money in an account if P dollars are deposited for t years at annual interest rate r, compounded n times per year. In Example 5, the number of compounding periods was four. What would happen if we let the number of compounding periods become larger and larger, so that we compounded the interest every day, then every hour, then every second, and so on? If we take this as far as it can go, we end up compounding the interest every moment. When this happens, we have an account with interest that is compounded continuously, and the amount of money in such an account depends on the number e. Here are the details.

Continuously Compounded Interest   If P dollars are deposited in an account with annual interest rate r, compounded continuously, then the amount of money in the account after t years is given by the formula

$$A(t) = Pe^{rt}$$

EXAMPLE 6   Suppose you deposit $500 in an account with an annual interest rate of 8% compounded continuously. Find an equation that gives the amount of money in the account after t years. Then find the amount of money in the account after 5 years.

Solution   Because the interest is compounded continuously, we use the formula $A(t) = Pe^{rt}$. Substituting $P = 500$ and $r = 0.08$ into this formula, we have

$$A(t) = 500e^{0.08t}$$

After 5 years, this account will contain

$$A(5) = 500e^{0.08 \cdot 5} = 500e^{0.4} \approx \$745.91$$

to the nearest cent. Compare this result with the answer to Example 5a.

6. Suppose you deposit $600 in an account with an annual interest rate of 6% compounded continuously. How much money is in the account after 5 years?

GETTING READY FOR CLASS

After reading through the preceding section, respond in your own words and in complete sentences.

A. What is an exponential function?

B. In an exponential function, explain why the base b cannot equal 1.

C. Explain continuously compounded interest.

D. What is the special number e in an exponential function?

Vocabulary Review

Choose the correct words to fill in the blanks below.

number e exponential function x-axis $A(t) = Pe^{rt}$ $A(t) = P\left(1 + \dfrac{r}{n}\right)^{nt}$

1. An _____ is any function that can be written in the form $f(x) = b^x$ where b is a positive real number other than 1.

2. The graphs of all exponential functions have two things in common: each crosses the y-axis at $(0, 1)$, and none can cross the _____.

3. The _____ is an irrational number and is approximately 2.7183.

4. An account with an initial deposit P and an annual interest rate r, compounded n times per year, will have a balance after t years given by the function _____.

5. An account with an initial deposit P and an annual interest rate r, compounded continuously, will have a balance after t years given by the function _____.

Problems

A Let $f(x) = 3^x$ and $g(x) = \left(\dfrac{1}{2}\right)^x$, and evaluate each of the following.

1. $g(0)$

2. $f(0)$

3. $g(-1)$

4. $g(-4)$

5. $f(-3)$

6. $f(-1)$

7. $f(2) + g(-2)$

8. $f(2) - g(-2)$

Let $f(x) = 4^x$ and $g(x) = \left(\dfrac{1}{3}\right)^x$. Evaluate each of the following.

9. $f(-1) + g(1)$

10. $f(2) + g(-2)$

11. $\dfrac{f(-2)}{g(1)}$

12. $f(3) - f(2)$

13. $\dfrac{f(3) - f(2)}{3 - 2}$

14. $\dfrac{g(3) - g(2)}{3 - 2}$

15. $\dfrac{f(-1) - f(-2)}{-1 - (-2)}$

16. $\dfrac{g(-2) - g(-4)}{-2 - (-4)}$

Let $f(x) = 3^x$ and $g(x) = \left(\dfrac{1}{2}\right)^x$. Simplify each difference quotient.

17. $\dfrac{f(x) - f(2)}{x - 2}$

18. $\dfrac{g(x) - g(2)}{x - 2}$

19. $\dfrac{f(x) - f(3)}{x - 3}$

20. $\dfrac{g(x) - g(3)}{x - 3}$

21. $\dfrac{f(x + h) - f(x)}{h}$

22. $\dfrac{g(x + h) - g(x)}{h}$

SCAN TO ACCESS

B Graph each of the following exponential functions.

23. $f(x) = 4^x$

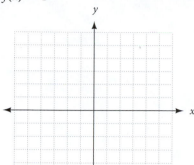

24. $f(x) = 2^{-x}$

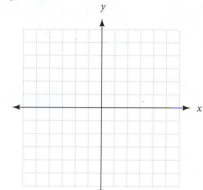

25. $f(x) = 3^{-x}$

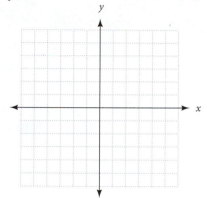

26. $f(x) = \left(\dfrac{1}{3}\right)^{-x}$

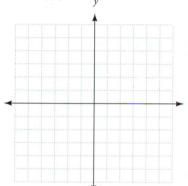

27. $g(x) = 2^{x+1}$

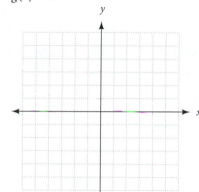

28. $g(x) = 2^{x-3}$

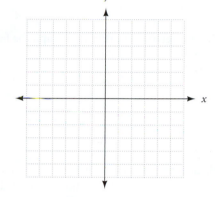

29. $g(x) = e^x$

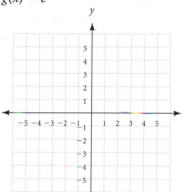

30. $g(x) = e^{-x}$

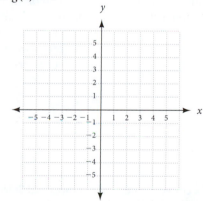

31. $h(x) = \left(\dfrac{1}{3}\right)^x$

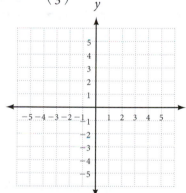

32. $h(x) = \left(\dfrac{1}{2}\right)^{-x}$

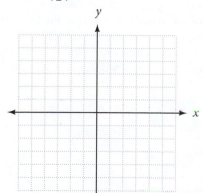

33. $h(x) = 3^{x+2}$

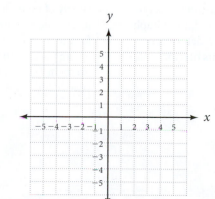

34. $h(x) = 2 \cdot 3^{-x}$

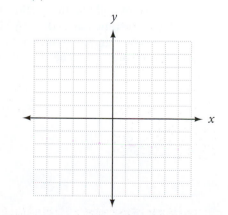

Graph each of the following functions on the same coordinate system for positive values of x only.

35. $y = 2x, y = x^2, y = 2^x$

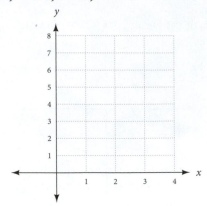

Based on your graph, which curve is growing faster?

36. $y = 3x, y = x^3, y = 3^x$

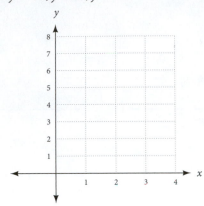

Based on your graph, which curve is growing faster?

37. On a graphing calculator, graph the family of curves $y = b^x$, $b = 2, 4, 6, 8$.

38. On a graphing calculator, graph the family of curves $y = b^x$, $b = \frac{1}{2}, \frac{1}{4}, \frac{1}{6}, \frac{1}{8}$.

Applying the Concepts

39. Bouncing Ball Suppose the ball mentioned in the introduction to this section is dropped from a height of 6 feet above the ground. Find an exponential function that gives the height h the ball will attain during the nth bounce. How high will it bounce on the fifth bounce?

40. Bouncing Ball A golf ball is manufactured so that if it is dropped from A feet above the ground onto a hard surface, the maximum height of each bounce will be one half of the height of the previous bounce. Find an exponential function that gives the height h the ball will attain during the nth bounce. If the ball is dropped from 10 feet above the ground onto a hard surface, how high will it bounce on the eighth bounce?

41. Exponential Decay Twinkies on the shelf of a convenience store lose their fresh tastiness over time. We say that the taste quality is 1 when the Twinkies are first put on the shelf at the store, and that the quality of tastiness declines according to the function $Q(t) = 0.85^t$ (t in days). Graph this function on a graphing calculator, and determine when the taste quality will be one half of its original value.

42. Exponential Growth Automobiles built before 1993 use Freon in their air conditioners. The federal government now prohibits the manufacture of Freon. Because the supply of Freon is decreasing, the price per pound is increasing exponentially. Current estimates put the formula for the price per pound of Freon at $p(t) = 1.89(1.25)^t$, where t is the number of years since 1990. Find the price of Freon in 1995 and 1990. How much will Freon cost in the year 2020?

43. Compound Interest Suppose you deposit $1,200 in an account with an annual interest rate of 6% compounded quarterly.

   a. Find a function that gives the amount of money in the account after t years.

   b. Find the amount of money in the account after 8 years.

   c. How many years will it take for the account to contain $2,400?

   d. If the interest were compounded continuously, how much money would the account contain after 8 years?

44. Compound Interest Suppose you deposit $500 in an account with an annual interest rate of 8% compounded monthly.

   a. Find a function that gives the amount of money in the account after t years.

   b. Find the amount of money in the account after 5 years.

   c. How many years will it take for the account to contain $1,000?

   d. If the interest were compounded continuously, how much money would the account contain after 5 years?

45. Student Loan Suppose you take out a student loan for $5,400. The loan will accrue interest at a rate of 5%, compounded quarterly, but you will not make any payments until after you graduate.

   a. Find a function that gives the amount of money you owe when you graduate after t years.

   b. Find the amount of money you owe if it takes you 5 years to graduate.

   c. How long did it take you to graduate if you owe $8,000? Round to the nearest year.

   d. If the interest were compounded continuously, how much money would you owe after 5 years?

46. Student Loan Suppose you take out a student loan for $2,300. The loan will accrue interest at a rate of 7%, compounded monthly, but you will not make any payments until after you graduate.

   a. Find a function that gives the amount of money you owe when you graduate after t years.

   b. Find the amount of money you owe if it takes you 6 years to graduate.

   c. How long did it take you to graduate if you owe $4,000? Round to the nearest year.

   d. If the interest were compounded continuously, how much money would you owe after 6 years?

47. Bacteria Growth Suppose it takes 12 hours for a certain strain of bacteria to reproduce by dividing in half. If 50 bacteria are present to begin with, then the total number present after x days will be $f(x) = 50 \cdot 4^x$. Find the total number present after 1 day, 2 days, and 3 days.

48. Bacteria Growth Suppose it takes 1 day for a certain strain of bacteria to reproduce by dividing in half. If 100 bacteria are present to begin with, then the total number present after x days will be $f(x) = 100 \cdot 2^x$. Find the total number present after 1 day, 2 days, 3 days, and 4 days. How many days must elapse before over 100,000 bacteria are present?

Declining-Balance Depreciation The declining-balance method of depreciation is an accounting method businesses use to deduct most of the cost of new equipment during the first few years of purchase. Unlike other methods, the declining-balance formula does not consider salvage value.

49. Value of a Crane The function

$$V(t) = 450,000 (1 - 0.30)^t$$

where V is value and t is time in years, can be used to find the value of a crane for the first 6 years of use.

a. What is the value of the crane after 3 years and 6 months?

b. State the domain of this function using interval notation.

c. Sketch the graph of this function.

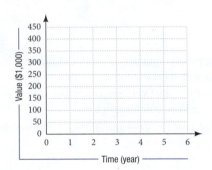

d. State the range of this function using interval notation.

e. After how many years will the crane be worth only $85,000?

50. Value of a Printing Press The function

$$V(t) = 375,000(1 - 0.25)^t$$

where V is value and t is time in years, can be used to find the value of a printing press during the first 7 years of use.

a. What is the value of the printing press after 4 years and 9 months?

b. State the domain of this function using interval notation.

c. Sketch the graph of this function.

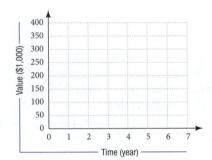

d. State the range of this function using interval notation.

e. After how many years will the printing press be worth only $65,000?

51. Value of a Painting Alana purchased a painting as an investment for $150. If the painting's value doubles every 3 years, then its value is given by the function

$$V(t) = 150 \cdot 2^{t/3} \text{ for } t \geq 0$$

where t is the number of years since it was purchased, and $V(t)$ is its value (in dollars) at that time. Graph this function.

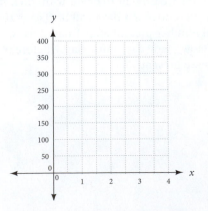

52. Value of a Painting Michael purchased a painting as an investment for $125. If the painting's value doubles every 5 years, then its value is given by the function

$$V(t) = 125 \cdot 2^{t/5} \text{ for } t \geq 0$$

where t is the number of years since it was purchased, and $V(t)$ is its value (in dollars) at that time. Graph this function.

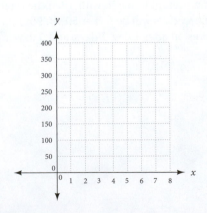

53. Cost Increase The cost of a can of Coca Cola in 1960 was $0.10. The exponential function that models the cost of a Coca Cola by year is given below, where t is the number of years since 1960.

$$C(t) = 0.10e^{0.0576t}$$

a. What was the expected cost of a can of Coca Cola in 1985?

b. What was the expected cost of a can of Coca Cola in 2000?

c. What was the expected cost of a can of Coca Cola in 2010?

d. What is the expected cost of a can of Coca Cola in 2050?

54. Airline Travel The number of airline passengers in 1990 was 466 million. The number of passengers traveling by airplane each year has increased exponentially according to the model, $P(t) = 466 \cdot 1.035^t$, where t is the number of years since 1990 (U.S. Census Bureau).

a. How many passengers traveled in 1997?

b. How many passengers will travel in 2015?

55. Bankruptcy Model In 1997, there were a total of 1,316,999 bankruptcies filed under the Bankruptcy Reform Act (Administrative Office of the U.S. Courts, Statistical Tables for the Federal Judiciary). The model for the number of bankruptcies filed is $B(t) = 0.798 \cdot 1.164^t$, where t is the number of years since 1994 and B is the number of bankruptcies filed in terms of millions. How close was the model in predicting the actual number of bankruptcies filed in 1997?

56. Value of a Car As a car ages, its value decreases. The value of a particular car with an original purchase price of $25,600 is modeled by the following function, where c is the value at time t. (Kelly Blue Book)

$$c(t) = 25,600(1 - 0.22)^t$$

a. What is the value of the car when it is 3 years old?

b. What is the total depreciation amount after 4 years?

57. Bacteria Decay You are conducting a biology experiment and begin with 5,000,000 cells, but some of those cells are dying each minute. The rate of death of the cells is modeled by the function $A(t) = A_0 \cdot e^{-.598t}$, where A_0 is the original number of cells, t is time in minutes, and A is the number of cells remaining after t minutes.

a. How may cells remain after 5 minutes?

b. How many cells remain after 10 minutes?

c. How many cells remain after 20 minutes?

58. Health Care In 1990, $699 billion were spent on health care expenditures. The amount of money, E, in billions spent on health care expenditures can be estimated using the function $E(t) = 78.16(1.11)^t$, where t is time in years since 1970 (U.S. Census Bureau).

a. How close was the estimate determined by the function in estimating the actual amount of money spent on health care expenditures in 1990?

b. What were the expected health care expenditures in 2008, 2009, and 2010?

Maintaining Your Skills

For each of the following relations, specify the domain and range; then indicate which are also functions.

59. $(-2, 6), (-2, 8), (2, 3)$

60. $(1, 2), (3, 4), (4, 1)$

State the domain for each of the following functions. Use interval notation.

61. $f(x) = \dfrac{-4}{x^2 + 2x - 35}$

62. $f(x) = \sqrt{3x + 1}$

If $f(x) = 2x^2 - 18$ and $g(x) = 2x - 6$, find

63. $f(0)$

64. $(g \circ f)(0)$

65. $\dfrac{g(x + h) - g(x)}{h}$

66. $\dfrac{g}{f(x)}$

The chart shows the increasing cost of Disneyland tickets from 1981 to 2011. Use it to answer the following questions.

67. Find the equation of the line that connects the first and the last points of the graph.

68. Use the equation found in problem 63 to find the cost of admission in 1990, 2000, and 2010.

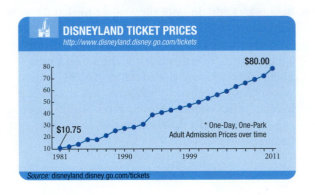

One Step Further

69. Reading Graphs The graphs of two exponential functions are given in Figures 8 and 9. Use the graphs to find the following:

   a. $f(0)$ **b.** $f(-1)$ **c.** $f(1)$ **d.** $g(0)$ **e.** $g(1)$ **f.** $g(-1)$ **g.** $f(g(0))$ **h.** $g(f(0))$

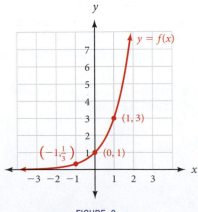

FIGURE 8

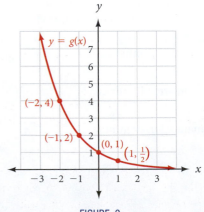

FIGURE 9

70. Analyzing Graphs The goal of this problem is to obtain a sense of the growth rate of an exponential function. We will compare the exponential function $y = 2^x$ to the familiar polynomial function $y = x^2$.

   a. On your calculator, graph $y = 2^x$ and $y = x^2$ using the window X: 0 to 3 and Y: 0 to 10. Which function appears to be growing faster as x grows larger?

   b. Show algebraically that both graphs contain the points (2, 4) and (4, 16). You may also want to confirm this on your graph by enlarging the window.

   c. On your calculator, graph $y = 2^x$ and $y = x^2$ with the window X: 0 to 15 and Y: 0 to 2,500. Which function dominates (grows the fastest) as x gets larger in the positive direction? Confirm your answer by evaluating $2^{(100)}$ and $(100)^2$.

Getting Ready for the Next Section

Solve each equation for y.

71. $x = 2y - 3$

72. $x = \dfrac{y + 7}{5}$

73. $x = y^2 - 3$

74. $x = (y + 4)^3$

75. $x = \dfrac{y - 4}{y - 2}$

76. $x = \dfrac{y + 5}{y - 3}$

77. $x = \sqrt{y - 3}$

78. $x = \sqrt{y} + 5$

Find the Mistake

Each problem below contains a mistake. Circle the mistake and write the correct number(s) or word(s) on the line provided.

1. The exponential function is any function written in the form $f(x) = x^b$. _____

2. The graph of $f(x) = 2^x$ gets very close to the y-axis but does not intersect it. _____

3. The special number e is a rational number. _____

4. To calculate annual interest rate compounded n times per year, use the formula $A(t) = Pe^{rt}$. _____

Navigation Skills: Prepare, Study, Achieve

The chapters in this course are organized such that each chapter builds on the previous chapters. You have already learned the tools to master the final topics in this book and be successful on the final exam. However, studying for the final exam may still seem like an overwhelming task. To ease that anxiety, begin now to lay out a study plan. Dedicate time in your day to do the following:

- Stay calm and maintain a positive attitude.
- Scan each section of the book to review headers, graphics, definitions, rules, properties, formulas, italicized words, and margin notes. Take new notes as you scan.
- Review your class notes and homework; review chapter summaries, and make your own outlines for each chapter.
- Rework problems in each section and from your difficult problems list.
- Make and review flashcards of definitions, properties, or formulas. Make sure you are understanding the topics in full. This will aid any memorization needed.
- Schedule time to meet with a study partner or group. Explain concepts to your group or out loud to yourself.

Another way to prepare is to visualize yourself being successful on the exam. Close your eyes and picture yourself arriving on exam day, receiving the test, staying calm, and working through difficult problems successfully. Now picture yourself receiving a high score on the exam. Complete this visualization multiple times before the day of the exam. If you picture yourself achieving success, you are more likely to be successful.

OBJECTIVES

A Find the equation of the inverse of a function.

B Sketch a graph of a function and its inverse.

KEY WORDS

inverse relation

one-to-one function

inverse function notation

Note Recall that a relation has one or more outputs for every one input; whereas, a function has only one output for every input. Previously, we have used the vertical line test to tell if a graph is a function or a relation.

Practice Problems

1. If $f(x) = 4x + 1$, find the equation for the inverse of f.

Answer

1. $y = \frac{x-1}{4}$

Rocket Man Bob Maddox plans to strap himself to four rockets and blast himself into space. While he's raising money to build these rockets, he spends his free time building small pulse jet engines for bicycles. For power, the engines pulse 70 times per second. A single engine on a standard bicycle can propel the rider to a top speed of 50 miles per hour in 7 seconds. A bicycle with two engines attached can reach a top speed of 73 miles per hour. The function $f(m) = \frac{15m}{22}$ converts miles per hour, m, to feet per second. Based on our past knowledge of unit analysis, we can convert the bicycle's top speed with two engines from 73 miles per hour to approximately 49.77 feet per second. After reading this section, we will be able to use the inverse function to convert the feet per second quantity back to miles per hour.

A Finding the Inverse Relation

Suppose the function $f(x)$ is given by

$$f(x) = \{(1, 4), (2, 5), (3, 6), (4, 7)\}$$

The inverse of $f(x)$ is obtained by reversing the order of the coordinates in each ordered pair in $f(x)$. The inverse of $f(x)$ is the relation given by

$$g(x) = \{(4, 1), (5, 2), (6, 3), (7, 4)\}$$

It is obvious that the domain of $f(x)$ is now the range of $g(x)$, and the range of $f(x)$ is now the domain of $g(x)$. Every function (or relation) has an inverse that is obtained from the original function by interchanging the components of each ordered pair.

Suppose a function $f(x)$ is defined with an equation instead of a list of ordered pairs. We can obtain the equation of the inverse of $f(x)$ by interchanging the role of x and y in the equation for $f(x)$.

EXAMPLE 1 If the function $f(x)$ is defined by $f(x) = 2x - 3$, find the equation that represents the inverse of $f(x)$.

Solution Because the inverse of $f(x)$ is obtained by interchanging the components of all the ordered pairs belonging to $f(x)$, and each ordered pair in $f(x)$ satisfies the equation $y = 2x - 3$, we simply exchange x and y in the equation $y = 2x - 3$ to get the formula for the inverse of $f(x)$.

$$x = 2y - 3$$

We now solve this equation for y in terms of x.

$$x + 3 = 2y$$

$$\frac{x + 3}{2} = y$$

$$y = \frac{x + 3}{2}$$

Chapter 8 Exponential and Logarithmic Functions

The last line gives the equation that defines the inverse of $f(x)$. Let's compare the graphs of $f(x)$ and its inverse as given here. (See Figure 1.)

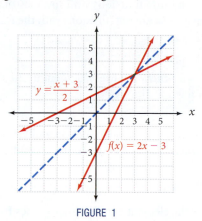

FIGURE 1

The graphs of $f(x)$ and its inverse have symmetry about the line $y = x$. This is a reasonable result since the one function was obtained from the other by interchanging x and y in the equation. The ordered pairs (a, b) and (b, a) always have symmetry about the line $y = x$.

B Graph a Function with its Inverse

EXAMPLE 2   Graph the function $y = x^2 - 2$ and its inverse. Give the equation for the inverse.

Solution   We can obtain the graph of the inverse of $y = x^2 - 2$ by graphing $y = x^2 - 2$ by the usual methods, and then reflecting the graph about the line $y = x$. This reflection is graphically the equivalent to switching the domain and range.

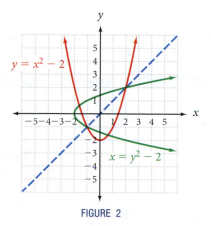

FIGURE 2

The equation that corresponds to the inverse of $y = x^2 - 2$ is obtained by interchanging x and y to get $x = y^2 - 2$.

We can solve the equation $x = y^2 - 2$ for y in terms of x as follows:

$$x = y^2 - 2$$
$$x + 2 = y^2$$
$$y = \pm\sqrt{x + 2}$$

Comparing the graphs from Examples 1 and 2, we observe that the inverse of a function is not always a function. In Example 1, both $f(x)$ and its inverse have graphs that are nonvertical straight lines and therefore both represent functions. In Example 2, the inverse of function $f(x)$ is not a function, since a vertical line crosses it in more than one place.

2. Graph $y = x^2 + 1$ and its inverse. Give the equation for the inverse.

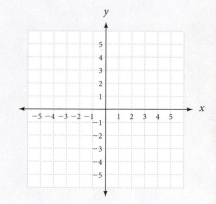

Answer

2. $y = \pm\sqrt{x - 1}$

One-to-One Functions

We can distinguish between those functions with inverses that are also functions and those functions with inverses that are not functions with the following definition:

> **DEFINITION** one-to-one functions
>
> A function is a *one-to-one function* if every element in the range comes from exactly one element in the domain.

This definition indicates that a one-to-one function will yield a set of ordered pairs in which no two different ordered pairs have the same second coordinates. For example, the function

$$f(x) = \{(2, 3), (-1, 3), (5, 8)\}$$

is not one-to-one because the element 3 in the range comes from both 2 and −1 in the domain. On the other hand, the function

$$g(x) = \{(5, 7), (3, -1), (4, 2)\}$$

is a one-to-one function because every element in the range comes from only one element in the domain.

Horizontal Line Test

If we have the graph of a function, then we can determine if the function is one-to-one with the following test. If a horizontal line crosses the graph of a function in more than one place, then the function is not a one-to-one function because the points at which the horizontal line crosses the graph will be points with the same y-coordinates, but different x-coordinates. Therefore, the function will have an element in the range (the y-coordinate) that comes from more than one element in the domain (the x-coordinates).

Of the functions we have covered previously, all the linear functions and exponential functions are one-to-one functions because no horizontal lines can be found that will cross their graphs in more than one place.

Functions Whose Inverses Are Also Functions

Because one-to-one functions do not repeat second coordinates, when we reverse the order of the ordered pairs in a one-to-one function, we obtain a relation in which no two ordered pairs have the same first coordinate — by definition, this relation must be a function. In other words, every one-to-one function has an inverse that is itself a function. Because of this, we can use function notation to represent that inverse.

> **PROPERTY** Inverse Function Notation
>
> If $y = f(x)$ is a one-to-one function, then the inverse of $f(x)$ is also a function and can be denoted by $y = f^{-1}(x)$.

Note The notation $f^{-1}(x)$ does not represent the reciprocal of $f(x)$. That is, the −1 in this notation is not an exponent. The notation $f^{-1}(x)$ is defined as representing the inverse function for a one-to-one function. In symbols,

$$f^{-1}(x) \neq \frac{1}{f(x)}$$

To illustrate, in Example 1 we found that the inverse of $f(x) = 2x - 3$ was the function $y = \frac{x+3}{2}$. We can write this inverse function with inverse function notation as

$$f^{-1}(x) = \frac{x+3}{2}$$

On the other hand, the inverse of the function in Example 2 is not itself a function, so we do not use the notation $f^{-1}(x)$ to represent it.

EXAMPLE 3 Find the inverse of $g(x) = \dfrac{x-4}{x-2}$.

Solution To find the inverse for g, we begin by replacing $g(x)$ with y to obtain

$$y = \frac{x-4}{x-2} \qquad \textcolor{green}{\text{Original function}}$$

To find an equation for the inverse, we exchange x and y.

$$x = \frac{y-4}{y-2} \qquad \textcolor{green}{\text{Inverse of the original function}}$$

To solve for y, we first multiply each side by $y-2$ to obtain

$$x(y-2) = y-4$$

$$xy - 2x = y - 4 \qquad \textcolor{green}{\text{Distributive property}}$$

$$xy - y = 2x - 4 \qquad \textcolor{green}{\text{Collect all terms containing } y \text{ on the left side.}}$$

$$y(x-1) = 2x - 4 \qquad \textcolor{green}{\text{Factor } y \text{ from each term on the left side.}}$$

$$y = \frac{2x-4}{x-1} \qquad \textcolor{green}{\text{Divide each side by } x-1.}$$

Because our original function is one-to-one, as verified by the graph in Figure 3, its inverse is also a function. Therefore, we can use inverse function notation to write

$$g^{-1}(x) = \frac{2x-4}{x-1}$$

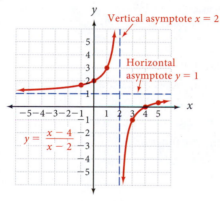

FIGURE 3

EXAMPLE 4 Graph the function $y = 2^x$ and its inverse $x = 2^y$.

Solution We graphed $y = 2^x$ in the preceding section. We simply reflect its graph about the line $y = x$ to obtain the graph of its inverse $x = 2^y$.

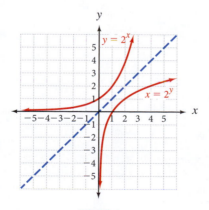

FIGURE 4

3. Find the inverse of $g(x) = \dfrac{x-3}{x+1}$.

4. Graph the function $y = 3^x$ and its inverse $x = 3^y$.

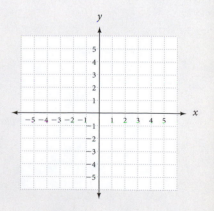

Answer

3. $g^{-1}(x) = \dfrac{3+x}{1-x}$

As you can see from the graph, $x = 2^y$ is a function. However, we have not developed the mathematical tools to solve this equation for y. Therefore, we are unable to use the inverse function notation to represent this function. In the next section, we will give a definition that solves this problem. For now, we simply leave the equation as $x = 2^y$.

Functions, Relations, and Inverses — A Summary

Here is a summary of some of the things we know about functions, relations, and their inverses:

1. Every function is a relation, but not every relation is a function.

2. Every function has an inverse, but only one-to-one functions have inverses that are also functions.

3. The domain of a function is the range of its inverse, and the range of a function is the domain of its inverse.

4. If $y = f(x)$ is a one-to-one function, then we can use the notation $y = f^{-1}(x)$ to represent its inverse function.

5. The graph of a function and its inverse have symmetry about the line $y = x$.

6. If (a, b) belongs to the function $f(x)$, then the point (b, a) belongs to its inverse.

GETTING READY FOR CLASS

After reading through the preceding section, respond in your own words and in complete sentences.

A. What is the inverse of a function?

B. What is the relationship between the graph of a function and the graph of its inverse?

C. What are one-to-one functions?

D. In words, explain inverse function notation.

Vocabulary Review

Choose the correct words to fill in the blanks below.

inverse   function   relation   range   one-to-one

1. Every function is a _____ , but not every relation is a function.

2. A function is a one-to-one function if every element in the _____ comes from exactly one element in the domain.

3. A horizontal line test can help determine if a function is _____ .

4. If $y = f(x)$ is a one-to-one function, then the _____ of $f(x)$ is also a function, given by $y = f^{-1}(x)$.

5. The graph of a _____ and its inverse have symmetry about the line $y = x$.

Problems

A For each of the following one-to-one functions, find the equation of the inverse. Write the inverse using the notation $f^{-1}(x)$.

1. $f(x) = 3x - 1$ **2.** $f(x) = 2x - 5$ **3.** $f(x) = x^3$ **4.** $f(x) = x^3 - 2$

5. $f(x) = \dfrac{x - 3}{x - 1}$ **6.** $f(x) = \dfrac{x - 2}{x - 3}$ **7.** $f(x) = \dfrac{x - 3}{4}$ **8.** $f(x) = \dfrac{x + 7}{2}$

9. $f(x) = \dfrac{1}{2}x - 3$ **10.** $f(x) = \dfrac{1}{3}x + 1$ **11.** $f(x) = \dfrac{2}{3}x - 3$ **12.** $f(x) = -\dfrac{1}{2}x + 4$

13. $f(x) = x^3 - 4$ **14.** $f(x) = -3x^3 + 2$ **15.** $f(x) = \dfrac{4x - 3}{2x + 1}$ **16.** $f(x) = \dfrac{3x - 5}{4x + 3}$

17. $f(x) = \dfrac{2x + 1}{3x + 1}$ **18.** $f(x) = \dfrac{3x + 2}{5x + 1}$

19. $f(x) = \sqrt[3]{3x - 2}$ **20.** $f(x) = \sqrt[3]{4x + 1}$

B For each of the following relations, sketch the graph of the relation and its inverse, and write an equation for the inverse.

21. $y = 2x - 1$

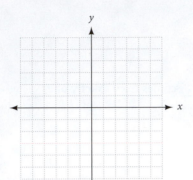

22. $y = 3x + 1$

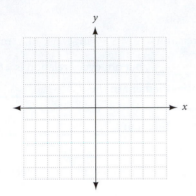

23. $y = x^2 - 3$

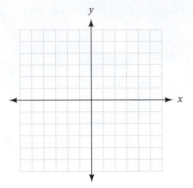

24. $y = x^2 + 1$

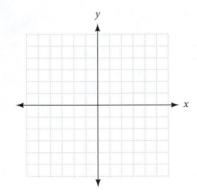

25. $y = x^2 - 2x - 3$

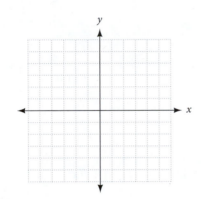

26. $y = x^2 + 2x - 3$

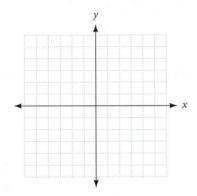

27. $y = 3^x$

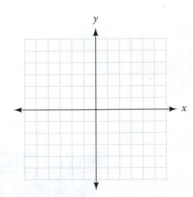

28. $y = \left(\dfrac{1}{2}\right)^x$

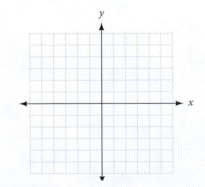

29. $y = 4$

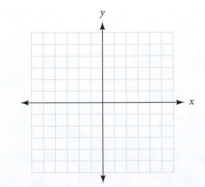

30. $y = -2$

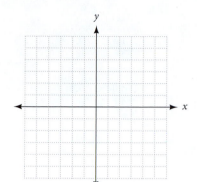

31. $y = \frac{1}{2}x^3$

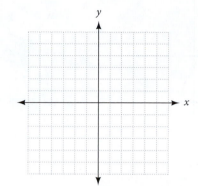

32. $y = x^3 - 2$

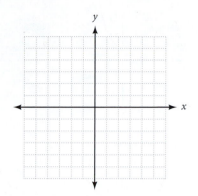

33. $y = \frac{1}{2}x + 2$

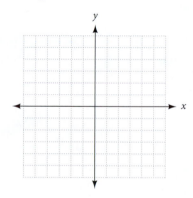

34. $y = \frac{1}{3}x - 1$

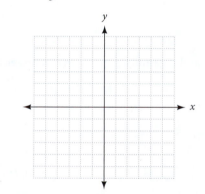

35. $y = \sqrt{x + 2}$

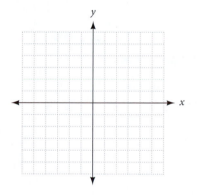

36. $y = \sqrt{x} + 2$

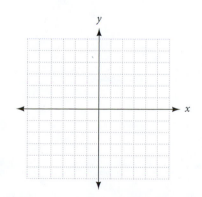

37. $y = \sqrt[3]{x} + 4$

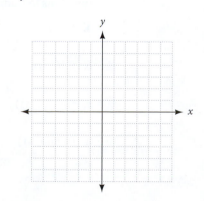

38. $y = \sqrt[3]{x} - 3$

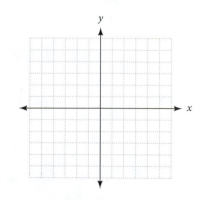

39. Determine if the following functions are one-to-one.

a.

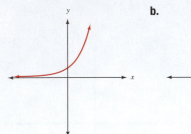

b.

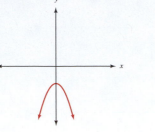

40. Could the following tables of values represent ordered pairs from one-to-one functions? Explain your answer.

a.

x	y
−2	5
−1	4
0	3
1	4
2	5

b.

x	y
1.5	0.1
2.0	0.2
2.5	0.3
3.0	0.4
3.5	0.5

c.

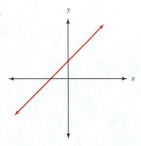

For the given function $f(x)$, find the inverse $g(x)$ and determine whether $g(x)$ represents a function.

41. $f(x) = \{(1, 4), (2, 5), (3, 7)\}$

42. $f(x) = \{(-1, 4), (0, 3), (1, 4)\}$

43. $f(x) = \{(a, A), (b, B), (c, C)\}$

44. $f(x) = \{(d, 4), (f, 6), (g, 7), (D, 4)\}$

45. If $f(x) = 3x - 2$, then $f^{-1}(x) = \dfrac{x + 2}{3}$. Use these two functions to find

   a. $f(2)$

   b. $f^{-1}(2)$

   c. $f(f^{-1}(2))$

   d. $f^{-1}(f(2))$

46. If $f(x) = \frac{1}{2}x + 5$, then $f^{-1}(x) = 2x - 10$. Use these two functions to find

   a. $f(-4)$

   b. $f^{-1}(-4)$

   c. $f(f^{-1}(-4))$

   d. $f^{-1}(f(-4))$

47. Let $f(x) = \dfrac{1}{x}$, and find $f^{-1}(x)$.

48. Let $f(x) = \dfrac{a}{x}$, and find $f^{-1}(x)$. (a is a real number constant.)

Applying the Concepts

49. Inverse Functions in Words Inverses of one-to-one functions may also be found by *inverse reasoning*. For example, to find the inverse of $f(x) = 3x + 2$, first list, in order, the operations done to variable x:

a. Multiply by 3.

b. Add 2.

Then, to find the inverse, simply apply the inverse operations, in reverse order, to the variable x. That is:

c. Subtract 2.

d. Divide by 3.

The inverse function then becomes $f^{-1}(x) = \frac{x-2}{3}$. Use this method of inverse reasoning to find the inverse of the function $f(x) = \frac{x}{7} - 2$.

50. Inverse Functions in Words Refer to the method of inverse reasoning explained in Problem 49. Use inverse reasoning to find the following inverses:

a. $f(x) = 2x + 7$

b. $f(x) = \sqrt{x} - 9$

c. $f(x) = x^3 - 4$

d. $f(x) = \sqrt{x^3 - 4}$

51. Reading Tables Evaluate each of the following functions using the functions defined by Tables 1 and 2.

a. $f(g(-3))$

b. $g(f(-6))$

c. $g(f(2))$

d. $f(g(3))$

e. $f(g(-2))$

f. $g(f(3))$

Table 1	
x	$f(x)$
-6	3
2	-3
3	-2
6	4

Table 2	
x	$g(x)$
-3	2
-2	3
3	-6
4	6

g. What can you conclude about the relationship between functions $f(x)$ and $g(x)$?

52. Reading Tables Use the functions defined in Tables 1 and 2 in Problem 51 to answer the following questions.

a. What are the domain and range of $f(x)$?

b. What are the domain and range of $g(x)$?

c. How are the domain and range of $f(x)$ related to the domain and range of $g(x)$?

d. Is $f(x)$ a one-to-one function?

e. Is $g(x)$ a one-to-one function?

53. Temperature The function $C(F) = \frac{5(F - 32)}{9}$ is used to convert the measurement of temperature from Fahrenheit to Celsius. In January 2010, the lowest recorded temperature was $-64°F$ in O'Brien, Alaska (USA Today).

a. Use this function to determine the temperature in Celsius.

b. Determine the inverse of this function.

c. Use the inverse function to convert $0°C$ to Fahrenheit.

54. Temperature The function $F(C) = \frac{9}{5}C + 32$ is used to convert the measurement of temperature from Celsius to Fahrenheit. The lowest temperature recorded in Honolulu, Hawaii was $11°C$ in 1902 (The Weather Channel).

a. Use this function to determine the temperature measured in Fahrenheit.

b. Determine the inverse of this function.

c. Use the inverse function to convert $32°F$ to Celsius.

55. Use the function given in problem 53 to find the temperature at which $C(F) = F$. What does this represent?

56. Use the function given in problem 54 to find the temperature at which $F(C) = C$. What does this represent?

57. Social Security A function that models the billions of dollars of Social Security payment (as shown in the chart) per year is $s(t) = 16t + 249.4$, where t is time in years since 1990 (U.S. Census Bureau).

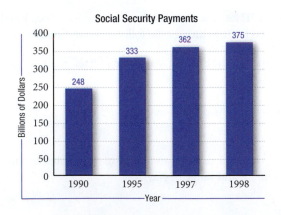

a. Use the model to estimate the amount of Social Security payments paid in 2010.

b. Write the inverse of the function.

c. Using the inverse function, estimate the year in which payments reached $507 billion.

58. Families The function for the percentage of one-parent families (as shown in the following chart) is $f(x) = 0.417x + 24$, when x is the time in years since 1990 (U.S. Census Bureau).

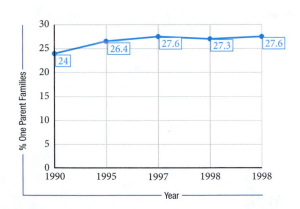

a. Use the function to predict the percentage of families with one parent in the year 2010.

b. Determine the inverse of the function, and estimate the year in which approximately 29% of the families were one-parent families.

59. Speed The fastest type of plane, a rocket plane, can travel at a speed of 4,520 miles per hour. The function $f(m) = \frac{22m}{15}$ converts miles per hour, m, to feet per second (World Book Encyclopedia).

  a. Use the function to convert the speed of the rocket plane to feet per second.

  b. Write the inverse of the function.

  c. Using the inverse function, convert 2 feet per second to miles per hour.

60. Speed A Lockheed SR-71A airplane set a world record (as reported by Air Force Armament Museum in 1996) with an absolute speed record of 2,193.167 miles per hour. The function $s(h) = 0.4468424h$ converts miles per hour, h, to meters per second, s (Air Force Armament Museum).

  a. What is the absolute speed of the Lockheed SR-71A in meters per second?

  b. What is the inverse of this function?

  c. Using the inverse function, determine the speed of an airplane in miles per hour that flies 150 meters per second.

Maintaining Your Skills

Solve each equation.

61. $(2x - 1)^2 = 25$

62. $(3x + 5)^2 = -12$

63. What number would you add to $x^2 - 10x$ to make it a perfect square trinomial?

64. What number would you add to $x^2 - 5x$ to make it a perfect square trinomial?

Solve by completing the square.

65. $x^2 - 10x + 8 = 0$

66. $x^2 - 5x + 4 = 0$

67. $3x^2 - 6x + 6 = 0$

68. $4x^2 - 16x - 8 = 0$

One Step Further

For each of the following functions, find $f^{-1}(x)$. Then show that $f(f^{-1}(x)) = x$.

69. $f(x) = 3x + 5$

70. $f(x) = 6 - 8x$

71. $f(x) = x^3 + 1$

72. $f(x) = x^3 - 8$

73. $f(x) = \dfrac{x - 4}{x - 2}$

74. $f(x) = \dfrac{x - 3}{x - 1}$

75. $f(x) = \dfrac{2x - 3}{x + 2}$

76. $f(x) = \dfrac{3x + 1}{2x - 1}$

77. Reading Graphs The graphs of a function and its inverse are shown in Figure 5. Use the graphs to find the following:

a. $f(0)$

b. $f(1)$

c. $f(2)$

d. $f^{-1}(1)$

e. $f^{-1}(2)$

f. $f^{-1}(5)$

g. $f^{-1}(f(2))$

h. $f(f^{-1}(5))$

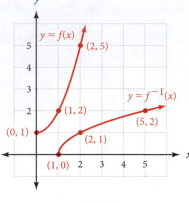

FIGURE 5

78. Domain From a Graph The function f is defined by the equation $f(x) = \sqrt{x + 4}$ for $x \geq -4$, meaning the domain is all x in the interval $[-4, \infty)$.

a. Graph the function f.

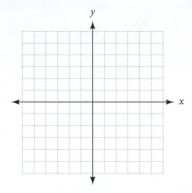

b. On the same set of axes, graph the line $y = x$ and the inverse function f^{-1}.

c. State the domain for the inverse function for f^{-1}.

d. Find the equation for $f^{-1}(x)$ and state its domain.

79. Currency Clark took a trip to Germany in November 2010. He had $1,000 USD of spending money for his trip. While in Germany he spent 694 EURO, the currency used for trade in Europe. The average conversion from USD to EURO at that time followed the function $E(D) = 0.73539D$, where E is the amount of EURO and D is the USD (ONADA FxConverter). How much money in EURO did Clark have left after his trip? USD?

80. Currency Lois left New York on October 21, 2010, to travel to Europe. On that day, she converted $500 USD in cash to EURO currency. The function for the conversion was $E(D) = 0.72088D$ where E is EURO and D is USD (ONADA FxConverter). On the date of her return trip, November 14, 2010, she had 86.50 EURO to convert to USD. On that date, the function for conversion was $E(D) = 0.73067D$.

a. How much money in EURO did Lois take with her to Europe?

b. How much money in USD did Lois return with to New York?

81. Assume that both $f(x)$ and $f^{-1}(x)$ are one-to-one functions. Given $f^{-1}(3) = -6$, solve the equation $f(3x + 2) = 3$.

82. Assume that both $f(x)$ and $f^{-1}(x)$ are one-to-one functions. Given $f(4) = -7$, solve the equation $f^{-1}(4x - 1) = 4$.

83. Let $f(x) = \dfrac{ax + b}{cx + d}$. Find $f^{-1}(x)$.

84. Let $f(x) = \dfrac{5x + 7}{6x - 5}$. Show that f is its own inverse.

85. Let $f(x) = \dfrac{ax + 5}{3x - a}$. Show that f is its own inverse.

86. Let $f(x) = \dfrac{ax + b}{cx - a}$. Show that f is its own inverse.

Getting Ready for the Next Section

Simplify.

87. 3^{-2}

88. 2^3

89. $\left(\dfrac{1}{2}\right)^4$

90. $\left(\dfrac{1}{3}\right)^{-3}$

Solve.

91. $2 = 3x$

92. $3 = 5x$

93. $4 = x^3$

94. $12 = x^2$

Fill in the blanks to make each statement true.

95. $8 = 2^{\square}$

96. $27 = 3^{\square}$

97. $10{,}000 = 10^{\square}$

98. $1{,}000 = 10^{\square}$

99. $81 = 3^{\square}$

100. $81 = 9^{\square}$

101. $6 = 6^{\square}$

102. $1 = 5^{\square}$

103. $27 = \left(\dfrac{1}{3}\right)^{\square}$

104. $16 = \left(\dfrac{1}{4}\right)^{\square}$

105. $\dfrac{1}{16} = 2^{\square}$

106. $\dfrac{1}{27} = 3^{\square}$

Find the Mistake

Each problem below contains a mistake. Circle the mistake and write the correct number(s) or word(s) on the line provided.

1. A function is one-to-one if every element in the domain comes from exactly one element in the range.

_____

2. If $y = f(x)$ is a one-to-one function, then the inverse of $f(x)$ is also a relation and can be denoted by $y = f^{-1}(x)$.

_____

3. The graph of a function and its inverse intersect the line $y = x$. _____

4. If (a, b) belongs to the function $f(x)$, then the point $(b, 0)$ belongs to its inverse. _____

OBJECTIVES

A Convert between logarithmic form and exponential form.

B Use the definition of logarithms to solve simple logarithmic equations.

C Sketch the graph of logarithmic functions.

D Simplify expressions involving logarithms.

E Solve applications involving logarithms.

KEY WORDS

logarithm

logarithmic function

exponential form

logarithmic form

exponential function

©iStockphoto.com/arindambanerjee

In January 2010, news organizations reported that an earthquake had rocked the tiny island nation of Haiti and caused massive destruction. They reported the strength of the earthquake as having measured 7.0 on the Richter scale. For comparison, Table 1 gives the Richter magnitude of a number of other earthquakes.

Table 1 Earthquakes		
Year	Earthquake	Richter Magnitude
1971	Los Angeles	6.6
1985	Mexico City	8.1
1989	San Francisco	7.1
1992	Kobe, Japan	7.2
1994	Northridge	6.6
1999	Armenia, Colombia	6.0
2004	Indian Ocean	9.1
2010	Haiti	7.0

Source: USGS

Although the sizes of the numbers in the table do not seem to be very different, the intensity of the earthquakes they measure can be very different. For example, the 1989 San Francisco earthquake was more than 10 times stronger than the 1999 earthquake in Colombia. The reason behind this is that the Richter scale is a *logarithmic scale*. In this section, we start our work with logarithms, which will give you an understanding of the Richter scale. Let's begin.

A Logarithmic Functions

As you know from your work in the previous sections, functions of the form

$$y = b^x \quad b > 0, b \neq 1$$

are called exponential functions. Because the equation of the inverse of a function can be obtained by exchanging x and y in the equation of the original function, the inverse of an exponential function must have the form

$$x = b^y \quad b > 0, b \neq 1$$

Now, this last equation is actually the equation of a logarithmic function, as the following definition indicates.

> **DEFINITION** logarithmic function
>
> The equation $y = \log_b x$ is read "y is the logarithm to the base b of x" and is equivalent to the equation
>
> *In symbols:* $x = b^y$ $b > 0, b \neq 1$
>
> *In words:* y is the number we raise b to in order to get x.

Notation When an equation is in the form $x = b^y$, it is said to be in *exponential form*. On the other hand, if an equation is in the form $y = \log_b x$, it is said to be in *logarithmic form*.

Here are some equivalent statements written in both forms:

Exponential Form		Logarithmic Form
$8 = 2^3$	$\Leftrightarrow$	$\log_2 8 = 3$
$25 = 5^2$	$\Leftrightarrow$	$\log_5 25 = 2$
$0.1 = 10^{-1}$	$\Leftrightarrow$	$\log_{10} 0.1 = -1$
$\frac{1}{8} = 2^{-3}$	$\Leftrightarrow$	$\log_2 \frac{1}{8} = -3$
$r = z^s$	$\Leftrightarrow$	$\log_z r = s$

B Logarithmic Equations

Practice Problems

EXAMPLE 1   Solve $\log_3 x = -2$.

Solution   In exponential form, the equation looks like this:

$$x = 3^{-2} \quad \text{or} \quad x = \frac{1}{9}$$

The solution is $\frac{1}{9}$.

1. Solve $\log_2 x = 3$.

Converting an equation in logarithmic form to its equivalent exponential form, as we did in Example 1, is the easiest way to isolate and solve for the variable x.

EXAMPLE 2   Solve $\log_x 4 = 3$.

Solution   Again, we use the definition of logarithms to write the expression in exponential form.

$$4 = x^3$$

Taking the cube root of both sides, we have

$$\sqrt[3]{4} = \sqrt[3]{x^3}$$
$$x = \sqrt[3]{4}$$

The solution is $\sqrt[3]{4}$.

2. Solve $\log_x 5 = 2$.

EXAMPLE 3   Solve $\log_8 4 = x$.

Solution   We write the expression again in exponential form.

$$4 = 8^x$$

Because both 4 and 8 can be written as powers of 2, we write them in terms of powers of 2.

$$2^2 = (2^3)^x$$
$$2^2 = 2^{3x}$$

The only way the left and right sides of this last line can be equal is if the exponents are equal — that is, if

$$2 = 3x$$
$$x = \frac{2}{3}$$

3. Solve $\log_9 27 = x$.

Answers

1. 8
2. $\sqrt{5}$
3. $\frac{3}{2}$

The solution is $\frac{2}{3}$. We check as follows:

$$\log_8 4 = \frac{2}{3} \Leftrightarrow 4 = 8^{2/3}$$
$$4 = (\sqrt[3]{8})^2$$
$$4 = 2^2$$
$$4 = 4$$

The solution checks in the original equation.

C Graphing Logarithmic Functions

Graphing logarithmic functions can be done using the graphs of exponential functions and the fact that the graphs of inverse functions have symmetry about the line $y = x$. Here's an example to illustrate:

4. Graph the function $f(x) = \log_3 x$.

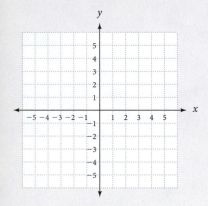

EXAMPLE 4   Graph the function $f(x) = \log_2 x$.

Solution   Let $y = f(x)$. The equation $y = \log_2 x$ is, by definition, equivalent to the exponential equation

$$x = 2^y$$

which is the equation of the inverse of the function

$$y = 2^x$$

The graph of $y = 2^x$ was given in Figure 2 of Section 8.1. We simply reflect the graph of $y = 2^x$ about the line $y = x$ to get the graph of $x = 2^y$, which is also the graph of $y = \log_2 x$. (See Figure 1.)

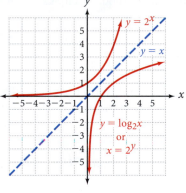

FIGURE 1

It is apparent from the graph that $y = \log_2 x$ is a function, because no vertical line will cross its graph in more than one place. The same is true for all logarithmic equations of the form $y = \log_b x$, where b is a positive number other than 1. Note also that the graph of $y = \log_b x$ will always appear to the right of the y-axis, meaning that x will always be positive in the equation $y = \log_b x$.

D Two Special Identities

If b is a positive real number other than 1, then each of the following is a consequence of the definition of a logarithm:

$$(1)\ b^{\log_b x} = x \text{ where } x > 0 \quad \text{and} \quad (2)\ \log_b b^x = x \text{ where } x \in \mathbb{R}.$$

The justifications for these identities are similar. Let's consider only the first one. Consider the expression

$$y = \log_b x$$

By definition, it is equivalent to

$$x = b^y$$

Note $x \in \mathbb{R}$ is read, "x is an element of the set of all real numbers."

Substituting $\log_b x$ for y in the last line gives us

$$x = b^{\log_b x}$$

The next examples in this section show how these two special properties can be used to simplify expressions involving logarithms.

EXAMPLE 5 Simplify the following logarithmic expressions.

a. $\log_2 8$ **b.** $\log_{10} 10,000$ **c.** $\log_b b$ **d.** $\log_b 1$ **e.** $\log_4 (\log_5 5)$

Solution

a. Substitute 2^3 for 8: $\log_2 8 = \log_2 2^3 = 3$

b. 10,000 can be written as 10^4: $\log_{10} 10,000 = \log_{10} 10^4 = 4$

c. Because $b^1 = b$, we have $\log_b b = \log_b b^1 = 1$

d. Because $1 = b^0$, we have $\log_b 1 = \log_b b^0 = 0$

e. Because $\log_5 5 = 1$, $\log_4 (\log_5 5) = \log_4 1 = 0$

5. Simplify.

  a. $\log_3 27$

  b. $\log_{10} 1,000$

  c. $\log_6 6$

  d. $\log_3 1$

  e. $\log_2 (\log_8 8)$

Here is a summary of what we know about the logarithmic function $f(x) = \log_b x$, where b is a positive real number other than zero:

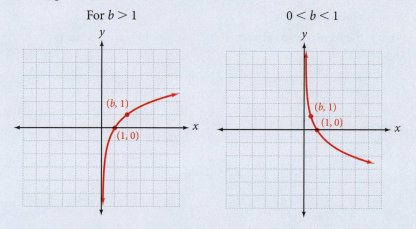

For $b > 1$ $0 < b < 1$

The function $f(x) = \log_b x$ is a one-to-one function. It has an x-intercept of $(1, 0)$, but it does not cross the y-axis. It also has a domain of $(0, \infty)$ and a range of $(-\infty, \infty)$.

Application

As we mentioned in the introduction to this section, one application of logarithms is in measuring the magnitude of an earthquake. If an earthquake has a shock wave T times greater than the smallest shock wave that can be measured on a seismograph, then the magnitude M of the earthquake, as measured on the Richter scale, is given by the formula

$$M = \log_{10} T$$

(When we talk about the size of a shock wave, we are talking about its amplitude. The amplitude of a wave is half the difference between its highest point and its lowest point.)

To illustrate the discussion, an earthquake that produces a shock wave that is 10,000 times greater than the smallest shock wave measurable on a seismograph will have a magnitude M on the Richter scale of

$$M = \log_{10} 10,000 = 4$$

Answers

5. a. 3 **b.** 3 **c.** 1 **d.** 0 **e.** 0

6. Repeat Example 6 if $M = 6$.

EXAMPLE 6   If an earthquake has a magnitude of $M = 5$ on the Richter scale, what can you say about the size of its shock wave?

Solution   To answer this question, we put $M = 5$ into the formula $M = \log_{10} T$ to obtain

$$5 = \log_{10} T$$

Writing this expression in exponential form, we have

$$T = 10^5 = 100,000$$

We can say that an earthquake that measures 5 on the Richter scale has a shock wave 100,000 times greater than the smallest shock wave measurable on a seismograph.

From Example 6 and the discussion that preceded it, we find that an earthquake of magnitude 5 has a shock wave that is 10 times greater than an earthquake of magnitude 4, because 100,000 is 10 times 10,000.

GETTING READY FOR CLASS

After reading through the preceding section, respond in your own words and in complete sentences.

A. What is a logarithm?

B. What is the relationship between $y = 2^x$ and $y = \log_2 x$? How are their graphs related?

C. Will the graph of $y = \log_b x$ ever appear in the second or third quadrants? Explain why or why not.

D. How would you use the identity $b^{\log_b x} = x$ to simplify an equation involving a logarithm?

Answer

6. 1,000,000

Vocabulary Review

Match the following statements written in exponential form with the statement written in the correct logarithmic form.

1. $64 = 4^3$

2. $9 = 3^2$

3. $0.01 = 10^{-1}$

4. $\frac{1}{16} = 2^{-4}$

5. $z = x^y$

a. $\log_{10} 0.01 = -1$

b. $\log_4 64 = 3$

c. $\log_2 \frac{1}{16} = -4$

d. $\log_3 9 = 2$

e. $\log_x z = y$

Problems

A Write each of the following equations in logarithmic form.

1. $2^4 = 16$

2. $3^2 = 9$

3. $125 = 5^3$

4. $16 = 4^2$

5. $0.01 = 10^{-2}$

6. $0.001 = 10^{-3}$

7. $2^{-5} = \frac{1}{32}$

8. $4^{-2} = \frac{1}{16}$

9. $\left(\frac{1}{2}\right)^{-3} = 8$

10. $\left(\frac{1}{3}\right)^{-2} = 9$

11. $27 = 3^3$

12. $81 = 3^4$

Write each of the following equations in exponential form.

13. $\log_{10} 100 = 2$

14. $\log_2 8 = 3$

15. $\log_2 64 = 6$

16. $\log_2 32 = 5$

17. $\log_8 1 = 0$

18. $\log_9 9 = 1$

19. $\log_{10} 0.001 = -3$

20. $\log_{10} 0.0001 = -4$

21. $\log_6 36 = 2$

22. $\log_7 49 = 2$

23. $\log_5 \frac{1}{25} = -2$

24. $\log_3 \frac{1}{81} = -4$

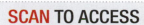

B Solve each of the following equations for x.

25. $\log_3 x = 2$

26. $\log_4 x = 3$

27. $\log_5 x = -3$

28. $\log_2 x = -4$

29. $\log_2 16 = x$

30. $\log_3 27 = x$

31. $\log_8 2 = x$

32. $\log_{25} 5 = x$

33. $\log_x 4 = 2$

34. $\log_x 16 = 4$

35. $\log_x 5 = 3$

36. $\log_x 8 = 2$

37. $\log_5 25 = x$

38. $\log_5 x = -2$

39. $\log_x 36 = 2$

40. $\log_x \dfrac{1}{25} = 2$

41. $\log_8 4 = x$

42. $\log_{16} 8 = x$

43. $\log_9 \dfrac{1}{3} = x$

44. $\log_{27} 9 = x$

45. $\log_8 x = -2$

46. $\log_{36} \dfrac{1}{6} = x$

47. $\log_{16} 8 = x$

48. $\log_9 \dfrac{1}{27} = x$

C Sketch the graph of each of the following logarithmic functions.

49. $f(x) = \log_3 x$

50. $f(x) = \log_{1/2} x$

51. $f(x) = \log_{1/3} x$

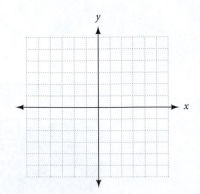

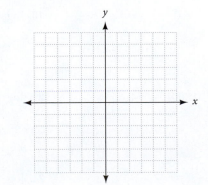

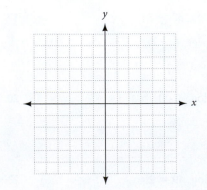

52. $f(x) = \log_4 x$

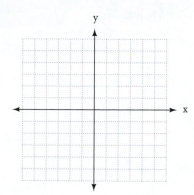

53. $f(x) = \log_5 x$

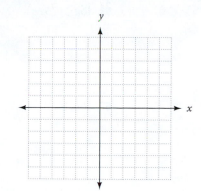

54. $f(x) = \log_{1/5} x$

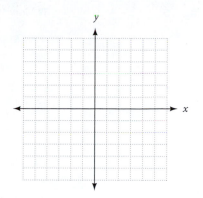

55. $f(x) = \log_{10} x$

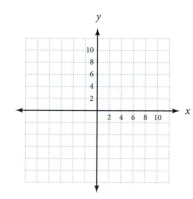

56. $f(x) = \log_{1/4} x$

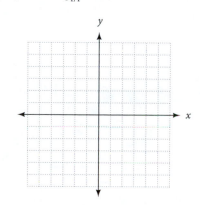

Each of the following graphs is a function of the form $f(x) = b^x$ or $f(x) = \log_b x$. Find the equation for each graph.

57.

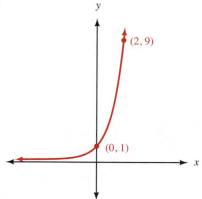

58.

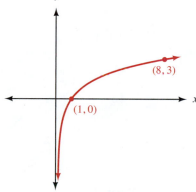

59.

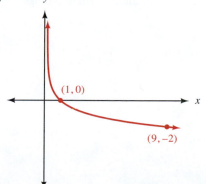

60.

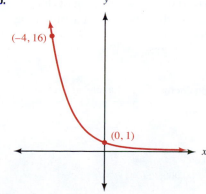

D Simplify each of the following.

61. $\log_2 16$

62. $\log_3 9$

63. $\log_{25} 125$

64. $\log_9 27$

65. $\log_{10} 1{,}000$

66. $\log_{10} 10{,}000$

67. $\log_3 3$

68. $\log_4 4$

69. $\log_5 1$

70. $\log_{10} 1$

71. $\log_{17} 1$

72. $\log_4 8$

73. $\log_{16} 4$

74. $\log_{10} 0.0001$

75. $\log_{100} 1000$

76. $\log_{32} 16$

77. $\log_3 (\log_2 8)$

78. $\log_5 (\log_{32} 2)$

79. $\log_{1/2} (\log_3 81)$

80. $\log_9 (\log_8 2)$

81. $\log_3 (\log_6 6)$

82. $\log_5 (\log_3 3)$

83. $\log_4 [\log_2 (\log_2 16)]$

84. $\log_4 [\log_3 (\log_2 8)]$

Applying the Concepts

85. Metric System The metric system uses logical and systematic prefixes for multiplication. For instance, to multiply a unit by 100, the prefix "hecto" is applied, so a hectometer is equal to 100 meters. For each of the prefixes in the following table find the logarithm, base 10, of the multiplying factor.

Prefix	Multiplying Factor	$\log_{10}$ (Multiplying Factor)
Nano	0.000 000 001	
Micro	0.000 001	
Deci	0.1	
Giga	1,000,000,000	
Peta	1,000,000,000,000,000	

86. Domain and Range Use the graphs of $y = 2^x$ and $y = \log_2 x$ shown in Figure 1 of this section to find the domain and range for each function. Explain how the domain and range found for $y = 2^x$ relate to the domain and range found for $y = \log_2 x$.

87. Magnitude of an Earthquake Find the magnitude M of an earthquake with a shock wave that measures $T = 100$ on a seismograph.

88. Magnitude of an Earthquake Find the magnitude M of an earthquake with a shock wave that measures $T = 100,000$ on a seismograph.

89. Shock Wave If an earthquake has a magnitude of 8 on the Richter scale, how many times greater is its shock wave than the smallest shock wave measurable on a seismograph?

90. Shock Wave If the 1999 Colombia earthquake had a magnitude of 6 on the Richter scale, how many times greater was its shock wave than the smallest shock wave measurable on a seismograph?

Earthquake The table below categorizes earthquake by the magnitude and identifies the average annual occurrence.

Earthquakes

Descriptor	Magnitude	Average Annual Occurrence
Great	≥ 8.0	1
Major	7 – 7.9	18
Strong	6 – 6.9	120
Moderate	5 – 5.9	800
Light	4 – 4.9	6,200
Minor	3 – 3.9	49,000
Very Minor	2 – 2.9	1,000 per day
Very Minor	1 – 1.9	8,000 per day

Source: USGS

91. What is the average number of earthquakes that occur per year when T is 100,000 or greater?

92. What is the average number of earthquakes that occur per year when T is 1,000,000 or greater?

Maintaining Your Skills

93. eBooks The chart shows the memory capacity on some consumer eReaders. If the average novel is 400 KB, find the number of novels that can fit on the following eReaders. Note: 1 MB = 1,024 KB and 1 GB = 1,024 MB.

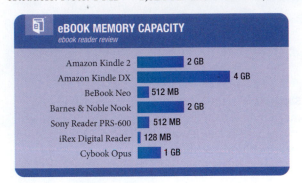

a. Amazon Kindle DX

b. Sony Ready PRS-600

c. Barnes and Noble Nook

94. iPads Using the chart, what percentage of iPad sales did not go to the seven states listed?

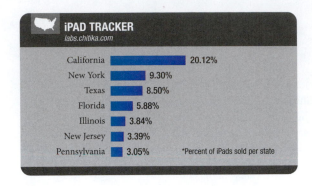

Getting Ready for the Next Section

Simplify.

95. $8^{2/3}$

96. $27^{2/3}$

97. $16^{3/4}$

98. $81^{3/4}$

99. $(-27)^{2/3}$

100. $16^{-3/4}$

101. $27^{-2/3}$

102. $(-32)^{4/5}$

Solve.

103. $(x+2)(x) = 2^3$

104. $(x+3)(x) = 2^2$

105. $\dfrac{x-2}{x+1} = 9$

106. $\dfrac{x+1}{x-4} = 25$

Write in exponential form.

107. $\log_2[(x+2)(x)] = 3$

108. $\log_4[x(x-6)] = 2$

109. $\log_3\left(\dfrac{x-2}{x+1}\right) = 4$

110. $\log_3\left(\dfrac{x-1}{x-4}\right) = 2$

Find the Mistake

Each problem below contains a mistake. Circle the mistake and write the correct number(s) or word(s) on the line provided.

1. The expression $y = \log_b x$ is read "y is the logarithm to the base x of b." _____

2. The expression $\frac{1}{16} = 2^{-4}$ is equivalent to the exponential expression $\log_2 \frac{1}{16} = -4$. _____

3. The graph of $y = \log_2 x$ is a function because no horizontal line will cross its graph in more than one place.
_____

4. A special identity for logarithms states that b raised to the power of $\log_b x$ is equal to b. _____

Landmark Review: Checking Your Progress

Let $f(x) = 3^x$ and $g(x) = \left(\frac{1}{3}\right)^x$, and evaluate each of the following.

1. $g(0)$

2. $f(-1)$

3. $f(-2) + g(2)$

4. $\dfrac{f(-3) - g(-2)}{-3-2}$

Graph each of the following exponential functions.

5. $f(x) = 3^x$

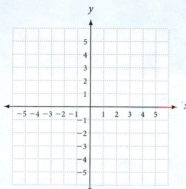

6. $g(x) = 4^{(x+1)}$

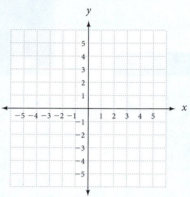

For each of the following functions, find the equation of the inverse.

7. $f(x) = 5x - 2$

8. $f(x) = 2x^3 - 2$

For each of the following relations, sketch the graph of the relation and its inverse, and write an equation for the inverse.

9. $y = 2x + 3$

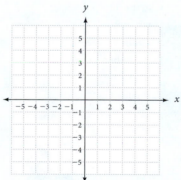

10. $y = 2^x$

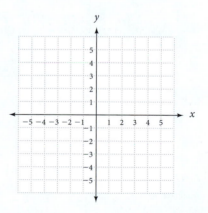

Write each of the following expressions in logarithmic form.

11. $4^2 = 16$

12. $2^4 = 16$

Write each of the following expressions in exponential form.

13. $\log(3)\frac{1}{9} = -2$

14. $\log(7)49 = 2$

Solve each of the following equations for x.

15. $\log(5)x = 2$

16. $\log(x)\frac{1}{64} = -2$

17. $\log(x)3 = 5$

18. $\log(25)\left(\frac{1}{5}\right) = x$

A Use the properties of logarithms to convert between expanded form and single logarithms.

B Use the properties of logarithms to solve equations that contain logarithms.

8.4 Properties of Logarithms

Image © sxc.hu, JWilsher, 2008

Recall the pulse jet engines we discussed earlier in the chapter. A rider of the jet-powered bicycle must wear ear protection while the engine is going. The sound of the pulse jet can reach 150 decibels, which is enough to pop a person's eardrums without the protection.

A decibel is a measurement of sound intensity, more specifically the difference in intensity between two sounds. This difference is equal to ten times the logarithm of the ratio of the two sounds. The following table and bar chart compare the intensity of sounds with which we are familiar:

Decibels	Comparable to
10	Calm breathing
20	Whisper
60	Normal conversation
80	City traffic
90	Lawn mower
110	Amplified music
120	Snowmobile
140	Fireworks
180	Airplane takeoff

Source: *hearingaid.know.com*

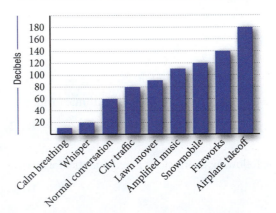

Chapter 8 Exponential and Logarithmic Functions

The precise definition for a *decibel* is

$$D = 10 \log_{10}\left(\frac{I}{I_0}\right)$$

where I is the intensity of the sound being measured, and I_0 is the intensity of the least audible sound. In this section, we will see that the preceding formula can also be written as

$$D = 10(\log_{10} I - \log_{10} I_0)$$

The rules we use to rewrite expressions containing logarithms are called the *properties of logarithms*. We will consider three of them here.

A Properties of Logarithms

For the following three properties, x, y, and b are all positive real numbers, $b \neq 1$, and r is any real number.

PROPERTY Product Property for Logarithms

$$\log_b(xy) = \log_b x + \log_b y$$

In words: The logarithm of a **product** is the **sum** of the logarithms.

PROPERTY Quotient Property for Logarithms

$$\log_b\left(\frac{x}{y}\right) = \log_b x - \log_b y$$

In words: The logarithm of a **quotient** is the **difference** of the logarithms.

PROPERTY Power Property for Logarithms

$$\log_b x^r = r \log_b x$$

In words: The logarithm of a number raised to a **power** is the **product** of the power and the logarithm of the number.

> **Note** Here are some common mistakes students make when working with properties of logarithms:
>
> $$\log_b x + \log_b y \neq \log_b(x + y)$$
> $$\log_b x - \log_b y \neq \log_b(x - y)$$
> $$r\log_b x \neq (\log_b x)^r$$
>
> Compare these errors with the correct properties to make sure you don't make the same mistakes.

Proof of the Product Property for Logarithms To prove the product property for logarithms, we simply apply the first identity for logarithms given in the preceding section.

$$b^{\log_b xy} = xy = (b^{\log_b x})(b^{\log_b y}) = b^{\log_b x + \log_b y}$$

Because the first and last expressions are equal and the bases are the same, the exponents $\log_b xy$ and $\log_b x + \log_b y$ must be equal. Therefore,

$$\log_b xy = \log_b x + \log_b y$$

The proofs of the quotient and power properties for logarithms proceed in much the same manner, so we will omit them here. The examples that follow show how the three properties can be used.

Practice Problems

1. Expand, using the properties of logarithms.

$$\log_3 \frac{5a}{b}$$

2. Expand $\log_{10} \dfrac{x^2}{\sqrt[3]{y}}$.

Note When evaluating logarithmic expressions, it is generally helpful to change any radicals to equivalent exponents, as we have done in Example 2. However, we recommend converting final answers from exponential form to radical form, unless instructed otherwise, as we did in Example 3.

3. Write as a single logarithm.

$$3\log_4 x + \log_4 y - 2\log_4 z$$

EXAMPLE 1   Expand using the properties of logarithms: $\log_5 \dfrac{3xy}{z}$.

Solution   Applying the quotient property for logarithms, we can write the quotient of $3xy$ and z in terms of a difference.

$$\log_5 \frac{3xy}{z} = \log_5 3xy - \log_5 z$$

Applying the product property for logarithms to the result $3xy$, we write it in terms of addition.

$$\log_5 \frac{3xy}{z} = \log_5 3 + \log_5 x + \log_5 y - \log_5 z$$

EXAMPLE 2   Expand using the properties of logarithms.

$$\log_2 \frac{x^4}{\sqrt{y} \cdot z^3}$$

Solution   We write $\sqrt{y}$ as $y^{1/2}$ and apply the properties.

$$\frac{\log_2 x^4}{\sqrt{y} \cdot z^3} = \log_2 \frac{x^4}{y^{1/2}z^3} \qquad \sqrt{y} = y^{1/2}$$

$$= \log_2 x^4 - \log_2(y^{1/2} \cdot z^3) \qquad \text{Quotient property for logarithms}$$

$$= \log_2 x^4 - (\log_2 y^{1/2} + \log_2 z^3) \qquad \text{Product property for logarithms}$$

$$= \log_2 x^4 - \log_2 y^{1/2} - \log_2 z^3 \qquad \text{Remove parentheses and distribute} -1.$$

$$= 4\log_2 x - \frac{1}{2}\log_2 y - 3\log_2 z \qquad \text{Power property for logarithms}$$

We can also use the three properties to write an expression in expanded form as just one logarithm.

EXAMPLE 3   Write the following as a single logarithm.

$$2\log_{10} a + 3\log_{10} b - \frac{1}{3}\log_{10} c$$

Solution   We begin by applying the power property for logarithms.

$$2\log_{10} a + 3\log_{10} b - \frac{1}{3}\log_{10} c$$

$$= \log_{10} a^2 + \log_{10} b^3 - \log_{10} c^{1/3} \qquad \text{Power property for logarithms}$$

$$= \log_{10}(a^2 \cdot b^3) - \log_{10} c^{1/3} \qquad \text{Product property for logarithms}$$

$$= \log_{10} \frac{a^2 b^3}{c^{1/3}} \qquad \text{Quotient property for logarithms}$$

$$= \log_{10} \frac{a^2 b^3}{\sqrt[3]{c}} \qquad c^{1/3} = \sqrt[3]{c}$$

When evaluating logarithmic expressions, it is generally helpful to change any radicals to an equivalent exponent, as we have done in Example 2. However, we recommend converting final answers from exponential form to radical form, unless instructed otherwise, as we did in Example 3.

Answers

1. $\log_3 5 + \log_3 a - \log_3 b$

2. $2\log_{10} x - \frac{1}{3}\log_{10} y$

3. $\log_4 \dfrac{x^3 y}{z^2}$

B Equations with Logarithms

The properties of logarithms along with the definition of logarithms are useful in solving equations that involve logarithms.

EXAMPLE 4   Solve $\log_2(x + 2) + \log_2 x = 3$.

4. Solve $\log_2 (x + 3) + \log_2 x = 2$.

Solution   Applying the product property for logarithms to the left side of the equation allows us to write it as a single logarithm.

$$\log_2(x + 2) + \log_2 x = 3$$

$$\log_2[(x + 2)(x)] = 3$$

The last line can be written in exponential form using the definition of logarithms:

$$(x + 2)(x) = 2^3$$

$$x^2 + 2x = 8$$

$$x^2 + 2x - 8 = 0$$

$$(x + 4)(x - 2) = 0$$

$$x + 4 = 0 \quad \text{or} \quad x - 2 = 0$$

$$x = -4 \qquad x = 2$$

In the previous section, we noted the fact that x in the expression $y = \log_b x$ cannot be a negative number. Because substitution of $x = -4$ into the original equation gives

$$\log_2(-2) + \log_2(-4) = 3$$

which contains logarithms of negative numbers, we cannot use -4 as a solution. The solution is 2.

GETTING READY FOR CLASS

After reading through the preceding section, respond in your own words and in complete sentences.

A. Explain why the following statement is false: "The logarithm of a product is the product of the logarithms."

B. Explain the quotient property for logarithms in words.

C. Explain the difference between $\log_b m + \log_b n$ and $\log_b(m + n)$. Are they equivalent?

D. Explain the difference between $\log_b(mn)$ and $(\log_b m)(\log_b n)$. Are they equivalent?

Answer

4. 1

Vocabulary Review

Choose the correct words to fill in the blanks below.

      quotient   decibel   power   product

1. The definition of _____ is $D = 10 \log_{10}\left(\dfrac{I}{I_0}\right)$.

2. The _____ property for logarithms states that $\log_b(xy) = \log_b x + \log_b y$.

3. The _____ property for logarithms states that $\log_b\left(\dfrac{x}{y}\right) = \log_b x - \log_b y$.

4. The _____ property for logarithms states that $\log_b x^r = r \log_b x$.

Problems

A Use the three properties of logarithms given in this section to expand each expression as much as possible.

1. $\log_3 4x$

2. $\log_2 5x$

3. $\log_6 \dfrac{5}{x}$

4. $\log_3 \dfrac{x}{5}$

5. $\log_2 y^5$

6. $\log_7 y^3$

7. $\log_9 \sqrt[3]{z}$

8. $\log_8 \sqrt{z}$

9. $\log_6 x^2 y^4$

10. $\log_{10} x^2 y^4$

11. $\log_5 \sqrt{x} \cdot y^4$

12. $\log_8 \sqrt[3]{xy^6}$

13. $\log_b \dfrac{xy}{z}$

14. $\log_b \dfrac{3x}{y}$

15. $\log_{10} \dfrac{4}{xy}$

16. $\log_{10} \dfrac{5}{4y}$

SCAN TO ACCESS

17. $\log_{10} \dfrac{x^2 y}{\sqrt{z}}$

18. $\log_{10} \dfrac{\sqrt{x} \cdot y}{z^3}$

19. $\log_{10} \dfrac{x^3 \sqrt{y}}{z^4}$

20. $\log_{10} \dfrac{x^4\sqrt[3]{y}}{\sqrt{z}}$

21. $\log_b \sqrt[3]{\dfrac{x^2y}{z^4}}$

22. $\log_b \sqrt[4]{\dfrac{x^4y^3}{z^5}}$

23. $\log_3 \sqrt[3]{\dfrac{x^2y}{z^6}}$

24. $\log_8 \sqrt[4]{\dfrac{x^5y^6}{z^3}}$

25. $\log_a \dfrac{4x^5}{9a^2}$

26. $\log_b \dfrac{16b^2}{25y^3}$

27. $\log_b \dfrac{9x^3}{64b^2}$

28. $\log_a \dfrac{27a^6}{16y^2}$

Write each expression as a single logarithm.

29. $\log_b x + \log_b z$

30. $\log_b x - \log_b z$

31. $2\log_3 x - 3\log_3 y$

32. $4\log_2 x + 5\log_2 y$

33. $\dfrac{1}{2}\log_{10} x + \dfrac{1}{3}\log_{10} y$

34. $\dfrac{1}{3}\log_{10} x - \dfrac{1}{4}\log_{10} y$

35. $3\log_2 x + \dfrac{1}{2}\log_2 y - \log_2 z$

36. $2\log_3 x + 3\log_3 y - \log_3 z$

37. $\dfrac{1}{2}\log_2 x - 3\log_2 y - 4\log_2 z$

38. $3\log_{10} x - \log_{10} y - \log_{10} z$

39. $\dfrac{3}{2}\log_{10} x - \dfrac{3}{4}\log_{10} y - \dfrac{4}{5}\log_{10} z$

40. $3\log_{10} x - \dfrac{4}{3}\log_{10} y - 5\log_{10} z$

41. $\dfrac{1}{2}\log_5 x + \dfrac{2}{3}\log_5 y - 4\log_5 z$

42. $\dfrac{1}{4}\log_7 x + 5\log_7 y - \dfrac{1}{3}\log_7 z$

43. $\log_3(x^2 - 16) - 2\log_3(x + 4)$

44. $\log_4(x^2 - x - 6) - \log_4(x^2 - 9)$

B Solve each of the following equations.

45. $\log_2 x + \log_2 3 = 1$

46. $\log_3 x + \log_3 3 = 1$

47. $\log_3 x - \log_3 2 = 2$

48. $\log_3 x + \log_3 2 = 2$

49. $\log_3 x + \log_3(x - 2) = 1$

50. $\log_6 x + \log_6(x - 1) = 1$

51. $\log_3(x + 3) - \log_3(x - 1) = 1$

52. $\log_4(x - 2) - \log_4(x + 1) = 1$

53. $\log_2 x + \log_2(x - 2) = 3$

54. $\log_4 x + \log_4(x + 6) = 2$

55. $\log_8 x + \log_8(x - 3) = \dfrac{2}{3}$

56. $\log_{27} x + \log_{27}(x + 8) = \dfrac{2}{3}$

57. $\log_3(x + 2) - \log_3 x = 1$

58. $\log_2(x + 3) - \log_2(x - 3) = 2$

59. $\log_2(x + 1) + \log_2(x + 2) = 1$

60. $\log_3 x + \log_3(x + 6) = 3$

61. $\log_9 \sqrt{x} + \log_9 \sqrt{2x + 3} = \dfrac{1}{2}$

62. $\log_8 \sqrt{x} + \log_8 \sqrt{5x + 2} = \dfrac{2}{3}$

63. $4\log_3 x - \log_3 x^2 = 6$

64. $9\log_4 x - \log_4 x^3 = 12$

65. $\log_5 \sqrt{x} + \log_5 \sqrt{6x + 5} = 1$

66. $\log_2 \sqrt{x} + \log_2 \sqrt{6x + 5} = 1$

Applying the Concepts

67. Decibel Formula Use the properties of logarithms to rewrite the decibel formula $D = 10 \log_{10}\left(\frac{I}{I_0}\right)$ as

$$D = 10(\log_{10} I - \log_{10} I_0).$$

68. Decibel Formula In the decibel formula $D = 10 \log_{10}\left(\frac{I}{I_0}\right)$, the threshold of hearing, I_0, is

$$I_0 = 10^{-12} \text{ watts/meter}^2$$

Substitute 10^{-12} for I_0 in the decibel formula, then show that it simplifies to

$$D = 10(\log_{10} I + 12)$$

69. Finding Logarithms If $\log_{10} 8 = 0.903$ and $\log_{10} 5 = 0.699$, find the following without using a calculator.

a. $\log_{10} 40$

b. $\log_{10} 320$

c. $\log_{10} 1,600$

70. Matching Match each expression in the first column with an equivalent expression in the second column.

a. $\log_2(ab)$ **i.** b

b. $\log_2\left(\frac{a}{b}\right)$ **ii.** 2

c. $\log_5 a^b$ **iii.** $\log_2 a + \log_2 b$

d. $\log_a b^a$ **iv.** $\log_2 a - \log_2 b$

e. $\log_a a^b$ **v.** $a \log_a b$

f. $\log_3 9$ **vi.** $b \log_5 a$

71. Henderson–Hasselbalch Formula Doctors use the Henderson–Hasselbalch formula to calculate the pH of a person's blood. pH is a measure of the acidity and/or the alkalinity of a solution. This formula is represented as

$$pH = 6.1 + \log_{10}\left(\frac{x}{y}\right)$$

where x is the base concentration and y is the acidic concentration. Rewrite the Henderson–Hasselbalch formula so that the logarithm of a quotient is not involved.

72. Henderson–Hasselbalch Formula Refer to the information in the preceding problem about the Henderson–Hasselbalch formula. If most people have a blood pH of 7.4, use the Henderson–Hasselbalch formula to find the ratio of $\frac{x}{y}$ for an average person.

73. Food Processing The formula $M = 0.21(\log_{10} a - \log_{10} b)$ is used in the food processing industry to find the number of minutes M of heat processing a certain food should undergo at 250°F to reduce the probability of survival of *Clostridium botulinum* spores. The letter a represents the number of spores per can before heating, and b represents the number of spores per can after heating. Find M if $a = 1$ and $b = 10^{-12}$. Then find M using the same values for a and b in the formula $M = 0.21 \log_{10} \frac{a}{b}$.

74. Acoustic Powers The formula $N = \log_{10} \frac{P_1}{P_2}$ is used in radio electronics to find the ratio of the acoustic powers of two electric circuits in terms of their electric powers. Find N if P_1 is 100 and P_2 is 1. Then use the same two values of P_1 and P_2 to find N in the formula $N = \log_{10} P_1 - \log_{10} P_2$.

Getting Ready for the Next Section

Simplify.

75. 5^0

76. 4^1

77. $\log_3 3$

78. $\log_5 5$

79. $\log_b b^4$

80. $\log_a a^k$

81. $4^{\log_4 x}$

82. $3^{\log_3 a}$

Use a calculator to find each of the following. Write your answer in scientific notation with the first number in each answer rounded to the nearest tenth.

83. $10^{-5.6}$

84. $10^{-4.1}$

Divide and round to the nearest whole number

85. $\dfrac{2.00 \times 10^8}{3.96 \times 10^6}$

86. $\dfrac{3.25 \times 10^{12}}{1.72 \times 10^{10}}$

Find the Mistake

Each problem below contains a mistake. Circle the mistake and write the correct number(s) or word(s) on the line provided.

1. The logarithm of a product is the difference of the logarithms. _____

2. The quotient property for logarithms states that $\log_b \frac{x}{y} = \log_b x + \log_b y$. _____

3. The logarithm of a number raised to a power is the product of the number and the logarithm of the power.

_____

4. To expand $\log_6 \frac{4xy}{z}$, apply the quotient property and the power property for logarithms. _____

8.5 Common and Natural Logarithms with Applications

Image © sxc.hu, sarahjmoon, 2007

OBJECTIVES

A Use a calculator to find common logarithms.

B Use a calculator to find a number given its common logarithm.

C Solve applications involving common logarithms.

D Simplify expressions containing natural logarithms.

KEY WORDS

common logarithm

characteristic

mantissa

antilogarithm

natural logarithm

Acid rain was first discovered in the 1960s by Gene Likens and his research team who studied the damage caused by acid rain to Hubbard Brook in New Hampshire. Acid rain is rain with a pH of 5.6 and below. pH is defined in terms of common logarithms—one of the topics we will present here. When you are finished with this section, you will have a more detailed knowledge of pH and acid rain.

Two kinds of logarithms occur more frequently than other logarithms. Logarithms with a base of 10 are very common because our number system is a base-10 number system. For this reason, we call base-10 logarithms *common logarithms*.

> **DEFINITION** common logarithms
>
> A *common logarithm* is a logarithm with a base of 10. Because common logarithms are used so frequently, it is customary, in order to save time, to omit notating the base. That is,
>
> $$\log_{10} x = \log x$$
>
> When the base is not shown, it is assumed to be 10.

A Common Logarithms

Common logarithms of powers of 10 are simple to evaluate. We need only recognize that $\log 10 = \log_{10} 10 = 1$ and apply the power property for logarithms $\log_b x^r = r \log_b x$.

$$
\begin{aligned}
\log 1{,}000 &= \log 10^3 &= 3\log 10 = &\quad 3(1) = &\quad 3 \\
\log 100 &= \log 10^2 &= 2\log 10 = &\quad 2(1) = &\quad 2 \\
\log 10 &= \log 10^1 &= 1\log 10 = &\quad 1(1) = &\quad 1 \\
\log 1 &= \log 10^0 &= 0\log 10 = &\quad 0(1) = &\quad 0 \\
\log 0.1 &= \log 10^{-1} &= -1\log 10 = &\quad -1(1) = &\quad -1 \\
\log 0.01 &= \log 10^{-2} &= -2\log 10 = &\quad -2(1) = &\quad -2 \\
\log 0.001 &= \log 10^{-3} &= -3\log 10 = &\quad -3(1) = &\quad -3
\end{aligned}
$$

To find common logarithms of numbers that are not powers of 10, we use a calculator with a ⌑log⌑ key.

Check the following logarithms to be sure you know how to use your calculator. (These answers have been rounded to the nearest ten-thousandth.)

$$\log 7.02 \approx 0.8463$$
$$\log 1.39 \approx 0.1430$$
$$\log 6.00 \approx 0.7782$$
$$\log 9.99 \approx 0.9996$$

Practice Problems

1. Find log 27,600.

EXAMPLE 1   Use a calculator to find log 2,760.
Solution

$$\log 2{,}760 \approx 3.4409$$

To work this problem on a scientific calculator, we simply enter the number 2,760 and press the key labeled log. On a graphing calculator we press the log key first, then 2,760.

The 3 in the answer is called the *characteristic,* and the decimal part of the logarithm is called the *mantissa.*

2. Find log 0.00391.

EXAMPLE 2   Find log 0.0391.
Solution $\log 0.0391 \approx -1.4078$

3. Find log 0.00952.

EXAMPLE 3   Find log 0.00523.
Solution $\log 0.00523 \approx -2.2815$

B Antilogorithms

4. Find x if $\log x = 3.9786$.

EXAMPLE 4   Find x if $\log x = 3.8774$.

Solution We are looking for the number whose logarithm is 3.8774. On a scientific calculator, we enter 3.8774 and press the key labeled 10^x. On a graphing calculator we press 10^x first, then 3.8774. The result is 7,540 to four significant digits. Here's why:

$$\text{If} \qquad \log x = 3.8774$$
$$\text{then} \qquad x = 10^{3.8774}$$
$$\approx 7{,}540$$

The number 7,540 is called the *antilogarithm* or just *antilog* of 3.8774. That is, 7,540 is the number whose logarithm is 3.8774.

5. Find x if $\log x = -1.5901$.

EXAMPLE 5   Find x if $\log x = -2.4179$.
Solution   Using the 10^x key, the result is 0.00382.

$$\text{If} \qquad \log x = -2.4179$$
$$\text{then} \qquad x = 10^{-2.4179}$$
$$\approx 0.00382$$

The antilog of -2.4179 is 0.00382. That is, the logarithm of 0.00382 is -2.4179.

C Applications of Logarithms

In Section 8.3, we found that the magnitude M of an earthquake that produces a shock wave T times larger than the smallest shock wave that can be measured on a seismograph is given by the formula

$$M = \log_{10} T$$

We can rewrite this formula using our shorthand notation for common logarithms as

$$M = \log T$$

Answers
1. 4.4409
2. −2.4078
3. −2.0214
4. 9,519
5. 0.0257

EXAMPLE 6 The San Francisco earthquake of 1906 is estimated to have measured 8.3 on the Richter scale. The San Fernando earthquake of 1971 measured 6.6 on the Richter scale. Find T for each earthquake, and then give some indication of how much stronger the 1906 earthquake was than the 1971 earthquake.

Solution For the 1906 earthquake:

If $\log T = 8.3$, then $T = 2.00 \times 10^8$.

For the 1971 earthquake:

If $\log T = 6.6$, then $T = 3.98 \times 10^6$.

Dividing the two values of T and rounding our answer to the nearest whole number, we have

$$\frac{2.00 \times 10^8}{3.98 \times 10^6} \approx 50$$

The shock wave for the 1906 earthquake was approximately 50 times larger than the shock wave for the 1971 earthquake.

In chemistry, the pH of a solution is the measure of the acidity of the solution. The definition for pH involves common logarithms. Here it is:

$$\text{pH} = -\log[\text{H}^+]$$

where $[\text{H}^+]$ is the concentration of the hydrogen ion in moles per liter. The range for pH is from 0 to 14. Pure water, a neutral solution, has a pH of 7. An acidic solution, such as vinegar, will have a pH less than 7, and an alkaline solution, such as ammonia, has a pH above 7.

EXAMPLE 7 Normal rainwater has a pH of 5.6. What is the concentration of the hydrogen ion in normal rainwater?

Solution Substituting 5.6 for pH in the formula $\text{pH} = -\log[\text{H}^+]$, we have

$5.6 = -\log[\text{H}^+]$	Substitution
$\log[\text{H}^+] = -5.6$	Isolate the logarithm.
$[\text{H}^+] = 10^{-5.6}$	Write in exponential form.
$\approx 2.5 \times 10^{-6}$ moles per liter	Answer in scientific notation.

EXAMPLE 8 The concentration of the hydrogen ion in a sample of acid rain known to kill fish is 3.2×10^{-5} mole per liter. Find the pH of this acid rain to the nearest tenth.

Solution Substituting 3.2×10^{-5} for $[\text{H}^+]$ in the formula $\text{pH} = -\log[\text{H}^+]$, we have

$\text{pH} = -\log[3.2 \times 10^{-5}]$	Substitution
$\approx -(-4.5)$	Evaluate the logarithm.
≈ 4.5	Simplify.

D Natural Logarithms

Recall that the number e is an irrational number with an approximate value of 2.7183. Here is a logarithm that uses the number e.

> **DEFINITION** natural logarithms
>
> A *natural logarithm* is a logarithm with a base of e. The natural logarithm of x is denoted by $\ln x$. That is,
>
> $$\ln x = \log_e x$$

6. Find T if an earthquake measures 5.5 on the Richter scale.

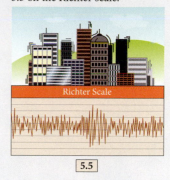

Richter Scale

5.5

THE ACID SCALE

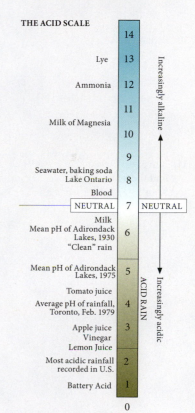

	14	Increasingly alkaline
Lye	13	
Ammonia	12	
	11	
Milk of Magnesia	10	
	9	
Seawater, baking soda Lake Ontario	8	
Blood		
NEUTRAL	7	NEUTRAL
Milk		
Mean pH of Adirondack Lakes, 1930 "Clean" rain	6	
Mean pH of Adirondack Lakes, 1975	5	ACID RAIN
Tomato juice		Increasingly acidic
Average pH of rainfall, Toronto, Feb. 1979	4	
Apple juice Vinegar Lemon Juice	3	
Most acidic rainfall recorded in U.S.	2	
Battery Acid	1	
	0	

7. Find the concentration of the hydrogen ion in a can of cola if the pH is 4.1.

8. The concentration of hydrogen ions in a sample of acid rain is 1.8×10^{-5}. Find the pH. Round to the nearest tenth.

Answers

6. 3.16×10^5
7. 7.9×10^{-5}
8. 4.7

We can assume that all our properties of exponents and logarithms hold for expressions with a base of e, because e is a real number. Here are some examples intended to make you more familiar with the number e and natural logarithms.

9. Simplify each expression.
  a. $\ln e^2$
  b. $\ln e^4$
  c. $\ln e^{-2}$
  d. $\ln e^x$

EXAMPLE 9   Simplify each of the following expressions.

a. $e^0 = 1$

b. $e^1 = e$

c. $\ln e = 1$ *In exponential form, $e^1 = e$.*

d. $\ln 1 = 0$ *In exponential form, $e^0 = 1$.*

e. $\ln e^3 = 3$

f. $\ln e^{-4} = -4$

g. $\ln e^t = t$

10. Expand $\ln Pe^{rt}$.

EXAMPLE 10   Use the properties of logarithms to expand the expression $\ln (Ae^{5t})$.

Solution   Because the properties of logarithms hold for natural logarithms, we have

$$\ln (Ae^{5t}) = \ln A + \ln e^{5t}$$
$$= \ln A + 5t \ln e$$
$$= \ln A + 5t \qquad \text{Because } \ln e = 1$$

11. If $\ln 5 = 1.6094$ and $\ln 7 = 1.9459$, find the following:
  a. $\ln 35$
  b. $\ln 0.2$
  c. $\ln 49$

EXAMPLE 11   If $\ln 2 = 0.6931$ and $\ln 3 = 1.0986$, find

  a. $\ln 6$ **b.** $\ln 0.5$ **c.** $\ln 8$

Solution

a. Because $6 = 2 \cdot 3$, we have

$$\ln 6 = \ln 2 \cdot 3$$
$$= \ln 2 + \ln 3$$
$$= 0.6931 + 1.0986$$
$$= 1.7917$$

b. Writing 0.5 as $\frac{1}{2}$ and applying the quotient property for logarithms gives us

$$\ln 0.5 = \ln \frac{1}{2}$$
$$= \ln 1 - \ln 2$$
$$= 0 - 0.6931$$
$$= -0.6931$$

c. Writing 8 as 2^3 and applying the power property for logarithms, we have

$$\ln 8 = \ln 2^3$$
$$= 3 \ln 2$$
$$= 3(0.6931)$$
$$= 2.0793$$

GETTING READY FOR CLASS

After reading through the preceding section, respond in your own words and in complete sentences.

A. What is a common logarithm?

B. What is a natural logarithm?

C. Is e a rational number? Explain.

D. Find $\ln e$, and explain how you arrived at your answer.

Answers
9. a. 2   **b.** 4   **c.** -2 **d.** x
10. $\ln P + rt$
11. a. 3.5553   **b.** -1.6094 **c.** 3.8918

Vocabulary Review

Choose the correct words to fill in the blanks below.

natural   common   characteristic   mantissa

1. A _____ logarithm is a logarithm with a base of 10 and denoted as $\log_{10} x = \log x$.

2. When a logarithm is calculated as a decimal, the whole number part of the logarithm is called the _____ and the decimal part is called the _____.

3. A _____ logarithm is a logarithm with a base of e and denoted as $\ln x = \log_e x$.

Problems

A Find the following logarithms.

1. $\log 378$

2. $\log 426$

3. $\log 37.8$

4. $\log 42{,}600$

5. $\log 3{,}780$

6. $\log 0.4260$

7. $\log 0.0378$

8. $\log 0.0426$

9. $\log 37{,}800$

10. $\log 4{,}900$

11. $\log 600$

12. $\log 900$

13. $\log 2{,}010$

14. $\log 10{,}200$

15. $\log 0.00971$

16. $\log 0.0312$

17. $\log 0.0314$

18. $\log 0.00052$

19. $\log 0.399$

20. $\log 0.111$

B Find x in the following equations.

21. $\log x = 2.8802$

22. $\log x = 4.8802$

23. $\log x = -2.1198$

24. $\log x = -3.1198$

25. $\log x = 3.1553$

26. $\log x = 5.5911$

27. $\log x = -5.3497$

28. $\log x = -1.5670$

SCAN TO ACCESS

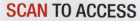

29. $\log x = -7.0372$

30. $\log x = -4.2000$

31. $\log x = 10$

32. $\log x = -1$

33. $\log x = -10$

34. $\log x = 1$

35. $\log x = 20$

36. $\log x = -20$

37. $\log x = -2$

38. $\log x = 4$

39. $\log x = \log_2 8$

40. $\log x = \log_3 9$

41. $\ln x = -1$

42. $\ln x = 4$

43. $\log x = 2 \log 5$

44. $\log x = -\log 4$

45. $\ln x = -3 \ln 2$

46. $\ln x = 5 \ln 3$

47. $\ln x = 4 \ln 3$

48. $\ln x = -2 \ln 5$

D Simplify each of the following expressions.

49. $\ln e$

50. $\ln 1$

51. $\ln e^5$

52. $\ln e^{-3}$

53. $\ln e^x$

54. $\ln e^y$

55. $\log 10{,}000$

56. $\log 0.001$

57. $\ln \dfrac{1}{e^3}$

58. $\ln \sqrt{e}$

59. $\log \sqrt{1000}$

60. $\log \sqrt[3]{10{,}000}$

Use the properties of logarithms to expand each of the following expressions.

61. $\ln (10e^{3t})$

62. $\ln (10e^{4t})$

63. $\ln (Ae^{-2t})$

64. $\ln (Ae^{-3t})$

65. $\log [100(1.01)^{3t}]$

66. $\log \left[\dfrac{1}{10} (1.5)^{t+2} \right]$

67. $\ln (Pe^{rt})$

68. $\ln \left(\dfrac{1}{2} e^{-kt} \right)$

69. $-\log (4.2 \times 10^{-3})$

70. $-\log (5.7 \times 10^{-10})$

71. $-\log (3.4 \times 10^4)$

72. $-\log (5.6 \times 10^8)$

If $\ln 2 = 0.6931$, $\ln 3 = 1.0986$, and $\ln 5 = 1.6094$, find each of the following.

73. $\ln 15$

74. $\ln 10$

75. $\ln \dfrac{1}{3}$

76. $\ln \dfrac{1}{5}$

77. $\ln 9$

78. $\ln 25$

79. $\ln 16$

80. $\ln 81$

Applying the Concepts

81. Atomic Bomb Tests The Bikini Atoll in the Pacific Ocean was used as a location for atomic bomb tests by the United States government in the 1950s. One such test resulted in an earthquake measurement of 5.0 on the Richter scale. Compare the 1906 San Francisco earthquake of estimated magnitude 8.3 on the Richter scale to this atomic bomb test. Use the shock wave T for purposes of comparison. The formula for determining the magnitude, M, of an earthquake on the Richter Scale is $M = \log_{10} T$, where T is the number of times the shockwave is greater than the smallest measurable shockwave.

82. Atomic Bomb Tests Today's nuclear weapons are 1,000 times more powerful than the atomic bombs tested in the Bikini Atoll mentioned in Problem 81. Use the shock wave T to determine the Richter scale measurement of a nuclear test today.

83. On March 11, 2011, one of the largest natural disasters in history occurred on the coast of Tohoku, Japan. An earthquake, which measured 9.0 on the Richter scale, and the resulting tsunami killed over 20,000 people. Below are the four largest earthquakes since 1900. For each, use the formula $T = 10^m$ to convert each magnitude to the level of energy of the quake.

Date	Location	Magnitude M	Energy Level T
5/22/1960	Chile	9.5	
3/28/1964	Alaska	9.2	
12/26/2004	Indonesia	9.1	
3/11/2011	Japan	9.0	

Source: *USGS*

84. For each of these earthquakes listed in problem 83, find how many times more energy they released compared to the 1906 San Francisco earthquake of magnitude 8.3.

85. Getting Close to *e* Use a calculator to complete the following table.

x	$(1 + x)^{1/x}$
1	
0.5	
0.1	
0.01	
0.001	
0.0001	
0.00001	

86. Getting Close to *e* Use a calculator to complete the following table.

x	$\left(1 + \dfrac{1}{x}\right)^x$
1	
10	
50	
100	
500	
1,000	
10,000	
1,000,000	

87. What number does the expression $(1 + x)^{1/x}$ seem to approach as x gets closer and closer to zero?

88. What number does the expression $\left(1 + \frac{1}{x}\right)^x$ seem to approach as x gets larger and larger?

89. University Enrollment The percentage of students enrolled in a university who are between the ages of 25 and 34 can be modeled by the formula $s = 5 \ln x$, where s is the percentage of students and x is the number of years since 1989. Predict the year in which approximately 15% of students enrolled in a university were between the ages of 25 and 34.

90. Memory A class of students take a test on the mathematics concept of solving quadratic equations. That class agrees to take a similar form of the test each month for the next 6 months to test their memory of the topic since instruction. The function of the average score earned each month on the test is $m(x) = 75 - 5 \ln(x + 1)$, where x represents time in months. Complete the table to indicate the average score earned by the class at each month.

Time x	Score m
0	
1	
2	
3	
4	
5	
6	

Use the following figure to solve Problems 91–94.

pH Scale

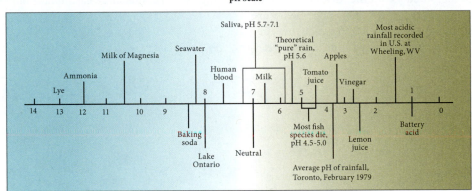

91. pH Find the pH of orange juice if the concentration of the hydrogen ion in the juice is $[\text{H}^+] = 6.50 \times 10^{-4}$.

92. pH Find the pH of milk if the concentration of the hydrogen ions in milk is $[\text{H}^+] = 1.88 \times 10^{-6}$.

93. pH Find the concentration of hydrogen ions in a glass of wine if the pH is 4.75.

94. pH Find the concentration of hydrogen ions in a bottle of vinegar if the pH is 5.75.

The Richter Scale Find the relative size T of the shock wave of earthquakes with the following magnitudes, as measured on the Richter scale.

95. 5.5

96. 6.6

97. 8.3

98. 8.7

99. Earthquake The chart below is a partial listing of earthquakes that were recorded in Canada during 2000. Complete the chart by computing the magnitude on the Richter Scale M or the number of times the associated shockwave is larger than the smallest measurable shockwave T.

Location	Date	Magnitude M	Shockwave T
Moresby Island	Jan. 23	4.0	
Vancouver Island	Apr. 30		1.99×10^5
Quebec City	June 29	3.2	
Mould Bay	Nov. 13	5.2	
St. Lawrence	Dec. 14		5.01×10^3

Source: *National Resources Canada, National Earthquake Hazards Program*

100. Earthquake On January 6, 2001, an earthquake with a magnitude of 7.7 on the Richter Scale hit southern India (*National Earthquake Information Center*). By what factor was this earthquake's shockwave greater than the smallest measurable shockwave?

Depreciation The annual rate of depreciation r on a car that is purchased for P dollars and is worth W dollars t years later can be found from the formula

$$\log(1 - r) = \frac{1}{t} \log \frac{W}{P}$$

101. Find the annual rate of depreciation on a car that is purchased for \$9,000 and sold 5 years later for \$4,500.

102. Find the annual rate of depreciation on a car that is purchased for \$9,000 and sold 4 years later for \$3,000.

Two cars depreciate in value according to the following depreciation tables. In each case, find the annual rate of depreciation.

103.

Age in years	Value in dollars
New	7,550
5	5,750

104.

Age in years	Value in dollars
New	7,550
3	5,750

Getting Ready for the Next Section

Solve.

105. $5(2x + 1) = 12$

106. $4(3x - 2) = 21$

107. $3^x = \dfrac{1}{27}$

108. $2^x = \dfrac{1}{8}$

Use a calculator to evaluate the following. Give answers to 4 decimal places.

109. $\dfrac{100{,}000}{32{,}000}$

110. $\dfrac{1.4982}{6.5681} + 3$

111. $\dfrac{1}{2}\left(\dfrac{-0.6931}{1.4289} + 3\right)$

112. $1 + \dfrac{0.04}{52}$

Use the power property to rewrite the following logarithms.

113. $\log 1.05^t$

114. $\log 1.033^t$

115. $\ln 1.045^{2t}$

116. $\ln 1.016^{10t}$

Use identities to simplify.

117. $\ln e^{0.05t}$

118. $\ln e^{-0.000121t}$

119. $\ln e^{-0.0042t}$

120. $\ln e^{0.0046t}$

Use a calculator to evaluate the following. Give answers to 2 decimal places.

121. $\dfrac{\ln 10}{0.0015}$

122. $\dfrac{\ln 2}{0.00125}$

123. $\dfrac{\ln 2}{\ln 1.043}$

124. $\dfrac{\ln 3}{\ln\left(1 + \dfrac{.08}{12}\right)}$

Find the Mistake

Each problem below contains a mistake. Circle the mistake and write the correct number(s) or word(s) on the line provided.

1. A common logarithm uses the expression $\log_{10} x$, which is equal to $\log x^{10}$. _____

2. When the base of a log is not shown, it is assumed to be 1. _____

3. When the logarithm of a whole number is approximated as a decimal, the decimal part of the answer is called the characteristic. _____

4. An antilogarithm is a logarithm with a base of e, denoted by $\ln x = \log_e x$. _____

OBJECTIVES

A   Solve exponential equations.

B   Use the change-of-base property to calculate logarithms.

C   Solve application problems involving logarithmic or exponential equations.

KEY WORDS

doubling time

exponential equation

change-of-base property

For items involved in exponential growth, the time it takes for a quantity to double is called the *doubling time*. For example, if you invest $5,000 in an account that pays 5% annual interest, compounded quarterly, you may want to know how long it will take for your money to double in value. You can find this doubling time if you can solve the equation

$$10,000 = 5,000 \, (1.0125)^{4t}$$

You will see as you progress through this section that logarithms are the key to solving equations of this type.

A Exponential Equations

Logarithms are very important in solving equations in which the variable appears as an exponent. The equation

$$5^x = 12$$

is an example of one such equation. Equations of this form are called *exponential equations*. Because the quantities 5^x and 12 are equal, so are their common logarithms.

We begin our solution by taking the logarithm of both sides.

$$\log 5^x = \log 12$$

We now apply the power property for logarithms, $\log x^r = r \log x$, to turn x from an exponent into a coefficient.

$$x \log 5 = \log 12$$

Dividing both sides by $\log 5$ gives us

$$x = \frac{\log 12}{\log 5}$$

If we want a decimal approximation to the solution, we can find $\log 12$ and $\log 5$ on a calculator and divide.

$$x \approx \frac{1.0792}{0.6990} \approx 1.5439$$

> **Note**   It is common to mistake the equation $y = x^2$ as an exponential equation. Yes, the equation has an exponent but it is not the variable appearing as such. The equation $y = 2^x$ is an exponential equation because the variable x appears as the exponent.

The complete problem looks like this:

$$5^x = 12$$

$$\log 5^x = \log 12$$

$$x \log 5 = \log 12$$

$$x = \frac{\log 12}{\log 5}$$

$$\approx \frac{1.0792}{0.6990}$$

$$\approx 1.5439$$

Here is another example of solving an exponential equation using logarithms.

EXAMPLE 1   Solve $25^{2x+1} = 15$.

Solution   Taking the logarithm of both sides and then writing the exponent $(2x + 1)$ as a coefficient, we proceed as follows:

$$25^{2x+1} = 15$$

$\log 25^{2x+1} = \log 15$	*Take the log of both sides.*
$(2x + 1)\log 25 = \log 15$	*Power property for logarithms*
$2x + 1 = \dfrac{\log 15}{\log 25}$	*Divide by log 25.*
$2x = \dfrac{\log 15}{\log 25} - 1$	*Add −1 to both sides.*
$x = \dfrac{1}{2}\left(\dfrac{\log 15}{\log 25} - 1\right)$	*Multiply both sides by $\frac{1}{2}$.*

Using a calculator, we can write a decimal approximation to the answer.

$$x \approx \frac{1}{2}\left(\frac{1.1761}{1.3979} - 1\right)$$

$$\approx \frac{1}{2}(0.8413 - 1)$$

$$\approx \frac{1}{2}(-0.1587)$$

$$\approx -0.079$$

B Change of Base

There is a fourth property of logarithms we have not yet considered. This last property allows us to change from one base to another and is therefore called the *change-of-base property*.

> **PROPERTY** Change-of-Base Property for Logarithms
>
> If a and b are both positive numbers other than 1, and if $x > 0$, then
>
> $$\log_a x = \frac{\log_b x}{\log_b a}$$
>
> $\uparrow$ $\uparrow$
>
> Base a Base b

The logarithm on the left side has a base of a, and both logarithms on the right side have a base of b. This allows us to change from base a to any other base b that is a positive number other than 1. Here is a proof of the change-of-base property for logarithms:

Practice Problems

1. Solve $12^{x+2} = 20$.

Answer

1. -0.7945 or -0.744 depending on whether you round your intermediate answers.

Proof   We begin by writing the identity

$$a^{\log_a x} = x$$

Taking the logarithm base b of both sides and writing the exponent $\log_a x$ as a coefficient, we have

$$\log_b a^{\log_a x} = \log_b x$$

$$(\log_a x)(\log_b a) = \log_b$$

Dividing both sides by $\log_b a$, we have the desired result:

$$\frac{(\log_a x)(\log_b a)}{\log_b a} = \frac{\log_b x}{\log_b a}$$

$$\log_a x = \frac{\log_b x}{\log_b a}$$

We can use this property to find logarithms we could not otherwise compute on our calculators — that is, logarithms with bases other than 10 or e. The next example illustrates the use of this property.

EXAMPLE 2   Calculate $\log_8 24$.

Solution   Because we do not have base-8 logarithms on our calculators, we can change this expression to an equivalent expression that contains only base-10 logarithms.

$$\log_8 24 = \frac{\log 24}{\log 8} \qquad \textcolor{green}{\textit{Change-of-base property}}$$

Don't be confused. We did not just drop the base, we changed to base 10. We could have written the last line like this:

$$\log_8 24 = \frac{\log_{10} 24}{\log_{10} 8}$$

From our calculators, we write

$$\log_8 24 \approx \frac{1.3802}{0.9031} \approx 1.5283$$

C Applications

If you invest P dollars in an account with an annual interest rate r that is compounded n times a year, then t years later the amount of money in that account will be

$$A = P\left(1 + \frac{r}{n}\right)^{nt}$$

EXAMPLE 3   How long does it take for $5,000 to double if it is deposited in an account that yields 5% interest compounded once a year?

Solution   Substituting $P = 5,000$, $r = 0.05$, $n = 1$, and $A = 10,000$ into our formula, we have

$$10,000 = 5,000(1 + 0.05)^t$$

$$10,000 = 5,000(1.05)^t$$

$$2 = (1.05)^t \qquad \textcolor{green}{\textit{Divide by 5,000.}}$$

This is an exponential equation. We solve by taking the logarithm of both sides.

$$\log 2 = \log(1.05)^t$$

$$\log 2 = t \log 1.05$$

$$t = \frac{\log 2}{\log 1.05} \qquad \textcolor{green}{\textit{Divide both sides by log 1.05.}}$$

$$\approx 14.2$$

It takes a little over 14 years for $5,000 to double if it earns 5% interest per year, compounded once a year.

2. Calculate $\log_6 14$.

3. How long does it take for $5,000 to double if it is invested in an account that pays 11% interest compounded once a year?

Answers

2. 1.4728 or 1.4729

3. 6.64 years

4. How long will it take the city in Example 4 to grow to 75,000 people from its current population of 32,000?

EXAMPLE 4   Suppose that the population of a small city is 32,000 in the beginning of 2010. The city council assumes that the population size t years later can be estimated by the equation

$$P = 32{,}000e^{0.05t}$$

Approximately when will the city have a population of 50,000?

Solution   We substitute 50,000 for P in the equation and solve for t.

$$50{,}000 = 32{,}000e^{0.05t}$$

$$1.5625 = e^{0.05t} \qquad \frac{50{,}000}{32{,}000} = 1.5625$$

To solve this equation for t, we can take the natural logarithm of each side.

$$\ln 1.5625 = \ln e^{0.05t}$$

$$= 0.05t \ln e \qquad \text{Power property for logarithms}$$

$$= 0.05t \qquad \text{Because } \ln e = 1$$

$$t = \frac{\ln 1.5625}{0.05} \qquad \text{Divide each side by 0.05.}$$

$$\approx 8.93 \text{ years}$$

We can estimate that the population will reach 50,000 toward the end of 2018.

USING TECHNOLOGY Graphing Calculators

We can evaluate many logarithmic expressions on a graphing calculator by using the fact that logarithmic functions and exponential functions are inverses.

5. Evaluate the logarithmic expression $\log_5 8$ from the graph of an exponential function.

EXAMPLE 5   Evaluate the logarithmic expression $\log_3 7$ from the graph of an exponential function.

Solution   First, we let $\log_3 7 = x$. Next, we write this expression in exponential form as $3^x = 7$. We can solve this equation graphically by finding the intersection of the graphs $Y_1 = 3^x$ and $Y_2 = 7$, as shown in Figure 1.

Using the calculator, we find the two graphs intersect at $(1.77, 7)$. Therefore, $\log_3 7 = 1.77$ to the nearest hundredth. We can check our work by evaluating the expression $3^{1.77}$ on our calculator with the key strokes

$$3 \; \boxed{\wedge} \; 1.77 \; \boxed{\text{ENTER}}$$

The result is 6.99 to the nearest hundredth, which seems reasonable since 1.77 is accurate to the nearest hundredth. To get a result closer to 7, we would need to find the intersection of the two graphs more accurately.

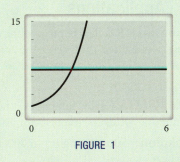

FIGURE 1

GETTING READY FOR CLASS

After reading through the preceding section, respond in your own words and in complete sentences.

A. What is an exponential equation?

B. How do logarithms help you solve exponential equations?

C. What is the change-of-base property?

D. Write an application modeled by the equation $A = 10{,}000\left(1 + \frac{0.08}{2}\right)^{(2)(5)}$.

Answers

4. Approximately 17.04 years

5. 1.29

Vocabulary Review

Choose the correct words to fill in the blanks below.

variable   logarithms   exponential   change-of-base   doubling time

1. When considering _____ growth, the time it takes for a quantity to double is called the _____.

2. An equation where the _____ appears as an exponent is called an exponential equation.

3. We use _____ to help solve exponential equations.

4. The _____ property states that $\log_a x = \dfrac{(\log_b s)}{(\log_b a)}$.

Problems

A Solve each exponential equation. Use a calculator to write the answer in decimal form. Round answers to four decimal places where appropriate.

1. $3^x = 5$

2. $4^x = 3$

3. $5^x = 3$

4. $3^x = 4$

5. $5^{-x} = 12$

6. $7^{-x} = 8$

7. $12^{-x} = 5$

8. $8^{-x} = 7$

9. $8^{x+1} = 4$

10. $9^{x+1} = 3$

11. $4^{x-1} = 4$

12. $3^{x-1} = 9$

13. $3^{2x+1} = 2$

14. $2^{2x+1} = 3$

15. $3^{1-2x} = 2$

16. $2^{1-2x} = 3$

17. $15^{3x-4} = 10$

18. $10^{3x-4} = 15$

19. $6^{5-2x} = 4$

20. $9^{7-3x} = 5$

21. $3^{-4x} = 81$

22. $2^{5x} = \dfrac{1}{16}$

23. $5^{3x-2} = 15$

24. $7^{4x+3} = 200$

25. $64^x = 16^{x-3}$

26. $81^x = 27^{x-4}$

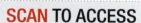

27. $8^a = 16^{2a+5}$

28. $9^y = 81^{2y-3}$

29. $16^{x+1} = 8^{3x-2}$

30. $32^{x+3} = 4^{3x-2}$

31. $27^{3b-5} = 81^{2b-3}$

32. $16^{3x-3} = 128^{2x+4}$

33. $100e^{3t} = 250$

34. $150e^{0.065t} = 400$

35. $1200\left(1 + \dfrac{0.072}{4}\right)^{4t} = 25000$

36. $2700\left(1 + \dfrac{0.086}{12}\right)^{12t} = 10000$

37. $50e^{-0.0742t} = 32$

38. $19e^{-0.000243t} = 12$

B Use the change-of-base property to calculate each of the following logarithms. Round answers to four decimal places, where appropriate.

39. $\log_8 16$

40. $\log_9 27$

41. $\log_{16} 8$

42. $\log_{27} 9$

43. $\log_7 15$

44. $\log_3 12$

45. $\log_{15} 7$

46. $\log_{12} 3$

47. $\log_8 240$

48. $\log_6 180$

49. $\log_4 321$

50. $\log_5 462$

Find a decimal approximation to each of the following natural logarithms.

51. $\ln 345$

52. $\ln 3,450$

53. $\ln 0.345$

54. $\ln 0.0345$

55. $\ln 10$

56. $\ln 100$

57. $\ln 45,000$

58. $\ln 450,000$

Applying the Concepts

Use the formula $A = P\left(1 + \dfrac{r}{n}\right)^{nt}$ to solve the following compound interest problems. Round all answers to two decimal places.

59. Compound Interest How long will it take for $500 to double if it is invested at 6% annual interest compounded 2 times a year?

60. Compound Interest How long will it take for $500 to double if it is invested at 6% annual interest compounded 12 times a year?

61. Compound Interest How long will it take for $1,000 to triple if it is invested at 12% annual interest compounded 6 times a year?

62. Compound Interest How long will it take for $1,000 to become $4,000 if it is invested at 12% annual interest compounded 6 times a year?

63. Doubling Time How long does it take for an amount of money P to double itself if it is invested at 8% interest compounded 4 times a year?

64. Tripling Time How long does it take for an amount of money P to triple itself if it is invested at 8% interest compounded 4 times a year?

65. Tripling Time If a $25 investment is worth $75 today, how long ago must that $25 have been invested at 6% interest compounded twice a year?

66. Doubling Time If a $25 investment is worth $50 today, how long ago must that $25 have been invested at 6% interest compounded twice a year?

Recall from Section 8.1 that if P dollars are invested in an account with annual interest rate r, compounded continuously, then the amount of money in the account after t years is given by the formula

$$A(t) = Pe^{rt}$$

67. Continuously Compounded Interest Repeat Problem 51 if the interest is compounded continuously.

68. Continuously Compounded Interest Repeat Problem 54 if the interest is compounded continuously.

69. Continuously Compounded Interest How long will it take $500 to triple if it is invested at 6% annual interest, compounded continuously? Round your answer to two decimal places.

70. Continuously Compounded Interest How long will it take $500 to triple if it is invested at 12% annual interest, compounded continuously? Round your answer to two decimal places.

71. Continuously Compounded Interest How long will it take for $1,000 to be worth $2,500 at 8% interest, compounded continuously? Round your answer to two decimal places.

72. Continuously Compounded Interest How long will it take for $1,000 to be worth $5,000 at 8% interest, compounded continuously? Round your answer to two decimal places.

73. Exponential Growth Suppose that the population in a small city is 32,000 at the beginning of 2005 and that the city council assumes that the population size t years later can be estimated by the equation

$$P(t) = 32,000e^{0.05t}$$

In what month and year will the city have a population of 64,000?

74. Exponential Growth Suppose the population of a city is given by the equation

$$P(t) = 100,000e^{0.05t}$$

where t is the number of years from the present time. How large is the population now? ("Now" corresponds to a certain value of t. Once you realize what that value of t is, the problem becomes very simple.)

75. Airline Travel The number of airline passengers in 1990 was 466 million. The number of passengers traveling by airplane each year has increased exponentially according to the model, $P(t) = 466 \cdot 1.035^t$, where t is the number of years since 1990 (U.S. Census Bureau). In what year is it predicted that 900 million passengers will travel by airline?

76. Bankruptcy Model In 1997, there were a total of 1,316,999 bankruptcies filed under the Bankruptcy Reform Act. The model for the number of bankruptcies filed is $B(t) = 0.798 \cdot 1.164^t$, where t is the number of years since 1994 and B is the number of bankruptcies filed in terms of millions (Administrative Office of the U.S. Courts, *Statistical Tables for the Federal Judiciary*). In what year is it predicted that 12 million bankruptcies will be filed?

77. Health Care In 1990, $699 billion was spent on health care expenditures. The amount of money, E, in billions spent on health care expenditures can be estimated using the function $E(t) = 78.16(1.11)^t$, where t is time in years since 1970 (*U.S. Census Bureau*). In what year was it estimated that $800 billion would be spent on health care expenditures?

78. Value of a Car As a car ages, its value decreases. The value of a particular car with an original purchase price of $25,600 is modeled by the function $c(t) = 25,600(1 - 0.22)^t$, where c is the value at time t (Kelly Blue Book). How old is the car when its value is $10,000? Round your answer to two decimal places.

79. Compound Interest In 1991, the average cost of attending a public university through graduation was $20,972 (U.S. Department of Education, National Center for Educational Statistics). If John's parents deposited that amount in an account in 1991 at an interest rate of 7% compounded semi-annually, how long will it take for the money to double? Round your answer to two decimal places.

80. Carbon Dating Scientists use Carbon-14 dating to find the age of fossils and other artifacts. The amount of Carbon-14 in an organism will yield information concerning its age. A formula used in Carbon-14 dating is $A(t) = A_0 \cdot 2^{-t/5600}$, where A_0 is the amount of carbon originally in the organism, t is time in years, and A is the amount of carbon remaining after t years. Determine the number of years since an organism died if it originally contained 1,000 gram of Carbon-14 and it currently contains 600 gram of Carbon-14. Round to the nearest year.

81. **Cost Increase** The cost of a can of Coca Cola in 1960 was $0.10. The function that models the cost of a Coca Cola by year is $C(t) = 0.10e^{0.0576t}$, where t is the number of years since 1960. In what year was it expected that a can of Coca Cola will cost $1.00?

82. **Online Banking Use** The number of households using online banking services has increased from 754,000 in 1995 to 12,980,000 in 2000. The formula $H(t) = 0.76e^{0.55t}$ models the number of households H in millions when time is t years since 1995 (Home Banking Report). In what year was it estimated that 50,000,000 households would use online banking services?

Newtons Law of Cooling states that the temperature of an object after t minutes is given by the formula $T(t) = A + (T_o - A)e^{-kt}$, where A is the ambient (outside) temperature, T_o is the initial temperature of the object, and k is a constant. Use this formula to solve the following problems.

83. Coffee which is 200°F is poured into a coffee cup held in a room with temperature 75°F.

  a. If the coffee cools to 180°F in 5 minutes, find k.

  b. What will the coffee temperature be after 15 minutes using your value for k?

  c. When will the temperature of the coffee be 100°F using your value for k?

84. A cold soda which is 37°F is put on a table in a room with temperature 80°F

  a. If the soda heats to 45°F in 4 minutes, find k.

  b. What will the soda temperature be after 10 minutes using your value for k?

  c. When will the temperature of the soda be 70°F using your value for k?

Maintaining Your Skills

Find the vertex for each of the following parabolas, and then indicate if it is the highest or lowest point on the graph.

85. $y = 2x^2 + 8x - 15$

86. $y = 3x^2 - 9x - 10$

87. $y = 12x - 4x^2$

88. $y = 18x - 6x^2$

89. **Maximum Height** An object is projected into the air with an initial upward velocity of 64 feet per second. Its height h at any time t is given by the formula $h = 64t - 16t^2$. Find the time at which the object reaches its maximum height. Then find the maximum height.

90. **Maximum Height** An object is projected into the air with an initial upward velocity of 64 feet per second from the top of a building 40 feet high. If the height h of the object t seconds after it is projected into the air is $h = 40 + 64t - 16t^2$, find the time at which the object reaches its maximum height. Then find the maximum height it attains.

Find the Mistake

Each problem below contains a mistake. Circle the mistake and write the correct number(s) or word(s) on the line provided.

1. The equation $5^x = 12$ is an example of a logarithmic equation. _____

2. The change-of-base property for logarithms states that $\log_a x = \dfrac{\log_b a}{\log_b x}$. _____

3. For the change-of-base property to work, base a and base b must both be negative numbers. _____

4. To calculate $\log_9 36$, change the base to 10 and find the product of log 36 and log 9. _____

Math in Real Life

As you can see from the various application problems in this chapter, there are a number of ways people use logarithms in real life. Most people don't realize how often we use math skills in our daily lives. The obvious skills are used doing things such as paying for groceries, calculating supplies for a home project, or cooking from a recipe. This project will help shed light on some lesser known ways of using math. Choose an occupation from the list below. Research what and how math skills are used in the chosen occupation. Where applicable, include math concepts you have encountered in this book. Present your findings to the class.

Accountant	Lawyer
Agriculturist	Manager
Biologist	Marketer
Carpenter	Nurse or Doctor
Chemist	Meteorologist
Computer Programmer	Politician
Engineer	Professor or Teacher
Geologist	Repair Technician
Graphic Designer	(e.g., plumber, electrician, or mechanic)

Bonus Assignment: Even if you are not using a specific math concept in your daily life, the process of learning it helps improve your overall critical thinking and problem solving skills. Explain why this is important.

Chapter 8 Summary

Exponential Functions [8.1]

Any function of the form

$$f(x) = b^x$$

where $b > 0$ and $b \neq 1$, is an *exponential function*.

EXAMPLES

1. For the exponential function $f(x) = 2^x$,
$$f(0) = 2^0 = 1$$
$$f(1) = 2^1 = 2$$
$$f(2) = 2^2 = 4$$
$$f(3) = 2^3 = 8$$

One-to-One Functions [8.2]

A function is a *one-to-one function* if every element in the range comes from exactly one element in the domain.

2. The function $f(x) = x^2$ is not one-to-one because 9, which is in the range, comes from both 3 and -3 in the domain.

Inverse Functions [8.2]

The *inverse* of a function is obtained by reversing the order of the coordinates of the ordered pairs belonging to the function. Only one-to-one functions have inverses that are also functions.

3. The inverse of $f(x) = 2x - 3$ is
$$f^{-1}(x) = \frac{x + 3}{2}$$

Definition of Logarithms [8.3]

If b is a positive number not equal to 1, then the expression

$$y = \log_b x$$

is equivalent to $x = b^y$; that is, in the expression $y = \log_b x$, y is the number to which we raise b in order to get x. Expressions written in the form $y = \log_b x$ are said to be in *logarithmic form*. Expressions like $x = b^y$ are in *exponential form*.

4. The definition allows us to write expressions like
$$y = \log_3 27$$
equivalently in exponential form as
$$3^y = 27$$
which makes it apparent that y is 3.

Two Special Identities [8.3]

For $b > 0$, $b \neq 1$, the following two expressions hold for all positive real numbers x:

1. $b^{\log_b x} = x$

2. $\log_b b^x = x$

5. Examples of the two special identities are
$$5^{\log_5 12} = 12$$
and
$$\log_8 8^3 = 3$$

Properties of Logarithms [8.4]

If x, y, and b are positive real numbers, $b \neq 1$, and r is any real number, then

1. $\log_b(xy) = \log_b x + \log_b y$ *Product property for logarithms*

2. $\log_b\left(\dfrac{x}{y}\right) = \log_b x - \log_b y$ *Quotient property for logarithms*

3. $\log_b x^r = r \log_b x$ *Power property for logarithms*

6. We can rewrite the expression
$$\log_{10} \frac{45^6}{273}$$
using the properties of logarithms, as
$$6 \log_{10} 45 - \log_{10} 273$$

Common Logarithms [8.5]

7. $\log_{10} 10{,}000 = \log 10{,}000$
$= \log 10^4$
$= 4$

Common logarithms are logarithms with a base of 10. To save time in writing, we omit the base when working with common logarithms; that is,

$$\log x = \log_{10} x$$

Natural Logarithms [8.5]

8. $\ln e = 1$
$\ln 1 = 0$

Natural logarithms, written $\ln x$, are logarithms with a base of e, where the number e is an irrational number (like the number π). A decimal approximation for e is 2.7183. All the properties of exponents and logarithms hold when the base is e.

Change of Base [8.6]

9. $\log_6 475 = \dfrac{\log 475}{\log 6}$

$\approx \dfrac{2.6767}{0.7782}$

≈ 3.44

If x, a, and b are positive real numbers, $a \neq 1$ and $b \neq 1$, then

$$\log_a x = \frac{\log_b x}{\log_b a}$$

COMMON MISTAKE

The most common mistakes that occur with logarithms come from trying to apply the properties of logarithms to situations in which they don't apply. For example, a very common mistake looks like this:

$$\frac{\log 3}{\log 2} = \log 3 - \log 2 \qquad \text{Mistake}$$

This is not a property of logarithms. To write the equation $\log 3 - \log 2$, we would have to start with

$$\log \frac{3}{2} \qquad NOT \qquad \frac{\log 3}{\log 2}$$

There is a difference.

Graph each exponential function. [8.1]

1. $f(x) = -5^x$

2. $g(x) = 3^{-x}$

Sketch the graph of each function and its inverse. Find $f^{-1}(x)$ for Problem 3. [8.2]

3. $f(x) = \frac{1}{3}x + \frac{4}{3}$

4. $f(x) = x^2 - 2$

Solve for x. [8.3]

5. $\log_4 x = 3$

6. $\log_x 16 = 4$

Graph each of the following. [8.3]

7. $y = \log_4 x$

8. $y = \log_{1/4} x$

Evaluate each of the following. Round to the nearest hundredth if necessary. [8.3, 8.4, 8.5]

9. $\log_{16} 4$

10. $\log_8 24$

11. $\log 57{,}300$

12. $\log 0.0507$

13. $\ln 27.4$

14. $\ln 0.024$

Use the properties of logarithms to expand each expression. [8.4]

15. $\log_3 \dfrac{5x^3}{y}$

16. $\log \dfrac{\sqrt{x}}{y^5 \sqrt[3]{z}}$

Write each expression as a single logarithm. [8.4]

17. $\dfrac{1}{2} \log_4 x + 3 \log_4 y$

18. $3 \log x - 4\log y + \dfrac{1}{2} \log z$

Use a calculator to find x. Round to the nearest thousandth. [8.5]

19. $\log x = 2.6532$

20. $\log x = -1.2518$

Solve for x. Round to the nearest thousandth if necessary. [8.4, 8.6]

21. $5 = 2^x$

22. $4^{2x-3} = 64$

23. $\log_4 x + \log_4 2 = 3$

24. $\log_2 x + \log_2 (x - 2) = 3$

25. Chemistry Use the formula $\text{pH} = -\log[\text{H}^+]$ to find the pH of a solution in which $[\text{H}^+] = 9.3 \times 10^{-6}$. [8.5]

26. Compound Interest If $400 is deposited in an account that earns 6% annual interest compounded quarterly, how much money will be in the account after 7 years? [8.1]

27. Compound Interest How long will it take $600 to become $2,400 if the $600 is deposited in an account that earns 9% annual interest compounded quarterly? [8.6]

Determine if the function is one-to-one. [8.2]

28. $f(x) = \{(3, 6), (4, 2), (5, -3), (6, 2)\}$

29. $g(x) = \{(-1, 2), (0, 4), (2, 8), (3, 10)\}$

30.

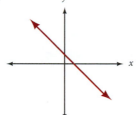

31.

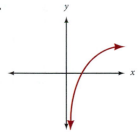

32.

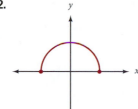

33.

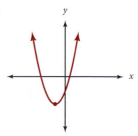

Simplify.

1. $-3 - 2[4 + 8(-1 - 5)]$ **2.** $4(3x + 1) - 7(2x - 2)$

3. $\dfrac{8}{3} \div \dfrac{1}{6} \cdot \dfrac{3}{2}$ **4.** $\dfrac{8}{3} - \dfrac{1}{6} + \dfrac{3}{2}$

5. $4\sqrt{32} - 3\sqrt{8} + 2\sqrt{18}$ **6.** $(\sqrt{5} + 3\sqrt{3})(2\sqrt{5} + \sqrt{3})$

7. $[(5 + i) - (2 - 3i)] - (2 - 5i)$ **8.** $\dfrac{(1 + i)^2}{2}$

9. $\log_{13}(\log_{17} 17)$ **10.** $\log_9[\log_3(\log_2 8)]$

11. $\dfrac{1 - \dfrac{1}{x - 2}}{1 + \dfrac{1}{x - 2}}$ **12.** $1 - \dfrac{x}{1 - \dfrac{1}{x}}$

Reduce the following to lowest terms.

13. $\dfrac{137}{274}$ **14.** $\dfrac{216}{168}$

Multiply.

15. $\left(4t^2 + \dfrac{1}{5}\right)\left(5t^2 - \dfrac{1}{4}\right)$ **16.** $(5x + 3)(x^2 - 5x + 3)$

Divide.

17. $\dfrac{x^2 - 6x + 3}{x - 1}$ **18.** $\dfrac{12x^2 + 23x - 3}{3x + 2}$

Add or subtract.

19. $\dfrac{6}{(x + 3)(x - 3)} - \dfrac{5}{(x + 3)(x - 2)}$

20. $\dfrac{3}{6x^2 - 5x - 4} + \dfrac{1}{6x^2 - 11x + 4}$

Solve.

21. $8 - 5(3x + 2) = 4$ **22.** $\dfrac{3}{4}(12x - 7) + \dfrac{1}{4} = 13$

23. $32x^2 = 4x + 3$

24. $4x - 4 = x^2$

25. $\dfrac{1}{x + 2} + \dfrac{1}{x - 5} = 1$

26. $\dfrac{1}{(x + 3)(x - 2)} + \dfrac{4}{(x + 2)(x - 2)} = \dfrac{-2}{(x + 2)(x + 3)}$

27. $x - 4\sqrt{x} - 5 = 0$

28. $2(3y + 1)^2 + (3y + 1) - 10 = 0$

29. $\log_4 x = 4$ **30.** $\log_x 0.2 = -1$

31. $\log_6(x + 2) - \log_6(x + 5) = 1$

32. $\log_3(x - 1) + \log_3(x - 6) = 2$

Solve each system.

33. $\begin{aligned} -x + 5y &= 19 \\ 2x - 9y &= -34 \end{aligned}$ **34.** $\begin{aligned} 4x + 7y &= 24 \\ -8x - 9y &= -48 \end{aligned}$

35. $\begin{aligned} x + 3y &= -12 \\ x - 3z &= -4 \\ y + 3z &= 0 \end{aligned}$ **36.** $\begin{aligned} 6x - y + 9z &= 0 \\ 6x + 6y + 9z &= 0 \\ 8x + y + z &= -11 \end{aligned}$

Graph each line.

37. $5x + 4y = 4$ **38.** $x = 2$

Write in symbols.

39. The difference of 8 and y is $-3x$.

40. The difference of $2a$ and $3b$ is less than their sum.

Write in scientific notation.

41. 0.00000129   **42.** 83,000,000

Factor completely.

43. $64x^3 + 27$ **44.** $48a^4 + 36a^2 + 6$

Specify the domain and range for the relation and state whether the relation is a function.

45. $\{(3, -2), (3, -4), (5, -1)\}$ **46.** $\{(1, 2), (2, 1), (3, 2)\}$

The illustration shows the yearly sales of different brands of cookies. Use the information to answer the following questions

47. Write the combined total for Nabisco Oreo and Oreo Double Stuff cookies in scientific notation.

48. What is the difference in sales between Private Label cookies and Chips Ahoy! cookies?

Graph each exponential function. [8.1]

1. $f(x) = -3^x$ **2.** $g(x) = 2^{-x}$

Sketch the graph of each function and its inverse. Find $f^{-1}(x)$ for Problem 3. [8.2]

3. $f(x) = \dfrac{1}{2}x + \dfrac{3}{2}$ **4.** $f(x) = x^2 - 3$

Solve for x. [8.3]

5. $\log_5 x = 2$ **6.** $\log_x 7 = 3$

Graph each of the following. [8.3]

7. $y = \log_3 x$ **8.** $y = \log_{1/3} x$

Evaluate each of the following. Round to the nearest hundredth if necessary. [8.3, 8.4, 8.5]

9. $\log_9 3$ **10.** $\log_7 35$

11. $\log 14{,}500$ **12.** $\log 0.0203$

13. $\ln 56.3$ **14.** $\ln 0.034$

Use the properties of logarithms to expand each expression. [8.4]

15. $\log_7 \dfrac{4x^3}{y}$ **16.** $\log \dfrac{\sqrt[3]{x}}{y^2 \sqrt{z}}$

Write each expression as a single logarithm. [8.4]

17. $\dfrac{1}{3} \log_2 x + 5 \log_2 y$ **18.** $\dfrac{1}{4} \log_3 x - \log_3 y + 3 \log_3 z$

Use a calculator to find x. Round to the nearest thousandth. [8.5]

19. $\log x = 5.3819$ **20.** $\log x = -2.4531$

Solve for x. Round to the nearest thousandth if necessary. [8.4, 8.6]

21. $4 = 3^x$ **22.** $9^{4x-1} = 27$

23. $\log_6 x + \log_6 3 = 1$ **24.** $\log_3 x + \log_3(x - 26) = 3$

25. Chemistry Use the formula $\text{pH} = -\log[\text{H}^+]$ to find the pH of a solution in which $[\text{H}^+] = 7.3 \times 10^{-6}$. [8.5]

26. Compound Interest If $300 is deposited in an account that earns 9% annual interest compounded twice a year, how much money will be in the account after 6 years? [8.1]

27. Compound Interest How long will it take $500 to become $2,500 if the $500 is deposited in an account that earns 6% annual interest compounded twice a year? [8.6]

Determine if the function is one-to-one. [8.2]

28. $f(x) = \{(4, 7), (5, 8), (7, 4), (8, 5)\}$

29. $g(x) = \{(1, 7), (2, 0), (3, 1), (4, 0)\}$

30.
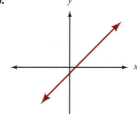

31.
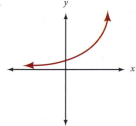

32.
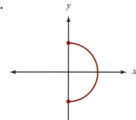

33.
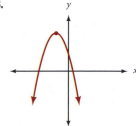

Conic Sections

The Mormon Tabernacle is a historic building in Salt Lake City, Utah that has a reputation for nearly perfect acoustics. It was originally built to host the semi-annual General Conference for The Church of Jesus Christ of Latter-Day Saints in 1867. Today, it is the home of the world-famous Mormon Tabernacle Choir. Built well before the invention of electronics, audiences of 7,000 often gathered to hear the sermons given there. The roof of the Tabernacle is a three-dimensional ellipse, which allows much of the sound from the pulpit to be concentrated and projected to every corner of the building. The roof also rests on piers outside the building. The absence of interior supports allows the sound waves to travel throughout the building without any obstacles.

An elliptical shape similar to the roof of the Mormon Tabernacle can be drawn using thumbtacks, string, and a pencil. As shown in the illustration, anchor the string using the tacks, then trace with the pencil. The tacks each represent one focus of the ellipse, a topic that will be expanded in the exercise set of Section 9.3.

An ellipse is an example of a conic section. This and other types of conics, including circles, parabolas, and hyperbolas, will be the focus of this chapter.

A Use the distance formula.

B Write the equation of a circle, given its center and radius.

C Find the center and radius of a circle from its equation, and then sketch the graph.

9.1 Circles

Because of their perfect symmetry, circles have been used for thousands of years in many disciplines, including art, science, and religion. The photograph above is of Stonehenge, a 4,500-year-old site in England. The arrangement of the stones is based on a circular plan that is thought to have both religious and astronomical significance.

As discussed in the chapter opener, conic sections include ellipses, circles, hyperbolas, and parabolas. They are called conic sections because each can be found by slicing a cone with a plane as shown in Figure 1. We begin our work with conic sections by studying circles. Before we find the general equation of a circle, we must first derive what is known as the *distance formula*.

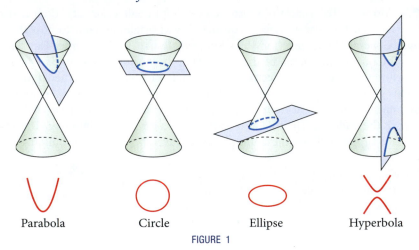

Parabola   Circle   Ellipse   Hyperbola

FIGURE 1

A The Distance Formula

Suppose (x_1, y_1) and (x_2, y_2) are any two points in the first quadrant. (Actually, we could choose the two points to be anywhere on the coordinate plane. It is just more convenient to have them in the first quadrant.) We can name the points P_1 and P_2, respectively, and draw the diagram shown in Figure 2.

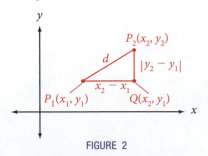

FIGURE 2

Notice the coordinates of point Q. The x-coordinate is x_2 because Q is directly below point P_2. The y-coordinate of Q is y_1 because Q is directly across from point P_1. It is evident from the diagram that the length of P_2Q is $|y_2 - y_1|$ and the length of P_1Q is $|x_2 - x_1|$. Using the Pythagorean theorem, we have

$$(P_1P_2)^2 = (P_1Q)^2 + (P_2Q)^2$$

or

$$d^2 = (x_2 - x_1)^2 + (y_2 - y_1)^2$$

Taking the square root of both sides, we have

$$d = \sqrt{(x_2 - x_1)^2 + (y_2 - y_1)^2}$$

We know this is the positive square root, because d is the distance from P_1 to P_2 and must therefore be non-negative. This formula is called the *distance formula*.

EXAMPLE 1 Find the distance between $(3, 5)$ and $(2, -1)$.

Solution If we let $(3, 5)$ be (x_1, y_1) and $(2, -1)$ be (x_2, y_2) and apply the distance formula, we have

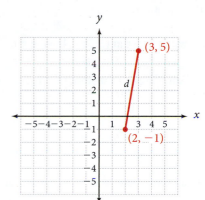

$$d = \sqrt{(2 - 3)^2 + (-1 - 5)^2}$$

$$= \sqrt{(-1)^2 + (-6)^2}$$

$$= \sqrt{1 + 36}$$

$$= \sqrt{37}$$

FIGURE 3

EXAMPLE 2 Find x if the distance from $(x, 5)$ to $(3, 4)$ is $\sqrt{2}$.

Solution Using the distance formula, we have

$\sqrt{2} = \sqrt{(x - 3)^2 + (5 - 4)^2}$	Distance formula
$2 = (x - 3)^2 + 1^2$	Square each side.
$2 = x^2 - 6x + 9 + 1$	Expand $(x - 3)^2$.
$0 = x^2 - 6x + 8$	Simplify.
$0 = (x - 4)(x - 2)$	Factor.
$x = 4 \quad$ or $\quad x = 2$	Set factors equal to 0.

The two solutions are 4 and 2, which indicates that two points, $(4, 5)$ and $(2, 5)$, are $\sqrt{2}$ units from $(3, 4)$.

B Circles

We can model circles very easily in algebra by using equations that are based on the distance formula.

> **THEOREM**   Circle Theorem
>
> The *equation of the circle* with center at (h, k) and radius r is given by
>
> $$(x - h)^2 + (y - k)^2 = r^2$$

Proof   By definition, all points on the circle are a distance r from the center (h, k). If we let (x, y) represent any point on the circle, then (x, y) is r units from (h, k). Applying the distance formula, we have

$$r = \sqrt{(x - h)^2 + (y - k)^2}$$

Squaring both sides of this equation gives the equation of the circle.

$$(x - h)^2 + (y - k)^2 = r^2$$

We can use the circle theorem to find the equation of a circle, given its center and radius, or to find its center and radius, given the equation.

EXAMPLE 3   Find the equation of the circle with center at $(-3, 2)$ having a radius of 5.

Solution   We have $(h, k) = (-3, 2)$ and $r = 5$. Applying our theorem for the equation of a circle yields

$$[x - (-3)]^2 + (y - 2)^2 = 5^2$$
$$(x + 3)^2 + (y - 2)^2 = 25$$

EXAMPLE 4   Give the equation of the circle with diameter 6 whose center is at the origin.

Solution   The coordinates of the center are $(0, 0)$, and the radius is 3 since the radius is equal to half the diameter.. The equation must be

$$(x - 0)^2 + (y - 0)^2 = 3^2$$
$$x^2 + y^2 = 9$$

We can see from Example 4 that the equation of any circle with its center at the origin and radius r will be

$$x^2 + y^2 = r^2$$

C Graphing Circles

EXAMPLE 5   Find the center and radius, and sketch the graph of the circle whose equation is

$$(x - 1)^2 + (y + 3)^2 = 4$$

Solution   Writing the equation in the form

$$(x - h)^2 + (y - k)^2 = r^2$$

we have

$$(x - 1)^2 + [y - (-3)]^2 = 2^2$$

Note   Notice in Example 3 how we placed the negative value of h in a set of parentheses. Pay special attention to the signs of h and k in the circle theorem to ensure clarity and accuracy.

3. Find the equation of the circle with center at $(4, -3)$ having a radius of 2.

4. Give the equation of the circle with center at the origin having a diameter of 10.

5. Find the center and radius of the circle whose equation is $(x - 3)^2 + (y - 4)^2 = 9$ and then sketch the graph.

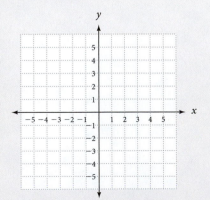

Answers

3. $(x - 4)^2 + (y + 3)^2 = 4$

4. $x^2 + y^2 = 25$

5. Center $= (3, 4)$; radius $= 3$

The center is at $(1, -3)$, and the radius is 2. Once you plot the center point, $(1, -3)$, plot four points that are r units, in this case, 2 units, above and below as well as to the right and left of the center. The graph shown in Figure 6 is the circle that passes through the equation $(x - 1)^2 + (y + 3)^2 = 4$.

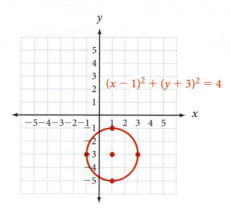

FIGURE 6

EXAMPLE 6 Sketch the graph of $x^2 + y^2 = 9$.

Solution Because the equation can be written in the form

$$(x - 0)^2 + (y - 0)^2 = 3^2$$

it must have its center at $(0, 0)$ and a radius of 3. The graph is shown in Figure 7.

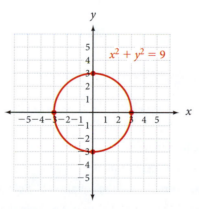

FIGURE 7

EXAMPLE 7 Sketch the graph of $x^2 + y^2 + 6x - 4y - 12 = 0$.

Solution To sketch the graph, we must find the center and radius of our circle. We can do so easily if the equation is in standard form. That is, if it has the form

$$(x - h)^2 + (y - k)^2 = r^2$$

To put our equation in standard form, we start by using the addition property of equality to group all the constant terms together on the right side of the equation. In this case, we add 12 to each side of the equation. We do this because we are going to add our own constants later to complete the square.

$$x^2 + y^2 + 6x - 4y = 12$$

Next, we group all the terms containing x together and all terms containing y together, and we leave some space at the end of each group for the numbers we will add when we complete the square on each group.

$$x^2 + 6x \qquad + y^2 - 4y \qquad = 12$$

6. Graph $x^2 + y^2 = 25$.

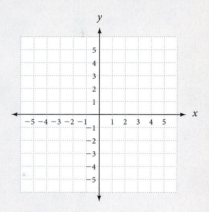

7. Graph $x^2 + y^2 - 6x + 4y - 3 = 0$.

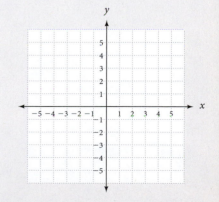

To complete the square on x, we add 9 to each side of the equation. To complete the square on y, we add 4 to each side of the equation. Remember, in order to complete the square, we add the square of half the coefficient of the linear term to both sides of the equation.

$$x^2 + 6x + 9 + y^2 - 4y + 4 = 12 + 9 + 4$$

The first three terms on the left side can be written as $(x + 3)^2$. Likewise, the last three terms on the left side simplify to $(y - 2)^2$. The right side simplifies to 25.

$$(x + 3)^2 + (y - 2)^2 = 25$$

Writing 25 as 5^2, we have our equation in standard form.

$$(x + 3)^2 + (y - 2)^2 = 5^2$$

From this last line, it is apparent that the center is at $(-3, 2)$ and the radius is 5. Using this information, we create the graph shown in Figure 8.

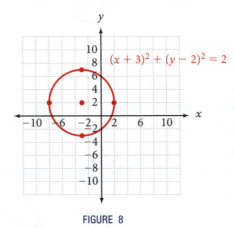

FIGURE 8

Vocabulary Review

Choose the correct words to fill in the blanks below.

center   circle   distance   radius

1. The equation of the _____ with center at (h, k) and radius r is given by $(x - h)^2 + (y - k)^2 = r^2$.

2. If d is the distance from P_1 to P_2 on a graph, then the _____ formula is given by
$d = \sqrt{(x_2 - x_1)^2 + (y_2 - y_1)^2}$.

3. A circle given by the equation $(x - 4)^2 + (y - 1)^2 = 36$ has a _____ at $(4, 1)$.

4. A circle given by the equation $(x + 4)^2 + (y - 3) = 25$ has a _____ of 5.

Problems

A Find the distance between the following points.

1. $(3, 7)$ and $(6, 3)$

2. $(4, 7)$ and $(8, 1)$

3. $(0, 9)$ and $(5, 0)$

4. $(-3, 0)$ and $(0, 4)$

5. $(3, -5)$ and $(-2, 1)$

6. $(-8, 9)$ and $(-3, -2)$

7. $(-1, -2)$ and $(-10, 5)$

8. $(-3, -8)$ and $(-1, 6)$

9. Find x so the distance between $(x, 2)$ and $(1, 5)$ is $\sqrt{13}$.

10. Find x so the distance between $(-2, 3)$ and $(x, 1)$ is 3.

11. Find x so the distance between $(x, 5)$ and $(3, 9)$ is 5.

12. Find y so the distance between $(-4, y)$ and $(2, 1)$ is 8.

13. Find x so the distance between $(x, 4)$ and $(2x + 1, 6)$ is 6.

14. Find y so the distance between $(3, y)$ and $(7, 3y - 1)$ is 6.

B Write the equation of the circle with the given center and radius.

15. Center $(3, -2)$; $r = 3$

16. Center $(-2, 4)$; $r = 1$

17. Center $(-5, -1)$; $r = \sqrt{5}$

18. Center $(-7, -6)$; $r = \sqrt{3}$

19. Center $(0, -5)$; $r = 1$

20. Center $(0, -1)$; $r = 7$

21. Center $(0, 0)$; $r = 2$

22. Center $(0, 0)$; $r = 5$

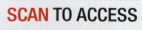

SCAN TO ACCESS

C Give the center and radius, and sketch the graph of each of the following circles.

23. $x^2 + y^2 = 4$

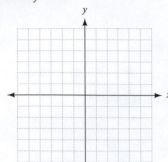

24. $x^2 + y^2 = 16$

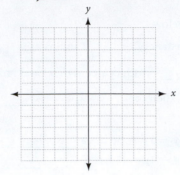

25. $(x - 1)^2 + (y - 3)^2 = 25$

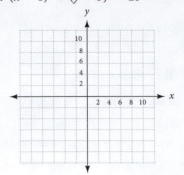

26. $(x - 4)^2 + (y - 1)^2 = 36$

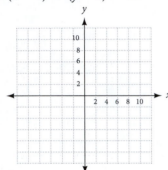

27. $(x + 2)^2 + (y - 4)^2 = 8$

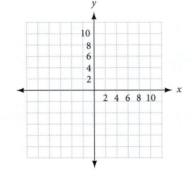

28. $(x - 3)^2 + (y + 1)^2 = 12$

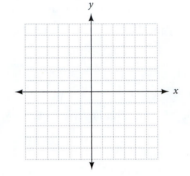

29. $(x + 2)^2 + (y - 4)^2 = 17$

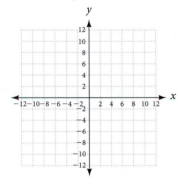

30. $x^2 + (y + 2)^2 = 11$

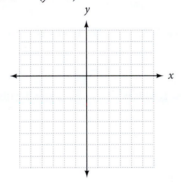

31. $x^2 + y^2 + 2x - 4y = 4$

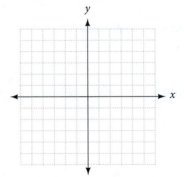

32. $x^2 + y^2 - 4x + 2y = 11$

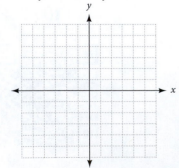

33. $x^2 + y^2 - 6y = 7$

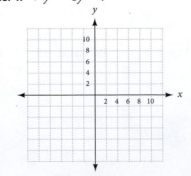

34. $x^2 + y^2 - 4y = 5$

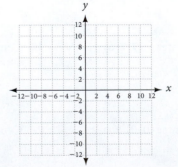

35. $x^2 + y^2 + 2x = 1$

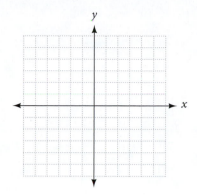

36. $x^2 + y^2 + 10x = 0$

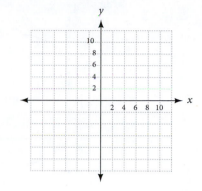

37. $x^2 + y^2 - 4x - 6y = -4$

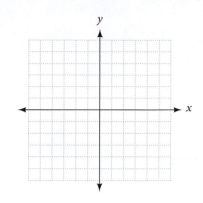

38. $x^2 + y^2 - 4x + 2y = 4$

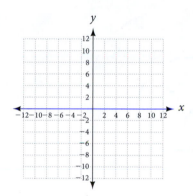

39. $x^2 + y^2 + 2x + y = \dfrac{11}{4}$

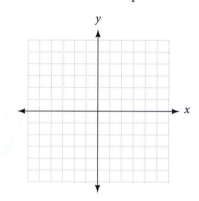

40. $x^2 + y^2 - 6x - y = -\dfrac{1}{4}$

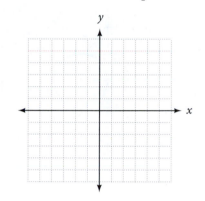

41. $4x^2 + 4y^2 - 4x + 8y = 11$

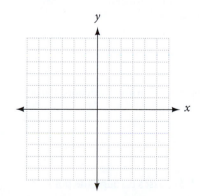

42. $36x^2 + 36y^2 - 24x - 12y = 31$

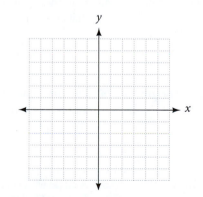

Each of the following circles passes through the origin. In each case, find the equation.

43.

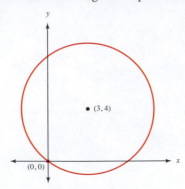

44.

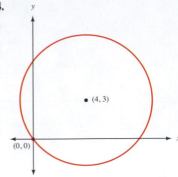

45. Find the equations of circles *A, B,* and *C* in the following diagram. The three points are the centers of the three circles.

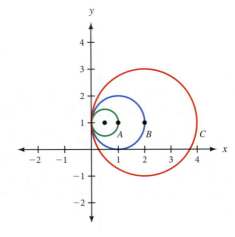

46. Each of the following circles passes through the origin. The centers are as shown. Find the equation of each circle.

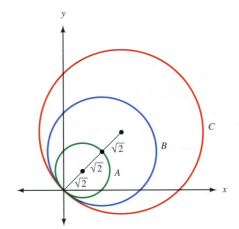

47. Find the equation of the circle with center at the origin that contains the point (3, 4).

48. Find the equation of the circle with center at the origin that contains the point (−5, 12).

49. Find the equation of the circle with center at the origin and *x*-intercepts 3 and −3.

50. Find the equation of the circle with *y*-intercepts 4 and −4 and center at the origin.

51. A circle with center at (−1, 3) passes through the point (4, 3). Find the equation.

52. A circle with center at (2, 5) passes through the point (−1, 4). Find the equation.

53. Find the equation of the circle with center at (−2, 5), which passes through the point (1, −3).

54. Find the equation of the circle with center at (4, −1), which passes through the point (6, −5).

55. Find the equation of the circle with center on the y-axis and y-intercepts at -2 and 6.

56. Find the equation of the circle with center on the x-axis and x-intercepts at -8 and 2.

57. Find the circumference and area of the circle

$$x^2 + (y - 3)^2 = 18$$

Leave your answer in terms of π.

58. Find the circumference and area of the circle

$$(x + 2)^2 + (y + 6)^2 = 12$$

Leave your answer in terms of π.

59. Find the circumference and area of the circle

$$x^2 + y^2 + 4x + 2y = 20$$

Leave your answer in terms of π.

60. Find the circumference and area of the circle

$$x^2 + y^2 - 6x + 2y = 6$$

Leave your answer in terms of π.

Applying the Concepts

61. Search Area A 3-year-old child has wandered away from home. The police have decided to search a circular area with a radius of 6 blocks. The child turns up at his grandmother's house, 5 blocks east and 3 blocks north of home. Was he found within the search area?

62. Placing a Bubble Fountain A circular garden pond with a diameter of 12 feet is to have a bubble fountain. The water from the bubble fountain falls in a circular pattern with a radius of 1.5 feet. If the center of the bubble fountain is placed 4 feet west and 3 feet north of the center of the pond, will all the water from the fountain fall inside the pond? What is the farthest distance from the center of the pond that water from the fountain will fall?

63. Ferris Wheel A giant Ferris wheel has a diameter of 240 feet and sits 12 feet above the ground. As shown in the diagram below, the wheel is 500 feet from the entrance to the park. The xy-coordinate system containing the wheel has its origin on the ground at the center of the entrance. Write an equation that models the shape of the wheel.

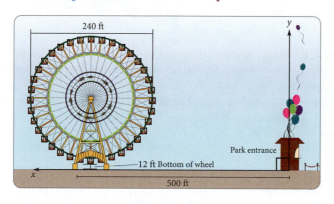

64. Magic Rings A magician is holding two rings that seem to lie in the same plane and intersect in two points. Each ring is 10 inches in diameter.

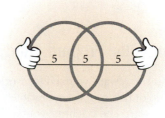

a. Find the equation of each ring if a coordinate system is placed with its origin at the center of the first ring and the x-axis contains the center of the second ring.

b. Find the equation of each ring if a coordinate system is placed with its origin at the center of the second ring and the x-axis contains the center of the first ring.

Getting Ready for the Next Section

Solve each equation.

65. $y^2 = 9$

66. $x^2 = 25$

67. $-y^2 = 4$

68. $-x^2 = 16$

69. $-x^2 = 9$

70. $y^2 = 100$

71. Divide $4x^2 + 9y^2$ by 36.

72. Divide $25x^2 + 4y^2$ by 100.

Find the x-coordinates of the x-intercepts and the y-coordinates of the y-intercepts of each equation.

73. $3x - 4y = 12$

74. $y = 3x^2 + 5x - 2$

75. If $\dfrac{x^2}{25} + \dfrac{y^2}{9} = 1$, find y when x is 3.

76. If $\dfrac{x^2}{25} + \dfrac{y^2}{9} = 1$, find y when x is -4.

Find the Mistake

Each problem below contains a mistake. Circle the mistake and write the correct number(s) or word(s) on the line provided.

1. The distance between two points on a graph can be found by using the formula $d = \sqrt{(x-a)^2 + (y-b)^2}$.

_____

2. The equation $(x-2)^2 + (y-7)^2 = 16$ gives the graph of a circle with a center at $(-2,-7)$ and a radius of 4. _____

3. A circle given by the equation $(x+4)^2 + (y-3)^2 = 64$ has a radius of 64. _____

4. The first step to graphing $x^2 + y^2 + 8x - 10y = 0$ is to use the distance formula.

_____

Navigation Skills: Prepare, Study, Achieve

It is important to reward yourself for working hard in this class. Recognizing your achievement will help foster success through the end of this course, and enable you to carry over the skills you've learned to future classes. What are some things you can do to reward and celebrate yourself? Also, for future classes, you could decide on these rewards at the beginning of the course. Schedule smaller short-term rewards to work towards as you complete each chapter. And then have a long-term reward that you achieve at the end of the course to celebrate your overall success. You deserve it. Congratulations!

9.2 Parabolas

©iStockphoto.com/Joe_Potato

At the 1997 Washington County Fair in Oregon, David Smith, Jr., The Bullet, was shot from a cannon. As a human cannonball, he reached a height of 70 feet, high enough to clear a ferris wheel, before landing in a net 160 feet from the cannon. If we sketch the graph of Smith's flight path, we get the shape of a vertical parabola. We will continue our study of parabolas in this section.

A Vertical Parabolas

As you recall from Section 7.5, parabolas that represent functions (vertical parabolas) have a standard form of

$$y = ax^2 + bx + c$$

By completing the square on x, these can be written as

$$y = a(x - h)^2 + k$$

which has a vertex of (h, k), opens upward if $a > 0$, and opens downward if $a < 0$. Also recall that the x-coordinate of the vertex occurs when $x = \frac{-b}{2a}$, so $h = \frac{-b}{2a}$. The line $x = \frac{-b}{2a}$ is called the *line of symmetry* of a parabola.

EXAMPLE 1 Graph $y = 2(x - 1)^2 - 3$.

Solution This parabola is already written in the form $y = a(x - h)^2 + k$, with $a = 2$, $h = 1$, and $k = -3$. So the vertex is $(1, -3)$ and the parabola opens upward. We need to find other points on the parabola near the vertex.

Completing the table:

x	$y = 2(x - 1)^2 - 3$
-1	$2(-1 - 1)^2 - 3 = 5$
0	$2(0 - 1)^2 - 3 = -1$
1	$2(1 - 1)^2 - 3 = -3$
2	$2(2 - 1)^2 - 3 = -1$
3	$2(3 - 1)^2 - 3 = 5$

Sketching the graph:

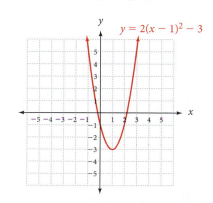

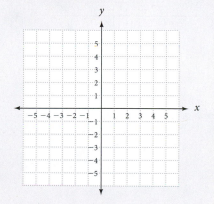

2. Graph $y = x^2 - 6x + 5$.

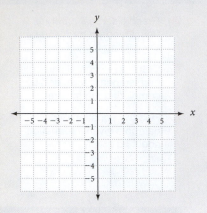

EXAMPLE 2   Graph $y = x^2 + 4x + 7$.

Solution

Method 1

We complete the square to convert the parabola to the form $y = a(x - h)^2 + k$.

$$y = (x^2 + 4x) + 7$$
$$= (x^2 + 4x + 4) + 7 - 4 \qquad \text{Add and subtract 4.}$$
$$= (x + 2)^2 + 3 \qquad \text{Complete the square.}$$

The vertex is $(-2, 3)$ and $a = 1$. Choosing points near the vertex to complete the table.

Completing the table:

x	$y = (x + 1)^2 + 3$
0	$(0 + 2)^2 + 3 = 7$
-1	$(-1 + 2)^2 + 3 = 4$
-2	$(-2 + 2)^2 + 3 = 3$
-3	$(-3 + 2)^2 + 3 = 4$
-4	$(-4 + 2)^2 + 3 = 7$

Sketching the graph:

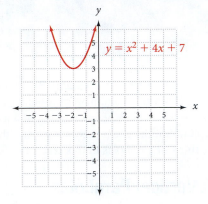

Method 2

Using the vertex formula from Chapter 7,

$$x = \frac{-b}{2a} = \frac{-4}{2(1)} = \frac{-4}{2} = -2$$

Now complete the table (given above) to graph the parabola. Using this method does not require completing the square.

---

3. Graph $y = -x^2 + 3x + 2$.

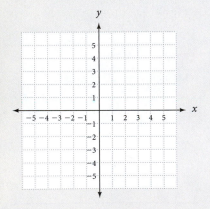

EXAMPLE 3   Graph $y = -2x^2 + 8x + 3$.

Solution

Using Method 2, apply the vertex formula.

$$x = \frac{-b}{2a} = \frac{-8}{2(-2)} = \frac{-8}{-4} = 2$$

Choosing points near $x = 2$ to complete the table.

Completing the table:

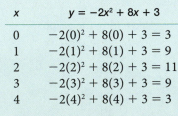

x	$y = -2x^2 + 8x + 3$
0	$-2(0)^2 + 8(0) + 3 = 3$
1	$-2(1)^2 + 8(1) + 3 = 9$
2	$-2(2)^2 + 8(2) + 3 = 11$
3	$-2(3)^2 + 8(3) + 3 = 9$
4	$-2(4)^2 + 8(4) + 3 = 3$

Sketching the graph:

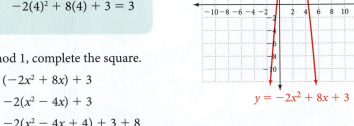

Using Method 1, complete the square.

$$y = (-2x^2 + 8x) + 3$$
$$= -2(x^2 - 4x) + 3$$
$$= -2(x^2 - 4x + 4) + 3 + 8$$
$$= -2(x - 2)^2 + 11$$

The vertex is $(2, 11)$ and $a = -2$. Then complete the above table.

B Horizontal Parabolas

Horizontal parabolas have a standard form of

$$x = ay^2 + by + c$$

By completing the square on y, these can be written as

$$x = a(y - k)^2 + h$$

which has a vertex of (h, k), opens to the right if $a > 0$, and opens to the left if $a < 0$.

The vertex formula can also be used to find the y-coordinate of the vertex, so $y = \frac{-b}{2a}$. The line $y = \frac{-b}{2a}$ is now a horizontal line of symmetry for the parabola.

EXAMPLE 4   Graph $x = -2(y + 3)^2 + 4$.

Solution   This parabola is already written in the form $x = a(y - k)^2 + h$, with $a = -2$, $h = 4$, and $k = -3$. So the vertex is $(4, 3)$ and the parabola opens to the left. Complete the table to find other points near the vertex, however now we are picking y-values near $y = -3$.

Completing the table:

x	y
$-2(-1 + 3)^2 + 4 = -4$	-1
$-2(-2 + 3)^2 + 4 = 2$	-2
$-2(-3 + 3)^2 + 4 = 4$	-3
$-2(-4 + 3)^2 + 4 = 2$	-4
$-2(-5 + 3)^2 + 4 = -4$	-5

Sketching the graph:

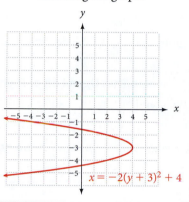

$x = -2(y + 3)^2 + 4$

4. Graph $x = -(y - 2)^2 + 3$.

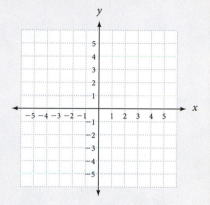

EXAMPLE 5   Graph $x = y^2 - 4y - 2$.

Solution   Again we have two methods to use.

Method 1

We complete the square to convert the parabola to the form $x = a(y - k)^2 + h$.

$$
\begin{aligned}
y &= (y^2 - 4y) - 2 \\
&= (y^2 - 4y + 4) - 2 - 4 \qquad \text{Add and subtract 4.} \\
&= (y - 2)^2 - 6 \qquad \text{Complete the square.}
\end{aligned}
$$

The vertex is $(-6, 2)$ and $a = 1$, so the parabola opens to the right. Choosing values of y near $y = 2$.

Completing the table:

x	y
$(0 - 2)^2 - 6 = -2$	0
$(1 - 2)^2 - 6 = -5$	1
$(2 - 2)^2 - 6 = 6$	2
$(3 - 2)^2 - 6 = -5$	3
$(4 - 2)^2 - 6 = -2$	4

Sketching the graph:

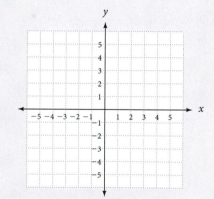

$x = y^2 - 4y - 2$

5. Graph $x = -y^2 + 2y + 4$.

Method 2

Using the vertex formula,

$$y = \frac{-b}{2a} = \frac{-(-4)}{2(1)} = \frac{4}{2} = 2$$

Now complete the table (given above) to graph the parabola. Again note that this method does not require us to complete the square.

6. Graph $x = 2y^2 - 4y - 1$.

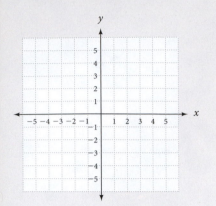

EXAMPLE 6   Graph $x = 3y^2 - 6y + 1$.

Solution   For this example, we simply show the graph of our equation alongside the graph of the equation $y = 3x^2 - 6x + 1$.

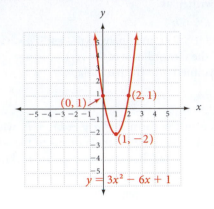

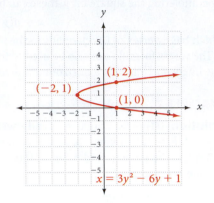

A summary of graphing horizontal parabolas is given below.

Parabolas That Open Left or Right

The graph of $x = ay^2 + by + c$ will have the following:

1. An x-coordinate for the x-intercept at $x = c$

2. y-coordinates for the y-intercepts (if they exist) at

$$y = \frac{-b \pm \sqrt{b^2 - 4ac}}{2a}$$

3. A vertex when $y = -\dfrac{b}{2a}$

4. A concavity opening to the right if $a > 0$, and a concavity opening to the left if $a < 0$

C Finding the Equation of a Parabola

7. A parabola has a vertex of $(2, -4)$ and x-intercepts of 0 and 4. Find its equation.

EXAMPLE 7   A parabola has a vertex of $(3, -2)$ and x-intercepts of 0 and 6. Find its equation.

Solution   For the parabola to have two x-intercepts, it must be a vertical parabola of the form $y = a(x - h)^2 + k$. Since $h = 3$ and $k = -2$, the equation looks like this:

$$y = a(x - 3)^2 - 2$$

Since the x-intercepts are 0 and 6, two other points on the parabola are $(0, 0)$ and $(6, 0)$. Substituting either of these points:

$(0, 0)$	$(6, 0)$
$0 = a(0 - 3)^2 - 2$	$0 = a(6 - 3)^2 - 2$
$0 = 9a - 2$	$0 = 9a - 2$
$2 = 9a$	$2 = 9a$
$a = \dfrac{2}{9}$	$a = \dfrac{2}{9}$

So the equation of the parabola is $y = \dfrac{2}{9}(x - 3)^2 - 2$.

Answer

7. $y = (x - 2)^2 - 4$

EXAMPLE 8   A parabola has a vertex of $(-4, 1)$ and y-intercepts of -7 and 9. Find its equation.

Solution   For the parabola to have two y-intercepts, it must be a horizontal parabola of the form $x = a(y - k)^2 + h$. Since $h = -4$ and $k = 1$, the equation looks like this:

$$x = a(y - 1)^2 - 4$$

Since the y-intercepts are -7 and 9, two other points on the parabola are $(0, -7)$ and $(0, 9)$. Substituting either of these points:

$(0, -7)$	$(0, 9)$
$0 = a(-7 - 1)^2 - 4$	$0 = a(9 - 1)^2 - 4$
$0 = 64a - 4$	$0 = 64a - 4$
$4 = 64a$	$4 = 64a$
$a = \dfrac{1}{16}$	$a = \dfrac{1}{16}$

So the equation of the parabola is $x = \dfrac{1}{16}(y - 1)^2 - 4$.

GETTING READY FOR CLASS

After reading through the preceding section, respond in your own words and in complete sentences.

A. How can you tell the difference between the equation of a horizontal parabola and the equation of a vertical parabola?

B. Do both methods of graphing parabolas require us to complete the square? Explain why or why not.

C. Besides coordinates of the vertex, what other information is useful in finding the equation of a parabola?

D. When graphing a parabola, how many points (including the vertex) do we usually plot?

8. A parabola has a vertex of $(3, -6)$ and y-intercepts of -7 and -5. Find its equation.

Answer

8. $x = -3(y + 6)^2 + 3$

Vocabulary Review

Choose the correct words to fill in the blanks below.

| left | right | horizontal | vertical | vertex | complete the square |

1. A _____ parabola that represents a function has a standard form of $y = ax^2 + bx + c$.

2. We can _____ on x in the equation $y = ax^2 + bx + c$ to get $y = a(x - h)^2 + k$, where (h, k) is the _____.

3. A _____ parabola is given by the equation in standard form of $x = ay^2 + by + c$.

4. A parabola with an equation in the form of $x = ay^2 + by + c$ opens to the _____ if $a > 0$.

5. A parabola with an equation in the form of $x = ay^2 + by + c$ opens to the _____ if $a < 0$.

Problems

A Graph the following vertical parabolas. Be sure to plot at least 5 points to get an accurate graph.

1. $y = 2(x + 1)^2 - 3$

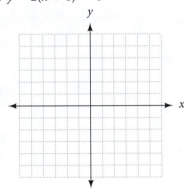

2. $y = 3(x + 2)^2 - 4$

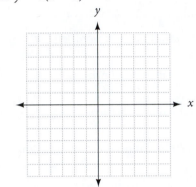

3. $y = -(x - 4)^2 + 2$

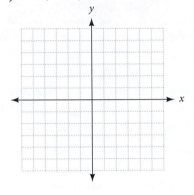

4. $y = -(x - 3)^2 + 5$

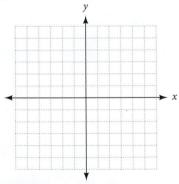

5. $y = 3(x - 2)^2 + 1$

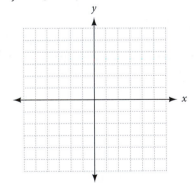

6. $y = 4(x - 2)^2 - 3$

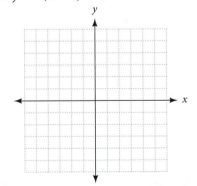

SCAN TO ACCESS

7. $y = x^2 + 3x - 5$

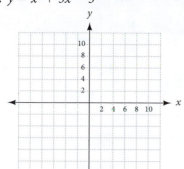

8. $y = x^2 - 4x + 1$

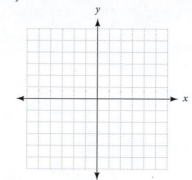

9. $y = -x^2 + 4x + 3$

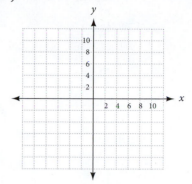

10. $y = -x^2 + 5x - 2$

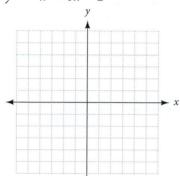

11. $y = 2x^2 + 8x + 5$

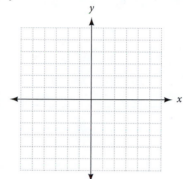

12. $y = 4x^2 + 8x + 7$

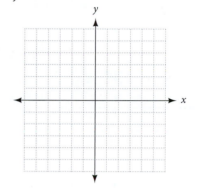

B Graph the following horizontal parabolas. Be sure to plot at least 5 points to get an accurate graph.

13. $x = 2(y + 2)^2 - 5$

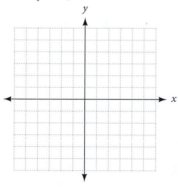

14. $x = 3(y - 1)^2 + 2$

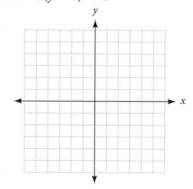

15. $x = -(y + 3)^2 + 3$

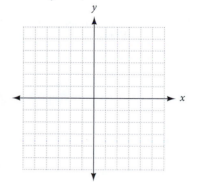

16. $x = -(y + 2)^2 + 4$

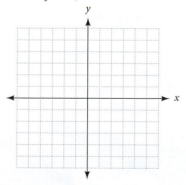

17. $x = -3(y - 2)^2 + 5$

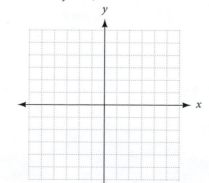

18. $x = -2(y - 4)^2 - 3$

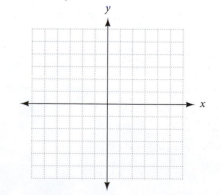

19. $x = y^2 + 4y + 7$

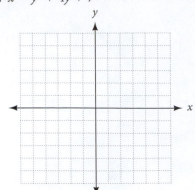

20. $x = y^2 - 6y + 5$

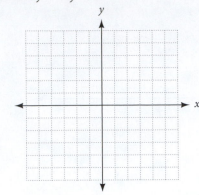

21. $x = -y^2 - 4y + 3$

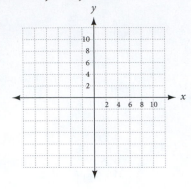

22. $x = -y^2 - 6y + 2$

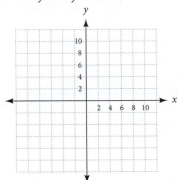

23. $x = -3y^2 + 6y - 4$

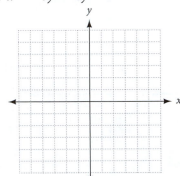

24. $x = -2y^2 + 6y - 3$

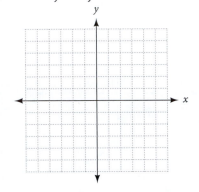

C Find the equation for the parabola satisfying the given information. Give your answer in the form $y = a(x - h)^2 + k$ or $x = a(y - k)^2 + h$.

25. Vertex $= (2, -1)$; x-intercepts $= (0, 0), (4, 0)$

26. Vertex $= (-2, 4)$; x-intercepts $= (1, 0), (3, 0)$

27. Vertex $= (-2, 3)$; x-intercepts $= (-3, 0), (-1, 0)$

28. Vertex $= (-2, -5)$; x-intercepts $= (-3, 0), (-1, 0)$

29. Vertex $= (-4, 1)$; y-intercepts $= (0, 0), (0, 2)$

30. Vertex $= (-3, 5)$; y-intercepts $= (0, 3), (0, 7)$

31. Vertex $= (3, -5)$; y-intercepts $= (0, -2), (0, -8)$

32. Vertex $= (4, 2)$; y-intercepts $= (0, 0), (0, 4)$

33. x-intercepts $= (0, 0), (6, 0)$; maximum value $= 8$

34. x-intercepts $= (3, 0), (5, 0)$; maximum value $= 12$

35. *x*-intercepts = $(-2, 0)$, $(4, 0)$; minimum value = -3

36. *x*-intercepts = $(-7, 0)$, $(-3, 0)$; minimum value = -5

37. *y*-intercepts = $(0, -1)$, $(0, 5)$; largest *x*-value = 4

38. *y*-intercepts = $(0, 0)$, $(0, 8)$; largest *x*-value = 7

39. *y*-intercepts = $(0, -6)$, $(0, -2)$; smallest *x*-value = -7

40. *y*-intercepts = $(0, -3)$, $(0, 5)$; smallest *x*-value = 2

Maintaining Your Skills

41. Laptops The chart shows the most popular laptop manufacturers for college students. If a college has 15,000 students, how many students have Apple computers? If a college has 35,000 students, how many students have Dell computers?

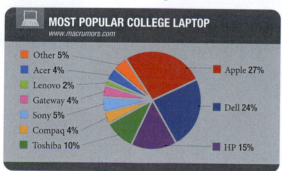

42. Baseball The chart shows attendance at Major League Baseball games over several years. Using the slope of the line, estimate the baseball game attendance for 2002 and 2006.

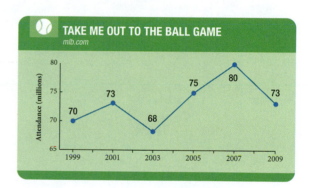

43. Human Cannonball In the introduction to this section, David Smith Jr. was shot out of a cannon, reaching a height of 70 feet and landing 160 feet away from the cannon in a net. Assume his path follows a parabolic arch.

a. If $(0, 0)$ represents the cannon position, find the coordinates of the vertex and the landing position.

b. Write a general form for the equation of the parabola

c. Using the vertex point, write the specific equation.

44. Human Cannonball Draw a graph illustrating the parabolic arc of the human cannonball from problem 43 assuming the cannon is at the origin.

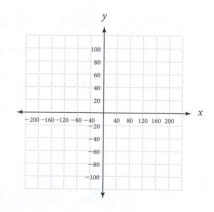

Getting Ready for the Next Section

Solve each equation.

45. $(x + 4)^2 + 3 = 12$ **46.** $(x - 3)^2 - 2 = 7$ **47.** $(y - 1)^2 - 6 = 19$ **48.** $(y + 2)^2 + 17 = 33$

Find the x-coordinates of the x-intercepts and the y-coordinates of the y-intercepts of each equation.

49. $\dfrac{x^2}{9} + \dfrac{y^2}{4} = 1$ **50.** $\dfrac{x^2}{4} + \dfrac{y^2}{9} = 1$ **51.** $\dfrac{x^2}{5} + \dfrac{y^2}{16} = 1$ **52.** $\dfrac{x^2}{25} + \dfrac{y^2}{15} = 1$

53. If $\dfrac{x^2}{9} + \dfrac{y^2}{16} = 1$, find y when $x = 1$.   **54.** If $\dfrac{x^2}{9} + \dfrac{y^2}{16} = 1$, find x when $y = 2$.

55. If $\dfrac{x^2}{25} + \dfrac{y^2}{4} = 1$, find x when $y = 1$.   **56.** If $\dfrac{x^2}{25} + \dfrac{y^2}{4} = 1$, find y when $x = 3$.

Find the Mistake

Each problem below contains a mistake. Circle the mistake and write the correct number(s) or word(s) on the line provided.

1. The line $x = -\dfrac{a}{2b}$ is called the line of symmetry of a parabola. _____

2. A parabola that has two x-intercepts must have the form $x = a(y - k)^2 + h$. _____

3. A parabola that has two y-intercepts must be a vertical parabola of the form $x = a(y - k)^2 + h$. _____

4. A parabola that has a vertex of $(4, 5)$ and x-intercepts of $(0, 2)$ and $(0, 6)$ will open up. _____

9.3 Ellipses

OBJECTIVES

A Graph an ellipse from its standard form.

B Use completing the square to convert an ellipse to standard form to graph the curve.

KEY WORDS

ellipse

x'y'-coordinate system

vertices

The photograph above shows Halley's comet as it passed close to Earth in 1986. Like the planets in our solar system, it orbits the sun in an elliptical path. While it takes the Earth one year to complete one orbit around the sun, it takes Halley's comet 76 years. The first known sighting of Halley's comet was in 239 BC. Its most famous appearance occurred in 1066 AD, when it was seen a few months before the Battle of Hastings and believed to be an omen.

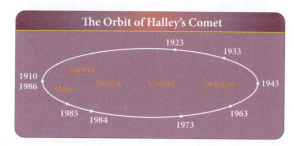

A Graphing Ellipses

This section is concerned with the graphs of ellipses. To begin, we will consider only those graphs that are centered about the origin.

Suppose we want to graph the equation

$$\frac{x^2}{25} + \frac{y^2}{9} = 1$$

We can find the *y*-intercepts by letting $x = 0$, and we can find the *x*-intercepts by letting $y = 0$.

When $x = 0$ When $y = 0$

$$\frac{0^2}{25} + \frac{y^2}{9} = 1 \qquad\qquad \frac{x^2}{25} + \frac{0^2}{9} = 1$$

$$y^2 = 9 \qquad\qquad\qquad x^2 = 25$$

$$y = \pm 3 \qquad\qquad\qquad x = \pm 5$$

The graph crosses the *y*-axis at $(0, 3)$ and $(0, -3)$ and the *x*-axis at $(5, 0)$ and $(-5, 0)$. Graphing these points and then connecting them with a smooth curve gives the graph shown in Figure 1.

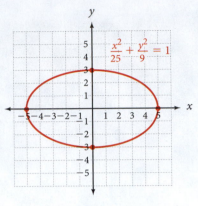

FIGURE 1

We can find other ordered pairs on the graph by substituting in values for x (or y) and then solving for y (or x). For example, if we let $x = 3$, then

$$\frac{3^2}{25} + \frac{y^2}{9} = 1$$

$$\frac{9}{25} + \frac{y^2}{9} = 1$$

$$0.36 + \frac{y^2}{9} = 1$$

$$\frac{y^2}{9} = 0.64 \qquad \text{Add } -0.36 \text{ to each side.}$$

$$y^2 = 5.76 \qquad \text{Multiply each side by 9.}$$

$$y = \pm 2.4 \qquad \text{Square root property for equations.}$$

This would give us the two ordered pairs $(3, -2.4)$ and $(3, 2.4)$.

A graph of the type shown in Figure 1 is called an *ellipse*. If we were to find some other ordered pairs that satisfy our original equation, we would find that their graphs lie on the ellipse. Also, the coordinates of any point on the ellipse will satisfy the equation. We can generalize these results as follows:

The Ellipse

The graph of any equation of the form

$$\frac{x^2}{a^2} + \frac{y^2}{b^2} = 1 \qquad \text{Standard form}$$

will be an *ellipse* centered at the origin. The ellipse will cross the x-axis at $(a, 0)$ and $(-a, 0)$. It will cross the y-axis at $(0, b)$ and $(0, -b)$. When a and b are equal, the ellipse will be a circle. Each of the points $(a, 0)$, $(-a, 0)$, $(0, b)$, and $(0, -b)$ is a *vertex* (intercept) of the graph.

The most convenient way to graph an ellipse is to locate the intercepts (vertices).

EXAMPLE 1   Sketch the graph of $\dfrac{x^2}{9} + \dfrac{y^2}{36} = 1$.

Solution   The ellipse is in standard form with $a = 3$ and $b = 6$. The graph crosses the x-axis at $(3, 0)$, $(-3, 0)$, and it crosses the y-axis at $(0, 6)$, $(0, -6)$.

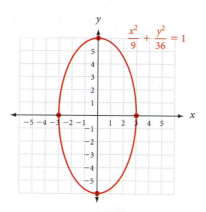

FIGURE 2

EXAMPLE 2   Sketch the graph of $4x^2 + 9y^2 = 36$.

Solution   To write the equation in the form

$$\frac{x^2}{a^2} + \frac{y^2}{b^2} = 1$$

we must divide both sides by 36.

$$\frac{4x^2}{36} + \frac{9y^2}{36} = \frac{36}{36}$$

$$\frac{x^2}{9} + \frac{y^2}{4} = 1$$

The graph crosses the x-axis at $(3, 0)$, $(-3, 0)$ and the y-axis at $(0, 2)$, $(0, -2)$. (See Figure 3.)

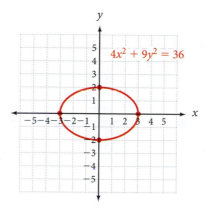

FIGURE 3

Practice Problems

1. Sketch the graph of $\dfrac{x^2}{16} + \dfrac{y^2}{25} = 1$.

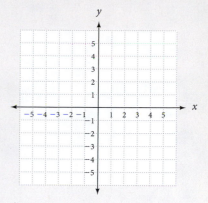

2. Graph $25x^2 + 4y^2 = 100$. (*Hint:* First, divide both sides by 100.)

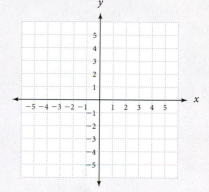

B Ellipses not Centered at the Origin

The following equation is that of an ellipse with its center at the point (4, 1):

$$\frac{(x-4)^2}{9} + \frac{(y-1)^2}{4} = 1$$

To see why the center is at (4, 1) we substitute x' (read "x prime") for $x - 4$ and y' for $y - 1$ in the equation.

If $x' = x - 4$

and $y' = y - 1$

the equation $\dfrac{(x-4)^2}{9} + \dfrac{(y-1)^2}{4} = 1$

becomes $\dfrac{(x')^2}{9} + \dfrac{(y')^2}{4} = 1$

This is the equation of an ellipse in a coordinate system with an x'-axis and a y'-axis. We call this new coordinate system the **$x'y'$-coordinate system**. The center of our ellipse is at the origin in the $x'y'$-coordinate system. The question is this: What are the coordinates of the center of this ellipse in the original xy-coordinate system? To answer this question, we go back to our original substitutions:

$$x' = x - 4$$
$$y' = y - 1$$

In the $x'y'$-coordinate system, the center of our ellipse is at $x' = 0, y' = 0$ (the origin of the $x'y'$ system). Substituting these numbers for x' and y', we have

$$0 = x - 4$$
$$0 = y - 1$$

Solving these equations for x and y will give us the coordinates of the center of our ellipse in the xy-coordinate system. As you can see, the solutions are $x = 4$ and $y = 1$. Therefore, in the xy-coordinate system, the center of our ellipse is at the point (4, 1). Figure 4 shows the graph.

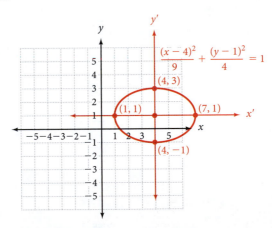

FIGURE 4

The coordinates of all points labeled in Figure 4 are given with respect to the xy-coordinate system. The x'- and y'-axes are shown simply for reference in our discussion. Note that the horizontal distance from the center to the vertices is 3—the square root of the denominator of the $(x - 4)^2$ term. Likewise, the vertical distance from the center to the other vertices is 2—the square root of the denominator of the $(y - 1)^2$ term.

We summarize the information on the previous page with the following:

An Ellipse with Center at (h, k)

The graph of the equation

$$\frac{(x-h)^2}{a^2} + \frac{(y-k)^2}{b^2} = 1$$

will be an *ellipse with center at (h, k)*. The vertices of the ellipse will be at the points $(h + a, k)$, $(h - a, k)$, $(h, k + b)$, and $(h, k - b)$.

EXAMPLE 3   Graph the ellipse $x^2 + 9y^2 + 4x - 54y + 76 = 0$.

Solution   To identify the coordinates of the center, we must complete the square on x and also on y. To begin, we rearrange the terms so that those containing x are together, those containing y are together, and the constant term is on the other side of the equal sign. Doing so gives us the following equation:

$$x^2 + 4x \quad + 9y^2 - 54y \quad = -76$$

Before we can complete the square on y, we must factor 9 from each term containing y.

$$x^2 + 4x \quad + 9(y^2 - 6y) \quad = -76$$

To complete the square on x, we add 4 to each side of the equation. To complete the square on y, we add 9 inside the parentheses. This increases the left side of the equation by 81 since each term within the parentheses is multiplied by 9. Therefore, we must add 81 to the right side of the equation also.

$$x^2 + 4x + 4 + 9(y^2 - 6y + 9) = -76 + 4 + 81$$

$$(x + 2)^2 + 9(y - 3)^2 = 9$$

To identify the distances to the vertices, we divide each term on both sides by 9.

$$\frac{(x + 2)^2}{9} + \frac{9(y - 3)^2}{9} = \frac{9}{9}$$

$$\frac{(x + 2)^2}{9} + \frac{(y - 3)^2}{1} = 1$$

The graph is an ellipse with center at $(-2, 3)$, as shown in Figure 5.

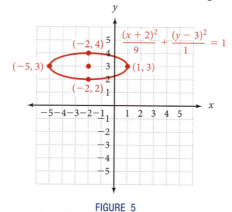

FIGURE 5

3. Graph the ellipse

$$16x^2 + y^2 - 128x - 6y + 249 = 0.$$

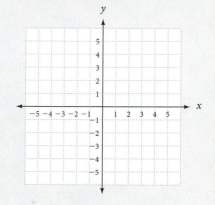

Vocabulary Review

Choose the correct words to fill in the blanks in the paragraph below.

vertices   center   ellipse   *x*-intercepts   *y*-intercepts

The graph of any equation of the form $\frac{x^2}{a^2} + \frac{y^2}{b^2} = 1$ will be an _____ centered at the origin. The

_____ for the ellipse will be $(a, 0)$ and $(-a, 0)$. The _____ will be $(0, b)$

and $(0, -b)$. The graph of the equation $\frac{(x - h)^2}{a^2} + \frac{(y - k)^2}{b^2} = 1$ will be an ellipse with _____ at (h, k).

The _____ of this ellipse will be at the points $(h + a, k)$, $(h - a, k)$, $(h, k + b)$, and $(h, k - b)$.

Problems

A Graph each of the following ellipses. Be sure to label both the *x*- and *y*-intercepts.

1. $\frac{x^2}{9} + \frac{y^2}{16} = 1$

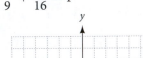

2. $\frac{x^2}{25} + \frac{y^2}{4} = 1$

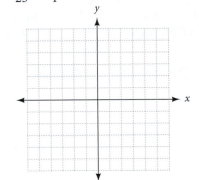

3. $\frac{x^2}{16} + \frac{y^2}{9} = 1$

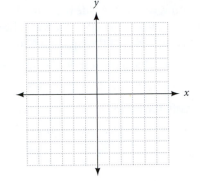

4. $\frac{x^2}{4} + \frac{y^2}{25} = 1$

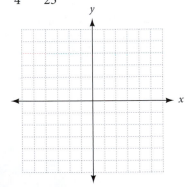

5. $\frac{x^2}{3} + \frac{y^2}{4} = 1$

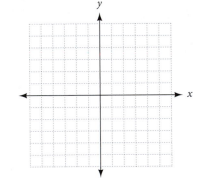

6. $\frac{x^2}{4} + \frac{y^2}{3} = 1$

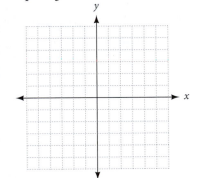

SCAN TO ACCESS

7. $4x^2 + 25y^2 = 100$

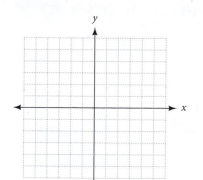

8. $4x^2 + 9y^2 = 36$

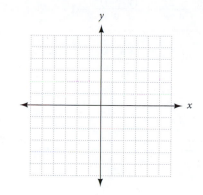

9. $x^2 + 8y^2 = 16$

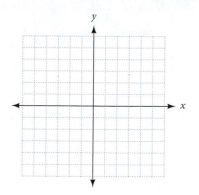

10. $12x^2 + y^2 = 36$

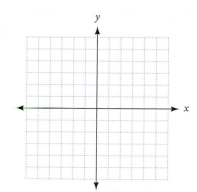

11. $3x^2 + 4y^2 = 12$

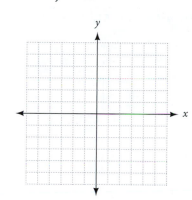

12. $6x^2 + y^2 = 36$

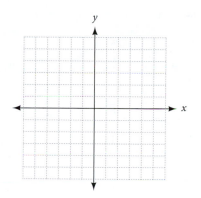

Find the x- and y-coordinates for the x- and y-intercepts, if they exist, for each of the following equations. Do not graph.

13. $0.4x^2 + 0.9y^2 = 3.6$

14. $1.6x^2 + 0.9y^2 = 14.4$

15. $\dfrac{x^2}{0.04} + \dfrac{y^2}{0.09} = 1$

16. $\dfrac{y^2}{0.16} + \dfrac{x^2}{0.25} = 1$

17. $\dfrac{25x^2}{9} + \dfrac{25y^2}{4} = 1$

18. $\dfrac{16x^2}{9} + \dfrac{16y^2}{25} = 1$

19. $\dfrac{4x^2}{9} + \dfrac{25y^2}{16} = 1$

20. $\dfrac{9x^2}{16} + \dfrac{4y^2}{25} = 1$

B Graph each of the following ellipses. In each case, label the coordinates of the center and the vertices.

21. $\dfrac{(x-4)^2}{4} + \dfrac{(y-2)^2}{9} = 1$

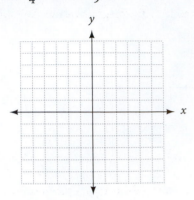

22. $\dfrac{(x-2)^2}{4} + \dfrac{(y-4)^2}{9} = 1$

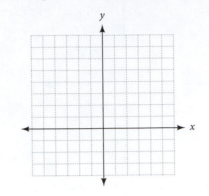

23. $4x^2 + y^2 - 4y - 12 = 0$

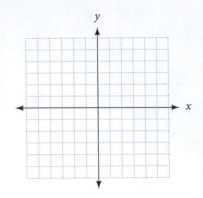

24. $4x^2 + y^2 - 24x - 4y + 36 = 0$

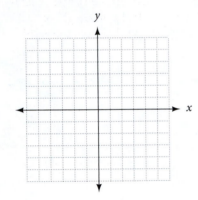

25. $x^2 + 9y^2 + 4x - 54y + 76 = 0$

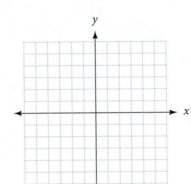

26. $4x^2 + y^2 - 16x + 2y + 13 = 0$

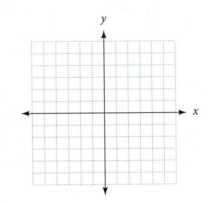

27. Find x when $y = 4$ in the equation $\dfrac{x^2}{25} + \dfrac{y^2}{16} = 1$.

28. Find y when $x = -5$ in the equation $\dfrac{x^2}{25} + \dfrac{y^2}{16} = 1$.

29. Find y when $x = 3$ in the equation $\dfrac{x^2}{25} + \dfrac{y^2}{16} = 1$.

30. Find y when $x = 4$ in the equation $\dfrac{x^2}{25} + \dfrac{y^2}{16} = 1$.

31. Find x when $y = -3$ in the equation $\dfrac{x^2}{25} + \dfrac{y^2}{16} = 1$.

32. Find x when $y = -2$ in the equation $\dfrac{x^2}{25} + \dfrac{y^2}{16} = 1$.

33. The longer line segment connecting opposite vertices of an ellipse is called the *major axis* of the ellipse. Give the length of the major axis of the ellipse you graphed in Problem 3.

34. The shorter line segment connecting opposite vertices of an ellipse is called the *minor axis* of the ellipse. Give the length of the minor axis of the ellipse you graphed in Problem 3.

Applying the Concepts

Some of the problems that follow use the major and minor axes mentioned in Problems 33 and 34. The diagram below shows the minor axis and the major axis for an ellipse.

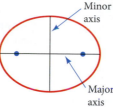

Minor axis

Major axis

In any ellipse, the length of the major axis is $2a$, and the length of the minor axis is $2b$ (these are the same a and b that appear in the general equations of an ellipse). Each of the two points shown on the major axis is a focus of the ellipse. If the distance from the center of the ellipse to each focus is c, then it is always true that $a^2 = b^2 + c^2$. You will need this information for some of the problems that follow.

35. The Colosseum The Colosseum in Rome seated 50,000 spectators around a central elliptical arena. The base of the Colosseum measured 615 feet long and 510 feet wide. Write an equation for the elliptical shape of the Colosseum.

510 ft wide

615 ft long

36. Archway A new theme park is planning an archway at its main entrance. The arch is to be in the form of a semi-ellipse with the major axis as the span. If the span is to be 40 feet and the height at the center is to be 10 feet, what is the equation of the ellipse? How far left and right of center could a 6-foot man walk upright under the arch?

10 feet

40 feet

37. Garden Trellis John is planning to build an arched trellis for the entrance to his botanical garden. If the arch is to be in the shape of the upper half of an ellipse that is 6 feet wide at the base and 9 feet high, what is the equation for the ellipse?

Butterfly Garden

9 ft

6 ft

38. The Ellipse President's Park, located between the White House and the Washington Monument in Washington, DC, is also called The Ellipse. The park is enclosed by an elliptical path with major axis 458 meters and minor axis 390 meters. What is the equation for the path around The Ellipse?

39. Elliptical Pool Table A children's science museum plans to build an elliptical pool table to demonstrate that a ball rolled from a particular point (focus) will always go into a hole located at another particular point (the other focus). The focus needs to be 1 foot from the vertex of the ellipse. If the table is to be 8 feet long, how wide should it be? *Hint:* The distance from the center to each focus point is represented by c and is found by using the equation $a^2 = b^2 + c^2$.

40. Elliptical Pool Table The eccentricity of the ellipse $\frac{x^2}{a^2} + \frac{y^2}{b^2}$ is defined by the formula

$$\sqrt{\frac{a^2 - b^2}{a^2}}$$

It measures how "circular" and an ellipse appears (circles have an eccentricity of 0). Find the eccentricity of the pool table from Problem 39.

41. The Oval Office One of the most famous oval shaped rooms is the Oval Office, the US President's official office at the White House. The oval office is approximately 35 feet long and 29 feet wide.

   a. Let (0, 0) represent the center of the oval office, with the length along the *x*-axis. Find the equation for the outer ellipse of the office.

   b. Approximately how far from the center are the foci positioned?

   c. Using the formula from problem 40, find the eccentricity of the oval office.

42. The Blue Room Another elliptical room at the White House is the Blue Room, which measures 40 feet long and 30 feet wide. Repeat problem 41 for the Blue Room.

Maintaining Your Skills

43. Disney Movies The chart shows the Disney movies that have made the most money. How much more money did *101 Dalmatians* make than *The Jungle Book*?

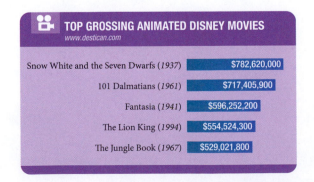

44. Farmer's Markets The chart shows the increasing number of farmer's markets in the United States. Using the equation of the line that connects the first and the last point, estimate the number of farmer's markets in 2007. Round to the nearest market.

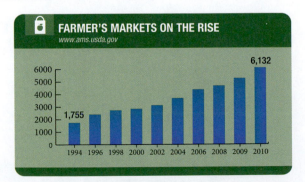

Getting Ready for the Next Section

Find the x-coordinates of the x-intercepts and the y-coordinates of the y-intercepts (if they exist) of each equation.

45. $\dfrac{x^2}{9} - \dfrac{y^2}{4} = 1$

46. $\dfrac{x^2}{4} - \dfrac{y^2}{9} = 1$

47. $\dfrac{y^2}{5} - \dfrac{x^2}{9} = 1$

48. $\dfrac{y^2}{9} - \dfrac{x^2}{5} = 1$

49. If $\dfrac{x^2}{9} - \dfrac{y^2}{16} = 1$, find x when $y = 1$.

50. If $\dfrac{y^2}{16} - \dfrac{x^2}{9} = 1$, find y when $x = 1$.

51. If $\dfrac{x^2}{25} - \dfrac{y^2}{4} = 1$, find y when $x = -7$.

52. If $\dfrac{y^2}{4} - \dfrac{x^2}{25} = 1$, find x when $y = -6$.

Find the Mistake

Each problem below contains a mistake. Circle the mistake and write the correct number(s) or word(s) on the line provided.

1. The equation $\dfrac{x^2}{a^2} + \dfrac{y^2}{b^2} = 1$ will give a graph in the shape of a parabola. _____

2. The graph of the equation $\dfrac{x^2}{a^2} + \dfrac{y^2}{b^2} = 1$ will be an ellipse centered at $(0, 1)$. _____

3. When a and b in the equation $\dfrac{x^2}{a^2} + \dfrac{y^2}{b^2} = 1$ are opposite, the graph will be a circle. _____

4. The graph of the equation $\dfrac{(x-h)^2}{a^2} + \dfrac{(y-k)^2}{b^2} = 1$ will have vertices at (h, k) and (a, b).

_____

Landmark Review: Checking Your Progress

Find the distance between the following points.

1. $(5, 2)$ and $(0, 2)$ **2.** $(3, 1)$ and $(1, 2)$ **3.** $(0, 4)$ and $(3, 3)$ **4.** $(0, 0)$ and $(4, 2)$

Write the equation of a circle with the given center and radius.

5. Center $(-4, 3)$; $r = 2$ **6.** Center $(0, 3)$; $r = 5$

Give the radius and center of each of the following circles.

7. $x^2 + y^2 = 9$ **8.** $(x + 3)^2 + (y - 2)^2 = 5$

Graph each of the following parabolas.

9. $y = (x + 3)^2 - 2$ **10.** $x = 2y^2 - 4y - 2$

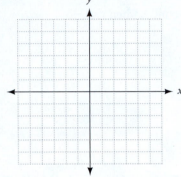 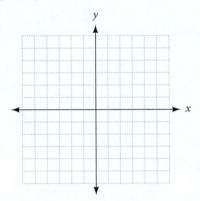

11. $\dfrac{x^2}{4} + \dfrac{y^2}{16} = 1$ **12.** $16x^2 + 9y^2 = 144$

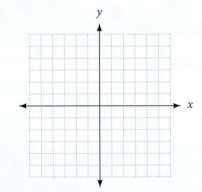 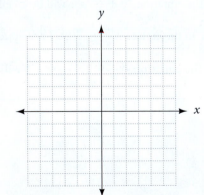

9.4 Hyperbolas

OBJECTIVES

A Graph a hyperbola from its standard form.

B Use completing the square to convert a hyperbola to standard form to graph the curve.

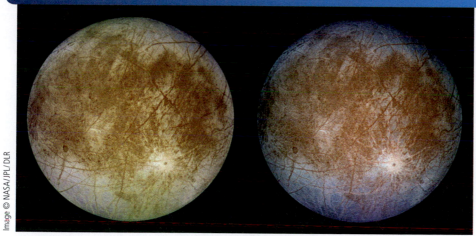

Image © NASA/JPL/DLR

The photo above shows Europa, one of Jupiter's moons, as it was photographed by the Galileo space probe in the late 1990s. To speed up the trip from Earth to Jupiter—nearly a billion miles—Galileo made use of the *slingshot effect*. This involves flying a hyperbolic path very close to a planet, so that gravity can be used to gain velocity as the space probe hooks around the planet (Figure 1). In this section we will work with equations that produce hyperbolas.

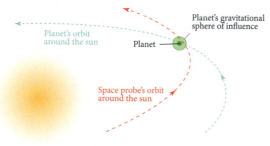

Planet's gravitational sphere of influence

Planet's orbit around the sun

Planet

Space probe's orbit around the sun

FIGURE 1

A Hyperbolas

Consider the following equation:

$$\frac{x^2}{9} - \frac{y^2}{4} = 1$$

If we were to find a number of ordered pairs that are solutions to this equation and connect their graphs with a smooth curve, we would have Figure 2.

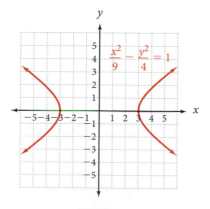

$$\frac{x^2}{9} - \frac{y^2}{4} = 1$$

FIGURE 2

This graph is an example of a *hyperbola*. Notice that the graph has x-intercepts at $(3, 0)$ and $(-3, 0)$. The graph has no y-intercepts and hence does not cross the y-axis. We can show this by substituting $x = 0$ into the equation.

$$\frac{0^2}{9} - \frac{y^2}{4} = 1 \qquad \text{Substitute 0 for } x.$$

$$-\frac{y^2}{4} = 1 \qquad \text{Simplify left side.}$$

$$y^2 = -4 \qquad \text{Multiply each side by } -4.$$

for which there is no real solution.

We want to produce reasonable sketches of hyperbolas without having to build extensive tables. We can produce the graphs we are after by using what are called *asymptotes* for our graphs. The discussion that follows is intended to give you some insight as to why these asymptotes exist. However, even if you don't understand this discussion completely, you will still be able to graph hyperbolas.

Asymptotes for Hyperbolas　　Let's solve the equation we graphed above for y.

$$\frac{x^2}{9} - \frac{y^2}{4} = 1 \qquad \text{Original equation}$$

$$-\frac{y^2}{4} = -\frac{x^2}{9} + 1 \qquad \text{Add } -\frac{x^2}{9} \text{ to each side.}$$

$$y^2 = \frac{4x^2}{9} - 4 \qquad \text{Multiply each side by } -4.$$

$$y = \pm\sqrt{\frac{4x^2}{9} - 4} \qquad \text{Square root property of equality}$$

To understand what comes next, you need to see that for very large values of x, the following expressions are almost the same:

$$\sqrt{\frac{4x^2}{9} - 4} \qquad \text{and} \qquad \sqrt{\frac{4x^2}{9}} = \frac{2}{3}x$$

This is because the 4 becomes insignificant compared with $\frac{4x^2}{9}$ for very large values of x. In fact, the larger x becomes, the closer these two expressions are to being equal. The table below is intended to help you see this fact.

x	$\sqrt{\dfrac{4x^2}{9} - 4}$	$\sqrt{\dfrac{4x^2}{9}} = \dfrac{2}{3}x$
1	undefined	0.67
10	6.35959	6.66667
100	66.63666	66.66667
1000	666.66367	666.66667
10000	6666.66637	6666.66667

Extending the idea presented above, we can say that, for very large values of x, the graphs of the equations

$$y = \pm\sqrt{\frac{4x^2}{9} - 4} \qquad \text{and} \qquad y = \pm\frac{2}{3}x$$

will be close to each other. Further, the larger x becomes, the closer the graphs are to one another. (Using a similar line of reasoning, we can draw the same conclusion for values of x on the other side of the origin, $-10, -100, -1{,}000$, and $-10{,}000$.) Believe it or not, this helps us find the shape of our hyperbola. We simply note that the graph of

$$\frac{x^2}{9} - \frac{y^2}{4} = 1$$

crosses the x-axis at -3 and 3, and that as x gets further and further from the origin, the graph looks more like the graph of $y = \frac{2}{3}x$ and $y = -\frac{2}{3}x$.

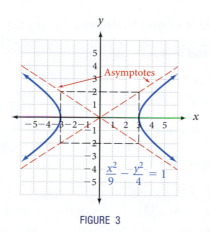

FIGURE 3

The lines $y = \frac{2}{3}x$ and $y = -\frac{2}{3}x$ are asymptotes for the graph of the hyperbola $\frac{x^2}{9} - \frac{y^2}{4} = 1$. The further we get from the origin, the closer the hyperbola is to these lines.

Asymptotes from a Rectangle In Figure 3, note the rectangle that has its sides parallel to the x- and y-axes and that passes through the x-intercepts and the points on the y-axis corresponding to the square roots of the number below y^2, $+2$ and -2. The lines that connect opposite corners of the rectangle are the *asymptotes* for graph of the hyperbola

$$\frac{x^2}{9} - \frac{y^2}{4} = 1$$

EXAMPLE 1 Graph the equation $\frac{y^2}{9} - \frac{x^2}{16} = 1$.

Solution In this case the y-intercepts are 3 and -3, and the x-intercepts do not exist. We can use the square roots of the number below x^2, however, to find the asymptotes associated with the graph. The sides of the rectangle used to draw the asymptotes must pass through 3 and -3 on the y-axis, and 4 and -4 on the x-axis. Figure 4 shows the rectangle, the asymptotes, and the hyperbola.

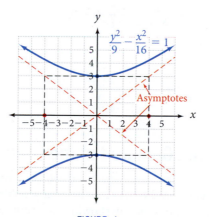

FIGURE 4

Practice Problems

1. Graph the equation
$$\frac{x^2}{25} - \frac{y^2}{9} = 1.$$

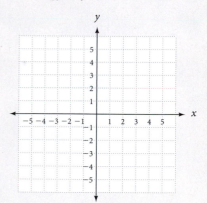

Here is a summary of what we have for hyperbolas:

Hyperbolas Centered at the Origin

The graph of the equation

$$\frac{x^2}{h^2} - \frac{y^2}{k^2} = 1$$

will be a *hyperbola centered at the origin.* The graph will have *x*-intercepts (*vertices*) at $(-a, 0)$ and $(a, 0)$.

The graph of the equation

$$\frac{y^2}{k^2} - \frac{x^2}{h^2} = 1$$

will be a *hyperbola centered at the origin.* The graph will have *y*-intercepts (*vertices*) at $(-b, 0)$ and $(b, 0)$.

As an aid in sketching either of these equations, the asymptotes can be found by graphing the lines $y = \frac{b}{a}x$ and $y = -\frac{b}{a}x$, or by drawing lines through opposite corners of the rectangle whose sides pass through $-a, a, -b,$ and b on the axes.

2. Graph the equation

$$25x^2 - 100y^2 = 100.$$

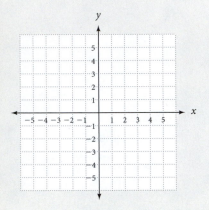

EXAMPLE 2   Graph the equation $64x^2 - 16y^2 = 64$.

Solution   Dividing by 64 to convert the equation to standard form.

$$\frac{64x^2}{64} - \frac{16y^2}{64} = \frac{64}{64}$$

$$\frac{x^2}{1} - \frac{y^2}{4} = 1$$

The graph crosses the *x*-axis at $(1, 0), (-1)$ which are the *x*-intercepts (vertices). The sides of the rectangle used to draw the asymptotes must pass through -1 and 1 on the *x*-axis, and -2 and 2 on the *y*-axis.

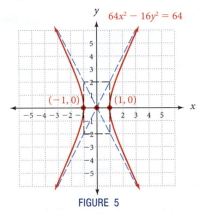

FIGURE 5

B Hyperbolas not Centered at the Origin

The ideas associated with graphing hyperbolas whose centers are not at the origin parallel the ideas presented in the last section about graphing ellipses whose centers have been moved off the origin. Without showing the justification for doing so, we state the following guidelines for graphing hyperbolas:

Note   Remember to be careful when determining the signs of *h* and *k* since each is preceded by a subtraction sign.

Hyperbolas with Centers at (h, k)

The graphs of the equations

$$\frac{(x - h)^2}{a^2} - \frac{(y - k)^2}{b^2} = 1 \quad \text{and} \quad \frac{(y - k)^2}{b^2} - \frac{(x - h)^2}{a^2} = 1$$

will be *hyperbolas with their centers at (h, k).* The vertices of the graph of the first equation will be at the points $(h + a, k)$ and $(h - a, k)$, and the vertices for the graph of the second equation will be at $(h, k + b)$ and $(h, k - b)$. In either case, the asymptotes can be found by connecting opposite corners of the rectangle that contains the four points $(h + a, k), (h - a, k), (h, k + b),$ and $(h, k - b)$.

EXAMPLE 3 Graph the hyperbola $4x^2 - y^2 + 4y - 20 = 0$.

Solution To identify the coordinates of the center of the hyperbola, we need to complete the square on y. (Because there is no linear term in x, we do not need to complete the square on x. The x-coordinate of the center will be $x = 0$.)

$$4x^2 - y^2 + 4y - 20 = 0$$

$$4x^2 - y^2 + 4y = 20 \qquad \text{Add 20 to each side.}$$

$$4x^2 - 1(y^2 - 4y) = 20 \qquad \text{Factor } -1 \text{ from each term containing } y.$$

To complete the square on y, we add 4 to the terms inside the parentheses. Doing so adds -4 to the left side of the equation because everything inside the parentheses is multiplied by -1. To keep from changing the equation we must add -4 to the right side also.

$$4x^2 - 1(y^2 - 4y + 4) = 20 - 4 \qquad \text{Add } -4 \text{ to each side.}$$

$$4x^2 - 1(y - 2)^2 = 16 \qquad y^2 - 4y + 4 = (y - 2)^2$$

$$\frac{4x^2}{16} - \frac{(y - 2)^2}{16} = \frac{16}{16} \qquad \text{Divide each side by 16.}$$

$$\frac{x^2}{4} - \frac{(y - 2)^2}{16} = 1 \qquad \text{Simplify each term.}$$

This is the equation of a hyperbola with center at $(0, 2)$. The graph opens to the right and left as shown in Figure 6.

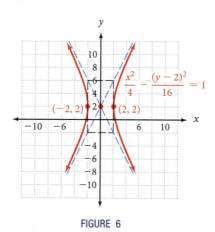

FIGURE 6

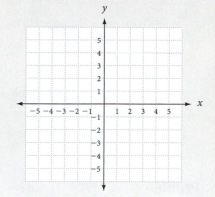

3. Graph the hyperbola

$$9y^2 - 36y - 4x^2 - 32x - 64 = 0.$$

GETTING READY FOR CLASS

After reading through the preceding section, respond in your own words and in complete sentences.

A. How do we find the vertices of a hyperbola?

B. How can you tell by looking at an equation if its graph will be an ellipse or a hyperbola?

C. How do you find the asymptotes of a hyperbola?

D. Are the points on the asymptotes of a hyperbola in the solution set of the equation of the hyperbola? Explain. (That is, are the asymptotes actually part of the graph?)

Vocabulary Review

Choose the correct words to fill in the blanks below.

origin   vertices   asymptotes   hyperbola

1. A _____ centered at the origin will have a graph of $\frac{x^2}{a^2} - \frac{y^2}{b^2} = 1$ or $\frac{y^2}{b^2} - \frac{x^2}{a^2} = 1$.

2. The further the graph of a hyperbola gets from its _____, the closer it gets to its asymptotes.

3. To help you draw the graph of a hyperbola centered at the origin, graph the lines of the _____ given by $y = \left(\frac{b}{a}\right)x$ and $y = -\left(\frac{b}{a}\right)x$.

4. A hyperbola that is not centered at the _____ will have a center at (h, k).

Problems

A Graph each of the following. Show the intercepts and the asymptotes in each case.

1. $\frac{x^2}{9} - \frac{y^2}{16} = 1$

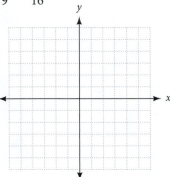

2. $\frac{x^2}{25} - \frac{y^2}{4} = 1$

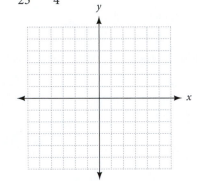

3. $\frac{x^2}{16} - \frac{y^2}{9} = 1$

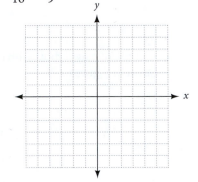

4. $\frac{x^2}{4} - \frac{y^2}{25} = 1$

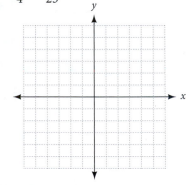

5. $\frac{y^2}{9} - \frac{x^2}{16} = 1$

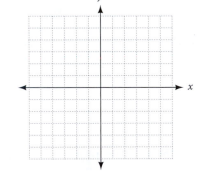

6. $\frac{y^2}{25} - \frac{x^2}{4} = 1$

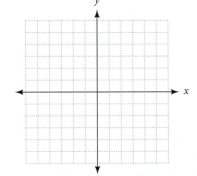

7. $\dfrac{y^2}{36} - \dfrac{x^2}{4} = 1$

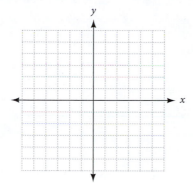

8. $\dfrac{y^2}{4} - \dfrac{x^2}{36} = 1$

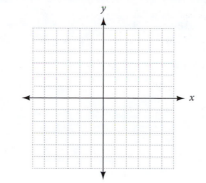

9. $x^2 - 4y^2 = 4$

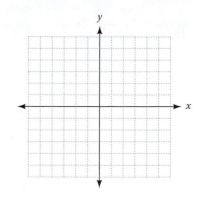

10. $y^2 - 4x^2 = 4$

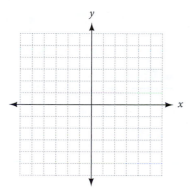

11. $16y^2 - 9x^2 = 144$

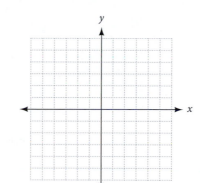

12. $4y^2 - 25x^2 = 100$

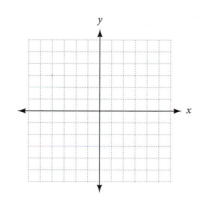

Find the x- and y-coordinates for the x- and y-intercepts, if they exist, for each of the following. Do not graph.

13. $0.4x^2 - 0.9y^2 = 3.6$

14. $1.6x^2 - 0.9y^2 = 14.4$

15. $\dfrac{x^2}{0.04} - \dfrac{y^2}{0.09} = 1$

16. $\dfrac{y^2}{0.16} - \dfrac{x^2}{0.25} = 1$

17. $\dfrac{25x^2}{9} - \dfrac{25y^2}{4} = 1$

18. $\dfrac{16x^2}{9} - \dfrac{16y^2}{25} = 1$

19. $\dfrac{25y^2}{16} - \dfrac{4x^2}{9} = 1$

20. $\dfrac{4y^2}{25} - \dfrac{9x^2}{16} = 1$

B Graph each of the following hyperbolas. In each case, label the coordinates of the center and the vertices and show the asymptotes.

21. $\dfrac{(x-2)^2}{16} - \dfrac{y^2}{4} = 1$

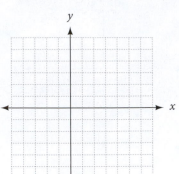

22. $\dfrac{(y-2)^2}{16} - \dfrac{x^2}{4} = 1$

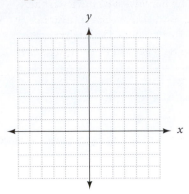

23. $9y^2 - x^2 - 4x + 54y + 68 = 0$

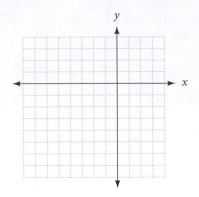

24. $4x^2 - y^2 - 24x + 4y + 28 = 0$

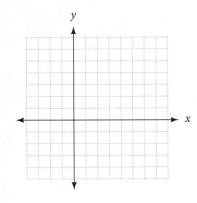

25. $4y^2 - 9x^2 - 16y + 72x - 164 = 0$

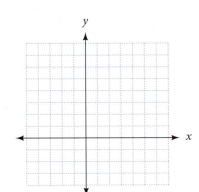

26. $4x^2 - y^2 - 16x - 2y + 11 = 0$

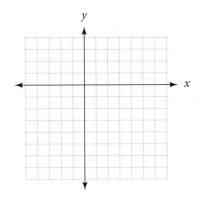

27. Find x when $y = 4$ in the equation $\dfrac{x^2}{25} - \dfrac{y^2}{16} = 1$.

28. Find y when $x = -5$ in the equation $\dfrac{x^2}{25} - \dfrac{y^2}{16} = 1$.

29. Find y when $x = 3$ in the equation $\dfrac{x^2}{25} - \dfrac{y^2}{16} = 1$.

30. Find y when $x = 6$ in the equation $\dfrac{x^2}{25} - \dfrac{y^2}{16} = 1$.

31. Find x when $y = -3$ in the equation $\dfrac{x^2}{25} - \dfrac{y^2}{16} = 1$.

32. Find x when $y = -2$ in the equation $\dfrac{x^2}{25} - \dfrac{y^2}{16} = 1$.

33. Find the equations of the asymptotes of the hyperbola.

$$\frac{x^2}{25} - \frac{y^2}{16} = 1$$

34. Find the equations of the asymptotes of the hyperbola.

$$\frac{y^2}{25} - \frac{x^2}{16} = 1$$

35. Find the equations of the asymptotes of the hyperbola.

$$\frac{(x+1)^2}{9} - \frac{(y-2)^2}{16} = 1$$

36. Find the equations of the asymptotes of the hyperbola.

$$\frac{(y-2)^2}{9} - \frac{(x+1)^2}{16} = 1$$

Applying the Concepts

37. Entering the Zoo A zoo is planning a new entrance. Visitors are to be "funneled" into the zoo between two tall brick fences. The bases of the fences will be in the shape of a hyperbola. The narrowest passage East and West between the fences will be 24 feet. The total North–South distance of the fences is to be 50 feet. Write an equation for the hyperbolic shape of the fences if the center of the hyperbola is to be placed at the origin of the coordinate system.

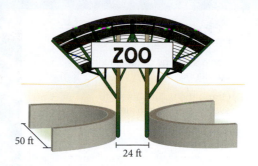

50 ft

24 ft

38. Make an argument and table similar to that preceding Example 1 of this section to show that the asymptotes of the hyperbola $\frac{y^2}{25} - \frac{x^2}{4} = 1$ are $y = \pm\frac{5}{2}x$.

Getting Ready for the Next Section

39. Which of the following are solutions to $x^2 + y^2 < 16$?

$(0, 0)$ $(4, 0)$ $(0, 5)$

40. Which of the following are solutions to $y \geq x^2 - 16$?

$(0, 0)$ $(-2, 0)$ $(0, -2)$

Expand and multiply.

41. $(2y + 4)^2$

42. $(y + 3)^2$

43. Solve $x - 2y = 4$ for x.

44. Solve $2x + 3y = 6$ for y.

Simplify.

45. $x^2 - 2(x^2 - 3)$

46. $x^2 + (x^2 - 4)$

47. $x^2 - 4(x^2 + 2)$

48. $y^2 - (2y^2 - 4)$

Factor.

49. $5y^2 + 16y + 12$

50. $3x^2 + 17x - 28$

51. $2y^2 - 8$

52. $2x^3 - 54$

Solve.

53. $y^2 = 4$

54. $x^2 = 25$

55. $-x^2 + 6 = 2$

56. $5y^2 + 16 + 12 = 0$

Find the Mistake

Each problem below contains a mistake. Circle the mistake and write the correct number(s) or word(s) on the line provided.

1. The graph of $\frac{x^2}{a^2} + \frac{y^2}{b^2} = 1$ will have y-intercepts at $(0, -b)$ and $(0, b)$. _____

2. Asymptotes for a hyperbola given by the equation $\frac{y^2}{b^2} + \frac{x^2}{a^2} = 1$ can be found by graphing the lines $x = \frac{b}{a}y$ and $x = -\frac{b}{a}y$. _____

3. The asymptotes of a hyperbola are lines that the graph of the hyperbola will intersect. _____

4. Hyperbolas given by the equation $\frac{(x-h)^2}{a^2} - \frac{(y-k)^2}{b^2} = 1$ or $\frac{(y-k)^2}{b^2} - \frac{(x-h)^2}{a^2} = 1$ will have centers at the origin. _____

OBJECTIVES

A Graph second-degree inequalities.

B Solve systems of nonlinear equations.

C Graph the solution sets to systems of inequalities.

KEY WORDS

second-degree inequalities

nonlinear system

nonlinear equation

boundary

The tarantula is a very large arachnid often kept as a pet. Venom vents through his two curved fangs, but is rarely hazardous to humans. As a defense mechanism, the tarantula's abdomen is covered with a layer of barbed hairs called urticating hairs. The tarantula uses his back pair of legs to flick these hairs off his abdomen towards a threatening target. Tarantula owners are advised to wear eye protection when handling the spider. One such owner learned this lesson the difficult way. The man was cleaning a stain on the glass wall of the tarantula's tank. He turned toward the spider, who hit him in the face with the hairs. A few of the hairs lodged in the man's cornea and began to irritate his eye. Doctors treated the man's condition with antibiotics after being unable to extract the hairs.

We can use this scenario to learn about systems of nonlinear equations. Suppose we can picture the man's eye as a circle given by the equation $x^2 + y^2 = 9$, and the projection of the hairs given by the equation $y = -\frac{1}{3}x + 1$. The solution for this system can be used to picture where the hairs lodged in the man's eye. In this section, we will learn how to solve systems of nonlinear equations, such as this one. But first, let's explore second-degree inequalities.

In Section 2.7, we graphed linear inequalities by first graphing the boundary and then choosing a test point not on the boundary to indicate the region used for the solution set. The problems in this section are very similar. We will use the same general methods for graphing second-degree inequalities in two variables that we used in Section 2.7.

A Second-Degree Inequalities

A *second-degree inequality* is an inequality that contains at least one squared variable.

EXAMPLE 1 Graph $x^2 + y^2 < 16$.

Solution   The boundary is $x^2 + y^2 = 16$, which is a circle with center at the origin and a radius of 4. Because the inequality sign is $<$, the boundary is not included in the solution set and must therefore be represented with a broken line. The graph of the boundary is shown in Figure 1.

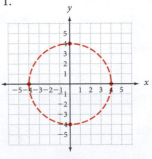

FIGURE 1

Practice Problems

1. Graph $x^2 + y^2 > 9$.

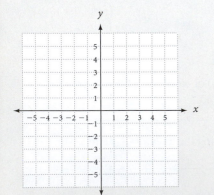

The solution set for $x^2 + y^2 < 16$ is either the region inside the circle or the region outside the circle. To see which region represents the solution set, we choose a convenient point not on the boundary and test it in the original inequality. The origin $(0, 0)$ is a convenient point. Because the origin satisfies the inequality $x^2 + y^2 < 16$, all points in the same region will also satisfy the inequality. The graph of the solution set is shown in Figure 2.

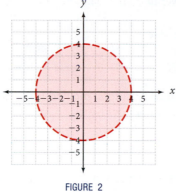

FIGURE 2

EXAMPLE 2 Graph the inequality $y \leq x^2 - 2$.

Solution The parabola $y = x^2 - 2$ is the boundary and is included in the solution set. Using $(0, 0)$ as the test point, we see that $0 \leq 0^2 - 2$ is a false statement, which means that the region containing $(0, 0)$ is not in the solution set. (See Figure 3.)

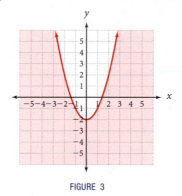

FIGURE 3

2. Graph $y \geq x^2 + 3$.

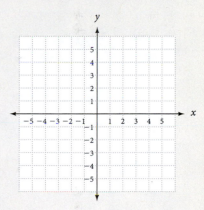

EXAMPLE 3 Graph $4y^2 - 9x^2 < 36$.

Solution The boundary is the hyperbola $4y^2 - 9x^2 = 36$ and is not included in the solution set. Testing $(0, 0)$ in the original inequality yields a true statement, which means that the region containing the origin is the solution set. (See Figure 4.)

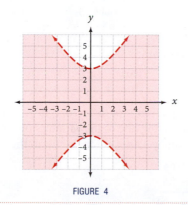

FIGURE 4

3. Graph $9x^2 - 4y^2 > 36$.

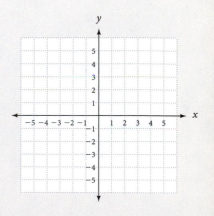

B Nonlinear Systems

In an earlier chapter, we worked with equations of linear systems. Now we will practice solving nonlinear systems, which include at least one nonlinear equation. A *nonlinear equation* is an equation whose graph is something other than a line, such as a circle, ellipse, hyperbola, or a parabola. In most cases, we can use the substitution method to solve nonlinear systems. However, we can use the elimination method if like variables are raised to the same power in both equations.

EXAMPLE 4   Solve the system.

$$x^2 + y^2 = 4$$
$$x - 2y = 4$$

Solution   In this case, the substitution method is the most convenient. Solving the second equation for x in terms of y, we have

$$x - 2y = 4$$
$$x = 2y + 4$$

We now substitute $2y + 4$ for x in the first equation in our original system and proceed to solve for y.

$$(2y + 4)^2 + y^2 = 4$$

$$4y^2 + 16y + 16 + y^2 = 4 \qquad \text{Expand } (2y + 4)^2.$$

$$5y^2 + 16y + 16 = 4 \qquad \text{Simplify left side.}$$

$$5y^2 + 16y + 12 = 0 \qquad \text{Add } -4 \text{ to each side.}$$

$$(5y + 6)(y + 2) = 0 \qquad \text{Factor.}$$

$$5y + 6 = 0 \quad \text{or} \quad y + 2 = 0 \qquad \text{Set factors equal to 0.}$$

$$y = -\frac{6}{5} \qquad\qquad y = -2 \qquad \text{Solve.}$$

These are the y-coordinates of the two solutions to the system. Substituting $y = -\frac{6}{5}$ into $x - 2y = 4$ and solving for x gives us $x = \frac{8}{5}$. Using $y = -2$ in the same equation yields $x = 0$. The two solutions to our system are $\left(\frac{8}{5}, -\frac{6}{5}\right)$ and $(0, -2)$. Although graphing the system is not necessary, it does help us visualize the situation. (See Figure 5.)

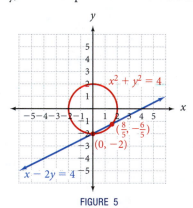

FIGURE 5

EXAMPLE 5   Solve the system.

$$16x^2 - 4y^2 = 64$$
$$x^2 + y^2 = 9$$

Solution   Because each equation is of the second degree in both x and y, it is easier to solve this system by eliminating one of the variables by addition. To eliminate y, we multiply the bottom equation by 4 and add the result to the top equation.

$$\begin{array}{r} 16x^2 - 4y^2 = 64 \\ 4x^2 + 4y^2 = 36 \\ \hline 20x^2 \qquad\quad = 100 \end{array}$$

---

4. Solve the system.

$$x^2 + y^2 = 9$$
$$x - y = 3$$

Note   We chose to solve the second equation for x because the absolute value of its coefficient is 1; whereas the coefficient of y is -2 and would require more steps to isolate.

5. Solve the system.

$$16x^2 - 4y^2 = 64$$
$$x^2 + y^2 = 4$$

Answers

4. $(3, 0), (0, -3)$

5. $(2, 0), (-2, 0)$

$$x^2 = 5$$
$$x = \pm\sqrt{5}$$

The x-coordinates of the points of intersection are $\sqrt{5}$ and $-\sqrt{5}$. We substitute each back into the second equation in the original system and solve for y.

$$\text{When} \rightarrow \qquad x = \sqrt{5}$$
$$(\sqrt{5})^2 + y^2 = 9$$
$$5 + y^2 = 9$$
$$y^2 = 4$$
$$y = \pm 2$$

$$\text{When} \rightarrow \qquad x = -\sqrt{5}$$
$$(-\sqrt{5})^2 + y^2 = 9$$
$$5 + y^2 = 9$$
$$y^2 = 4$$
$$y = \pm 2$$

The four points of intersection are $(\sqrt{5}, 2)$, $(\sqrt{5}, -2)$, $(\sqrt{5}, 2)$, and $(\sqrt{5}, -2)$. Graphically the situation is as shown in Figure 6.

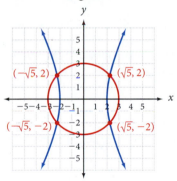

FIGURE 6

EXAMPLE 6 Solve the system.

$$x^2 - 2y = 2$$
$$y = x^2 - 3$$

Solution We can solve this system using the substitution method. Replacing y in the first equation with $x^2 - 3$ from the second equation, we have

$$x^2 - 2(x^2 - 3) = 2$$
$$-x^2 + 6 = 2$$
$$x^2 = 4$$
$$x = \pm 2$$

Using either 2 or -2 in the equation $y = x^2 - 3$ gives us $y = 1$. The system has two solutions: $(2, 1)$ and $(-2, 1)$.

EXAMPLE 7 The sum of the squares of two numbers is 34. The difference of their squares is 16. Find the two numbers.

Solution Let x and y be the two numbers. The sum of their squares is $x^2 + y^2$, and the difference of their squares is $x^2 - y^2$. (We can assume here that x^2 is the larger number.) The system of equations that describes the situation is

$$x^2 + y^2 = 34$$
$$x^2 - y^2 = 16$$

6. Solve the system.

$$x^2 + y^2 = 4$$
$$y = x^2 - 4$$

7. One number is two less than the square of another number. The sum of the squares of the two numbers is 58. Find the two numbers.

Answers

6. $(2, 0), (-2, 0), (\sqrt{3}, -1),$
$(-\sqrt{3}, -1)$

7. 3, 7 and -3, 7

We can eliminate y by simply adding the two equations. The result of doing so is

$$2x^2 = 50$$
$$x^2 = 25$$
$$x = \pm 5$$

Substituting $x = 5$ into either equation in the system gives $y = \pm 3$. Using $x = -5$ gives the same results, $y = \pm 3$. The four pairs of numbers that are solutions to the original problem are

$$(5, 3) \qquad (-5, 3) \qquad (5, -3) \qquad (-5, -3)$$

We now turn our attention to systems of inequalities. To solve a system of inequalities by graphing, we simply graph each inequality on the same set of axes. The solution set for the system is the region common to both graphs — the intersection of the individual solution sets. Each ordered pair in the common region (the part we shade) is a solution for the nonlinear system of inequalities.

C Graphing Systems of Inequalities

8. Graph the solution set for the system.

$$x^2 + y^2 \geq 9$$

$$\frac{x^2}{4} + \frac{y^2}{25} \leq 1$$

EXAMPLE 8   Graph the solution set for the system.

$$x^2 + y^2 \leq 9$$

$$\frac{x^2}{4} + \frac{y^2}{25} \geq 1$$

Solution   The boundary for the top inequality is a circle with center at the origin and a radius of 3. The solution set lies inside the boundary. The boundary for the second inequality is an ellipse. In this case, the solution set lies outside the boundary. (See Figure 7.)

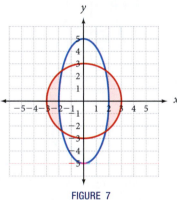

FIGURE 7

The solution set is the intersection of the two individual solution sets.

Vocabulary Review

Choose the correct words to fill in the blanks below.

    nonlinear   solid   broken   solution set   boundary   second-degree

1. A _____ inequality is an inequality that contains at least one squared variable.

2. When graphing the inequality $x^2 + y^2 < 4$, the _____ is not included in the solution set and is represented with a _____ line.

3. The boundary for the equation $y \leq x^2 + 6$ is included in the _____ and is represented by a _____ line.

4. A _____ system of equalities produces graphs for circles, ellipses, hyperbolas, and parabolas.

Problems

A Graph each of the following inequalities.

1. $x^2 + y^2 \leq 49$

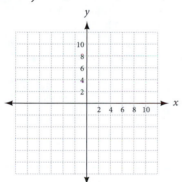

2. $x^2 + y^2 < 49$

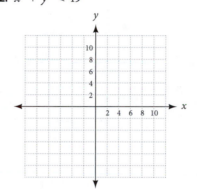

3. $(x - 2)^2 + (y + 3)^2 < 16$

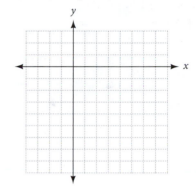

4. $(x + 3)^2 + (y - 2)^2 \geq 25$

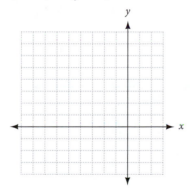

5. $y < x^2 - 6x + 7$

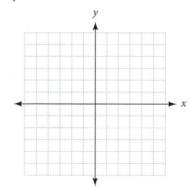

6. $y \geq x^2 + 2x - 8$

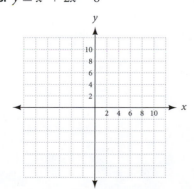

7. $\dfrac{x^2}{25} - \dfrac{y^2}{9} \geq 1$

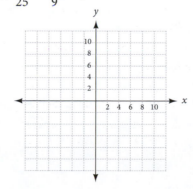

8. $\dfrac{x^2}{25} - \dfrac{y^2}{9} \leq 1$

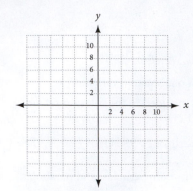

9. $4x^2 + 25y^2 \leq 100$

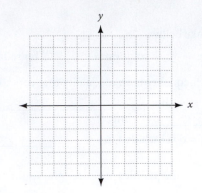

10. $25x^2 - 4y^2 > 100$

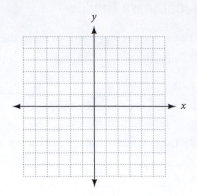

11. $\dfrac{(x+2)^2}{25} + \dfrac{(y-1)^2}{9} \leq 1$

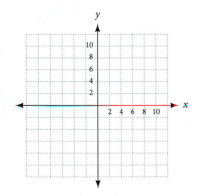

12. $\dfrac{(x-1)^2}{16} - \dfrac{(y+1)^2}{16} < 1$

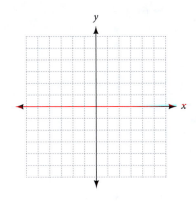

13. $16x^2 - 9y^2 \geq 144$

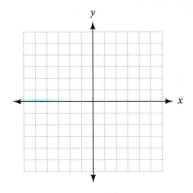

14. $16y^2 - 9x^2 < 144$

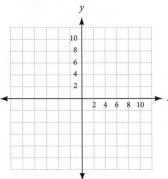

15. $9x^2 + 4y^2 + 36x - 8y + 4 < 0$

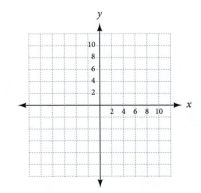

16. $9x^2 - 4y^2 + 36x + 8y \geq 4$

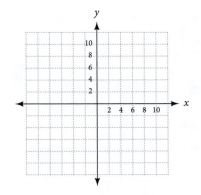

17. $9y^2 - x^2 + 18y + 2x > 1$

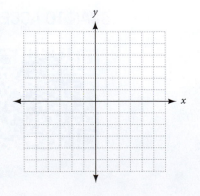

18. $x^2 + y^2 - 6x - 4y \leq 12$

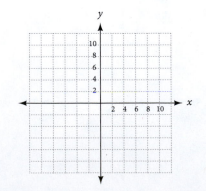

B Solve each of the following systems of equations.

19. $x^2 + y^2 = 9$
$2x + y = 3$

20. $x^2 + y^2 = 9$
$x + 2y = 3$

21. $x^2 + y^2 = 16$
$x + 2y = 8$

22. $x^2 + y^2 = 16$
$x - 2y = 8$

23. $x^2 + y^2 = 25$
$x^2 - y^2 = 25$

24. $x^2 + y^2 = 4$
$2x^2 - y^2 = 5$

25. $x^2 + y^2 = 9$
$y = x^2 - 3$

26. $x^2 + y^2 = 4$
$y = x^2 - 2$

27. $x^2 + y^2 = 16$
$y = x^2 - 4$

28. $x^2 + y^2 = 1$
$y = x^2 - 1$

29. $3x + 2y = 10$
$y = x^2 - 5$

30. $4x + 2y = 10$
$y = x^2 - 10$

31. $y = x^2 + 2x - 3$
$y = -x + 1$

32. $y = -x^2 - 2x + 3$
$y = x - 1$

33. $y = x^2 - 6x + 5$
$y = x - 5$

34. $y = x^2 - 2x - 4$
$y = x - 4$

35. $4x^2 - 9y^2 = 36$
$4x^2 + 9y^2 = 36$

36. $4x^2 + 25y^2 = 100$
$4x^2 - 25y^2 = 100$

37. $x - y = 4$
$x^2 + y^2 = 16$

38. $x + y = 2$
$x^2 - y^2 = 4$

39. $2x^2 - y = 1$
$x^2 + y = 7$

40. $x^2 + y^2 = 52$
$y = x + 2$

41. $y = x^2 - 3$
$y = x^2 - 2x - 1$

42. $y = 8 - 2x^2$
$y = x^2 - 1$

43. $4x^2 + 5y^2 = 40$
$4x^2 - 5y^2 = 40$

44. $x + 2y^2 = -4$
$3x - 4y^2 = -3$

C Graph the solution sets to the following systems.

45. $x^2 + y^2 < 9$
$y \geq x^2 - 1$

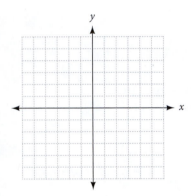

46. $x^2 + y^2 \leq 16$
$y < x^2 + 2$

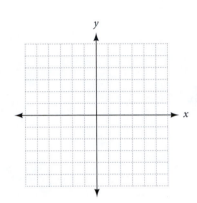

47. $\dfrac{x^2}{9} + \dfrac{y^2}{25} \leq 1$

$\dfrac{x^2}{4} - \dfrac{y^2}{9} > 1$

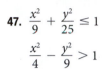
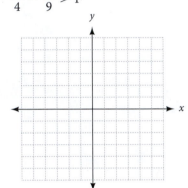

48. $\dfrac{x^2}{4} + \dfrac{y^2}{16} \geq 1$

$\dfrac{x^2}{9} - \dfrac{y^2}{25} < 1$

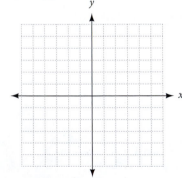

49. $4x^2 + 9y^2 \leq 36$
$y > x^2 + 2$

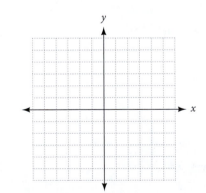

50. $9x^2 + 4y^2 \geq 36$
$y < x^2 + 1$

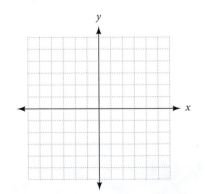

51. $x + y \leq 3$
$x - 3y \leq 3$

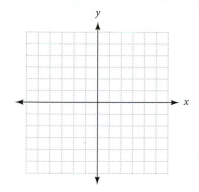

52. $x - y \leq 4$
$x + 2y \leq 4$

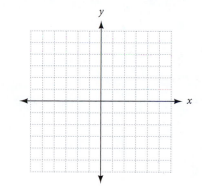

53. $x + y \leq 2$
$-x + y \leq 2$

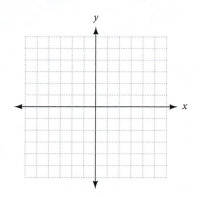

54. $x - y \leq 3$
$-x - y \leq 3$

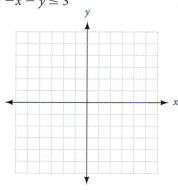

55. $x + y \leq 4$
$x \geq 0$

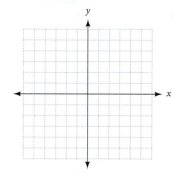

56. $x - y \leq 2$
$x \geq 0$

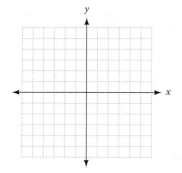

57. $2x + 3y \leq 6$
$x \geq 0$

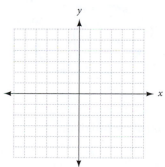

58. $x + 2y \leq 10$
$3x + 2y \leq 12$

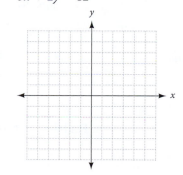

59. $x^2 + y^2 \leq 25$

$\dfrac{x^2}{9} - \dfrac{y^2}{16} > 1$

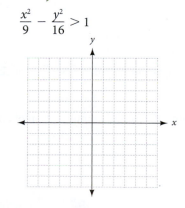

60. $\dfrac{x^2}{16} + \dfrac{y^2}{25} < 1$

$\dfrac{y^2}{4} - \dfrac{x^2}{1} \geq 1$

61. $x + y \leq 2$

$y > x^2$

62. $\dfrac{x^2}{9} + \dfrac{y^2}{16} \leq 1$

$\dfrac{x^2}{16} + \dfrac{y^2}{9} > 1$

Applying the Concepts

63. Number Problem The sum of the squares of two numbers is 89. The difference of their squares is 39. Find the numbers.

64. Number Problem The difference of the squares of two numbers is 35. The sum of their squares is 37. Find the numbers.

65. Consider the equations for the three circles below.

Circle A
$(x + 8)^2 + y^2 = 64$

Circle B
$x^2 + y^2 = 64$

Circle C
$(x - 8)^2 + y^2 = 64$

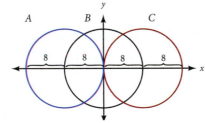

a. Find the points of intersection of circles A and B.

b. Find the points of intersection of circles B and C.

66. A magician is holding two rings that seem to lie in the same plane and intersect in two points. Each ring is 10 inches in diameter. If a coordinate system is placed with its origin at the center of the first ring and the x-axis contains the center of the second ring, then the equations are as follows:

First Ring
$x^2 + y^2 = 25$

Second Ring
$(x - 5)^2 + y^2 = 25$

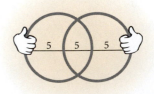

Find the points of intersection of the two rings.

Maintaining Your Skills

Let $f(x) = 3x - 7$ and $g(x) = x^2 - 4$. Find the following values.

67. $f(-4)$

68. $g(-3)$

69. $f(g(-4))$

70. $g(f(2))$

71. $f(2) - g(2)$

72. $g(3) - f(3)$

73. $\dfrac{g(4) - g(2)}{4 - 2}$

74. $\dfrac{f(3) - f(1)}{3 - 1}$

75. $\dfrac{f(g(3)) - f(g(1))}{3 - 1}$

76. $\dfrac{g(f(3)) - g(f(1))}{3 - 1}$

77. $\dfrac{g(f(-1)) - g(f(-4))}{-1 - (-4)}$

78. $\dfrac{f(g(-2)) - f(g(-5))}{-2 - (-5)}$

Find the Mistake

Each problem below contains a mistake. Circle the mistake and write the correct number(s) or word(s) on the line provided.

1. A second-degree inequality contains at least two squared variables. _____

2. Graphing the inequality $y \geq x^2 - 5$ will include the boundary but not the origin as part of the solution set.

_____

3. To solve the following system, we first solve the first equation for x and use the substitution method.
$$x^2 - y^2 = 36$$
$$x^2 + y^2 = -4$$

4. The graph of the solution set for the following system will include both boundaries.
$$x^2 + y^2 < 4$$
$$y \geq x^2$$

A Access to the internet for research

B Your choice of paper and tape, modeling clay, or other simple building materials

Regular Solids

In this course, we have worked plentifully with the cube, finding its dimensions, surface area, and volume. The cube, also called a hexahedron, is part of a larger group of solids called platonic solids. This group includes a tetrahedron, an octahedron, a dodecahedron, and an icosahedron. For this project, we would also like you to discuss the math behind the construction of these polyhedrons by answering the following questions:

1. What polygon makes up each polyhedron?

2. What is important about the vertices of each polyhedron?

3. How would you find the surface area, volume, side lengths, and slant lengths (if applicable) for each polyhedron?

4. What is the difference between a platonic solid and an Archimedean solid?

5. Can these solids, either platonic or Archimedean, be found in nature? Explain.

The next part of this project is to create a model of a platonic or Archimedean solid. You may use whatever materials you choose, whether it be paper and tape, modeling clay, or everyday supplies found at home. Once your model is complete, compose five application problems based on your three-dimensional solid.

Finally, the table below pairs a second-degree inequality with a group of 3 or 4 linear inequalities. When the linear inequalities intersect, they will represent one face of a platonic solid. This pairing creates a system of inequalities. Graph the solution set for each system.

	Linear Inequalities	Second-Degree Inequality
Graph 1	a. $x + y < -2$ b. $y > x + 6$ c. $y > -x + -6$ d. $y + 2 < x$	$x^2 + y^2 \leq 36$
Graph 2	a. $y \leq -2$ b. $y \geq \dfrac{2\sqrt{3} + 1}{2} x + 2\sqrt{3} - 1$ c. $y \geq \dfrac{2\sqrt[2]{3} + 1}{2} x + 2\sqrt{3} - 1$	$x^{2/9} + y^{2/25} < 1$

Chapter 9 Summary

Distance Formula [9.1]

The distance between the two points (x_1, y_1) and (x_2, y_2) is given by the formula

$$d = \sqrt{(x_2 - x_1)^2 + (y_2 - y_1)^2}$$

EXAMPLES

1. The distance between $(5, 2)$ and $(-1, 1)$ is

$$d = \sqrt{(5 + 1)^2 + (2 - 1)^2} = \sqrt{37}$$

Circles [9.1]

The graph of any equation of the form

$$(x - a)^2 + (y - b)^2 = r^2$$

will be a circle having its center at (a, b) and a radius of r.

2. The graph of the circle

$$(x - 3)^2 + (y + 2)^2 = 25$$

will have its center at $(3, -2)$ and the radius will be 5.

Horizontal Parabolas [9.2]

Any equation that can be written in the form $y = a(x - h)^2 + k$ is a parabola with the vertex (h, k), opening up if $a > 0$ and down if $a < 0$.

3. The graph of the parabola

$$y = 2(x - 1)^2 + 3$$

has a vertex at $(1, 3)$ and is opening upward.

Vertical Parabolas [9.2]

Any equation that can be written in the form $x = a(y - k)^2 + h$ is a parabola with vertex (h, k), opening to the right if $a > 0$ and to the left if $a < 0$.

4. The graph of the parabola

$$x = -3(y + 2)^2 + 1$$

has a vertex at $(1, -2)$ and is opening to the left.

Ellipses [9.3]

Any equation that can be put in the form

$$\frac{x^2}{a^2} + \frac{y^2}{b^2} = 1$$

will have an ellipse for its graph. The x-intercepts will be at a and $-a$, and the y-intercepts will be at b and $-b$.

5. The ellipse

$$\frac{x^2}{9} + \frac{y^2}{4} = 1$$

will cross the x-axis at 3 and -3 and will cross the y-axis at 2 and -2.

Hyperbolas [9.4]

The graph of an equation that can be put in either of the forms

$$\frac{x^2}{a^2} - \frac{y^2}{b^2} = 1 \quad \text{or} \quad \frac{y^2}{a^2} - \frac{x^2}{b^2} = 1$$

will be a hyperbola. The x-intercepts, for the first equation, will be at a and $-a$. The y-intercepts, for the second equation, will be at a and $-a$. Two straight lines, called *asymptotes,* are associated with the graph of every hyperbola. Although the asymptotes are not part of the hyperbola, they are useful in sketching the graph.

6. The hyperbola

$$\frac{x^2}{4} - \frac{y^2}{9} = 1$$

will cross the x-axis at 2 and -2. It will not cross the y-axis.

Second-Degree Inequalities in Two Variables [9.5]

7. The graph of the inequality

$$x^2 + y^2 < 9$$

is all points inside the circle with center at the origin and radius 3. The circle itself is not part of the solution and therefore is shown with a broken curve.

We graph second-degree inequalities in two variables in much the same way that we graphed linear inequalities; that is, we begin by graphing the boundary, using a solid curve if the boundary is included in the solution (this happens when the inequality symbol is $\geq$ or $\leq$) or a broken curve if the boundary is not included in the solution (when the inequality symbol is $>$ or $<$). After we have graphed the boundary, we choose a test point that is not on the boundary and try it in the original inequality. A true statement indicates we are in the region of the solution. A false statement indicates we are not in the region of the solution.

Systems of Nonlinear Equations [9.5]

8. We can solve the system

$$x^2 + y^2 = 4$$
$$x = 2y + 4$$

by substituting $2y + 4$ from the second equation for x in the first equation.

$$(2y + 4)^2 + y^2 = 4$$

$$4y^2 + 16y + 16 + y^2 = 4$$

$$5y^2 + 16y + 12 = 0$$

$$(5y + 6)(y + 2) = 0$$

$$y = -\frac{6}{5} \quad \text{or} \quad y = -2$$

Substituting these values of y into the second equation in our system gives $x = \frac{8}{5}$ and $x = 0$. The solutions are $\left(\frac{8}{5}, -\frac{6}{5}\right)$ and $(0, -2)$.

A system of nonlinear equations is two equations, at least one of which is not linear, considered at the same time. The solution set for the system consists of all ordered pairs that satisfy both equations. In most cases we use the substitution method to solve these systems; however, the elimination method can be used if like variables are raised to the same power in both equations. It is sometimes helpful to graph each equation in the system on the same set of axes to anticipate the number and approximate positions of the solutions.

Find the distance between the points. [9.1]

1. $(4, -6)$ and $(-1, 6)$ **2.** $(-4, 3)$ and $(2, -5)$

3. Find y so that $(-5, y)$ is $5\sqrt{2}$ units from $(2, 4)$. [9.1]

4. Give the equation of the circle with center at $(-4, 2)$ and radius 6. [9.1]

5. Give the equation of the circle with center at the origin that contains the point $(9, 12)$. [9.1]

Find the center and radius of the circle. [9.1]

6. $x^2 + (y + 5)^2 = 9$ **7.** $x^2 + y^2 - 6x + 10y = 2$

Graph each of the following. [9.1, 9.2, 9.3]

8. $(x - 3)^2 + (y + 1)^2 = 16$ **9.** $\dfrac{x^2}{81} + \dfrac{y^2}{36} = 1$

10. $9x^2 - y^2 = 81$

11. $36x^2 + 144x + 4y^2 - 24y + 36 = 0$

12. $(x - 3)^2 + (y + 1)^2 \geq 16$ **13.** $y > x^2 - 3$

Solve the following systems. [9.5]

14. $x^2 + y^2 = 25$ **15.** $x^2 + y^2 = \dfrac{25}{4}$
     $15x + 3y = 15$ $y = x^2 - \dfrac{5}{2}$

16. Graph the solution set to the system. [9.5]
$$x^2 + y^2 > 25$$
$$y \leq -x^2 + 6$$

Match each equation with its graph. [9.1, 9.2]

17. $x^2 + y^2 = 81$ **18.** $9x^2 + y^2 = 81$

19. $x^2 - y^2 = 9$ **20.** $x^2 + 9y^2 = 81$

A.

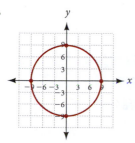

B.

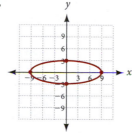

C.

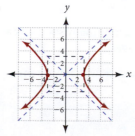

D.

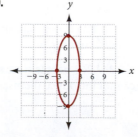

Find an equation for each graph. [9.1, 9.2]

21.

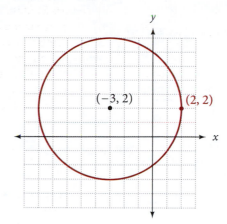

22.

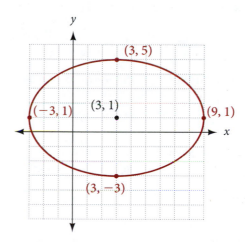

23.

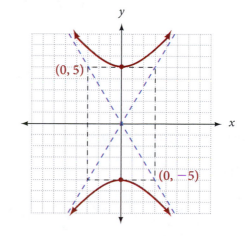

Simplify.

1. $\dfrac{-3-5}{6-7}$

2. $\dfrac{3(-2)+4(-5)}{-8-5}$

3. $\dfrac{19a^4b^{-7}}{38a^{-3}b^4}$

4. $\dfrac{x^{5/3}y^2}{x^{1/6}y^{5/6}}$

5. $\dfrac{36x^3y^2 - 12x^2y^3 + 18x^4y}{6x^2y}$

6. $\dfrac{y^2 + 7y + 12}{y^2 - 9}$

7. $\log_6 36$

8. $\log_2 8$

Factor completely.

9. $ab^3 - b^3 + 4a - 4$

10. $7x^2 - 20x - 3$

Solve.

11. $4 + 2(3x - 8) - 7x = 3$

12. $6 - 7(2x + 1) = -8$

13. $(x - 2)(x + 5) = 8$

14. $1 - \dfrac{3}{x} = \dfrac{10}{x^2}$

15. $t - 6 = \sqrt{t + 14}$

16. $\sqrt{6x - 3} = -3$

17. $(7x + 2)^2 = -48$

18. $(x - 6)^2 = -5$

Solve and graph the solution on the number line.

19. $|3x + 6| + 1 > 7$

20. $|4x - 7| - 3 < 10$

Graph on a rectangular coordinate system.

21. $y < -\dfrac{1}{2}x + 1$

22. $2x - y < -2$

23. $y = -(x + 1)^2 + 3$

24. $y = (x - 2)^2 - 3$

Multiply.

25. $\dfrac{y^2 - 3y}{2y^2 + 12y + 10} \cdot \dfrac{2y + 10}{y^2 - 5y + 6}$

26. $\dfrac{x^2 - 9}{x^2 - 2x - 15} \cdot \dfrac{x^2 - 3x - 10}{x^3 + x^2 - 2x}$

27. $(3 - 2i)(1 + 5i)$

28. $(5 + i)(6 - 8i)$

Rationalize the denominator.

29. $\dfrac{3}{\sqrt{5} - \sqrt{2}}$

30. $\dfrac{3\sqrt{10}}{2\sqrt{10} + 1}$

Find the inverse, $f^{-1}(x)$.

31. $f(x) = \dfrac{1 - x}{3}$

32. $f(x) = 5x + 2$

33. Find x to the nearest hundredth if $\log x = 3.2164$.

34. Find $\log_{17} 13$ to the nearest hundredth.

35. Solve $2mn - 1 = nx + 4$ for x.

36. Solve $S = 3x^2 + 5xy - 2y$ for y.

37. Find the slope of the line through $(5, 4)$ and $(-6, 2)$.

38. Find the slope and y-intercept for $5x - 6y = 18$.

39. If $f(x) = -\dfrac{5}{4}x + 2$, find $f(16)$.

40. If $C(t) = 60\left(\dfrac{1}{2}\right)^{t/15}$, find $C(15)$ and $C(30)$.

41. Find an equation that has solutions $x = \dfrac{1}{2}$ and $x = 4$.

42. Find an equation with solutions $t = -\dfrac{4}{3}$ and $t = \dfrac{1}{8}$.

43. If $f(x) = 2x + 5$ and $g(x) = 3 - x$, find $(f \circ g)(x)$.

44. If $f(x) = 9 - x^2$ and $g(x) = 3x + 2$, find $(g \circ f)(x)$.

45. Add -7 to the difference of -3 and 2.

46. Subtract 8 from the sum of -7 and 4.

47. Direct Variation y varies directly with x. If y is 36 when x is 12, find y when x is 7.

48. Inverse Variation y varies inversely with the square of x. If y is 2 when x is 4, find y when x is 5.

49. Mixture How much 40% alcohol solution and 90% alcohol solution must be mixed to get 50 gallons of 85% alcohol solution?

50. Mixture How many gallons of 10% alcohol solution and 70% alcohol solution must be mixed to get 8 gallons of 55% alcohol solution?

The chart shows the extent to which Americans say they know a foreign language. Use the information to answer the following questions. Round to the nearest person.

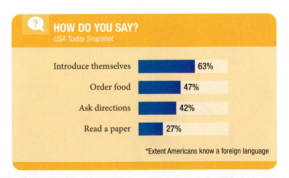

HOW DO YOU SAY?
USA Today Snapshot

Introduce themselves 63%
Order food 47%
Ask directions 42%
Read a paper 27%

*Extent Americans know a foreign language

51. If 425,382 people were surveyed, how many say they can ask directions in another language?

52. If 5,347,265 people were surveyed, how many say they cannot introduce themselves?

Find the distance between the points. [9.1]

1. $(0, -3)$ and $(-4, -6)$ **2.** $(-2, 4)$ and $(6, 0)$

3. Find y so that $(4, y)$ is $2\sqrt{5}$ units from $(6, 3)$. [9.1]

4. Give the equation of the circle with center at $(3, -1)$ and radius 4. [9.1]

5. Give the equation of the circle with center at the origin that contains the point $(-12, 5)$. [9.1]

Find the center and radius of the circle. [9.1]

6. $x^2 + (y - 1)^2 = 36$ **7.** $x^2 + y^2 - 4x + 8y = -11$

Graph each of the following. [9.1, 9.2, 9.3]

8. $(x + 2)^2 + (y - 2)^2 = 9$ **9.** $\dfrac{x^2}{100} + \dfrac{y^2}{16} = 1$

10. $4x^2 - y^2 = 64$

11. $25x^2 - 100x + 4y^2 + 8y + 4 = 0$

12. $(x - 2)^2 + (y - 2)^2 \le 9$ **13.** $y < -x^2 + 4$

Solve the following systems. [9.5]

14. $x^2 + y^2 = 16$ **15.** $x^2 + y^2 = 9$
    $2x - 2y = 8$ $y = x^2 - 7$

16. Graph the solution set to the system. [9.5]
$$x^2 + y^2 \ge 1$$
$$y > x^2 - 1$$

Match each equation with its graph. [9.1, 9.2]

17. $x^2 + y^2 = 64$ **18.** $x^2 - 4y^2 = 4$

19. $4x^2 + y^2 = 64$ **20.** $x^2 + 4y^2 = 64$

A.

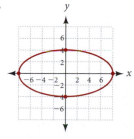

B.

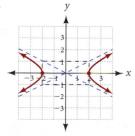

C.

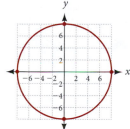

D.

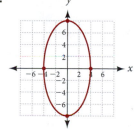

Find an equation for each graph. [9.1, 9.2]

21.

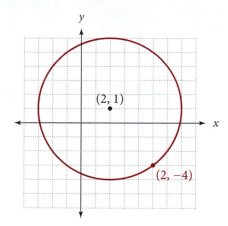

22.

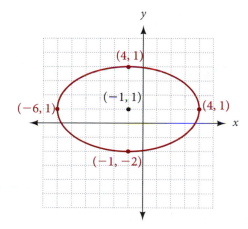

23.

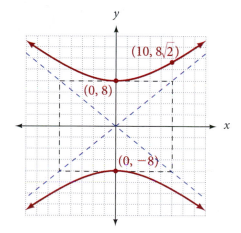

Answers to Odd-Numbered Problems

Chapter 0

Exercise Set 0.1

Vocabulary Review **1.** real number line **2.** origin
3. coordinate **4.** numerator, denominator **5.** equivalent
6. absolute value **7.** opposites **8.** factor **9.** prime
10. composite **11.** reciprocals

Problems 1–7.

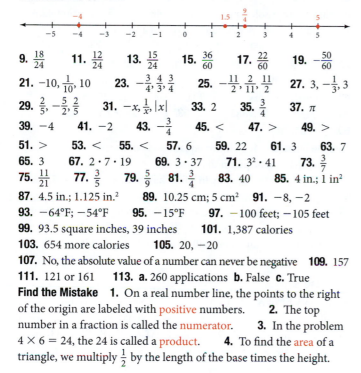

9. $\frac{18}{24}$ **11.** $\frac{12}{24}$ **13.** $\frac{15}{24}$ **15.** $\frac{36}{60}$ **17.** $\frac{22}{60}$ **19.** $-\frac{50}{60}$

21. $-10, \frac{1}{10}, 10$ **23.** $-\frac{3}{4}, \frac{4}{3}, \frac{3}{4}$ **25.** $-\frac{11}{2}, \frac{2}{11}, \frac{11}{2}$ **27.** $3, -\frac{1}{3}, 3$

29. $\frac{2}{5}, -\frac{5}{2}, \frac{2}{5}$ **31.** $-x, \frac{1}{x}, |x|$ **33.** 2 **35.** $\frac{3}{4}$ **37.** π

39. -4 **41.** -2 **43.** $-\frac{3}{4}$ **45.** $<$ **47.** $>$ **49.** $>$

51. $>$ **53.** $<$ **55.** $<$ **57.** 6 **59.** 22 **61.** 3 **63.** 7

65. 3 **67.** $2 \cdot 7 \cdot 19$ **69.** $3 \cdot 37$ **71.** $3^2 \cdot 41$ **73.** $\frac{3}{7}$

75. $\frac{11}{21}$ **77.** $\frac{3}{5}$ **79.** $\frac{5}{9}$ **81.** $\frac{3}{4}$ **83.** 40 **85.** 4 in.; 1 in²

87. 4.5 in.; 1.125 in.² **89.** 10.25 cm; 5 cm² **91.** $-8, -2$

93. $-64°F; -54°F$ **95.** $-15°F$ **97.** -100 feet; -105 feet

99. 93.5 square inches, 39 inches **101.** 1,387 calories

103. 654 more calories **105.** $20, -20$

107. No, the absolute value of a number can never be negative **109.** 157

111. 121 or 161 **113. a.** 260 applications **b.** False **c.** True

Find the Mistake **1.** On a real number line, the points to the right of the origin are labeled with positive numbers. **2.** The top number in a fraction is called the numerator. **3.** In the problem $4 \times 6 = 24$, the 24 is called a product. **4.** To find the area of a triangle, we multiply $\frac{1}{2}$ by the length of the base times the height.

Exercise Set 0.2

Vocabulary Review **1.** e **2.** b **3.** d **4.** g **5.** c
6. f **7.** i **8.** h **9.** a

Problems **1.** Commutative property of addition
3. Commutative property of multiplication
5. Additive inverse property **7.** Commutative property of addition
9. Associative and commutative properties of multiplication
11. Commutative and associative properties of addition
13. Distributive property **15.** Multiplicative inverse property
17. $6 + x$ **19.** $a + 8$ **21.** $15y$ **23.** x **25.** a
27. x **29.** $3x + 18$ **31.** $12x + 8$ **33.** $15a + 10b$
35. $\frac{4}{3}x + 2$ **37.** $2 + y$ **39.** $40t + 8$ **41.** $9x + 3y - 6z$
43. $3x + 7y$ **45.** $6x + 7y$ **47.** $3x + 1$ **49.** $2x - 1$
51. $x + 2$ **53.** $a - 3$ **55.** $x + 24$ **57.** $3x - 2y$
59. $3x + 8y$ **61.** $8x + 5y$ **63.** $15x + 10$ **65.** $8y + 32$
67. $15t + 9$ **69.** $28x + 11$ **71.** $\frac{7}{15}$ **73.** $\frac{29}{35}$ **75.** $\frac{35}{144}$
77. $\frac{949}{1260}$ **79.** $\frac{47}{105}$ **81.** 15 **83.** $14a + 7$ **85.** $12x + 2$
87. $24a + 15$ **89.** $8x + 13$ **91.** $17x + 14y$ **93.** $17b + 9a$
95. $x + \frac{x}{4} = 15, 4x + x = 60, 5x = 60, x = 12$
97. 767 **99.** 1.1% **101. a.** From 2006 − 2008 **b.** Negative
c. 2002 **103.** $x^2 + 8x$ **105.** $xy - 4x$ **107.** False
109. False **111.** True **113.** False **115.** False

Find the Mistake **1.** The order of the numbers in a sum does not affect the result. **2.** The statement $2 \cdot 4 = 4 \cdot 2$ is an example of the commutative property of multiplication. **3.** The least common denominator for the fractions $\frac{2}{3}$ and $\frac{1}{7}$ is 21.
4. Reciprocals multiply to one.

Landmark Review **1.** $\frac{9}{15}$ **2.** $\frac{10}{15}$ **3.** $\frac{30}{15}$ **4.** $\frac{15}{15}$
5. $-5, \frac{1}{5}, 5$ **6.** $\frac{1}{2}, -2, \frac{1}{2}$ **7.** 14 **8.** -3 **9.** $\frac{2}{3}$ **10.** $\frac{4}{5}$
11. $\frac{3}{5}$ **12.** $\frac{3}{4}$ **13.** $\frac{7}{15}$ **14.** $\frac{19}{48}$ **15.** $\frac{31}{40}$ **16.** $\frac{5}{4}$
17. $7x + 8$ **18.** $5y - 1$ **19.** $19a + 13$ **20.** $12x + 10y + 5$

Exercise Set 0.3

Vocabulary Review **1.** positive **2.** larger **3.** difference
4. negative **5.** quotient **6.** undefined **7.** algebraic expression
Problems **1.** 4 **3.** -4 **5.** -10 **7.** -4 **9.** $\frac{19}{12} = 1\frac{7}{12}$
11. $-\frac{32}{105}$ **13.** -8 **15.** -12 **17.** $-7x$ **19.** 13
21. -14 **23.** $6a$ **25.** -15 **27.** 15 **29.** -24
31. $-10x$ **33.** x **35.** y **37.** $-8x + 6$ **39.** $-3a + 4$
41. 39 **43.** 11 **45.** -5 **47.** 11 **49.** 7 **51.** -44
53. 2 **55.** $\frac{4}{3}$ **57.** 0 **59.** Undefined **61.** $-\frac{2}{3}$
63. 0 **65.** $14x + 12$ **67.** $7m - 15$ **69.** $-2x + 9$
71. $7y + 10$ **73.** $-20x + 5$ **75.** $-11x + 10$
77. $4x + 13$ **79.** Undefined **81.** 0 **83.** $-\frac{2}{3}$ **85.** 32
87. 64 **89.** $-\frac{1}{18}$ **91.** $\frac{5}{3}$ **93.** 11 **95.** 12

97.

		Sum	Difference	Product	Quotient
a	b	$a + b$	$a - b$	ab	$\frac{a}{b}$
3	12	15	-9	36	$\frac{1}{4}$
-3	12	9	-15	-36	$-\frac{1}{4}$
3	-12	-9	15	-36	$-\frac{1}{4}$
-3	-12	-15	9	36	$\frac{1}{4}$

99.

x	$3(5x - 2)$	$15x - 6$	$15x - 2$
-2	-36	-36	-32
-1	-21	-21	-17
0	-6	-6	-2
1	9	9	13
2	24	24	28

101. a. 1 **b.** $\frac{3}{2}$ **c.** -1 **d.** 135 **103.** 1.5282 **105.** -0.0794
107. -0.0714 **109.** 3.4 **111.** 1.6 **113.** 1,200 **115.** 190
117. a. 0.068 stores/mi² **b.** 0.034 stores/mi² **c.** Georgia
119. a. $-0:0:56$ **b.** $-0:0:03$ **c.** $-0:01:49$ **121.** 3.8 times
123. 1,027 acres

Find the Mistake **1.** $-3(-5) = 15$ **2.** $-6 + 7 = 1$
3. $-4 + 2 = -2$ **4.** $-8 \div 4 = -2$ **5.** The value of the expression $2x + 5$ when $x = -3$ is -1. **6.** The value of the expression $x^2 - 4x$ when $x = 2$ is -4.

Exercise Set 0.4

Vocabulary Review **1.** c **2.** b **3.** f **4.** h **5.** g **6.** e
7. i **8.** d **9.** a **Order of Operations:** 2, 4, 3, 1
Problems **1.** 16 **3.** -16 **5.** -0.027 **7.** 32 **9.** $\frac{1}{8}$
11. $\frac{25}{36}$ **13.** $\frac{1}{10,000}$ **15.** $\frac{25}{36}$ **17.** $\frac{9}{49}$ **19.** x^9 **21.** 64
23. $-6a^6$ **25.** $16x^4$ **27.** $\frac{1}{9}$ **29.** $-\frac{1}{32}$ **31.** $\frac{16}{9}$ **33.** 17
35. $80x^3y^6$ **37.** $72x^3y^3$ **39.** $15xy$ **41.** $12x^7y^6$
43. $54a^6b^2c^4$ **45.** $36x^4y^6$ **47.** $2x^3$ **49.** $4x$ **51.** $-4xy^4$
53. $-2x^2y^2$ **55.** $\frac{1}{x^{10}}$ **57.** a^{10} **59.** $\frac{1}{t^6}$ **61.** x^{12} **63.** x^{18}
65. $\frac{1}{x^{22}}$ **67.** $\frac{a^3b^7}{4}$ **69.** $\frac{y^{38}}{x^{16}}$ **71.** $\frac{16y^{16}}{x^8}$ **73.** x^4y^6 **75.** 1
77. 1 **79.** 3.78×10^5 **81.** 4.9×10^3 **83.** 3.7×10^{-4}
85. 4.95×10^{-3} **87.** 5,340 **89.** 7,800,000 **91.** 0.00344
93. 0.49 **95.** 8×10^4 **97.** 2×10^9 **99.** 2.5×10^{-6}
101. 1.8×10^{-7} **103.** 50 **105.** 7.9×10^{-5}
107. a. 19 **b.** 27 **c.** 27 **109. a.** 16 **b.** 12 **c.** 18
111. a. 144 **b.** 74 **c.** 144 **113. a.** 23 **b.** 41 **c.** 65
115. 6.3×10^8 **117.** 1.003×10^{19} miles
119. a. 4.204×10^6 **b.** 7.995×10^6 **c.** 6.761×10^6
121. 2.478×10^{13} miles **123.** 70,000 years
125. a. 8.0×10^{11} **b.** \$4,444
127.

Number of Bytes		
Unit	Exponential Form	Scientific Notation
Kilobyte	$2^{10} = 1,024$	1.024×10^3
Megabyte	$2^{20} \approx 1,048,000$	1.048×10^6
Gigabyte	$2^{30} \approx 1,074,000,000$	1.074×10^9
Terabyte	$2^{40} \approx 1,099,500,000,000$	1.100×10^{12}

129. x^5 **131.** y^6 **133.** x^4 **135.** x^{-9b-2} **137.** x^{16m+5}
139. x^{-3b-8} **141.** x^{10m-15} **143.** x^{6a+18}
Find the Mistake **1.** To find the volume of a cube with a side length of 3 cm, we raise 3 to the third power. **2.** The product of $x^4 \cdot x^5$ written with a single exponent is x^9. **3.** To write 3×10^{-4} in expanded form, move the decimal point 4 places to the left. **4.** According to the order of operations, we begin to simplify the expression $10 - 3 + 5^3 \cdot 2$ by evaluating the exponent.

Chapter 0 Review

1. $\frac{24}{36}$ **2.** $\frac{28}{36}$ **3.** $<$ **4.** $<$ **5.** $>$ **6.** $<$ **7.** 3
8. -5 **9.** -8 **10.** 9 **11.** $\frac{7}{13}$ **12.** $\frac{13}{31}$ **13.** $2x + 15$
14. $7x + 2$ **15.** $9x + 8y$ **16.** $4a + 6b + 4$ **17.** $4x + 2y - 8$
18. $10x + 2$ **19.** $\frac{157}{168}$ **20.** $\frac{43}{48}$ **21.** $-2 + 4x$
22. $-5x + 10$ **23.** $-5x$ **24.** -11 **25.** $\frac{1}{4x^4}$
26. $27x^6y^6$ **27.** $\frac{27}{8}$ **28.** $-4x^5y^5$ **29.** $-21x^7y^3$ **30.** $\frac{6a^3}{b^3}$
31. $-\frac{4a^5}{5b}$ **32.** $4a^4b^8$ **33.** 4.82×10^5 **34.** 7.28×10^6
35. 4.21×10^{-3} **36.** 5.26×10^{-2} **37.** 0.00629 **38.** 3,290
39. 0.0631 **40.** 48,200,000 **41.** 3.0×10^{-3} **42.** 2.0×10^{-4}

Chapter 0 Test

1. $\frac{9}{18}$ **2.** $\frac{15}{18}$ **3.** $<$ **4.** $<$ **5.** $>$ **6.** $<$ **7.** 7
8. 1 **9.** 4 **10.** 5 **11.** $\frac{8}{13}$ **12.** $\frac{6}{17}$ **13.** $12x + 2$
14. $10x + 2$ **15.** $10x + 6y$ **16.** $3a + 10b + 7$
17. $2y + 3x + 9$ **18.** $10y - 12x$ **19.** $\frac{7}{6}$ **20.** $\frac{197}{204}$

21. $2x + 11$ **22.** $-2x + 2$ **23.** $-x + 3$ **24.** 15
25. $-\frac{1}{27x^{12}}$ **26.** $64x^6y^3$ **27.** $\frac{81}{25}$ **28.** $35x^5y^2$ **29.** $\frac{18y^7}{x}$
30. $\frac{6a^3}{b^2}$ **31.** $\frac{3b^5}{4a^2}$ **32.** $27x^9y^3$ **33.** 1.253×10^7 **34.** 5.2×10^{-3}
35. 6.32×10^3 **36.** 3.4×10^{-4} **37.** 0.00526 **38.** 490,000
39. 0.00063 **40.** 78,000 **41.** 1.3×10^1 **42.** 1.7×10^8

Chapter 1

Exercise Set 1.1

Vocabulary Review **1.** linear equation in one variable **2.** solution
3. equivalent **4.** addition **5.** multiplication **6.** identity
Problems **1.** A solution **3.** Not a solution **5.** Not a solution
7. A solution **9.** A solution **11.** Not a solution **13.** 3
15. $-\frac{4}{3}$ **17.** 7,000 **19.** -3 **21.** $-\frac{9}{2}$ **23.** $-\frac{7}{640}$
25. $\frac{7}{10}$ **27.** 24 **29.** $\frac{46}{15}$ **31.** 2 **33.** 5 **35.** 4
37. 6,000 **39.** No solution **41.** No solution
43. All real numbers **45.** 30,000 feet **47.** 32.36 psi
49. 206 feet **51.** 883 psi **53.** $\frac{1}{2}$ **55.** 62.5 **57.** 0
59. $\frac{5}{4}$ **61.** 13
Find the Mistake **1.** A linear equation in one variable is any equation that can be put in the form $ax + b = c$. **2.** Using the addition property of equality to solve the equation $4x - 7 = 5$, we start by adding 7 to both sides of the equation. **3.** The first step to solving linear equations in one variable is to use the distributive property to separate terms. **4.** An equation is called an identity if the left side is always identically equal to the right side.

Exercise Set 1.2

Vocabulary Review **1.** formula **2.** area **3.** perimeter
4. variable **5.** rate equation **6.** average speed
Problems **1.** -3 **3.** 0 **5.** $\frac{3}{2}$ **7.** 4 **9.** $\frac{8}{5}$ **11.** $-\frac{7}{640}$
13. 675 **15. a.** 3,400 **b.** 3,400 **17. a.** 23 **b.** 23 **19.** $c = 2$
21. $k = 3$ **23.** $k = 2,400$ **25.** $\frac{25}{3}$ **27.** $\frac{14}{3}$ **29.** $r = \frac{d}{t}$
31. $t = \frac{d}{r + c}$ **33.** $l = \frac{A}{w}$ **35.** $t = \frac{I}{pr}$ **37.** $T = \frac{PV}{nR}$
39. $x = \frac{y - b}{m}$ **41.** $F = \frac{9}{5}C + 32$ **43.** $v = \frac{h - 16t^2}{t}$ **45.** $d = \frac{A - a}{n - 1}$
47. $y = -\frac{2}{3}x + 2$ **49.** $y = \frac{3}{5}x + 3$ **51.** $y = \frac{1}{3}x + 2$ **53.** $x = \frac{5}{a - b}$
55. $h = \frac{S - \pi r^2}{2\pi r}$ **57.** $x = \frac{4}{3}y - 4$ **59.** $x = -\frac{10}{a - c}$ **61.** $y = \frac{1}{2}x + \frac{3}{2}$
63. $y = -2x - 5$ **65.** $y = -\frac{2}{3}x + 1$ **67.** $y = -\frac{1}{2}x + \frac{7}{2}$
69. a. $y = 4x - 1$ **b.** $y = -\frac{1}{2}x$ **c.** $y = -3$ **71.** $y = -\frac{1}{4}x + 2$
73. $y = \frac{3}{5}x - 3$ **75. a.** $-\frac{15}{4}$ **b.** -7 **c.** $y = \frac{4}{5}x + 4$ **d.** $x = \frac{5}{4}y - 5$
77. $\frac{9}{5}$ tons **79.** 6 miles per hour **81.** 42 miles per hour
83. 6.8 feet per second **85.** 13,330 KB **87.** 35.7 million grams
89. 8 **91.** Shar: 128.4 beats per min, Sara: 140.4 beats per min
93. $2x - 3$ **95.** $x + y = 180$ **97.** 30 **99.** 8.5
101. 6,000 **103.** $x = -\frac{a}{b}y + a$ **105.** $a = \frac{bc}{b - c}, b \neq c$
Find the Mistake **1.** A formula is an equation that has more than one variable. **2.** Solving the formula $y - 3 = -2(x - 5)$ for y, you get $y = -2x + 13$. **3.** The value of y in the formula $4y + 2x = 20$ when $x = 8$ is $y = 1$. **4.** The rate equation is given by distance = rate $\cdot$ time.

Exercise Set 1.3

Vocabulary Review   1. vertex   **2.** right, straight   **3.** acute, obtuse
4. complementary   **5.** supplementary   **6.** isosceles, equilateral
Blueprint for Problem Solving: 2, 4, 1, 3, 6, 5
Problems   1. 10 feet by 20 feet   **3.** 7 feet   **5.** 5 inches
7. 4 meters   **9.** $92.00   **11.** $200.00   **13.** $86.47
15. $81.41 million   **17.** 20°, 160°   **19. a.** 20.4°, 69.6°   **b.** 38.4°, 141.6°
21. 27°, 72°, 81°   **23.** 102°, 44°, 34°   **25.** 43°, 43°, 94°
27. $6,000 at 8%; $3,000 at 9%   **29.** $5,000 at 12%; $10,000 at 10%
31. $4,000 at 8%; 2,000 at 9%   **33.** 30 fathers, 45 sons   **35.** $54
37.

t	0	$\frac{1}{4}$	1	$\frac{7}{4}$	2
h	0	7	16	7	0

39.

Speed (miles per hour)	Distance (miles)
20	10
30	15
40	20
50	25
60	30
70	35

41.

Time (hours)	Distance Upstream (miles)	Distance Downstream (miles)
1	6	14
2	12	28
3	18	42
4	24	56
5	30	70
6	36	84

43.

	Hot Coffee Sales
Year	Sales (billions of dollars)
2005	7
2006	7.5
2007	8
2008	8.6
2009	9.2

45.

w (ft)	l (ft)	A (ft²)
2	22	44
4	20	80
6	18	108
8	16	128
10	14	140
12	12	144

47.

Age (years)	Maximum Heart Rate (beats per minute)
18	202
19	201
20	200
21	199
22	198
23	197

49.

Resting Heart Rate (beats per minute)	Training Heart Rate (beats per minute)
60	144
62	144.8
64	145.6
68	147.2
70	148
72	148.8

51.

53.
(number line from -3)

55. -5 **57.** 6

Find the Mistake   1. False, $6w - 8 = 22$ **2.** True
3. False, $0.155x = 7,218.75$ **4.** True
Landmark Review   1. $x = 3$ **2.** $y = -3$ **3.** $x = -\frac{7}{10}$
4. $x = 1$ **5.** $y = 5$ **6.** $y = 10$ **7.** $y = \frac{10}{3}$ **8.** $y = -\frac{5}{3}$
9. $\frac{-3x + 5}{4}$ **10.** $c = \frac{d}{t} - r$ **11.** $r = \frac{A - p}{pt}$ **12.** $x = \frac{3}{5}y + 3$
13.

t	0	$\frac{1}{4}$	1	$\frac{5}{4}$	2
h	0	15	48	55	64

14.

t	5	7	9	11
h	14	10	7.78	6.36

Exercise Set 1.4

Vocabulary Review   1. linear inequality in one variable
2. addition property for inequalities   **3.** set   **4.** interval
5. multiplication property for inequalities

Problems

1. (number line with bracket at $\frac{3}{2}$)

3. (number line with parenthesis at 4)

5. (number line with bracket at -5)

7. (number line with parenthesis at 4)

9. (number line with bracket at -6)

11. (number line with bracket at 4)

13. (number line with parenthesis at -3)

15. (number line with bracket at -1)

17. (number line with bracket at -3)

19. (number line with bracket at $\frac{7}{2}$)

21. (number line with parenthesis at 6)

23. (number line with bracket at -52)

25. (number line with parenthesis at $\frac{54}{7}$)

27. (number line with bracket at -32)

29. (number line with parenthesis at 40)

31. $(-\infty, -2]$ **33.** $[1, \infty)$ **35.** $(-\infty, 3)$ **37.** $(-\infty, -1]$

39. $[-17, \infty)$ **41.** $(-\infty, 10]$ **43.** $(1, \infty)$ **45.** $(435, \infty)$

47. $(-\infty, 900]$ **49. a.** 1 **b.** 16 **c.** No **d.** $x > 16$

51. $x \geq 6$; the width is at least 6 meters.

53. $x > 6$; the shortest side is even and greater than 6 inches.

55. $t \geq 100$ **57.** Lose money if they sell less than 200 tickets. Make a profit if they sell more than 200 tickets.

59. a. $p \leq \$2.00$ **b.** $p < \$1.00$ **c.** $p > \$1.25$ **d.** $p \geq \$1.75$

61. a. 1983 and earlier **b.** 1991 and later **63.** 15.6 years

65. 19.6 years **67.** $x \geq 2$ **69.** $x < 4$ **71.** $x < -1$

73. $x > -3$ **75.** $x < \frac{c-b}{a}$ **77.** $x < -\frac{a}{b}y + a$

79. $x < \frac{d-b}{a-c}$ **81.** $x < \frac{abc - acd}{bcd - abd}$

Find the Mistake **1.** False; The graph of the solution for the inequality $-16 \leq 4x$ will use a bracket to show that -4 is included in the solution. **2.** True **3.** False; $x > 3$ is a solution to $-\frac{2}{3}x - 5 < -7$. **4.** False; The interval notation for the solution set to $5 > \frac{1}{2} + 3(4x - 1)$ begins with a parentheses and ends with a parentheses, $\left(-\infty, \frac{5}{8}\right)$.

Exercise Set 1.5

Vocabulary Review **1.** set, elements **2.** subset **3.** empty set **4.** union **5.** intersection **6.** set-builder **7.** compound **8.** continued **9.** closed, open

Problems **1.** True **3.** False **5.** True **7.** True

9. $\{0, 1, 2, 3, 4, 5, 6\}$ **11.** $\{2, 4\}$ **13.** $\{0, 1, 2, 3, 4, 5, 6\}$

15. $\{0, 2\}$ **17.** $\{0, 1, 2, 3, 4, 5, 6, 7\}$ **19.** $\{1, 2, 4, 5\}$

21. a. $\{0, -1, -2\}$ **b.** $\{0, 1, 2\}$ **c.** $\{0, 1, 2, -1, -2\}$ **d.** $\{0\}$

23. a. $\{4, 5\}$ **b.** $\{-4, -5\}$ **c.** $\{-4, -5, 4, 5\}$ **d.** $\varnothing$

25.

27. $(-\infty, 1)$ **29.** $[3, \infty)$ **31.** $(-\infty, \infty)$

33. $(-\infty, -1) \cup (5, \infty)$

35. $(1, 5)$

37. $(0, 7]$

39. $(-\infty, 2) \cup (4, \infty)$

41. $(-1, 3)$

43. $(-3, -2]$

45. $[-3, 2)$

47. $[3, 7]$

49. $\varnothing$

51. $[4, 6]$

53. $(-4, 2)$

55. $(-3, 3)$

57. $[0, 6)$

59. $\left[-\frac{2}{3}, 0\right]$

61. $\left[-\frac{3}{2}, \frac{17}{2}\right]$

63. $\left[\frac{3}{8}, \frac{3}{2}\right]$

65. $(-\infty, -7] \cup [-3, \infty)$

67. $(-\infty, -1] \cup \left[\frac{3}{5}, \infty\right)$

69. $(-\infty, -10) \cup (6, \infty)$

71. $(-\infty, -3) \cup [2, \infty)$

73. $-2 < x \leq 4$ **75.** $x < -4$ or $x \geq 1$ **77. a.** $35° \leq C \leq 45°$ **b.** $20° \leq C \leq 30°$ **c.** $-25° \leq C \leq -10°$ **d.** $-20° \leq C \leq -5°$

79. Adults: $0.61 \leq r \leq 0.83$; 61% to 83%; Juveniles: $0.06 \leq r \leq 0.20$; 6% to 20% **81.** -3 **83.** 15 **85.** No solution

87. -1 **89.** $\frac{-c-b}{a} < x < \frac{c-b}{a}$ **91.** $\frac{-cd-b}{a} < x < \frac{cd-b}{a}$

Find the Mistake **1.** The notation $x \in A$ is read "x is an element of set A." **2.** The union of two sets A and B is written $A \cup B$. **3.** The interval $[-6, 8]$ for the continued inequality $-6 \leq x \leq 8$ is called a closed interval. **4.** To solve the compound inequality $-2 \leq 4x - 3 \leq 9$, add 3 to the first two parts of the inequality.

Exercise Set 1.6

Vocabulary Review **1.** negative **2.** positive **3.** $\Leftrightarrow$ **4.** same

Problems **1.** $-4, 4$ **3.** $-2, 2$ **5.** $\varnothing$ **7.** $-1, 1$ **9.** $\varnothing$

11. $\frac{17}{3}, \frac{7}{3}$ **13.** $-\frac{5}{2}, \frac{5}{6}$ **15.** $-1, 5$ **17.** $\varnothing$ **19.** $20, -4$

21. $-4, 8$ **23.** $1, 4$ **25.** $-\frac{1}{7}, \frac{9}{7}$ **27.** $-3, 12$ **29.** $\frac{2}{3}, -\frac{10}{3}$

31. $\varnothing$ **33.** $\frac{3}{2}, -1$ **35.** $5, 25$ **37.** $-30, 26$ **39.** $-12, 28$

41. $-2, 0$ **43.** $-\frac{1}{2}, \frac{7}{6}$ **45.** $0, 15$ **47.** $-\frac{23}{7}, -\frac{11}{7}$

49. $-5, \frac{3}{5}$ **51.** $1, \frac{1}{9}$ **53.** $-\frac{1}{2}$ **55.** 0 **57.** $-\frac{1}{6}, -\frac{7}{4}$

59. All real numbers **61.** All real numbers **63.** $-\frac{3}{10}, \frac{3}{2}$

65. $-\frac{1}{10}, -\frac{3}{5}$ **67. a.** $\frac{5}{4}$ **b.** $\frac{5}{4}$ **c.** 2 **d.** $\frac{1}{2}, 2$ **e.** $\frac{1}{3}, 4$

69. 1987 and 1995 **71.** $x < 4$ **73.** $-\frac{11}{3} \leq a$ **75.** $t \leq -\frac{3}{2}$

77. $x = a \pm b$ **79.** $x = \frac{-b \pm c}{a}$ **81.** $x = -\frac{a}{b}y \pm a$

Find the Mistake **1.** Some absolute value equations have two solutions. **2.** There is one solution to the absolute value equation $|2x - 3| = |2x + 4|$. **3.** There is a solution of all real numbers for the absolute equation $|y - 7| = |7 - y|$. **4.** The absolute value equation $|6a - 4| = -9$ has no solution.

Exercise Set 1.7

Vocabulary Review **1.** e **2.** f **3.** b **4.** d **5.** c **6.** a

Problems

1.  $-3 < x < 3, (-3, 3)$

3. $x \le -2$ or $x \ge 2$, $(-\infty, -2] \cup [2, \infty)$

5. $-3 < x < 3, (-3, 3)$

7. $t < -7$ or $t > 7$, $(-\infty, -7] \cup [7, \infty)$

9. $\varnothing$

11. All real numbers, $\mathbb{R}$

13. $-4 < x < 10, (-4, 10)$

15. $a \le -9$ or $a \ge -1$, $(-\infty, -9] \cup [-1, \infty)$

17. $2 < x < 8, (2, 8)$

19. $\varnothing$

21. $-1 < x < 5, (-1, 5)$

23. $y \le -5$ or $y \ge -1$, $(-\infty, -5] \cup [-1, \infty)$

25. $k \le -5$ or $k \ge 2$, $(-\infty, -5] \cup [2, \infty)$

27. $-1 < x < 7, (-1, 7)$

29. $a \le -2$ or $a \ge 1$, $(-\infty, -2] \cup [1, \infty)$

31. $-6 < x < \frac{8}{3}, \left(-6, \frac{8}{3}\right)$

33. $[-2, 8]$ **35.** $\left(-2, \frac{4}{3}\right)$ **37.** $(-\infty, -5] \cup [-3, \infty)$

39. $\left(-\infty, -\frac{7}{2}\right) \cup \left(-\frac{3}{2}, \infty\right)$ **41.** $\left[-1, \frac{11}{5}\right]$ **43.** $\left(\frac{5}{3}, 3\right)$

45. $x < 2$ or $x > 8$ **47.** $x \le -3$ or $x \ge 12$ **49.** $x < 2$ or $x > 6$

51. $x < -6$ or $x > 18$ **53.** $0.99 < x < 1.01$

55. $x \le -\frac{3}{5}$ or $x \ge -\frac{2}{5}$ **57.** $\frac{5}{9} \le x \le \frac{7}{9}$ **59.** $x < -\frac{2}{3}$ or $x > 0$

61. $x \le \frac{2}{3}$ or $x \ge 2$ **63.** $-\frac{1}{6} \le x \le \frac{3}{2}$ **65.** $-0.05 < x < 0.25$

67. $\frac{1}{12} < x < \frac{5}{12}$ **69.** $|x| \le 4$ **71.** $|x - 5| \le 1$

73. a. 3 **b.** $-2, \frac{4}{5}$ **c.** no **d.** $x < -2$ or $x > \frac{4}{5}$

75. $-23 < v - 455 < 23$; $432 < v < 478$; blue

77. $x < a + b$ **79.** $x > \frac{b + c}{a}$ **81.** $x \le \frac{c - b}{a}$

83. $x > \frac{abc - a}{b}$ or $x < \frac{-abc - a}{b}$ **85.** $-2a < x < a$

87. $-a^2 \le x \le 3a^2$

Find the Mistake **1.** False; The expression $|b| > 9$ is read "*b* is greater than 9 units from zero on the number line."

2. True **3.** False; When an absolute value expression is greater than a negative value, the solution set will be all real numbers.

4. True

Chapter 1 Review

1. 11 **2.** $-\frac{5}{3}$ **3.** $\frac{10}{7}$ **4.** 5 **5.** $\frac{1}{3}$ **6.** -3 **7.** -7

8. -14 **9.** $y = \frac{1}{2}x - 2$ **10.** $x = \frac{dy - by}{a - c}$ **11.** $y = 5x - 5$

12. $y = 2x - 5$ **13.** $y = \frac{2}{3}x + 6$ **14.** $y = 3x + 7$

15. 34 inches by 17 inches **16.** 5 meters **17.** 25°, 155°

18. 36°, 54° **19.** \$71.72 **20.** \$500 at 9%, \$2,500 at 11%

21. $(-4, \infty)$ **22.** $(-\infty, -6]$ **23.** $[4, \infty)$ **24.** $(-16, \infty)$

25. $\{3\}$ **26.** $\{3, 4\}$ **27.** $3 \le x \le 18$ **28.** $x > 8$ or $x < 5$

29. 28, 36 **30.** $-2, -\frac{22}{3}$ **31.** $\frac{9}{2}, -2$ **32.** $\varnothing$

33. $2 \le y$ or $y \le \frac{2}{3}$ **34.** $-\frac{16}{5} \le x \le 2$ **35.** All real numbers

36. $\varnothing$

Chapter 1 Cumulative Review

1. 36 **2.** 5 **3.** -5 **4.** -3 **5.** 2 **6.** 54 **7.** 18

8. 45 **9.** -2 **10.** -6 **11.** 2 **12.** $\frac{5}{3}$ **13.** $\frac{3}{4}$

14. $\frac{1}{3}$ **15.** $\frac{2}{3}$ **16.** $\frac{271}{168}$ **17.** 9 **18.** -25

19. $\{1, 3, 5, 6, 7, 8\}$ **20.** $\{3\}$ **21.** $\{9, 12\}$ **22.** $\{5, 6, 7\}$

23. $2x + 14$ **24.** $12x - 1$ **25.** $x + 6 + \frac{9}{x}$ **26.** $30x + 10y$

27. 2 **28.** $\frac{2}{15}$ **29.** 6 **30.** $0, \frac{1}{3}$ **31.** $-2, 2$

32. $-\frac{21}{5}, 3$ **33.** $t = 100$ **34.** 3 **35.** $m = \frac{y - b}{x}$

36. $r = \frac{S - a}{S}$ **37.** $F = \frac{9}{5}C + 32$ **38.** $x = \frac{5}{a - b}$

39. 17°, 73°; $5x + 5 = 90$ **40.** 8 feet by 24 feet; $8x = 64$

41. $(-1, \infty)$ **42.** $[-2, 4]$ **43.** $x < -2$ or $x > \frac{8}{3}$

44. $\frac{1}{4} < t < \frac{5}{4}$ **45.** 2.8 hours per day **46.** 0.9 hours per day

Chapter 1 Test

1. -1 **2.** $-\frac{2}{5}$ **3.** $\frac{3}{2}$ **4.** $\frac{1}{5}$ **5.** 4 **6.** 0 **7.** -7

8. 2 **9.** $m = \frac{y - b}{x}$ **10.** $F = \frac{9}{5}C + 32$ **11.** $y = 7x - 2$

12. $y = \frac{4}{5}x - 2$ **13.** $y = \frac{1}{5}x - 7$ **14.** $y = 12x - 1$

15. 9 inches by 27 inches **16.** 2 meters **17.** 42°, 138°

18. 36°, 54° **19.** \$53.30 **20.** \$3,000 at 13%, \$4,000 at 15%

21. $[-4, \infty)$ **22.** $[-12, \infty)$ **23.** $(-\infty, 4)$ **24.** $(19, \infty)$

25. 10, 14 **26.** $-21, -7$ **27.** $-6, 0$ **28.** $\varnothing$

29. $-\frac{5}{2} < y < -1$ **30.** $x \le -2$ or $x \ge 1$

31. All real numbers **32.** $\varnothing$

Chapter 2

Vocabulary Review 2.1

1. x-axis, y-axis **2.** x-coordinate, y-coordinate **3.** origin
4. x-intercept **5.** y-intercept **6.** vertical **7.** horizontal

Exercise Set 2.1

1.

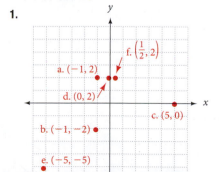

3. A. $(4, 1)$ **B.** $(-4, 3)$ **C.** $(-2, -5)$ **D.** $(2, -2)$ **E.** $(0, 5)$
F. $(-4, 0)$ **G.** $(1, 0)$

5. b   **7.** b   **9.** $y = x + 3$ **11.** $y = |x| - 3$

13. a.

b.
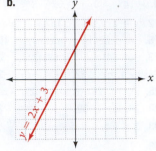

c.
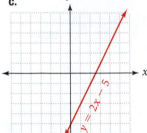

15. a.
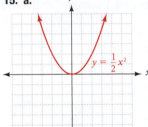

b.
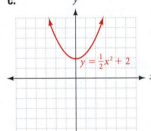

c.
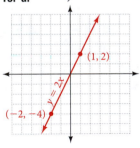

17.
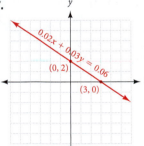

19. a.
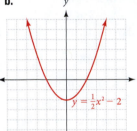

b.
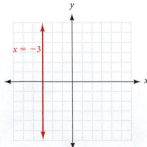

c.
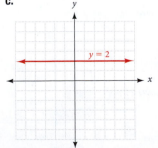

21. a.
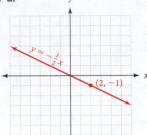

b.
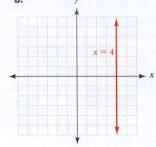

c.
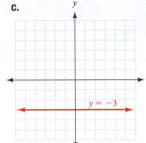

23.
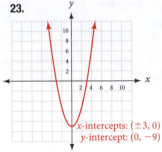

x-intercepts: $(\pm 3, 0)$
y-intercept: $(0, -9)$

25.
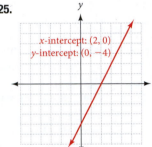

x-intercept: $(2, 0)$
y-intercept: $(0, -4)$

27.
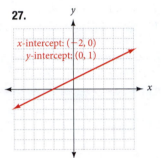

x-intercept: $(-2, 0)$
y-intercept: $(0, 1)$

29.
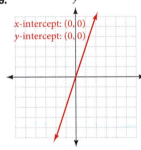

x-intercept: $(0, 0)$
y-intercept: $(0, 0)$

31.
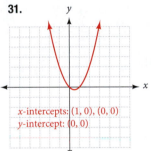

x-intercepts: $(1, 0), (0, 0)$
y-intercept: $(0, 0)$

33.
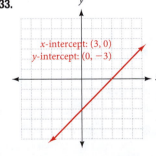

x-intercept: $(3, 0)$
y-intercept: $(0, -3)$

35. a. -7 **b.** -4 **c.** $-\dfrac{4}{3}$

d.
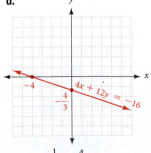

e. $y = -\dfrac{1}{3}x - \dfrac{4}{3}$

37. a. Yes **b.** No **c.** Yes

39.

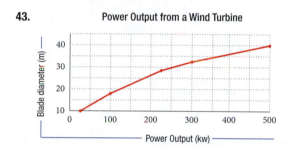

Weekly Wages

41.

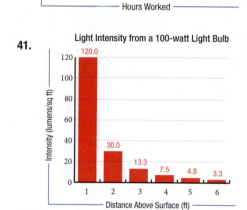

Light Intensity from a 100-watt Light Bulb

43.

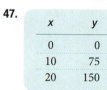

Power Output from a Wind Turbine

45. a. 60 **b.** 70 **c.** 10 **d.** 6:30 & 7:00 **e.** about 22 minutes

47.

x	y
0	0
10	75
20	150

49.

x	y
0	0
$\frac{1}{2}$	3.75
1	7.5

Find the Mistake **1.** False; On a rectangular coordinate system, the vertical axis is called the *y*-axis. **2.** True **3.** True **4.** True

Exercise Set 2.2

Vocabulary Review **1.** domain, range **2.** left, right
3. function **4.** relation **5.** vertical line test
Problems

1.

x	y
−2	−8
0	0
2	8
5	20

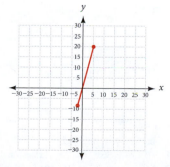

3.

x	y
0	100
5	80
10	60
15	40
20	20
25	0

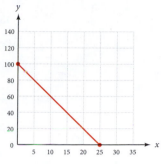

5.

x	y
0	1600
40	1200
80	800
120	400
160	0

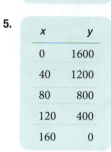

7.

x	y
0	40
25	102.5
50	165
75	227.5
100	290

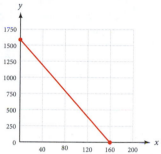

9. Domain = { 1, 3, 5, 7 }; Range = { 2, 4, 6, 8 }; A function
11. Domain = { 0, 1, 2, 3 }; Range = { 4, 5, 6 }; A function
13. Domain = { a, b, c, d }; Range = { 3, 4, 5 }; A function
15. Domain = { a }; Range = { 1, 2, 3, 4 }; Not a function
17. Domain = { AAPL, DELL, MSFT, FB, GRPN };
Range = { $562, $12, $29, $32 }
19. Domain = { Atlanta, Chicago, Boston, Houston };
Range = { Braves, White Sox, Cubs, Red Sox, Astros }
21. Yes **23.** No **25.** No **27.** Yes **29.** Yes **31.** Yes
33. Domain = { x | −5 ≤ x ≤ 5 }, [−5, 5];
Range = { y | 0 ≤ y ≤ 5 }, [0, 5]
35. Domain = { x | −5 ≤ x ≤ 3 }, [−5, 3];
Range = { y | y = 3 }, [3, 3]

37.

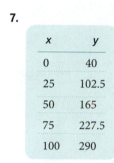

$y = x^2 − 1$

39.

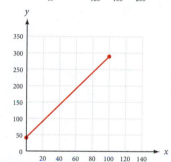

$y = x^2 + 4$

Domain = All real numbers;   Domain = All real numbers;
Range = { y | y ≥ −1 }; A function   Range = { y | y ≥ 4 }; A function

41.

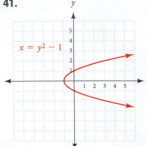

Domain = $\{x \mid x \geq -1\}$;
Range = All real numbers;
Not a function

43.

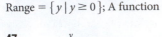

Domain = All real numbers;
Range = $\{y \mid y \geq 0\}$; A function

45.

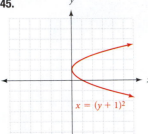

Domain = $\{x \mid x \geq 0\}$;
Range = All real numbers;
Not a function

47.

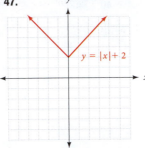

Domain = All real numbers;
Range = $\{y \mid y \geq 2\}$; A function

49. Domain = $\{2001, 2003, 2005, 2007, 2009\}$;
Range = $\{5, 10, 40, 80, 160\}$

51. a. $y = 8.5x$ for $10 \leq x \leq 40$
b.

Table 4 Weekly Wages		
Hours Worked	**Function Rule**	**Gross Pay ($)**
x	$y = 8.5x$	y
10	$y = 8.5(10) = 85$	85
20	$y = 8.5(20) = 170$	170
30	$y = 8.5(30) = 255$	255
40	$y = 8.5(40) = 340$	340

c.

d. Domain = $\{x \mid 10 \leq x \leq 40\}$; Range = $\{y \mid 85 \leq y \leq 340\}$
e. Minimum = $85; Maximum = $340

53. a. Domain = $\{2008, 2009, 2010, 2011, 2012\}$;
Range = $\{5.8, 5.2\text{-}5.5, 7.5, 9.7, 12.1\}$ **b.** 2012 **c.** 2009
55. a. III **b.** I **c.** II **d.** IV **57. a.** 6 **b.** 7.5 **59. a.** 27 **b.** 6
61. 1 **63.** -3 **65.** $x = 0$ **67.** $-\frac{6}{5}$ **69.** $-\frac{35}{32}$
Find the Mistake **1.** The set of all inputs for a function is called the
domain. **2.** For the relation $(1, 2), (3, 4), (5, 6), (7, 8)$, the range
is $\{2, 4, 6, 8\}$. **3.** Graphing the relation $y = x^2 - 3$ will show that
the domain is all real numbers. **4.** If a vertical line crosses the
graph of a relation in more than one place, the relation cannot be a
function.

Exercise Set 2.3
Vocabulary Review **1.** independent **2.** dependent **3.** $f(x)$
4. function f
Problems **1.** -1 **3.** -11 **5.** 2 **7.** 4 **9.** $a^2 + 3a + 4$
11. $2a + 7$ **13.** 1 **15.** -9 **17.** 8 **19.** 0 **21.** $3a^2 - 4a + 1$
23. $3a^2 + 8a + 5$ **25.** 4 **27.** 0 **29.** 2 **31.** 4
33. $x = \frac{1}{2}$ **35.** 24 **37.** -1 **39.** $2x^2 - 19x + 12$ **41.** 99
43. 28 **45.** 225 **47.** $\frac{3}{10}$ **49.** $\frac{2}{5}$ **51.** Undefined **53.** $\frac{1}{2}$
55. a. $a^2 - 7$ **b.** $a^2 - 6a + 5$ **c.** $x^2 - 2$ **d.** $x^2 + 4x$
e. $a^2 + 2ab + b^2 - 4$ **f.** $x^2 + 2xh + h^2 - 4$

57.

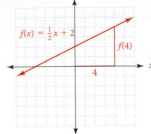

59. $x = 4$

61.

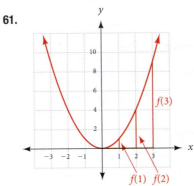

63. $V(3) = 300$, the painting is worth $300 in 3 years; $V(6) = 600$,
the painting is worth $600 in 6 years.
65. a. True **b.** False **c.** True **d.** False **e.** True
67. a. $5,625 **b.** $1,500 **c.** $\{t \mid 0 \leq t \leq 5\}$
d.
e.

$\{V(t) \mid 1,500 \leq V(t) \leq 18,000\}$ **f.** About 2.42 years

69. a.

d (feet)	$f(d)$ (KWH/year)
0	0
5	332
10	1328
15	2988
20	5312

Source: build_itsolar.com/Projects/Wind/Wind.htm

b. $0 \leq f(d) \leq 5312$
71. $-0.1x^2 + 35x$ **73.** $4x^2 - 7x + 3$ **75.** $-0.1x^2 + 27x - 500$
77. $2x^2 + 8x + 8$ **79.** 10 **81.** $2x^2 - 17x + 33$ **83.** $24x - 6$

Find the Mistake   1. To evaluate the function $f(x) = 10x + 3$ for $f(1)$ we begin by replacing x with 1.   **2.** For the average speed function $s(t) = \frac{60}{t}$, finding $s(6)$ shows an average speed of 10 miles per hour.   **3.** If the volume of a ball is given by $V(r) = \frac{4}{3}\pi r^3$ where r is the radius, we see that the volume increases as the radius of the ball increases.   **4.** If $f(x) = x^2$ and $g(x) = 2x + 4$, evaluating $f(g(2))$ is the same as finding $g(2)$ and using that value in $f(x) = x^2$ for x.

Exercise Set 2.4

Vocabulary Review   1. c   **2.** e   **3.** d   **4.** a   **5.** b

Problems   1. $6x + 2$ **3.** $-2x + 8$ **5.** $8x^2 + 14x - 15$
7. $\frac{2x + 5}{4x - 3}$ **9.** $3x - 5 + \frac{5}{x}$ **11.** $\frac{5}{x} - 3x + 5$ **13.** $15 - \frac{25}{x}$
15. $\frac{5}{3x^2 - 5x}$ **17.** $4x - 7$ **19.** $3x^2 - 10x + 8$ **21.** $-2x + 3$
23. $3x^2 - 11x + 10$ **25.** $9x^3 - 48x^2 + 85x - 50$ **27.** $x - 2$
29. $\frac{1}{x - 2}$ **31.** $3x^2 - 7x + 3$ **33.** $6x^2 - 22x + 20, \mathbb{R}$
35. $\frac{4x - 7}{3x^2 - 11x + 10}, D = \left(-\infty, \frac{5}{3}\right) \cup \left(\frac{5}{3}, 2\right) \cup (2, \infty)$
37. $9x^2 - 30x + 25, D = $ All real numbers
39. $8x^2 - 26x + 21, D = $ All real numbers   **41.** 15   **43.** 98   **45.** $\frac{3}{2}$
47. 1   **49.** 40   **51.** 147   **53. a.** 81   **b.** 29   **c.** $(x + 4)^2$ **d.** $x^2 + 4$
55. a. -2 **b.** -1 **c.** $16x^2 + 4x - 2$ **d.** $4x^2 + 12x - 1$
57. a. 1   **b.** $\frac{1}{3}$ **c.** $\frac{9}{x^2}$ **d.** $\frac{3}{x^2}$
59. $(f \circ g)(x) = 5\left(\frac{x + 4}{5}\right) - 4$ **61.** $(f \circ g)(x) = 3\left(\frac{x}{3} - 2\right) + 6$
$\qquad = x + 4 - 4$ $\qquad\qquad = x - 6 + 6$
$\qquad = x$ $\qquad\qquad\qquad = x$

$(g \circ f)(x) = \frac{(5x - 4) + 4}{5}$ $\qquad (g \circ f)(x) = \frac{3x + 6}{3} - 2$
$\qquad = \frac{5x}{5}$ $\qquad\qquad\qquad = x + 2 - 2$
$\qquad = x$ $\qquad\qquad\qquad = x$

63. -1 **65.** $-\frac{4}{3}$ **67.** $\frac{2}{5}$ **69.** -2 **71.** $\frac{2}{7}$
73. a. $R(x) = 11.5x - 0.05x^2$ **b.** $C(x) = 2x + 200$
   c. $P(x) = -0.05x^2 + 9.5x - 200$ **d.** $\overline{C}(x) = 2 + \frac{200}{x}$
75. a. $M(x) = 220 - x$ **b.** $M(24) = 196$ **c.** 142   **d.** 135   **e.** 128
77. a. Coal: $f(x) = 94.80x$, Wind: $g(x) = 97.00x$
   b.   **c.** $h(x) = 3.20x$

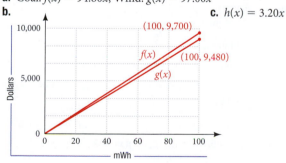

79. a. Coal: $f(x) = 136.20x$, Wind: $g(x) = 97.00x$
   b.   **c.** $h(x) = 39.20x$

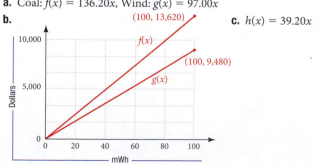

81. 12   **83.** 28   **85.** $-\frac{7}{4}$ **87.** $w = \frac{P - 2l}{2}$
89.

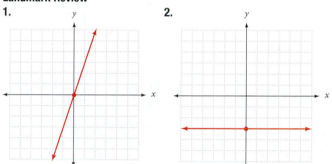

$[-6, \infty)$

91. $(-\infty, 6)$

93. 6, 2   **95.** $\varnothing$

Find the Mistake   1. The function $(f + g)(x)$ is the sum of the functions $f(x)$ and $g(x)$.   **2.** The function $\left(\frac{f}{g}\right)(x) = \frac{f(x)}{g(x)}$.
3. If $f(x) = 3x + 4$ and $g(x) = x - 3$, the function $(f + g)(x) = (3x + 4) + (x - 3)$ or $4x + 1$.
4. For $(f \circ g)(x)$ the numbers in the domain of the composite function must be numbers in the domain of f.

Landmark Review

1.   **2.**

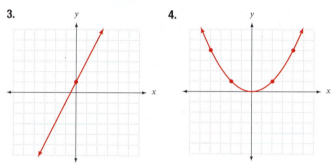

3.   **4.**

5. Domain = { 4, 3, 1, 7 }; Range = { 3, 2, 9 }; A function
6. Domain = { 0, 1, 3 }; Range = { 3, 4, 2 }; Not a function
7. Domain = { a, 4, b, 5 }; Range = { 0, 3, 1 }; A function
8. Domain = { 1 }; Range = { 3, 5, 2, 0 }; Not a function   **9.** 5
10. 12   **11.** $3a - 1$ **12.** $a^2 + 9a + 20$ **13.** $5x - 4$
14. $x - 6$ **15.** $6x^2 - 7x - 5$ **16.** $\frac{3x - 5}{2x + 1}$ **17.** -3 **18.** 16

Exercise Set 2.5

Vocabulary Review   1. rise, run   **2.** slope   **3.** positive, negative
4. parallel   **5.** perpendicular
Problems   1. $\frac{3}{2}$ **3.** Undefined slope   **5.** $\frac{2}{3}$
7.   **9.**

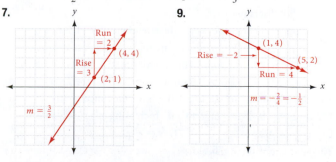

11.

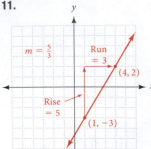

13.

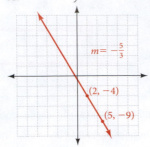

15.

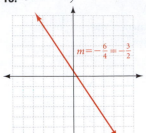

17.

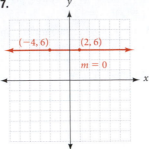

19.

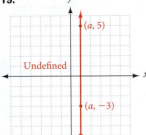

21. $a = 5$ **23.** $b = 2$
25. $x = 3$ **27.** $x = -4$

29.

x	y
0	2
3	0

$m = -\frac{2}{3}$

31.

x	f(x)
0	-5
3	-3

$m = \frac{2}{3}$

33.

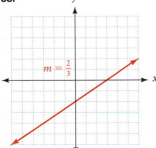

35. $\frac{1}{5}$ **37.** 0   **39.** -1 **41.** $-\frac{3}{2}$ **43. a.** Yes **b.** No
45. 17.5 mph   **47.** 120 ft/sec   **49.** 10   **51.** 5   **53.** -2
55. 12   **57. a.** 10 minutes **b.** 20 minutes **c.** 20°C per minute
d. 10°C per minute **e.** 1st minute
59. 66.5 million users per year. Between 2004 and 2010 the number
of active Facebook users has increased an average of 66.5
million users per year.
61. a. 300,000 people per year. The attendance for Major League
Baseball increased by 300,000 people over the last 10 years.
b. -3.5 million people. The attendance for Major League
Baseball decreased by 3.5 million people over the 2 year period.
63. a. 10.7 **b.** 12.5 **c.** 12.5 **d.** 28.6 **e.** Longer
65. a. 2.15 **b.** 2.20 **c.** 2.40 **d.** Increasing at a steady rate
67. $-\frac{2}{3}$ **69.** $-2x + 3$ **71.** $y = -\frac{2}{3}x + 2$
73. $y = -\frac{2}{3}x + 1$ **75.** 1

Find the Mistake   1. A line that rises from left to right has a
positive slope.   **2.** All horizontal lines have a slope that equals
zero.   **3.** To find the slope of a line between two points (x_1, y_1) and
(x_2, y_2) divide the expression $y_2 - y_1$ by the expression $x_2 - x_1$.
4. The graphs of two non-vertical lines are perpendicular if and only
if the product of their slopes is -1.

Exercise Set 2.6
Vocabulary Review   1. slope-intercept   **2.** coefficient,
constant term   **3.** point slope
Problems   1. $f(x) = -4x - 3$ **3.** $f(x) = -\frac{2}{3}x$
5. $f(x) = -\frac{2}{3}x + \frac{1}{4}$ **7. a.** 3 **b.** $-\frac{1}{3}$ **9. a.** -3 **b.** $\frac{1}{3}$
11. a. $-\frac{2}{5}$ **b.** $\frac{5}{2}$ **13. a.** 0 **b.** Undefined

15.

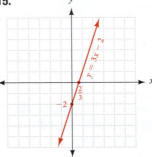

Slope = 3, y-intercept = -2,
perpendicular slope = $-\frac{1}{3}$

17.

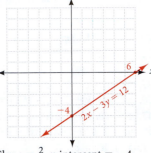

Slope = $\frac{2}{3}$, y-intercept = -4,
perpendicular slope = $-\frac{3}{2}$

19.

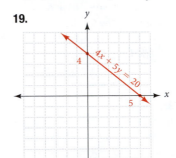

Slope = $-\frac{4}{5}$, y-intercept = 4,
perpendicular slope = $\frac{5}{4}$

21. Slope = $\frac{1}{2}$, y-intercept = -4, $f(x) = \frac{1}{2}x - 4$
23. Slope = $-\frac{2}{3}$, y-intercept = 3, $f(x) = -\frac{2}{3}x + 3$
25. $f(x) = 2x - 1$ **27.** $f(x) = -\frac{1}{2}x - 1$ **29.** $f(x) = -3x + 1$
31. $f(x) = \frac{2}{3}x + \frac{14}{3}$ **33.** $f(x) = -\frac{1}{4}x - \frac{13}{4}$ **35.** $f(x) = -\frac{2}{3}x - \frac{8}{3}$
37. $3x + 5y = -1$ **39.** $x - 12y = -8$ **41.** $6x - 5y = 3$
43. $x - y = 0$ **45.** $(0, -4), (2, 0); f(x) = 2x - 4$
47. $(-2, 0), (0, 4); f(x) = 2x + 4$
49. a. $x: \frac{10}{3}, y: -5$ **b.** $(4, 1)$, Answers may vary **c.** $f(x) = \frac{3}{2}x - 5$ **d.** No
51. a. 2 **b.** $\frac{3}{2}$ **c.** -3 **d.**   **e.** $y = 2x - 3$

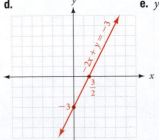

53. a. $y = \frac{1}{2}x$

$m = \frac{1}{2}$
$b = 0$
x-intercept $= 0$

b. $x = 3$

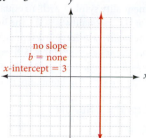

no slope
$b =$ none
x-intercept $= 3$

c. $y = -2$

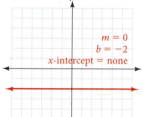

$m = 0$
$b = -2$
x-intercept $=$ none

55. $f(x) = 3x + 7$

57. $y = -\frac{5}{2}x - 13$

59. $f(x) = \frac{1}{4}x + \frac{1}{4}$

61. $f(x) = -\frac{2}{3}x + 2$

63. b. $86°$

65. a. $190,000 **b.** $19 **c.** $6.50

67. $y = 0.19x - 361.6$

69. $V(t) = -350t + 1800$

71. a. $P = \frac{95}{6}d - \frac{400}{3}$ **b.** 658.33 kw   **c.** 33.68 m

73. $(0, 0), (4, 0)$ **75.** $(0, 0), (2, 0)$

Find the Mistake   **1.** The equation of a line with slope $-\frac{2}{3}$ and y-intercept 8 is given by $y = -\frac{2}{3}x + 8$.   **2.** If you are given a point and the slope of a line, you can find the equation of the line using point-slope form.   **3.** The standard form for the equation of a line is $ax + by = c$.   **4.** The graphs of two non-vertical lines are perpendicular if and only if the product of their slopes is -1.

Exercise Set 2.7
Vocabulary Review   **1.** inequality   **2.** solid   **3.** broken
4. equals   **5.** same   **6.** opposite

Problems

1.

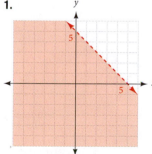

3.

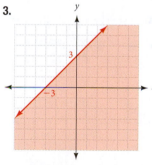

5.

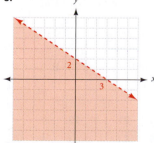

7.

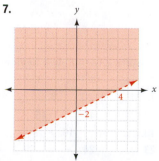

9.

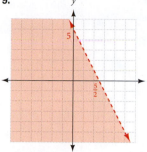

11.

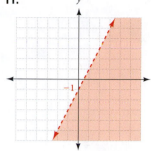

13.

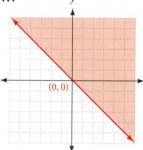

15.

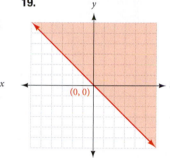

17.

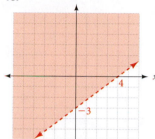

(0, 0)

19.

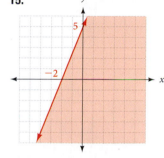

(0, 0)

21. $y > -x + 4$ **23.** $y \leq \frac{1}{2}x + 2$ **25.** $x \geq 0$ **27.** $x > -2$

29.

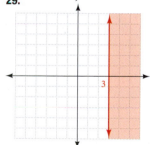

31.

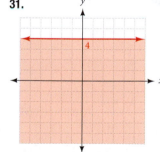

33.

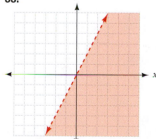

35.

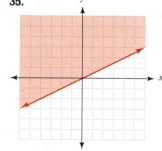

37.
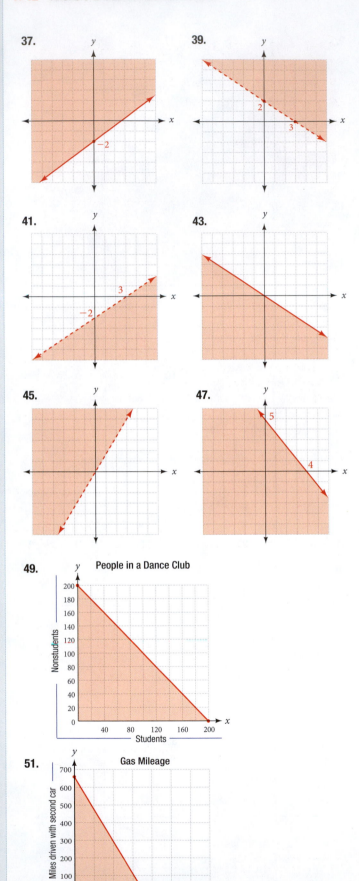

39.

41.

43.

45.

47.

49.

People in a Dance Club

51.

Gas Mileage

Find the Mistake   **1.** The equation found when replacing an inequality symbol with an equals sign represents the boundary for the inequality.   **2.** The boundary for the solution set for the inequality $y > 5x + 3$ is represented by a broken line.
3. The solution set for $y \leq 8$ includes all points below the boundary $y = 8$.   **4.** Use $(0, 0)$ as a convenient test point to find the solution set of an inequality that does not pass through the origin.

Chapter 2 Review

1.
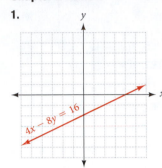

2.
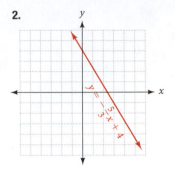

3.
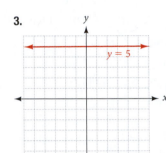

4. Domain $= \{2, 3\}$;
Range $= \{0, -4, 1\}$;
Not a function

5. Domain $= \{3, 4, 5\}$;
Range $= \{0, 2, -3\}$;
A function

6. 3   **7.** 0   **8.** 1   **9.** 9   **10.** 0   **11.** 40   **12.** 13   **13.** 1
14. $\frac{8}{5}$ **15.** 29   **16.** $-\frac{1}{2}$ **17.** $-\frac{5}{2}$ **18.** 3   **19.** 5
20. $-\frac{7}{12}$ **21.** -1 **22.** $y = -5x + 4$ **23.** $y = -7x - 3$
24. $m = 4, b = -3$ **25.** $m = \frac{4}{5}, b = -2$ **26.** $y = 3x - 4$
27. $y = \frac{3}{2}x + 8$ **28.** $y = -\frac{4}{5}x + \frac{12}{5}$ **29.** $y = -x - 2$
30. $y = -\frac{1}{3}x + \frac{7}{3}$ **31.** $y = -3x + 4$ **32.** $y = \frac{3}{2}x + 6$

33.
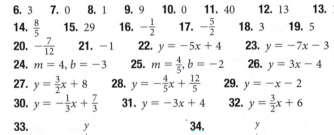

34.
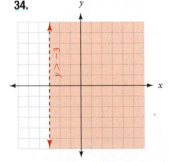

Chapter 2 Cumulative Review

1. -36 **2.** -4 **3.** 43   **4.** 169   **5.** 18   **6.** 54
7. 34   **8.** 33   **9.** 33   **10.** -11 **11.** $\frac{6}{7}$ **12.** $\frac{7}{13}$
13. $\frac{49}{72}$ **14.** $\frac{1}{96}$ **15.** $\frac{1}{2}$ **16.** $\frac{53}{9}$ **17.** $8x - 10y$
18. $2x - 9y$ **19.** -18 **20.** $\frac{7}{6}$ **21.** 4   **22.** -8
23. $3, \frac{1}{3}$ **24.** $-\frac{1}{2}$ **25.** $y = -4x + 11$ **26.** $y = 6x - 13$
27. $-2 \leq x \leq 7$ **28.** $x > \frac{5}{3}$ or $x < -5$

29.

30.
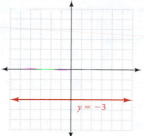

31.
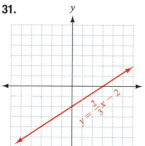

32.
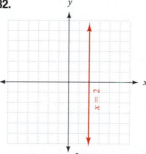

33. -1 **34.** 8 **35.** No slope   **36.** $y = -\frac{3}{5}x + 7$
37. $m = \frac{1}{2}, b = -2$ **38.** $y = 2x + 6$ **39.** $y = \frac{1}{2}x + 4$
40. $y = \frac{5}{7}x - 13$ **41.** 3 **42.** 1 **43.** 104 **44.** $x^2 + 3x - 3$
45. Domain $= \{-2, -5, 1\}$; Range $= \{3, 4\}$; A function
46. 15% **47.** $\$16.40$ **48.** 58.5% more   **49.** 13.6% more

Chapter 2 Test

1. $3; -9; 3$
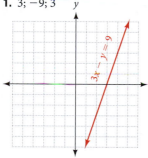

2. $8; 4; -\frac{1}{2}$
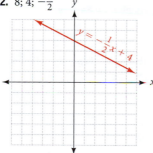

3. $10; -6; \frac{3}{5}$
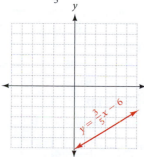

4. -4; None; No slope
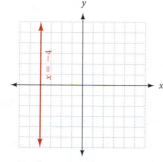

5. Domain $= \{0, 2, 6\}$; Range $= \{3, 5, -4\}$; A function
6. Domain $=$ All real numbers; Range $= \{y \mid y \geq 2\}$; A function
7. Domain $= \{0, -6\}$; Range $= \{5, 7\}$; Not a function
8. Domain $=$ All real numbers; Range $=$ All real numbers; A function
9. Domain $= \{x \mid -2 \leq x \leq 2\}$; Range $= \{y \mid -1 \leq y \leq 2\}$
10. Domain $= \{x \mid -2 \leq x \leq 2\}$; Range $= \{y \mid -3 \leq y \leq 3\}$
11. -16 **12.** 11 **13.** -12 **14.** -9 **15.** 2

16. -1 **17.** 3 **18.** 1 **19.** $m = -\frac{2}{3}$ **20.** $m = 3$
21. $y = -\frac{3}{2}x + 3$ **22.** $y = \frac{1}{3}x + 1$ **23.** $y = -3x - 17$
24. $y = \frac{2}{3}x + 4$ **25.** $y = \frac{3}{2}x + 7$ **26.** $y = 2x - 6$
27. $y \geq -\frac{2}{3}x + 2$ **28.** $y < 2x + 4$

29.
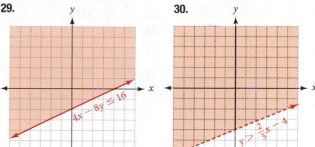

30.

Chapter 3

Exercise Set 3.1
Vocabulary Review   **1.** system   **2.** intersect   **3.** inconsistent
4. dependent   **5.** solution set   **6.** $1, 0$, infinite
7. consistent, inconsistent, consistent
8. independent, independent, dependent
Problems

1.
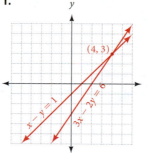

3.

5.
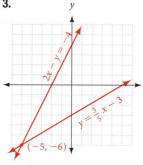

7.
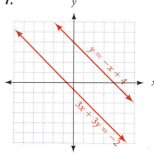

Inconsistent (no solution)

9.
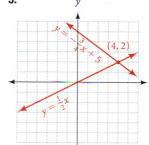

11.

13.

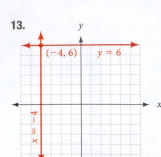

15. $(4, 7)$ **17.** $(3, 17)$
19. $\left(\frac{3}{2}, 2\right)$ **21.** $(-1, 3)$
23. $(1, 1)$ **25.** $(2, -3)$
27. Dependent (lines coincide)
29. Inconsistent (no solution)
31. $(10, 12)$ **33.** $(2, 1)$ **35.** $(-1, 0)$
37. Dependent (lines coincide)
39. $(10, 24)$ **41.** $\left(-\frac{32}{7}, -\frac{50}{21}\right)$
43. $(4, 8)$ **45.** $\left(\frac{1}{5}, 1\right)$

47. $(1, 0)$ **49.** $\left(\frac{13}{11}, -\frac{16}{11}\right)$ **51.** $(-1, -2)$ **53.** $\left(-5, \frac{3}{4}\right)$
55. $(-4, 5)$ **57.** $\left(-\frac{17}{16}, -\frac{19}{8}\right)$ **59.** $\left(-\frac{11}{7}, -\frac{20}{7}\right)$ **61.** $\left(2, \frac{4}{3}\right)$
63. $(-12, -12)$ **65.** Inconsistent (no solution)   **67.** $y = 5, z = 2$
69. $\left(\frac{3}{2}, \frac{3}{8}\right)$ **71.** $\left(-4, -\frac{8}{3}\right)$ **73.** $\left(-\frac{10}{103}, \frac{14}{103}\right)$
75. a. 350 miles   **b.** Car   **c.** Truck   **d.** We are only working with positive numbers.
77. a. \$20,000   **b.** $f(t) = 1{,}200t - 20{,}000$ **c.** 16.67 years
79. a. Charter jet: $f(x) = 2{,}600$; Business class: $g(x) = 1{,}300x$
b.

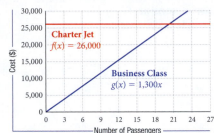

c. Business class
d. More than 20

81. a. Hybrid: $c(x) = 22{,}600 + 0.095x$, Ex: $c(x) = 18{,}710 + 0.14x$
b. 86,444.44 miles
83. -10 **85.** $3y + 2z$ **87.** $y = 1$ **89.** $z = 3$
91. $10x - 2z$ **93.** $9x + 3y - 6z$
Find the Mistake   **1.** When two lines are parallel, the equations of those lines are inconsistent.   **2.** Equations are said to be dependent if their lines coincide.   **3.** False, To solve the following system, it would be easiest to add $3y$ to both sides of second equation to isolate x.   **4.** No solution, system inconsistent, lines are parallel.

Exercise Set 3.2
Vocabulary Review   **1.** triples   **2.** dependent   **3.** infinite
4. inconsistent
Problems   1. $(1, 2, 1)$ **3.** $(2, 1, 3)$ **5.** $(2, 0, 1)$
7. $\left(\frac{1}{2}, \frac{2}{3}, -\frac{1}{2}\right)$ **9.** Inconsistent (no solution)   **11.** $(4, -3, -5)$
13. Dependent (parallel planes)   **15.** $(4, -5, -3)$
17. Dependent (parallel planes)   **19.** $\left(\frac{1}{2}, 1, 2\right)$ **21.** $\left(\frac{1}{2}, \frac{1}{3}, \frac{1}{4}\right)$
23. $\left(\frac{10}{3}, -\frac{5}{3}, -\frac{1}{3}\right)$ **25.** $\left(\frac{1}{4}, -\frac{1}{3}, \frac{1}{8}\right)$ **27.** $(6, 8, 12)$
29. $(-141, -210, -104)$ **31.** 4 amp, 3 amp, 1 amp
33. Suzie: 3 dozen; Minnie: 16 dozen; Bunny: 12 dozen
35. $2 + 3x$ **37.** $2x + 5y$ **39.** 6   **41.** $(-1, 5)$
43. 3, 7, 10   **45.** 3 ft, 7 ft, 10 ft
Find the Mistake   1. The solution to the following system is $(3, -3, -2)$.
2. It is easiest to solve the following system by eliminating z from each equation.   **3.** The following system has no unique solution, dependent solution.   **4.** The following system has solution: $(4, 3, 1)$.

Exercise Set 3.3
Vocabulary Review   1. matrix   **2.** rectangular array; elements
3. dimensions   **4.** coefficient; constant   **5.** augmented
Problems   1. $(2, 3)$ **3.** $(-1, -2)$ **5.** $(7, 1)$ **7.** $(-3, 4)$
9. $(0, -9)$ **11.** $(-4, 3)$ **13.** $(-5, 7)$ **15.** $(0, -8)$ **17.** $(8, 4)$
19. $(1, 2, 1)$ **21.** $(2, 0, 1)$ **23.** $(0, 1, -2)$ **25.** $(0, -2, 4)$
27. $(2, 1, 1)$ **29.** $(1, 1, 2)$ **31.** $(4, 1, 5)$ **33.** $(1, 3, 1)$ **35.** $(-4, 3, 5)$
37. $(0, 3, -1)$ **39.** $(1, 3, -4)$ **41.** $\left(4, \frac{10}{3}\right)$ **43.** $(6, 4)$

49.

51.

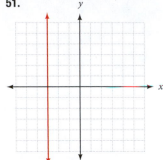

Find the Mistake   1. The dimensions of the matrix $[5 \quad 2 \quad -1]$ is 1×3.   **2.** When transforming an augmented matrix into an equivalent system, we can interchange any two rows of a matrix, multiply any row by a nonzero constant.   **3.** from left to right: coefficient matrix, constant matrix, augmented matrix.
4. The correct system is $x - z = 4$
$\qquad\qquad\qquad\qquad y = 8$
$\qquad\qquad\qquad x + y + z = 5$

Exercise Set 3.4
Vocabulary Review   1. 2×2 **2.** coefficients   **3.** 3×3
4. minor   **5.** sign array   **6.** constant
Problems   1. 3   **3.** 5   **5.** -2 **7.** -1 **9.** 0   **11.** 10   **13.** 2
15. -3 **17.** -2 **19.** -3 **21.** -2 **23.** $-2, 5$ **25.** $-8, 4$
27. $-3, 3$ **29.** $(3, 1)$ **31.** Inconsistent system; $\varnothing$
33. $\left(-\frac{15}{43}, -\frac{27}{43}\right)$ **35.** $\left(\frac{60}{43}, \frac{46}{43}\right)$ **37.** $(2, 0)$ **39.** $\left(\frac{474}{323}, \frac{40}{323}\right)$
41. 3   **43.** 0   **45.** 3   **47.** 8   **49.** 6   **51.** -228
53. 27   **55.** -57 **57.** $(3, -1, 2)$ **59.** $\left(\frac{1}{2}, \frac{5}{2}, 1\right)$
61. No unique solution   **63.** $\left(-\frac{10}{91}, -\frac{9}{13}, \frac{107}{91}\right)$
65. $\left(\frac{83}{18}, -\frac{7}{9}, -\frac{17}{18}\right)$ **67.** $\left(\frac{111}{53}, \frac{57}{53}, \frac{80}{53}\right)$ **69.** $\left(\frac{31}{4}, \frac{7}{2}, \frac{9}{4}\right)$
71. $\left(\frac{71}{13}, -\frac{12}{13}, \frac{24}{13}\right)$ **73.** $(3, 1, 2)$ **75.** $y - mx = b$; $y = mx + b$
77. a. $y = 0.3x + 3.4$ **b.** $y = 0.3(2) + 3.4$; 4 billion
79. a. $I = 767.5x + 21{,}363$ **b.** \$24,433
81. 14.4 million   **83.** 1,540 heart transplants   **85.** $x = 50$ items
87. 1986   **89.** 4   **91.** 4   **93.** 171   **95.** $(2, 1, 3, 1)$
97. $\left(\frac{1}{a+2}, \frac{1}{a+2}, \frac{1}{a+2}\right)$ **99.** $3x + 2$ **101.** $-\frac{160}{9}$
103. 320   **105.** $2x + 5y$ **107.** 6   **109.** $y = 5, z = 2$

Find the Mistake   1. The 2×2 determinant $\begin{vmatrix} 4 & 9 \\ -3 & 6 \end{vmatrix}$ is equal to $4(6) - 9(-3) = 51$.   **2.** The second step when evaluating a 3×3 determinant by expansion of minors is to write the product of each element in the row or column chosen with its minor.
3. For a three-variable system, Cramer's rule states that there is no unique solution to the system if $D = 0$.

4. When using Cramer's rule to solve the following system of

$$x + 0y + 2z = 4$$

equations $2x + 3y - z = 7$, the value for the determinant D_y is

$$0x + y + 3z = 2$$

found by evaluating $\begin{vmatrix} 1 & 4 & -2 \\ 2 & 7 & -1 \\ 0 & 2 & -3 \end{vmatrix}$.

Landmark Review

1. $(1, 2)$

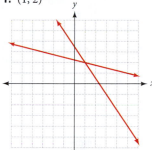

2. $(0, -4)$

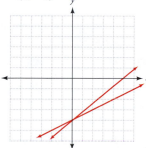

3. $(2, 3)$

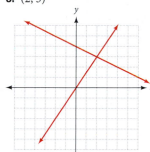

4. Inconsistent (no solution)

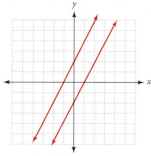

5. $(3, 4)$ **6.** $(5, 1)$ **7.** $(1, 1)$ **8.** $(3, 2)$ **9.** $(1, 1)$
10. Dependent (lines coincide)   **11.** $(5, -7)$ **12.** $(-1, -2)$
13. $(-2, 2, 2)$ **14.** $(-1, 2, 1)$ **15.** $(7, 3, -5)$ **16.** $\left(\frac{1}{4}, \frac{1}{2}, \frac{1}{2}\right)$

Exercise Set 3.5

Vocabulary Review   **1.** known   **2.** unknown   **3.** variables
4. system   **5.** sentences   **6.** solution
Problems   **1.** 5, 13   **3.** 10, 16   **5.** 1, 3, 4   **7.** 3 and 23
9. 15 and 24   **11.** 225 adult and 700 children's tickets
13. $12,000 at 6%, $8,000 at 7%   **15.** $4,000 at 6%, $8,000 at 7.5%
17. $200 at 6%, $1,400 at 8%, $600 at 9%
19. 6 gallons of 20%, 3 gallons of 50%
21. 5 gallons of 20%, 10 gallons of 14%
23. 12.5 lbs of oats, 12.5 lbs of nuts
25. Speed of boat: 9 miles/hour, speed of current: 3 miles/hour
27. Airplane: 270 miles per hour, wind: 30 miles per hour
29. 12 nickels, 8 dimes   **31.** 3 of each   **33.** 110 nickels
35. 14 nickels and 10 dimes   **37.** $x = -200p + 700$; when
$p = \$3, x = 100$ items   **39.** $L = 23$ inches; $W = 6$ inches;
41. $h = -16t^2 + 64t + 80$ **43.** 18 preregistered, 11 on-site
45. No   **47.** $(4, 0)$ **49.** $x > 435$
Find the Mistake   **1.** $a + s = 950$; $2.5a + 1.5s = 1875$
2. $0.08x + 0.06y = 900$ **3.** $x + y + z = 1700$;
$0.06x + 0.07y + 0.08z = 123$ **4.** $n + d + q = 55$; $2q = d$

Exercise Set 3.6

Vocabulary Review   **1.** common   **2.** shaded   **3.** second
4. first   **5.** fourth   **6.** third
Problems

1.

3.
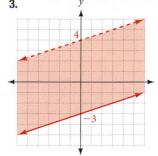

5.
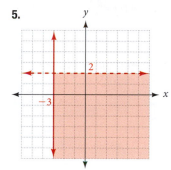

7.
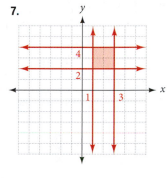

9.
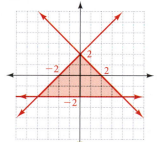

11.
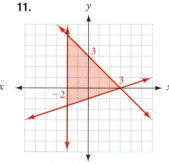

13.
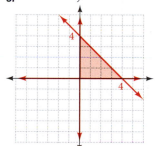

15.
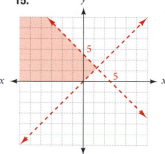

17.
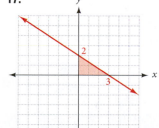

19.
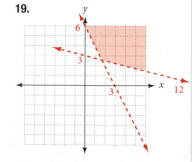

21. $x + y \leq 4; -x + y < 4$ **23.** $x + y \geq 4; -x + y < 4$

25. a.

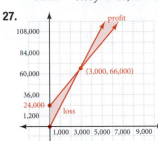

$0.55x + 0.65y \leq 40; x \geq 2y; x > 15, y \geq 0$

b. 10 65-cent stamps

27.

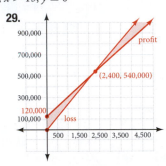

3,000 items

29.

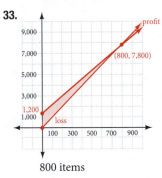

2,400 items

31.

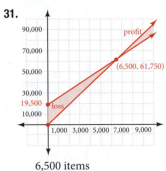

6,500 items

33.

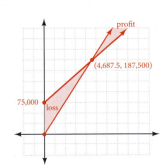

800 items

35. $p = \$420; 1,680$ units   **37.** $p = \$500; 4,500$ units

39. $p = \$600; 9,600$ units

41. a. $C(x) = 75,000 + 24x,$
   $R(x) = 40x,$
   $P(x) = 16x - 75,000$

b. $C(2,000) = \$123,000,$
   $R(2,000) = \$80,000,$
   $P(2,000) = -\$43,000$

c. $C(8,000) = \$267,000,$
   $R(8,000) = \$320,000,$
   $P(8,000) = \$53,000$

d. 4,688 items

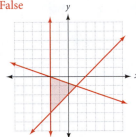

43.

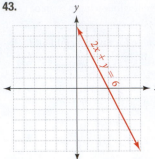

x-intercept $= 3$; y-intercept $= 6$;
slope $= -2$

45.

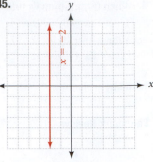

x-intercept $= -2$;
no y-intercept; no slope

47. $y = -\frac{3}{7}x + \frac{5}{7}$ **49.** $x = 4$

51. Domain $= \mathbb{R}$; Range $= \{y \mid y \geq -9\}$; A function

53. -4 **55.** 4 **57.** $3x + 2$ **59.** $\dfrac{1}{x - 2}$

Find the Mistake   **1.** The solution set for the system of inequalities $x \leq 0$ and $y \geq 0$ is all the points in the second quadrant only.

2. The solution set for the system of inequalities $x \leq 0$ and $y \leq 0$ is all the points in the third quadrant.

3. False

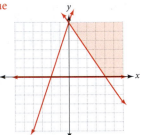

4. True

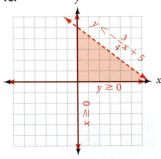

Chapter 3 Review

1. $(-3, 5)$ **2.** $(0, -2)$ **3.** $(2, 5)$ **4.** $(4, 2)$ **5.** $\left(\frac{1}{2}, 1, 1\right)$

6. $(1, -2, 3)$ **7.** 2 **8.** -2 **9.** $\left(\frac{34}{11}, \frac{3}{11}\right)$ **10.** 3 and 13

11. $\$1,750$ at 12%, $\$3,500$ at 10%

12. 543 adult tickets, 362 children's tickets

13. Boat: 7 mph; Current: 3 mph

14. 2 quarters, 5 dimes, 15 nickels

15.

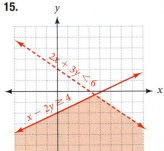

16.

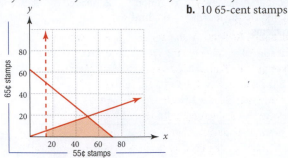

17. $y \leq \frac{2}{5}x + 2, y < -\frac{3}{2}x + 3$ **18.** $y > x + 2, y \geq -\frac{3}{2}x - 3$

Chapter 3 Cumulative Review

1. -40 **2.** 16 **3.** 1 **4.** -49 **5.** -4 **6.** -35
7. -46 **8.** -16 **9.** $21x - 17$ **10.** $2x + 37$
11. $-4x - 15y$ **12.** $2x - 3y$ **13.** 1 **14.** 0 **15.** $\varnothing$
16. $\varnothing$ **17.** $y = -2x + 2$ **18.** $y = \frac{3}{4}x + 4$ **19.** $x = \frac{-4}{a-b}$
20. $x = \frac{-1}{a-c}$ **21.** $[-3, \infty)$ **22.** $\left(-\infty, -\frac{4}{3}\right] \cup [2, \infty)$
23. $\{0, 1, 3, 5, 7, 10, 15\}$ **24.** $\{5\}$ **25.** $(-4, -5)$
26. Lines are parallel; $\varnothing$ **27.** Lines coincide   **28.** $(0, -2, 4)$
29.

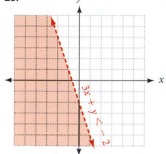

30. -9
31.

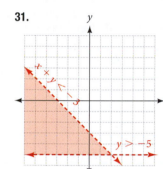

32.

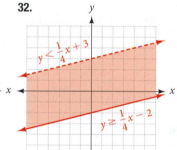

33. -1 **34.** $m = \frac{2}{7}, b = -2$ **35.** $y = -\frac{1}{2}x + 13$
36. $y = -\frac{3}{4}x - 3$ **37.** $y = 2x - 16$ **38.** -4 **39.** 4
40. $-x^2 + x - 4$ **41.** $-x^2 + 2x + 2$ **42.** 1 **43.** 2
44. -1 **45.** 8

Chapter 3 Test

1. $\left(\frac{25}{16}, -\frac{11}{8}\right)$ **2.** $(4, 9)$ **3.** $(6, 7)$ **4.** $(5, 3)$ **5.** $\left(1, \frac{2}{3}, \frac{2}{3}\right)$
6. $(-2, 2, 1)$ **7.** 2 **8.** 0 **9.** $\left(\frac{47}{26}, \frac{29}{26}\right)$ **10.** $4, 4$
11. $\$1,200$ at 13%, $\$3,600$ at 17%
12. 730 adult tickets, 160 children's tickets
13. Boat: 6 mph; Current: 3 mph   **14.** 3 quarters, 1 dime, 4 nickels
15.

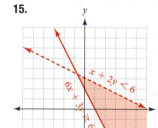

16.

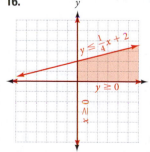

17. $y < \frac{4}{3}x + 1, y \le -\frac{4}{3}x + 1$ **18.** $y < \frac{4}{3}x + 1, y \ge -\frac{4}{3}x + 1$

Chapter 4

Exercise Set 4.1

Vocabulary Review   **1.** monomial   **2.** coefficient
3. binomial   **4.** polynomial   **5.** degree   **6.** similar or like
Problems   **1.** Trinomial, degree 2, leading coefficient 5
3. Binomial, degree 1, leading coefficient 3
5. Trinomial, degree 2, leading coefficient 8
7. Polynomial, degree 3, leading coefficient 4
9. Monomial, degree 0, leading coefficient $-\frac{3}{4}$
11. Trinomial, degree 3, leading coefficient 6   **13.** $7x + 1$
15. $2x^2 + 7x - 15$ **17.** $12a^2 - 7ab - 10b^2$ **19.** $x^2 - 13x + 3$
21. $\frac{1}{4}x^2 - \frac{7}{12}x - \frac{1}{4}$ **23.** $-y^3 - y^2 - 4y + 7$
25. $2x^3 + x^2 - 3x - 17$ **27.** $\frac{1}{14}x^2 + \frac{1}{7}xy + \frac{5}{7}y^2$
29. $-3a^3 + 6a^2b - 5ab^2$ **31.** $-3x$ **33.** $3x^2 - 12xy$
35. $17x^5 - 12$ **37.** $14a^2 - 2ab + 8b^2$ **39.** $2 - x$
41. $10x - 5$ **43.** $9x - 35$ **45.** $9y - 4x$ **47.** $9a + 2$
49. -2 **51. a.** 208 **b.** 103 **53. a.** 51 **b.** -15
55. a. 110 **b.** -120 **57. a.** $-5,000$ **b.** $3,000$
59. $3(43) + 2(23) + 54 = 229$ milligrams   **61.** 240 feet; 240 feet
63. $P(x) = -300 + 40x - 0.5x^2; P(60) = \$300$
65. $P(x) = -800 + 3.5x - 0.002x^2; P(1,000) = \$700$
67. $A = (10 + 2x)(40 + 2x); x = 5, A = 1,000\,\text{ft}^2; x = 8, A = 1,456\,\text{ft}^2$
69. $A = (9 - 2x)(12 - 2x)x; x = 2, A = 80\,\text{in}^2$ **71.** $2x^2 + 7x - 15$
73. $6x^3 - 11x^2y + 11xy^2 - 12y^3$ **75.** $-12x^4$ **77.** $20x^5$
79. a^6 **81.** 650 **83.** $x - 5$ **85.** $x - 7$ **87.** $5x - 1$
89. $x - 1$ **91.** $3x + 8y$ **93.** $x^2 - 4$
95. a. 5 **b.** -4 **c.** -3 **d.** 3 **e.** 4 **f.** 0 **g.** -4 **h.** 4
i. 0 **j.** -4 **k.** 5 **l.** -5 **m.** ± 2 **n.** ± 3 **o.** ± 2
Find the Mistake   **1.** The numerical coefficient for the monomial
$4x^2y^3z^5$ is 4.   **2.** The degree of the polynomial $\frac{1}{2}x^6 + 2x^2 - 4$ is 6.
3. Similar terms of a polynomial have the same variable parts, but
may have different coefficients.   **4.** The sum of the polynomials
$2x^2 + 4x - 3$ and $3x + 5$ is $2x^2 + 7x + 2$.

Exercise Set 4.2

Vocabulary Review   **1.** distributive   **2.** second   **3.** first
4. binomial   **5.** difference
Problems   **1.** $12x^3 - 10x^2 + 8x$ **3.** $-3a^5 + 18a^4 - 21a^2$
5. $2a^5b - 2a^3b^2 + 2a^2b^4$ **7.** $x^2 - 2x - 15$ **9.** $6x^4 - 19x^2 + 15$
11. $x^3 + 9x^2 + 23x + 15$ **13.** $a^3 - b^3$ **15.** $8x^3 + y^3$
17. $2a^3 - a^2b - ab^2 - 3b^3$ **19.** $x^2 + x - 6$ **21.** $6a^2 + 13a + 6$
23. $20 - 2t - 6t^2$ **25.** $x^6 - 2x^3 - 15$ **27.** $20x^2 - 9xy - 18y^2$
29. $18t^2 - \frac{2}{9}$ **31. a.** $2x - 4$ **b.** $6x - 2$ **c.** $8x^2 - 2x - 3$
33. $25x^2 + 20xy + 4y^2$ **35.** $25 - 30t^3 + 9t^6$ **37.** $4a^2 - 9b^2$
39. $9r^4 - 49s^2$ **41.** $y^2 + 3y + \frac{9}{4}$ **43.** $a^2 - a + \frac{1}{4}$
45. $x^2 + \frac{1}{2}x + \frac{1}{16}$ **47.** $t^2 + \frac{2}{3}t + \frac{1}{9}$ **49.** $\frac{1}{9}x^2 - \frac{4}{25}$
51. $x^3 - 6x^2 + 12x - 8$ **53.** $x^3 - \frac{3}{2}x^2 + \frac{3}{4}x - \frac{1}{8}$
55. $3x^3 - 18x^2 + 33x - 18$ **57.** $a^2b^2 + b^2 + 8a^2 + 8$
59. $3x^2 + 12x + 14$ **61.** $24x$ **63.** $x^2 + 4x - 5$
65. $4a^2 - 30a + 56$ **67.** $32a^2 + 20a - 18$
69. a. $2^4 - 3^4 = -65$ **b.** $(2 - 3)^4 = 1$ **c.** $(2^2 + 3^2)(2 + 3)(2 - 3) = -65$
71. $[6 + 0.05(x - 400)] = 6 + 0.05(592 - 400) = \$15.60$
73. $R(p) = 900p - 300p^2; R(x) = 3x - \frac{x^2}{300}; \$672$

75. $R(p) = 350p - 10p^2$; $R(x) = 35x - \frac{x^2}{10}$; $1,852.50

77. $P(x) = -\frac{x^2}{10} + 30x - 500$; $P(60) = \$940$

79. $A = 100 + 400r + 600r^2 + 400r^3 + 100r^4$ **81.** $8.64

83. 20 years ago: \$1,100.85; 30 years ago: \$509.91   **85.** $8a^2$

87. -48 **89.** $2a^3b$ **91.** $-3b^2$ **93.** $-y^4$ **95.** $(1, 2, 3)$

97. $(1, 3, 1)$ **99.** $(x + y)^2 + (x + y) - 20 = x^2 + 2xy + y^2 + x + y - 20$

101. $x^{2n} - 5x^n + 6$ **103.** $10x^{2n} + 13x^n - 3$

105. $x^{2n} + 10x^n + 25$ **107.** $x^{3n} + 1$

Find the Mistake   1. To multiply binomials using the FOIL method, find the sum of the first, the outside, the inside, and the last terms.   **2.** To square the binomial $(a + b)$, find the sum of the square of the first term, twice the product of the two inside terms, and the square of the last term.   **3.** The product found by expanding and multiplying $(2x + 4)^2$ is $4x^2 + 16x + 16$.
4. Multiplying two binomials that differ only in the sign between their two terms will result in the difference of two squares.

Exercise Set 4.3

Vocabulary Review   1. factor   **2.** coefficient   **3.** similar
4. grouping

Problems   1. $5x^2(2x - 3)$ **3.** $9y^3(y^3 + 2)$ **5.** $3ab(3a - 2b)$

7. $7xy^2(3y^2 + x)$ **9.** $3(a^2 - 7a + 11)$ **11.** $4x(x^2 - 4x + 5)$

13. $10x^2y^2(x^2 + 2xy + 3y^2)$ **15.** $xy(-x + y - xy)$

17. $2xy^2z(2x^2 - 4xz + 3z^2)$ **19.** $5abc(4abc - 6b + 5ac)$

21. $(a - 2b)(5x - 3y)$ **23.** $3(x + y)^2(x^2 - 2y^2)$

25. $(x + 5)(2x^2 + 7x + 8)$ **27.** $(x + 1)(3y + 2a)$

29. $(xy + 1)(x + 3)$ **31.** $(x - 2)(3y^2 + 4)$ **33.** $(x - a)(x - b)$

35. $(b + 5)(a - 1)$ **37.** $(b^2 + 1)(a^4 - 5)$

39. $(x + 3)(x^2 + 4)$ **41.** $(x + 2)(x^2 + 25)$

43. $(2x + 3)(x^2 + 4)$ **45.** $(x + 3)(4x^2 + 9)$ **47.** 6

49. $P(1 + r) + P(1 + r)r = (1 + r)(P + Pr)$
$\qquad = (1 + r)P(1 + r)$
$\qquad = P(1 + r)^2$

51. \$28.50

53. $y = -3.5x + 7,104.5$; 69.5 million   **55.** $3x^2(x^2 - 3xy - 6y^2)$

57. $(x - 3)(2x^2 - 4x - 3)$ **59.** $3x^2 + 5x - 2$ **61.** $3x^2 - 5x + 2$

63. $x^2 + 5x + 6$ **65.** $6y^2 + y - 35$ **67.** $20 - 19a + 3a^2$ **69.** $35 - 4x^2$

71.

Two Numbers a and b	Their Product ab	Their Sum $a + b$
1, −24	−24	−23
−1, 24	−24	23
2, −12	−24	−10
−2, 12	−24	10
3, −8	−24	−5
−3, 8	−24	5
4, −6	−24	−2
−4, 6	−24	2

73. 9, 4

75. 8, −5

77. −12, 4

Find the Mistake   1. The greatest common factor for a polynomial is the largest monomial that is a factor of each term of the polynomial.
2. The greatest common factor for the polynomial $9x^3 + 18x^2 + 36x$ is $9x$.   **3.** The greatest common factor for the polynomial $x^5y^3z - x^3y^2 + 6y^4z^2$ is y^2.   **4.** To begin to factor the polynomial $12 - 4y^2 - 5x^3 + x^3y^5$ by grouping, we can factor 4 from the first two terms and x^3 from the last two terms.

Landmark Review   1. Binomial degree 1, leading coefficient 5
2. Monomial degree 0, leading coefficient 14
3. Trinomial degree 2, leading coefficient 9
4. Trinomial degree 2, leading coefficient 14   **5.** $-2x - 7$
6. $5x^2 + 11x + 1$ **7.** $8y^2 + x^2 + 4x - 3$ **8.** $15x^3 - 9x^2 + 7y - 13$
9. $4a^3 + 8a^2 - 10a$ **10.** $40x^2y + 24xy^2 + 16x^2y^2$
11. $x^2 - 8x - x + 8$ **12.** $xy + 6x - 2y - 12$ **13.** $x^3 - 7x^2 - 4x + 28$
14. $x^3 - 8x^2 + 6x - 48$ **15.** $3a^2 + 11a + 10$ **16.** $6y^2 - 7y + 2$
17. $5x(3x^2 + 1)$ **18.** $6xy(4xy + 3)$ **19.** $7xy(2x^2y + 3xy + 5)$
20. $3x^2y^2(3x^3y + 2y + 6x)$ **21.** $(x + 3)(4y^2 + 5a)$
22. $(2b + 6)(3a - 1)$ **23.** $(4 - 2b)(a^2 + 2c^2)$ **24.** $(2x + 6)(3x^2 + 3)$

Exercise Set 4.4

Vocabulary Review   1. coefficient   **2.** common   **3.** trial and error   **4.** grouping   **List of Steps:** product, sum, middle, factor

Problems   1. $(x + 3)(x + 4)$ **3.** $(x + 3)(x - 4)$

5. $(y + 3)(y - 2)$ **7.** $(2 - x)(8 + x)$ **9.** $(2 + x)(6 + x)$

11. $(4 + x)(4 - x)$ **13.** $(x + 2y)(x + y)$ **15.** $(a + 6b)(a - 3b)$

17. $(x - 8a)(x + 6a)$ **19.** $(x - 6b)^2$ **21.** $3(a - 2)(a - 5)$

23. $4x(x - 5)(x + 1)$ **25.** $3(x - 3y)(x + y)$

27. $2a^3(a^2 + 2ab + 2b^2)$ **29.** $10x^2y^2(x + 3y)(x - y)$

31. $(2x - 3)(x + 5)$ **33.** $(2x - 5)(x + 3)$ **37.** $(2x - 5)(x - 3)$

39. Prime   **41.** $(2 + 3a)(1 + 2a)$ **43.** $15(4y + 3)(y - 1)$

45. $x^2(3x - 2)(2x + 1)$ **47.** $10r(2r - 3)^2$ **49.** $(4x + y)(x - 3y)$

51. $(2x - 3a)(5x + 6a)$ **53.** $(3a + 4b)(6a - 7b)$

55. $200(1 + 2t)(3 - 2t)$ **57.** $y^2(3y - 2)(3y + 5)$

59. $2a^2(3 + 2a)(4 - 3a)$ **61.** $2x^2y^2(4x + 3y)(x - y)$

63. $100(3x^2 + 1)(x^2 + 3)$ **65.** $(5a^2 + 3)(4a^2 + 5)$

67. $3(3 + 4r^2)(1 + r^2)$ **69.** $(x + 5)(2x + 3)(x + 2)$

71. $(2x + 3)(x + 5)(x + 2)$ **73.** $(3x - 5)(x + 4)(x - 3)$

75. $(2x - 3)(3x - 4)(x - 2)$ **77.** $(3x - 5)(4x + 9)(x + 3)$

79. $(3x + 2)(2x - 5)(5x - 2)$ **81.** $(5x + 3)(4x + 7)(2x + 3)$

83. $9x^2 - 25y^2$ **85.** $a + 250$ **87.** $12x - 35$ **89.** $9x + 8$

91. $7x + 8$ **93.** $y = 2(2x - 1)(x + 5)$; $y = 0$ when $x = \frac{1}{2}$ or
$\qquad x = -5$, $y = 42$ when $x = 2$

95. $4x^2 - 9$ **97.** $4x^2 - 12x + 9$ **99.** $8x^3 - 27$ **101.** $\frac{5}{8}$

103. x^3 **105.** $4x^2$ **107.** $\frac{1}{2}$ **109.** x^2 **111.** $3x$

113. $2y$ **115.** $(2x^3 + 5y^2)(4x^3 + 3y^2)$ **117.** $(3x - 5)(x + 100)$

119. $\left(\frac{1}{4}x + 1\right)\left(\frac{1}{3}x + 2\right)$ **121.** $(2x + 0.5)(x + 0.5)$

Find the Mistake   1. True   **2.** True   **3.** False, the leading coefficient of the polynomial $4x^4 - 2x^3y^2 - 12x^2y^3$ is 4.   **4.** True

Exercise Set 4.5

Vocabulary Review   1. c   **2.** a   **3.** b   **4.** e   **5.** d

Problems   1. $(x - 3)^2$ **3.** $(a - 6)^2$ **5.** $(5 - t)^2$

7. $\left(\frac{1}{3}x + 3\right)^2$ **9.** $(2y^2 - 3)^2$ **11.** $(4a + 5b)^2$

13. $\left(\frac{1}{5} + \frac{1}{4}t^2\right)^2$ **15.** $\left(y + \frac{3}{2}\right)^2$ **17.** $\left(a - \frac{1}{2}\right)^2$ **19.** $\left(x - \frac{1}{4}\right)^2$

21. $\left(t + \frac{1}{3}\right)^2$ **23.** $4(2x - 3)^2$ **25.** $3a(5a + 1)^2$

27. $(x + 2 + 3)^2 = (x + 5)^2$ **29.** $(x + 3)(x - 3)$

31. $(7x + 8y)(7x - 8y)$ **33.** $\left(2a + \frac{1}{2}\right)\left(2a - \frac{1}{2}\right)$

35. $\left(x + \frac{3}{5}\right)\left(x - \frac{3}{5}\right)$ **37.** $(3x + 4y)(3x - 4y)$

39. $10(5 + t)(5 - t)$ **41.** $(x^2 + 9)(x + 3)(x - 3)$

43. $(3x^3 + 1)(3x^3 - 1)$ **45.** $(4a^2 + 9)(2a + 3)(2a - 3)$

47. $\left(\frac{1}{9} + \frac{y^2}{4}\right)\left(\frac{1}{3} + \frac{y}{2}\right)\left(\frac{1}{3} - \frac{y}{2}\right)$ **49.** $\left(\frac{x^2}{4} + \frac{4}{9}\right)\left(\frac{x}{2} + \frac{2}{3}\right)\left(\frac{x}{2} - \frac{2}{3}\right)$

51. $\left(a^2 + \frac{9}{16}\right)\left(a + \frac{3}{4}\right)\left(a - \frac{3}{4}\right)$

53. $(x - y)(x + y)(x^2 + xy + y^2)(x^2 - xy + y^2)$

55. $2a(a - 2)(a + 2)(a^2 + 2a + 4)(a^2 - 2a + 4)$ **57.** $(x + 1)(x - 5)$

59. $y(y + 8)$ **61.** $(x - 5 + y)(x - 5 - y)$

63. $(a + 4 + b)(a + 4 - b)$ **65.** $(x + y + a)(x + y - a)$

67. $(x + 3)(x + 2)(x - 2)$ **69.** $(x + 2)(x + 5)(x - 5)$

71. $(2x + 3)(x + 2)(x - 2)$ **73.** $(x + 3)(2x + 3)(2x - 3)$

75. $(2x - 15)(2x + 5)$ **77.** $(a - 3 - 4b)(a - 3 + 4b)$

79. $(a - 3 - 4b)(a - 3 + 4b)$ **81.** $(x + 4)(x - 3)^2$

83. $(x - y)(x^2 + xy + y^2)$ **85.** $(a + 2)(a^2 - 2a + 4)$

87. $(3 + x)(9 - 3x + x^2)$ **89.** $(y - 1)(y^2 + y + 1)$

91. $10(r - 5)(r^2 + 5r + 25)$ **93.** $(4 + 3a)(16 - 12a + 9a^2)$

95. $(2x - 3y)(4x^2 + 6xy + 9y^2)$ **97.** $\left(t + \frac{1}{3}\right)\left(t^2 - \frac{1}{3}t + \frac{1}{9}\right)$

99. $\left(3x - \frac{1}{3}\right)\left(9x^2 + x + \frac{1}{9}\right)$ **101.** $(4a + 5b)(16a^2 - 20ab + 25b^2)$

103. 30 and -30 **105.** 81   **107.** $y(y^2 + 25)$

109. $2ab^3(b^2 + 4b + 1)$ **111.** $(2x - 3)(2x + a)$

113. $(3x - 2)(5a + 4)$ **115.** $(x + 2)(x - 2)$ **117.** $(A + 5)(A - 5)$

119. $(x - 3)^2$ **121.** $(x - 4y)^2$ **123.** $(3a - 4)(2a - 1)$

125. $(6x + 5)(2x - 7)$ **127.** $(x + 2)(x^2 - 2x + 4)$

129. $(2x - 3)(4x^2 + 6x + 9)$ **131.** $\left(-\frac{15}{43}, -\frac{27}{43}\right)$ **133.** $(1, 3, 1)$

135. $(a - b + 3)(a + b - 3)$ **137.** $(x - y - 8)(x + y + 2)$

139. $k = 144$ **141.** $k = \pm 126$

Find the Mistake   **1.** The left side of the formula $a^2 + 2ab + b^2 = (a + b)^2$ is called a perfect square trinomial.

2. The difference of two squares $(a^2 - b^2)$ factors into $(a + b)(a - b)$.   **3.** The difference of two cubes $(a^3 - b^3)$ factors into $(a - b)(a^2 + ab + b^2)$.   **4.** The sum of two cubes $(27x^3 + y^8)$ factors into $(3x + y^2)(9x^2 - 3xy^2 + y^4)$.

Exercise Set 4.6

Vocabulary Review   List of Steps: polynomial, greatest, squares, cubes, trinomial, binomial, grouping, common

Problems   1. $(x + 9)(x - 9)$ **3.** $(x - 3)(x + 5)$

5. $(x + 2)(x + 3)^2$ **7.** $(x^2 + 2)(y^2 + 1)$ **9.** $2ab(a^2 + 3a + 1)$

11. Does not factor, prime   **13.** $3(2a + 5)(2a - 5)$

15. $(3x - 2y)^2$ **17.** $(5 - t)^2$ **19.** $4x(x^2 + 4y^2)$

21. $2y(y + 5)^2$ **23.** $a^4(a + 2b)(a^2 - 2ab + 4b^2)$

25. $(t + 3 + x)(t + 3 - x)$ **27.** $(x + 5)(x + 3)(x - 3)$

29. $5(a + b)^2$ **31.** Does not factor, prime

33. $3(x + 2y)(x + 3y)$ **35.** $\left(3a + \frac{1}{3}\right)^2$ **37.** $(x - 3)(x - 7)^2$

39. $(x + 8)(x - 8)$ **41.** $(2 - 5x)(4 + 3x)$ **43.** $a^5(7a + 3)(7a - 3)$

45. $\left(r + \frac{1}{5}\right)\left(r - \frac{1}{5}\right)$ **47.** Does not factor, prime

49. $100(x - 3)(x + 2)$ **51.** $a(5a + 3)(5a + 1)$

53. $(3x^2 + 1)(x^2 - 5)$ **55.** $3a^2b(2a - 1)(4a^2 + 2a + 1)$

57. $(4 - r)(16 + 4r + r^2)$ **59.** $5x^2(2x + 3)(2x - 3)$

61. $100(2t + 3)(2t - 3)$ **63.** $2x^3(4x - 5)(2x - 3)$

65. $(y + 1)(y - 1)(y^2 - y + 1)(y^2 + y + 1)$ **67.** $2(5 + a)(5 - a)$

69. $3x^2y^2(2x + 3y)^2$ **71.** $(x - 2 + y)(x - 2 - y)$

73. $\left(a - \frac{2}{3}b\right)^2$ **75.** $\left(x - \frac{2}{5}y\right)^2$ **77.** $\left(a - \frac{5}{6}b\right)^2$

79. $\left(x - \frac{4}{5}y\right)^2$ **81.** $(2x - 3)(x - 5)(x + 2)$

83. $(x - 4)^3(x - 3)$ **85.** $2(y - 3)(y^2 + 3y + 9)$

87. $2(a - 4b)(a^2 + 4ab + 16b^2)$ **89.** $2(x + 6y)(x^2 - 6xy + 36y^2)$

91. 60 geese; 48 ducks   **93.** 150 oranges; 144 apples

95. $2x^2 + 2x + 1$ **97.** $t^2 - 4t + 3$ **99.** $(x - 6)(x + 4)$

101. $x(2x + 1)(x - 3)$ **103.** $(x + 2)(x - 3)(x + 3)$

105. $(x + 2)(x^2 - 5)$ **107.** 6   **109.** $-\frac{1}{2}$

Find the Mistake   1. The first step to factoring a polynomial is to factor out the greatest common factor.   **2.** If a polynomial is a perfect square trinomial, we can factor it into the square of a binomial.   **3.** If the polynomial has more than three terms, factor it by grouping.   **4.** A sum of two squares will not factor.

Exercise Set 4.7

Vocabulary Review   1. quadratic   **2.** constants   **3.** linear   **4.** zero-factor   **5.** standard   **6.** right

Problems   1. $6, -1$ **3.** $0, 2, 3$ **5.** $\frac{1}{3}, -4$ **7.** $\frac{2}{3}, \frac{3}{2}$

9. $5, -5$ **11.** $0, -3, 7$ **13.** $-4, \frac{5}{2}$ **15.** $0, \frac{4}{3}$ **17.** $-\frac{1}{5}, \frac{1}{3}$

19. $-10, 0$ **21.** $-5, 1$ **23.** $1, 2$ **25.** $-2, 3$ **27.** $-2, \frac{1}{4}$

29. $-2, \frac{5}{3}$ **31.** $-1, 9$ **33.** $0, -3$ **35.** $-4, -2$ **37.** $\frac{1}{2}, 5$

39. $\frac{3}{2}, -6$ **41.** $9, 2$ **43.** $-3, -2, 2$ **45.** $-2, -5, 5$

47. $0, -\frac{4}{3}, \frac{4}{3}$ **49.** $-\frac{3}{2}, -2, 2$ **51.** $-3, -\frac{3}{2}, \frac{3}{2}$ **53.** $-\frac{3}{2}$

55. $-2, 8$ **57.** -3 **59.** $-7, 1$ **61.** $0, 5$ **63.** $-1, 8$

65. $0, 1$ **67.** $\frac{3}{2}$ **69.** 9 and 11, or -11 and -9 **71.** 8 and 6

73. 8 and 10   **75.** 7 and 9   **77.** $l = 14$ cm, $w = 9$ cm

79. $b = 16$ m, $h = 10$ m   **81.** $b = 9$ in., $h = 14$ in.   **83.** $x = 5$

85. 6, 8, 10   **87.** 2 meters, 8 meters   **89.** 18 cm, 4 cm

91. $t = 0$ or $t = 2$, 0 and 2 seconds   **93.** 2 yards   **95.** 350 square feet

97. 12 feet   **99.** 0.5 inches   **101.** 1 and 2 seconds

103. 0 and $\frac{3}{2}$ seconds   **105.** 2 and 3 seconds   **107.** $(1, 2)$

109. $(15, 12)$

111.   **113.**

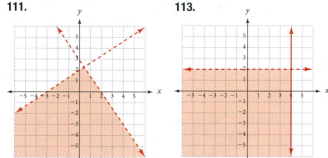

Find the Mistake   1. The quadratic term for the equation $4x^2 + 4x - 3$ is $4x^2$.   **2.** The quadratic equation $x^2 - 5x = 6$ written in standard form is $x^2 - 5x - 6 = 0$.   **3.** The constant term for the equation $\frac{1}{10}y^2 - \frac{5}{2} = 0$ is $-\frac{5}{2}$.   **4.** The sum of the two cubes $(27x^3 + y^6)$ factors into $(3x + y^2)(9x^2 - 3xy^2 + y^4)$.

Chapter 4 Review

1. $\frac{2}{3}x^3 - \frac{1}{3}x^2 - \frac{7}{3}x - \frac{8}{3}$ **2.** $-21x - 18$ **3.** $R(x) = 24x - 0.2x^2$

4. $P(x) = -0.2x^2 + 21x - 25$ **5.** \$400   **6.** \$325   **7.** \$75

8. $-3x^2 - 4x + 15$ **9.** $4x^3 - 2x^2 - 14x - 3$

10. $16a^2 - 24a + 9$ **11.** $4x^2 - 25$ **12.** $2x^3 + 3x^2 - 20x$

13. $15x^2 - \frac{1}{15}$ **14.** $(x - 5)(x + 2)$ **15.** $2(x^2 + 4)(3x^2 - 5)$

16. $(2x - 3y)(2x + 3y)(4x^2 + 9y^2)$ **17.** $(3a - 4x)(2b + y)$

18. $\left(y - \frac{1}{4}\right)\left(y^2 + \frac{1}{4}y + \frac{1}{16}\right)$ **19.** $2x^2y^3(x - 2y)(x + 5y)$

20. $(2a - b - 4)(2a - b + 4)$ **21.** $(3 - x)(3 + x)(9 + x^2)$

22. $7, \frac{2}{3}$ **23.** $0, \frac{1}{4}$ **24.** $-5, 4$ **25.** $-4, -3, 4$ **26.** $-7, 3$

27. $-8, 2$ **28.** 7   **29.** 3 inches   **30.** 6 centimeters

Chapter 4 Cumulative Review

1. -27 **2.** $-\frac{1}{27}$ **3.** 58 **4.** 161 **5.** 48 **6.** 121
7. $\frac{1}{44}$ **8.** -1 **9.** 12.24 **10.** -12.26 **11.** 10 **12.** 6
13. $17x - 1$ **14.** $10x - 21y$ **15.** $-x + 3$ **16.** $2a - 6$
17. $7x^3 + x^2 + 15$ **18.** $\frac{41}{15}x^2 - \frac{8}{5}x$ **19.** $32x^{10}y^5$ **20.** $-20a^3b^5$
21. $\frac{3x}{y^8}$ **22.** $\frac{125b}{a^2}$ **23.** $8x^2 + 6x - 5$ **24.** $18x^3 + 8x^2 + 14x$
25. $x^2 - \frac{3}{2}x + \frac{9}{16}$ **26.** $x^3 + \frac{9}{4}x^2 + \frac{27}{16}x + \frac{27}{64}$ **27.** 1.6×10^8
28. 3×10^6 **29.** $(x + a)(x - b)$ **30.** $(4a + 7)(2a - 1)$
31. $(x + 3)(x - 5)$ **32.** $\left(t + \frac{1}{3}\right)\left(t^3 - \frac{1}{3}t + \frac{1}{9}\right)$ **33.** 3
34. $-9, -5$ **35.** -30 **36.** 1 **37.** $-\frac{7}{4}, 0, \frac{7}{4}$ **38.** -4
39. $(1, 3)$ **40.** $(0, -2)$ **41.** $-\frac{1}{7}$ **42.** $\frac{2}{19}$ **43.** -6
44. $\frac{13}{2}$ **45.** 11 **46.** 2 **47.** $b = 3$ feet, $h = 8$ feet
48. $30°, 60°, 90°$ **49.** 74 **50.** 6

Chapter 4 Test

1. $\frac{6}{5}x^3 - \frac{6}{5}x^2 - \frac{8}{5}x - \frac{6}{5}$ **2.** $-14x - 58$ **3.** $R(x) = 36x - 0.3x^2$
4. $P(x) = -0.3x^2 + 32x - 50$ **5.** $\$600$ **6.** $\$450$ **7.** $\$150$
8. $-5x^2 - 31x + 28$ **9.** $6x^3 + 14x^2 - 27x + 10$
10. $9a^8 - 42a^4 + 49$ **11.** $4x^2 - 9$ **12.** $3x^3 - 17x^2 - 28x$
13. $14x^2 - \frac{1}{14}$ **14.** $(x - 1)(x - 5)$ **15.** $3(5x^2 - 4)(x^2 + 3)$
16. $(3x + 2)(3x - 2)(9x^2 + 4y^2)$ **17.** $(6x - y)(a + 3b^2)$
18. $\left(y - \frac{1}{3}\right)\left(y^2 + \frac{1}{3}y + \frac{1}{9}\right)$ **19.** $3x^2y^4(x - 3y)(x + 8y)$
20. $(a - 6 - b)(a + 6 - b)$ **21.** $(2 - x)(2 + x)(4 + x^2)$
22. $-10, -\frac{1}{2}$ **23.** $0, 3$ **24.** $-6, 3$ **25.** $-5, -3, 3$ **26.** $-3, 5$
27. $-5, 4$ **28.** 3 **29.** 5 inches   **30.** 6 centimeters

Chapter 5

Exercise Set 5.1

Vocabulary Review   **1.** rational   **2.** nonzero   **3.** function
4. domain   **5.** quotient
Problems   **1.** $\frac{x - 4}{6}$ **3.** $(a^2 + 9)(a + 3)$ **5.** $\frac{2y + 3}{y + 1}$ **7.** $\frac{x - 2}{x - 1}$
9. $\frac{x - 3}{x + 2}$ **11.** $\frac{x^2 - x + 1}{x - 1}$ **13.** $\frac{-4a}{3}$ **15.** $\frac{b - 1}{b + 1}$ **17.** $\frac{7x - 3}{7x + 5}$
19. $\frac{4x + 3}{4x - 3}$ **21.** $\frac{x + 5}{2x - 7}$ **23.** $-\frac{a + 5x}{3a + 5x}$ **25.** $\frac{a^2 - ab + b^2}{a - b}$
27. $\frac{2x - 2}{x}$ **29.** $\frac{x + 3}{y - 4}$ **31.** $x + 2$ **33.** $-\frac{3x + 2y}{3x + y}$
35. $\frac{x^2 + 2x + 4}{x + 2}$ **37.** $\frac{4x^2 + 6x + 9}{2x + 3}$ **39.** $\frac{1}{x^2 - 2x + 4}$ **41.** -1
43. $-(y + 6)$ **45.** $-\frac{3a + 1}{3a - 1}$ **47.** 3 **49.** $x + a$ **51.** $x + a + 3$
53. $\{x \mid x \neq 1\}$ **55.** $\{x \mid x \neq 2\}$ **57.** $\{t \mid t \neq 4, t \neq -4\}$
59. $g(0) = -3, g(-3) = 0, g(3) = 3, g(-1) = -1, g(1) =$ undefined
61. $h(0) = -3, h(-3) = 3, h(3) = 0, h(-1)$ is undefined, $h(1) = -1$
63. a. 4 **b.** 4 **65. a.** 5 **b.** 5 **67. a.** $x + a$ **b.** $2x + h$
69. a. $x + a$ **b.** $2x + h$ **71. a.** $x + a - 3$ **b.** $2x + h - 3$
73. a. $2x + 2a - 3$ **b.** $4x + 2h + 3$
75.

Weeks	Weight (lb)
x	$W(x)$
0	200
1	194.3
4	184
12	173.3
24	168

77. a. $C(n) = 1200 + 540n$
b. $\overline{C}(n) = \dfrac{1200 + 540n}{n}$
c. $\$660$ **d.** $\$620$ **e.** Decreasing
79. $\frac{2}{3}$ **81.** $20x^2y^2$
83. $(x + 2)(x - 2)$ **85.** $x^2(x - y)$
87. a. 2 **b.** -4 **c.** Undefined
d. 2 **e.** 1 **f.** -6 **g.** 4

Find the Mistake   **1.** The rational expression $\frac{8x^2 + 2x - 1}{16x^2 - 6x - 7}$ reduced to lowest terms is $\frac{4x - 1}{8x - 7}$.   **2.** To reduce the rational expression $\frac{15x^2 - 16x + 4}{5x^2 - 17x + 6}$ to lowest terms, we must divide the numerator and denominator by the factor $5x - 2$.   **3.** The rational function $f(x) = \frac{x^3 - 4x^2 + 4x + 16}{x^2 - 6x + 8}$ is undefined when $f(4)$ and $f(2)$.   **4.** To find the difference quotient for $f(x) = x^2 + x - 7$, we use $f(a) = a^2 + a - 7$.

Exercise Set 5.2

Vocabulary Review   **1.** product, numerators   **2.** quotient, reciprocal
Problems   **1.** $\frac{1}{6}$ **3.** $\frac{9}{4}$ **5.** $\frac{1}{2}$ **7.** $\frac{15y}{x^2}$ **9.** $\frac{b}{a}$ **11.** $\frac{2y^5}{z^3}$
13. $\frac{x + 3}{x + 2}$ **15.** $y + 1$ **17.** $\frac{3(x + 4)}{x - 2}$ **19.** $\frac{y^2}{xy + 1}$ **21.** $\frac{x^2 + 9}{x^2 - 9}$
23. $-\frac{20}{9}$ **25.** 1 **27.** $-\frac{x + 2y}{x - 2y}$ **29.** $\frac{9t^2 - 6t + 4}{4t^2 - 2t + 1}$
31. $\frac{x + 3}{x + 4}$ **33.** $\frac{a - b}{5}$ **35.** $\frac{5c - 1}{3c - 2}$ **37.** $-\frac{x + 2y}{x - 2y}$
39. 2 **41.** $x(x - 1)(x + 1)$ **43.** $\frac{(a + 4b)(a - 3b)}{(a - 4b)(a + 5b)}$
45. $\frac{2y - 1}{2y - 3}$ **47.** $\frac{(y - 2)(y + 1)}{(y + 2)(y - 1)}$ **49.** $\frac{x - 1}{x + 1}$ **51.** $\frac{x - 2}{x + 3}$
53. $\frac{w(y - 1)}{w - x}$ **55.** $\frac{(m + 2)(x + y)}{(2x + y)^2}$ **57.** $\frac{(1 - 2d)(d + c)}{d - c}$
59. $\frac{(r - s)^2(r + s)}{(r^2 + s^2)^2}$ **61.** $3x$ **63.** $2(x + 5)$ **65.** $x - 2$
67. $-(y - 4)$ or $4 - y$ **69.** $(a - 5)(a + 1)$
71. a. $\frac{5}{21}$ **b.** $\frac{5x + 3}{25x^2 + 15x + 9}$ **c.** $\frac{5x - 3}{25x^2 + 15x + 9}$ **d.** $\frac{5x + 3}{5x - 3}$
73. a. $\frac{(x + 2)^2}{(x - 1)^2}$ **b.** $(x - 3)^2$ **c.** $\frac{1}{(x - 3)^2}$
75. a.

Number of Copies	Price per Copy ($)
x	$p(x)$
1	20.33
10	9.33
20	6.40
50	4.00
100	3.05

b. $\$200$
c. $\$305$
d. $\dfrac{2x(x + 60)}{x + 5}$

77. Venus: 7,110 miles; Saturn: 75,050 miles   **79.** $\frac{2}{3}$
81. $\frac{47}{105}$ **83.** $x - 7$ **85.** $-4x + 10$ **87.** $(x + 1)(x - 1)$
89. $2(x + 5)$ **91.** $(a - b)(a^2 + ab + b^2)$ **93.** $(2x + 3)(2x - 3)$
95. $\frac{x^4 - x^2y^2 + y^4}{x^2 + y^2}$ **97.** $\frac{(a - 1)(a + 5)}{3a^2 - 2a + 1}$ **99.** $-(p^2 - pq + q^2)$

Find the Mistake   **1.** Multiplying $\frac{x^2 - 4}{x - 2} \cdot \frac{x^2 + 2x - 8}{x^2 + 6x + 8}$ we get $x - 2$.
2. Multiplying $\frac{2x - 6}{4x^2 - 4x - 24} \cdot \frac{6x^2 + 2x - 20}{3x - 5}$ we get 1.   **3.** To divide the rational expression $\frac{5x^2 + 21x - 20}{15x^2 - 17x + 4}$ by $\frac{x^2 - 25}{3x^2 - 4x + 1}$ we multiply by the reciprocal of the expression that follows the division symbol.   **4.** Dividing $\frac{4x^2 + 27xy + 18y^2}{x^2 + 7xy + 6y^2} \div \frac{16x + 12y}{x^2 - y^2}$ to get $\frac{x - y}{4}$.

Exercise Set 5.3

Vocabulary Review   **1.** common   **2.** denominator   **3.** different
List of Steps: equivalent, numerators, lowest
Problems   **1.** 1 **3.** -1 **5.** $\frac{1}{x + y}$ **7.** 1 **9.** $\frac{a^2 + 2a - 3}{a^3}$
11. 1 **13.** $\frac{5}{4}$ **15.** $\frac{1}{3}$ **17.** $\frac{41}{24}$ **19.** $\frac{19}{144}$ **21.** $\frac{31}{24}$ **23.** $\frac{2}{3}$

25. a. $\frac{1}{16}$ **b.** $\frac{9}{4}$ **c.** $\frac{13}{24}$ **d.** $\frac{5x+15}{(x-3)^2}$ **e.** $\frac{x+3}{5}$ **f.** $\frac{x-2}{x-3}$

27. $f(x) - g(x) = \frac{1}{2}$ **29.** $f(x) - g(x) = \frac{1}{5}$ **31.** $\frac{x+3}{2(x+1)}$

33. $\frac{a-b}{a^2+ab+b^2}$ **35.** $\frac{x-3}{2x-1}$ **37.** $\frac{2y-3}{4y^2+6y+9}$

39. $\frac{2(2x-3)}{(x-3)(x-2)}$ **41.** $\frac{1}{2t-7}$ **43.** $\frac{4}{(a-3)(a+1)}$

45. $\frac{-4x^2}{(2x+1)(2x-1)(4x^2+2x+1)}$ **47.** $\frac{2}{(2x+3)(4x+3)}$ **49.** $\frac{a}{(a+4)(a+5)}$

51. $\frac{x+1}{(x-2)(x+3)}$ **53.** $\frac{x-1}{(x+1)(x+2)}$ **55.** $\frac{1}{(x+2)(x+1)}$

57. $\frac{1}{(x+2)(x+3)}$ **59.** $\frac{4x+5}{2x+1}$ **61.** $\frac{22-5t}{4-t}$ **63.** $\frac{2x^2+3x-4}{2x+3}$

65. $\frac{2x-3}{2x}$ **67.** $\frac{1}{2}$ **69.** $\frac{3}{x+4}$ **71.** $\frac{(2x+1)(x+5)}{(x-2)(x+1)(x+3)}$

73. $\frac{51}{10}$ **75.** 1 hour, 20 minutes **77.** $x + \frac{4}{x} = \frac{x^2+4}{x}$

79. $\frac{1}{x} + \frac{1}{x+1} = \frac{2x+1}{x(x+1)}$ **81.** $\frac{6}{5}$ **83.** $\frac{1}{9}$ **85.** $x + 2$

87. $3 - x$ **89.** $(x+2)(x-2)$ **91.** $(x-3)(x^2+3x+9)$ **93.** $\frac{x-1}{x+3}$

95. $\frac{a^2+ab+a-v}{u+v}$ **97.** $\frac{6x+5}{(4x-1)(3x-4)}$ **99.** $\frac{y(y^2+1)}{(y+1)^2(y-1)^2}$

Find the Mistake **1.** If we add the rational expressions $\frac{x+5}{3x+9}$ and $\frac{4}{x^2-9}$ and reduce to the lowest terms, we get $\frac{x-1}{3x-9}$.

2. The least common denominator for the rational expressions $\frac{2x+3}{3x^2+6x-45}$ and $\frac{x+7}{3x^2+14x-5}$ is $(3x-1)(x+5)(3x-9)$.

3. If we subtract $\frac{5}{2x^2+x-3}$ from $\frac{2}{x^2-1}$, we get $\frac{-1}{(x+1)(2x+3)}$.

4. To subtract $\frac{x-4}{x^2-x-20}$ from $\frac{2x+5}{x^2-4x-5}$, make the expressions equivalent by first using LCD $(x+1)(x-5)(x+4)$.

Exercise Set 5.4

Vocabulary Review   complex, quotient, numerator, LCD, reciprocal, denominator

Problems **1.** $\frac{9}{8}$ **3.** $\frac{2}{15}$ **5.** $\frac{119}{20}$ **7.** $\frac{1}{x+1}$ **9.** $\frac{a+1}{a-1}$

11. $\frac{y-x}{y+x}$ **13.** $\frac{1}{(x+5)(x-2)}$ **15.** $\frac{1}{a^2-a+1}$ **17.** $\frac{x+3}{x+2}$

19. $\frac{a+3}{a-2}$ **21.** $\frac{x-3}{x+3}$ **23.** $\frac{a-1}{a+1}$ **25.** $-\frac{x}{3}$ **27.** $\frac{y^2+1}{2y}$

29. $\frac{-x^2+x-1}{x-1}$ **31.** $\frac{5}{3}$ **33.** $\frac{2x-1}{2x+3}$ **35.** $-\frac{1}{x(x+h)}$

37. $-\frac{1}{x^2+8x}$ **39.** $-\frac{12x+36}{x^2(x+6)^2}$ **41.** $\frac{3c+4a-2b}{5}$

43. $\frac{(t-4)(t+1)}{(t+6)(t-3)}$ **45.** $\frac{(5b-1)(b+5)}{2(2b-11)}$ **47.** $-\frac{3}{2x+14}$

49. $2m - 9$ **51.** $-\frac{2}{x}$ **53.** $\frac{f(x)}{g(x)} = \frac{x-3}{x}$ **55.** $\frac{f(x)}{g(x)} = \frac{x+4}{x+2}$

57. a. $\frac{-4}{ax}$ **b.** $\frac{-1}{(x+1)(a+1)}$ **c.** $-\frac{a+x}{a^2x^2}$

59. a. As v approaches 0, the denominator approaches 1 **b.** $v = \frac{fs}{h} - s$

61. $xy - 2x$ **63.** $3x - 18$ **65.** ab **67.** $2t^2 - 3s$

69. $(y+5)(y-5)$ **71.** $x(a+b)$ **73.** 2 **75.** $-2, 6$

Find the Mistake **1.** Simplifying $\frac{\frac{2}{m}+\frac{4}{n}}{\frac{2}{m}-\frac{6}{n}}$ gives us $-\frac{2m+n}{3m-n}$.

2. If $f(x) = \frac{x+2}{x^2-16}$ and $g(x) = \frac{x^2-x-6}{x+4}$, then $\frac{f(x)}{g(x)} = \frac{1}{(x-4)(x-3)}$.

3. Simplifying $5 - \frac{2}{x+\frac{1}{4}}$ gives us $\frac{20x-3}{4x+1}$.

4. The first step to simplifying $\frac{1-36x^{-2}}{1-9x^{-1}+18x^{-2}}$ is to write the expression with only positive exponents.

Landmark Review **1.** $\frac{a+3}{2}$ **2.** $\frac{1}{y-3}$ **3.** $\frac{3x(x-5)}{x+2}$ **4.** $\frac{x+4}{5x-2}$

5. $\frac{a}{2}$ **6.** $\frac{x(x+4)}{x-1}$ **7.** $\frac{y(y-3)}{2(y+2)(y-5)}$ **8.** $2x^2+6$ **9.** 1

10. $\frac{4}{y-3}$ **11.** $\frac{a-3}{a-4}$ **12.** $x-4$ **13.** $\frac{x+4}{x+1}$ **14.** $\frac{x-2}{(x+3)(x+2)}$

15. $\frac{a-3}{a-4}$ **16.** $\frac{x+2y}{x+3y}$ **17.** $\frac{1}{(x+3)(4x-1)}$ **18.** $-\frac{2y}{y^2+1}$

Exercise Set 5.5

Vocabulary Review **1.** rational **2.** extraneous **3.** function **4.** asymptote

Problems **1.** $-\frac{35}{3}$ **3.** $-\frac{18}{5}$ **5.** $\frac{36}{11}$ **7.** 2 **9.** 5 **11.** 2

13. $-3, 4$ **15.** $1, -\frac{4}{3}$ **17.** No solution **19.** -4 **21.** $-\frac{1}{2}, \frac{5}{3}$

23. $\frac{2}{3}$ **25.** -4 **27.** -6 **29.** -5 **31.** $\frac{53}{17}$ **33.** 2

35. No solution **37.** $\frac{22}{3}$ **39.** 18 **41.** No solution

43. a. $-\frac{9}{5}, 5$ **b.** $\frac{9}{2}$ **c.** No solution **45. a.** $\frac{1}{3}$ **b.** 3 **c.** 9 **d.** 4 **e.** $\frac{1}{3}, 3$

47. a. $\frac{6}{(x-4)(x+3)}$ **b.** $\frac{x-3}{x-4}$ **c.** 5 **49.** $x = \frac{ab}{a-b}$

51. $y = \frac{x-3}{x-1}$ **53.** $y = \frac{1-x}{3x-2}$

55.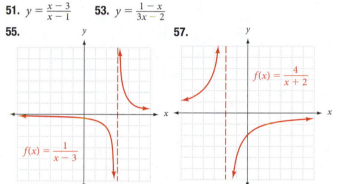

57.

59.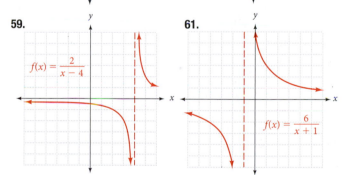

61.

63. Vertical asymptote: $x = \frac{5}{3}$; Domain: $\left\{ x \mid x \neq \frac{5}{3} \right\}$

65. Vertical asymptote: $x = \frac{3}{2}$; Domain: $\left\{ x \mid x \neq \frac{3}{2} \right\}$

67. Vertical asymptote: $x = \pm 3$; Domain: $\left\{ x \mid x \neq \pm 3 \right\}$

69. Vertical asymptote: $x = 1, -3$; Domain: $\left\{ x \mid x \neq 1, -3 \right\}$

71. Vertical asymptote: $x = 1, 2$; Domain: $\left\{ x \mid x \neq 1, 2 \right\}$

73. Vertical asymptote: $x = -\frac{1}{2}, 3$; Domain: $\left\{ x \mid x \neq -\frac{1}{2}, 3 \right\}$

75. $\frac{24}{5}$ feet **77.** 2,358 **79.** 3 **81.** 12.3 **83.** 9, -1

85. 60 **87.** $-\frac{2}{3}, -\frac{5}{2}, \frac{5}{2}$ **89.** $-2, 2, 3$ **91.** $x = -\frac{a}{5}$

93. $v = \frac{16t^2+s}{t}$ **95.** $f = \frac{pg}{g-p}$

Find the Mistake **1.** When we solve the rational equation $\frac{1}{x-3} - \frac{5}{x+3} = \frac{x}{x^2-9}$, we get $x = \frac{18}{5}$. **2.** To solve the rational equation $\frac{x+5}{x^2-5x} = \frac{20}{x^2-25}$, we must use the LCD $x(x-5)(x+5)$.

3. The graph for $f(x) = \frac{3}{x-1}$ will come very close but not touch the line $x = 1$. **4.** The graph for $f(x) = \frac{2}{x+2}$ has a vertical asymptote of $x = -2$.

Exercise Set 5.6

Vocabulary Review   1. blueprint **2.** LCD **3.** table **4.** asymptote

Problems   1. $\frac{1}{5}$ and $\frac{3}{5}$ **3.** 3 or $\frac{1}{3}$ **5.** 3, 4   **7.** 3

9. a-b.

	d	r	t
Upstream	1.5	$5 - x$	$\frac{1.5}{5-x}$
Downstream	3	$5 + x$	$\frac{3}{5+x}$

c. They are the same. $\frac{1.5}{5-x} = \frac{3}{5+x}$

d. The speed of the current is $\frac{5}{3}$ mph.   **11.** 6 mph

13. a-b.

	d	r	t
Train A	150	$x + 15$	$\frac{150}{x+15}$
Train B	120	x	$\frac{120}{x}$

c. They are the same. $\frac{150}{x+15} = \frac{120}{x}$

d. The speed of train A is 75 mph, and for train B it is 60 mph

15. 540 mph   **17.** 54 mph   **19.** 16 hours   **21.** 15 hours

23. 5.25 minutes   **25.** 1.52 minutes

27. $10 = \frac{1}{3}\left[\left(x + \frac{2}{3}x\right) + \frac{1}{3}\left(x + \frac{2}{3}x\right)\right]; x = \frac{27}{2}$

29. a. 30 grams **b.** 3.25 moles

31.

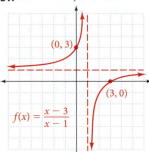

33.

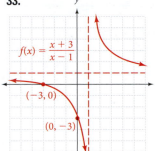

35.

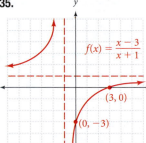

37.

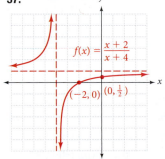

39. 121,215 fans   **41.** 2   **43.** $-2x^2y^2$ **45.** 185.12

47. $x - 3$ **49.** $4x^3 - 8x^2$ **51.** $4x^3 - 6x - 20$

53. $-3x + 9$ **55.** $(x + a)(x - a)$ **57.** $(x - 7y)(x + y)$

Find the Mistake   1. The equation to find the numbers is $\frac{1}{x} + \frac{1}{3x} = 8$.

2. The rate equation $t = \frac{d}{r}$, is very useful for many applications of rational equations.   **3.** To find the speed of the hikers, you must recognize that their travel distance is the same.   **4.** To find how long it takes to finish together, you would use the equation $\frac{1}{4} - \frac{1}{3} = \frac{1}{x}$.

Exercise Set 5.7

Vocabulary Review   1. average   **2.** monomial   **3.** factor
4. long division   **5.** estimate   **6.** numerator

Problems   1. $2x^2 - 4x + 3$ **3.** $-2x^2 - 3x + 4$

5. $2y^2 + \frac{5}{2} - \frac{3}{2y^2}$ **7.** $-\frac{5}{2}x + 4 + \frac{3}{x}$ **9.** $4ab^3 + 6a^2b$

11. $-xy + 2y^2 + 3xy^2$ **13.** $x + 2$ **15.** $a - 3$ **17.** $5x + 6y$

19. $x^2 + xy + y^2$ **21.** $(y^2 + 4)(y + 2)$ **23.** $(x + 2)(x + 5)$

25. $(2x + 3)(2x - 3)$ **27.** $x - 7 + \frac{7}{x+2}$ **29.** $2x + 5 + \frac{2}{3x-4}$

31. $y^2 - 3y - 13$ **33.** $x - 3$ **35.** $3y^2 + 6y + 8 + \frac{37}{2y-4}$

37. $a^3 + 2a^2 + 4a + 6 + \frac{17}{a-2}$ **39.** $y^3 + 2y^2 + 4y + 8$

41. $\frac{p(x)}{q(x)} = 2x^2 - 5x + 1 + \frac{4}{x+1}$ **43.** $\frac{p(x)}{q(x)} = x^2 - 2x + 1$

45. $(x + 3)(x + 2)(x + 1)$ **47.** $(x + 3)(x + 4)(x - 2)$ **49.** Yes

51. -7; same   **53. a.** $(x - 2)(x^2 - x + 3)$ **b.** $(x - 5)(x^3 - x + 1)$

55. b. $p(x) = (x - 4)(x + 2)(x + 3)$

57. a.

x	1	5	10	15	20
C(x)	2.15	2.75	3.50	4.25	5.00

b. $\overline{C}(x) = \frac{2}{x} + 0.15$

c.

x	1	5	10	15	20
$\overline{C}(x)$	2.15	0.55	0.35	0.28	0.25

d. It decreases.

e.

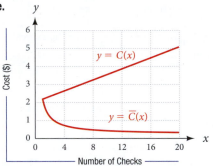

f. $y = C(x)$: domain $= \{x \mid 1 \le x \le 20\}$;
range $= \{y \mid 2.15 \le y \le 5.00\}$
$y = \overline{C}(x)$: domain $= \{x \mid 1 \le x \le 20\}$;
range $= \{y \mid 0.25 \le y \le 2.15\}$

59. a. $T(100) = 11.95$, $T(400) = 32.95$, $T(500) = 39.95$
b. $\frac{4.95 + .07m}{m}$ **c.** $\overline{T}(100) = 0.1195$, $\overline{T}(400) = 0.082$, $\overline{T}(500) = 0.0799$

61. 1,487,340 people   **63.** $\frac{2}{3a}$ **65.** $(x - 3)(x + 2)$ **67.** 1

69. $\frac{3-x}{x+3}$ **71.** No solution   **73.** 196   **75.** 4   **77.** 1.6

79. 3   **81.** 2,400

Find the Mistake   1. Dividing the polynomial $15x^3 + 6x^2 + 9x$ by the monomial $3x$ is the same as multiplying by $\frac{1}{3x}$.

2. Dividing $\frac{4x^3 + 12x^2 - 20x}{4x}$, we get $x^2 + 3x - 5$.   **3.** When using long division to divide $\frac{2x^3 - 9x^2 + 19x - 15}{2x - 3}$, we first estimate $2x^3 \div 2x = x^2$.

4. Dividing $3x^3 - 10x^2 + 7$ by $x - 3$ leaves us with a remainder $= -2$.

Exercise Set 5.8

Vocabulary Review   1. d   **2.** b   **3.** c   **4.** j   **5.** f   **6.** l
7. g   **8.** h   **9.** i   **10.** a   **11.** k   **12.** e
Problems   1. 30   **3.** −6   **5.** 40   **7.** $\frac{81}{5}$ **9.** 64   **11.** 108
13. 300   **15.** ±2   **17.** 1600   **19.** ±8   **21.** $\frac{50}{3}$ pounds
23. a. $T = 4P$ **b.** T

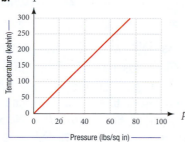

   c. 70 pounds per square inch
25. 12 pounds per square inch
27. a. $f = \frac{80}{d}$ **b.** f

c. An f-stop of 8

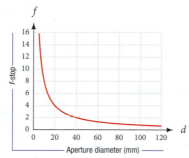

29. $\frac{1504}{15}$ square inches   **31.** 1.5 ohms
33. a. $P = 0.21\sqrt{L}$
   b. y

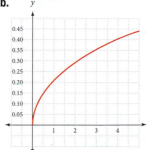

35. $0.6M - 42$
37. $16x^3 - 40x^2 + 33x - 9$
39. $4x^2 - 3x$
41. $6x^2 - 2x - 4$
43. 11

   c. 3.15

Find the Mistake   1. If y varies directly with x and $y = 9$ when $x = 27$, then $x = \frac{1}{3}$ when $y = \frac{1}{9}$.   **2.** If s is directly proportional to the square root of t, and $t = 100$ when $s = 40$, then $s = 4\sqrt{8}$ when $t = 8$.   **3.** If z varies inversely with the square of t and $z = 16$ when $t = \frac{1}{4}$, then $z = 9$.   **4.** If w is directly proportional to l and inversely proportional to p when $w = 18$, $l = 9$ and $p = 3$, then $w = 2.4$ when $l = 4$ and $p = 10$.

Chapter 5 Review

1. $x - y$ **2.** $\frac{x-2}{x+1}$ **3.** 201.9 feet/min   **4. a.** 3 **b.** 3
5. a. $2x + 2 + h$ **b.** $x + a + 2$ **6.** $3(x + 3)$ **7.** $5(x + 3)$
8. $\frac{2x+4}{x^2+2x+4}$ **9.** $\frac{93}{140}$ **10.** $\frac{25}{24}$ **11.** $\frac{1}{a-9}$ **12.** $\frac{3(4x-3)}{(x-1)(2x-1)}$
13. $\frac{2x}{(x+1)(x+3)}$ **14.** $\frac{x+1}{(x-1)(x+3)}$ **15.** $\frac{4a+7}{4a+9}$
16. $\frac{x+4}{x+3}$ **17.** $\frac{5}{8}$ **18.** $-\frac{5}{3}, 2$ **19.** $-\frac{5}{3}, 3$ **20.** 5, 6
21. 6   **22.** 11 mph   **23.** 30 mins   **24.** $2x^2y - 5x^3 + 7y^3$
25. $3x^2 + 2x + 4 + \frac{8}{3x-5}$ **26.** $\frac{3}{2}$ **27.** 112

Chapter 5 Cumulative Review

1. $-\frac{125}{27}$ **2.** $\frac{49}{64}$ **3.** −40   **4.** 93   **5.** 9   **6.** 12
7. $\frac{1}{11}$ **8.** 1   **9.** $-\frac{11}{5}$ **10.** $2a - 9b > 2a + 9b$ **11.** 141
12. −139   **13.** 94   **14.** 90   **15.** $-27x - 20$ **16.** $20x$
17. $\frac{1}{(y+3)(y+2)}$ **18.** $\frac{1}{y-x}$ **19.** $20t^4 - \frac{1}{20}$ **20.** $\frac{x-9}{x+3}$
21. $3x^{5n} - 2x^{2n}$ **22.** $2a^4 + 2a^3 + 2a^2 + 3a + 3 - \frac{5}{a-1}$ **23.** x^5
24. x^7y^2 **25.** $x + 5$ **26.** $\frac{1}{a^2+a+1}$ **27.** $y = -\frac{3}{5}x + 7$
28. $y = -1$ **29.** $2^2 \cdot 3^2 \cdot 11$ **30.** $(x-5)(x+8)$
31. $(x - 6 + y)(x + 6 + y)$ **32.** $\left(y - \frac{4}{5}x\right)\left(y^2 + \frac{4}{5}xy + \frac{16}{25}x^2\right)$
33. 14   **34.** −5   **35.** −1   **36.** $\frac{3}{2}$ **37.** $\frac{31}{3}$ **38.** $-\frac{2}{3}, 6$
39. $(1, 3)$ **40.** $\left(\frac{57}{31}, \frac{8}{31}\right)$ **41.** $x \geq -6$ **42.** $2 < x < 6$
43.   **44.**

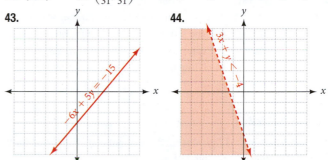

45. 4   **46.** 10   **47.** 7   **48.** 3

Chapter 5 Test

1. $x + y$ **2.** $\frac{2x-3}{x-1}$ **3.** 4.1 feet/second   **4. a.** 8 **b.** 8
5. a. $2x + h + 1$ **b.** $x + a + 1$ **6.** $3(a + 2)$ **7.** $4(a - 2)$
8. $\frac{x+1}{2}$ **9.** $\frac{27}{50}$ **10.** $\frac{17}{32}$ **11.** $\frac{1}{a+5}$ **12.** $\frac{2(4x+1)}{(x+1)(2x-1)}$
13. $\frac{x}{(x+4)(x+3)}$ **14.** $\frac{x+2}{(x-5)(x+3)}$ **15.** $\frac{2a+11}{2a+13}$ **16.** $\frac{x-5}{x+2}$
17. $-\frac{10}{19}$ **18.** ∅   **19.** $-\frac{12}{67}$ **20.** 7, 3   **21.** −2
22. 7 mph   **23.** 10 hours   **24.** $2x^2 + 11xy - 2y^2$
25. $x^2 - 3x + 2 + \frac{2}{4x+3}$ **26.** $\frac{1}{14}$ **27.** $\frac{25}{4}$

Chapter 6

Exercise Set 6.1

Vocabulary Review   1. positive   **2.** cube   **3.** radical   **4.** radicand
5. index   **6.** root   **7.** rational   **8.** exponent   **9.** ratio
Problems   1. 12   **3.** Not a real number   **5.** −7   **7.** −3
9. 2   **11.** Not a real number   **13.** 0.2   **15.** 0.2   **17.** $6a^4$
19. $3a^4$ **21.** xy^2 **23.** $2x^2y$ **25.** $2a^3b^5$ **27.** $-3x^3y^4$ **29.** $x^{4/5}$
31. $n^{2/3}$ **33.** $a^{3/6} = a^{1/2}$ **35.** $m^{5/8}$ **37.** 6   **39.** −3   **41.** 2
43. −2   **45.** 2   **47.** $\frac{9}{5}$ **49.** $\frac{4}{5}$ **51.** $\frac{2}{3}$ **53.** 9   **55.** 125
57. 8   **59.** 9   **61.** $\frac{1}{3}$ **63.** $\frac{1}{27}$ **65.** $\frac{6}{5}$ **67.** $\frac{8}{27}$ **69.** 7   **71.** $\frac{3}{4}$
73. $x^{4/5}$ **75.** a **77.** $\frac{1}{x^{2/5}}$ **79.** $x^{1/6}$ **81.** $a^{2/9}b^{1/6}c^{1/12}$ **83.** $x^{9/25}y^{1/2}z^{1/5}$
85. $\frac{b^{7/4}}{a^{1/8}}$ **87.** $y^{3/10}$ **89.** $\frac{1}{x^2y}$ **91.** $\frac{1}{a^2b^4}$ **93.** $\frac{s^{1/2}}{r^{20}}$ **95.** $10b^3$
97. $(9^{1/2} + 4^{1/2})^2 \stackrel{?}{=} 9 + 4$ **99.** $(a^{1/2})^{1/2} \stackrel{?}{=} a^{1/4}$ **101.** 25 mph
       $(3 + 2)^2 \stackrel{?}{=} 13$ $a^{1/4} = a^{1/4}$ **103.** 1.618
       $5^2 \stackrel{?}{=} 13$
       $25 \neq 13$
105. a. 420 picometers **b.** 594.0 pico meters **c.** 5.94×10^{-10} meters

107.

Pyramid Name	b	h	Slant height (ft)
Temple of Kukulkan	181 ft	98 ft	133
Temple IV at Tecal	118 ft	187 ft	196
Pyramid of the Sun	733 ft	246 ft	441
Great Pyramid of Khufu	755 ft	455 ft	591
Luxor Hotel	600 ft	350 ft	461
Transamerica Pyramid	175 ft	850 ft	854

Source: skyscrapercity.com

109. Transamerica   **111.** 5   **113.** 6   **115.** $4x^2y$ **117.** $5y$
119. 3   **121.** 2   **123.** $2ab$ **125.** $5x^2y$ **127.** 25
129. $48x^4y^2$ **131.** $4x^6y^6$ **133.** $2xy^4$
Find the Mistake   **1.** $\sqrt{x}$ is the positive square number we square
to get x.   **2.** If we raise the fifth root of x to the fifth power, we get
x.   **3.** For the radical expression $\sqrt[4]{16}$, the 4 is the index and the
16 is the radicand.   **4.** Using the rational exponent theorem, we
can say that $16^{-3/2} = (16^{1/2})^{-3} = \frac{1}{64}$.

Exercise Set 6.2

Vocabulary Review   **1.** product   **2.** quotient   **3.** simplified form
4. absolute value   **5.** spiral of roots
Problems   **1.** $2\sqrt{2}$ **3.** $7\sqrt{2}$ **5.** $12\sqrt{2}$ **7.** $4\sqrt{5}$
9. $4\sqrt{3}$ **11.** $15\sqrt{3}$ **13.** $3\sqrt[3]{2}$ **15.** $4\sqrt[3]{2}$ **17.** $6\sqrt[3]{2}$
19. $2\sqrt[5]{2}$ **21.** $3x\sqrt{2x}$ **23.** $2y\sqrt[4]{2y^3}$ **25.** $2xy^2\sqrt[3]{5xy}$
27. $4abc^2\sqrt{3b}$ **29.** $2bc\sqrt[3]{6a^2c}$ **31.** $2xy^2\sqrt[5]{2x^3y^2}$
33. $3xy^2z\sqrt[5]{x^2}$ **35.** $5xy^3\sqrt[3]{2x^2y}$ **37.** $-3x^3y^4\sqrt[4]{xy}$ **39.** $2\sqrt{3}$
41. $\sqrt{-20}$; not real number   **43.** $\frac{\sqrt{11}}{2}$ **45.** $\frac{2\sqrt{3}}{3}$
47. $\frac{5\sqrt{6}}{6}$ **49.** $\frac{\sqrt{2}}{2}$ **51.** $\frac{\sqrt{5}}{5}$ **53.** $2\sqrt[3]{4}$ **55.** $\frac{2\sqrt[3]{3}}{3}$
57. $\frac{\sqrt[4]{24x^2}}{2x}$ **59.** $\frac{\sqrt[4]{8y^3}}{y}$ **61.** $\frac{\sqrt[3]{36xy^2}}{3y}$ **63.** $\frac{\sqrt[3]{6xy^2}}{3y}$ **65.** $\frac{\sqrt[4]{2x}}{2x}$
67. $\frac{\sqrt[4]{18x^2}}{2x}$ **69.** $\frac{3x\sqrt{15xy}}{5y}$ **71.** $\frac{5xy\sqrt{6xz}}{2z}$ **73.** $\frac{2ab\sqrt[3]{6ac^2}}{3c}$
75. $\frac{2xy^2\sqrt[3]{3z^2}}{3z}$ **77.** $5|x|$ **79.** $3|xy|\sqrt{3x}$ **81.** $|x-5|$
83. $|2x+3|$ **85.** $2|a(a+2)|$ **87.** $2|x|\sqrt{x-2}$
89. $\sqrt{9+16} \overset{?}{=} \sqrt{9}+\sqrt{16}$; $\sqrt{25} \overset{?}{=} 3+4$; $5 \neq 7$
91. $5\sqrt{13}$ feet   **93. a.** 89.4 miles **b.** 126.5 miles **c.** 154.9 miles
97. $a_{10} = \sqrt{11}$; $a_{100} = \sqrt{101}$ **99.** $7x$ **101.** $27xy^2$ **103.** $\frac{5}{6}x$
105. $-7(x+1)^2$ **107.** $3\sqrt{2}$ **109.** $5y\sqrt{3xy}$ **111.** $2a\sqrt[3]{ab^2}$
113. $2x^2y\sqrt[4]{2y}$

Find the Mistake   **1.** False: To write the radical expression $\sqrt{4x^7y^2}$
in simplified form, we apply the product property for radicals.
2. False: The radical expression in statement 1 written in simplified
form is $2x^3y\sqrt{x}$.   **3.** False: To rationalize the denominator of the
radical expression $\sqrt{\frac{8x^4y}{3x}}$, we use the quotient property for radicals and
then multiply the numerator and denominator by $\sqrt{3x}$.   **4.** False:
For the spiral of roots, each new line segment drawn to the end of a
diagonal is at a right angle to the previous line segment.

Exercise Set 6.3

Vocabulary Review   simplified form, distributive, similar, radicand,
radicals
Problems   **1.** $7\sqrt{5}$ **3.** $-x\sqrt{7}$ **5.** $\sqrt[3]{10}$ **7.** $9\sqrt[5]{6}$
9. 0   **11.** $\sqrt{5}$ **13.** $-32\sqrt{2}$ **15.** $-3x\sqrt{2}$
17. $5\sqrt{5y}+3\sqrt{2y}$ **19.** $-2\sqrt[3]{2}$ **21.** $8x\sqrt[3]{xy^2}$ **23.** $3a^2b\sqrt{3ab}$
25. $2ab\sqrt[3]{3a^2b}+3a^2b\sqrt[3]{3a^2b}$ **27.** $11ab\sqrt[3]{3a^2b}$ **29.** $10xy\sqrt[4]{3y}$
31. $\sqrt{2}$ **33.** $\frac{8\sqrt{5}}{15}$ **35.** $\frac{(x-1)\sqrt{x}}{x}$ **37.** $\frac{3\sqrt{2}}{2}$ **39.** $\frac{5\sqrt{6}}{6}$
41. $\frac{8\sqrt[3]{25}}{5}$ **43.** $\frac{7\sqrt{5}}{10}$ **45.** $\sqrt{12} \approx 3.464$; $2\sqrt{3} \approx 2(1.732) = 3.464$
47. $\sqrt{8}+\sqrt{18} \approx 2.828+4.243 = 7.071$; $\sqrt{50} \approx 7.071$; $\sqrt{26} \approx 5.099$
49. $8\sqrt{2x}$ **51.** 5   **59.** $\sqrt{2}$
61. a. $\sqrt{2}:1 \approx 1.414:1$ **b.** $5:\sqrt{2}$ **c.** 5:4   **63.** 6
65. $4x^2+3xy-y^2$ **67.** x^2+6x+9 **69.** x^2-4 **71.** $6\sqrt{2}$
73. 6   **75.** $9x$ **77.** $48x$ **79.** $\frac{\sqrt{6}}{2}$ **81.** $\frac{\sqrt{3}}{2}$
Find the Mistake   **1.** False: When adding or subtracting radical
expressions, put each in simplified form and apply the
distributive property.   **2.** True   **3.** False: When combining
$3\sqrt{5}+5\sqrt{3}-2\sqrt{5}$, the first and last radical are similar, so they
can be combined to get $5\sqrt{3}+\sqrt{5}$.   **4.** False: Combining
$4\sqrt{8}+3\sqrt{32}, 20\sqrt{2}$.

Exercise Set 6.4

Vocabulary Review   **1.** polynomials   **2.** conjugate   **3.** radical
4. denominator
Problems   **1.** $3\sqrt{2}$ **3.** $10\sqrt{21}$ **5.** 720   **7.** 54
9. $\sqrt{6}-9$ **11.** $24+6\sqrt[3]{4}$ **13.** $7+2\sqrt{6}$
15. $x+2\sqrt{x}-15$ **17.** $34+20\sqrt{3}$ **19.** $19+8\sqrt{3}$
21. $x-6\sqrt{x}+9$ **23.** $4a-12\sqrt{ab}+9b$ **25.** $x+4\sqrt{x}-4$
27. $x-6\sqrt{x-5}+4$ **29.** 1   **31.** $a-49$ **33.** $25-x$
35. $x-8$ **37.** $10+6\sqrt{3}$ **39.** $30\sqrt{3}-54$ **41.** $\frac{\sqrt{3}+1}{2}$
43. $\frac{5-\sqrt{5}}{4}$ **45.** $\frac{x+3\sqrt{x}}{x-9}$ **47.** $\frac{10+3\sqrt{5}}{11}$ **49.** $\frac{3\sqrt{x}+3\sqrt{y}}{x-y}$
51. $2+\sqrt{3}$ **53.** $\frac{11-4\sqrt{7}}{3}$ **55.** $\frac{a+2\sqrt{ab}+b}{a-b}$ **57.** $\frac{x+4\sqrt{x}+4}{x-4}$
59. $\frac{5-\sqrt{21}}{4}$ **61.** $\frac{\sqrt{x}-3x+2}{1-x}$ **63.** $\frac{6x-\sqrt{x}-1}{1-9x}$ **67.** $10\sqrt{3}$
69. $x+6\sqrt{x}+9$ **71.** 75   **73.** $x^2+2x\sqrt{3}+3$
75. $\frac{5\sqrt{2}}{4}$ second; $\frac{5}{2}$ second   **81.** $t^2+10t+25$ **83.** x **85.** 7
87. $-4, -3$ **89.** $-6, -3$ **91.** $-5, -2$ **93.** Yes   **95.** No
97. North America: 24.5 million; Asia: 7 million; Africa: 500,000
Find the Mistake   **1.** To multiply $\sqrt{5}(\sqrt{7}+6\sqrt{5})$, we use the
distributive property to get our answer $30+\sqrt{35}$.
2. To multiply $(2\sqrt{2}+\sqrt{7})(4\sqrt{2}-3\sqrt{7})$, we apply the FOIL
method to get $-5-2\sqrt{14}$.   **3.** To divide 4 by $(2\sqrt{5}-2)$, we
rationalize the denominator by multiplying the numerator and
the denominator by $2\sqrt{5}+2$.   **4.** Dividing $\frac{(2\sqrt{3}-\sqrt{2})}{(4\sqrt{3}+\sqrt{2})}$ we
get $\frac{(13-3\sqrt{6})}{23}$.

Landmark Review   **1.** 9   **2.** -9 **3.** Not a real number
4. -3 **5.** x^2y^4 **6.** 8   **7.** $\frac{5}{4}$ **8.** 19   **9.** $2\sqrt{41}$
10. $3x^2y^3\sqrt{2x}$ **11.** $3x^2y^2z^3\sqrt[5]{2x^2z^3}$ **12.** $\frac{4\sqrt{5}}{5}$ **13.** $28\sqrt{6}$
14. $\frac{5\sqrt{2}}{6}$ **15.** $2\sqrt{3}$ **16.** 480   **17.** $30+11\sqrt{6}$
18. $6-2\sqrt{5}$ **19.** $\frac{1+\sqrt{3}}{2}$ **20.** $\frac{-7-3\sqrt{5}}{2}$

Exercise Set 6.5

Vocabulary Review   **1.** radical   **2.** solutions   **3.** extraneous
4. original   **5.** negative

Problems   **1.** 4   **3.** No solution   **5.** 5   **7.** No solution
9. $\frac{39}{2}$ **11.** No solution   **13.** 5   **15.** 3   **17.** $-\frac{32}{3}$
19. 3, 4   **21.** $-1, -2$ **23.** -1 **25.** No solution   **27.** 7
29. 0, 3   **31.** -4 **33.** 8   **35.** 0   **37.** 9   **39.** 0   **41.** 8
43. No solution   **45.** 0   **47.** 6   **49.** 3   **51.** 1   **53.** $\frac{16}{3}$
55. $h = 100 - 16t^2$ **57.** $\frac{392}{121} \approx 3.24$ feet   **59.** $l = \frac{10}{3}(\sqrt{x+1})$
61. 5 meters   **63.** 2,500 meters   **65.** $0 \le y \le 100$

67.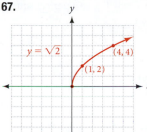

69.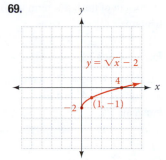

71.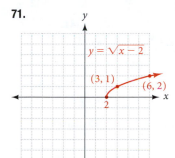

73.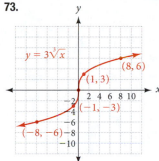

75.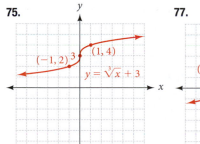

77.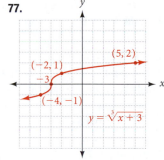

79. 5   **81.** $2\sqrt{3}$ **83.** -1 **85.** 1   **87.** 4   **89.** 2
91. $10 - 2x$ **93.** $2 - 3x$ **95.** $6 + 7x - 20x^2$
97. $8x - 12x^2$ **99.** $4 + 12x + 9x^2$ **101.** $4 - 9x^2$
Find the Mistake   **1.** Solving $\sqrt{6x+1} = -5$ gives us an
extraneous solution.   **2.** Solving $\sqrt[4]{3x-8} = 2$ gives us a solution
of $x = 8$.   **3.** Solving $\sqrt{x+8} = x+6$ gives a solution of $x = -4$
because $x = -7$ gives us a false statement when checked in the
original equation.   **4.** The graphs of $y = \sqrt{x-1}$ and $y = \sqrt{2x-6}$
intersect at $(5, 2)$.

Exercise Set 6.6

Vocabulary Review   **1.** number i **2.** complex number   **3.** zero
4. equal   **5.** imaginary   **6.** conjugates

Problems   **1.** $6i$ **3.** $-5i$ **5.** $6i\sqrt{2}$ **7.** $-2i\sqrt{3}$
9. 1   **11.** -1 **13.** $-i$ **15.** 1   **17.** $x = 3, y = -1$
19. $x = -2, y = -\frac{1}{2}$ **21.** $x = -8, y = -5$ **23.** $x = 7, y = \frac{1}{2}$
25. $x = \frac{3}{7}, y = \frac{2}{5}$ **27.** $5 + 9i$ **29.** $5 - i$ **31.** $2 - 4i$
33. $1 - 6i$ **35.** $2 + 2i$ **37.** $-1 - 7i$ **39.** $6 + 8i$ **41.** $2 - 24i$
43. $-15 + 12i$ **45.** $18 + 24i$ **47.** $10 + 11i$ **49.** $21 + 23i$
51. $-2 + 2i$ **53.** $2 - 11i$ **55.** $-21 + 20i$ **57.** $-2i$
59. $-7 - 24i$ **61.** 5   **63.** 40   **65.** 13   **67.** 164   **69.** 7
71. $-3 - 2i$ **73.** $-2 + 5i$ **75.** $\frac{8}{13} + \frac{12}{13}i$ **77.** $-\frac{18}{13} - \frac{12}{13}i$
79. $-\frac{5}{13} + \frac{12}{13}i$ **81.** $\frac{13}{15} - \frac{2}{5}i$ **83.** $-11 - 7i$ ohms
85. $-\frac{3}{2}$ **87.** $-3, \frac{1}{2}$ **89.** $\frac{5}{4}$ or $\frac{4}{5}$ **91.** $m = 0.23$, \$8.90

Find the Mistake   **1.** The number i is such that $i = \sqrt{-1}$ is the
same as saying $i^2 = -1$.   **2.** The form $a + bi$ is called standard
form for complex numbers and the bi is the imaginary part of
the number.   **3.** The product of the complex numbers $(5 - 3i)$
and $(2 + 4i)$ is $22 + 14i$.   **4.** The product of the two complex
conjugates $6 + 4i$ and $6 - 4i$ is $6^2 + 4^2 = 36 + 16 = 52$.

Chapter 6 Review

1. $\frac{1}{9}$ **2.** $\frac{7}{12}$ **3.** $x^{13/15}$ **4.** ab^4 **5.** $7x^4y^6$ **6.** $3x^2y^4$
7. $\frac{7b^4}{3a}$ **8.** $x^{1-n}y^{2-n^3}$ **9.** $5xy^3\sqrt{2xy}$ **10.** $3xy^2\sqrt{5x^2y}$
11. $\frac{\sqrt{35}}{7}$ **12.** $\frac{2ab^2\sqrt{15ab}}{5c}$ **13.** $3\sqrt{5} + 6$
14. $\frac{x - 2\sqrt{5x} + 5}{x - 5}$ **15.** $\frac{3+x}{x^{1/3}}$ **16.** $\frac{6\sqrt{x^4-6}}{x^4-6}$ **17.** $-4\sqrt{3}$
18. $0.5ab\sqrt{5b^2}$ **19.** $6a^4 - 3a^6$ **20.** $9a^{8/3} - 12a^{4/3} + 4$
21. $x + 2\sqrt{x} - 15$ **22.** $58 - 8\sqrt{15}$ **23.** 9   **24.** 12   **25.** 17

26.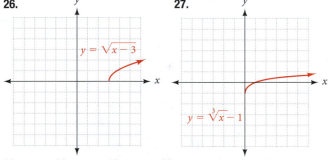

27.

28. C   **29.** A   **30.** D   **31.** B   **32.** E   **33.** F
34. $x = -1; y = 3$ **35.** $4 + 12i$ **36.** $18 + i$
37. $16 - 30i$ **38.** $\frac{15 + 8i}{17}$ **39.** $i^{39} = (i^4)^9 \cdot i^3 = 1 \cdot i^3 = -i$

Chapter 6 Cumulative Review

1. 5   **2.** 14   **3.** -4 **4.** $2\sqrt[4]{8}$ **5.** 14   **6.** $\frac{1}{6}$
7. $-\frac{5}{2}$ **8.** 45   **9.** $\frac{8}{9}$ **10.** $3x - 5y$ **11.** $\frac{x-1}{x+3}$ **12.** $\frac{a^5b^6}{4}$
13. $3x - 2\sqrt{x} - 5$ **14.** $50 + 4i$ **15.** $\frac{11}{13} + \frac{3i}{13}$
16. $-5 - 2\sqrt{5}$ **17.** $28x^2 + 10xy - 2y^2$ **18.** $(4x + 3)(x - 7)$
19. $\frac{20x^6}{y^8}$ **20.** $\frac{x - 2\sqrt{xy} - y}{x - y}$ **21.** -5 **22.** 5
23. $\frac{-4 \pm \sqrt{7}}{3}$ **24.** $\frac{1}{2}, -4$ **25.** 16   **26.** $-2, 0, \frac{2}{3}$
27. $-5, 1$ **28.** 3   **29.** $\frac{1}{4}, 2$ **30.** 4
31. $20 \le x \le 30$ **32.** $x < -\frac{5}{7}$ or $x > \frac{1}{7}$

33. $(0, 5)$ **34.** $(1, -5)$

35.

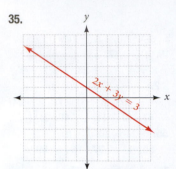

36.

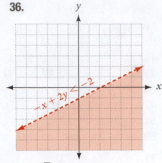

37. 4 **38.** 11 **39.** $2\sqrt[4]{3}$ **40.** $\frac{5\sqrt{7}+15}{-2}$

41. $(3a-4b)(3a+4b)(9a^2+16b^2)$ **42.** $2(3a^2-7)(4a^2+2)$

43. $y=-\frac{3}{4}$ **44.** $z=-54$ **45.** $2b+2$ **46.** 4

47. 20 **48.** 2 **49.** \$80,557,200 **50.** \$438,632,000

Chapter 6 Test

1. $\frac{1}{8}$ **2.** $\frac{8}{9}$ **3.** $\frac{1}{x^{11/21}}$ **4.** $\frac{a^{7/15}}{b^{12}}$ **5.** $5x^3y^{10}$ **6.** $3xy^5$

7. $\frac{2}{7a^2b^3}$ **8.** $x^{n-n^2}y^{n^3-1}$ **9.** $3x^2y\sqrt{3xy}$ **10.** $4y^2\sqrt[3]{2x^2y}$

11. $\frac{\sqrt{15}}{5}$ **12.** $\frac{2ab^3\sqrt{14bc}}{7c}$ **13.** $7\sqrt{2}-7$ **14.** $\frac{x+2\sqrt{3x}+3}{x-3}$

15. $\frac{5-x\sqrt{x}}{x}$ **16.** $\frac{17\sqrt{x^3-17}}{x^3-17}$ **17.** 0 **18.** $3a^2b\sqrt[3]{4}$

19. $4a^4-12a^2$ **20.** $4a^5-12a^{5/2}+9$ **21.** $x+4\sqrt{x}-5$

22. $14-4\sqrt{6}$ **23.** 6, 2 does not check **24.** $-\frac{3}{2}$ **25.** 10

26.

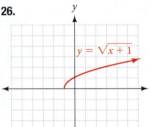

$y=\sqrt{x+1}$

27.

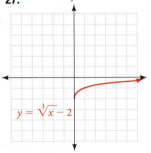

$y=\sqrt[3]{x}-2$

28. E **29.** F **30.** A **31.** B **32.** D

33. C **34.** $x=-2; y=1$ **35.** $7i$ **36.** $13+i$

37. $-27-36i$ **38.** $\frac{5-12i}{13}$ **39.** $i^{36}=(i^2)^{18}=(-1)^{18}=1$

Chapter 7

Exercise Set 7.1

Vocabulary Review 2, 1, 4, 3, 5

Problems 1. ± 5 **3.** $\pm 3i$ **5.** $\pm\frac{\sqrt{3}}{2}$ **7.** $\pm 2i\sqrt{3}$

9. $\pm\frac{3\sqrt{5}}{2}$ **11.** $-2, 3$ **13.** $\frac{-3\pm 3i}{2}$ **15.** $\frac{-2\pm 2i\sqrt{2}}{5}$

17. $-4\pm 3i\sqrt{3}$ **19.** $\frac{3\pm 2i}{2}$ **21.** 36, 6 **23.** 4, 2

25. 25, 5 **27.** $\frac{25}{4}, \frac{5}{2}$ **29.** $\frac{49}{4}, \frac{7}{2}$ **31.** $\frac{1}{16}, \frac{1}{4}$ **33.** $\frac{1}{9}, \frac{1}{3}$

35. $-6, 2$ **37.** $-3, -9$ **39.** $1\pm 2i$ **41.** $4\pm\sqrt{15}$

43. $\frac{5\pm\sqrt{37}}{2}$ **45.** $1\pm\sqrt{5}$ **47.** $\frac{4\pm\sqrt{13}}{3}$ **49.** $\frac{3\pm i\sqrt{71}}{8}$

51. $\frac{-2\pm\sqrt{7}}{3}$ **53.** $\frac{5\pm\sqrt{47}}{2}$ **55.** $\frac{5\pm i\sqrt{19}}{4}$ **57.** $\frac{15\pm 2\sqrt{3}}{9}$

59. a. No **b.** $\pm 3i$ **61. a.** 0, 6 **b.** 0, 6 **63. a.** $-7, 5$ **b.** $-7, 5$

65. No **67. a.** $\frac{7}{5}$ **b.** 3 **c.** $\frac{7\pm 2\sqrt{2}}{5}$ **d.** $\frac{71}{5}$ **e.** 3 **69.** $-1, 5$

71. $\frac{-3\pm\sqrt{29}}{2}$ **73.** $\frac{7\pm\sqrt{85}}{2}$ **75.** None **77.** $\frac{-3\pm\sqrt{41}}{4}$

79. $\frac{-1\pm 2\sqrt{3}}{2}$ **81.** $\frac{\sqrt{3}}{2}$ meter, 1 meter **83.** $\sqrt{2}$ inches

85. 781 feet **87.** 7.3% to the nearest tenth **89.** $20\sqrt{2}\approx 28$ feet

91. 169 **93.** 49 **95.** $\frac{85}{12}$ **97.** 1 **99.** $(3t-2)(9t^2+6t+4)$

101. $2(x+3)(x^2-3x+9)$ **103. a.** 1,006 **b.** 346

Find the Mistake 1. The square root property for equations says that if $a^2=b$, then $a=\sqrt{b}$ or $a=-\sqrt{b}$. **2.** Solving $(4x+1)^2=16$ gives us two solutions, $\frac{3}{4}$ or $-\frac{3}{4}$. **3.** The first step to solving a quadratic equation by completing the square is to put the equation in the form $ax^2+bx=c$. **4.** The method of completing the square says that adding the square of half the coefficient of the linear term to both sides will give a perfect square trinomial.

Exercise Set 7.2

Vocabulary Review 1. standard form **2.** quadratic **3.** rational **4.** factored

Problems 1. $-3, -2$ **3.** $2\pm\sqrt{3}$ **5.** 1, 2 **7.** $\frac{2\pm i\sqrt{14}}{3}$

9. 0, 5 **11.** $0, -\frac{4}{3}$ **13.** $\frac{3\pm\sqrt{5}}{4}$ **15.** $-3\pm\sqrt{17}$

17. $\frac{-1\pm i\sqrt{5}}{2}$ **19.** 1 **21.** $\frac{1\pm i\sqrt{47}}{6}$ **23.** $4\pm\sqrt{2}$ **25.** $\frac{1}{2}, 1$

27. $-\frac{1}{2}, 3$ **29.** $\frac{3+\sqrt{5}}{2}$ **31.** $\frac{-1\pm i\sqrt{7}}{2}$ **33.** $1\pm\sqrt{2}$

35. $\frac{-3\pm\sqrt{5}}{2}$ **37.** 3, -5 **39.** $1\pm\sqrt{10}$ **41.** $\frac{3\pm\sqrt{17}}{4}$

43. $2, -1\pm i\sqrt{3}$ **45.** $-\frac{3}{2}, \frac{3\pm 3i\sqrt{3}}{4}$ **47.** $\frac{1}{5}, \frac{-1\pm i\sqrt{3}}{10}$

49. $0, \frac{-1\pm i\sqrt{5}}{2}$ **51.** $0, 1\pm i$ **53.** $0, \frac{-1\pm i\sqrt{2}}{3}$

55. a and b **57. a.** $\frac{5}{3}, 0$ **b.** $\frac{5}{3}, 0$ **59.** No, $2\pm i\sqrt{3}$

61. Yes **63.** 0, 4 **65.** 0, 0.4 **67.** $-2\pm\sqrt{2}$

69. $\frac{1\pm\sqrt{13}}{2}$ **71.** $\pm\sqrt{6}$ **73.** 2 seconds **75.** 20 or 60 items

77. 0.49 centimeter (8.86 cm is impossible)

79. a. $l+w=10, lw=15$ **b.** 8.16 yards, 1.84 yards **c.** Because either dimension (long or short) may be considered the length.

81. 169 **83.** 0 **85.** ± 12 **87.** x^2-x-6 **89.** $x^3-4x^2-3x+18$

Find the Mistake 1. Using the quadratic formula to solve a quadratic equation gives the solutions $x=\frac{-b\pm\sqrt{b^2-4ac}}{2a}$.
2. Using the quadratic formula to solve $x^2+10x=11, a=1$, $b=10$, and $c=-11$. **3.** To solve $x^2-10x=-24$ using the quadratic formula, begin by writing the equation in standard form.
4. Solving $3x^2-2x-4=0$ using the quadratic formula, we get $\frac{1\pm\sqrt{13}}{7}$ for our answer.

Exercise Set 7.3

Vocabulary Review 1. discriminant **2.** quadratic **3.** negative **4.** zero **5.** rational **6.** irrational **7.** multiplicity

Problems 1. $D=16$, two rational **3.** $D=0$, one rational

5. $D=5$, two irrational **7.** $D=17$, two irrational

9. $D=36$, two rational **11.** $D=116$, two irrational **13.** ± 10

15. ± 12 **17.** 9 **19.** -16 **21.** $\pm 2\sqrt{6}$ **23.** 0, 12

25. $x^2-7x+10=0$ **27.** $t^2-3t-18=0$

29. $y^3-4y^2-4y+16=0$ **31.** $2x^2-7x+3=0$

33. $4t^2-9t-9=0$ **35.** $6x^3-5x^2-54x+45=0$

37. $10a^2-a-3=0$ **39.** $9x^3-9x^2-4x+4=0$

41. $x^4-13x^2+36=0$ **43.** $x^2-7=0$ **45.** $x^2+25=0$

47. $x^2-2x+2=0$ **49.** $x^2+4x+13=0$

51. $x^2+8x+25=0$ **53.** $16x^2+16x+29=0$

55. $x^2+2x-5=0$ **57.** $9x^2+24x+14=0$

59. $49x^2 - 42x + 14 = 0$ **61.** $x^3 + 7x^2 - 5x - 75 = 0$
63. $x^4 - 18x^2 + 81 = 0$ **65.** $-3, -2, -1$ **67.** $-4, -3, 2$
69. $1 \pm i$ **71.** $5, 4 \pm 3i$ **73.** $x^2 + 4x - 5$ **75.** $4a^2 - 30a + 56$
77. $32a^2 + 20a - 18$ **79.** $\pm\frac{1}{2}$ **81.** No solution   **83.** 1
85. $-2, 4$ **87.** $-2, \frac{1}{4}$ **89.** $-\frac{1}{4}, \frac{1}{3}$

91. a. Four and forty-seven hundredths   **b.** Three and eighty-five hundredths

Find the Mistake   1. The equation $3x^2 + 2x + 5 = 0$ will have two complex solutions.   **2.** The equation $5x^2 - 10x + 5 = 0$ will have one rational solution.   **3.** The equation $9x^2 + 8x - 3 = 0$ will have two irrational solutions.   **4.** The equation $6x^2 - 13x + 2 = 0$ will have two rational solutions.

Landmark Review   1. $x = \pm 3$ **2.** $y = \pm\frac{\sqrt{3}}{2}$ **3.** $x = 1, -7$
4. $y = -2, 8$ **5.** $x = -9, 1$ **6.** $y = -6, -4$ **7.** $x = -8, 2$
8. $x = -1, 4$ **9.** $x = \pm\frac{\sqrt{37} - 5}{2}$ **10.** $x = \pm\frac{i\sqrt{31} - 1}{2}$
11. $y = 1 \pm \sqrt{7}$ **12.** $x = \frac{-3 \pm \sqrt{21}}{2}$ **13.** $D = 36$, 2 rational
14. $D = 24$, 2 irrational   **15.** $D = -15$, 2 complex
16. $D = 1$, 2 rational   **17.** $x^2 + 8x + 12 = 0$
18. $x^3 - 5x^2 - 9x + 45 = 0$

Exercise Set 7.4

Vocabulary Review   1. quadratic   **2.** radical, squaring
3. extraneous
Problems 1. $1, 2$ **3.** $-8, -\frac{5}{2}$ **5.** $\pm 3, \pm i\sqrt{3}$ **7.** $\pm 2i, \pm i\sqrt{5}$
9. $\frac{7}{2}, 4$ **11.** $-\frac{9}{8}, \frac{1}{2}$ **13.** $\pm\frac{\sqrt{30}}{6}, \pm i$ **15.** $\pm\frac{\sqrt{21}}{3}, \pm\frac{i\sqrt{21}}{3}$
17. $4, 25$ **19.** 25 **21.** $\frac{25}{9}$ **23.** $27, 38$ **25.** $4, 12$
27. $t = \frac{v \pm \sqrt{v^2 + 64h}}{32}$ **29.** $x = \frac{-4 \pm 2\sqrt{4 - k}}{k}$ **31.** $x = -y$
33. $t = \frac{1 + \sqrt{1 + h}}{4}$ **35.** 1.30 sec, 1.52 sec, 1.71 sec
37. a. $f(x) = -\frac{23}{4000}(x^2 - 400x)$ **b.** $f(x) = -\frac{77}{21025}(x^2 - 290x)$
   c. $f(x) = -\frac{16}{625}(x^2 - 250x)$ **d.** $f(x) = -\frac{984}{75,625}(x^2 - 275x)$
39. $\frac{x(x + 2)}{2}$
41. a. $l + 2w = 160$ **b.** $A(w) = -2w^2 + 160w$
   c.

w (yd)	A (yd²)
5	750
10	1,400
15	1,950
20	2,400
25	2,750

   d. 3200 square yards
43. -2 **45.** $1,322.52$
47. $-\frac{7}{640}$ **49.** $1, 5$
51. $-3, 1$ **53.** $\frac{3}{2} \pm \frac{1}{2}i$
55. $9, 3$ **57.** $1, 1$

Find the Mistake   1. False: The equation $(x + 4)^2 - 3(x + 4) - 7 = 0$ is quadratic in form.   **2.** False: Solutions for the equation $(y + 2)^2 - 3(y + 2) - 18 = 0$ are $y = 4$ and $y = -5$.   **3.** True
4. False: The equation $x + \sqrt{x} - 12 = 0$ has one extraneous solution and one real solution.

Exercise Set 7.5

Vocabulary Review   1. y-intercept   **2.** x-intercepts   **3.** vertex
4. $x = -\frac{b}{2a}$ **5.** up   **6.** down   **7.** (h, k) **8.** $y = -\frac{b}{2a}$
Problems

1.

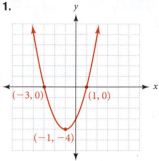

x-intercepts $= -3, 1$;
vertex $= (-1, -4)$

3.

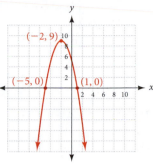

x-intercepts $= -5, 1$;
vertex $= (-2, 9)$

5.

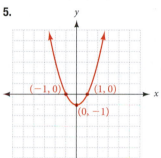

x-intercepts $= -1, 1$;
vertex $= (0, -1)$

7.

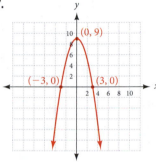

x-intercepts $= -3, 3$;
vertex $= (0, 9)$

9.

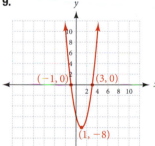

x-intercepts $= -1, 3$;
vertex $= (1, -8)$

11.

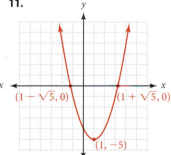

x-intercepts $= 1 - \sqrt{5}, 1 + \sqrt{5}$;
vertex $= (1, -5)$

13.

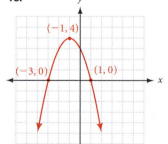

x-intercepts $= -3, 1$;
vertex $= (-1, 4)$

15.

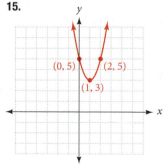

17.

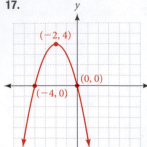

19.

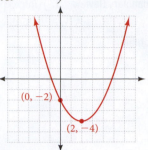

21.

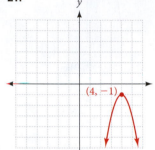

23.

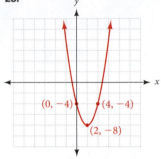

Vertex = $(2, -8)$

25.

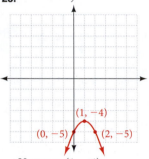

Vertex = $(1, -4)$

27.

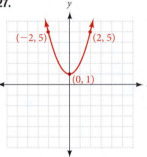

Vertex = $(0, 1)$

29.

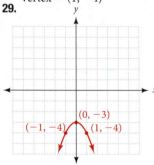

Vertex = $(0, -3)$

31.

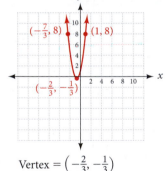

Vertex = $\left(-\frac{2}{3}, -\frac{1}{3}\right)$

33.

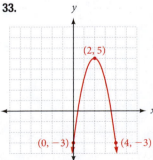

Vertex = $(2, 5)$

35. $(3, -4)$ lowest
37. $(1, 9)$ highest
39. $(2, 16)$ highest
41. $(-4, 16)$ highest
43. 875 patterns; maximum profit $731.25
45. The ball is in her hand when $h(t) = 0$, which means $t = 0$ or $t = 2$ seconds. Maximum height is $h(1) = 16$ feet.

47. Maximum $R = \$3,600$ when $p = \$6.00$
49. Maximum $R = \$7,225$ when $p = \$8.50$
51. $y = -\frac{1}{135}(x - 90)^2 + 60$

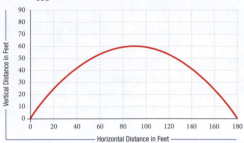

53. $-2, 4$ **55.** $-\frac{1}{2}, \frac{2}{3}$ **57.** 3   **59.** $-2, 2$

Find the Mistake   **1.** The graph of $f(x) = -x^2 - 2x + 3$ is concave down.   **2.** The graph of $f(x) = \frac{1}{4}(x + 1)^2 - 4$ is concave up.
3. The graph of $x = -2(y + 1)^2 - 1$ opens to the left.
4. The graph of $x = y^2 - 6x + 4$ opens to the right.

Exercise Set 7.6

Vocabulary Review   **1.** inequalities   **2.** constants
3. real number line, factors   **4.** same   **5.** opposite

1.

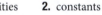

3.

5.

7.

9.

11.

13.

15.

17.

19.

All real numbers

21.

No solution

23.

25.

27.
-4 0 2

29.
-4 1

31.
-6 $\frac{8}{3}$

33.
2 6

35.
-3 2 4

37.
2 3 4

39.
5 6

41.
-3 -1

43. a. $(-2, 2)$ **b.** $(-\infty, -2) \cup (2, \infty)$ **c.** $x = -2, 2$

45. a. $(-2, 5)$ **b.** $(-\infty, -2) \cup (5, \infty)$ **c.** $x = -2, 5$

47. a. $(-\infty, -1) \cup (1, 3)$ **b.** $(-1, 1) \cup (3, \infty)$ **c.** $x = -1, 1, 3$

49. $\{x \mid x \neq -1, 4\}$ **51.** $\{x \mid x \leq 0 \text{ or } x \geq 7\}$

53. $\{x \mid x \leq -\frac{1}{2} \text{ or } x \geq 3\}$ **55.** $\{x \mid x \leq \frac{2}{5} \text{ or } x \geq 3\}$

57. $x \geq 4$; the width is at least 4 inches

59. $5 \leq p \leq 8$; she should charge at least \$5 but no more than \$8 for each radio

61. \$300, \$1,800,000 **63.** \$30, \$900 **65.** 1.5625

67. 0.6549 **69.** $\frac{2}{3}$ **71.** 6

73. **75.**

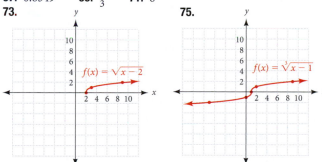

$f(x) = \sqrt{x - 2}$ $f(x) = \sqrt[3]{x - 1}$

Find the Mistake **1.** The solution to the inequality $4x^2 - 31x > -21$ is $\left(-\infty, \frac{3}{4}\right) \cup (7, \infty)$. **2.** The solution to the inequality $x(x + 5)(x - 6) \geq 0$ is $[-5, 0] \cup [6, \infty)$.

3. The solution to the inequality $\frac{x + 8}{x - 3} \geq 0$ is $(-\infty, -8) \cup (3, \infty)$.

4. For the inequality $\frac{4}{x + 5} - \frac{3}{x + 4} > 0$ lies within $(-5, -4) \cup (-1, \infty)$.

Chapter 7 Review

1. $-3, \frac{1}{3}$ **2.** $\frac{3 \pm 3i\sqrt{2}}{2}$ **3.** $-3, 9$ **4.** $-6, 3$

5. $\frac{4}{5}, \frac{2 \pm 2i\sqrt{3}}{5}$ **6.** $1 \pm i\sqrt{5}$ **7.** $r = 4 \pm \frac{\sqrt{A}}{6}$ **8.** $4 \pm \sqrt{3}$

9. 6 cm, $6\sqrt{3}$ cm **10.** $\frac{1}{2}$ second, $\frac{5}{2}$ seconds **11.** 2 or 300 watches

12. 27 **13.** Two rational solutions **14.** $5x^2 - 12x - 9 = 0$

15. $x^3 - 3x^2 - 16x + 48 = 0$ **16.** $\pm \frac{3i}{2}, \pm \sqrt{5}$ **17.** $-1, \frac{10}{9}$

18. 9 **19.** $t = \frac{7 \pm 2\sqrt{18 - 8h}}{8}$

20. $(2, -10)$ **21.** $(-1, 5)$

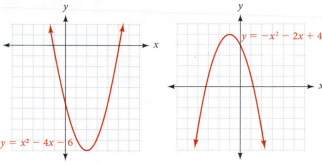

$y = x^2 - 4x - 6$ $y = -x^2 - 2x + 4$

22. $y = 3(x - 4)^2 - 3$ **23.** \$102.50

24. -5 2 $-5 \leq x \leq 2$

25. $-\frac{1}{3}$ 3 $x < -\frac{1}{3} \text{ or } x > 3$

26. a. $x < -3 \text{ or } -1 < x < 2$ **b.** $-3 < x < -1 \text{ or } x > 2$ **c.** $x = -3 \text{ or } x = -1 \text{ or } x = 2$

Chapter 7 Cumulative Review

1. 4 **2.** 78 **3.** $-\frac{125}{8}$ **4.** $\frac{9}{16}$ **5.** $-13x + 13$ **6.** $2x + 13$

7. xy^7 **8.** $\frac{a^{21}}{b^3}$ **9.** $3\sqrt[3]{2}$ **10.** $4x^2\sqrt{2x}$ **11.** $-\frac{3}{16}$ **12.** $\frac{4}{25}$

13. $-\frac{3}{11}$ **14.** $-\frac{25}{4}$ **15.** $x - 7y$ **16.** $x - 5$ **17.** $1 + 4i$

18. $\frac{9}{13} + \frac{7}{13}i$ **19.** 4 **20.** $\frac{51}{4}$ **21.** $-3, 3$ **22.** $\varnothing$ **23.** $-3, -5$

24. $a = 4$ **25.** $\frac{3 \pm 2\sqrt{2}}{2}$ **26.** $0, \frac{9 \pm \sqrt{65}}{8}$ **27.** 1, 4

28. $\frac{-1 \pm i\sqrt{35}}{9}$ **29.** $-4, -3$ **30.** 9 **31.** 9 **32.** $-125, 216$

33. $-36 \leq x \leq -18$ **34.** $x \leq -1, x \geq -\frac{1}{3}$

-36 -18 | -1 $-\frac{1}{3}$

35. $-3 < x < 6$ **36.** $x < -2 \text{ or } x \geq 4$

-3 6 | -2 4

37. Lines parallel $\varnothing$ **38.** Lines parallel $\varnothing$

39. **40.**

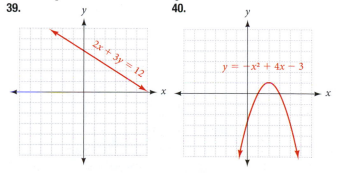

$2x + 3y = 12$ $y = -x^2 + 4x - 3$

41. $y = \frac{3}{2}x + \frac{5}{2}$ **42.** $y = 4x - 2$ **43.** $(x - 3 - y)(x - 3 + y)$

44. $(x + 3)(x - 4)$ **45.** $3\sqrt[3]{2}$ **46.** $-\frac{3 + \sqrt{15}}{2}$ **47.** $\{1, 4\}$

48. $\{2, 8\}$ **49.** 30°, 50°, 100° **50.** \$3,500 at 2%; \$2,500 at 3%

51. $-\frac{25}{9}$ **52.** 4 **53.** 328,290 people **54.** 38,372 people

Chapter 7 Test

1. $\frac{9}{2}, -\frac{3}{2}$ **2.** $\frac{-1 \pm 2i\sqrt{3}}{4}$ **3.** $-4 \pm 5i$ **4.** $-5 \pm i\sqrt{2}$

5. $\frac{3}{4}, \frac{-3 \pm 3i\sqrt{3}}{8}$ **6.** $\frac{1 \pm i\sqrt{7}}{2}$ **7.** $r = 7 \pm \frac{\sqrt{A}}{7}$ **8.** $3 \pm i\sqrt{2}$

9. 4 m, $4\sqrt{3}$ m   **10.** $\frac{1}{4}$ second, $\frac{7}{4}$ seconds

11. 5 or 100 letter openers   **12.** 8   **13.** Two complex solutions

14. $2x^2 + 9x - 35 = 0$ **15.** $x^3 - 5x^2 - 9x + 45$ **16.** $\pm\frac{3}{4}i, \pm\sqrt{3}$

17. $2, \frac{1}{4}$ **18.** $\frac{1}{9}, 4$ **19.** $t = \frac{5 \pm \sqrt{16h + 25}}{16}$

20. $(1, 0)$ **21.** $(2, 2)$

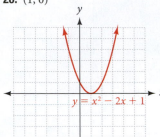

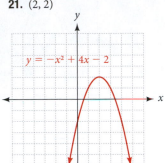

22. $y = -(x - 2)^2 + 4$ **23.** $645

24. $-1 \le x \le 4$

25. $x < -2$ or $x > \frac{1}{2}$

26. a. $-2 < x < -1$ or $x > 1$ **b.** $x < -2$ or $-1 < x < 1$
    c. $x = -2$ or $x = -1$ or $x = 1$

Chapter 8

Exercise Set 8.1

Vocabulary Review   **1.** exponential function   **2.** x–axis

3. number e **4.** $A(t) = P\left(1 + \frac{r}{n}\right)^{nt}$ **5.** $A(t) = Pe^{rt}$

Problems   **1.** 1   **3.** 2   **5.** $\frac{1}{27}$ **7.** 13   **9.** $\frac{7}{12}$ **11.** $\frac{3}{16}$

13. 48   **15.** $-\frac{35}{4}$ **17.** $\frac{3^x - 9}{x - 2}$ **19.** $\frac{3^x - 27}{x - 3}$ **21.** $\frac{3^x(3^h - 1)}{h}$

23.

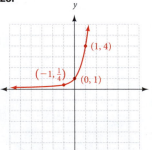

25.

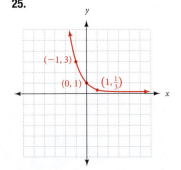

27.

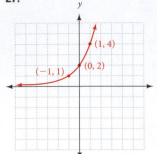

29.

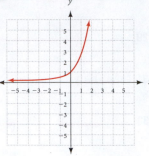

31.

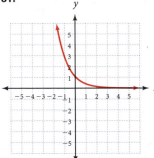

33.

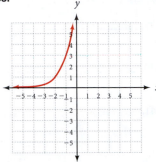

35. $y = 2x$

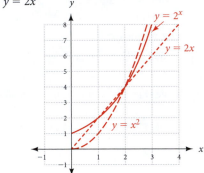

39. $h(n) = 6 \cdot \left(\frac{2}{3}\right)^n$; 5th bounce: $6\left(\frac{2}{3}\right)^5 \approx 0.79$ feet   **41.** 4.27 days

43. a. $A(t) = 1{,}200\left(1 + \frac{.06}{4}\right)^{4t}$ **b.** $1,932.39   **c.** About 11.64 years
    d. $1,939.29

45. a. $A(t) = 5{,}400\left(1 + \frac{.05}{4}\right)^{4t}$ **b.** $6,923.00   **c.** 8 years   **d.** $6,933.74

47. $f(1) = 200, f(2) = 800, f(3) = 3{,}200$

49. a. $129,138.48   **b.** $[0, 6]$
   c.

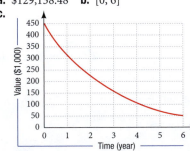

   d. $[52{,}942.05, 450{,}000]$
   e. After approximately 4 years and 8 months

51.

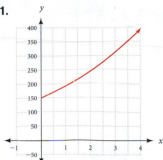

53. a. \$0.42 **b.** \$1.00 **c.** \$1.78 **d.** \$17.84

55. 1,258,525 bankruptcies, which is 58,474 less than the actual number.

57. a. 251,437 cells **b.** 12,644 cells **c.** 32 cells

59. $D = \{-2, 2\}$, $R = \{6, 8, 3\}$, Not a function

61. $(-\infty, -7) \cup (-7, 5) \cup (5, \infty)$ **63.** -18 **65.** 2

67. $y = 2.308x - 4562.06$

69. a. 1 **b.** $\frac{1}{3}$ **c.** 3 **d.** 1 **e.** $\frac{1}{2}$ **f.** 2 **g.** 3 **h.** $\frac{1}{2}$

71. $y = \frac{x+3}{2}$ **73.** $y = \pm\sqrt{x+3}$ **75.** $y = \frac{2x-4}{x-1}$ **77.** $y = x^2 + 3$

Find the Mistake **1.** The exponential function is any function written in the form $f(x) = b^x$. **2.** The graph of $f(x) = 2^x$ gets very close to the x-axis but does not intersect it. **3.** The special number e is an irrational number. **4.** To calculate annual interest rate compounded n times per year, use the formula $A(t) = P\left(1 + \frac{r}{n}\right)^{nt}$.

Exercise Set 8.2

Vocabulary Review **1.** relation **2.** range **3.** one-to-one **4.** inverse **5.** function

Problems **1.** $f^{-1}(x) = \frac{x+1}{3}$ **3.** $f^{-1}(x) = \sqrt[3]{x}$

5. $f^{-1}(x) = \frac{x-3}{x-1}$ **7.** $f^{-1}(x) = 4x + 3$ **9.** $f^{-1}(x) = 2x + 6$

11. $f^{-1}(x) = \frac{3}{2}x + \frac{9}{2}$ **13.** $f^{-1}(x) = \sqrt[3]{x+4}$ **15.** $f^{-1}(x) = \frac{x+3}{4-2x}$

17. $f^{-1}(x) = \frac{1-x}{3x-2}$ **19.** $f^{-1}(x) = \frac{x^3+2}{3}$

21.

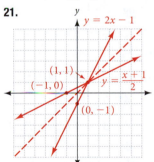

23.

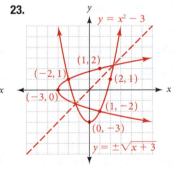

25.

27.

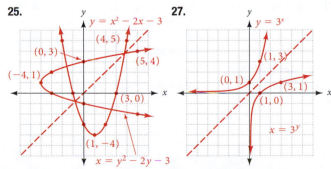

29.

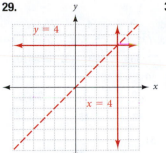

31.

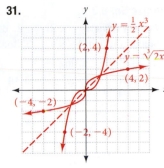

33.

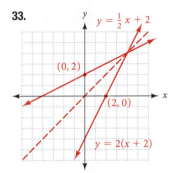

35.

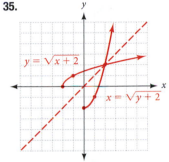

37.

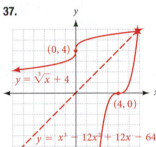
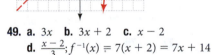

39. a. Yes **b.** No **c.** Yes

41. $g = \{(4, 1), (5, 2), (7, 3)\}$; Yes

43. $g = \{(A, a), (B, b), (C, c)\}$; Yes

45. a. 4 **b.** $\frac{4}{3}$ **c.** 2 **d.** 2

47. $f^{-1}(x) = \frac{1}{x}$

49. a. $3x$ **b.** $3x + 2$ **c.** $x - 2$
d. $\frac{x-2}{3}$; $f^{-1}(x) = 7(x + 2) = 7x + 14$

51. a. -3 **b.** -6 **c.** 2 **d.** 3 **e.** -2 **f.** 3 **g.** inverses

53. a. $-53.3°C$ **b.** $C^{-1}(F) = \frac{9}{5}F + 32$ **c.** $32°F$ **55.** -40

57. a. 569.4 **b.** $s^{-1}(t) = \frac{t - 249.4}{16}$ **c.** 2006

59. a. 6629.33 ft/s **b.** $f^{-1}(m) = \frac{15m}{22}$ **c.** 1.36 mph

61. $-2, 3$ **63.** 25 **65.** $5 \pm \sqrt{17}$ **67.** $1 \pm i$ **69.** $f^{-1}(x) = \frac{x-5}{3}$

71. $f^{-1}(x) = \sqrt[3]{x-1}$ **73.** $f^{-1}(x) = \frac{2x-4}{x-1}$ **75.** $f^{-1}(x) = \frac{2x+3}{2-x}$

77. a. 1 **b.** 2 **c.** 5 **d.** 0 **e.** 1 **f.** 2 **g.** 2 **h.** 5

79. 41.39 EURO, \$56.28 **81.** $x = -\frac{8}{3}$ **83.** $f^{-1}(x) = \frac{b - dx}{cx - a}$

85. $f^{-1}(x) = \frac{ax + 5}{3x - a}$ **87.** $\frac{1}{9}$ **89.** $\frac{1}{16}$ **91.** $\frac{2}{3}$ **93.** $\sqrt[3]{4}$

95. 3 **97.** 4 **99.** 4 **101.** 1 **103.** -3 **105.** -4

Find the Mistake **1.** A function is a one-to-one function if every element in the range comes from exactly one element in the domain. **2.** If $y = f(x)$ is a one-to-one function, then the inverse of $f(x)$ is also a function and can be denoted by $y = f^{-1}(x)$. **3.** The graph of a function and its inverse have symmetry about the line $y = x$. **4.** If (a, b) belongs to the function $f(x)$, then the point (b, a) belongs to its inverse.

Exercise Set 8.3

Vocabulary Review   **1.** b   **2.** d   **3.** a   **4.** c   **5.** e

Problems   **1.** $\log_2 16 = 4$ **3.** $\log_5 125 = 3$ **5.** $\log_{10} 0.01 = -2$

7. $\log_2 \frac{1}{32} = -5$ **9.** $\log_{1/2} 8 = -3$ **11.** $\log_3 27 = 3$ **13.** $10^2 = 100$

15. $2^6 = 64$ **17.** $8^0 = 1$ **19.** $10^{-3} = 0.001$ **21.** $6^2 = 36$

23. $5^{-2} = \frac{1}{25}$ **25.** 9   **27.** $\frac{1}{125}$ **29.** 4   **31.** $\frac{1}{3}$ **33.** 2

35. $\sqrt[3]{5}$ **37.** 2   **39.** 6   **41.** $\frac{2}{3}$ **43.** $-\frac{1}{2}$ **45.** $\frac{1}{64}$ **47.** $\frac{3}{4}$

49.   **51.**

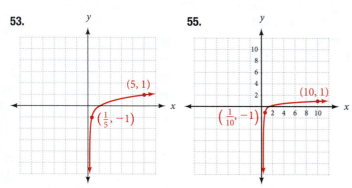

53.   **55.**

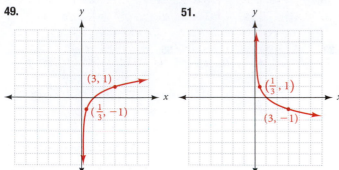

57. $f(x) = 3^x$ **59.** $f(x) = \log_{1/3} x$ **61.** 4   **63.** $\frac{3}{2}$ **65.** 3   **67.** 1

69. 0   **71.** 0   **73.** $\frac{1}{2}$ **75.** $\frac{3}{2}$ **77.** 1   **79.** -2 **81.** 0   **83.** $\frac{1}{2}$

85.

Prefix	Multiplying Factor	$\log_{10}$ (Multiplying Factor)
Nano	0.000 000 001	-9
Micro	0.000 001	-6
Deci	0.1	-1
Giga	1,000,000,000	9
Peta	1,000,000,000,000,000	15

87. 2   **89.** 10^8 times as large   **91.** 939

93. a. 10,486   **b.** 1,311   **c.** 5,243   **95.** 4   **97.** 8   **99.** 9

101. $\frac{1}{9}$ **103.** $-4, 2$ **105.** $-\frac{11}{8}$ **107.** $2^3 = (x+2)(x)$

109. $3^4 = \frac{x-2}{x+1}$

Find the Mistake   **1.** The expression $y = \log_b x$ is read "y is the logarithm to the base b of x."   **2.** The expression $\frac{1}{16} = 2^{-4}$ is equivalent to the logarithmic expression $\log_2 \frac{1}{16} = -4$.   **3.** The graph of $y = \log_2 x$ is a function because no vertical line will cross its graph in more than one place.   **4.** A special identity for logarithms states that b raised to the power of $\log_b x$ is equal to x.

Landmark Review

1. 1   **2.** $\frac{1}{3}$ **3.** $\frac{2}{9}$ **4.** $-\frac{2}{27}$

5.   **6.**

7. $\frac{x+2}{5}$ **8.** $\frac{\sqrt[3]{4x+8}}{2}$

9.   **11.**

11. $\log_2 16 = 4$ **12.** $\log_4 16 = 2$ **13.** $3^{-2} = \frac{1}{9}$ **14.** $7^2 = 49$

15. 25   **16.** 8   **17.** $\sqrt[5]{3}$ **18.** $-\frac{1}{2}$

Exercise Set 8.4

Vocabulary Review   **1.** decibel   **2.** product   **3.** quotient   **4.** power

Problems   **1.** $\log_3 4 + \log_3 x$ **3.** $\log_6 5 - \log_6 x$ **5.** $5 \log_2 y$

7. $\frac{1}{3} \log_9 z$ **9.** $2 \log_6 x + 4 \log_6 y$ **11.** $\frac{1}{2} \log_5 x + 4 \log_5 y$

13. $\log_b x + \log_b y - \log_b z$ **15.** $\log_{10} 4 - \log_{10} x - \log_{10} y$

17. $2\log_{10} x + \log_{10} y - \frac{1}{2}\log_{10} z$ **19.** $3\log_{10} x + \frac{1}{2}\log_{10} y - 4\log_{10} z$

21. $\frac{2}{3}\log_b x + \frac{1}{3}\log_b y - \frac{4}{3}\log_b z$ **23.** $\frac{2}{3}\log_3 x + \frac{1}{3}\log_3 y - 2\log_3 z$

25. $2\log_a 2 + 5\log_a x - 2\log_a 3 - 2$ **27.** $2\log_b 3 + 3\log_b x - 6\log_b 2 - 2$

29. $\log_b xz$ **31.** $\log_3 \frac{x^2}{y^3}$ **33.** $\log_{10}\left(\sqrt{x}\sqrt[3]{y}\right)$ **35.** $\log_2\left(\frac{x^3\sqrt{y}}{z}\right)$

37. $\log_2\left(\frac{\sqrt{x}}{y^3 z^4}\right)$ **39.** $\log_{10}\left(\frac{x^{3/2}}{y^{3/4}z^{4/5}}\right)$ **41.** $\log_5\left(\frac{\sqrt{x}\cdot\sqrt[3]{y^2}}{z^4}\right)$

43. $\log_3\left(\frac{x-4}{x+4}\right)$ **45.** $\frac{2}{3}$ **47.** 18   **49.** 3   **51.** 3   **53.** 4

55. 4   **57.** 1   **59.** 0   **61.** $\frac{3}{2}$ **63.** 27   **65.** $\frac{5}{3}$

69. a. 1.602   **b.** 2.505   **c.** 3.204   **71.** $\text{pH} = 6.1 + \log_{10} x - \log_{10} y$

73. 2.52   **75.** 1   **77.** 1   **79.** 4   **81.** x **83.** 2.5×10^{-6} **85.** 51

Find the Mistake   **1.** The logarithm of a product is the sum of the logarithms.   **2.** The quotient property for logarithms states that $\log_b \frac{x}{y} = \log_b x - \log_b y$.   **3.** The logarithm of a number raised to a power is the product of the power and the logarithm of the number.   **4.** To expand $\log_6 \frac{4xy}{z}$, apply the quotient property and the product property for logarithms.

Exercise Set 8.5

Vocabulary Review **1.** common **2.** characteristic, mantissa
3. natural
Problems **1.** 2.5775 **3.** 1.5775 **5.** 3.5775 **7.** -1.4225
9. 4.5775 **11.** 2.7782 **13.** 3.3032 **15.** -2.0128
17. -1.5031 **19.** -0.3990 **21.** 759 **23.** 0.00759
25. 1,430 **27.** 0.00000447 **29.** 0.0000000918 **31.** 10^{10}
33. 10^{-10} **35.** 10^{20} **37.** $\frac{1}{100}$ **39.** 1,000
41. 0.3679 or $\frac{1}{e}$ **43.** 25 **45.** $\frac{1}{8}$ **47.** 81 **49.** 1
51. 5 **53.** x **55.** 4 **57.** -3 **59.** $\frac{3}{2}$
61. $3t + \ln 10$ **63.** $-2t + \ln A$ **65.** $2 + 3t \log 1.01$
67. $rt + \ln P$ **69.** $3 - \log 4.2$ **71.** $-4 - \log 3.4$
73. 2.7080 **75.** -1.0986 **77.** 2.1972
79. 2.7724 **81.** San Francisco was approx. 2,000 times greater.

83.

Date	Location	Magnitude M	Energy Level T
5/22/1960	Chile	9.5	3.16×10^9
3/28/1964	Alaska	9.2	1.58×10^9
12/26/2004	Indonesia	9.1	1.26×10^9
3/11/2011	Japan	9.0	10^9

85.

x	$(1 + x)^{1/x}$
1	2
0.5	2.25
0.1	2.5937
0.01	2.7048
0.001	2.7169
0.0001	2.7181
0.00001	2.7183

87. e **89.** 2009
91. Approximately 3.19
93. 1.78×10^{-5}
95. 3.16×10^5
97. 2.00×10^8

99.

Location	Date	Magnitude M	Shockwave T
Moresby Island	Jan. 23	4.0	1.00×10^4
Vancouver Island	Apr. 30	5.3	1.99×10^5
Quebec City	June 29	3.2	1.58×10^3
Mould Bay	Nov. 13	5.2	1.58×10^5
St. Lawrence	Dec. 14	3.7	5.01×10^3

SOURCE: *National Resources Canada, National Earthquake Hazards Program.*

101. 12.9% **103.** 5.3% **105.** $\frac{7}{10}$ **107.** -3
109. 3.1250 **111.** 1.2575 **113.** $t \log 1.05$ **115.** $2t \ln 1.045$
117. $0.05t$ **119.** $-0.0042t$ **121.** 1535.06 **123.** 16.46

Find the Mistake **1.** A common logarithm uses the expression
$\log_{10} x$, which is equal to $\log x$. **2.** When the base of a log is not
shown, it is assumed to be 10. **3.** When the logarithm of a whole
number is approximated as a decimal, the decimal part of the answer
is called the mantissa. **4.** A natural logarithm is a logarithm with
a base of e, denoted by $\ln x = \log_e x$.

Exercise Set 8.6

Vocabulary Review **1.** exponential, doubling time **2.** variable
3. logarithms **4.** change-of-base
Problems **1.** 1.4650 **3.** 0.6826 **5.** -1.5440 **7.** -0.6477
9. $-\frac{1}{3}$ **11.** 2 **13.** -0.1845 **15.** 0.1845 **17.** 1.6168
19. 2.1131 **21.** -1 **23.** 1.2275 **25.** -6 **27.** -4 **29.** 2
31. 3 **33.** 0.3054 **35.** 42.5528 **37.** 6.0147 **39.** $\frac{4}{3}$
41. $\frac{3}{4}$ **43.** 1.3917 **45.** 0.7186 **47.** 2.6356 **49.** 4.1632
51. 5.8435 **53.** -1.0642 **55.** 2.3026 **57.** 10.7144
59. 11.72 years **61.** 9.25 years **63.** 8.75 years **65.** 18.58 years
67. 11.55 years **69.** 18.31 years **71.** 11.45 years
73. October 2018 **75.** 2009 **77.** 1992 **79.** 10.07 years
81. 2000 **83. a.** $K \approx 0.035$ **b.** 148.9°F **c.** 45.98 minutes
85. $(-2, -23)$, lowest **87.** $\left(\frac{3}{2}, 9\right)$, highest **89.** 2 seconds, 64 feet

Find the Mistake **1.** The equation $5^x = 12$ is an example of
exponential equation. **2.** The change-of-base property for
logarithms states that $\log_a x = \dfrac{\log_b x}{\log_b a}$. **3.** For the change-of-base
property to work, base a and base b must both be positive numbers.
4. To calculate $\log_9 36$, change the base to 10 and find the quotient
of log 36 and log 9.

Chapter 8 Review

1. 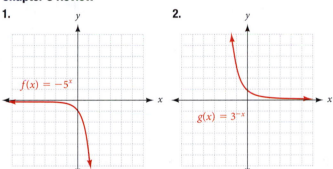 $f(x) = -5^x$

2. $g(x) = 3^{-x}$

3. 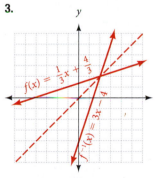 $f(x) = \frac{1}{3}x + \frac{4}{3}$ $f^{-1}(x) = 3x - 4$
$f^{-1}(x) = 3x - 4$

4. 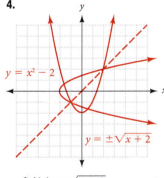 $y = x^2 - 2$ $y = \pm\sqrt{x + 2}$
$f^{-1}(x) = \sqrt{x + 2}$

5. 64 **6.** 2

7. 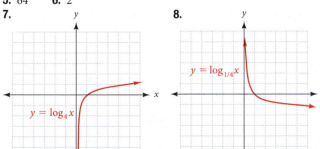 $y = \log_4 x$

8. $y = \log_{1/4} x$

9. $\frac{1}{2}$ **10.** 1.53   **11.** 4.76   **12.** -1.30 **13.** 3.31   **14.** -3.73

15. $\log_3 5 + 3\log_3 x - \log_3 y$ **16.** $\frac{1}{2}\log x - 5\log y - \frac{1}{3}\log z$

17. $\log_4 y^3 \sqrt{x}$ **18.** $\log \frac{x^3 \sqrt{z}}{y^4}$ **19.** 449.987   **20.** 0.0560

21. 2.322   **22.** 3   **23.** 32   **24.** 4, -2 does not check
25. 5.03   **26.** $606.89   **27.** 15.6 years   **28.** No   **29.** Yes
30. Yes   **31.** Yes   **32.** No   **33.** No

Chapter 8 Cumulative Review

1. 85   **2.** $-2x + 18$ **3.** 24   **4.** 4   **5.** $16\sqrt{2}$
6. $19 + 7\sqrt{15}$ **7.** $1 + 9i$ **8.** i **9.** 0   **10.** 0

11. $\frac{x+3}{x+1}$ **12.** $\frac{x^2-x+1}{1-x}$ **13.** $\frac{1}{2}$ **14.** $\frac{9}{7}$ **15.** $20t^4 - \frac{1}{20}$

16. $5x^3 - 22x^2 + 9$ **17.** $x - 5 - \frac{2}{x-1}$ **18.** $4x + 5 - \frac{13}{3x+2}$

19. $\frac{1}{(x-3)(x-2)}$ **20.** $\frac{8x-2}{(3x-4)(2x+1)(2x-1)}$ **21.** $-\frac{2}{5}$

22. 2   **23.** $-\frac{1}{4}, \frac{3}{8}$ **24.** 2   **25.** $\frac{5 \pm \sqrt{53}}{2}$ **26.** $-\frac{10}{7}$ **27.** 25

28. $-\frac{7}{6}, \frac{1}{3}$ **29.** 256   **30.** 5   **31.** $\varnothing$ **32.** $\frac{7+\sqrt{61}}{2}$

33. $(1, 4)$ **34.** $(6, 0)$ **35.** $\left(0, -4, \frac{4}{3}\right)$ **36.** $\left(-\frac{3}{2}, 0, 1\right)$

37.   **38.**

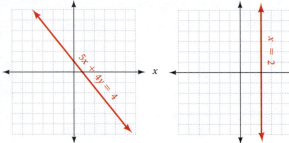

39. $8 - y = -3x$ **40.** $2a - 3b < 2a + 3b$ **41.** 1.29×10^{-6}

42. 8.3×10^7 **43.** $(4x + 3)(16x^2 - 12x + 9)$ **44.** $6(4a^2 + 1)(2a^2 + 1)$

45. Domain $= \{3, 5\}$; Range $= \{-4, -2, -1\}$; Not a function

46. Domain $= \{1, 2, 3\}$; Range $= \{1, 2\}$; A function

47. 4.34×10^8 **48.** 265.7 million

Chapter 8 Test

1.   **2.**

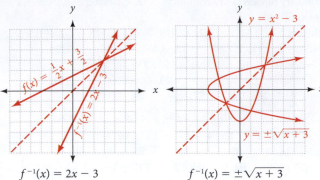

3.   **4.**

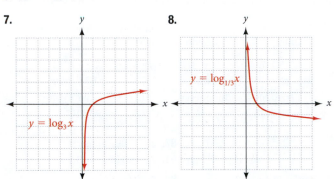

$f^{-1}(x) = 2x - 3$ $f^{-1}(x) = \pm\sqrt{x + 3}$

5. 25   **6.** $\sqrt[3]{7}$

7.   **8.**

9. $\frac{1}{2}$ **10.** 1.83   **11.** 4.16   **12.** -1.69 **13.** 4.03   **14.** -3.38

15. $2\log_7 2 + 3\log_7 x - \log_7 y$ **16.** $\frac{1}{3}\log x - 2\log y - \frac{1}{2}\log z$

17. $\log_2 x^{1/3} y^5$ **18.** $\log_3 \frac{x^{1/4} z^3}{y}$ **19.** 240,935.059   **20.** 0.004

21. 1.262   **22.** $\frac{5}{8}$ **23.** 2   **24.** 27   **25.** 5.14

26. $508.76   **27.** 27.88 years   **28.** Yes   **29.** No   **30.** Yes

31. Yes   **32.** No   **33.** No

Chapter 9

Exercise Set 9.1

Vocabulary Review   **1.** circle   **2.** distance   **3.** center
4. radius
Problems   **1.** 5   **3.** $\sqrt{106}$ **5.** $\sqrt{61}$ **7.** $\sqrt{130}$
9. 3 or -1 **11.** 0 or 6   **13.** $x = -1 \pm 4\sqrt{2}$
15. $(x - 3)^2 + (y + 2)^2 = 9$ **17.** $(x + 5)^2 + (y + 1)^2 = 5$
19. $x^2 + (y + 5)^2 = 1$ **21.** $x^2 + y^2 = 4$

23.

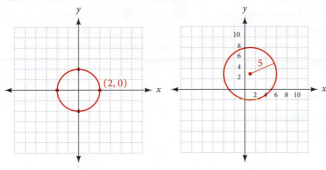

center = (0, 0); radius = 2

25.

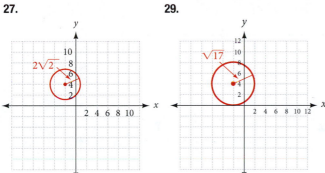

center = (1, 3); radius = 5

27.

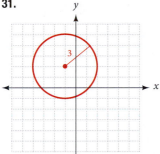

center = (−2, 4);
radius = $2\sqrt{2}$

29.

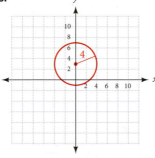

center = (−2, 4);
radius = $\sqrt{17}$

31.

center = (−1, 2); radius = 3

33.

center = (0, 3); radius = 4

35.

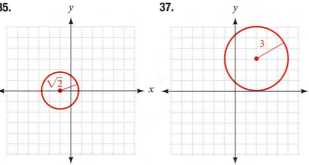

center = (−1, 0); radius = $\sqrt{2}$

37.

center = (2, 3); radius = 3

39.

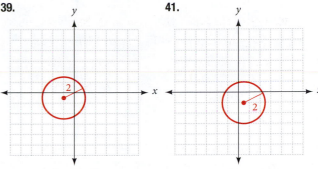

center = $\left(-1, -\frac{1}{2}\right)$; radius = 2

41.

center = $\left(\frac{1}{2}, -1\right)$; radius = 2

43. $(x - 3)^2 + (y - 4)^2 = 25$

45. $\left(x - \frac{1}{2}\right)^2 + (y - 1)^2 = \frac{1}{4}$; $(x - 1)^2 + (y - 1)^2 = 1$;
$(x - 2)^2 + (y - 1)^2 = 4$

47. $x^2 + y^2 = 25$ **49.** $x^2 + y^2 = 9$ **51.** $(x + 1)^2 + (y - 3)^2 = 25$

53. $(x + 2)^2 + (y - 5)^2 = 73$ **55.** $x^2 + (y - 2)^2 = 16$

57. $C = 6\pi\sqrt{2}$, $A = 18\pi$ **59.** $C = 10\pi$, $A = 25\pi$ **61.** Yes

63. $(x - 500)^2 + (y - 132)^2 = 120^2$ **65.** $y = \pm3$

67. $y = \pm2i$, no real solution **69.** $xi = \pm3$, no real solution

71. $\frac{x^2}{9} + \frac{y^2}{4}$ **73.** x-intercept 4, y-intercept −3 **75.** $\pm\frac{12}{5}$

Find the Mistake **1.** The distance between two points on a graph
can be found by using the formula $d = \sqrt{(x_2 - x_1)^2 + (y_2 - y_1)^2}$.
2. The equation $(x - 2)^2 + (y - 7)^2 = 16$ gives the graph of a circle
with a center at (2, 7) and a radius of 4. **3.** A circle given by the
equation $(x + 4)^2 + (y - 3)^2 = 64$ has a radius of 8. **4.** The first
step to graphing $x^2 + y^2 + 8x - 10y = 16$ is to use the addition
property of equality to group similar terms.

Exercise Set 9.2

Vocabulary Review **1.** vertical **2.** complete the square, vertex
3. horizontal **4.** right **5.** left

Problems

1.

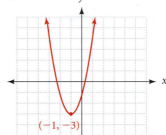

3.

(4, 2)

5.

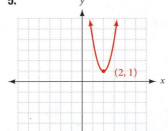

(2, 1)

7.

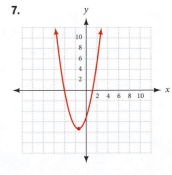

9.

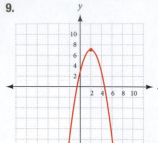

11.

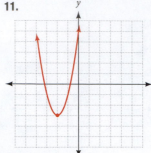

Find the Mistake **1.** The line $x = -\frac{b}{2a}$ is called the line of symmetry a parabola. **2.** A parabola that has two x-intercepts must have the form $y = a(x - k)^2 + h$. **3.** A parabola that has two y-intercepts must be a horizontal parabola of the form $x = a(y - k)^2 + h$. **4.** A parabola that has a vertex of $(4, 5)$ and x-intercepts of $(0, 2)$ and $(0, 6)$ will open down.

Exercise Set 9.3

Vocabulary Review ellipse, x-intercepts, y-intercepts, center, vertices
Problems

13.

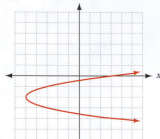

15.

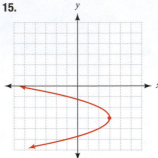

1.

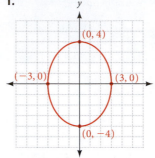

3.

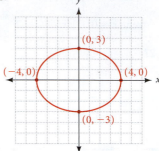

17.

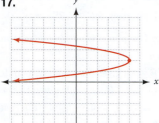

19.

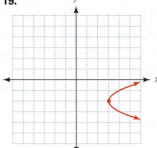

5.

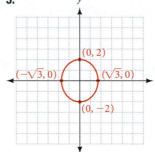

7.

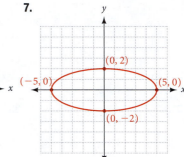

21.

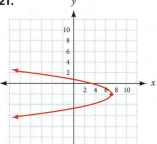

23.

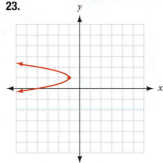

9.

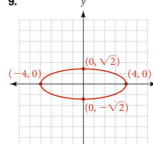

11.

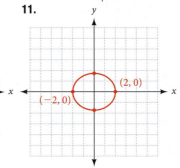

13. x-intercepts $= \pm 3$; y-intercepts $= \pm 2$

15. x-intercepts $= \pm 0.2$; y-intercepts $= \pm 0.3$

17. x-intercepts $= \pm\frac{3}{5}$; y-intercepts $= \pm\frac{2}{5}$

19. y-intercepts $= \pm\frac{3}{2}$; x-intercepts $= \pm\frac{4}{5}$

25. $y = \frac{1}{4}(x - 2)^2 - 1$ **27.** $y = -3(x + 2)^2 + 3$

29. $x = 4(y - 1)^2 - 4$ **31.** $x = -\frac{1}{3}(y + 5)^3 + 3$

33. $y = -\frac{8}{9}(x - 3)^2 + 8$ **35.** $y = \frac{1}{3}(x - 1)^2 - 3$

37. $x = -\frac{4}{9}(y - 2)^2 + 4$ **39.** $x = \frac{7}{4}(y + 4)^2 - 7$

41. Apple: 4,050 students; Dell: 8,400 students

43. **a.** Vertex $(80, 70)$, landing $(160, 0)$

   b. $y = ax(160 - x)$ or $y = -ax^2 + 160ax$

   c. $y = \frac{7}{640}x(160 - x)$ or $y = -\frac{7}{640}x^2 + \frac{7}{4}x$ **45.** $-7, -1$

47. $-4, 6$ **49.** x-intercepts $= -3, 3$, y-intercepts $= -2, 2$

51. x-intercepts $= -\sqrt{5}, \sqrt{5}$, y-intercepts $= -4, 4$ **53.** $\frac{\pm 8\sqrt{2}}{3}$

55. $\frac{\pm 5\sqrt{3}}{2}$

21.

23.

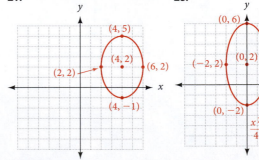

25.

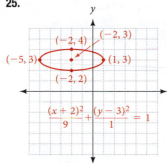

27. $x = 0$ **29.** $y = \pm\frac{16}{5}$

31. $x = \pm\frac{5\sqrt{7}}{4}$ **33.** 8

35. $\frac{x^2}{307.5^2} + \frac{y^2}{255^2} = 1$

37. $\frac{x^2}{9} + \frac{y^2}{81} = 1$

39. About 5.3 feet wide.

41. a. $\frac{4x^2}{35^2} + \frac{4y^2}{29^2} = 1$

   b. 19.6 feet **c.** ≈ 0.544

43. $188,384,100 **45.** x-intercepts $= \pm3$; no y-intercepts

47. y-intercepts $= \pm\sqrt{5}$; no x-intercepts **49.** $\frac{\pm3\sqrt{7}}{4}$ **51.** $\frac{\pm4\sqrt{6}}{5}$

Find the Mistake **1.** The equation $\frac{x^2}{a^2} + \frac{y^2}{b^2} = 1$ will give a graph in the shape of a ~~ellipse~~. **2.** The graph of the equation $\frac{x^2}{a^2} + \frac{y^2}{b^2} = 1$ will be an ellipse ~~centered at the origin~~. **3.** When a and b in the equation $\frac{x^2}{a^2} + \frac{y^2}{b^2} = 1$ are ~~equal~~, the graph will be a circle. **4.** The graph of the equation $\frac{(x-h)^2}{a^2} + \frac{(y-k)^2}{b^2} = 1$ will have vertices at $(h+a, k), (h-a, k), (h, k+b),$ and $(h, k-b)$.

Landmark Review **1.** 5 **2.** $\sqrt{5}$ **3.** $\sqrt{10}$ **4.** $2\sqrt{5}$

5. $(x+4)^2 + (y-3)^2 = 4$ **6.** $x^2 + (y-3)^2 = 25$

7. Center $(0, 0)$; $r = 3$ **8.** Center $(-3, 2)$; $r = \sqrt{5}$

9.

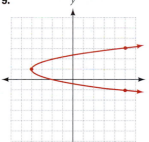

10.

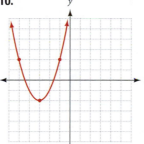

11.

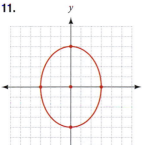

12.

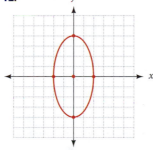

Exercise Set 9.4

Vocabulary Review **1.** hyperbola **2.** vertices

3. asymptotes **4.** origin

Problems

1.

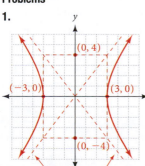

3.

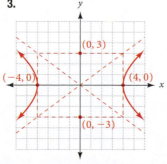

5.

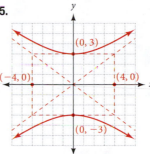

7.

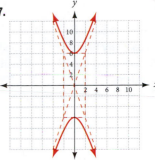

9.

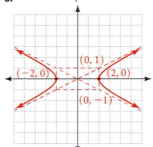

11.

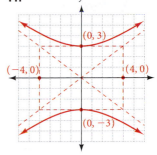

13. x-intercepts $= \pm3$; no y-intercepts

15. x-intercepts $= \pm0.2$; no y-intercepts

17. x-intercepts $= \pm\frac{3}{5}$; no y-intercepts

19. y-intercepts $= \pm\frac{4}{5}$; no x-intercepts

21.

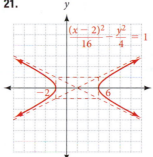

23.

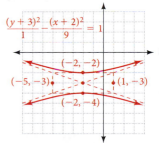

25.

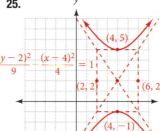

27. $x = 0$ **29.** No solution

31. $x = \pm\frac{25}{4}$ **33.** $y = \pm\frac{4}{5}$

35. $y = \frac{4}{3}x + \frac{10}{3}, y = -\frac{4}{3}x + \frac{2}{3}$

37. $\frac{x^2}{144} - \frac{y^2}{625} = 1$ **39.** $(0, 0)$

41. $4y^2 + 16y + 16$

43. $x = 2y + 4$ **45.** $-x^2 + 6$

47. $-3x^2 - 8$ **49.** $(5y + 6)(y + 2)$

51. $2(y + 2)(y - 2)$ **53.** $y = \pm2$ **55.** $x = \pm2$

Find the Mistake **1.** The graph of $\frac{x^2}{a^2} + \frac{y^2}{b^2} = 1$ will have ~~x-intercepts of $(-a, 0)$ and $(a, 0)$~~. **2.** Asymptotes for a hyperbola given by the equation $\frac{y^2}{b^2} + \frac{x^2}{a^2} = 1$ can be found by graphing the lines $y = \frac{b}{a}x$ and $y = -\frac{b}{a}x$. **3.** The asymptotes of a hyperbola are lines that the graph of the hyperbola ~~will approach but never cross~~. **4.** Hyperbolas given by the equation $\frac{(x-h)^2}{a^2} - \frac{(y-k)^2}{b^2} = 1$ or $\frac{(y-k)^2}{b^2} - \frac{(x-h)^2}{a^2} = 1$ will have centers at (h, k).

Set 9.5

Vocabulary Review   **1.** second-degree   **2.** boundary, broken
3. solution set, solid   **4.** nonlinear

Problems

1.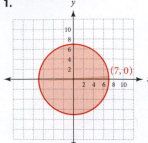

3.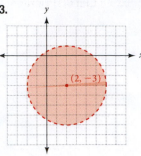

5.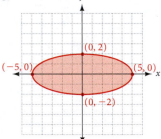

7.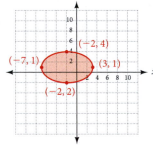

9.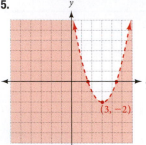

11.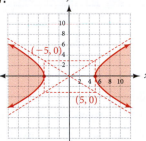

13.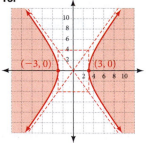

15.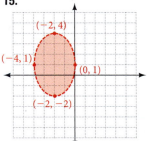

17.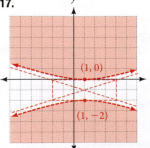

19. $(0, 3), \left(\frac{12}{5}, -\frac{9}{5}\right)$

21. $(0, 4), \left(\frac{16}{5}, \frac{12}{5}\right)$

23. $(5, 0), (-5, 0)$

25. $(0, -3), (\sqrt{5}, 2), (-\sqrt{5}, 2)$

27. $(0, -4), (\sqrt{7}, 3), (-\sqrt{7}, 3)$

29. $(-4, 11), \left(\frac{5}{2}, \frac{5}{4}\right)$

31. $(-4, 5), (1, 0)$

33. $(2, -3), (5, 0)$ **35.** $(3, 0), (-3, 0)$ **37.** $(4, 0), (0, -4)$

39. $\left(\frac{2\sqrt{6}}{3}, \frac{13}{3}\right), \left(-\frac{2\sqrt{6}}{3}, \frac{13}{3}\right)$

41. $(1, -2)$ **43.** $(\sqrt{10}, 0), (-\sqrt{10}, 0)$

45.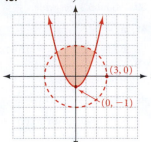

47.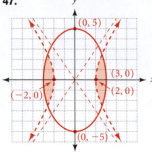

49.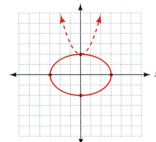

51.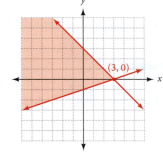

53.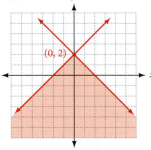

55.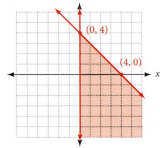

57.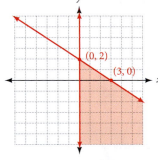

59.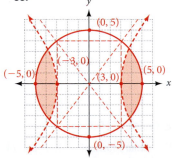

61.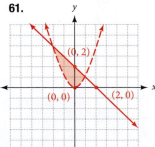

63. 8 and 5, -8 and -5, 8 and -5, -8 and 5

65. a. $(-4, 4\sqrt{3})$ and $(-4, -4\sqrt{3})$ **b.** $(4, 4\sqrt{3})$ and $(4, -4\sqrt{3})$

67. -19 **69.** 29   **71.** -1

73. 6   **75.** 12   **77.** -87

Find the Mistake **1.** A second-degree inequality contains at least one squared variable. **2.** Graphing the inequality $y \geq x^2 - 5$ will include the boundary and the origin as part of the solution set. **3.** To solve the following system, we first add the two equations to eliminate the variable y. **4.** The graph of the solution set for the following system will include the boundary of the second equation but not the first.

Chapter 9 Review

1. 13 **2.** 10 **3.** 3, 5 **4.** $(x + 4)^2 + (y - 2)^2 = 36$
5. $x^2 + y^2 = 15^2$ **6.** Center $= (0, -5)$; Radius $= 3$
7. Center $= (3, -5)$; Radius $= 6$

8.
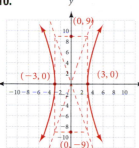

9.
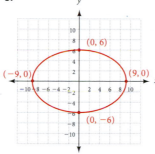

10.
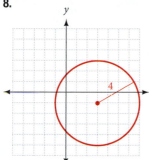

11.
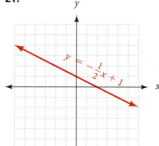

12.
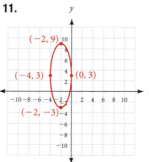

13.
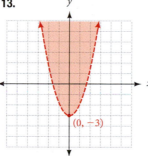

14. $(0, 5), \left(\frac{25}{13}, -\frac{60}{13}\right)$

15. $\left(0, -\frac{5}{2}\right), \left(2, \frac{3}{2}\right), \left(-2, \frac{3}{2}\right)$

16.
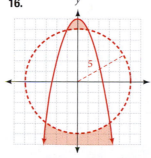

17. A **18.** D **19.** C **20.** B **21.** $(x + 3)^2 + (y - 2)^2 = 25$
22. $\frac{(x - 3)^2}{36} + \frac{(y - 1)^2}{16} = 1$ **23.** $\frac{y^2}{25} - \frac{x^2}{9} = 1$

Chapter 9 Cumulative Review

1. 8 **2.** 2 **3.** $\frac{a^7}{2b^{11}}$ **4.** $x^{3/2}y^{7/6}$ **5.** $6xy - 2y^2 + 3x^2$
6. $\frac{y + 4}{y - 3}$ **7.** 2 **8.** 3 **9.** $(b^3 + 4)(a - 1)$ **10.** $(7x + 1)(x - 3)$
11. -15 **12.** $\frac{1}{2}$ **13.** $-6, 3$ **14.** $-2, 5$
15. 11, 2 does not check **16.** $\varnothing$ **17.** $\frac{-2 \pm 4i\sqrt{3}}{7}$ **18.** $6 \pm i\sqrt{5}$
19. $x < -4$ or $x > 0$
20. $-\frac{3}{2} < x < 5$

21.
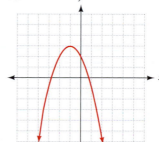

22.
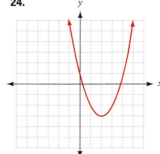

23.
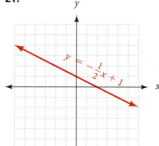

24.
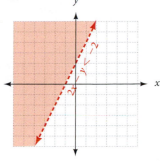

25. $\frac{y}{(y + 1)(y - 2)}$ **26.** $\frac{(x + 3)}{x(x - 1)}$ **27.** $13 + 13i$ **28.** $38 - 34i$
29. $\sqrt{5} + \sqrt{2}$ **30.** $\frac{20 - \sqrt{10}}{13}$ **31.** $f^{-1}(x) = 1 - 3x$
32. $f^{-1}(x) = \frac{x - 2}{5}$ **33.** 1,645.89 **34.** 0.91 **35.** $x = \frac{2mn - 5}{n}$
36. $y = \frac{S - 3x^2}{5x - 2}$ **37.** $\frac{2}{11}$ **38.** Slope $= \frac{5}{6}$; y-intercept $= -3$
39. -18 **40.** $C(15) = 30$; $C(30) = 15$ **41.** $2x^2 - 9x + 4 = 0$
42. $24t^2 + 29t - 4 = 0$ **43.** $-2x + 11$ **44.** $-3x^2 + 29$
45. -12 **46.** -11 **47.** 21 **48.** $\frac{32}{25}$
49. 5 gal of 40%; 45 gal of 90% **50.** 2 gal of 10%; 6 gal of 70%
51. 178,660 people **52.** 1,978,488 people

Chapter 9 Test

1. 5　　**2.** $4\sqrt{5}$　　**3.** 7, −1　　**4.** $(x - 3)^2 + (y + 1)^2 = 16$

5. $x^2 + y^2 = 169$　　**6.** Center = (0, 1); Radius = 6

7. Center = (2, −4); Radius = 3

8.

9.

10.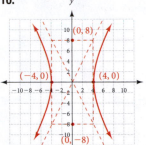

11.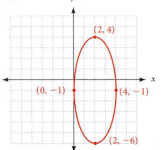

12.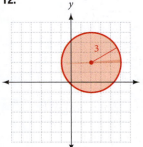

13.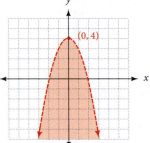

14. (0, −4), (4, 0)　　**15.** $(\pm 2\sqrt{2}, 1), (\pm\sqrt{5}, -2)$

16.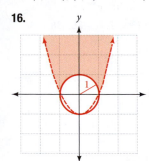

17. A　　**18.** D　　**19.** B　　**20.** C

21. $(x - 2)^2 + (y - 1)^2 = 25$

22. $\dfrac{(x - 1)^2}{25} + \dfrac{(y - 1)^2}{9} = 1$

23. $\dfrac{y^2}{64} - \dfrac{y^2}{100} = 1$

Index